Geometric Formulas

Conversion between Radians and Degrees: π radians $= 180°$

Triangle	**Circle**	**Sector of Circle**
$A = \frac{1}{2}bh$	$A = \pi r^2$	$A = \frac{1}{2}r^2\theta$ (θ in radians)
$= \frac{1}{2}ab\sin\theta$	$C = 2\pi r$	$s = r\theta$ (θ in radians)

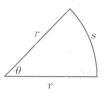

Sphere	**Cylinder**	**Cone**
$V = \frac{4}{3}\pi r^3 \quad A = 4\pi r^2$	$V = \pi r^2 h$	$V = \frac{1}{3}\pi r^2 h$

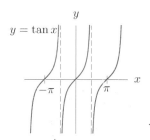

Trigonometric Functions

$$\sin\theta = \frac{y}{r}$$

$$\cos\theta = \frac{x}{r}$$

$$\tan\theta = \frac{y}{x}$$

$$\tan\theta = \frac{\sin\theta}{\cos\theta}$$

$$\cos^2\theta + \sin^2\theta = 1$$

$$\sin(A\pm B) = \sin A\cos B \pm \cos A\sin B$$

$$\cos(A\pm B) = \cos A\cos B \mp \sin A\sin B$$

$$\sin(2A) = 2\sin A\cos A$$

$$\cos(2A) = 2\cos^2 A - 1 = 1 - 2\sin^2 A$$

The Binomial Theorem

$$(x+y)^n = x^n + nx^{n-1}y + \frac{n(n-1)}{1\cdot 2}x^{n-2}y^2 + \frac{n(n-1)(n-2)}{1\cdot 2\cdot 3}x^{n-3}y^3 + \cdots + nxy^{n-1} + y^n$$

$$(x-y)^n = x^n - nx^{n-1}y + \frac{n(n-1)}{1\cdot 2}x^{n-2}y^2 - \frac{n(n-1)(n-2)}{1\cdot 2\cdot 3}x^{n-3}y^3 + \cdots \pm nxy^{n-1} \mp y^n$$

CALCULUS

Fourth Edition

Produced by the Calculus Consortium and initially funded by a National Science Foundation Grant.

Deborah Hughes-Hallett
University of Arizona

William G. McCallum
University of Arizona

Andrew M. Gleason
Harvard University

Brad G. Osgood
Stanford University

Daniel E. Flath
Macalester College

Douglas Quinney
University of Keele

Patti Frazer Lock
St. Lawrence University

Karen Rhea
University of Michigan

David O. Lomen
University of Arizona

Jeff Tecosky-Feldman
Haverford College

David Lovelock
University of Arizona

Thomas W. Tucker
Colgate University

with the assistance of

Otto K. Bretscher
Colby College

Sheldon P. Gordon
SUNY at Farmingdale

Andrew Pasquale
Chelmsford High School

Joe B. Thrash
University of Southern Mississippi

Coordinated by
Elliot J. Marks

WILEY

John Wiley & Sons, Inc.

ACQUISITIONS EDITOR	Michael Boezi
ASSOCIATE PUBLISHER	Laurie Rosatone
DEVELOPMENTAL EDITOR	Anne Scanlan-Rohrer/Two Ravens Editorial
ASSISTANT EDITOR	Kelly Boyle
EXECUTIVE MARKETING MANAGER	Julie Z. Lindstrom
MEDIA EDITOR	Lisa Schnettler
SENIOR PRODUCTION EDITOR	Ken Santor
COVER DESIGNER	Madelyn Lesure
COVER AND CHAPTER OPENING PHOTO	©Pete Turner/The Image Bank/Getty Images

Problems from Calculus: The Analysis of Functions, by Peter D. Taylor (Toronto: Wall & Emerson, Inc., 1992). Reprinted with permission of the publisher.

This book was set in Times Roman by the Consortium using TeX, Mathematica, and the package AsTeX, which was written by Alex Kasman. It was printed and bound by Von Hoffmann Press. The cover was printed by Von Hoffmann Press.

This book is printed on acid-free paper.

The paper in this book was manufactured by a mill whose forest management programs include sustained yield harvesting of its timberlands. Sustained yield harvesting principles ensure that the numbers of trees cut each year does not exceed the amount of new growth.

To order books or for customer service please, call 1(800)-CALL-WILEY (225-5945).

This material is based upon work supported by the National Science Foundation under Grant No. DUE-9352905. Opinions expressed are those of the authors and not necessarily those of the Foundation.

ISBN 0-471-48482-2

Printed in the United States of America

10 9 8 7 6 5 4 3 2 1

PREFACE

Calculus is one of the greatest achievements of the human intellect. Inspired by problems in astronomy, Newton and Leibniz developed the ideas of calculus 300 years ago. Since then, each century has demonstrated the power of calculus to illuminate questions in mathematics, the physical sciences, engineering, and the social and biological sciences.

Calculus has been so successful because of its extraordinary power to reduce complicated problems to simple rules and procedures. Therein lies the danger in teaching calculus: it is possible to teach the subject as nothing but the rules and procedures—thereby losing sight of both the mathematics and of its practical value. This edition of *Calculus* continues our effort to refocus the teaching of calculus on concepts as well as procedures.

A Focused Vision: Conceptual Understanding

Our goal is to provide students with a clear understanding of the ideas of calculus as a solid foundation for subsequent courses in mathematics and other disciplines. When we designed this curriculum we started with a clean slate. We increased the emphasis on some topics, such as differential equations, and decreased the emphasis on others after discussions with mathematicians, engineers, physicists, chemists, biologists, and economists. We focused on key concepts, emphasizing depth of understanding rather than breadth of coverage.

The Fourth Edition: Flexible and Focused

Since our First Edition struck a new balance between concepts, modeling, and skills in 1992, we have helped widen the range of choices for calculus instructors. Under our approach, which we called the "Rule of Four," ideas are presented graphically, numerically, symbolically, and verbally. Our problems probe student understanding in areas often taken for granted. The influence of these problems, praised for their creativity and variety, has extended far beyond the users of our textbook.

The text retains the hallmarks of our earlier editions: The Rule of Four, an emphasis on modeling, exposition that students can read, and a flexible approach to technology. Working in collaboration, consortium members and colleagues in mathematics and client disciplines have shaped a flexible curriculum that balances new and traditional topics.

In adding new topics, we have kept in mind two competing considerations: a textbook's role as a reference suitable for differing audiences, and the danger of encyclopedic clutter that overwhelms students and obscures the main vista. Our treatment of new topics is guided by the Rule of Four and our focus on conceptual understanding.

Choice of Paths: Lean or Expanded

For those who prefer the lean topic list of earlier editions, we have kept clear the main conceptual paths. For example,

- The Key Concept chapters on the derivative and the definite integral (Chapters 2 and 5) can be covered at the outset of the course, right after Chapter 1.
- Limits and Continuity (Sections 1.7 and 1.8) can be covered in depth before the introduction of the derivative (Sections 2.1 and 2.2), or after.
- Approximating Functions Using Series (Chapter 10) can be covered before, or without, Chapter 9.
- In Chapter 4 (Using the Derivative), instructors can select freely from Sections 4.4–4.8.
- Chapter 8 (Using the Definite Integral) contains a wide range of applications. Instructors can select one or two to do in detail.

The Development of Mathematical Thinking

The first stage in the development of mathematical thinking is the acquisition of a clear intuitive picture of the central ideas. In the next stage, the student learns to reason with the intuitive ideas and explain the reasoning clearly in plain English. After this foundation has been laid, there is a choice of direction. All students benefit from both theory and modeling, but the balance may differ for different groups. Some students, such as mathematics majors, may prefer more theory, while others, such as science and engineering majors, may prefer more modeling.

The Development of Mathematical Skills

To use calculus effectively, students need skill in both symbolic manipulation and the use of technology. The exact proportions of each may vary widely, depending on the preparation of the student and the wishes of the instructor. The book is adaptable to many different combinations.

The book does not require any specific software or technology. It has been used with graphing calculators, graphing software, and computer algebra systems. Any technology with the ability to graph functions and perform numerical integration will suffice. Students are expected to use their own judgment to determine where technology is useful.

Chapter-by-Chapter Description of Changes in the Fourth Edition:

Throughout the chapters, problems and exercises have been reorganized and hundreds of new problems and exercises have been added.

- Chapter 1: The chapter now ends with a section on limits, formerly Section 2.2. This allows a fuller discussion of continuity, both on an interval and at a point.

- Chapter 2: Some material has been reorganized. The material on limits and the discussion of continuity formerly in Section 2.7 has been moved to the end of Chapter 1.

- Chapter 3: There is a new section applying the chain rule to the derivatives of inverse functions. The section on hyperbolic functions has been moved to this chapter from Chapter 4. Section 3.3 on the derivatives of exponential functions has been extensively rewritten. The section on the chain rule has been expanded. Most of the theory on differentiable functions, including the Mean Value Theorem, has been moved from Section 4.7 to the end of this chapter. The second order error behavior of the tangent line approximation is addressed in greater detail, as a preview to Taylor polynomials.

- Chapter 4: There is a new section on rates and related rates, with dozens of new problems and exercises. The treatment of l'Hopital's rule has been moved from Chapter 3 to this chapter and expanded. The material on parametric equations has been moved to this chapter from Chapter 3 and expanded to included concavity of parametric curves.

- Chapter 5: This chapter has been reorganized and rewritten to clarify the Fundamental Theorem of Calculus, viewed as the integral of a rate of change giving the total change. There is also new material on area between curves.

- Chapter 6: There are no major changes in this chapter.

- Chapter 7: There are no major changes in this chapter.

- Chapter 8: There is a new section on area, arc length, and slope for curves in polar coordinates, including the former Appendix B on polar coordinates. Many of the problems and exercises in this section are new.

- Chapter 9: The chapter begins with a new section on sequences, including recursively defined sequences and a more detailed discussion of convergence for bounded sequences. Nearly all the problems and exercises in this section are new. The role of p-series in the integral comparison test is delineated. In Section 9.4, the limit comparison test has been added to the array of tests for convergence of a series, and there is a discussion of conditional convergence. The last section on functions defined by power series has been expanded, with more analysis of behavior at the endpoints of the interval of convergence. There is a new project proving the convergence of the sequence $s_n = (1 + 1/n)^n$, using the Bounded Monotone Convergence Theorem.

- Chapter 10: Section 10.4 on the Lagrange error bound for Taylor polynomials has been revised.
- Chapter 11: There is a new project on the 2003 SARS outbreak.
- Appendix: Polar Coordinates have been moved to Chapter 8.

Content

This content represents our vision of how calculus can be taught. It is flexible enough to accommodate individual course needs and requirements. Topics can easily be added or deleted, or the order changed.

Chapter 1: A Library of Functions

This chapter introduces all the elementary functions to be used in the book. Although the functions are probably familiar, the graphical, numerical, verbal, and modeling approach to them may be new. We introduce exponential functions at the earliest possible stage, since they are fundamental to the understanding of real-world processes. The chapter concludes with sections on limits and continuity.

Chapter 2: Key Concept—The Derivative

The purpose of this chapter is to give the student a practical understanding of the definition of the derivative and its interpretation as an instantaneous rate of change. The power rule is introduced; other rules are introduced in Chapter 3.

Chapter 3: Short-Cuts to Differentiation

The derivatives of all the functions in Chapter 1 are introduced, as well as the rules for differentiating products, quotients, composite, inverse, hyperbolic, and implicitly defined functions.

Chapter 4: Using the Derivative

The aim of this chapter is to enable the student to use the derivative in solving problems, including optimization, graphing, rates, parametric equations, and indeterminate forms. It is not necessary to cover all the sections in this chapter.

Chapter 5: Key Concept—The Definite Integral

The purpose of this chapter is to give the student a practical understanding of the definite integral as a limit of Riemann sums and to bring out the connection between the derivative and the definite integral in the Fundamental Theorem of Calculus.

Chapter 6: Constructing Antiderivatives

This chapter focuses on going backward from a derivative to the original function, first graphically and numerically, then analytically. It introduces the Second Fundamental Theorem of Calculus and the concept of a differential equation.

Chapter 7: Integration

This chapter includes several techniques of integration, including substitution, parts, partial fractions, and trigonometric substitutions; others are included in the table of integrals. There are discussions of numerical methods and of improper integrals.

Chapter 8: Using the Definite Integral

This chapter emphasizes the idea of subdividing a quantity to produce Riemann sums which, in the limit, yield a definite integral. It shows how the integral is used in geometry, physics, economics, and probability; polar coordinates are introduced. It is not necessary to cover all the sections in this chapter.

Chapter 9: Sequences and Series

This chapter focuses on sequences, series of constants and convergence. It includes the integral, ratio, comparison, limit comparison, and alternating series tests. It also introduces geometric series and general power series, including their intervals of convergence.

Chapter 10: Approximating Functions

This chapter introduces Taylor Series and Fourier Series using the idea of approximating functions by simpler functions.

Chapter 11: Differential Equations

This chapter introduces differential equations. The emphasis is on qualitative solutions, modeling, and interpretation.

Appendices

There are appendices on roots, accuracy, and bounds; complex numbers; and Newton's Method.

Supplementary Materials and Additional Resources

Supplements for the instructor can be obtained by sending a request on your institutional letterhead to Mathematics Marketing Manager, John Wiley & Sons, Inc., 111 River Street, Hoboken, NJ 07030, or by contacting your Wiley representative. The following supplementary materials are available for this edition:

- **Instructor's Manual** containing teaching tips, calculator programs, overhead transparency masters.
- **Test Bank** containing test questions and solutions arranged by section.
- **Instructor's Solution Manual** with complete solutions to all problems.
- **Student's Solution Manual** with complete solutions to half the odd-numbered problems.
- **Student's Study Guide** with additional study aids for students that are tied directly to the book.
- **Additional Material**, elaborating specially marked points in the text, as well as password protected electronic versions of the instructor ancillaries, can be found on the web at www.wiley.com/college/hugheshallett.

ConcepTests

ConcepTests, modeled on the pioneering work of Harvard physicist Eric Mazur, are questions designed to promote active learning during class, particularly (but not exclusively) in large lectures. Our evaluation data show students taught with ConcepTest outperformed students taught by traditional lecture methods 73% versus 17% on conceptual questions, and 63% versus 54% on computational problems. ConcepTests arranged by section are available from your Wiley representative.

Faculty Resource Network

A peer-to-peer network of academic faculty dedicated to the effective use of technology in the classroom, this group can help you apply innovative classroom techniques and implement specific software packages. Visit www.facultyresourcenetwork.com or speak to your Wiley representative.

eGrade Plus

eGrade Plus is a powerful online tool that provides a completely integrated suite of teaching and learning resources in one easy-to-use Web site. *eGrade Plus* includes an online version of the text, with fully integrated electronic versions of all student supplements, including the Student Solutions Manual, the Student Study Guide, and an Algebra and Trigonometry Refresher. Instructors will have additional access to electronic versions of the Instructor's Manual, the Instructor's Solutions Manual, additional projects, as well as other valuable resources. *eGrade Plus* also offers an online assessment system with full gradebook capabilities, which contains over 1000 skill-building questions from the Exercise sections in each chapter. Please view our online demo at www.wiley.com/college/egradeplus. Here you will find additional information about the features and benefits of eGrade Plus, how to request a "test drive" of eGrade Plus, and how to adopt it for class use.

AP Teacher's Guide

The AP Guide, written by experienced AP teachers, provides day-by-day syllabi for AB and BC Calculus, sample multiple choice questions, a listing of the past 25 years of AP free-response questions by chapter of the text, teaching tips, and labs to encourage student exploration of concepts.

Acknowledgements

First and foremost, we want to express our appreciation to the National Science Foundation for their faith in our ability to produce a revitalized calculus curriculum and, in particular, to our program officers, Louise Raphael, John Kenelly, John Bradley, and James Lightbourne. We also want to thank the members of our Advisory Board, Benita Albert, Lida Barrett, Simon Bernau, Robert Davis, M. Lavinia DeConge-Watson, John Dossey, Ron Douglas, Eli Fromm, William Haver, Seymour Parter, John Prados, and Stephen Rodi.

In addition, a host of other people around the country and abroad deserve our thanks for their contributions to shaping this edition. They include: Huriye Arikan, Ralph Baierlein, Ruth Baruth, Frank Beatrous, Alon Ben-David, Ebo Bentil, Paul Blanchard, Michael Boezi, Kelly Boyle, Melkana Brakalova, David Bressoud, Stephen Boyd, Morton Brown, Lucille Buonocore, Rick Cangelosi, Jo Cannon, Ray Cannon, Marilyn Carlson, Phil Cheifetz, C.K. Cheung, Scott Clark, Eric Connally, Caspar Curjel, Ehud de Shalit, Ann Davidian, Bob Decker, Dennis De Turck, Tom Dick, Srdjan Divac, Tevian Dray, Jeff Edmunds, Wade Ellis, Johann Engelbrecht, Bill Faris, Paul Feehan, Hermann Flaschka, Sol Friedberg, Paul Froeschl, Amanda Galtman, Lynn Garner, Tom Gearhart, Chris Goff, Joshua Green, Cristi Guevara, John Hagood, Gregory Hartman, Dudley Herschbach, Rob Indik, Adrian Iovita, Pallavi Jayawant, Millie Johnson, Alex Kasman, Matthias Kawski, Mike Klucznik, Donna Krawczyk, Mark Kunka, Ted Laetsch, Stephane Lafortune, Brigitte Lahme, Carl Leinbach, Carl Leinert, David Levermore, Don Lewis, Jim Lewis, Julie Lindstrom, Guadalupe Lozano, Bin Lu, Chris Lucas, John Lucas, Dan Madden, Alex Mallozzi, Corinne Manogue, Lisa A. Mantini, Nancy Marcus, Eric Mazur, Dan McGee, Tom McMahon, Ansie Meiring, Dave Meredith, Andrew Metrick, Lang Moore, Jerry Morris, Alan Newell, Steve Olson, John Orr, Arnie Ostebee, Igor Padure, Howard Penn, Amy Radunskaya, Wayne Raskind, Jeffrey Rauch, Janet Ray, Maria Robinson, Laurie Rosatone, Tenzin Chokey Rubling, Ann Ryu, Ayse Sahin, Nataliya Sandler, Ken Santor, Anne Scanlan-Rohrer, Pat Shure, Scott Pilzer, Lynne Small, David Smith, Seung Hye Song, Bob Speiser, Howard Stone, Steve Strogatz, "Suds" Sudholz, Al Taylor, Peter Taylor, Ralph Teixeira, Sally Thomas, Alan Tucker, Jerry Uhl, Doug Ulmer, Ignatios Vakalis, Bill Vélez, Aaron Wootton, Xianbao Xu, Kathy Yoshiwara, Lee Zia, and Paul Zorn.

Reports from the following reviewers were most helpful for the third edition:

Jennifer Aikens, Dean Allison, Cathy Anderson, Richard I. Avery, Janet Beery, Hamid Behmard, Victor Belfi, Tom Bengtson , David Borthwick, Sean Bradley, Carl T. Brezovec, Paul Wayne Britt, Denise Brown, Dave Bush, Don A. Campbell, Jack Carter, Stan Chadick, Ruth Charney, Jon Clauss, Ann Commito, Lawerence J. Cohen, Mary V. Connolly, Randall Crist, Rick Dalrymple, John Davis, Mike Dellens, Matt DeLong, Sylvia Desjardins, Meighan I. Dillon, Dawn Draus, Steven R. Dunbar, Jim Edmondson, Karen Edwards, Murray Eisenberg, Sue Esch, Phil Farmer, Richard E. Faulkenberry, George F. Feissner, Nelson Fong, Paula Castagna, Ben Fusaro, Kris Green, Michael Green, Leah Griffith, Cheryl Grood, David L. Gross, Brenda Guthrie, John Hagood, William Hamill, John Haverhals, Elizabeth Hentges, Curtis Herink, Jim Holmes, John Hoover, Rick Hough, Thomas Hunter, Kathy Ivey, Craig A. Jensen, Sue Jensen, Jerry Johnson, Nancy L. Johnson, Guarav Khanna, Helmut Knaust, Robert Kowalczyk, Anne Landry, James W. Lea, Phyllis Leonard, Tom Linton, Harold Lund, Juan Manfredi, William O. Martin, Alex Martsinkovsky, Anita Maxwell, Barbara McGee, Mary Ann McLoughlin, Douglas B. Meade, Richard Mercer, Raquel Mesa, C.R. Messer, Mark A. Mills, Yair Minsky, Derek Mitchem, Jeff Mock, Doug Mupasiri, Douglas Nelson, Dave Nolan, Sandy Norman, David A. Olson, Amos Ong, Grace Orzech, Wes Ostertag, Jeffrey S. Palmer, Barry Perratt, Peter Philliou, Shirley Pomeranz, G.J. Porter, Rebecca Porter, Vrunda Prabhu, Harald Proppe, Margie Reinhart, Mike Robillard, Donna Fields Rochon, Peter Rojas, Ned I. Rosen, Peter Rosnick, Tom Rousseau, Brenda Rudin, Mark D. Schlatter, Howard Schwesky, Patrick Shanahan, Michael Sheard, Angela B. Shiflet, Mehrdad Simkani, Lynne Small, Katy Snyder, Edward J. Soares, Ernie Solheid, Richard Stout, Robert Styer, James Sullivan, Ted Sundstrom, Stuart Swain, Ray Tennant, Buren Thomas, George Tintera, Dave Trautman, Mark Turner, Marian Van Vleet, Mike Veatch, Rachel Westlake, Bill Waltmann, Michael Ward, Steve Waters, Jack Winn, John Wood, Lisa Woodrum, William Wunderlich, Christos Xenophontos, W. Xie, Deane Yang, Kathie

Yoder, Bruce Yoshiwara, Andrzej Zarach, Ed Zeidman, and Peter Zekert.

Reports from the following reviewers were most helpful for the fourth edition:

Patricia Beaulieu, Phil Blau, Russell D. Blyth, Ray Cannon, Paula Castagna, Annette Cooper, Randall Crist, James D'Archangelo, Michael Divinia, Diane Doyle, Jim Edmondson, Leslie O. Foged, Scott R. Fulton, Lynn E. Garner, Sandy Grabiner, Steve Harris, Jeff Hirst, Vlajko Kocic, Kenneth Levasseur, Greg Marks, Kevin McDougal, Wilbur Mellema, Dan Oberlin, Tom Rousseau, Jeffrey Scroggs, Ernie S. Solheid, Barry S. Spieler, Diana Staats, Steve Szydlik, Mark Turner, Tony Vance, Gordon Woodward, and William Wunderlich.

Deborah Hughes-Hallett	Patti Frazer Lock	Douglas Quinney
Andrew M. Gleason	David O. Lomen	Karen R. Rhea
William G. McCallum	David Lovelock	Jeff Tecosky-Feldman
Daniel E. Flath	Brad G. Osgood	Thomas W. Tucker

To Students: How to Learn from this Book

- This book may be different from other math textbooks that you have used, so it may be helpful to know about some of the differences in advance. This book emphasizes at every stage the *meaning* (in practical, graphical or numerical terms) of the symbols you are using. There is much less emphasis on "plug-and-chug" and using formulas, and much more emphasis on the interpretation of these formulas than you may expect. You will often be asked to explain your ideas in words or to explain an answer using graphs.

- The book contains the main ideas of calculus in plain English. Your success in using this book will depend on your reading, questioning, and thinking hard about the ideas presented. Although you may not have done this with other books, you should plan on reading the text in detail, not just the worked examples.

- There are very few examples in the text that are exactly like the homework problems. This means that you can't just look at a homework problem and search for a similar–looking "worked out" example. Success with the homework will come by grappling with the ideas of calculus.

- Many of the problems that we have included in the book are open-ended. This means that there may be more than one approach and more than one solution, depending on your analysis. Many times, solving a problem relies on common sense ideas that are not stated in the problem but which you will know from everyday life.

- Some problems in this book assume that you have access to a graphing calculator or computer. There are many situations where you may not be able to find an exact solution to a problem, but you can use a calculator or computer to get a reasonable approximation.

- This book attempts to give equal weight to four methods for describing functions: graphical (a picture), numerical (a table of values) algebraic (a formula), and verbal. Sometimes you may find it easier to translate a problem given in one form into another. The best idea is to be flexible about your approach: if one way of looking at a problem doesn't work, try another.

- Students using this book have found discussing these problems in small groups very helpful. There are a great many problems which are not cut-and-dried; it can help to attack them with the other perspectives your colleagues can provide. If group work is not feasible, see if your instructor can organize a discussion session in which additional problems can be worked on.

- You are probably wondering what you'll get from the book. The answer is, if you put in a solid effort, you will get a real understanding of one of the most important accomplishments of the millennium—calculus—as well as a real sense of the power of mathematics in the age of technology.

Table of Contents

8 USING THE DEFINITE INTEGRAL   367

9 SEQUENCES AND SERIES   437

10 APPROXIMATING FUNCTIONS USING SERIES   477

11 DIFFERENTIAL EQUATIONS   521

APPENDIX   603

READY REFERENCE   623

ANSWERS TO ODD NUMBERED PROBLEMS   634

INDEX   665

Chapter One

A LIBRARY OF FUNCTIONS

Functions are truly fundamental to mathematics. In everyday language we say, "The fuel needed to launch a rocket is a function of its payload" or "The patient's blood pressure is a function of the drugs prescribed." In each case, the word *function* expresses the idea that knowledge of one fact tells us another. In mathematics, the most important functions are those in which knowledge of one number tells us another number. If we know the length of the side of a square, its area is determined. If the circumference of a circle is known, its radius is determined.

Calculus starts with the study of functions. This chapter lays the foundation for calculus by surveying the behavior of the most common functions, including exponential, logarithmic, and trigonometric functions. We see functions represented by graphs, tables, and formulas and investigate what limits and continuity can tell us about functions.

1.1 FUNCTIONS AND CHANGE

In mathematics, a *function* is used to represent the dependence of one quantity upon another.

Let's look at an example. In December 2000, the temperatures in Chicago were unusually low over winter vacation. The daily high temperatures for December 19–28 are given in Table 1.1.

Table 1.1   *Daily high temperature in Chicago, December 19–28, 2000*

Date (December 2000)	19	20	21	22	23	24	25	26	27	28
High temperature (°F)	20	17	19	7	20	11	17	19	17	20

Although you may not have thought of something so unpredictable as temperature as being a function, the temperature *is* a function of date, because each day gives rise to one and only one high temperature. There is no formula for temperature (otherwise we would not need the weather bureau), but nevertheless the temperature does satisfy the definition of a function: Each date, t, has a unique high temperature, H, associated with it.

We define a function as follows:

A **function** is a rule that takes certain numbers as inputs and assigns to each a definite output number. The set of all input numbers is called the **domain** of the function and the set of resulting output numbers is called the **range** of the function.

The input is called the *independent variable* and the output is called the *dependent variable*. In the temperature example, the domain is the set of dates $t = \{19, 20, 21, 22, 23, 24, 25, 26, 27, 28\}$ and the range is the set of temperatures $H = \{7, 11, 17, 19, 20\}$. We call the function f and write $H = f(t)$. Notice that a function may have identical outputs for different inputs (December 20, 25, and 27, for example).

Some quantities, such as date, are *discrete*, meaning they take only certain isolated values (dates must be integers). Other quantities, such as time, are *continuous* as they can be any number. For a continuous variable, domains and ranges are often written using interval notation:

The set of numbers t such that $a \leq t \leq b$ is written $[a, b]$.

The set of numbers t such that $a < t < b$ is written (a, b).

The Rule of Four: Tables, Graphs, Formulas, and Words

Functions can be represented by tables, graphs, formulas, and descriptions in words. For example, the function giving the daily high temperatures in Chicago can be represented by the graph in Figure 1.1, as well as by Table 1.1.

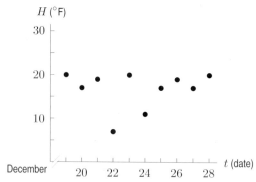

Figure 1.1: Chicago temperatures, December 2000

As another example of a function, consider the snow tree cricket. Surprisingly enough, all such crickets chirp at essentially the same rate if they are at the same temperature. That means that the chirp rate is a function of temperature. In other words, if we know the temperature, we can determine the chirp rate. Even more surprisingly, the chirp rate, C, in chirps per minute, increases steadily with the temperature, T, in degrees Fahrenheit, and can be computed by the formula

$$C = 4T - 160$$

to a fair degree of accuracy. We write $C = f(T)$ to express the fact that we think of C as a function of T and that we have named this function f. The graph of this function is in Figure 1.2.

Figure 1.2: Cricket chirp rate versus temperature

Examples of Domain and Range

If the domain of a function is not specified, we usually take it to be the largest possible set of real numbers. For example, we usually think of the domain of the function $f(x) = x^2$ as all real numbers. However, the domain of the function $g(x) = 1/x$ is all real numbers except zero, since we cannot divide by zero.

Sometimes we restrict the domain to be smaller than the largest possible set of real numbers. For example, if the function $f(x) = x^2$ is used to represent the area of a square of side x, we restrict the domain to nonnegative values of x.

Example 1   The function $C = f(T)$ gives chirp rate as a function of temperature. We restrict this function to temperatures for which the predicted chirp rate is positive, and up to the highest temperature ever recorded at a weather station, $136°$F. What is the domain of this function f?

Solution   If we consider the equation

$$C = 4T - 160$$

simply as a mathematical relationship between two variables C and T, any T value is possible. However, if we think of it as a relationship between cricket chirps and temperature, then C cannot be less than 0. Since $C = 0$ leads to $0 = 4T - 160$, and so $T = 40°$F, we see that T cannot be less than $40°$F. (See Figure 1.2.) In addition, we are told that the function is not defined for temperatures above $136°$. Thus, for the function $C = f(T)$ we have

$$\text{Domain} = \text{All } T \text{ values between } 40°F \text{ and } 136°F$$
$$= \text{All } T \text{ values with } 40 \leq T \leq 136$$
$$= [40, 136].$$

Example 2   Find the range of the function f, given the domain from Example 1. In other words, find all possible values of the chirp rate, C, in the equation $C = f(T)$.

Solution   Again, if we consider $C = 4T - 160$ simply as a mathematical relationship, its range is all real C values. However, when thinking of the meaning of $C = f(T)$ for crickets, we see that the function predicts cricket chirps per minute between 0 (at $T = 40°$F) and 384 (at $T = 136°$F). Hence,

$$\text{Range} = \text{All } C \text{ values from } 0 \text{ to } 384$$
$$= \text{All } C \text{ values with } 0 \leq C \leq 384$$
$$= [0, 384].$$

In using the temperature to predict the chirp rate, we thought of the temperature as the *independent variable* and the chirp rate as the *dependent variable*. However, we could do this backward, and calculate the temperature from the chirp rate. From this point of view, the temperature is dependent on the chirp rate. Thus, which variable is dependent and which is independent may depend on your viewpoint.

Linear Functions

The chirp-rate function, $C = f(T)$, is an example of a *linear function*. A function is linear if its slope, or rate of change, is the same at every point. The rate of change of a function that is not linear may vary from point to point.

Olympic and World Records

During the early years of the Olympics, the height of the men's winning pole vault increased approximately 8 inches every four years. Table 1.2 shows that the height started at 130 inches in 1900, and increased by the equivalent of 2 inches a year. So the height was a linear function of time from 1900 to 1912. If y is the winning height in inches and t is the number of years since 1900, we can write

$$y = f(t) = 130 + 2t.$$

Since $y = f(t)$ increases with t, we say that f is an *increasing function*. The coefficient 2 tells us the rate, in inches per year, at which the height increases.

Table 1.2 *Men's Olympic pole vault winning height (approximate)*

Year	1900	1904	1908	1912
Height (inches)	130	138	146	154

This rate of increase is the *slope* of the line in Figure 1.3. The slope is given by the ratio

$$\text{Slope} = \frac{\text{Rise}}{\text{Run}} = \frac{146 - 138}{8 - 4} = \frac{8}{4} = 2 \text{ inches/year.}$$

Calculating the slope (rise/run) using any other two points on the line gives the same value.

What about the constant 130? This represents the initial height in 1900, when $t = 0$. Geometrically, 130 is the *intercept* on the vertical axis.

You may wonder whether the linear trend continues beyond 1912. Not surprisingly, it doesn't exactly. The formula $y = 130 + 2t$ predicts that the height in the 2000 Olympics would be 330 inches or 27 feet 6 inches, which is considerably higher than the actual value of 19 feet 4.27 inches. There is clearly a danger in *extrapolating* too far from the given data. You should also observe that the data in Table 1.2 is discrete, because it is given only at specific points (every four years). However, we have treated the variable t as though it were continuous, because the function $y = 130 + 2t$ makes

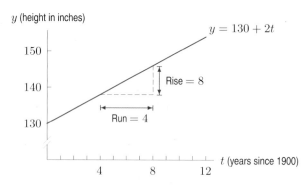

Figure 1.3: Olympic pole vault records

sense for all values of t. The graph in Figure 1.3 is of the continuous function because it is a solid line, rather than four separate points representing the years in which the Olympics were held.

As the pole vault heights have increased over the years, the time to run the mile has decreased. If y is the world record time to run the mile, in seconds, and t is the number of years since 1900, then records show that, approximately,

$$y = g(t) = 260 - 0.4t.$$

The 260 tells us that the world record was 260 seconds in 1900 (at $t = 0$). The slope, -0.4, tells us that the world record decreased by about 0.4 seconds per year. We say that g is a *decreasing function*.

Difference Quotients and Delta Notation

We use the symbol Δ (the Greek letter capital delta) to mean "change in," so Δx means change in x and Δy means change in y.

The slope of a linear function $y = f(x)$ can be calculated from values of the function at two points, given by x_1 and x_2, using the formula

$$m = \frac{\text{Rise}}{\text{Run}} = \frac{\Delta y}{\Delta x} = \frac{f(x_2) - f(x_1)}{x_2 - x_1}.$$

The quantity $(f(x_2) - f(x_1))/(x_2 - x_1)$ is called a *difference quotient* because it is the quotient of two differences. (See Figure 1.4). Since $m = \Delta y/\Delta x$, the units of m are y-units over x-units.

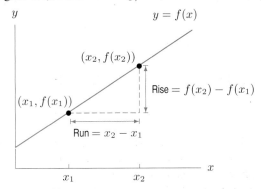

Figure 1.4: Difference quotient $= \dfrac{f(x_2) - f(x_1)}{x_2 - x_1}$

Families of Linear Functions

A **linear function** has the form

$$y = f(x) = b + mx.$$

Its graph is a line such that
- m is the **slope**, or rate of change of y with respect to x.
- b is the **vertical intercept**, or value of y when x is zero.

Notice that if the slope, m, is zero, we have $y = b$, a horizontal line.

To recognize that a table of x and y values comes from a linear function, $y = b + mx$, look for differences in y-values that are constant for equally spaced x-values.

Formulas such as $f(x) = b + mx$, in which the constants m and b can take on various values, give a *family of functions*. All the functions in a family share certain properties—in this case, all the graphs are straight lines. The constants m and b are called *parameters*; their meaning is shown in Figures 1.5 and 1.6. Notice the greater the magnitude of m, the steeper the line.

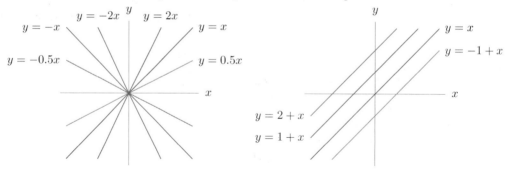

Figure 1.5: The family $y = mx$
(with $b = 0$)

Figure 1.6: The family $y = b + x$
(with $m = 1$)

Increasing versus Decreasing Functions

The terms increasing and decreasing can be applied to other functions, not just linear ones. See Figure 1.7. In general,

> A function f is **increasing** if the values of $f(x)$ increase as x increases.
> A function f is **decreasing** if the values of $f(x)$ decrease as x increases.
>
> The graph of an *increasing* function *climbs* as we move from left to right.
> The graph of a *decreasing* function *falls* as we move from left to right.

Figure 1.7: Increasing and decreasing functions

Proportionality

A common functional relationship occurs when one quantity is *proportional* to another. For example, the area, A, of a circle is proportional to the square of the radius, r, because

$$A = f(r) = \pi r^2.$$

> We say y is (directly) **proportional** to x if there is a nonzero constant k such that
>
> $$y = kx.$$
> This k is called the constant of proportionality.

We also say that one quantity is *inversely proportional* to another if one is proportional to the reciprocal of the other. For example, the speed, v, at which you make a 50-mile trip is inversely proportional to the time, t, taken, because v is proportional to $1/t$:

$$v = 50 \left(\frac{1}{t} \right) = \frac{50}{t}.$$

Exercises and Problems for Section 1.1

Exercises

1. The population of a city, P, in millions, is a function of t, the number of years since 1950, so $P = f(t)$. Explain the meaning of the statement $f(35) = 12$ in terms of the population of this city.

2. The pollutant PCB (polychlorinated biphenyl) affects the thickness of pelican eggs. Thinking of the thickness, T, of the eggs, in mm, as a function of the concentration, P, of PCBs in ppm (parts per million), we have $T = f(P)$. Explain the meaning of $f(200)$ in terms of thickness of pelican eggs and concentration of PCBs.

3. The value of a car, $V = f(a)$, in thousands of dollars, is a function of the age of the car, a, in years.

   (a) Interpret the statement $f(5) = 6$
   (b) Sketch a possible graph of V against a. Is f an increasing or decreasing function? Explain.
   (c) Explain the significance of the horizontal and vertical intercepts in terms of the value of the car.

For Exercises 4–6, find the equation of the line that passes through the given points.

4. $(0, 0)$ and $(1, 1)$ 5. $(0, 2)$ and $(2, 3)$

6. $(-2, 1)$ and $(2, 3)$

For Exercises 7–9, determine the slope and the y-intercept of the line whose equation is given.

7. $7y + 12x - 2 = 0$ 8. $-4y + 2x + 8 = 0$

9. $12x = 6y + 4$

10. Match the graphs in Figure 1.8 with the following equations. (Note that the x and y scales may be unequal.)

    (a) $y = x - 5$ (b) $-3x + 4 = y$
    (c) $5 = y$ (d) $y = -4x - 5$
    (e) $y = x + 6$ (f) $y = x/2$

(I)   (II)   (III)

(IV)   (V)   (VI)

Figure 1.8

11. Match the graphs in Figure 1.9 with the following equations. (Note that the x and y scales may be unequal.)

    (a) $y = -2.72x$ (b) $y = 0.01 + 0.001x$
    (c) $y = 27.9 - 0.1x$ (d) $y = 0.1x - 27.9$
    (e) $y = -5.7 - 200x$ (f) $y = x/3.14$

(I)   (II)   (III)

(IV)   (V)   (VI)

Figure 1.9

For Exercises 12–14, use the facts that parallel lines have equal slopes and that the slopes of perpendicular lines are negative reciprocals of one another.

12. Find the equation of the line through the point $(2, 1)$ which is perpendicular to the line $y = 5x - 3$.

13. Find the equations of the lines through the point $(1, 5)$ that are parallel to and perpendicular to the line with equation $y + 4x = 7$.

14. Find the equations of the lines through the point (a, b) that are parallel and perpendicular to the line $y = mx + c$, assuming $m \neq 0$.

For Exercises 15–17, give the approximate domain and range of each function. Assume the entire graph is shown.

15.

16.

17.

Find domain and range in Exercises 18–19.

18. $y = x^2 + 2$

19. $y = \dfrac{1}{x^2 + 2}$

20. If $f(t) = \sqrt{t^2 - 16}$, find all values of t for which $f(t)$ is a real number. Solve $f(t) = 3$.

21. If $g(x) = (4 - x^2)/(x^2 + x)$, find the domain of $g(x)$. Solve $g(x) = 0$.

In Exercises 22–24, write a formula representing the function.

22. The strength, S, of a beam is proportional to the square of its thickness, h.

23. The energy, E, expended by a swimming dolphin is proportional to the cube of the speed, v, of the dolphin.

24. The number of animal species, N, of a certain body length, l, is inversely proportional to the square of l.

Problems

25. Match each story to the most appropriate graph in Figure 1.10. Write a story for the remaining graph.

   (a) During my journey, I had a flat tire. Then after fixing the flat tire, I had to speed up to avoid being late.

   (b) My car broke down and I parked it at the road side.

   (c) As soon as I dropped off the package, I turned around and drove home.

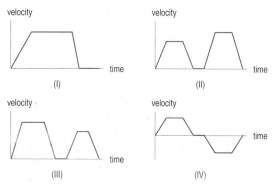

Figure 1.10

26. A car starts out slowly and then goes faster and faster until a tire blows out. Sketch a possible graph of the distance the car has traveled as a function of time.

27. A flight from Dulles Airport in Washington, DC, to LaGuardia Airport in New York City has to circle LaGuardia several times before being allowed to land. Plot a graph of the distance of the plane from Washington, DC, against time, from the moment of takeoff until landing.

28. You drive at a constant speed from Chicago to Detroit, a distance of 275 miles. About 120 miles from Chicago you pass through Kalamazoo, Michigan. Sketch a graph of your distance from Kalamazoo as a function of time.

29. The monthly charge for a waste collection service is $32 for 100 kg of waste and is $48 for 180 kg of waste.

   (a) Find a linear formula for the cost, C, of waste collection as a function of the number of kilograms of waste, w.

   (b) What is the slope of the line found in part (a)? Give units and interpret your answer in terms of the cost of waste collection.

   (c) What is the vertical intercept of the line found in part (a)? Give units with your answer and interpret it in terms of the cost of waste collection.

30. Residents of the town of Maple Grove who are connected to the municipal water supply are billed a fixed amount yearly plus a charge for each cubic foot of water used. A household using 1000 cubic feet was billed $90, while one using 1600 cubic feet was billed $105.

   (a) What is the charge per cubic foot?

   (b) Write an equation for the total cost of a resident's water as a function of cubic feet of water used.

   (c) How many cubic feet of water used would lead to a bill of $130?

31. The graph of Fahrenheit temperature, $°F$, as a function of Celsius temperature, $°C$, is a line. You know that $212°F$ and $100°C$ both represent the temperature at which water boils. Similarly, $32°F$ and $0°C$ both represent water's freezing point.

   (a) What is the slope of the graph?

   (b) What is the equation of the line?

   (c) Use the equation to find what Fahrenheit temperature corresponds to $20°C$.

   (d) What temperature is the same number of degrees in both Celsius and Fahrenheit?

32. An object is put outside on a cold day at time $t = 0$. Its temperature, $H = f(t)$, in $°C$, is graphed in Figure 1.11.

   (a) What does the statement $f(30) = 10$ mean in terms of temperature? Include units for 30 and for 10 in your answer.

   (b) Explain what the vertical intercept, a, and the horizontal intercept, b, represent in terms of temperature of the object and time outside.

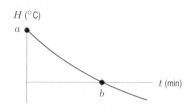

Figure 1.11

33. For tax purposes, you may have to report the value of your assets, such as cars or refrigerators. The value you report drops with time. "Straight-line depreciation" assumes that the value is a linear function of time. If a $950 refrigerator depreciates completely in seven years, find a formula for its value as a function of time.

34. The table gives the average weight, w, in pounds, of American men in their sixties for various heights, h, in inches.[1]

   (a) How do you know that the data in this table could represent a linear function?

   (b) Find weight, w, as a linear function of height, h. What is the slope of the line? What are the units for the slope?

   (c) Find height, h, as a linear function of weight, w. What is the slope of the line? What are the units for the slope?

h (inches)	68	69	70	71	72	73	74	75
w (pounds)	166	171	176	181	186	191	196	201

35. The demand function for a certain product, $q = D(p)$, is linear, where p is the price per item in dollars and q is the quantity demanded. If p increases by $5, market research shows that q drops by two items. In addition, 100 items are purchased if the price is $550.

   (a) Find a formula for

       (i) q as a linear function of p

       (ii) p as a linear function of q

   (b) Draw a graph with q on the horizontal axis.

36. When a cold yam is put into a hot oven to bake, the temperature of the yam rises. The rate, R (in degrees per minute), at which the temperature of the yam rises is governed by Newton's Law of Heating, which says that the rate is proportional to the temperature difference between the yam and the oven. If the oven is at $350°$F and the temperature of the yam is $H°$F,

   (a) Write a formula giving R as a function of H.

   (b) Sketch the graph of R against H.

37. For small changes in temperature, the formula for the expansion of a metal rod under a change in temperature is:

$$l - l_0 = al_0(t - t_0),$$

where l is the length of the object at temperature t, and l_0 is the length at temperature t_0, and a is a constant which depends on the type of metal.

   (a) Express l as a linear function of t. Find the slope and vertical intercept in terms of l_0, t_0, and a.

   (b) A rod is 100 cm long at $60°$F and made of a metal with $a = 10^{-5}$. Write an equation giving the length of this rod at temperature t.

   (c) What does the sign of the slope tell you about the expansion of a metal under a change in temperature?

38. A body of mass m is falling downward with velocity v. Newton's Second Law of Motion, $F = ma$, says that the net downward force, F, on the body is proportional to its downward acceleration, a. The net force, F, consists of the force due to gravity, F_g, which acts downward, minus the air resistance, F_r, which acts upward. The force due to gravity is mg, where g is a constant. Assume the air resistance is proportional to the velocity of the body.

   (a) Write an expression for the net force, F, as a function of the velocity, v.

   (b) Write a formula giving a as a function of v.

   (c) Sketch a against v.

39. When Galileo was formulating the laws of motion, he considered the motion of a body starting from rest and falling under gravity. He originally thought that the velocity of such a falling body was proportional to the distance it had fallen. What do the experimental data in Table 1.3 tell you about Galileo's hypothesis? What alternative hypothesis is suggested by the two sets of data in Table 1.3 and Table 1.4?

Table 1.3

Distance (ft)	0	1	2	3	4
Velocity (ft/sec)	0	8	11.3	13.9	16

Table 1.4

Time (sec)	0	1	2	3	4
Velocity (ft/sec)	0	32	64	96	128

1.2 EXPONENTIAL FUNCTIONS

Population Growth

The population of Mexico in the early 1980s is given in Table 1.5. To see how the population is growing, we look at the increase in population in the third column. If the population had been growing linearly, all the numbers in the third column would be the same.

---

[1] Adapted from "Average Weight of Americans by Height and Age", *The World Almanac* (New Jersey: Funk and Wagnalls, 1992), p. 956.

Table 1.5 *Population of Mexico (estimated), 1980–1986*

Year	Population (millions)	Change in population (millions)
1980	67.38	
		1.75
1981	69.13	
		1.80
1982	70.93	
		1.84
1983	72.77	
		1.89
1984	74.66	
		1.94
1985	76.60	
		1.99
1986	78.59	

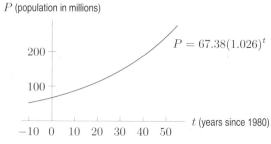

Figure 1.12: Population of Mexico (estimated): Exponential growth

Suppose we divide each year's population by the previous year's population. For example,

$$\frac{\text{Population in 1981}}{\text{Population in 1980}} = \frac{69.13 \text{ million}}{67.38 \text{ million}} = 1.026$$

$$\frac{\text{Population in 1982}}{\text{Population in 1981}} = \frac{70.93 \text{ million}}{69.13 \text{ million}} = 1.026.$$

The fact that both calculations give 1.026 shows the population grew by about 2.6% between 1980 and 1981 *and* between 1981 and 1982. Similar calculations for other years show that the population grew by a factor of about 1.026, or 2.6%, every year. Whenever we have a constant growth factor (here 1.026), we have exponential growth. The population t years after 1980 is given by the *exponential* function

$$P = 67.38(1.026)^t.$$

If we assume that the formula holds for 50 years, the population graph has the shape shown in Figure 1.12. Since the population is growing faster and faster as time goes on, the graph is bending upward; we say it is *concave up*. Even exponential functions which climb slowly at first, such as this one, eventually climb extremely quickly.

> To recognize that a table of t and P values comes from an exponential function $P = P_0 a^t$, look for ratios of P values that are constant for equally spaced t values.

Concavity

We have used the term concave up[2] to describe the graph in Figure 1.12. In words:

> The graph of a function is **concave up** if it bends upward as we move left to right; it is **concave down** if it bends downward. (See Figure 1.13.) A line is neither concave up nor concave down.

Figure 1.13: Concavity of a graph

---

[2]In Chapter 2 we consider concavity in more depth.

Elimination of a Drug from the Body

Now we look at a quantity which is decreasing exponentially instead of increasing. When a patient is given medication, the drug enters the bloodstream. As the drug passes through the liver and kidneys, it is metabolized and eliminated at a rate that depends on the particular drug. For the antibiotic ampicillin, approximately 40% of the drug is eliminated every hour. A typical dose of ampicillin is 250 mg. Suppose $Q = f(t)$, where Q is the quantity of ampicillin, in mg, in the bloodstream at time t hours since the drug was given. At $t = 0$, we have $Q = 250$. Since every hour the amount remaining is 60% of the previous amount, we have

$$f(0) = 250$$
$$f(1) = 250(0.6)$$
$$f(2) = (250(0.6))(0.6) = 250(0.6)^2,$$

and after t hours,

$$Q = f(t) = 250(0.6)^t.$$

This is an *exponential decay function*. Some values of the function are in Table 1.6; its graph is in Figure 1.14.

Notice the way the function in Figure 1.14 is decreasing. Each hour a smaller quantity of the drug is removed than in the previous hour. This is because as time passes, there is less of the drug in the body to be removed. Compare this to the exponential growth in Figure 1.12, where each step upward is larger than the previous one. Notice, however, that both graphs are concave up.

Table 1.6

t (hours)	Q (mg)
0	250
1	150
2	90
3	54
4	32.4
5	19.4

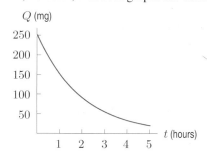

Figure 1.14: Drug elimination: Exponential decay

The General Exponential Function

We say P is an **exponential function** of t with base a if

$$P = P_0 a^t,$$

where P_0 is the initial quantity (when $t = 0$) and a is the factor by which P changes when t increases by 1.
If $a > 1$, we have exponential growth; if $0 < a < 1$, we have exponential decay.

Provided $a > 0$, the largest possible domain for the exponential function is all real numbers. The reason we do not want $a < 0$ is that, for example, we cannot define $a^{1/2}$ if $a < 0$. Also, we do not usually have $a = 1$, since $P = P_0 1^t = P_0$ is then a constant function.

The value of a is closely related to the percent growth (or decay) rate. For example, if $a = 1.03$, then P is growing at 3%; if $a = 0.94$, then P is decaying at 6%.

Example 1   Suppose that $Q = f(t)$ is an exponential function of t. If $f(20) = 88.2$ and $f(23) = 91.4$:

(a)   Find the base.   (b)   Find the growth rate.   (c)   Evaluate $f(25)$.

Solution   (a) Let $Q = Q_0 a^t$.

Substituting $t = 20, Q = 88.2$ and $t = 23, Q = 91.4$ gives two equations for Q_0 and a:

$$88.2 = Q_0 a^{20} \quad \text{and} \quad 91.4 = Q_0 a^{23}.$$

Dividing the two equations enables us to eliminate Q_0:

$$\frac{91.4}{88.2} = \frac{Q_0 a^{23}}{Q_0 a^{20}} = a^3.$$

Solving for the base, a, gives

$$a = \left(\frac{91.4}{88.2}\right)^{1/3} = 1.012.$$

(b) Since $a = 1.012$, the growth rate is $0.012 = 1.2\%$.

(c) We want to evaluate $f(25) = Q_0 a^{25} = Q_0(1.012)^{25}$. First we find Q_0 from the equation

$$88.2 = Q_0(1.012)^{20}.$$

Solving gives $Q_0 = 69.5$. Thus,

$$f(25) = 69.5(1.012)^{25} = 93.6.$$

Half-Life and Doubling Time

Radioactive substances, such as uranium, decay exponentially. A certain percentage of the mass disintegrates in a given unit of time; the time it takes for half the mass to decay is called the *half-life* of the substance.

A well-known radioactive substance is carbon-14, which is used to date organic objects. When a piece of wood or bone was part of a living organism, it accumulated small amounts of radioactive carbon-14. Once the organism dies, it no longer picks up carbon-14. Using the half-life of carbon-14 (about 5730 years), we can estimate the age of the object. We use the following definitions:

The **half-life** of an exponentially decaying quantity is the time required for the quantity to be reduced by a factor of one half.

The **doubling time** of an exponentially increasing quantity is the time required for the quantity to double.

The Family of Exponential Functions

The formula $P = P_0 a^t$ gives a family of exponential functions with positive parameters P_0 (the initial quantity) and a (the base, or growth/decay factor). The base tells us whether the function is increasing ($a > 1$) or decreasing ($0 < a < 1$). Since a is the factor by which P changes when t is increased by 1, large values of a mean fast growth; values of a near 0 mean fast decay. (See Figures 1.15 and 1.16.) All members of the family $P = P_0 a^t$ are concave up.

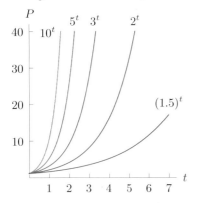

Figure 1.15: Exponential growth: $P = a^t$, for $a > 1$

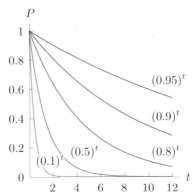

Figure 1.16: Exponential decay: $P = a^t$, for $0 < a < 1$

Example 2   Figure 1.17 is the graph of three exponential functions. What can you say about the values of the six constants, a, b, c, d, p, q?

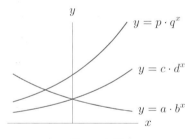

Figure 1.17

Solution   All the constants are positive. Since a, c, p represent y-intercepts, we see that $a = c$ because these graphs intersect on the y-axis. In addition, $a = c < p$, since $y = p \cdot q^x$ crosses the y-axis above the other two.

Since $y = a \cdot b^x$ is decreasing, we have $0 < b < 1$. The other functions are increasing, so $1 < d$ and $1 < q$.

Exponential Functions with Base e

The most frequently used base for an exponential function is the famous number $e = 2.71828\ldots$. This base is used so often that you will find an e^x button on most scientific calculators. At first glance, this is all somewhat mysterious. Why is it convenient to use the base $2.71828\ldots$? The full answer to that question must wait until Chapter 3, where we show that many calculus formulas come out neatly when e is used as the base. We often use the following result:

Any **exponential growth** function can be written, for some $a > 1$ and $k > 0$, in the form

$$P = P_0 a^t \quad \text{or} \quad P = P_0 e^{kt}$$

and any **exponential decay** function can be written, for some $0 < a < 1$ and $k > 0$, as

$$Q = Q_0 a^t \quad \text{or} \quad Q = Q_0 e^{-kt},$$

where P_0 and Q_0 are the initial quantities.

We say that P and Q are growing or decaying at a *continuous*[3] *rate* of k. (For example, $k = 0.02$ corresponds to a continuous rate of 2%.)

Example 3   Convert the functions $P = e^{0.5t}$ and $Q = 5e^{-0.2t}$ into the form $y = y_0 a^t$. Use the results to explain the shape of the graphs in Figures 1.18 and 1.19.

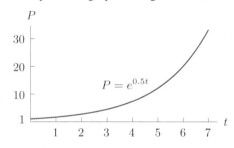

Figure 1.18: An exponential growth function

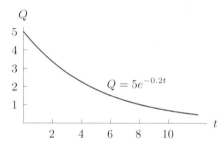

Figure 1.19: An exponential decay function

[3]The reason that k is called the continuous rate is explored in detail in Chapter 11.

Solution   We have

$$P = e^{0.5t} = (e^{0.5})^t = (1.65)^t.$$

Thus, P is an exponential growth function with $P_0 = 1$ and $a = 1.65$. The function is increasing and its graph is concave up, similar to those in Figure 1.15. Also,

$$Q = 5e^{-0.2t} = 5(e^{-0.2})^t = 5(0.819)^t,$$

so Q is an exponential decay function with $Q_0 = 5$ and $a = 0.819$. The function is decreasing and its graph is concave up, similar to those in Figure 1.16.

Example 4   The quantity, Q, of a drug in a patient's body at time t is represented for positive constants S and k by the function $Q = S(1 - e^{-kt})$. For $t \geq 0$, describe how Q changes with time. What does S represent? How does the graph of Q relate to the exponential decay graph in Figure 1.19?

Solution   The graph of Q is shown in Figure 1.20. Initially none of the drug is present, but the quantity increases with time. Since the graph is concave down, the quantity increases at a decreasing rate. This is realistic because as the quantity of the drug in the body increases, so does the rate at which the body excretes the drug. Thus, we expect the quantity to level off. Figure 1.20 shows that S is the saturation level.

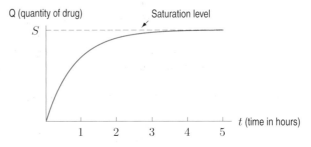

Figure 1.20: Buildup of the quantity of a drug in body

The graph in Figure 1.20 looks like an exponential decay function, but upside down. Since $Q = S - Se^{-kt}$, we have

$$S - Q = Se^{-kt}.$$

Thus, the difference between the saturation level, S, and the quantity in the blood, Q, is decaying exponentially. The graph in Figure 1.20 can be obtained from the graph in Figure 1.19 by reflection about the t-axis and moving up by a distance S.

Exercises and Problems for Section 1.2

Exercises

The functions in Exercises 1–4 represent exponential growth or decay. What is the initial quantity? What is the growth rate? State if the growth rate is continuous.

1. $P = 5(1.07)^t$

2. $P = 7.7(0.92)^t$

3. $P = 3.2e^{0.03t}$

4. $P = 15e^{-0.06t}$

5. A town has a population of 1000 people at time $t = 0$. In each of the following cases, write a formula for the population, P, of the town as a function of year t.

(a) The population increases by 50 people a year.
(b) The population increases by 5% a year.

6. An air-freshener starts with 30 grams and evaporates. In each of the following cases, write a formula for the quantity, Q grams, of air-freshener remaining t days after the start and sketch a graph of the function. The decrease is:

(a) 2 grams a day   (b) 12% a day

In Exercises 7–10, decide whether the graph is concave up, concave down, or neither.

7.

8.

9.

10.

11. Identify the x-intervals on which the function graphed in Figure 1.21 is:

    (a) Increasing and concave up
    (b) Increasing and concave down
    (c) Decreasing and concave up
    (d) Decreasing and concave down

Figure 1.21

Problems

12. In 1999, the world's population reached 6 billion and was increasing at a rate of 1.3% per year. Assume that this growth rate remains constant. (In fact, the growth rate has decreased since 1987.)

    (a) Write a formula for the world population (in billions) as a function of the number of years since 1999.
    (b) Use your formula to estimate the population of the world in the year 2020.
    (c) Sketch a graph of world population as a function of years since 1999. Use the graph to estimate the doubling time of the population of the world.

13. A photocopy machine can reduce copies to 80% of their original size. By copying an already reduced copy, further reductions can be made.

    (a) If a page is reduced to 80%, what percent enlargement is needed to return it to its original size?
    (b) Estimate the number of times in succession that a page must be copied to make the final copy less than 15% of the size of the original.

14. **(a)** Niki invested $10,000 in the stock market. The investment was a loser, declining in value 10% per year each year for 10 years. How much was the investment worth after 10 years?
    (b) After 10 years, the stock began to gain value at 10% per year. After how long will the investment regain its initial value ($10,000)?

15. When a new product is advertised, more and more people try it. However, the rate at which new people try it slows as time goes on.

    (a) Graph the total number of people who have tried such a product against time.
    (b) What do you know about the concavity of the graph?

16. Sketch reasonable graphs for the following. Pay particular attention to the concavity of the graphs.

    (a) The total revenue generated by a car rental business, plotted against the amount spent on advertising.
    (b) The temperature of a cup of hot coffee standing in a room, plotted as a function of time.

Give a possible formula for the functions in Problems 17–20.

17.

18.

19.

20.

In Problems 21–22, let $f(t) = Q_0 a^t = Q_0(1 + r)^t$.
    (a) Find the base, a.
    (b) Find the percentage growth rate, r.

21. $f(5) = 75.94$ and $f(7) = 170.86$

22. $f(0.02) = 25.02$ and $f(0.05) = 25.06$

Write the functions in Problems 23–26 in the form $P = P_0 a^t$. Which represent exponential growth and which represent exponential decay?

23. $P = 15e^{0.25t}$ **24.** $P = 2e^{-0.5t}$

25. $P = P_0 e^{0.2t}$ **26.** $P = 7e^{-\pi t}$

27. **(a)** A population, P, grows at a continuous rate of 2% a year and starts at 1 million. Write P in the form $P = P_0 e^{kt}$, with P_0, k constants.
    (b) Plot the population in part (a) against time.

28. When the Olympic Games were held outside Mexico City in 1968, there was much discussion about the effect the high altitude (7340 feet) would have on the athletes. Assuming air pressure decays exponentially by 0.4% every 100 feet, by what percentage is air pressure reduced by moving from sea level to Mexico City?

29. During 1988, Nicaragua's inflation rate averaged 1.3% a day. This means that, on average, prices went up by 1.3% from one day to the next.

    (a) By what percentage did Nicaraguan prices increase in June of 1988?
    (b) What was Nicaragua's annual inflation rate during 1988?

30. **(a)** The half-life of radium-226 is 1620 years. Write a formula for the quantity, Q, of radium left after t years, if the initial quantity is Q_0.
    (b) What percentage of the original amount of radium is left after 500 years?

31. In the early 1960s, radioactive strontium-90 was released during atmospheric testing of nuclear weapons and got into the bones of people alive at the time. If the half-life of strontium-90 is 29 years, what fraction of the strontium-90 absorbed in 1960 remained in people's bones in 1990?

32. A certain region has a population of 10,000,000 and an annual growth rate of 2%. Estimate the doubling time by guessing and checking.

33. Estimate graphically the doubling time of the exponentially growing population shown in Figure 1.22. Check that the doubling time is independent of where you start on the graph. Show algebraically that if $P = P_0 a^t$ doubles between time t and time $t + d$, then d is the same number for any t.

Figure 1.22

34. Aircrafts require longer takeoff distances, called takeoff rolls, at high altitude airports because of diminished air density. The table shows how the takeoff roll for a certain light airplane depends on the airport elevation. (Takeoff rolls are also strongly influenced by air temperature; the data shown assume a temperature of $0°$ C.) Determine a formula for this particular aircraft that gives the takeoff roll as an exponential function of airport elevation.

Elevation (ft)	Sea level	1000	2000	3000	4000
Takeoff roll (ft)	670	734	805	882	967

35. Each of the functions g, h, k in Table 1.7 is increasing, but each increases in a different way. Which of the graphs in Figure 1.23 best fits each function?

Table 1.7

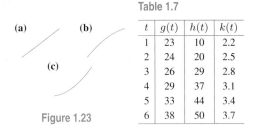

Figure 1.23

t	$g(t)$	$h(t)$	$k(t)$
1	23	10	2.2
2	24	20	2.5
3	26	29	2.8
4	29	37	3.1
5	33	44	3.4
6	38	50	3.7

36. **(a)** Which (if any) of the functions in the following table could be linear? Find formulas for those functions.
    (b) Which (if any) of these functions could be exponential? Find formulas for those functions.

x	$f(x)$	$g(x)$	$h(x)$
-2	12	16	37
-1	17	24	34
0	20	36	31
1	21	54	28
2	18	81	25

37. Match the functions $h(s)$, $f(s)$, and $g(s)$, whose values are in Table 1.8, with the formulas

$$y = a(1.1)^s, \quad y = b(1.05)^s, \quad y = c(1.03)^s,$$

assuming a, b, and c are constants. Note that the function values have been rounded to two decimal places.

Table 1.8

s	$h(s)$	s	$f(s)$	s	$g(s)$
2	1.06	1	2.20	3	3.47
3	1.09	2	2.42	4	3.65
4	1.13	3	2.66	5	3.83
5	1.16	4	2.93	6	4.02
6	1.19	5	3.22	7	4.22

38. Table 1.9 shows some values of a linear function f and an exponential function g. Find exact values (not decimal approximations) for each of the missing entries.

Table 1.9

x	0	1	2	3	4
$f(x)$	10	?	20	?	?
$g(x)$	10	?	20	?	?

39. The median price, P, of a home rose from \$50,000 in 1970 to \$100,000 in 1990. Let t be the number of years since 1970.

(a) Assume the increase in housing prices has been linear. Give an equation for the line representing price, P, in terms of t. Use this equation to complete column (a) of Table 1.10. Use units of \$1000.

(b) If instead the housing prices have been rising exponentially, find an equation of the form $P = P_0 a^t$ to represent housing prices. Complete column (b) of Table 1.10.

(c) On the same set of axes, sketch the functions represented in column (a) and column (b) of Table 1.10.

(d) Which model for the price growth do you think is more realistic?

Table 1.10

t	(a) Linear growth price in \$1000 units	(b) Exponential growth price in \$1000 units
0	50	50
10		
20	100	100
30		
40		

1.3   NEW FUNCTIONS FROM OLD

Shifts and Stretches

The graph of a constant multiple of a given function is easy to visualize: each y-value is stretched or shrunk by that multiple. For example, consider the function $f(x)$ and its multiples $y = 3f(x)$ and $y = -2f(x)$. Their graphs are shown in Figure 1.24. The factor 3 in the function $y = 3f(x)$ stretches each $f(x)$ value by multiplying it by 3; the factor -2 in the function $y = -2f(x)$ stretches $f(x)$ by multiplying by 2 and reflects it about the x-axis. You can think of the multiples of a given function as a family of functions.

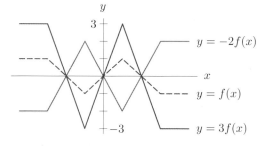

Figure 1.24: Multiples of the function $f(x)$

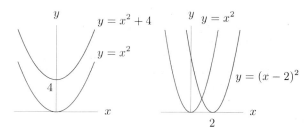

Figure 1.25: Graphs of $y = x^2$ with $y = x^2 + 4$ and $y = (x - 2)^2$

It is also easy to create families of functions by shifting graphs. For example, $y - 4 = x^2$ is the same as $y = x^2 + 4$, which is the graph of $y = x^2$ shifted up by 4. Similarly, $y = (x - 2)^2$ is the graph of $y = x^2$ shifted right by 2. (See Figure 1.25.)

- Multiplying a function by a constant, c, stretches the graph vertically (if $c > 1$) or shrinks the graph vertically (if $0 < c < 1$). A negative sign (if $c < 0$) reflects the graph about the x-axis, in addition to shrinking or stretching.
- Replacing y by $(y - k)$ moves a graph up by k (down if k is negative).
- Replacing x by $(x - h)$ moves a graph to the right by h (to the left if h is negative).

Composite Functions

If oil is spilled from a tanker, the area of the oil slick grows with time. Suppose that the oil slick is always a perfect circle. Then the area, A, of the oil slick is a function of its radius, r:

$$A = f(r) = \pi r^2.$$

The radius is also a function of time, because the radius increases as more oil spills. Thus, the area, being a function of the radius, is also a function of time. If, for example, the radius is given by

$$r = g(t) = 1 + t,$$

then the area is given as a function of time by substitution:

$$A = \pi r^2 = \pi(1 + t)^2.$$

We are thinking of A as a *composite function* or a "function of a function," which is written

$$A = \underbrace{f(g(t))}_{\substack{\text{Composite function;} \\ f \text{ is outside function,} \\ g \text{ is inside function}}} = \pi(g(t))^2 = \pi(1 + t)^2.$$

To calculate A using the formula $\pi(1 + t)^2$, the first step is to find $1 + t$, and the second step is to square and multiply by π. The first step corresponds to the inside function $g(t) = 1 + t$, and the second step corresponds to the outside function $f(r) = \pi r^2$.

Example 1   If $f(x) = x^2$ and $g(x) = x + 1$, find each of the following:
(a) $f(g(2))$ (b) $g(f(2))$ (c) $f(g(x))$ (d) $g(f(x))$

Solution   (a) Since $g(2) = 3$, we have $f(g(2)) = f(3) = 9$.
(b) Since $f(2) = 4$, we have $g(f(2)) = g(4) = 5$. Notice that $f(g(2)) \neq g(f(2))$.
(c) $f(g(x)) = f(x + 1) = (x + 1)^2$.
(d) $g(f(x)) = g(x^2) = x^2 + 1$. Again, notice that $f(g(x)) \neq g(f(x))$.

Example 2   Express each of the following functions as a composition:

(a) $h(t) = (1 + t^3)^{27}$ (b) $k(y) = e^{-y^2}$ (c) $l(y) = -(e^y)^2$

Solution   In each case think about how you would calculate a value of the function. The first stage of the calculation gives you the inside function, and the second stage gives you the outside function.
(a) For $(1 + t^3)^{27}$, the first stage is cubing and adding 1, so an inside function is $g(t) = 1 + t^3$. The second stage is taking the 27th power, so an outside function is $f(y) = y^{27}$. Then

$$f(g(t)) = f(1 + t^3) = (1 + t^3)^{27}.$$

In fact, there are lots of different answers: $g(t) = t^3$ and $f(y) = (1 + y)^{27}$ is another possibility.
(b) To calculate e^{-y^2} we square y, take its negative, and then take e to that power. So if $g(y) = -y^2$ and $f(z) = e^z$, then we have

$$f(g(y)) = e^{-y^2}.$$

(c) To calculate $-(e^y)^2$, we find e^y, square it, and take the negative. Using the same definitions of f and g as in part (b), the composition is

$$g(f(y)) = -(e^y)^2.$$

Since parts (b) and (c) give different answers, we see the order in which functions are composed is important.

Odd and Even Functions: Symmetry

There is a certain symmetry apparent in the graphs of $f(x) = x^2$ and $g(x) = x^3$ in Figure 1.26. For each point (x, x^2) on the graph of f, the point $(-x, x^2)$ is also on the graph; for each point (x, x^3) on the graph of g, the point $(-x, -x^3)$ is also on the graph. The graph of $f(x) = x^2$ is symmetric about the y-axis, whereas the graph of $g(x) = x^3$ is symmetric about the origin. The graph of any polynomial involving only even powers of x has symmetry about the y-axis, while polynomials with only odd powers of x are symmetric about the origin. Consequently, any functions with these symmetry properties are called *even* and *odd*, respectively.

Figure 1.26: Symmetry of even and odd functions

For any function f,
f is an **even** function if $f(-x) = f(x)$ for all x.
f is an **odd** function if $f(-x) = -f(x)$ for all x.

For example, $g(x) = e^{x^2}$ is even and $h(x) = x^{1/3}$ is odd. However, many functions do not have any symmetry and are neither even nor odd.

Inverse Functions

On August 18, 1989, the runner Arturo Barrios of Mexico set a world record for the 10,000-meter race. His times, in seconds, at 2000-meter intervals are recorded in Table 1.11, where $t = f(d)$ is the number of seconds Barrios took to complete the first d meters of the race. For example, Barrios ran the first 4000 meters in 650.10 seconds, so $f(4000) = 650.10$. The function f was useful to athletes planning to compete with Barrios.

Let us now change our point of view and ask for distances rather than times. If we ask how far Barrios ran during the first 650.10 seconds of his race, the answer is clearly 4000 meters. Going backward in this way from numbers of seconds to numbers of meters gives f^{-1}, the *inverse function*[4] of f. We write $f^{-1}(650.10) = 4000$. Thus, $f^{-1}(t)$ is the number of meters that Barrios ran during the first t seconds of his race. See Table 1.12 which contains values of f^{-1}.

The independent variable for f is the dependent variable for f^{-1}, and vice versa. The domains and ranges of f and f^{-1} are also interchanged. The domain of f is all distances d such that $0 \le d \le 10000$, which is the range of f^{-1}. The range of f is all times t, such that $0 \le t \le 1628.23$, which is the domain of f^{-1}.

Table 1.11   *Barrios's running time*

d (meters)	$t = f(d)$ (seconds)
0	0.00
2000	325.90
4000	650.10
6000	975.50
8000	1307.00
10000	1628.23

Table 1.12   *Distance run by Barrios*

t (seconds)	$d = f^{-1}(t)$ (meters)
0.00	0
325.90	2000
650.10	4000
975.50	6000
1307.00	8000
1628.23	10000

[4]The notation f^{-1} represents the inverse function, which is not the same as the reciprocal, $1/f$.

Which Functions Have Inverses?

If a function has an inverse, we say it is *invertible*. Let's look at a function which is not invertible. Consider the flight of the Mercury spacecraft *Freedom 7*, which carried Alan Shepard, Jr. into space in May 1961. Shepard was the first American to journey into space. After launch, his spacecraft rose to an altitude of 116 miles, and then came down into the sea. The function $f(t)$ giving the altitude in miles t minutes after lift-off does not have an inverse. To see why not, try to decide on a value for $f^{-1}(100)$, which should be the time when the altitude of the spacecraft was 100 miles. However, there are two such times, one when the spacecraft was ascending and one when it was descending. (See Figure 1.27.)

The reason the altitude function does not have an inverse is that the altitude has the same value for two different times. The reason the Barrios time function did have an inverse is that each running time, t, corresponds to a unique distance, d.

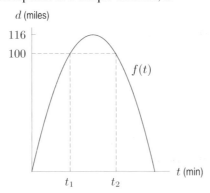

Figure 1.27: Two times, t_1 and t_2, at which altitude of spacecraft is 100 miles

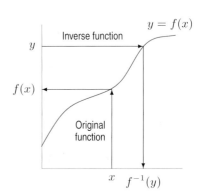

Figure 1.28: A function which has an inverse

Figure 1.28 suggests when an inverse exists. The original function, f, takes us from an x-value to a y-value, as shown in Figure 1.28. Since having an inverse means there is a function going from a y-value to an x-value, the crucial question is whether we can get back. In other words, does each y-value correspond to a unique x-value? If so, there's an inverse; if not, there is not. This principle may be stated geometrically, as follows:

> A function has an inverse if (and only if) its graph intersects any horizontal line at most once.

For example, the function $f(x) = x^2$ does not have an inverse because many horizontal lines intersect the parabola twice.

Definition of an Inverse Function

If the function f is invertible, its inverse is defined as follows:

$$f^{-1}(y) = x \quad \text{means} \quad y = f(x).$$

Formulas for Inverse Functions

If a function is defined by a formula, it is sometimes possible to find a formula for the inverse function. In Section 1.1, we looked at the snow tree cricket, whose chirp rate, C, in chirps per minute, is approximated at the temperature, T, in degrees Fahrenheit, by the formula

$$C = f(T) = 4T - 160.$$

So far we have used this formula to predict the chirp rate from the temperature. But it is also possible to use this formula backward to calculate the temperature from the chirp rate.

Example 3 Find the formula for the function giving temperature in terms of the number of cricket chirps per minute; that is, find the inverse function f^{-1} such that

$$T = f^{-1}(C).$$

Solution Since C is an increasing function, f is invertible. We know $C = 4T - 160$. We solve for T, giving

$$T = \frac{C}{4} + 40,$$

so

$$f^{-1}(C) = \frac{C}{4} + 40.$$

Graphs of Inverse Functions

The function $f(x) = x^3$ is increasing everywhere and so has an inverse. To find the inverse, we solve

$$y = x^3$$

for x, giving

$$x = y^{1/3}.$$

The inverse function is

$$f^{-1}(y) = y^{1/3}$$

or, if we want to call the independent variable x,

$$f^{-1}(x) = x^{1/3}.$$

The graphs of $y = x^3$ and $y = x^{1/3}$ are shown in Figure 1.29. Notice that these graphs are the reflections of one another about the line $y = x$. For example, $(8, 2)$ is on the graph of $y = x^{1/3}$ because $2 = 8^{1/3}$, and $(2, 8)$ is on the graph of $y = x^3$ because $8 = 2^3$. The points $(8, 2)$ and $(2, 8)$ are reflections of one another about the line $y = x$. In general, if the x- and y-axes have the same scales:

> The graph of f^{-1} is the reflection of the graph of f about the line $y = x$.

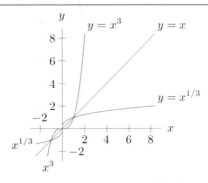

Figure 1.29: Graphs of inverse functions, $y = x^3$ and $y = x^{1/3}$, are reflections about the line $y = x$

Exercises and Problems for Section 1.3

Exercises

For the functions f and g in Exercises 1–4, find

(a) $f(g(1))$ (b) $g(f(1))$ (c) $f(g(x))$
(d) $g(f(x))$ (e) $f(t)g(t)$

1. $f(x) = x^2, g(x) = x + 1$
2. $f(x) = \sqrt{x + 4}, g(x) = x^2$
3. $f(x) = e^x, g(x) = x^2$
4. $f(x) = 1/x, g(x) = 3x + 4$

In Exercises 5–8, use Figure 1.30 to graph the functions.

Figure 1.30

5. $n(t) = m(t) + 2$ **6.** $p(t) = m(t - 1)$

7. $k(t) = m(t + 1.5)$

8. $w(t) = m(t - 0.5) - 2.5$

In Exercises 9–13, use Figure 1.31 to graph the function.

Figure 1.31

9. $f(-x)$ **10.** $f(x) + 5$

11. $f(x + 5)$ **12.** $5f(x)$

13. $f(5x)$

14. For $g(x) = x^2 + 2x + 3$, find and simplify:

 (a) $g(2 + h)$ **(b)** $g(2)$

 (c) $g(2 + h) - g(2)$

15. If $f(x) = x^2 + 1$, find and simplify:

 (a) $f(t + 1)$ **(b)** $f(t^2 + 1)$ **(c)** $f(2)$

 (d) $2f(t)$ **(e)** $[f(t)]^2 + 1$

16. For $f(n) = 3n^2 - 2$ and $g(n) = n + 1$, find and simplify:

 (a) $f(n) + g(n)$

 (b) $f(n)g(n)$

 (c) The domain of $f(n)/g(n)$

 (d) $f(g(n))$

 (e) $g(f(n))$

Simplify the quantities in Exercises 17–20 using $m(z) = z^2$.

17. $m(z + 1) - m(z)$ **18.** $m(z + h) - m(z)$

19. $m(z) - m(z - h)$ **20.** $m(z + h) - m(z - h)$

21. Let p be the price of an item and q be the number of items sold at that price, where $q = f(p)$. What do the following quantities mean in terms of prices and quantities sold?

 (a) $f(25)$ **(b)** $f^{-1}(30)$

22. Let $C = f(A)$ be the cost, in dollars, of building a store of area A square feet. In terms of cost and square feet, what do the following quantities represent?

 (a) $f(10,000)$ **(b)** $f^{-1}(20,000)$

23. Let $f(x)$ be the temperature ($^\circ$F) when the column of mercury in a particular thermometer is x inches long. What is the meaning of $f^{-1}(75)$ in practical terms?

24. Let $m = f(A)$ be the minimum annual gross income, in thousands of dollars, needed to obtain a 30-year home mortgage loan of A thousand dollars at an interest rate of 9%. What do the following quantities represent in terms of the income needed for a loan?

 (a) $f(100)$ **(b)** $f^{-1}(75)$

For Exercises 25–26, decide if the function $y = f(x)$ is invertible.

25.   **26.**

27. (a) Use Figure 1.32 to estimate $f^{-1}(2)$.

 (b) Sketch a graph of f^{-1} on the same axes.

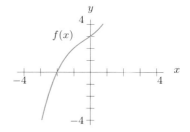

Figure 1.32

Find possible formulas for the graphs in Exercises 28–29 using shifts of x^2 or x^3.

28.   **29.**

          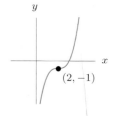

Are the functions in Exercises 30–37 even, odd, or neither?

30. $f(x) = x^6 + x^3 + 1$ **31.** $f(x) = x^3 + x^2 + x$

32. $f(x) = x^4 - x^2 + 3$ **33.** $f(x) = x^3 + 1$

34. $f(x) = 2x$ **35.** $f(x) = e^{x^2 - 1}$

36. $f(x) = x(x^2 - 1)$ **37.** $f(x) = e^x - x$

Problems

38. (a) Write an equation for a graph obtained by vertically stretching the graph of $y = x^2$ by a factor of 2, followed by a vertical upward shift of 1 unit. Sketch it.

(b) What is the equation if the order of the transformations (stretching and shifting) in part (a) is interchanged?

(c) Are the two graphs the same? Explain the effect of reversing the order of transformations.

For Problems 39–42, decide if the function f is invertible.

39. $f(t)$ is the number of customers in Macy's department store at t minutes past noon on December 18, 2000.

40. $f(n)$ is the number of students in your calculus class whose birthday is on the nth day of the year.

41. $f(x)$ is the volume in liters of x kg of water at $4°C$.

42. $f(w)$ is the cost of mailing a letter weighing w grams.

For Problems 43–48, use the graphs in Figure 1.33.

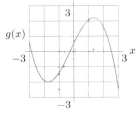

Figure 1.33

43. Estimate $f(g(1))$.

44. Estimate $g(f(2))$.

45. Estimate $f(f(1))$.

46. Graph $f(g(x))$.

47. Graph $g(f(x))$.

48. Graph $f(f(x))$.

For Problems 49–50, determine functions f and g such that $h(x) = f(g(x))$. [Note: There is more than one correct answer. Do not choose $f(x) = x$ or $g(x) = x$.]

49. $h(x) = (x + 1)^3$

50. $h(x) = x^3 + 1$

51. A tree of height y meters has, on average, B branches, where $B = y - 1$. Each branch has, on average, n leaves where $n = 2B^2 - B$. Find the average number of leaves of a tree as a function of height.

52. The cost of producing q articles is given by the function $C = f(q) = 100 + 2q$.

(a) Find a formula for the inverse function.

(b) Explain in practical terms what the inverse function tells you.

53. A kilogram weighs about 2.2 pounds.

(a) Write a formula for the function, f, which gives an object's mass in kilograms, k, as a function of its weight in pounds, p.

(b) Find a formula for the inverse function of f. What does this inverse function tell you, in practical terms?

54. Complete the following table with values for the functions f, g, and h, given that:

(a) f is an even function.

(b) g is an odd function.

(c) h is the composition $h(x) = g(f(x))$.

x	$f(x)$	$g(x)$	$h(x)$
−3	0	0	
−2	2	2	
−1	2	2	
0	0	0	
1			
2			
3			

1.4 LOGARITHMIC FUNCTIONS

In Section 1.2, we approximated the population of Mexico (in millions) by the function

$$P = f(t) = 67.38(1.026)^t,$$

where t is the number of years since 1980. Now suppose that instead of calculating the population at time t, we ask when the population will reach 200 million. We want to find the value of t for which

$$200 = f(t) = 67.38(1.026)^t.$$

We use logarithms to solve for a variable in an exponent.

Logarithms to Base 10 and to Base e

We define the *logarithm* function, $\log_{10} x$, to be the inverse of the exponential function, 10^x, as follows:

> The **logarithm** to base 10 of x, written $\log_{10} \mathbf{x}$, is the power of 10 we need to get x. In other words,
>
> $$\log_{10} x = c \quad \text{means} \quad 10^c = x.$$
>
> We often write $\log x$ in place of $\log_{10} x$.

The other frequently used base is e. The logarithm to base e is called the *natural logarithm* of x, written $\ln x$ and defined to be the inverse function of e^x, as follows:

> The **natural logarithm** of x, written $\ln x$, is the power of e needed to get x. In other words,
>
> $$\ln x = c \quad \text{means} \quad e^c = x.$$

Values of $\log x$ are in Table 1.13. Because no power of 10 gives 0, $\log 0$ is undefined. The graph of $y = \log x$ is shown in Figure 1.34. The domain of $y = \log x$ is positive real numbers; the range is all real numbers. In contrast, the inverse function $y = 10^x$ has domain all real numbers and range all positive real numbers. The graph of $y = \log x$ has a vertical asymptote at $x = 0$, whereas $y = 10^x$ has a horizontal asymptote at $y = 0$.

One big difference between $y = 10^x$ and $y = \log x$ is that the exponential function grows extremely quickly whereas the log function grows extremely slowly. However, $\log x$ does go to infinity, albeit slowly, as x increases. Since $y = \log x$ and $y = 10^x$ are inverse functions, the graphs of the two functions are reflections of one another about the line $y = x$, provided the scales along the x- and y-axes are equal.

Table 1.13   *Values for $\log x$ and 10^x*

x	$\log x$	x	10^x
0	undefined	0	1
1	0	1	10
2	0.3	2	100
3	0.5	3	10^3
4	0.6	4	10^4
⋮	⋮	⋮	⋮
10	1	10	10^{10}

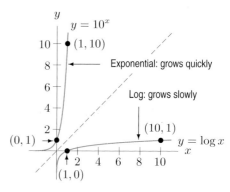

Figure 1.34: Graphs of $\log x$ and 10^x

The graph of $y = \ln x$ in Figure 1.35 has roughly the same shape as the graph of $y = \log x$. The x-intercept is $x = 1$, since $\ln 1 = 0$. The graph of $y = \ln x$ also climbs very slowly as x increases.

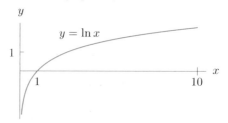

Figure 1.35: Graph of the natural logarithm

The following properties of logarithms may be deduced from the properties of exponents:

Properties of Logarithms

Note that $\log x$ and $\ln x$ are not defined when x is negative or 0.

1. $\log(AB) = \log A + \log B$
2. $\log\left(\dfrac{A}{B}\right) = \log A - \log B$
3. $\log(A^p) = p \log A$
4. $\log(10^x) = x$
5. $10^{\log x} = x$

1. $\ln(AB) = \ln A + \ln B$
2. $\ln\left(\dfrac{A}{B}\right) = \ln A - \ln B$
3. $\ln(A^p) = p \ln A$
4. $\ln e^x = x$
5. $e^{\ln x} = x$

In addition, $\log 1 = 0$ because $10^0 = 1$, and $\ln 1 = 0$ because $e^0 = 1$.

Solving Equations Using Logarithms

Logs are frequently useful when we have to solve for unknown exponents, as in the next examples.

Example 1   Find t such that $2^t = 7$.

Solution   First, notice that we expect t to be between 2 and 3 (because $2^2 = 4$ and $2^3 = 8$). To calculate t, we take logs to base 10. (Natural logs could also be used.)

$$\log(2^t) = \log 7.$$

Then use the third property of logs, which says $\log(2^t) = t \log 2$, and get:

$$t \log 2 = \log 7.$$

Using a calculator to find the logs gives

$$t = \frac{\log 7}{\log 2} \approx 2.81.$$

Example 2   Find when the population of Mexico reaches 200 million by solving $200 = 67.38(1.026)^t$.

Solution   Dividing both sides of the equation by 67.38, we get

$$\frac{200}{67.38} = (1.026)^t.$$

Now take logs of both sides:

$$\log\left(\frac{200}{67.38}\right) = \log(1.026^t).$$

Using the fact that $\log(A^t) = t \log A$, we get

$$\log\left(\frac{200}{67.38}\right) = t \log(1.026).$$

Solving this equation using a calculator to find the logs, we get

$$t = \frac{\log(200/67.38)}{\log(1.026)} \approx 42.4 \text{ years}$$

which is between $t = 42$ and $t = 43$. This value of t corresponds to the year 2022.

Example 3   The release of chlorofluorocarbons used in air conditioners and in household sprays (hair spray, shaving cream, etc.) destroys the ozone in the upper atmosphere. Currently, the amount of ozone, Q, is decaying exponentially at a continuous rate of 0.25% per year. What is the half-life of ozone?

Solution   We want to find how long it takes for half the ozone to disappear. If Q_0 is the initial quantity of ozone, then

$$Q = Q_0 e^{-0.0025t}.$$

We want to find T, the value of t making $Q = Q_0/2$, that is,

$$Q_0 e^{-0.0025T} = \frac{Q_0}{2}.$$

Dividing by Q_0 and then taking natural logs yields

$$\ln\left(e^{-0.0025T}\right) = -0.0025T = \ln\left(\frac{1}{2}\right) \approx -0.6931,$$

so

$$T \approx 277 \text{ years}.$$

The half-life of ozone is about 277 years.

In Example 3 the decay rate was given. However, in many situations where we expect to find exponential growth or decay, the rate is not given. To find it, we must know the quantity at two different times and then solve for the growth or decay rate, as in the next example.

Example 4   The population of Kenya was 19.5 million in 1984 and 21.2 million in 1986. Assuming it increases exponentially, find a formula for the population of Kenya as a function of time.

Solution   If we measure the population, P, in millions and time, t, in years since 1984, we can say

$$P = P_0 e^{kt} = 19.5 e^{kt},$$

where $P_0 = 19.5$ is the initial value of P. We find k by using the fact that $P = 21.2$ when $t = 2$, so

$$21.2 = 19.5 e^{k \cdot 2}.$$

To find k, we divide both sides by 19.5, giving

$$\frac{21.2}{19.5} = 1.087 = e^{2k}.$$

Now take natural logs of both sides:

$$\ln(1.087) = \ln(e^{2k}).$$

Using a calculator and the fact that $\ln(e^{2k}) = 2k$, this becomes

$$0.0834 = 2k.$$

So

$$k \approx 0.042,$$

and therefore

$$P = 19.5 e^{0.042t}.$$

Since $k = 0.042 = 4.2\%$, the population of Kenya was growing at a continuous rate of 4.2% per year.

In Example 4 we chose to use e for the base of the exponential function representing Kenya's population, making clear that the continuous growth rate was 4.2%. If we had wanted to emphasize the annual growth rate, we could have expressed the exponential function in the form $P = P_0 a^t$.

Example 5   Give a formula for the inverse of the following function (that is, solve for t in terms of P):

$$P = f(t) = 67.38(1.026)^t.$$

Solution   We want a formula expressing t as a function of P. Take logs:

$$\log P = \log(67.38(1.026)^t).$$

Since $\log(AB) = \log A + \log B$, we have

$$\log P = \log 67.38 + \log((1.026)^t).$$

Now use $\log(A^t) = t \log A$:

$$\log P = \log 67.38 + t \log 1.026.$$

Solve for t in two steps, using a calculator at the final stage:

$$t \log 1.026 = \log P - \log 67.38$$

$$t = \frac{\log P}{\log 1.026} - \frac{\log 67.38}{\log 1.026} \approx 89.7 \log P - 164.0.$$

Thus,

$$f^{-1}(P) = 89.7 \log P - 164.0.$$

Note that

$$f^{-1}(200) = 89.7(\log 200) - 164.0 \approx (89.7)(2.301) - 164.0 \approx 42.4,$$

which agrees with the result of Example 2.

Exercises and Problems for Section 1.4

Exercises

Simplify the expressions in Exercises 1–6 completely.

1. $e^{\ln(1/2)}$

2. $10^{\log(AB)}$

3. $5e^{\ln(A^2)}$

4. $\ln(e^{2AB})$

5. $\ln(1/e) + \ln(AB)$

6. $2\ln(e^A) + 3\ln B^e$

For Exercises 7–17, solve for x using logs.

7. $3^x = 11$

8. $17^x = 2$

9. $20 = 50(1.04)^x$

10. $4 \cdot 3^x = 7 \cdot 5^x$

11. $2^x = e^{x+1}$

12. $50 = 600e^{-0.4x}$

13. $2e^{3x} = 4e^{5x}$

14. $7^{x+2} = e^{17x}$

15. $10^{x+3} = 5e^{7-x}$

16. $2x - 1 = e^{\ln x^2}$

17. $9^x = 2e^{x^2}$

For Exercises 18–23, solve for t. Assume a and b are positive constants and k is nonzero.

18. $a = b^t$

19. $P = P_0 \, a^t$

20. $Q = Q_0 \, a^{nt}$

21. $P_0 \, a^t = Q_0 \, b^t$

22. $a = be^t$

23. $P = P_0 \, e^{kt}$

In Exercises 24–27, put the functions in the form $P = P_0 e^{kt}$.

24. $P = 15(1.5)^t$

25. $P = 10(1.7)^t$

26. $P = 174(0.9)^t$

27. $P = 4(0.55)^t$

Find the inverse function in Exercises 28–30.

28. $p(t) = (1.04)^t$

29. $f(t) = 50e^{0.1t}$

30. $f(t) = 1 + \ln t$

Problems

31. Without a calculator or computer, match the functions e^x, $\ln x$, x^2, and $x^{1/2}$ to their graphs in Figure 1.36.

Figure 1.36

32. The population of a region is growing exponentially. There were 40,000,000 people in 1980 ($t = 0$) and 56,000,000 in 1990. Find an expression for the population at any time t, in years. What population would you predict for the year 2000? What is the doubling time?

33. What is the doubling time of prices which are increasing by 5% a year?

34. If the size of a bacteria colony doubles in 5 hours, how long will it take for the number of bacteria to triple?

35. One hundred kilograms of a radioactive substance decay to 40 kg in 10 years. How much remains after 20 years?

36. Find the half-life of a radioactive substance that is reduced by 30% in 20 hours.

37. The sales at Borders bookstores went from $78 million in 1991 to $412 million in 1994. Find an exponential function to model the sales as a function of years since 1991. What is the continuous percent growth rate, per year, of sales?

38. Owing to an innovative rural public health program, infant mortality in Senegal, West Africa, is being reduced at a rate of 10% per year. How long will it take for infant mortality to be reduced by 50%?

39. Find the equation of the line l in Figure 1.37.

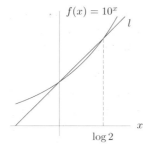

Figure 1.37

40. At time t hours after taking the cough suppressant hydrocodone bitartrate, the amount, A, in mg, remaining in the body is given by $A = 10(0.82)^t$.

(a) What was the initial amount taken?

(b) What percent of the drug leaves the body each hour?

(c) How much of the drug is left in the body 6 hours after the dose is administered?

(d) How long is it until only 1 mg of the drug remains in the body?

41. A cup of coffee contains 100 mg of caffeine, which leaves the body at a continuous rate of 17% per hour.

(a) Write a formula for the amount, A mg, of caffeine in the body t hours after drinking a cup of coffee.

(b) Graph the function from part (a). Use the graph to estimate the half-life of caffeine.

(c) Use logarithms to find the half-life of caffeine.

42. In 1980, there were about 170 million vehicles (cars and trucks) and about 227 million people in the United States. The number of vehicles has been growing at 4% a year, while the population has been growing at 1% a year. When was there, on average, one vehicle per person?

43. The air in a factory is being filtered so that the quantity of a pollutant, P (in mg/liter), is decreasing according to the function $P = P_0 e^{-kt}$, where t is time in hours. If 10% of the pollution is removed in the first five hours:

(a) What percentage of the pollution is left after 10 hours?

(b) How long is it before the pollution is reduced by 50%?

(c) Plot a graph of pollution against time. Show the results of your calculations on the graph.

(d) Explain why the quantity of pollutant might decrease in this way.

44. Air pressure, P, decreases exponentially with the height, h, in meters above sea level:

$$P = P_0 e^{-0.00012h}$$

where P_0 is the air pressure at sea level.

(a) At the top of Mount McKinley, height 6198 meters (about 20,330 feet), what is the air pressure, as a percent of the pressure at sea level?

(b) The maximum cruising altitude of an ordinary commercial jet is around 12,000 meters (about 39,000 feet). At that height, what is the air pressure, as a percent of the sea level value?

45. The half-life of radioactive strontium-90 is 29 years. In 1960, radioactive strontium-90 was released into the atmosphere during testing of nuclear weapons, and was absorbed into people's bones. How many years does it take until only 10% of the original amount absorbed remains?

46. A picture supposedly painted by Vermeer (1632–1675) contains 99.5% of its carbon-14 (half-life 5730 years). From this information decide whether the picture is a fake. Explain your reasoning.

1.5 TRIGONOMETRIC FUNCTIONS

Trigonometry originated as part of the study of triangles. The name *tri-gon-o-metry* means the measurement of three-cornered figures, and the first definitions of the trigonometric functions were in terms of triangles. However, the trigonometric functions can also be defined using the unit circle, a definition that makes them periodic, or repeating. Many naturally occurring processes are also periodic. The water level in a tidal basin, the blood pressure in a heart, an alternating current, and the position of the air molecules transmitting a musical note all fluctuate regularly. Such phenomena can be represented by trigonometric functions.

We use the three trigonometric functions found on a calculator: the sine, the cosine, and the tangent.

Radians

There are two commonly used ways to represent the input of the trigonometric functions: radians and degrees. The formulas of calculus, as you will see, are neater in radians than in degrees.

> An angle of 1 **radian** is defined to be the angle at the center of a unit circle which cuts off an arc of length 1, measured counterclockwise. (See Figure 1.38(a).) A unit circle has radius 1.

An angle of 2 radians cuts off an arc of length 2 on a unit circle. A negative angle, such as $-1/2$ radians, cuts off an arc of length $1/2$, but measured clockwise. (See Figure 1.38(b).)

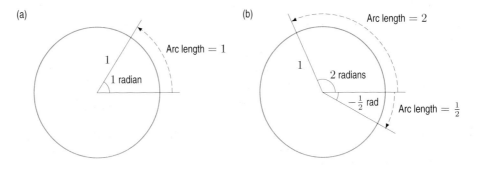

Figure 1.38: Radians defined using unit circle

It is useful to think of angles as rotations, since then we can make sense of angles larger than $360°$; for example, an angle of $720°$ represents two complete rotations counterclockwise. Since one full rotation of $360°$ cuts off an arc of length 2π, the circumference of the unit circle, it follows that

$$360° = 2\pi \text{ radians}, \quad \text{so} \quad 180° = \pi \text{ radians}.$$

In other words, 1 radian $= 180°/\pi$, so one radian is about $60°$. The word radians is often dropped, so if an angle or rotation is referred to without units, it is understood to be in radians.

Radians are useful for computing the length of an arc in any circle. If the circle has radius r and the arc cuts off an angle θ, as in Figure 1.39, then we have the following relation:

> Arc length $= s = r\theta$.

Figure 1.39: Arc length of a sector of a circle

The Sine and Cosine Functions

The two basic trigonometric functions—the sine and cosine—are defined using a unit circle. In Figure 1.40, an angle of t radians is measured counterclockwise around the circle from the point $(1, 0)$. If P has coordinates (x, y), we define

$$\cos t = x \quad \text{and} \quad \sin t = y.$$

We assume that the angles are *always* in radians unless specified otherwise.

Since the equation of the unit circle is $x^2 + y^2 = 1$, we have the following fundamental identity

$$\cos^2 t + \sin^2 t = 1.$$

As t increases and P moves around the circle, the values of $\sin t$ and $\cos t$ oscillate between 1 and -1, and eventually repeat as P moves through points where it has been before. If t is negative, the angle is measured clockwise around the circle.

Amplitude, Period, and Phase

The graphs of sine and cosine are shown in Figure 1.41. Notice that sine is an odd function, and cosine is even. The maximum and minimum values of sine and cosine are $+1$ and -1, because those are the maximum and minimum values of y and x on the unit circle. After the point P has moved around the complete circle once, the values of $\cos t$ and $\sin t$ start to repeat; we say the functions are *periodic*.

> For any periodic function of time, the
> - **Amplitude** is half the distance between the maximum and minimum values (if it exists).
> - **Period** is the smallest time needed for the function to execute one complete cycle.

The amplitude of $\cos t$ and $\sin t$ is 1, and the period is 2π. Why 2π? Because that's the value of t when the point P has gone exactly once around the circle. (Remember that $360° = 2\pi$ radians.)

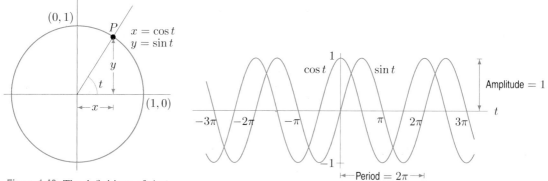

Figure 1.40: The definitions of $\sin t$ and $\cos t$

Figure 1.41: Graphs of $\cos t$ and $\sin t$

In Figure 1.41, we see that the sine and cosine graphs are exactly the same shape, only shifted horizontally. Since the cosine graph is the sine graph shifted $\pi/2$ to the left,

$$\cos t = \sin(t + \pi/2).$$

Equivalently, the sine graph is the cosine graph shifted $\pi/2$ to the right, so

$$\sin t = \cos(t - \pi/2).$$

We say that the *phase difference* or *phase shift*[5] between $\sin t$ and $\cos t$ is $\pi/2$.

Functions whose graphs are the shape of a sine or cosine curve are called *sinusoidal* functions.

> To describe arbitrary amplitudes and periods of sinusoidal functions, we use functions of the form
>
> $$f(t) = A\sin(Bt) \qquad \text{and} \qquad g(t) = A\cos(Bt),$$
>
> where $|A|$ is the amplitude and $2\pi/|B|$ is the period.
> The graph of a sinusoidal function is shifted horizontally by a distance $|h|$ when t is replaced by $t - h$ or $t + h$.
> Functions of the form $f(t) = A\sin(Bt) + C$ and $g(t) = A\cos(Bt) + C$ have graphs which are shifted vertically and oscillate about the value C.

Example 1 Find and show on a graph the amplitude and period of the functions

(a) $y = 5\sin(2t)$ 　　　　　(b) $y = -5\sin\left(\dfrac{t}{2}\right)$ 　　　　　(c) $y = 1 + 2\sin t$

Solution
(a) From Figure 1.42, you can see that the amplitude of $y = 5\sin(2t)$ is 5 because the factor of 5 stretches the oscillations up to 5 and down to -5. The period of $y = \sin(2t)$ is π, because when t changes from 0 to π, the quantity $2t$ changes from 0 to 2π, so the sine function goes through one complete oscillation.

(b) Figure 1.43 shows that the amplitude of $y = -5\sin(t/2)$ is again 5, because the negative sign reflects the oscillations in the t-axis, but does not change how far up or down they go. The period of $y = -5\sin(t/2)$ is 4π because when t changes from 0 to 4π, the quantity $t/2$ changes from 0 to 2π, so the sine function goes through one complete oscillation.

(c) The 1 shifts the graph $y = 2\sin t$ up by 1. Since $y = 2\sin t$ has an amplitude of 2 and a period of 2π, the graph of $y = 1 + 2\sin t$ goes up to 3 and down to -1, and has a period of 2π. (See Figure 1.44.) Thus, $y = 1 + 2\sin t$ also has amplitude 2.

Figure 1.42: Amplitude = 5, period = π

Figure 1.43: Amplitude = 5, period = 4π

Figure 1.44: Amplitude = 2, period = 2π

Example 2 Find possible formulas for the following sinusoidal functions.

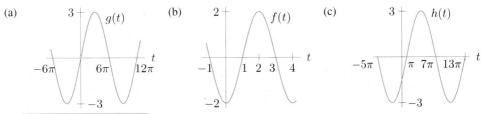

(a) $g(t)$

(b) $f(t)$

(c) $h(t)$

[5]Phase shift is defined in Section 11.10 on page 585.

Solution

(a) This function looks like a sine function of amplitude 3, so $g(t) = 3\sin(Bt)$. Since the function executes one full oscillation between $t = 0$ and $t = 12\pi$, when t changes by 12π, the quantity Bt changes by 2π. This means $B \cdot 12\pi = 2\pi$, so $B = 1/6$. Therefore, $g(t) = 3\sin(t/6)$ has the graph shown.

(b) This function looks like an upside down cosine function with amplitude 2, so $f(t) = -2\cos(Bt)$. The function completes one oscillation between $t = 0$ and $t = 4$. Thus, when t changes by 4, the quantity Bt changes by 2π, so $B \cdot 4 = 2\pi$, or $B = \pi/2$. Therefore, $f(t) = -2\cos(\pi t/2)$ has the graph shown.

(c) This function looks like the function $g(t)$ in part (a), but shifted a distance of π to the right. Since $g(t) = 3\sin(t/6)$, we replace t by $(t - \pi)$ to obtain $h(t) = 3\sin[(t - \pi)/6]$.

Example 3   On February 10, 1990, high tide in Boston was at midnight. The water level at high tide was 9.9 feet; later, at low tide, it was 0.1 feet. Assuming the next high tide is at exactly 12 noon and that the height of the water is given by a sine or cosine curve, find a formula for the water level in Boston as a function of time.

Solution   Let y be the water level in feet, and let t be the time measured in hours from midnight. The oscillations have amplitude 4.9 feet ($= (9.9 - 0.1)/2$) and period 12, so $12B = 2\pi$ and $B = \pi/6$. Since the water is highest at midnight, when $t = 0$, the oscillations are best represented by a cosine function. (See Figure 1.45.) We can say

$$\text{Height above average} = 4.9\cos\left(\frac{\pi}{6}t\right).$$

Since the average water level was 5 feet ($= (9.9 + 0.1)/2$), we shift the cosine up by adding 5:

$$y = 5 + 4.9\cos\left(\frac{\pi}{6}t\right).$$

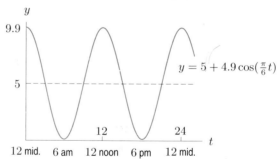

Figure 1.45: Function approximating the tide in Boston on February 10, 1990

Example 4   Of course, there's something wrong with the assumption in Example 3 that the next high tide is at noon. If so, the high tide would always be at noon or midnight, instead of progressing slowly through the day, as in fact it does. The interval between successive high tides actually averages about 12 hours 24 minutes. Using this, give a more accurate formula for the height of the water as a function of time.

Solution   The period is 12 hours 24 minutes $= 12.4$ hours, so $B = 2\pi/12.4$, giving

$$y = 5 + 4.9\cos\left(\frac{2\pi}{12.4}t\right) = 5 + 4.9\cos(0.507t).$$

Example 5   Use the information from Example 4 to write a formula for the water level in Boston on a day when the high tide is at 2 pm.

Solution   When the high tide is at midnight

$$y = 5 + 4.9\cos(0.507t).$$

Since 2 pm is 14 hours after midnight, we replace t by $(t - 14)$. Therefore, on a day when the high tide is at 2 pm,

$$y = 5 + 4.9\cos(0.507(t - 14)).$$

The Tangent Function

If t is any number with $\cos t \neq 0$, we define the tangent function as follows

$$\tan t = \frac{\sin t}{\cos t}.$$

Figure 1.40 on page 30 shows the geometrical meaning of the tangent function: $\tan t$ is the slope of the line through the origin $(0, 0)$ and the point $P = (\cos t, \sin t)$ on the unit circle.

The tangent function is undefined wherever $\cos t = 0$, namely, at $t = \pm\pi/2, \pm 3\pi/2, \ldots$, and it has a vertical asymptote at each of these points. The function $\tan t$ is positive where $\sin t$ and $\cos t$ have the same sign. The graph of the tangent is shown in Figure 1.46.

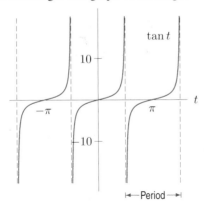

Figure 1.46: The tangent function

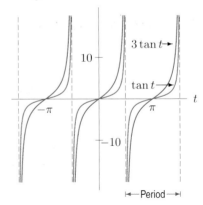

Figure 1.47: Multiple of tangent

The tangent function has period π, because it repeats every π units. Does it make sense to talk about the amplitude of the tangent function? Not if we're thinking of the amplitude as a measure of the size of the oscillation, because the tangent becomes infinitely large near each vertical asymptote. We can still multiply the tangent by a constant, but that constant no longer represents an amplitude. (See Figure 1.47.)

The Inverse Trigonometric Functions

On occasion, you may need to find a number with a given sine. For example, you might want to find x such that

$$\sin x = 0$$

or such that

$$\sin x = 0.3.$$

The first of these equations has solutions $x = 0, \pm\pi, \pm 2\pi, \ldots$. The second equation also has infinitely many solutions. Using a calculator, we get

$$x \approx 0.305, 2.84, 0.305 \pm 2\pi, 2.84 \pm 2\pi, \ldots.$$

For each equation, we pick out the solution between $-\pi/2$ and $\pi/2$ as the preferred solution. For example, the preferred solution to $\sin x = 0$ is $x = 0$, and the preferred solution to $\sin x = 0.3$ is $x = 0.305$. We define the inverse sine, written "arcsin" or "$\sin^{-1}$," as the function which gives the preferred solution.

For $-1 \le y \le 1$,
$$\arcsin y = x$$
means
$$\sin x = y \quad \text{with} \quad -\frac{\pi}{2} \le x \le \frac{\pi}{2}.$$

Thus the arcsine is the inverse function to the piece of the sine function having domain $[-\pi/2, \pi/2]$. (See Table 1.14 and Figure 1.48.) On a calculator, the arcsine function[6] is usually denoted by $\boxed{\sin^{-1}}$.

Table 1.14   *Values of* $\sin x$ *and* $\sin^{-1} x$

x	$\sin x$	x	$\sin^{-1} x$
$-\frac{\pi}{2}$	-1.000	-1.000	$-\frac{\pi}{2}$
-1.0	-0.841	-0.841	-1.0
-0.5	-0.479	-0.479	-0.5
0.0	0.000	0.000	0.0
0.5	0.479	0.479	0.5
1.0	0.841	0.841	1.0
$\frac{\pi}{2}$	1.000	1.000	$\frac{\pi}{2}$

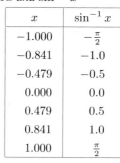

Figure 1.48: The arcsine function

The inverse tangent, written "arctan" or "$\tan^{-1}$," is the inverse function for the piece of the tangent function having the domain $-\pi/2 < x < \pi/2$. On a calculator, the inverse tangent is usually denoted by $\boxed{\tan^{-1}}$. The graph of the arctangent is shown in Figure 1.50.

For any y,
$$\arctan y = x$$
means
$$\tan x = y \quad \text{with} \quad -\frac{\pi}{2} < x < \frac{\pi}{2}.$$

The inverse cosine function, written "arccos" or "$\cos^{-1}$," is discussed in Problem 47. The range of the arccosine function is $0 \le x \le \pi$.

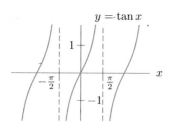

Figure 1.49: The tangent function

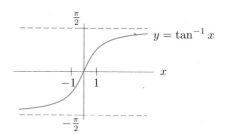

Figure 1.50: The arctangent function

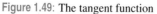

[6]Note that $\sin^{-1} x = \arcsin x$ is not the same as $(\sin x)^{-1} = 1/\sin x$.

Exercises and Problems for Section 1.5

Exercises

For Exercises 1–9, draw the angle using a ray through the origin, and determine whether the sine, cosine, and tangent of that angle are positive, negative, zero, or undefined.

1. $\frac{3\pi}{2}$ **2.** 2π **3.** $\frac{\pi}{4}$

4. 3π **5.** $\frac{\pi}{6}$ **6.** $\frac{4\pi}{3}$

7. $\frac{-4\pi}{3}$ **8.** 4 **9.** -1

Given that $\sin(\pi/12) = 0.259$ and $\cos(\pi/5) = 0.809$, compute (without using the trigonometric functions on your calculator) the quantities in Exercises 10–12. You may want to draw a picture showing the angles involved, and then check your answers on a calculator.

10. $\cos\left(-\frac{\pi}{5}\right)$ **11.** $\sin\frac{\pi}{5}$ **12.** $\cos\frac{\pi}{12}$

13. Consider the function $y = 5 + \cos(3x)$.

    (a) What is its amplitude?
    (b) What is its period?
    (c) Sketch its graph.

Find the period and amplitude in Exercises 14–17.

14. $y = 7\sin(3t)$ **15.** $z = 3\cos(u/4) + 5$

16. $w = 8 - 4\sin(2x + \pi)$ **17.** $r = 0.1\sin(\pi t) + 2$

18. Without a calculator or computer, match the formulas with the graphs in Figure 1.51.

    (a) $y = 2\cos(t - \pi/2)$ **(b)** $y = 2\cos t$
    (c) $y = 2\cos(t + \pi/2)$

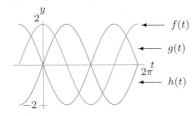

Figure 1.51

For Exercises 19–28, find a possible formula for each graph.

19.

20.

21.

22.

23.

24.

25.

26.

27.

28.
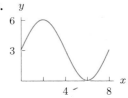

In Exercises 29–33, find a solution to the equation if possible. Give the answer in exact form and in decimal form.

29. $2 = 5\sin(3x)$ **30.** $1 = 8\cos(2x + 1) - 3$

31. $8 = 4\tan(5x)$ **32.** $1 = 8\tan(2x + 1) - 3$

33. $8 = 4\sin(5x)$

Problems

34. A compact disk spins at a rate of 200 to 500 revolutions per minute. What are the equivalent rates measured in radians per second?

35. When a car's engine makes less than about 200 revolutions per minute, it stalls. What is the period of the rotation of the engine when it is about to stall?

36. What is the difference between $\sin x^2$, $\sin^2 x$, and $\sin(\sin x)$? Express each of the three as a composition. (Note: $\sin^2 x$ is another way of writing $(\sin x)^2$.)

37. On the graph of $y = \sin x$, points P and Q are at consecutive lowest and highest points. Find the slope of the line through P and Q.

38. The Bay of Fundy in Canada has the largest tides in the world. The difference between low and high water levels is 15 meters (nearly 50 feet). At a particular point the depth of the water, y meters, is given as a function of time, t, in hours since midnight by

$$y = D + A \cos\left(B(t - C)\right).$$

(a) What is the physical meaning of D?
(b) What is the value of A?
(c) What is the value of B? Assume the time between successive high tides is 12.4 hours.
(d) What is the physical meaning of C?

39. A mass is oscillating on the end of a spring. The distance, y, of the mass from its equilibrium position is given by

$$y = y_0 \cos(2\pi\omega t).$$

Here y is in centimeters, t is time in seconds, and y_0 and ω are positive constants.

(a) What is the meaning of y_0 in terms of oscillations?
(b) How many oscillations are completed in 1 second?

40. In an electrical outlet, the voltage, V, in volts, is given as a function of time, t, in seconds, by the formula

$$V = V_0 \sin(120\pi t).$$

(a) What does V_0 represent in terms of voltage?
(b) What is the period of this function?
(c) How many oscillations are completed in 1 second?

41. In a US household, the voltage in volts in an electric outlet is given by

$$V = 156 \sin(120\pi t),$$

where t is in seconds. However, in a European house, the voltage is given (in the same units) by

$$V = 339 \sin(100\pi t).$$

Compare the voltages in the two regions, considering the maximum voltage and number of cycles (oscillations) per second.

42. A baseball hit at an angle of θ to the horizontal with initial velocity v_0 has horizontal range, R, given by

$$R = \frac{v_0^2}{g} \sin(2\theta).$$

Here g is the acceleration due to gravity. Sketch R as a function of θ for $0 \leq \theta \leq \pi/2$. What angle gives the maximum range? What is the maximum range?

43. A population of animals oscillates sinusoidally between a low of 700 on January 1 and a high of 900 on July 1.

(a) Graph the population against time.
(b) Find a formula for the population as a function of time, t, in months since the start of the year.

44. The desert temperature, H, oscillates daily between $40°$ F at 5 am and $80°$ F at 5 pm. Write a possible formula for H in terms of t, measured in hours from 5 am.

45. The point P is rotating around a circle of radius 5 shown in Figure 1.52. The angle θ, in radians, is given as a function of time, t, by the graph in Figure 1.53.

(a) Estimate the coordinates of P when $t = 1.5$.
(b) Describe in words the motion of the point P on the circle.

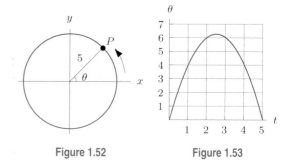

Figure 1.52   **Figure 1.53**

46. Find the area of the trapezoidal cross-section of the irrigation canal shown in Figure 1.54.

Figure 1.54

47. This problem introduces the arccosine function, or inverse cosine, denoted by $\boxed{\cos^{-1}}$ on most calculators.

(a) Using a calculator set in radians, make a table of values, to two decimal places, of $g(x) = \arccos x$, for $x = -1, -0.8, -0.6, \ldots, 0, \ldots, 0.6, 0.8, 1$.
(b) Sketch the graph of $g(x) = \arccos x$.
(c) Why is the domain of the arccosine the same as the domain of the arcsine?
(d) Why is the range of the arccosine *not* the same as the range of the arcsine? To answer this, look at how the domain of the original sine function was restricted to construct the arcsine. Why can't the domain of the cosine be restricted in exactly the same way to construct the arccosine?

1.6 POWERS, POLYNOMIALS, AND RATIONAL FUNCTIONS

Power Functions

A *power function* is a function in which the dependent variable is proportional to a power of the independent variable:

> A **power function** has the form
>
> $$f(x) = kx^p, \qquad \text{where } k \text{ and } p \text{ are constant.}$$

For example, the volume, V, of a sphere of radius r is given by

$$V = g(r) = \frac{4}{3}\pi r^3.$$

As another example, the gravitational force, F, on a unit mass at a distance r from the center of the earth is given by Newton's Law of Gravitation, which says that, for some positive constant k,

$$F = \frac{k}{r^2} \qquad \text{or} \qquad F = kr^{-2}.$$

We consider the graphs of the power functions x^n, with n a positive integer. Figures 1.55 and 1.56 show that the graphs fall into two groups: odd and even powers. For n greater than 1, the odd powers have a "seat" at the origin and are increasing everywhere else. The even powers are first decreasing and then increasing. For large x, the higher the power of x, the faster the function climbs.

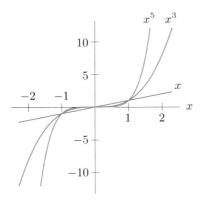

Figure 1.55: Odd powers of x: "Seat" shaped for $n > 1$

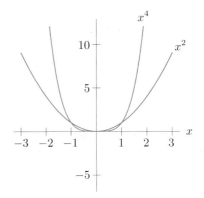

Figure 1.56: Even powers of x: $\bigcup$-shaped

Exponentials and Power Functions: Which Dominate?

In everyday language, the word exponential is often used to imply very fast growth. But do exponential functions always grow faster than power functions? To determine what happens "in the long run," we often want to know which functions *dominate* as x gets arbitrarily large.

Let's consider $y = 2^x$ and $y = x^3$. The close-up view in Figure 1.57(a) shows that between $x = 2$ and $x = 4$, the graph of $y = 2^x$ lies below the graph of $y = x^3$. The far-away view in Figure 1.57(b) shows that the exponential function $y = 2^x$ eventually overtakes $y = x^3$. Figure 1.57(c), which gives a very far–away view, shows that, for large x, the value of x^3 is insignificant compared to 2^x. Indeed, 2^x is growing so much faster than x^3 that the graph of 2^x appears almost vertical in comparison to the more leisurely climb of x^3.

We say that Figure 1.57(a) gives a *local* view of the functions' behavior, whereas Figure 1.57(c) gives a *global* view.

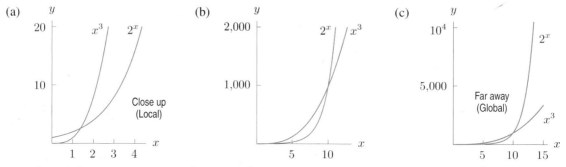

Figure 1.57: Comparison of $y = 2^x$ and $y = x^3$: Notice that $y = 2^x$ eventually dominates $y = x^3$

In fact, *every* exponential growth function eventually dominates *every* power function. Although an exponential function may be below a power function for some values of x, if we look at large enough x-values, a^x (with $a > 1$) will eventually dominate x^n, no matter what n is.

Polynomials

Polynomials are the sums of power functions with nonnegative integer exponents:

$$y = p(x) = a_n x^n + a_{n-1} x^{n-1} + \cdots + a_1 x + a_0.$$

Here n is a nonnegative integer called the *degree* of the polynomial, and $a_n, a_{n-1}, \ldots, a_1, a_0$ are constants, with leading coefficient $a_n \neq 0$. An example of a polynomial of degree $n = 3$ is

$$y = p(x) = 2x^3 - x^2 - 5x - 7.$$

In this case $a_3 = 2, a_2 = -1, a_1 = -5$, and $a_0 = -7$. The shape of the graph of a polynomial depends on its degree; typical graphs are shown in Figure 1.58. These graphs correspond to a positive coefficient for x^n; a negative leading coefficient turns the graph upside down. Notice that the quadratic "turns around" once, the cubic "turns around" twice, and the quartic (fourth degree) "turns around" three times. An nth degree polynomial "turns around" at most $n - 1$ times (where n is a positive integer), but there may be fewer turns.

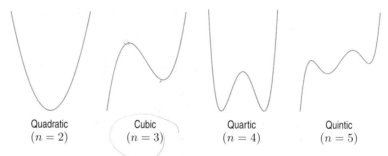

| Quadratic | Cubic | Quartic | Quintic |
| $(n = 2)$ | $(n = 3)$ | $(n = 4)$ | $(n = 5)$ |

Figure 1.58: Graphs of typical polynomials of degree n

Example 1   Find possible formulas for the polynomials whose graphs are in Figure 1.59.

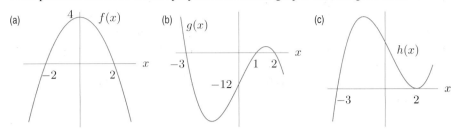

Figure 1.59: Graphs of polynomials

Solution

(a) This graph appears to be a parabola, turned upside down, and moved up by 4, so

$$f(x) = -x^2 + 4.$$

The negative sign turns the parabola upside down and the $+4$ moves it up by 4. Notice that this formula does give the correct x-intercepts since $0 = -x^2 + 4$ has solutions $x = \pm 2$. These values of x are called *zeros* of f.

We can also solve this problem by looking at the x-intercepts first, which tell us that $f(x)$ has factors of $(x + 2)$ and $(x - 2)$. So

$$f(x) = k(x + 2)(x - 2).$$

To find k, use the fact that the graph has a y-intercept of 4, so $f(0) = 4$, giving

$$4 = k(0 + 2)(0 - 2),$$

or $k = -1$. Therefore, $f(x) = -(x + 2)(x - 2)$, which multiplies out to $-x^2 + 4$.

Note that $f(x) = 4 - x^4/4$ also has the same basic shape, but is flatter near $x = 0$. There are many possible answers to these questions.

(b) This looks like a cubic with factors $(x + 3)$, $(x - 1)$, and $(x - 2)$, one for each intercept:

$$g(x) = k(x + 3)(x - 1)(x - 2).$$

Since the y-intercept is -12, we have

$$-12 = k(0 + 3)(0 - 1)(0 - 2).$$

So $k = -2$, and we get the cubic polynomial

$$g(x) = -2(x + 3)(x - 1)(x - 2).$$

(c) This also looks like a cubic with zeros at $x = 2$ and $x = -3$. Notice that at $x = 2$ the graph of $h(x)$ touches the x-axis but does not cross it, whereas at $x = -3$ the graph crosses the x-axis. We say that $x = 2$ is a *double zero*, but that $x = -3$ is a single zero.

To find a formula for $h(x)$, imagine the graph of $h(x)$ to be slightly lower down, so that the graph has one x-intercept near $x = -3$ and two near $x = 2$, say at $x = 1.9$ and $x = 2.1$. Then a formula would be

$$h(x) \approx k(x + 3)(x - 1.9)(x - 2.1).$$

Now move the graph back to its original position. The zeros at $x = 1.9$ and $x = 2.1$ move toward $x = 2$, giving

$$h(x) = k(x + 3)(x - 2)(x - 2) = k(x + 3)(x - 2)^2.$$

The double zero leads to a repeated factor, $(x - 2)^2$. Notice that when $x > 2$, the factor $(x - 2)^2$ is positive, and when $x < 2$, the factor $(x - 2)^2$ is still positive. This reflects the fact that $h(x)$ does not change sign near $x = 2$. Compare this with the behavior near the single zero at $x = -3$, where h does change sign.

We cannot find k, as no coordinates are given for points off of the x-axis. Any positive value of k stretches the graph vertically but does not change the zeros, so any positive k works.

Example 2   Using a calculator or computer, graph $y = x^4$ and $y = x^4 - 15x^2 - 15x$ for $-4 \leq x \leq 4$ and for $-20 \leq x \leq 20$. Set the y range to $-100 \leq y \leq 100$ for the first domain, and to $-100 \leq y \leq 200{,}000$ for the second. What do you observe?

Solution   From the graphs in Figure 1.60 we see that close up $(-4 \leq x \leq 4)$ the graphs look different; from far away, however, they are almost indistinguishable. The reason is that the leading terms (those with the highest power of x) are the same, namely x^4, and for large values of x, the leading term dominates the other terms.

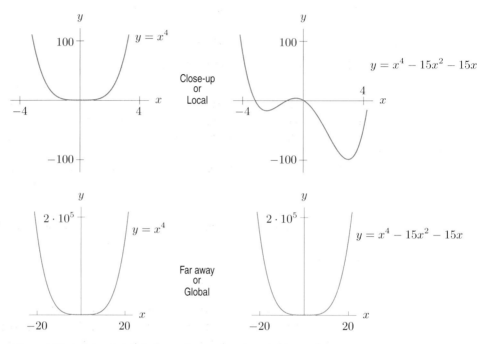

Figure 1.60: Local and global views of $y = x^4$ and $y = x^4 - 15x^2 - 15x$

Rational Functions

Rational functions are ratios of polynomials, p and q:

$$f(x) = \frac{p(x)}{q(x)}.$$

Example 3   Look at a graph and explain the behavior of $y = \dfrac{1}{x^2 + 4}$.

Solution   The function is even, so the graph is symmetric about the y-axis. As x gets larger, the denominator gets larger, making the value of the function closer to 0. Thus the graph gets arbitrarily close to the x-axis as x increases without bound. See Figure 1.61.

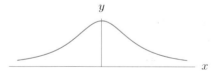

Figure 1.61: Graph of $y = \frac{1}{x^2+4}$

In the previous example, we say that $y = 0$ (i.e. the x-axis) is a *horizontal asymptote*. Writing "$\rightarrow$" to mean "tends to," we have $y \rightarrow 0$ as $x \rightarrow \infty$ and $y \rightarrow 0$ as $x \rightarrow -\infty$.

> If the graph of $y = f(x)$ approaches a horizontal line $y = L$ as $x \rightarrow \infty$ or $x \rightarrow -\infty$, then the line $y = L$ is called a **horizontal asymptote**.[7] This occurs when
>
> $$f(x) \rightarrow L \quad \text{as} \quad x \rightarrow \infty \qquad \text{or} \qquad f(x) \rightarrow L \quad \text{as} \quad x \rightarrow -\infty.$$
>
> If the graph of $y = f(x)$ approaches the vertical line $x = K$ as $x \rightarrow K$ from one side or the other, that is, if
>
> $$y \rightarrow \infty \quad \text{or} \quad y \rightarrow -\infty \quad \text{when} \quad x \rightarrow K,$$
>
> then the line $x = K$ is called a **vertical asymptote**.

The graphs of rational functions may have vertical asymptotes where the denominator is zero. For example, the function in Example 3 has no vertical asymptotes as the denominator is never zero. The function in Example 4 has two vertical asymptotes corresponding to the two zeros in the denominator.

Rational functions have horizontal asymptotes if $f(x)$ approaches a finite number as $x \rightarrow \infty$ or $x \rightarrow -\infty$. We call the behavior of a function as $x \rightarrow \pm\infty$ its *end behavior*.

Example 4   Look at a graph and explain the behavior of $y = \dfrac{3x^2 - 12}{x^2 - 1}$, including end behavior.

Solution   Factoring gives

$$y = \frac{3x^2 - 12}{x^2 - 1} = \frac{3(x+2)(x-2)}{(x+1)(x-1)}$$

so $x = \pm 1$ are vertical asymptotes. If $y = 0$, then $3(x+2)(x-2) = 0$ or $x = \pm 2$; these are the x-intercepts. Note that zeros of the denominator give rise to the vertical asymptotes, whereas zeros of the numerator give rise to x-intercepts. Substituting $x = 0$ gives $y = 12$; this is the y-intercept. The function is even, so the graph is symmetric about the y-axis.

To see what happens as $x \rightarrow \pm\infty$, look at the y-values in Table 1.15. Clearly y is getting closer to 3 as x gets large positively or negatively. Alternatively, realize that as $x \rightarrow \pm\infty$, only the highest powers of x matter. For large x, the 12 and the 1 are insignificant compared to x^2, so

$$y = \frac{3x^2 - 12}{x^2 - 1} \approx \frac{3x^2}{x^2} = 3 \quad \text{for large } x.$$

So $y \rightarrow 3$ as $x \rightarrow \pm\infty$, and therefore the horizontal asymptote is $y = 3$. See Figure 1.62. Since, for $x > 1$, the denominator is positive and the numerator is less than the denominator, the graph lies *below* its asymptote. (Why doesn't the graph lie below $y = 3$ when $-1 < x < 1$?)

Table 1.15   *Values of* $y = \frac{3x^2-12}{x^2-1}$

x	$y = \dfrac{3x^2-12}{x^2-1}$
± 10	2.909091
± 100	2.999100
± 1000	2.999991

Figure 1.62: Graph of the function $y = \frac{3x^2-12}{x^2-1}$

---

[7]We are assuming that $f(x)$ gets arbitrarily close to L as $x \rightarrow \infty$.

Exercises and Problems for Section 1.6

Exercises

In Exercises 1–2, which function dominates as $x \to \infty$?

1. $10 \cdot 2^x$ or $72{,}000x^{12}$ **2.** $0.25\sqrt{x}$ or $25{,}000x^{-3}$

For Problems 3–4, what happens to the value of the function as $x \to \infty$ and as $x \to -\infty$?

3. $y = 0.25x^3 + 3$ **4.** $y = 2 \cdot 10^{4x}$

5. Each of the graphs in Figure 1.63 is of a polynomial. The windows are large enough to show global behavior.

    (a) What is the minimum possible degree of the polynomial?

    (b) Is the leading coefficient of the polynomial positive or negative?

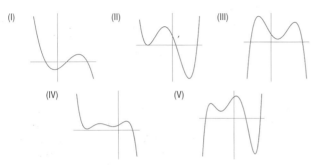

(I)   (II)   (III)

(IV)   (V)

Figure 1.63

6. For each function, fill in the blanks in the statements:

$f(x) \to$ _____ as $x \to -\infty$,

$f(x) \to$ _____ as $x \to +\infty$.

(a) $f(x) = 17 + 5x^2 - 12x^3 - 5x^4$

(b) $f(x) = \dfrac{3x^2 - 5x + 2}{2x^2 - 8}$

(c) $f(x) = e^x$

Find possible formulas for the graphs in Exercises 7–10.

7.   **8.**

9.   **10.**

11. Use Figure 1.64 to graph each of the following. Label any intercepts or asymptotes that can be determined.

    (a) $y = f(x) + 3$ **(b)** $y = 2f(x)$

    (c) $y = f(x + 4)$ **(d)** $y = 4 - f(x)$

Figure 1.64

Problems

Find cubic polynomials for the graphs in Exercises 12–13.

12.   **13.**

14. Which of the functions I–III meet each of the following descriptions? There may be more than one function for each description, or none at all.

    (a) Horizontal asymptote of $y = 1$.

    (b) The x-axis is a horizontal asymptote.

    (c) Symmetric about the y-axis.

    (d) An odd function.

    (e) Vertical asymptotes at $x = \pm 1$.

I. $y = \dfrac{x - 1}{x^2 + 1}$ II. $y = \dfrac{x^2 - 1}{x^2 + 1}$ III. $y = \dfrac{x^2 + 1}{x^2 - 1}$

15. The DuBois formula relates a person's surface area, s, in m^2, to weight w, in kg, and height h, in cm, by

$$s = 0.01 w^{0.25} h^{0.75}.$$

    (a) What is the surface area of a person who weighs 65 kg and is 160 cm tall?

    (b) What is the weight of a person whose height is 180 cm and who has a surface area of $1.5\ m^2$?

    (c) For people of fixed weight 70 kg, solve for h as a function of s. Simplify your answer.

16. A box of fixed volume V has a square base with side length x. Write a formula for the height, h, of the box in terms of x and V. Sketch a graph of h versus x.

17. According to *Car and Driver*, an Alfa Romeo going at 70 mph requires 177 feet to stop. Assuming that the stopping distance is proportional to the square of velocity, find the stopping distances required by an Alfa Romeo going at 35 mph and at 140 mph (its top speed).

18. Water is flowing down a cylindrical pipe of radius r.

    (a) Write a formula for the volume, V, of water that emerges from the end of the pipe in one second if the water is flowing at a rate of

        (i)   3 cm/sec   (ii) k cm/sec

    (b) Graph your answer to part (a)(ii) as a function of

        (i) r, assuming k is constant

        (ii) k, assuming r is constant

19. Poiseuille's Law gives the rate of flow, R, of a gas through a cylindrical pipe in terms of the radius of the pipe, r, for a fixed drop in pressure between the two ends of the pipe.

    (a) Find a formula for Poiseuille's Law, given that the rate of flow is proportional to the fourth power of the radius.

    (b) If $R = 400$ cm^3/sec in a pipe of radius 3 cm for a certain gas, find a formula for the rate of flow of that gas through a pipe of radius r cm.

    (c) What is the rate of flow of the same gas through a pipe with a 5 cm radius?

20. The height of an object above the ground at time t is given by

$$s = v_0 t - \frac{g}{2}t^2,$$

where v_0 is the initial velocity and g is the acceleration due to gravity.

    (a) At what height is the object initially?

    (b) How long is the object in the air before it hits the ground?

    (c) When will the object reach its maximum height?

    (d) What is that maximum height?

21. A pomegranate is thrown from ground level straight up into the air at time $t = 0$ with velocity 64 feet per second. Its height at time t seconds is $f(t) = -16t^2 + 64t$. Find the time it hits the ground and the time it reaches its highest point. What is the maximum height?

22. **(a)** If $f(x) = ax^2 + bx + c$, what can you say about the values of a, b, and c if:

        (i) $(1, 1)$ is on the graph of $f(x)$?

        (ii) $(1, 1)$ is the vertex of the graph of $f(x)$? [Hint: The axis of symmetry is $x = -b/(2a)$.]

        (iii)   The y intercept of the graph is $(0, 6)$?

    (b) Find a quadratic function satisfying all three conditions.

23. Values of three functions are given in Table 1.16, rounded to two decimal places. One function is of the form $y = ab^t$, one is of the form $y = ct^2$, and one is of the form $y = kt^3$. Which function is which?

Table 1.16

t	$f(t)$	t	$g(t)$	t	$h(t)$
2.0	4.40	1.0	3.00	0.0	2.04
2.2	5.32	1.2	5.18	1.0	3.06
2.4	6.34	1.4	8.23	2.0	4.59
2.6	7.44	1.6	12.29	3.0	6.89
2.8	8.62	1.8	17.50	4.0	10.33
3.0	9.90	2.0	24.00	5.0	15.49

24. Values of three functions are given in Table 1.17, rounded to two decimal places. Two are power functions and one is an exponential. One of the power functions is a quadratic and one a cubic. Which one is exponential? Which one is quadratic? Which one is cubic?

Table 1.17

x	$f(x)$	x	$g(x)$	x	$k(x)$
8.4	5.93	5.0	3.12	0.6	3.24
9.0	7.29	5.5	3.74	1.0	9.01
9.6	8.85	6.0	4.49	1.4	17.66
10.2	10.61	6.5	5.39	1.8	29.19
10.8	12.60	7.0	6.47	2.2	43.61
11.4	14.82	7.5	7.76	2.6	60.91

25. A cubic polynomial with positive leading coefficient is shown in Figure 1.65 for $-10 \le x \le 10$ and $-10 \le y \le 10$. What can be concluded about the total number of zeros of this function? What can you say about the location of each of the zeros? Explain.

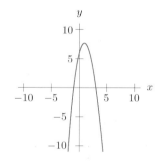

Figure 1.65

26. How many distinct roots can a polynomial of degree 5 have? (List all possibilities.) Sketch a possible graph for each case.

27. Suppose $f(x) = (x + a_1)^2$ and $g(x) = (x + a_2)^2$. Suppose $c > 0$ and $a_1 = -2a_2 = 2c$.

    (a) Solve $f(x) = g(x)$. Your answer should involve c but no other constants.

    (b) Sketch $f(x)$ and $g(x)$ on the same axes, labeling intercepts.

    (c) As c increases, what happens to the x- and y-coordinates of the point of intersection?

28. A rational function $y = f(x)$ is graphed in Figure 1.66. If $f(x) = g(x)/h(x)$ with $g(x)$ and $h(x)$ both quadratic functions, give possible formulas for $g(x)$ and $h(x)$.

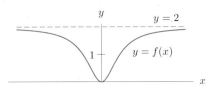

Figure 1.66

29. Match the following functions with the graphs in Figure 1.67. Assume $0 < b < a$.

(a) $y = \dfrac{a}{x} - x$

(b) $y = \dfrac{(x-a)(x+a)}{x}$

(c) $y = \dfrac{(x-a)(x^2+a)}{x^2}$

(d) $y = \dfrac{(x-a)(x+a)}{(x-b)(x+b)}$

Figure 1.67

30. Find a calculator window in which the graphs of $f(x) = x^3 + 1000x^2 + 1000$ and $g(x) = x^3 - 1000x^2 - 1000$ appear indistinguishable.

31. Use a graphing calculator or a computer to graph $y = x^4$ and $y = 3^x$. Determine approximate domains and ranges that give each of the graphs in Figure 1.68.

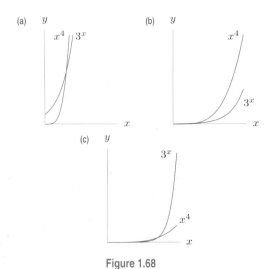

Figure 1.68

32. Newton's Second Law of Motion, $F = ma$, tells us that the net force, F, on a train of mass m is proportional to its acceleration, a. Suppose that the only forces are those of the engine, which exerts a constant force, F_E, in the direction of motion, and the wind resistance, which exerts a force proportional to the square of the train's velocity, v, but in the opposite direction.

(a) Write a formula giving a as a function of v.
(b) Sketch a graph of a against v.

1.7 INTRODUCTION TO CONTINUITY

This section introduces the idea of *continuity* on an interval and at a point. This leads to the concept of limit, which is investigated in the next section.

Continuity of a Function on an Interval: Graphical Viewpoint

Roughly speaking, a function is said to be *continuous* on an interval if its graph has no breaks, jumps, or holes in that interval. Continuity is important because, as we shall see, continuous functions have many desirable properties.

For example, to locate the zeros of a function, we often look for intervals where the function changes sign. In the case of the function $f(x) = 3x^3 - x^2 + 2x - 1$, for instance, we expect[8] to find a zero between 0 and 1 because $f(0) = -1$ and $f(1) = 3$. (See Figure 1.69.) To be sure that

---

[8]This is due to the Intermediate Value Theorem, which is discussed on page 46.

$f(x)$ has a zero there, we need to know that the graph of the function has no breaks or jumps in it. Otherwise the graph could jump across the x-axis, changing sign but not creating a zero. For example, $f(x) = 1/x$ has opposite signs at $x = -1$ and $x = 1$, but no zeros for $-1 \le x \le 1$ because of the break at $x = 0$. (See Figure 1.70.) To be certain that a function has a zero in an interval on which it changes sign, we need to know that the function is defined and continuous in that interval.

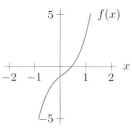

Figure 1.69: The graph of $f(x) = 3x^3 - x^2 + 2x - 1$

Figure 1.70: No zero although $f(-1)$ and $f(1)$ have opposite signs

Figure 1.71: Cost of mailing a letter

A continuous function has a graph which can be drawn without lifting the pencil from the paper.

Example: The function $f(x) = 3x^3 - x^2 + 2x - 1$ is continuous on any interval. (See Figure 1.69.)

Example: The function $f(x) = 1/x$ is not defined at $x = 0$. It is continuous on any interval not containing the origin. (See Figure 1.70.)

Example: Suppose $p(x)$ is the price of mailing a first-class letter weighing x ounces. It costs 34¢ for one ounce or less, 55¢ between the first and second ounces, and so on. So the graph (in Figure 1.71) is a series of steps. This function is not continuous on any open interval containing a positive integer because the graph jumps at these points.

Which Functions are Continuous?

Requiring a function to be continuous on an interval is not asking very much, as any function whose graph is an unbroken curve over the interval is continuous. For example, exponential functions, polynomials, and the sine and cosine are continuous on every interval. Rational functions are continuous on any interval in which their denominators are not zero. Functions created by adding, multiplying, or composing continuous functions are also continuous.

The Intermediate Value Theorem

Continuity tells us about the values taken by a function. In particular, a continuous function cannot skip values. For example, the function in the next example must have a zero because its graph cannot skip over the x-axis.

Example 1   What do the values in Table 1.18 tell you about the zeros of $f(x) = \cos x - 2x^2$?

Table 1.18

x	$f(x)$
0	1.00
0.2	0.90
0.4	0.60
0.6	0.11
0.8	-0.58
1.0	-1.46

Figure 1.72: Zeros occur where the graph of a continuous function crosses the horizontal axis

Solution   Since $f(x)$ is the difference of two continuous functions, it is continuous. We conclude that $f(x)$ has at least one zero in the interval $0.6 < x < 0.8$, since $f(x)$ changes from positive to negative on that interval. The graph of $f(x)$ in Figure 1.72 suggests that there is only one zero in the interval $0 \le x \le 1$, but we cannot be sure of this from the graph or the table of values.

In the previous example, we concluded that $f(x) = \cos x - 2x^2$ has a zero between $x = 0$ and $x = 1$ because $f(x)$ is positive at $x = 0$ and negative at $x = 1$. More generally, an intuitive notion of continuity tells us that, as we follow the graph of a continuous function f from some point $(a, f(a))$ to another point $(b, f(b))$, then f takes on all intermediate values between $f(a)$ and $f(b)$. (See Figure 1.73.) This is:

Theorem 1.1: Intermediate Value Theorem

Suppose f is continuous on a closed interval $[a, b]$. If k is any number between $f(a)$ and $f(b)$, then there is at least one number c in $[a, b]$ such that $f(c) = k$.

A proof of this result needs the formal definition of continuity in Section 2.6 and is given at www.wiley.com/college/hugheshallett. The key idea is to find successively smaller subintervals of $[a, b]$ on which f changes from less than k to more than k. These subintervals converge on the number c.

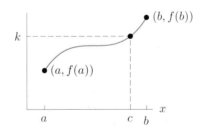

Figure 1.73: The Intermediate Value Theorem

Continuity of a Function at a Point: Numerical Viewpoint

A function is continuous if nearby values of the independent variable give nearby values of the function. In practical work, continuity is important because it means that small errors in the independent variable lead to small errors in the value of the function.

Example: Suppose that $f(x) = x^2$ and that we want to compute $f(\pi)$. Knowing f is continuous tells us that taking $x = 3.14$ should give a good approximation to $f(\pi)$, and that we can get as accurate an approximation to $f(\pi)$ as we want by using enough decimals of π.

Example: If $p(x)$ is the cost of mailing a letter weighing x ounces, then $p(0.99) = p(1) = 37¢$, whereas $p(1.01) = 60¢$, because as soon as we get over 1 ounce, the price jumps up to 60¢. So a small difference in the weight of a letter can lead to a significant difference in its mailing cost. Hence p is not continuous at $x = 1$.

In other words, if $f(x)$ is continuous at $x = c$, the values of $f(x)$ approach $f(c)$ as x approaches c. In Section 1.8, we discuss the concept of a *limit*, which allows us to define more precisely what it means for the values of $f(x)$ to approach $f(c)$ as x approaches c.

Example 2   Investigate the continuity of $f(x) = x^2$ at $x = 2$.

Solution   From Table 1.19, it appears that the values of $f(x) = x^2$ approach $f(2) = 4$ as x approaches 2. Thus f appears to be continuous at $x = 2$. Continuity at a point describes behavior of a function *near* a point, as well as *at* the point.

Table 1.19   *Values of x^2 near $x = 2$*

x	1.9	1.99	1.999	2.001	2.01	2.1
x^2	3.61	3.96	3.996	4.004	4.04	4.41

Exercises and Problems for Section 1.7

Exercises

Are the functions in Exercises 1–10 continuous on the given intervals?

1. $2x + x^{2/3}$ on $[-1, 1]$

2. $2x + x^{-1}$ on $[-1, 1]$

3. $\dfrac{1}{x - 2}$ on $[-1, 1]$

4. $\dfrac{1}{x - 2}$ on $[0, 3]$

5. $\dfrac{1}{\sqrt{2x - 5}}$ on $[3, 4]$

6. $\dfrac{x}{x^2 + 2}$ on $[-2, 2]$

7. $\dfrac{1}{\cos x}$ on $[0, \pi]$

8. $\dfrac{1}{\sin x}$ on $[-\frac{\pi}{2}, \frac{\pi}{2}]$

9. $\dfrac{e^x}{e^x - 1}$ on $[-1, 1]$

10. $\dfrac{e^{\sin \theta}}{\cos \theta}$ on $[-\frac{\pi}{4}, \frac{\pi}{4}]$

In Exercises 11–14, show that there is a number c, with $0 \le c \le 1$, such that $f(c) = 0$.

11. $f(x) = x^3 + x^2 - 1$

12. $f(x) = e^x - 3x$

13. $f(x) = x - \cos x$

14. $f(x) = 2^x - 1/x$

Problems

15. An electrical circuit switches instantaneously from a 6 volt battery to a 12 volt battery 7 seconds after being turned on. Graph the battery voltage against time. Give formulas for the function represented by your graph. What can you say about the continuity of this function?

16. A car is coasting down a hill at a constant speed. A truck collides with the rear of the car, causing it to lurch ahead. Graph the car's speed from a time shortly before impact to a time shortly after impact. Graph the distance from the top of the hill for this time period. What can you say about the continuity of each of these functions?

17. Find k so that the following function is continuous on any interval:

$$f(x) = \begin{cases} kx & 0 \le x < 2 \\ 3x^2 & 2 \le x. \end{cases}$$

18. Is the following function continuous on $[-1, 1]$?

$$f(x) = \begin{cases} \dfrac{x}{|x|} & x \ne 0 \\ 0 & x = 0 \end{cases}$$

19. If possible, choose k so that the following function is continuous on any interval:

$$f(x) = \begin{cases} \dfrac{5x^3 - 10x^2}{x - 2} & x \ne 2 \\ k & x = 2. \end{cases}$$

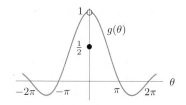

20. Discuss the continuity of the function g graphed in Figure 1.74 and defined as follows:

$$g(\theta) = \begin{cases} \dfrac{\sin \theta}{\theta} & \text{for } \theta \ne 0 \\ 1/2 & \text{for } \theta = 0. \end{cases}$$

Figure 1.74

21. Sketch the graphs of three different functions that are continuous on $0 \leq x \leq 1$ and that have the values given in the table. The first function is to have exactly one zero in $[0, 1]$, the second is to have at least two zeros in the interval $[0.6, 0.8]$, and the third is to have at least two zeros in the interval $[0, 0.6]$.

x	0	0.2	0.4	0.6	0.8	1.0
$f(x)$	1.00	0.90	0.60	0.11	−0.58	−1.46

22. (a) Sketch the graph of a continuous function f with *all* of the following properties:
     (i) $f(0) = 2$
     (ii) $f(x)$ is decreasing for $0 \leq x \leq 3$
     (iii) $f(x)$ is increasing for $3 < x \leq 5$
     (iv) $f(x)$ is decreasing for $x > 5$
     (v) $f(x) \to 9$ as $x \to \infty$
   (b) Is it possible that the graph of f is concave down for all $x > 6$? Explain.

23. A 0.6 ml dose of a drug is injected into a patient steadily for half a second. At the end of this time, the quantity, Q, of the drug in the body starts to decay exponentially at a continuous rate of 0.2% per second. Using formulas, express Q as a continuous function of time, t in seconds.

24. Use a computer or calculator to sketch the functions

$$y(x) = \sin x \quad \text{and} \quad z_k(x) = ke^{-x}$$

for $k = 1, 2, 4, 6, 8, 10$. In each case find the smallest positive solution of the equation $y(x) = z_k(x)$. Now define a new function f by

$$f(k) = \{\text{Smallest positive solution of } y(x) = z_k(x)\}.$$

Explain why the function $f(k)$ is not continuous on the interval $0 \leq k \leq 10$.

1.8   LIMITS

The concept of *limit* is the underpinning of calculus. In Section 1.7, we said that a function f is continuous at $x = c$ if the values of $f(x)$ approach $f(c)$ as x approaches c. In this section, we define a limit, which makes precise what we mean by approaching.

The Idea of a Limit

We first introduce some notation:

> We write $\lim\limits_{x \to c} f(x) = L$ if the values of $f(x)$ approach L as x approaches c.

How should we find L, or even know whether such a number exists? We will look for trends in the values of $f(x)$ as x gets closer to c, but $x \neq c$. A graph from a calculator or computer often helps.

Example 1   Use a graph to estimate $\lim\limits_{\theta \to 0} \left(\dfrac{\sin \theta}{\theta} \right)$. (Use radians.)

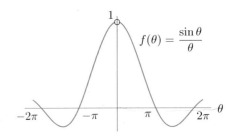

Figure 1.75: Find the limit as $\theta \to 0$

Solution   Figure 1.75 shows that as θ approaches 0 from either side, the value of $\sin\theta/\theta$ appears to approach 1, suggesting that $\lim_{\theta \to 0}(\sin\theta/\theta) = 1$. Zooming in on the graph near $\theta = 0$ provides further support for this conclusion. Notice that $\sin\theta/\theta$ is undefined at $\theta = 0$.

Figure 1.75 strongly suggests that $\lim_{\theta \to 0}(\sin\theta/\theta) = 1$, but to be sure we need to be more precise about words like "approach" and "close".

Definition of Limit

By the beginning of the 19th century, calculus had proved its worth, and there was no doubt about the correctness of its answers. However, it was not until the work of the French mathematician Augustin Cauchy (1789–1857) that calculus was put on a rigorous footing. Cauchy gave a formal definition of the limit, similar to the following:

> A function f is defined on an interval around c, except perhaps at the point $x = c$. We define the **limit** of the function $f(x)$ as x approaches c, written $\lim_{x \to c} f(x)$, to be a number L (if one exists) such that $f(x)$ is as close to L as we want whenever x is sufficiently close to c (but $x \neq c$). If L exists, we write
> $$\lim_{x \to c} f(x) = L.$$

Shortly, we see how "as close as we want" and "sufficiently close" are expressed using inequalities. First, we look at $\lim_{\theta \to 0}(\sin\theta/\theta)$ more closely (see Example 1).

Example 2   By graphing $y = (\sin\theta)/\theta$ in an appropriate window, find how close θ should be to 0 in order to make $(\sin\theta)/\theta$ within 0.01 of 1.

Solution   Since we want $(\sin\theta)/\theta$ to be within 0.01 of 1, we set the y-range on the graphing window to go from 0.99 to 1.01. Our first attempt with $-0.5 \leq \theta \leq 0.5$ yields the graph in Figure 1.76. Since we want the y-values to stay within the range $0.99 < y < 1.01$, we do not want the graph to leave the window through the top or bottom. By trial and error, we find that changing the θ-range to $-0.2 \leq \theta \leq 0.2$ gives the graph in Figure 1.77. Thus, the graph suggests that $(\sin\theta)/\theta$ is within 0.01 of 1 whenever θ is within 0.2 of 0. Proving this requires an analytical argument, not just graphs from a calculator.

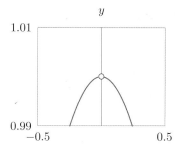

Figure 1.76: $(\sin\theta)/\theta$ with $-0.5 \leq \theta \leq 0.5$

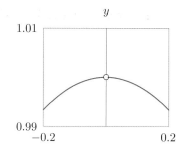

Figure 1.77: $(\sin\theta)/\theta$ with $-0.2 \leq \theta \leq 0.2$

When we say "$f(x)$ is close to L," we measure closeness by the distance between $f(x)$ and L, expressed using absolute values:

$$|f(x) - L| = \text{Distance between } f(x) \text{ and } L.$$

When we say "as close to L as we want," we use ϵ (the Greek letter epsilon) to specify how close. We write

$$|f(x) - L| < \epsilon$$

to indicate that we want the distance between $f(x)$ and L to be less than ϵ. In Example 2 we used $\epsilon = 0.01$. Similarly, we interpret "x is sufficiently close to c" as specifying a distance between x and c:

$$|x - c| < \delta,$$

where δ (the Greek letter delta) tells us how close x should be to c. In Example 2 we found $\delta = 0.2$.

If $\lim\limits_{x \to c} f(x) = L$, we know that no matter how narrow the horizontal band determined by ϵ in Figure 1.78, there is always a δ which makes the graph stay within that band, for $c - \delta < x < c + \delta$.

Thus we restate the definition of a limit, using symbols:

Definition of Limit

We define $\lim\limits_{x \to c} f(x)$ to be the number L (if one exists) such that for every $\epsilon > 0$ (as small as we want), there is a $\delta > 0$ (sufficiently small) such that if $|x - c| < \delta$ and $x \neq c$, then $|f(x) - L| < \epsilon$.

We have arrived at a formal definition of limit. Let's see if it agrees with our intuition.

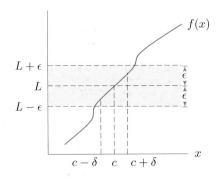

Figure 1.78: What the definition of the limit means graphically

---

Example 3   Use the definition of limit to show that $\lim\limits_{x \to 3} 2x = 6$.

Solution   We must show how, given any $\epsilon > 0$, we can find a $\delta > 0$ such that

$$\text{If } |x - 3| < \delta \text{ and } x \neq 3, \text{ then } |2x - 6| < \epsilon.$$

Since $|2x - 6| = 2|x - 3|$, to get $|2x - 6| < \epsilon$ we require that $|x - 3| < \epsilon/2$. Thus we take $\delta = \epsilon/2$.

It is important to understand that the ϵ, δ definition does not make it easier to calculate limits; rather the ϵ, δ definition makes it possible to put calculus on a rigorous foundation. From this foundation, we can prove the following properties. See Problems 49–51.

Theorem 1.2: Properties of Limits

Assuming all the limits on the right hand side exist:

1. If b is a constant, then $\lim\limits_{x \to c} (bf(x)) = b \left(\lim\limits_{x \to c} f(x) \right)$.

2. $\lim\limits_{x \to c} (f(x) + g(x)) = \lim\limits_{x \to c} f(x) + \lim\limits_{x \to c} g(x)$.

3. $\lim\limits_{x \to c} (f(x)g(x)) = \left(\lim\limits_{x \to c} f(x) \right) \left(\lim\limits_{x \to c} g(x) \right)$.

4. $\lim\limits_{x \to c} \dfrac{f(x)}{g(x)} = \dfrac{\lim_{x \to c} f(x)}{\lim_{x \to c} g(x)}$, provided $\lim\limits_{x \to c} g(x) \neq 0$.

5. For any constant k, $\lim\limits_{x \to c} k = k$.

6. $\lim\limits_{x \to c} x = c$.

These properties underlie many limit calculations, though we may not acknowledge them explicitly.

Example 4   Explain how the limit properties are used in the following calculation:

$$\lim_{x \to 3} \frac{x^2 + 5x}{x + 9} = \frac{3^2 + 5 \cdot 3}{3 + 9} = 2.$$

Solution   We calculate this limit in stages, using the limit properties to justify each step:

$$\lim_{x \to 3} \frac{x^2 + 5x}{x + 9} = \frac{\lim\limits_{x \to 3} (x^2 + 5x)}{\lim\limits_{x \to 3} (x + 9)} \qquad \text{(Property 4, since } \lim_{x \to 3}(x + 9) \neq 0\text{)}$$

$$= \frac{\lim\limits_{x \to 3} (x^2) + \lim\limits_{x \to 3} (5x)}{\lim\limits_{x \to 3} x + \lim\limits_{x \to 3} 9} \qquad \text{(Property 2)}$$

$$= \frac{\left(\lim\limits_{x \to 3} x \right)^2 + 5 \left(\lim\limits_{x \to 3} x \right)}{\lim\limits_{x \to 3} x + \lim\limits_{x \to 3} 9} \qquad \text{(Properties 1 and 3)}$$

$$= \frac{3^2 + 5 \cdot 3}{3 + 9} = 2. \qquad \text{(Properties 5 and 6)}$$

One- and Two-Sided Limits

When we write

$$\lim_{x \to 2} f(x),$$

we mean the number that $f(x)$ approaches as x approaches 2 *from both sides*. We examine values of $f(x)$ as x approaches 2 through values greater than 2 (such as 2.1, 2.01, 2.003) and values less than 2 (such as 1.9, 1.99, 1.994). If we want x to approach 2 only through values greater than 2, we write

$$\lim_{x \to 2^+} f(x)$$

for the number that $f(x)$ approaches (assuming such a number exists). Similarly,

$$\lim_{x \to 2^-} f(x)$$

denotes the number (if it exists) obtained by letting x approach 2 through values less than 2. We call $\lim_{x \to 2^+} f(x)$ a *right-hand limit* and $\lim_{x \to 2^-} f(x)$ a *left-hand limit*. Problems 24 and 25 ask for formal definitions of left and right-hand limits.

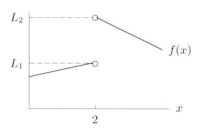

Figure 1.79: Left- and right-hand limits
at $x = 2$

For the function graphed in Figure 1.79, we have

$$\lim_{x \to 2^-} f(x) = L_1 \qquad \lim_{x \to 2^+} f(x) = L_2.$$

If the left- and right-hand limits were equal, that is, if $L_1 = L_2$, then it can be proved that $\lim_{x \to 2} f(x)$ exists and $\lim_{x \to 2} f(x) = L_1 = L_2$. Since $L_1 \neq L_2$ in Figure 1.79, we see that $\lim_{x \to 2} f(x)$ does not exist in this case.

When Limits Do Not Exist

Whenever there is no number L such that $\lim_{x \to c} f(x) = L$, we say $\lim_{x \to c} f(x)$ does not exist. Here are three examples in which limits fail to exist.

Example 5   Explain why $\lim_{x \to 2} \dfrac{|x - 2|}{x - 2}$ does not exist.

Solution   Figure 1.80 shows the problem: The right-hand limit and the left-hand limit are different. For $x > 2$, we have $|x - 2| = x - 2$, so as x approaches 2 from the right,

$$\lim_{x \to 2^+} \frac{|x - 2|}{x - 2} = \lim_{x \to 2^+} \frac{x - 2}{x - 2} = \lim_{x \to 2^+} 1 = 1.$$

Similarly, if $x < 2$, then $|x - 2| = 2 - x$ so

$$\lim_{x \to 2^-} \frac{|x - 2|}{x - 2} = \lim_{x \to 2^-} \frac{2 - x}{x - 2} = \lim_{x \to 2^-} (-1) = -1.$$

So if $\lim_{x \to 2} \dfrac{|x - 2|}{x - 2} = L$ then L would have to be both 1 and -1. Since L cannot have two different values, the limit does not exist.

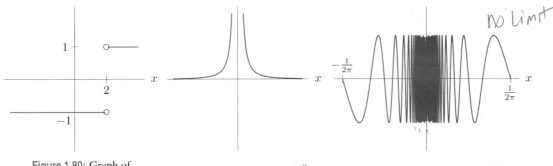

Figure 1.80: Graph of $|x-2|/(x-2)$

Figure 1.81: Graph of $1/x^2$

Figure 1.82: Graph of $\sin(1/x)$

Example 6   Explain why $\lim\limits_{x\to 0}\dfrac{1}{x^2}$ does not exist.

Solution   As x approaches zero, $1/x^2$ becomes arbitrarily large, so it cannot approach any finite number L. See Figure 1.81. Therefore we say $1/x^2$ has no limit as $x\to 0$.

If $\lim\limits_{x\to a} f(x)$ does not exist because $f(x)$ gets arbitrarily large on both sides of a, we also say $\lim\limits_{x\to a} f(x)=\infty$. So in Example 6 we could say $\lim\limits_{x\to 0} 1/x^2=\infty$. This behavior may also be described as "diverging to infinity."

Example 7   Explain why $\lim\limits_{x\to 0}\sin\left(\dfrac{1}{x}\right)$ does not exist.

Solution   The sine function has values between -1 and 1. The graph in Figure 1.82 oscillates more and more rapidly as $x\to 0$. There are x-values approaching 0 where $\sin(1/x)=-1$. There are also x-values approaching 0 where $\sin(1/x)=1$. So if the limit existed, it would have to be both -1 and 1. Thus, the limit does not exist.

Limits at Infinity

Sometimes we want to know what happens to $f(x)$ as x gets large, that is, the end behavior of f.

If $f(x)$ gets as close to a number L as we please when x gets sufficiently large, then we write

$$\lim_{x\to\infty} f(x)=L.$$

Similarly, if $f(x)$ approaches L when x is negative and has a sufficiently large absolute value, then we write

$$\lim_{x\to-\infty} f(x)=L.$$

The symbol ∞ does not represent a number. Writing $x\to\infty$ means that we consider arbitrarily large values of x. If the limit of $f(x)$ as $x\to\infty$ or $x\to-\infty$ is L, we say that the graph of f has $y=L$ as a *horizontal asymptote*. Problem 26 asks for a formal definition of $\lim_{x\to\infty} f(x)$.

Example 8 Investigate $\displaystyle\lim_{x\to\infty}\frac{1}{x}$ and $\displaystyle\lim_{x\to-\infty}\frac{1}{x}$.

Solution A graph of $f(x) = 1/x$ in a large window shows $1/x$ approaching zero as x increases in either the positive or the negative direction (See Figure 1.83). This is as we would expect, since dividing 1 by larger and larger numbers yields answers which are smaller and smaller. This suggests that

$$\lim_{x\to\infty}\frac{1}{x} = \lim_{x\to-\infty}\frac{1}{x} = 0,$$

and that $f(x) = 1/x$ has $y = 0$ as a horizontal asymptote as $x \to \pm\infty$.

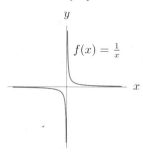

Figure 1.83: The end behavior of $f(x) = 1/x$

Definition of Continuity

We can now give a precise definition of continuity using limits.

> The function f is **continuous** at $x = c$ if f is defined at $x = c$ and if
>
> $$\lim_{x\to c} f(x) = f(c).$$
>
> In other words, $f(x)$ is as close as we want to $f(c)$ provided x is close enough to c. The function is **continuous on an interval** $[a, b]$ if it is continuous at every point in the interval.[9]

Constant functions and $f(x) = x$ are continuous for all x. Using the continuity of sums and products, we can show that any polynomial is continuous. Proving that $\sin x$, $\cos x$, and e^x are continuous is more difficult. The following theorem, based on the properties of limits on page 51, makes it easier to decide whether certain combinations of functions are continuous.

> ## Theorem 1.3: Continuity of Sums, Products, and Quotients of Functions
>
> Suppose that f and g are continuous on an interval and that b is a constant. Then, on that same interval,
> 1. $bf(x)$ is continuous
> 2. $f(x) + g(x)$ is continuous
> 3. $f(x)g(x)$ is continuous
> 4. $f(x)/g(x)$ is continuous, provided $g(x) \neq 0$ on the interval.

We prove the third of these properties.

---

[9]If c is an endpoint of the interval, we define continuity at $x = c$ using one-sided limits at c.

Proof Let c be any point in the interval. We must show that $\lim_{x \to c}(f(x)g(x)) = f(c)g(c)$. Since $f(x)$ and $g(x)$ are continuous, we know that $\lim_{x \to c} f(x) = f(c)$ and $\lim_{x \to c} g(x) = g(c)$. So, by the first property of limits in Theorem 1.2,

$$\lim_{x \to c}(f(x)g(x)) = \left(\lim_{x \to c} f(x)\right)\left(\lim_{x \to c} g(x)\right) = f(c)g(c).$$

Since c was chosen arbitrarily, we have shown that $f(x)g(x)$ is continuous at every point in the interval.

Theorem 1.4: Continuity of Composite Functions

If f and g are continuous, and if the composite function $f(g(x))$ is defined on an interval, then $f(g(x))$ is continuous on that interval.

Assuming the continuity of $\sin x$ and e^x, Theorem 1.4 shows us, for example, that $\sin(e^x)$ and $e^{\sin x}$ are both continuous.

Although we now have a formal definition of continuity, some properties of continuous functions, such as the Intermediate Value Theorem, can be difficult to prove. See www.wiley.com/college/hugheshallett.

Exercises and Problems for Section 1.8

Exercises

1. Use Figure 1.84 to give approximate values for the following limits (if they exist).

  (a) $\lim_{x \to -2} f(x)$ **(b)** $\lim_{x \to 0} f(x)$

  (c) $\lim_{x \to 2} f(x)$ **(d)** $\lim_{x \to 4} f(x)$

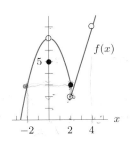

Figure 1.84

2. Using Figures 1.85 and 1.86, estimate

  (a) $\lim_{x \to 1^-} (f(x) + g(x))$ **(b)** $\lim_{x \to 1^+} (f(x) + 2g(x))$

  (c) $\lim_{x \to 1^-} f(x)g(x)$ **(d)** $\lim_{x \to 1^+} \dfrac{f(x)}{g(x)}$

Figure 1.85

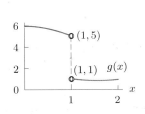

Figure 1.86

Use a graph to estimate each of the limits in Exercises 3–6. Use radians unless degrees are indicated by $\theta°$.

3. $\lim_{\theta \to 0} \dfrac{\sin(2\theta)}{\theta}$ **4.** $\lim_{\theta \to 0} \dfrac{\cos\theta - 1}{\theta}$

5. $\lim_{\theta \to 0} \dfrac{\sin\theta°}{\theta°}$ **6.** $\lim_{\theta \to 0} \dfrac{\theta}{\tan(3\theta)}$

For the functions in Exercises 7–14, do the following:

  (a) Make a table of values of $f(x)$ for $x = 0.1, 0.01, 0.001, 0.0001, -0.1, -0.01, -0.001$, and -0.0001.

  (b) Make a conjecture about the value of $\lim_{x \to 0} f(x)$.

  (c) Graph the function to see if it is consistent with your answers to parts (a) and (b).

  (d) Find an interval for x near 0 such that the difference between your conjectured limit and the value of the function is less than 0.01. (In other words, find a window of height 0.02 such that the graph exits the sides of the window and not the top or bottom of the window.)

7. $f(x) = 3x + 1$ **8.** $f(x) = x^2 - 1$

9. $f(x) = \sin 2x$ **10.** $f(x) = \sin 3x$

11. $f(x) = \dfrac{\sin 2x}{x}$ **12.** $f(x) = \dfrac{\sin 3x}{x}$

13. $f(x) = \dfrac{e^x - 1}{x}$ **14.** $f(x) = \dfrac{e^{2x} - 1}{x}$

For the functions in Exercises 15–18, do the following:

(a) Make a table of values of $f(x)$ for $x = a+0.1, a+0.01$, $a + 0.001$, $a + 0.0001$, $a - 0.1$, $a - 0.01$, $a - 0.001$, and $a - 0.0001$.

(b) Make a conjecture about the value of $\lim_{x \to a} f(x)$.

(c) Graph the function to see if it is consistent with your answers to parts (a) and (b).

(d) Find an interval for x containing a such that the difference between your conjectured limit and the value of the function is less than 0.01 on that interval. (In other words, find a window of height 0.02 such that the graph exits the sides of the window and not the top or bottom of the window.)

15. $f(x) = \dfrac{x^2 - 4}{x - 2}, \quad a = 2$

16. $f(x) = \dfrac{x^2 - 9}{x - 3}, \quad a = 3$

17. $f(x) = \dfrac{\sin x - 1}{x - \pi/2}, \quad a = \dfrac{\pi}{2}$

18. $f(x) = \dfrac{e^{2x-2} - 1}{x - 1}, \quad a = 1$

For the functions in Exercises 19–21, use algebra to evaluate the limits $\lim_{x \to a+} f(x)$, $\lim_{x \to a-} f(x)$, and $\lim_{x \to a} f(x)$ if they exist. Sketch a graph to confirm your answers.

19. $a = 4, \quad f(x) = \dfrac{|x - 4|}{x - 4}$

20. $a = 2, \quad f(x) = \dfrac{|x - 2|}{x}$

21. $a = 3, \quad f(x) = \begin{cases} x^2 - 2, & 0 < x < 3 \\ 2, & x = 3 \\ 2x + 1, & 3 < x \end{cases}$

22. Estimate how close θ should be to 0 to make $(\sin \theta)/\theta$ stay within 0.001 of 1.

23. Write the definition of the following statement both in words and in symbols:
$$\lim_{h \to a} g(h) = K.$$

Problems

In Problems 24–26, modify the definition of limit on page 50 to give a definition of each of the following.

24. A right-hand limit

25. A left-hand limit

26. $\lim_{x \to \infty} f(x) = L$

27. If $p(x)$ is the function on page 45 giving the price of mailing a first-class letter, explain why $\lim_{x \to 1} p(x)$ does not exist.

28. Investigate $\lim_{h \to 0} (1 + h)^{1/h}$ numerically.

29. The notation $\lim_{x \to 0+}$ means that we only consider values of x greater than 0. Estimate the limit
$$\lim_{x \to 0+} x^x,$$
either by evaluating x^x for smaller and smaller positive values of x (say $x = 0.1, 0.01, 0.001, \ldots$) or by zooming in on the graph of $y = x^x$ near $x = 0$.

Assuming that limits as $x \to \infty$ have the properties listed for limits as $x \to c$ on page 51, use algebraic manipulations to evaluate $\lim_{x \to \infty}$ for the functions in Problems 30–36.

30. $f(x) = \dfrac{x + 3}{2 - x}$

31. $f(x) = \dfrac{x^2 + 2x - 1}{3 + 3x^2}$

32. $f(x) = \dfrac{x^2 + 4}{x + 3}$

33. $f(x) = \dfrac{2x^3 - 16x^2}{4x^2 + 3x^3}$

34. $f(x) = \dfrac{x^4 + 3x}{x^4 + 2x^5}$

35. $f(x) = \dfrac{3e^x + 2}{2e^x + 3}$

36. $f(x) = \dfrac{2e^{-x} + 3}{3e^{-x} + 2}$

In Problems 37–44, find a value of the constant k such that the limit exists.

37. $\lim_{x \to 4} \dfrac{x^2 - k^2}{x - 4}$

38. $\lim_{x \to 1} \dfrac{x^2 - kx + 4}{x - 1}$

39. $\lim_{x \to -2} \dfrac{x^2 + 4x + k}{x + 2}$

40. $\lim_{x \to \infty} \dfrac{x^2 + 3x + 5}{4x + 1 + x^k}$

41. $\lim_{x \to -\infty} \dfrac{e^{2x} - 5}{e^{kx} + 3}$

42. $\lim_{x \to \infty} \dfrac{x^3 - 6}{x^k + 3}$

43. $\lim_{x \to \infty} \dfrac{3^{kx} + 6}{3^{2x} + 4}$

44. $\lim_{x \to -\infty} \dfrac{3^{kx} + 6}{3^{2x} + 4}$

For each value of ϵ in Problems 45–46, find a positive value of δ such that the graph of the function leaves the window $a - \delta < x < a + \delta, b - \epsilon < y < b + \epsilon$ by the sides and not through the top or bottom.

45. $f(x) = -2x + 3; a = 0; b = 3; \epsilon = 0.2, 0.1, 0.02,$ $0.01, 0.002, 0.001.$

46. $g(x) = -x^3 + 2; a = 0; b = 2; \epsilon = 0.1, 0.01, 0.001.$

47. Show that $\lim_{x \to 0} (-2x + 3) = 3$. [Hint: Use Problem 45.]

48. Consider the function $f(x) = \sin(1/x)$.

    (a) Find a sequence of x-values that approach 0 such that $\sin(1/x) = 0$.
[Hint: Use the fact that $\sin(\pi) = \sin(2\pi) = \sin(3\pi) = \ldots = \sin(n\pi) = 0$.]

    (b) Find a sequence of x-values that approach 0 such that $\sin(1/x) = 1$.
[Hint: Use the fact that $\sin(n\pi/2) = 1$ if $n = 1, 5, 9, \ldots$.]

    (c) Find a sequence of x-values that approach 0 such that $\sin(1/x) = -1$.

    (d) Explain why your answers to any two of parts (a)–(c) show that $\lim_{x \to 0} \sin(1/x)$ does not exist.

49. This problem suggests a proof of the first property of limits on page 51: $\lim_{x \to c} bf(x) = b \lim_{x \to c} f(x)$.

    (a) First, prove the property in the case $b = 0$.

    (b) Now suppose that $b \neq 0$. Let $\epsilon > 0$. Show that if $|f(x) - L| < \epsilon/|b|$, then $|bf(x) - bL| < \epsilon$.

    (c) Finally, prove that if $\lim_{x \to c} f(x) = L$ then $\lim_{x \to c} bf(x) = bL$. [Hint: Choose δ so that if $|x-c| < \delta$, then $|f(x) - L| < \epsilon/|b|$.]

50. Prove the second property of limits: $\lim_{x \to c} (f(x) + g(x)) = \lim_{x \to c} f(x) + \lim_{x \to c} g(x)$. Assume that the limits on the right exist.

51. This problem suggests a proof of the third property of limits (assuming the limits on the right exist):

$$\lim_{x \to c} (f(x)g(x)) = \left(\lim_{x \to c} f(x) \right) \left(\lim_{x \to c} g(x) \right)$$

Let $L_1 = \lim_{x \to c} f(x)$ and $L_2 = \lim_{x \to c} g(x)$.

    (a) First, show that if $\lim_{x \to c} f(x) = \lim_{x \to c} g(x) = 0$, then $\lim_{x \to c} (f(x)g(x)) = 0$.

    (b) Show algebraically that
$f(x)g(x) = (f(x) - L_1)(g(x) - L_2) + L_1 g(x) + L_2 f(x) - L_1 L_2$.

    (c) Use the second limit property (see Problem 50) to explain why
$\lim_{x \to c} (f(x) - L_1) = \lim_{x \to c} (g(x) - L_2) = 0$.

    (d) Use parts (a) and (c) to explain why $\lim_{x \to c} (f(x) - L_1)(g(x) - L_2) = 0$.

    (e) Finally, use parts (b) and (d) and the first and second limit properties to show that
$$\lim_{x \to c} (f(x)g(x)) = \left(\lim_{x \to c} f(x) \right) \left(\lim_{x \to c} g(x) \right).$$

52. Show $f(x) = x$ is continuous everywhere.

53. Use Problem 52 to show that for any positive integer n, the function x^n is continuous everywhere.

54. Use Theorem 1.2 on page 51 to explain why if f and g are continuous on an interval, then so are $f + g$, fg, and f/g (assuming $g(x) \neq 0$ on the interval).

CHAPTER SUMMARY (see also Ready Reference at the end of the book)

- **Function terminology**
  Domain/range, increasing/decreasing, concavity, zeros (roots), even/odd, end behavior, asymptotes.

- **Linear functions**
  Slope, vertical intercept. Grow by equal amounts in equal times.

- **Exponential functions**
  Exponential growth and decay, with base e, growth rate, continuous growth rate, doubling time, half life. Grow by equal percentages in equal times.

- **Logarithmic functions**
  Log base 10, natural logarithm.

- **Trigonometric functions**
  Sine and cosine, tangent, amplitude, period, arcsine, arctangent.

- **Power functions**

- **Polynomials and rational functions**

- **New functions from old**
  Inverse functions, composition of functions, shifting, stretching, shrinking.

- **Working with functions**
  Find a formula for a linear, exponential, power, logarithmic, or trigonometric function, given graph, table of values, or verbal description. Find vertical and horizontal asymptotes. End behavior.

- **Comparisons between functions**
  Exponential functions dominate power and linear functions.

- **Continuity**
  Interpret graphically and numerically. Intermediate Value Theorem.

- **Limits**
  Graphical interpretation, ϵ-δ definition, properties, one-sided limits, limits to infinity.

REVIEW EXERCISES AND PROBLEMS FOR CHAPTER ONE

Exercises

1. The entire graph of $f(x)$ is shown in Figure 1.87.

  (a) What is the domain of $f(x)$?
  (b) What is the range of $f(x)$?
  (c) List all zeros of $f(x)$.
  (d) List all intervals on which $f(x)$ is decreasing.
  (e) Is $f(x)$ concave up or concave down at $x = 6$?
  (f) What is $f(4)$?
  (g) Is this function invertible? Explain.

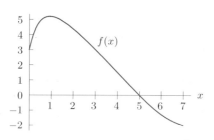

Figure 1.87

For Exercises 2–5, solve for x using logs.

2. $10 = 4^x$

3. $25 = 2(5)^x$

4. $2 \cdot 5^x = 11 \cdot 7^x$

5. $7 = 5e^{0.2x}$

Find the amplitudes and periods in Exercises 6–7.

6. $y = 5\sin(x/3)$ **7.** $y = 4 - 2\cos(5x)$

Find possible formulas for the graphs in Exercises 8–21.

8.

9.

10.

11.

12.

13.

14.

15.

16.

17.

18.

19.

20.

21.

For Exercises 22–23, find functions f and g such that $h(x) = f(g(x))$. [Note: Do not choose $f(x) = x$ or $g(x) = x$.]

22. $h(x) = \ln(x^3)$ **23.** $h(x) = (\ln x)^3$

State whether the functions in Exercises 24–27 are continuous on the interval $[-1, 1]$.

24. $f(x) = |x|$ **25.** $g(x) = \dfrac{|x|}{x}$

26. $h(\theta) = \theta\sin\theta$ **27.** $f(t) = \dfrac{\sin t}{t^2}$

For the functions in Exercises 28–31, do the following:

(a) Make a table of values of $f(x)$ for $x = a+0.1, a+0.01,$ $a + 0.001,\ a + 0.0001,\ a - 0.1,\ a - 0.01,\ a - 0.001,$ and $a - 0.0001$.

(b) Make a conjecture about the value of $\lim\limits_{x \to a} f(x)$.

(c) Graph the function to see if it is consistent with your answers to parts (a) and (b).

(d) Find an interval for x containing a such that the difference between your conjectured limit and the value of the function is less than 0.01 on that interval. (In other words, find a window of height 0.02 such that the graph exits the sides of the window and not the top or bottom of the window.)

28. $f(x) = \dfrac{\cos 2x - 1 + 2x^2}{x^3}, \quad a = 0$

29. $f(x) = \dfrac{\cos 3x - 1 + 4.5x^2}{x^3}, \quad a = 0$

30. $f(x) = \dfrac{\sin 5x - 1}{x - \pi/2}, \quad a = \dfrac{\pi}{2}$

31. $f(x) = \dfrac{e^{0.5x-1} - 1}{x - 2}, \quad a = 2$

For the functions in Exercises 32–33, use algebra to evaluate the limits $\lim\limits_{x \to a+} f(x)$, $\lim\limits_{x \to a-} f(x)$, and $\lim\limits_{x \to a} f(x)$ if they exist. Sketch a graph to confirm your answers.

32. $a = 3, \quad f(x) = \dfrac{x^3|2x - 6|}{x - 3}$

33. $a = 0, \quad f(x) = \begin{cases} e^x, & -1 < x < 0 \\ 1, & x = 0 \\ \cos x, & 0 < x < 1 \end{cases}$

Problems

34. The yield, Y, of an apple orchard (in bushels) as a function of the amount, a, of fertilizer (in pounds) used on the orchard is shown in Figure 1.88.

(a) Describe the effect of the amount of fertilizer on the yield of the orchard.

(b) What is the vertical intercept? Explain what it means in terms of apples and fertilizer.

(c) What is the horizontal intercept? Explain what it means in terms of apples and fertilizer.

(d) What is the range of this function for $0 \le a \le 80$?

(e) Is the function increasing or decreasing at $a = 60$?

(f) Is the graph concave up or down near $a = 40$?

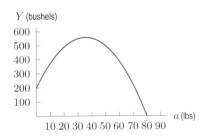

Figure 1.88

35. A rock is dropped from a window and falls to the ground below. The height, s (in meters), of the rock above ground is a function of the time, t (in seconds), since the rock was dropped, so $s = f(t)$.

(a) Sketch a possible graph of s as a function of t.

(b) Explain what the statement $f(7) = 12$ tells us about the rock's fall.

(c) The graph drawn as the answer for part (a) should have a horizontal and vertical intercept. Interpret each intercept in terms of the rock's fall.

36. A culture of 100 bacteria doubles after 2 hours. How long will it take for the number of bacteria to reach 3,200?

37. A controversial 1992 Danish study[10] reported that men's average sperm count has decreased from 113 million per milliliter in 1940 to 66 million per milliliter in 1990.

(a) Express the average sperm count, S, as a linear function of the number of years, t, since 1940.

(b) A man's fertility is affected if his sperm count drops below about 20 million per milliliter. If the linear model found in part (a) is accurate, in what year will the average male sperm count fall below this level?

38. Different kinds of the same element (called different *isotopes*) can have very different half-lives. With t in years, the decay of plutonium-240 is described by the formula

$$Q = Q_0 e^{-0.00011t},$$

whereas the decay of plutonium-242 is described by

$$Q = Q_0 e^{-0.0000018t}.$$

Find the half-lives of plutonium-240 and plutonium-242.

39. A culture of bacteria originally numbers 500. After 2 hours there are 1500 bacteria in the culture. Assuming exponential growth, how many are there after 6 hours?

40. The population of the United States was 226.5 million in 1980 and 248.7 million in 1990. Assuming exponential growth, in what year is the population expected to go over 300 million?

[10]"Investigating the Next Silent Spring", *US News and World Report*, p. 50-52, (March 11, 1996).

41. One of the main contaminants of a nuclear accident, such as that at Chernobyl, is strontium-90, which decays exponentially at a continuous rate of approximately 2.47% per year. After the Chernobyl disaster, it was suggested that it would be about 100 years before the region would again be safe for human habitation. What percent of the original strontium-90 would still remain then?

42. In the early 1920s, Germany had tremendously high inflation, called hyperinflation. Photographs of the time show people going to the store with wheelbarrows full of money. If a loaf of bread cost 1/4 RM in 1919 and 2,400,000 RM in 1922, what was the average yearly inflation rate between 1919 and 1922?

43. An airplane uses a fixed amount of fuel for takeoff, a (different) fixed amount for landing, and a third fixed amount per mile when it is in the air. How does the total quantity of fuel required depend on the length of the trip? Write a formula for the function involved. Explain the meaning of the constants in your formula.

44. Each planet moves around the sun in an elliptical orbit. The orbital period, T, of a planet is the time it takes the planet to go once around the sun. The semimajor axis of each planet's orbit is the average of the largest and the smallest distances between the planet and the sun. Johannes Kepler (1571-1630) discovered that the period of a planet is proportional to the $\frac{3}{2}$ power of its semimajor axis. What is the orbiting period (in days) of Mercury, the closest planet to the sun, with a semimajor axis of 58 million km? What is the period (in years) of Pluto, the farthest planet, with a semimajor axis of 6000 million km? The semimajor axis of the earth is 150 million km. [Hint: What is the earth's period?]

45. A closed cylindrical can of fixed volume V has radius r.

   (a) Find the surface area, S, as a function of r.
   (b) What happens to the value of S as $r \to \infty$?
   (c) Sketch a graph of S against r, if $V = 10$ cm^3.

46. **(a)** Consider the functions graphed in Figure 1.89(a). Find the coordinates of C.
   (b) Consider the functions in Figure 1.89(b). Find the coordinates of C in terms of b.

47. The depth of water in a tank oscillates sinusoidally once every 6 hours. If the smallest depth is 5.5 feet and the largest depth is 8.5 feet, find a possible formula for the depth in terms of time in hours.

48. The visitors' guide to St. Petersburg, Florida, contains the chart shown in Figure 1.90 to advertise their good weather. Fit a trigonometric function approximately to the data, where H is temperature in degrees Fahrenheit, and the independent variable is time in months. In order to do this, you will need to estimate the amplitude and period of the data, and when the maximum occurs. (There are many possible answers to this problem, depending on how you read the graph.)

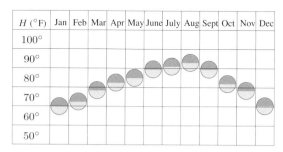

Figure 1.90: "St. Petersburg...where we're famous for our wonderful weather and year-round sunshine." (Reprinted with permission)

49. Figure 1.91 is the graph[11] of the function f, where $f(t)$ is the number (in millions) of motor vehicles registered in the world in the year t. (In 1988, one-third of the registered vehicles in the world were in the United States.)

   (a) Is f invertible? Explain.
   (b) What is the meaning of $f^{-1}(400)$ in practical terms? Evaluate $f^{-1}(400)$.
   (c) Sketch the graph of f^{-1}.

Figure 1.89

Figure 1.91

---

[11] From D. Blerics and P. Walzer, "Energy for Motor Vehicles," *Scientific American*, September 1990.

50. Each of the functions in the table is increasing over its domain, but each increases in a different way. Match the functions f, g, h to the graphs in Figure 1.92.

x	$f(x)$	x	$g(x)$	x	$h(x)$
1	1	3.0	1	10	1
2	2	3.2	2	20	2
4	3	3.4	3	28	3
7	4	3.6	4	34	4
11	5	3.8	5	39	5
16	6	4.0	6	43	6
22	7	4.2	7	46.5	7
29	8	4.4	8	49	8
37	9	4.6	9	51	9
47	10	4.8	10	52	10

(a)   (b)   (c)

Figure 1.92

51. (a) Use a graphing calculator or computer to estimate the period of $2\sin\theta + 3\cos(2\theta)$.
   (b) Explain your answer, given that the period of $\sin\theta$ is 2π and the period of $\cos(2\theta)$ is π.

52. Glucose is fed by intravenous injection at a constant rate, k, into a patient's bloodstream. Once there, the glucose is removed at a rate proportional to the amount of glucose present. If R is the net rate at which the quantity, G, of glucose in the blood is increasing:

   (a) Write a formula giving R as a function of G.
   (b) Sketch a graph of R against G.

53. A catalyst in a chemical reaction is a substance which speeds up the reaction but which does not itself change. If the product of a reaction is itself a catalyst, the reaction is said to be autocatalytic. The rate, r, of an autocatalytic reaction is proportional to the quantity of the original material remaining times the quantity of product, p, produced. The initial quantity of the original material is A and the amount remaining is $A - p$.

   (a) Express r as a function of p.
   (b) What is the value of p when the reaction is proceeding fastest?

54. What approximate domains and ranges make the graphs of $y = x^2$ and $y = 0.01e^{0.01x}$ look like each of the graphs in Figure 1.93?

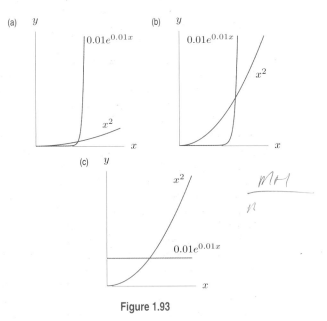

Figure 1.93

For each value of ϵ in Problems 55–56, find a positive value of δ such that the graph of the function leaves the window $a - \delta < x < a + \delta$, $b - \epsilon < y < b + \epsilon$ by the sides and not through the top or bottom.

55. $h(x) = \sin x$, $a = b = 0$, $\epsilon = 0.1, 0.05, 0.0007$.

56. $k(x) = \cos x$, $a = 0$, $b = 1$, $\epsilon = 0.1, 0.001, 0.00001$.

CAS Challenge Problems

57. (a) Factor $f(x) = x^4 + bx^3 - cx^3 - a^2x^2 - bcx^2 - a^2bx + a^2cx + a^2bc$ using a computer algebra system.
   (b) Assuming a, b, c are constants with $0 < a < b < c$, use your answer to part (a) to make a hand sketch of the graph of f. Explain how you know its shape.

58. (a) Using a computer algebra system, factor $f(x) = -x^5 + 11x^4 - 46x^3 + 90x^2 - 81x + 27$.
   (b) Use your answer to part (a) to make a hand sketch of the graph of f. Explain how you know its shape.

59. Let $f(x) = e^{6x} + e^{5x} - 2e^{4x} - 10e^{3x} - 8e^{2x} + 16e^x + 16$.

   (a) What happens to the value of $f(x)$ as $x \to \infty$? As $x \to -\infty$? Explain your answer.
   (b) Using a computer algebra system, factor $f(x)$ and predict the number of zeros of the function $f(x)$.
   (c) What are the exact values of the zeros? What is the relationship between successive zeros?

60. Let $f(x) = x^2 - x$.

   (a) Find the polynomials $f(f(x))$ and $f(f(f(x)))$ in expanded form.
   (b) What do you expect to be the degree of the polynomial $f(f(f(f(f(f(x))))))$? Explain.

61. (a) Use a computer algebra system to rewrite the rational function

$$f(x) = \frac{x^3 - 30}{x - 3}$$

in the form

$$f(x) = p(x) + \frac{r(x)}{q(x)},$$

where $p(x)$, $q(x)$, $r(x)$ are polynomials and the degree of $r(x)$ is less than the degree of $q(x)$.

(b) What is the vertical asymptote of f? Use your answer to part (a) to write the formula for a function whose graph looks like the graph of f for x near the vertical asymptote.

(c) Use your answer to part (a) to write the formula for a function whose graph looks like the graph of f for $x \to \infty$ and $x \to -\infty$.

(d) Using graphs, confirm the asymptote you found in part (b) and the formula you found in part (c).

We say that a function can be written as a polynomial in $\sin x$ (or $\cos x$) if it is of the form $p(\sin x)$ (or $p(\cos x)$) for some polynomial $p(x)$. For example, $\cos 2x$ can be written as a polynomial in $\sin x$ because $\cos(2x) = 1 - 2\sin^2 x = p(\sin x)$, where $p(x) = 1 - 2x^2$.

62. Use the trigonometric capabilities of your computer algebra system to express $\sin(5x)$ as a polynomial in $\sin x$.

63. Use the trigonometric capabilities of your computer algebra system to express $\cos(4x)$ as a polynomial in

(a) $\sin x$
(b) $\cos x$.

CHECK YOUR UNDERSTANDING

Are the statements in Problems 1–30 true or false? Give an explanation for your answer.

1. For any two points in the plane, there is a linear function whose graph passes through them.

2. The graph of $f(x) = 100(10^x)$ is a horizontal shift of the graph of $g(x) = 10^x$.

3. The graph of $f(x) = \ln x$ is concave down.

4. The graph of $g(x) = \log(x - 1)$ crosses the x-axis at $x = 1$.

5. Every polynomial of odd degree has at least one zero.

6. The function $y = 2 + 3e^{-t}$ has a y-intercept of $y = 3$.

7. The function $y = 5 - 3e^{-4t}$ has a horizontal asymptote of $y = 5$.

8. If $y = f(x)$ is a linear function, then increasing x by 1 unit changes the corresponding y by m units, where m is the slope.

9. If $y = f(x)$ is an exponential function and if increasing x by 1 increases y by a factor of 5, then increasing x by 2 increases y by a factor of 10.

10. If $y = Ab^x$ and increasing x by 1 increases y by a factor of 3, then increasing x by 2 increases y by a factor of 9.

11. The function $f(\theta) = \cos\theta - \sin\theta$ is increasing on $0 \le \theta \le \pi/2$.

12. The function $f(t) = \sin(0.05\pi t)$ has period 0.05.

13. If t is in seconds, $g(t) = \cos(200\pi t)$ executes 100 cycles in one second.

14. The function $f(\theta) = \tan(\theta - \pi/2)$ is not defined at $\theta = \pi/2, 3\pi/2, 5\pi/2 \ldots$.

15. The function $f(x) = \sin(x^2)$ is periodic, with period 2π.

16. The function $g(\theta) = e^{\sin\theta}$ is periodic.

17. If $f(x)$ is a periodic function with period k, then $f(g(x))$ is periodic with period k for every function $g(x)$.

18. If $g(x)$ is a periodic function, then $f(g(x))$ is periodic for every function $f(x)$.

19. The function $f(x) = e^{-x^2}$ is decreasing for all x.

20. The inverse function of $y = \log x$ is $y = 1/\log x$.

21. If f is an increasing function, then f^{-1} is an increasing function.

22. The function $f(x) = |\sin x|$ is even.

23. If a function is even, then it does not have an inverse.

24. If a function is odd, then it does not have an inverse function.

25. If a and b are positive constants, $b \ne 1$, then $y = a + ab^x$ has a horizontal asymptote.

26. If a and b are positive constants, then $y = \ln(ax + b)$ has no vertical asymptote.

27. The function $y = 20/(1 + 2e^{-kt})$ with $k > 0$, has a horizontal asymptote at $y = 20$.

28. If $g(x)$ is an even function then $f(g(x))$ is even for every function $f(x)$.

29. If $f(x)$ is an even function then $f(g(x))$ is even for every function $g(x)$.

30. If $\lim_{h \to 0} f(h) = L$, then $f(0.0001)$ is closer to L than is $f(0.01)$.

In Problems 31–36, give an example of a function with the specified properties. Express your answer using formulas.

31. Continuous on $[0, 1]$ but not continuous on $[1, 3]$.

32. Increasing but not continuous on $[0, 10]$.

33. Has a vertical asymptote at $x = -7\pi$.

34. Has exactly 17 vertical asymptotes.

35. Has a vertical asymptote which is crossed by a horizontal asymptote.

36. Two functions $f(x)$ and $g(x)$ such that moving the graph of f to the left 2 units gives the graph of g and moving the graph of f up 3 also gives the graph of g.

Suppose f is an increasing function and g is a decreasing function. In Problems 37–40, give an example for f and g for which the statement is true, or say why such an example is impossible.

37. $f(x) + g(x)$ is decreasing for all x.

38. $f(x) - g(x)$ is decreasing for all x.

39. $f(x)g(x)$ is decreasing for all x.

40. $f(g(x))$ is increasing for all x.

Are the statements in Problems 41–49 true or false? If a statement is true, explain how you know. If a statement is false, give a counterexample.

41. Every rational function that is not a polynomial has a vertical asymptote.

42. If a function is increasing on an interval, then it is concave up on that interval.

43. If y is a linear function of x, then the ratio y/x is constant for all points on the graph at which $x \neq 0$.

44. An exponential function can be decreasing.

45. If $y = f(x)$ is a linear function, then increasing x by 2 units adds $m + 2$ units to the corresponding y, where m is the slope.

46. If a function is not continuous at a point, then it is not defined at that point.

47. If f is continuous on the interval $[0, 10]$ and $f(x) = 0$ and $f(10) = 100$, then $f(c)$ cannot be negative for c in $[0, 10]$.

48. If $f(x)$ is not continuous on the interval $[a, b]$, then $f(x)$ must omit at least one value between $f(a)$ and $f(b)$.

49. There is a function which is both even and odd.

Suppose that $\lim_{x \to 3} f(x) = 7$. Are the statements in Problems 50–56 true or false? If a statement is true, explain how you know. If a statement is false, give a counterexample.

50. $\lim_{x \to 3}(x f(x)) = 21$.

51. If $g(3) = 4$, then $\lim_{x \to 3}(f(x)g(x)) = 28$.

52. If $\lim_{x \to 3} g(x) = 5$, then $\lim_{x \to 3}(f(x) + g(x)) = 12$.

53. If $\lim_{x \to 3}(f(x) + g(x)) = 12$, then $\lim_{x \to 3} g(x) = 5$.

54. $f(2.99)$ is closer to 7 than $f(2.9)$ is.

55. If $f(3.1) > 0$, then $f(3.01) > 0$.

56. If $\lim_{x \to 3} g(x)$ does not exist, then $\lim_{x \to 3}(f(x)g(x))$ does not exist.

Which of the statements in Problems 57–61 are true about every function $f(x)$ such that $\lim_{x \to c} f(x) = L$? Give a reason for your answer.

57. If $f(x)$ is within 10^{-3} of L, then x is within 10^{-3} of c.

58. There is a positive ϵ such that, provided x is within 10^{-3} of c, and $x \neq c$, we can be sure $f(x)$ is within ϵ of L.

59. For any positive ϵ, we can find a positive δ such that, provided x is within δ of c, and $x \neq c$, we can be sure that $f(x)$ is within ϵ of L.

60. For each $\epsilon > 0$, there is a $\delta > 0$ such that if x is not within δ of c, then $f(x)$ is not within ϵ of L.

61. For each $\epsilon > 0$, there is some $\delta > 0$ such that if $f(x)$ is within ϵ of L, then we can be sure that x is within δ of c.

62. Which of the following statements is a direct consequence of the statement: "If f and g are continuous at $x = a$ and $g(a) \neq 0$ then f/g is continuous at $x = a$"?

   (a) If f and g are continuous at $x = a$ and $f(a) \neq 0$ then g/f is continuous at $x = a$.
   (b) If f and g are continuous at $x = a$ and $g(a) = 0$, then f/g is not continuous at $x = a$.
   (c) If f, g, are continuous at $x = a$, but f/g is not continuous at $x = a$, then $g(a) = 0$.
   (d) If f and f/g are continuous at $x = a$ and $g(a) \neq 0$, then g is continuous at $x = a$.

PROJECTS FOR CHAPTER ONE

1. Matching Functions to Data

From the data in Table 1.20, determine a possible formula for each function.[12] Write an explanation of your reasoning.

Table 1.20

x	$f(x)$	$g(x)$	$h(x)$	$F(x)$	$G(x)$	$H(x)$
−5	−10	20	25	0.958924	0.544021	2.958924
−4.5	−9	19	20.25	0.97753	−0.412118	2.97753
−4	−8	18	16	0.756802	−0.989358	2.756802
−3.5	−7	17	12.25	0.350783	−0.656987	2.350783
−3	−6	16	9	−0.14112	0.279415	1.85888
−2.5	−5	15	6.25	−0.598472	0.958924	1.401528
−2	−4	14	4	−0.909297	0.756802	1.090703
−1.5	−3	13	2.25	−0.997495	−0.14112	1.002505
−1	−2	12	1	−0.841471	−0.909297	1.158529
−0.5	−1	11	0.25	−0.479426	−0.841471	1.520574
0	0	10	0	0	0	2
0.5	1	9	0.25	0.479426	0.841471	2.479426
1	2	8	1	0.841471	0.909297	2.841471
1.5	3	7	2.25	0.997495	0.14112	2.997495
2	4	6	4	0.909297	−0.756802	2.909297
2.5	5	5	6.25	0.598472	−0.958924	2.598472
3	6	4	9	0.14112	−0.279415	2.14112
3.5	7	3	12.25	−0.350783	0.656987	1.649217
4	8	2	16	−0.756802	0.989358	1.243198
4.5	9	1	20.25	−0.97753	0.412118	1.02247
5	10	0	25	−0.958924	−0.544021	1.041076

2. Compound Interest

The newspaper article below is from *The New York Times*, May 27, 1990. Fill in the three blanks. (For the first blank, assume that daily compounding is essentially the same as continuous compounding. For the last blank, assume the interest has been compounded yearly, and give your answer in dollars. Ignore the occurrence of leap years.)

213 Years After Loan, Uncle Sam Is Dunned

By LISA BELKIN

Special to The New York Times

SAN ANTONIO, May 26 — More than 200 years ago, a wealthy Pennsylvania merchant named Jacob DeHaven lent $450,000 to the Continental Congress to rescue the troops at Valley Forge. That loan was apparently never repaid.

So Mr. DeHaven's descendants are taking the United States Government to court to collect what they believe they are owed.

The total: ____ in today's dollars if the interest is compounded daily at 6 percent, the going rate at the time. If compounded yearly, the bill is only ____.

Family Is Flexible

The descendants say that they are willing to be flexible about the amount of a settlement and that they might even accept a heartfelt thank you or perhaps a DeHaven statue. But they also note that interest is accumulating at ____ a second.

---

[12] Based on a problem by Lee Zia

Chapter Two

KEY CONCEPT: THE DERIVATIVE

We begin this chapter by investigating the problem of speed: How can we measure the speed of a moving object at a given instant in time? Or, more fundamentally, what do we mean by the term *speed*? We give a definition of speed that has wide-ranging implications — not just for the speed problem, but for measuring the rate of change of any quantity. Our journey will lead us to the key concept of *derivative*, which forms the basis for our study of calculus.

The derivative can be interpreted geometrically as the slope of a curve, and physically as a rate of change. Because derivatives can be used to represent everything from fluctuations in interest rates to the rates at which fish populations vary and gas molecules move, they have applications throughout the sciences.

2.1 HOW DO WE MEASURE SPEED?

The speed of an object at an instant in time is surprisingly difficult to define precisely. Consider the statement "At the instant it crossed the finish line, the horse was traveling at 42 mph." How can such a claim be substantiated? A photograph taken at that instant will show the horse motionless—it is no help at all. There is some paradox in trying to study the horse's motion at a particular instant in time, since by focusing on a single instant we stop the motion!

Problems of motion were of central concern to Zeno and other philosophers as early as the fifth century B.C. The modern approach, made famous by Newton's calculus, is to stop looking for a simple notion of speed at an instant, and instead to look at speed over small time intervals containing the instant. This method sidesteps the philosophical problems mentioned earlier but introduces new ones of its own.

We illustrate the ideas discussed above by an idealized example, called a thought experiment. It is idealized in the sense that we assume that we can make measurements of distance and time as accurately as we wish.

A Thought Experiment: Average and Instantaneous Velocity

We look at the speed of a small object (say, a grapefruit) that is thrown straight upward into the air at $t = 0$ seconds. The grapefruit leaves the thrower's hand at high speed, slows down until it reaches its maximum height, and then speeds up in the downward direction and finally, "Splat!" (See Figure 2.1.)

Suppose that we want to determine the speed, say, at $t = 1$ second. Table 2.1 gives the height, y, of the grapefruit above the ground as a function of time. During the first second the grapefruit travels $90 - 6 = 84$ feet, and during the second second it travels only $142 - 90 = 52$ feet. Hence the grapefruit traveled faster over the first interval, $0 \leq t \leq 1$, than the second interval, $1 \leq t \leq 2$.

Figure 2.1: The grapefruit's path is straight up and down

Table 2.1   *Height of the grapefruit above the ground*

t (sec)	0	1	2	3	4	5	6
y (feet)	6	90	142	162	150	106	30

Velocity versus Speed

From now on, we will distinguish between velocity and speed. Suppose an object moves along a line. We pick one direction to be positive and say that the *velocity* is positive if it is in the same direction, and negative if it is in the opposite direction. For the grapefruit, upward is positive and downward is negative. (See Figure 2.1.) *Speed* is the magnitude of the velocity and so is always positive or zero.

If $s(t)$ is the position of an object at time t, then the **average velocity** of the object over the interval $a \leq t \leq b$ is

$$\text{Average velocity} = \frac{\text{Change in position}}{\text{Change in time}} = \frac{s(b) - s(a)}{b - a}.$$

In words, the **average velocity** of an object over an interval is the net change in position during the interval divided by the change in time.

Example 1   Compute the average velocity of the grapefruit over the interval $4 \leq t \leq 5$. What is the significance of the sign of your answer?

Solution   During this interval, the grapefruit moves $(106 - 150) = -44$ feet. Therefore the average velocity is -44 ft/sec. The negative sign means the height is decreasing and the grapefruit is moving downward.

Example 2   Compute the average velocity of the grapefruit over the interval $1 \leq t \leq 3$.

Solution   Average velocity $= (162 - 90)/(3 - 1) = 72/2 = 36$ ft/sec.

The average velocity is a useful concept since it gives a rough idea of the behavior of the grapefruit: if two grapefruits are hurled into the air, and one has an average velocity of 10 ft/sec over the interval $0 \leq t \leq 1$ while the second has an average velocity of 100 ft/sec over the same interval, the second one is moving faster.

But average velocity over an interval does not solve the problem of measuring the velocity of the grapefruit at *exactly* $t = 1$ second. To get closer to an answer to that question, we have to look at what happens near $t = 1$ in more detail. The data[1] in Figure 2.2 shows the average velocity over small intervals on either side of $t = 1$.

Notice that the average velocity before $t = 1$ is slightly more than the average velocity after $t = 1$. We expect to define the velocity *at* $t = 1$ to be between these two average velocities. As the size of the interval shrinks, the values of the velocity before $t = 1$ and the velocity after $t = 1$ get closer together. In the smallest interval in Figure 2.2, both velocities are 68.0 ft/sec (to one decimal place), so we define the velocity at $t = 1$ to be 68.0 ft/sec (to one decimal place).

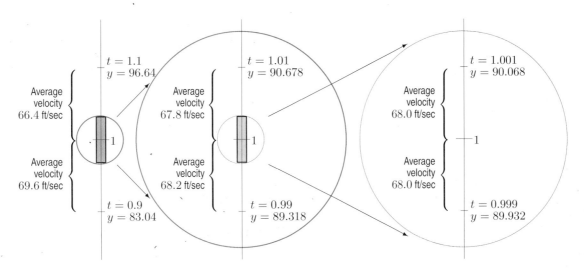

Figure 2.2: Average velocities over intervals on either side of $t = 1$: Showing successively smaller intervals

Of course, if we calculate to more decimal places, the average velocities before and after $t = 1$ would no longer agree. To calculate the velocity at $t = 1$ to more decimal places of accuracy, we take smaller and smaller intervals on either side of $t = 1$ until the average velocities agree to the number of decimal places we want. In this way, we can estimate the velocity at $t = 1$ to any accuracy.

---

[1]The data is in fact calculated from the formula $y = 6 + 100t - 16t^2$.

Defining Instantaneous Velocity Using Limit Notation

When we take smaller and smaller intervals, it turns out that the average velocities get closer and closer to 68 ft/sec. It seems natural, then, to define *instantaneous velocity* at the instant $t = 1$ to be 68 ft/sec. Its definition depends on our being convinced that smaller and smaller intervals give average speeds that come arbitrarily close to 68; that is, the average speeds approach 68 as a limit.

Notice how we have replaced the original difficulty of computing velocity at a point by a search for an argument to convince ourselves that the average velocities approach a limit as the time intervals shrink in size. Showing that the limit is exactly 68 requires the precise definition of limit given in Section 1.8.

To define instantaneous velocity at an arbitrary point $t = a$, we use the same method as for $t = 1$. On small intervals of size h around $t = a$, we calculate

$$\text{Average velocity} = \frac{s(a + h) - s(a)}{h}.$$

The instantaneous velocity is the number that the average velocities approach as the intervals decrease in size, that is, as h becomes smaller. So we make the following definition:

Let $s(t)$ be the position at time t. Then the **instantaneous velocity** at $t = a$ is defined as

$$\begin{array}{c}\text{Instantaneous velocity} \\ \text{at } t = a\end{array} = \lim_{h \to 0} \frac{s(a + h) - s(a)}{h}.$$

In words, the **instantaneous velocity** of an object at time $t = a$ is given by the limit of the average velocity over an interval, as the interval shrinks around a.

This limit refers to the number that the average velocities approach as the intervals shrink. To estimate the limit, we look at intervals of smaller and smaller, but never zero, length.

Visualizing Velocity: Slope of Curve

Now we visualize velocity using a graph of height. The cornerstone of the idea is the fact that, on a very small scale, most functions look almost like straight lines. Imagine taking the graph of a function near a point and "zooming in" to get a close-up view. (See Figure 2.3.) The more we zoom in, the more the curve appears to be a straight line. We call the slope of this line the *slope of the curve* at the point.

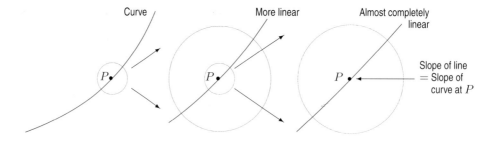

Figure 2.3: Estimating the slope of the curve at the point by "zooming in"

To visualize the instantaneous velocity, we think about how we calculated it. We took average velocities over small intervals containing at $t = 1$. Two such velocities are represented by the slopes of the lines in Figure 2.4. As the length of the interval shrinks, the slope of the line gets closer to the slope of the curve at $t = 1$.

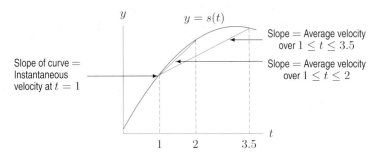

Figure 2.4: Average velocities over small intervals

> The **instantaneous velocity** is the slope of the curve at a point.

Let's go back to the grapefruit. Figure 2.5 shows the height of the grapefruit plotted against time. (Note that this is not a picture of the grapefruit's path, which is straight up and down.)

How can we visualize the average velocity on this graph? Suppose $y = s(t)$. We consider the interval $1 \leq t \leq 2$ and the expression

$$\text{Average velocity} = \frac{\text{Change in position}}{\text{Change in time}} = \frac{s(2) - s(1)}{2 - 1} = \frac{142 - 90}{1} = 52 \text{ ft/sec.}$$

Now $s(2) - s(1)$ is the change in position over the interval, and it is marked vertically in Figure 2.5. The 1 in the denominator is the time elapsed and is marked horizontally in Figure 2.5. Therefore,

$$\text{Average velocity} = \frac{\text{Change in position}}{\text{Change in time}} = \text{Slope of line joining } B \text{ and } C.$$

(See Figure 2.5.) A similar argument shows the following:

> The **average velocity** over any time interval $a \leq t \leq b$ is the slope of the line joining the points on the graph of $s(t)$ corresponding to $t = a$ and $t = b$.

Figure 2.5 shows how the grapefruit's velocity varies during its journey. At points A and B the curve has a large positive slope, indicating that the grapefruit is traveling up rapidly. Point D

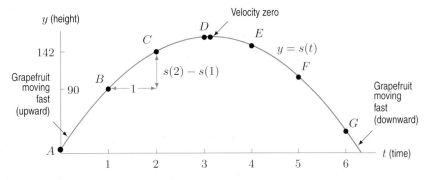

Figure 2.5: The height, y, of the grapefruit at time t

is almost at the top: the grapefruit is slowing down. At the peak, the slope of the curve is zero: the fruit has slowed to zero velocity for an instant in preparation for its return to earth. At point E the curve has a small negative slope, indicating a slow velocity of descent. Finally, the slope of the curve at point G is large and negative, indicating a large downward velocity that is responsible for the "Splat."

Using Limits to Compute the Instantaneous Velocity

Suppose we want to calculate the instantaneous velocity for $s(t) = t^2$ at $t = 3$. We must find:

$$\lim_{h \to 0} \frac{s(3+h) - s(3)}{h} = \lim_{h \to 0} \frac{(3+h)^2 - 9}{h}.$$

We show two possible approaches.

Example 3   Estimate $\lim\limits_{h \to 0} \dfrac{(3+h)^2 - 9}{h}$ numerically.

Solution   The limit is the value approached by this expression as h approaches 0. The values in Table 2.2 seem to be converging to 6 as $h \to 0$. So it is a reasonable guess that

$$\lim_{h \to 0} \frac{(3+h)^2 - 9}{h} = 6.$$

However, we cannot be sure that the limit is *exactly* 6 by looking at the table. To calculate the limit exactly requires algebra.

Table 2.2   *Values of $\left((3+h)^2 - 9\right)/h$ near $h = 0$*

h	-0.1	-0.01	-0.001	0.001	0.01	0.1
$\left((3+h)^2 - 9\right)/h$	5.9	5.99	5.999	6.001	6.01	6.1

Example 4   Use algebra to find $\lim\limits_{h \to 0} \dfrac{(3+h)^2 - 9}{h}$.

Solution   Expanding the numerator gives

$$\frac{(3+h)^2 - 9}{h} = \frac{9 + 6h + h^2 - 9}{h} = \frac{6h + h^2}{h}.$$

Since taking the limit as $h \to 0$ means looking at values of h near, but not equal, to 0, we can cancel h, giving

$$\lim_{h \to 0} \frac{(3+h)^2 - 9}{h} = \lim_{h \to 0} (6 + h).$$

As h approaches 0, the values of $(6 + h)$ approach 6, so

$$\lim_{h \to 0} \frac{(3+h)^2 - 9}{h} = \lim_{h \to 0} (6 + h) = 6.$$

Exercises and Problems for Section 2.1

Exercises

1. The distance, s, a car has traveled on a trip is shown in the table as a function of the time, t, since the trip started. Find the average velocity between $t = 2$ and $t = 5$.

t (hours)	0	1	2	3	4	5
s (km)	0	45	135	220	300	400

Slope	Point
-3	
-1	
0	
$1/2$	
1	
2	

Figure 2.6

2. In a time of t seconds, a particle moves a distance of s meters from its starting point, where $s = 3t^2$.

(a) Find the average velocity between $t = 1$ and $t = 1 + h$ if:

(i) $h = 0.1$, (ii) $h = 0.01$, (iii) $h = 0.001$.

(b) Use your answers to part (a) to estimate the instantaneous velocity of the particle at time $t = 1$.

3. In a time of t seconds, a particle moves a distance of s meters from its starting point, where $s = 4t^2 + 3$.

(a) Find the average velocity between $t = 1$ and $t = 1 + h$ if:

(i) $h = 0.1$, (ii) $h = 0.01$, (iii) $h = 0.001$.

(b) Use your answers to part (a) to estimate the instantaneous velocity of the particle at time $t = 1$.

4. In a time of t seconds, a particle moves a distance of s meters from its starting point, where $s = \sin(2t)$.

(a) Find the average velocity between $t = 1$ and $t = 1 + h$ if:

(i) $h = 0.1$, (ii) $h = 0.01$, (iii) $h = 0.001$.

(b) Use your answers to part (a) to estimate the instantaneous velocity of the particle at time $t = 1$.

5. Match the points labeled on the curve in Figure 2.6 with the given slopes.

6. For the function shown in Figure 2.7, at what labeled points is the slope of the graph positive? Negative? At which labeled point does the graph have the greatest (i.e., most positive) slope? The least slope (i.e., negative and with the largest magnitude)?

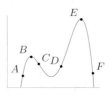

Figure 2.7

Estimate the limits in Exercises 7–10 by substituting smaller and smaller values of h. For trigonometric functions, use radians. Give answers to one decimal place.

7. $\displaystyle\lim_{h \to 0} \frac{(3+h)^3 - 27}{h}$

8. $\displaystyle\lim_{h \to 0} \frac{\cos h - 1}{h}$

9. $\displaystyle\lim_{h \to 0} \frac{7^h - 1}{h}$

10. $\displaystyle\lim_{h \to 0} \frac{e^{1+h} - e}{h}$

Problems

11. A car is driven at a constant speed. Sketch a graph of the distance the car has traveled as a function of time.

12. A car is driven at an increasing speed. Sketch a graph of the distance the car has traveled as a function of time.

13. A car starts at a high speed, and its speed then decreases slowly. Sketch a graph of the distance the car has traveled as a function of time.

14. For the graph $y = f(x)$ in Figure 2.8, arrange the following numbers from smallest to largest:

- The slope of the graph at A.
- The slope of the graph at B.
- The slope of the graph at C.
- The slope of the line AB.
- The number 0.
- The number 1.

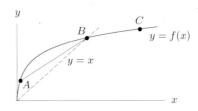

Figure 2.8

15. The graph of $f(t)$ in Figure 2.9 gives the position of a particle at time t. List the following quantities in order, smallest to largest.

- A, average velocity between $t = 1$ and $t = 3$,
- B, average velocity between $t = 5$ and $t = 6$,
- C, instantaneous velocity at $t = 1$,
- D, instantaneous velocity at $t = 3$,
- E, instantaneous velocity at $t = 5$,
- F, instantaneous velocity at $t = 6$.

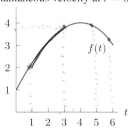

Figure 2.9

16. A particle moves at varying velocity along a line and $s = f(t)$ represents the particle's distance from a point as a function of time, t. Sketch a possible graph for f if the average velocity of the particle between $t = 2$ and $t = 6$ is the same as the instantaneous velocity at $t = 5$.

17. Find the average velocity over the interval $0 \le t \le 0.2$, and estimate the velocity at $t = 0.2$ of a car whose position, s, is given by the following table.

t (sec)	0	0.2	0.4	0.6	0.8	1.0
s (ft)	0	0.5	1.8	3.8	6.5	9.6

Use algebra to evaluate the limits in Problems 18–21.

18. $\displaystyle\lim_{h \to 0} \frac{(2+h)^2 - 4}{h}$

19. $\displaystyle\lim_{h \to 0} \frac{(1+h)^3 - 1}{h}$

20. $\displaystyle\lim_{h \to 0} \frac{3(2+h)^2 - 12}{h}$

21. $\displaystyle\lim_{h \to 0} \frac{(3+h)^2 - (3-h)^2}{2h}$

2.2   THE DERIVATIVE AT A POINT

Average Rate of Change

In Section 2.1, we looked at the change in height divided by the change in time, which tells us

$$\text{Average rate of change of height with respect to time} = \frac{s(a+h) - s(a)}{h}.$$

This ratio is called the *difference quotient*. Now we apply the same analysis to any function f, not necessarily a function of time. We say:

$$\boxed{\text{Average rate of change of } f \text{ over the interval from } a \text{ to } a+h = \frac{f(a+h) - f(a)}{h}.}$$

The numerator, $f(a+h) - f(a)$, measures the change in the value of f over the interval from a to $a+h$. So, the difference quotient is the change in f divided by the change in x. See Figure 2.10.

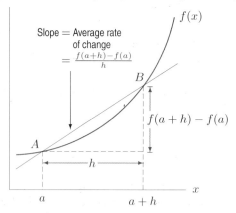

Figure 2.10: Visualizing the average rate of change of f

Although the interval is no longer necessarily a time interval, we still talk about the *average rate of change* of f over the interval. If we want to emphasize the independent variable, we talk about the average rate of change of f *with respect to x*.

Average Rate of Change versus Absolute Change

The average rate of change of a function over an interval is not the same as the absolute change. Absolute change is just the difference in the values of f at the ends of the interval:

$$f(a+h) - f(a).$$

The average rate of change is the absolute change divided by the size of the interval:

$$\frac{f(a+h) - f(a)}{h}.$$

The average rate of change tells how quickly (or slowly) the function changes from one end of the interval to the other, relative to the size of the interval. It is often more useful to know the rate of change than the absolute change. For example, if someone offers you a $100 salary, you will want to know how long to work to make that money. Just knowing the absolute change in your money, $100, is not enough, but knowing the rate of change (i.e., $100 divided by the time it takes to make it) helps you decide whether or not to accept the salary.

Blowing Up a Balloon

Consider the function which gives the radius of a sphere in terms of its volume. For example, think of blowing air into a balloon. You've probably noticed that a balloon seems to blow up faster at the start and then slows down as you blow more air into it. What you're seeing is variation in the rate of change of the radius with respect to volume.

Example 1   The volume, V, of a sphere of radius r is given by $V = 4\pi r^3/3$. Solving for r in terms of V gives

$$r = f(V) = \left(\frac{3V}{4\pi}\right)^{1/3}.$$

Calculate the average rate of change of r with respect to V over the intervals $0.5 \leq V \leq 1$ and $1 \leq V \leq 1.5$.

Solution   Using the formula for the average rate of change gives

$$\text{Average rate of change of radius for } 0.5 \leq V \leq 1 = \frac{f(1) - f(0.5)}{0.5} = 2\left(\left(\frac{3}{4\pi}\right)^{1/3} - \left(\frac{1.5}{4\pi}\right)^{1/3}\right) \approx 0.26.$$

$$\text{Average rate of change of radius for } 1 \leq V \leq 1.5 = \frac{f(1.5) - f(1)}{0.5} = 2\left(\left(\frac{4.5}{4\pi}\right)^{1/3} - \left(\frac{3}{4\pi}\right)^{1/3}\right) \approx 0.18.$$

So we see that the rate decreases as the volume increases.

Instantaneous Rate of Change: The Derivative

We define the *instantaneous rate of change* of a function at a point in the same way that we defined instantaneous velocity: we look at the average rate of change over smaller and smaller intervals. This instantaneous rate of change is called the *derivative of f at a*, denoted by $f'(a)$.

The **derivative of f at a**, written $f'(a)$, is defined as

$$\text{Rate of change} \atop \text{of } f \text{ at } a} = f'(a) = \lim_{h \to 0} \frac{f(a+h) - f(a)}{h}.$$

If the limit exists, then f is said to be **differentiable at a**.

To emphasize that $f'(a)$ is the rate of change of $f(x)$ as the variable x changes, we call $f'(a)$ the derivative of f *with respect to* x at $x = a$. When the function $y = s(t)$ represents the position of an object, the derivative $s'(t)$ is the velocity.

Example 2   By choosing small values for h, estimate the instantaneous rate of change of the radius, r, of a sphere with respect to change in volume at $V = 1$.

Solution   The formula for $r = f(V)$ is given in Example 1. With $h = 0.01$ and $h = -0.01$, we have the difference quotients

$$\frac{f(1.01) - f(1)}{0.01} \approx 0.2061 \qquad \text{and} \qquad \frac{f(0.99) - f(1)}{-0.01} \approx 0.2075.$$

With $h = 0.001$ and $h = -0.001$,

$$\frac{f(1.001) - f(1)}{0.001} \approx 0.2067 \qquad \text{and} \qquad \frac{f(0.999) - f(1)}{-0.001} \approx 0.2069.$$

The values of these difference quotients suggest that the limit is between 0.2067 and 0.2069. We conclude that the value is about 0.207; taking smaller h values confirms this. So we say

$$f'(1) = {\text{Instantaneous rate of change of radius} \atop \text{with respect to volume at } V = 1} \approx 0.207.$$

In this example we found an approximation to the instantaneous rate of change, or derivative, by substituting in smaller and smaller values of h. Now we see how to visualize the derivative.

Visualizing the Derivative: Slope of Curve and Slope of Tangent

As with velocity, we can visualize the derivative $f'(a)$ as the slope of the graph of f at $x = a$. In addition, there is another way to think of $f'(a)$. Consider the difference quotient $(f(a + h) - f(a))/h$. The numerator, $f(a + h) - f(a)$, is the vertical distance marked in Figure 2.11 and h is the horizontal distance, so

$$\text{Average rate of change of } f = \frac{f(a+h) - f(a)}{h} = \text{Slope of line } AB.$$

As h becomes smaller, the line AB approaches the tangent line to the curve at A. (See Figure 2.12.) We say

$$\text{Instantaneous rate of change} \atop \text{of } f \text{ at } a} = \lim_{h \to 0} \frac{f(a+h) - f(a)}{h} = \text{Slope of tangent at } A.$$

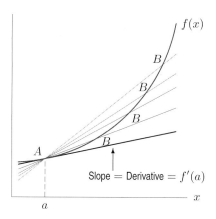

Figure 2.11: Visualizing the average rate of change of f

Figure 2.12: Visualizing the instantaneous rate of change of f

> The derivative at point A can be interpreted as:
> - The slope of the curve at A.
> - The slope of the tangent line to the curve at A.

The slope interpretation is often useful in gaining rough information about the derivative, as the following examples show.

Example 3   Is the derivative of $\sin x$ at $x = \pi$ positive or negative?

Solution   Looking at a graph of $\sin x$ in Figure 2.13 (remember, x is in radians), we see that a tangent line drawn at $x = \pi$ has negative slope. So the derivative at this point is negative.

Figure 2.13: Tangent line to $\sin x$ at $x = \pi$

Recall that if we zoom in on the graph of a function $y = f(x)$ at the point $x = a$, we usually find that the graph looks like a straight line with slope $f'(a)$.

Example 4   By zooming in on the point $(0, 0)$ on the graph of the sine function, estimate the value of the derivative of $\sin x$ at $x = 0$, with x in radians.

Solution   Figure 2.14 shows graphs of $\sin x$ with smaller and smaller scales. On the interval $-0.1 \le x \le 0.1$, the graph looks like a straight line of slope 1. Thus, the derivative of $\sin x$ at $x = 0$ is about 1.

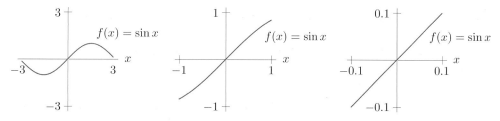

Figure 2.14: Zooming in on the graph of $\sin x$ near $x = 0$ shows the derivative is about 1 at $x = 0$

Later we will show that the derivative of $\sin x$ at $x = 0$ is exactly 1. (See page 136 in Section 3.5.) From now on we will assume that this is so.

Example 5   Use the tangent line at $x = 0$ to estimate values of $\sin x$ near $x = 0$.

Solution   In the previous example we see that near $x = 0$, the graph of $y = \sin x$ looks like the straight line $y = x$; we can use this line to estimate values of $\sin x$ when x is close to 0. For example, the point on the straight line $y = x$ with x coordinate 0.32 is $(0.32, 0.32)$. Since the line is close to the graph of $y = \sin x$, we estimate that $\sin 0.32 \approx 0.32$. (See Figure 2.15.) Checking on a calculator, we find that $\sin 0.32 \approx 0.3146$, so our estimate is quite close. Notice that the graph suggests that the real value of $\sin 0.32$ is slightly less than 0.32.

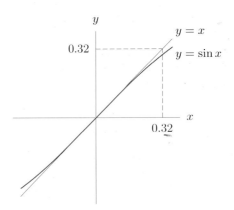

Figure 2.15: Approximating $y = \sin x$ by the line $y = x$

Why Do We Use Radians and Not Degrees?

After Example 4 we stated that the derivative of $\sin x$ at $x = 0$ is 1, when x is in radians. This is the reason we choose to use radians. If we had done Example 4 in degrees, the derivative of $\sin x$ would have turned out to be a much messier number. (See Problem 20, page 80.)

Estimating the Derivative of an Exponential Function

Example 6   Estimate the value of the derivative of $f(x) = 2^x$ at $x = 0$ graphically and numerically.

Solution   Graphically: Figure 2.16 indicates that the graph is concave up. Assuming this, the slope at A is between the slope of BA and the slope of AC. Since

$$\text{Slope of line } BA = \frac{(2^0 - 2^{-1})}{(0 - (-1))} = \frac{1}{2} \quad \text{and} \quad \text{Slope of line } AC = \frac{(2^1 - 2^0)}{(1 - 0)} = 1,$$

we know that at $x = 0$ the derivative of 2^x is between $1/2$ and 1.

Numerically: To estimate the derivative at $x = 0$, we look at values of the difference quotient

$$\frac{f(0 + h) - f(0)}{h} = \frac{2^h - 2^0}{h} = \frac{2^h - 1}{h}$$

for small h. Table 2.3 shows some values of 2^h together with values of the difference quotients. (See Problem 26 on page 80 for what happens for very small values of h.)

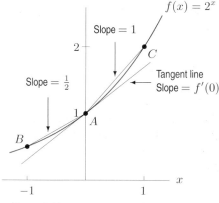

Figure 2.16: Graph of $y = 2^x$ showing the derivative at $x = 0$

Table 2.3 *Numerical values for difference quotient of 2^x at $x = 0$*

h	2^h	Difference quotient: $\dfrac{2^h - 1}{h}$
-0.0003	0.999792078	0.693075
-0.0002	0.999861380	0.693099
-0.0001	0.999930688	0.693123
0	1	
0.0001	1.00006932	0.693171
0.0002	1.00013864	0.693195
0.0003	1.00020797	0.693219

The concavity of the curve tells us that difference quotients calculated with negative h's are smaller than the derivative, and those calculated with positive h's are larger. From Table 2.3 we see that the derivative is between 0.693123 and 0.693171. To three decimal places, $f'(0) = 0.693$.

Example 7   Find an approximate equation for the tangent line to $f(x) = 2^x$ at $x = 0$.

Solution   From the previous example, we know the slope of the tangent line is about 0.693. Since the tangent line has y-intercept 1, its equation is

$$y = 0.693x + 1.$$

Computing the Derivative of $1/x$

The graph of $f(x) = 1/x$ in Figure 2.17 leads us to expect that $f'(2)$ is negative. To compute $f'(2)$ exactly, we use algebra.

Figure 2.17: Tangent line to $f(x) = 1/x$ at $x = 2$

Example 8   Find the derivative of $f(x) = 1/x$ at the point $x = 2$.

Solution   The derivative is the limit of the difference quotient, so we look at

$$f'(2) = \lim_{h \to 0} \frac{f(2 + h) - f(2)}{h}.$$

Using the formula for f and simplifying gives

$$f'(2) = \lim_{h \to 0} \frac{1}{h} \left(\frac{1}{2 + h} - \frac{1}{2} \right) = \lim_{h \to 0} \left(\frac{2 - (2 + h)}{2h(2 + h)} \right) = \lim_{h \to 0} \frac{-h}{2h(2 + h)}.$$

Since the limit only examines values of h close to, but not equal to, zero, we can cancel h. We get

$$f'(2) = \lim_{h \to 0} \frac{-1}{2(2 + h)} = -\frac{1}{4}.$$

Thus, $f'(2) = -1/4$. The slope of the tangent line in Figure 2.17 is $-1/4$.

Exercises and Problems for Section 2.2

Exercises

1. The table shows values of $f(x) = x^3$ near $x = 2$ (to three decimal places). Use it to estimate $f'(2)$.

x	1.998	1.999	2.000	2.001	2.002
x^3	7.976	7.988	8.000	8.012	8.024

2. (a) Make a table of values rounded to two decimal places for the function $f(x) = e^x$ for $x = 1, 1.5, 2, 2.5,$ and 3. Then use the table to answer parts (b) and (c).
  (b) Find the average rate of change of $f(x)$ between $x = 1$ and $x = 3$.
  (c) Use average rates of change to approximate the instantaneous rate of change of $f(x)$ at $x = 2$.

3. (a) Make a table of values, rounded to two decimal places, for $f(x) = \log x$ (that is, log base 10) with $x = 1, 1.5, 2, 2.5, 3$. Then use this table to answer parts (b) and (c).
  (b) Find the average rate of change of $f(x)$ between $x = 1$ and $x = 3$.
  (c) Use average rates of change to approximate the instantaneous rate of change of $f(x)$ at $x = 2$.

4. (a) Let $f(x) = x^2$. Explain what Table 2.4 tells us about $f'(1)$.
  (b) Find $f'(1)$ exactly.
  (c) If x changes by 0.1 near $x = 1$, what does $f'(1)$ tell us about how $f(x)$ changes? Illustrate your answer with a sketch.

Table 2.4

x	x^2	Difference in successive x^2 values
0.998	0.996004	
		0.001997
0.999	0.998001	
		0.001999
1.000	1.000000	
		0.002001
1.001	1.002001	
		0.002003
1.002	1.004004	

5. Graph $f(x) = \sin x$, and use the graph to decide whether the derivative of $f(x)$ at $x = 3\pi$ is positive or negative.

6. For the function $f(x) = \log x$, estimate $f'(1)$. From the graph of $f(x)$, would you expect your estimate to be greater than or less than $f'(1)$?

7. Estimate $f'(2)$ for $f(x) = 3^x$. Explain your reasoning.

8. Figure 2.18 shows the graph of f. Match the derivatives in the table with the points a, b, c, d, e.

x	$f'(x)$
d	0
b	0.5
c	2
a	-0.5
e	-2

Figure 2.18

9. Label points $A, B, C, D, E,$ and F on the graph of $y = f(x)$ in Figure 2.19.

  (a) Point A is a point on the curve where the derivative is negative.
  (b) Point B is a point on the curve where the value of the function is negative.
  (c) Point C is a point on the curve where the derivative is largest.
  (d) Point D is a point on the curve where the derivative is zero.
  (e) Points E and F are different points on the curve where the derivative is about the same.

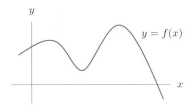

Figure 2.19

10. The graph of $y = f(x)$ is shown in Figure 2.20. Which is larger in each of the following pairs?

Figure 2.20

  (a) Average rate of change: Between $x = 1$ and $x = 3$? Or between $x = 3$ and $x = 5$?
  (b) $f(2)$ or $f(5)$?
  (c) $f'(1)$ or $f'(4)$?

Problems

11. Suppose that $f(x)$ is a function with $f(100) = 35$ and $f'(100) = 3$. Estimate $f(102)$.

12. The function in Figure 2.21 has $f(4) = 25$ and $f'(4) = 1.5$. Find the coordinates of the points A, B, C.

Figure 2.21

13. Use Figure 2.22 to fill in the blanks in the following statements about the function g at point B.

(a) $g(\underline{}) = \underline{}$ (b) $g'(\underline{}) = \underline{}$

Figure 2.22

14. Show how to represent the following on Figure 2.23.

(a) $f(4)$ (b) $f(4) - f(2)$

(c) $\dfrac{f(5) - f(2)}{5 - 2}$ (d) $f'(3)$

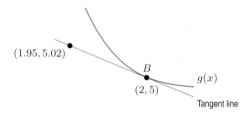

Figure 2.23

15. For each of the following pairs of numbers, use Figure 2.23 to decide which is larger. Explain your answer.

(a) $f(3)$ or $f(4)$?

(b) $f(3) - f(2)$ or $f(2) - f(1)$?

(c) $\dfrac{f(2) - f(1)}{2 - 1}$ or $\dfrac{f(3) - f(1)}{3 - 1}$?

(d) $f'(1)$ or $f'(4)$?

16. With the function f given by Figure 2.23, arrange the following quantities in ascending order:

$$0, \quad f'(2), \quad f'(3), \quad f(3) - f(2)$$

17. Suppose $y = f(x)$ graphed in Figure 2.23 represents the cost of manufacturing x kilograms of a chemical. Then $f(x)/x$ represents the average cost of producing 1 kilogram when x kilograms are made. This problem asks you to visualize these averages graphically.

(a) Show how to represent $f(4)/4$ as the slope of a line.

(b) Which is larger, $f(3)/3$ or $f(4)/4$?

18. On a copy of Figure 2.24, mark lengths that represent the quantities in parts (a) – (d). (Pick any convenient x, and assume $h > 0$.)

(a) $f(x)$ (b) $f(x+h)$ (c) $f(x+h) - f(x)$ (d) h

(e) Using your answers to parts (a)–(d), show how the quantity $\dfrac{f(x+h) - f(x)}{h}$ can be represented as the slope of a line on the graph.

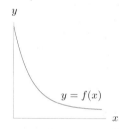

Figure 2.24

19. Consider the function shown in Figure 2.25.

(a) Write an expression involving f for the slope of the line joining A and B.

(b) Draw the tangent line at C. Compare its slope to the slope of the line in part (a).

(c) Are there any other points on the curve at which the slope of the tangent line is the same as the slope of the tangent line at C? If so, mark them on the graph. If not, why not?

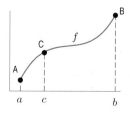

Figure 2.25

20. **(a)** Estimate $f'(0)$ if $f(x) = \sin x$, with x in degrees.

(b) In Example 4 on page 75, we found that the derivative of $\sin x$ at $x = 0$ was 1. Why do we get a different result here? (This problem shows why radians are almost always used in calculus.)

21. Estimate the derivative of $f(x) = x^x$ at $x = 2$.

22. For $y = f(x) = 3x^{3/2} - x$, use your calculator to construct a graph of $y = f(x)$, for $0 \le x \le 2$. From your graph, estimate $f'(0)$ and $f'(1)$.

23. Let $f(x) = \ln(\cos x)$. Use your calculator to approximate the instantaneous rate of change of f at the point $x = 1$. Do the same thing for $x = \pi/4$. (Note: Be sure that your calculator is set in radians.)

24. The population, P, of China, in billions, can be approximated by the function

$$P = 1.15(1.014)^t,$$

where t is the number of years since the start of 1993. According to this model, how fast is the population growing at the start of 1993 and at the start of 1995? Give your answers in millions of people per year.

25. **(a)** Graph $f(x) = \frac{1}{2}x^2$ and $g(x) = f(x) + 3$ on the same set of axes. What can you say about the slopes of the tangent lines to the two graphs at the point $x = 0$? $x = 2$? Any point $x = x_0$?

(b) Explain why adding a constant value, C, to any function does not change the value of the slope of its graph at any point. [Hint: Let $g(x) = f(x) + C$, and calculate the difference quotients for f and g.]

26. Suppose Table 2.3 on page 77 is continued with smaller values of h. A particular calculator gives the results in Table 2.5. (Your calculator may give slightly different results.) Comment on the values of the difference quotient in Table 2.5. In particular, why is the last value of $(2^h - 1)/h$ zero? What do you expect the calculated value of $(2^h - 1)/h$ to be when $h = 10^{-20}$?

Table 2.5 *Questionable values of difference quotients of 2^x near $x = 0$*

h	Difference quotient: $(2^h - 1)/h$
10^{-4}	0.6931712
10^{-6}	0.693147
10^{-8}	0.6931
10^{-10}	0.69
10^{-12}	0

Use algebra to evaluate the limits in Problems 27–32.

27. $\displaystyle\lim_{h \to 0} \frac{(-3 + h)^2 - 9}{h}$

28. $\displaystyle\lim_{h \to 0} \frac{(2 - h)^3 - 8}{h}$

29. $\displaystyle\lim_{h \to 0} \frac{1/(1 + h) - 1}{h}$

30. $\displaystyle\lim_{h \to 0} \frac{1/(1 + h)^2 - 1}{h}$

31. $\displaystyle\lim_{h \to 0} \frac{\sqrt{4 + h} - 2}{h}$ [Hint: Multiply by $\sqrt{4 + h} + 2$ in numerator and denominator.]

32. $\displaystyle\lim_{h \to 0} \frac{1/\sqrt{4 + h} - 1/2}{h}$

Find the derivatives in Problems 33–38 algebraically.

33. $f(x) = 5x^2$ at $x = 10$

34. $f(x) = x^3$ at $x = -2$

35. $g(t) = t^2 + t$ at $t = -1$

36. $f(x) = x^3 + 5$ at $x = 1$

37. $g(x) = 1/x$ at $x = 2$

38. $g(z) = z^{-2}$, find $g'(2)$

For Problems 39–42, find the equation of the line tangent to the function at the given point.

39. $f(x) = 5x^2$ at $x = 10$

40. $f(x) = x^3$ at $x = -2$

41. $f(x) = x$ at $x = 20$

42. $f(x) = 1/x^2$ at $(1, 1)$

2.3   THE DERIVATIVE FUNCTION

In the previous section we looked at the derivative of a function at a fixed point. Now we consider what happens at a variety of points. The derivative generally takes on different values at different points and is itself a function.

First, remember that the derivative of a function at a point tells us the rate at which the value of the function is changing at that point. Geometrically, we can think of the derivative as the slope of the curve or of the tangent line at the point.

Example 1   Estimate the derivative of the function $f(x)$ graphed in Figure 2.26 at $x = -2, -1, 0, 1, 2, 3, 4, 5$.

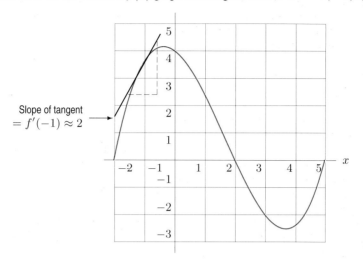

Figure 2.26: Estimating the derivative graphically as the slope of the tangent line

Solution   From the graph we estimate the derivative at any point by placing a straightedge so that it forms the tangent line at that point, and then using the grid squares to estimate the slope of the straightedge. For example, the tangent at $x = -1$ is drawn in Figure 2.26, and has a slope of about 2, so $f'(-1) \approx 2$. Notice that the slope at $x = -2$ is positive and fairly large; the slope at $x = -1$ is positive but smaller. At $x = 0$, the slope is negative, by $x = 1$ it has become more negative, and so on. Some estimates of the derivative are listed in Table 2.6. You should check these values. Are they reasonable? Is the derivative positive where you expect? Negative?

Table 2.6   *Estimated values of derivative of function in Figure 2.26*

x	-2	-1	0	1	2	3	4	5
$f'(x)$	6	2	-1	-2	-2	-1	1	4

Notice that for every x-value, there's a corresponding value of the derivative. Therefore, the derivative is itself a function of x.

For any function f, we define the **derivative function**, f', by

$$f'(x) = \text{Rate of change of } f \text{ at } x = \lim_{h \to 0} \frac{f(x+h) - f(x)}{h}.$$

For every x-value for which this limit exists, we say f is *differentiable at* that x-value. If the limit exists for all x in the domain of f, we say f is *differentiable everywhere*. Most functions we meet are differentiable at every point in their domain, except perhaps for a few isolated points.

The Derivative Function: Graphically

Example 2   Sketch the graph of the derivative of the function shown in Figure 2.26.

Solution   We plot the values of this derivative given in Table 2.6. We obtain Figure 2.27, which shows a graph of the derivative (the black curve), along with the original function (color).

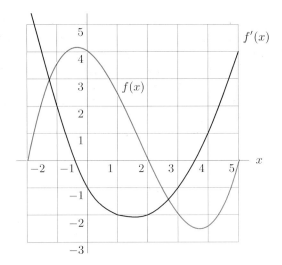

Figure 2.27: Function (colored) and derivative (black) from Example 1

You should check that this graph of f' makes sense. Where the values of f' are positive, f is increasing ($x < -0.3$ or $x > 3.8$) and where f' is negative, f is decreasing. Notice that at the points where f has large positive slope, such as $x = -2$, the graph of the derivative is far above the x-axis, as it should be, since the value of the derivative is large there. At points where the slope is gentler, such as $x = -1$, the graph of f' is closer to the x-axis, since the derivative is smaller.

What Does the Derivative Tell Us Graphically?

Where f' is positive, the tangent line to f is sloping up; where f' is negative, the tangent line to f is sloping down. If $f' = 0$ everywhere, then the tangent line to f is horizontal everywhere, and f is constant. We see that the sign of f' tells us whether f is increasing or decreasing.

> If $f' > 0$ on an interval, then f is *increasing* over that interval.
> If $f' < 0$ on an interval, then f is *decreasing* over that interval.

Moreover, the magnitude of the derivative gives us the magnitude of the rate of change; so if f' is large (positive or negative), then the graph of f is steep (up or down), whereas if f' is small the graph of f slopes gently. With this in mind, we can learn about the behavior of a function from the behavior of its derivative.

The Derivative Function: Numerically

If we are given values of a function instead of its graph, we can estimate values of the derivative.

Example 3   Table 2.7 gives values of $c(t)$, the concentration (μg/cm^3) of a drug in the bloodstream at time t (min). Construct a table of estimated values for $c'(t)$, the rate of change of $c(t)$ with respect to time.

Table 2.7 *Concentration as a function of time*

t (min)	0	0.1	0.2	0.3	0.4	0.5	0.6	0.7	0.8	0.9	1.0
$c(t)$ (μg/cm^3)	0.84	0.89	0.94	0.98	1.00	1.00	0.97	0.90	0.79	0.63	0.41

Solution   We estimate values of c' using the values in the table. To do this, we have to assume that the data points are close enough together that the concentration does not change wildly between them. From the table, we see that the concentration is increasing between $t = 0$ and $t = 0.4$, so we expect a positive derivative there. However, the increase is quite slow, so we expect the derivative to be small. The concentration does not change between 0.4 and 0.5, so we expect the derivative to be roughly 0 there. From $t = 0.5$ to $t = 1.0$, the concentration starts to decrease, and the rate of decrease gets larger and larger, so we expect the derivative to be negative and of greater and greater magnitude.

Using the data in the table, we estimate the derivative using the difference quotient:

$$c'(t) \approx \frac{c(t + h) - c(t)}{h}.$$

Since the data points are 0.1 apart, we use $h = 0.1$, giving

$$c'(0) \approx \frac{c(0.1) - c(0)}{0.1} = \frac{0.89 - 0.84}{0.1} = 0.5 \ \mu\text{g/cm}^3/\text{min}$$

$$c'(0.1) \approx \frac{c(0.2) - c(0.1)}{0.1} = \frac{0.94 - 0.89}{0.1} = 0.5 \ \mu\text{g/cm}^3/\text{min}$$

$$c'(0.2) \approx \frac{c(0.3) - c(0.2)}{0.1} = \frac{0.98 - 0.94}{0.1} = 0.4 \ \mu\text{g/cm}^3/\text{min}$$

$$c'(0.3) \approx \frac{c(0.4) - c(0.3)}{0.1} = \frac{1.00 - 0.98}{0.1} = 0.2 \ \mu\text{g/cm}^3/\text{min}$$

$$c'(0.4) \approx \frac{c(0.5) - c(0.4)}{0.1} = \frac{1.00 - 1.00}{0.1} = 0.0 \ \mu\text{g/cm}^3/\text{min}$$

and so on. These values are tabulated in Table 2.8. Notice that the derivative has small positive values up until $t = 0.4$, where it is roughly 0, and then it gets more and more negative, as we expected. The slopes are shown on the graph of $c(t)$ in Figure 2.28.

Table 2.8 *Estimated derivative of concentration*

t	$c'(t)$
0	0.5
0.1	0.5
0.2	0.4
0.3	0.2
0.4	0.0
0.5	−0.3
0.6	−0.7
0.7	−1.1
0.8	−1.6
0.9	−2.2

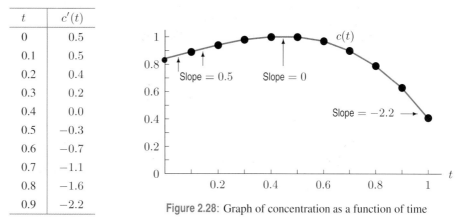

Figure 2.28: Graph of concentration as a function of time

Improving Numerical Estimates for the Derivative

In the previous example, the estimate for the derivative at 0.2 used the interval to the right; we found the average rate of change between $t = 0.2$ and $t = 0.3$. However, we could equally well have gone to the left and used the rate of change between $t = 0.1$ and $t = 0.2$ to approximate the derivative at 0.2. For a more accurate result, we could average these slopes and say

$$c'(0.2) \approx \frac{1}{2} \left(\begin{matrix} \text{Slope to left} \\ \text{of 0.2} \end{matrix} + \begin{matrix} \text{Slope to right} \\ \text{of 0.2} \end{matrix} \right) = \frac{0.5 + 0.4}{2} = 0.45.$$

In general, averaging the slopes leads to a more accurate answer.

Derivative Function: From a Formula

If we are given a formula for f, can we come up with a formula for f'? We often can, as shown in the next example. Indeed, much of the power of calculus depends on our ability to find formulas for the derivatives of all the functions we described earlier. This is done systematically in Chapter 3.

Derivative of a Constant Function

The graph of a constant function $f(x) = k$ is a horizontal line, with a slope of 0 everywhere. Therefore, its derivative is 0 everywhere. (See Figure 2.29.)

$$\boxed{\text{If } f(x) = k, \text{ then } f'(x) = 0.}$$

Figure 2.29: A constant function

Derivative of a Linear Function

We already know that the slope of a straight line is constant. This tells us that the derivative of a linear function is constant.

$$\boxed{\text{If } f(x) = b + mx, \text{ then } f'(x) = \text{Slope} = m.}$$

Derivative of a Power Function

Example 4   Find a formula for the derivative of $f(x) = x^2$.

Solution   Before computing the formula for $f'(x)$ algebraically, let's try to guess the formula by looking for a pattern in the values of $f'(x)$. Table 2.9 contains values of $f(x) = x^2$ (rounded to three decimals), which we can use to estimate the values of $f'(1)$, $f'(2)$, and $f'(3)$.

Table 2.9   *Values of $f(x) = x^2$ near $x = 1$, $x = 2$, $x = 3$ (rounded to three decimals)*

x	x^2	x	x^2	x	x^2
0.999	0.998	1.999	3.996	2.999	8.994
1.000	1.000	2.000	4.000	3.000	9.000
1.001	1.002	2.001	4.004	3.001	9.006
1.002	1.004	2.002	4.008	3.002	9.012

Near $x = 1$, the value of x^2 increases by about 0.002 each time x increases by 0.001, so

$$f'(1) \approx \frac{0.002}{0.001} = 2.$$

Similarly, near $x = 2$ and $x = 3$, the value of x^2 increases by about 0.004 and 0.006, respectively, when x increases by 0.001. So

$$f'(2) \approx \frac{0.004}{0.001} = 4 \quad \text{and} \quad f'(3) \approx \frac{0.006}{0.001} = 6.$$

Knowing the value of f' at specific points can never tell us the formula for f', but it certainly can be suggestive: Knowing $f'(1) \approx 2$, $f'(2) \approx 4$, $f'(3) \approx 6$ suggests that $f'(x) = 2x$.

The derivative is calculated by forming the difference quotient and taking the limit as h goes to zero. The difference quotient is

$$\frac{f(x+h) - f(x)}{h} = \frac{(x+h)^2 - x^2}{h} = \frac{x^2 + 2xh + h^2 - x^2}{h} = \frac{2xh + h^2}{h}.$$

Since h never actually reaches zero, we can cancel it in the last expression to get $2x + h$. The limit of this as h goes to zero is $2x$, so

$$f'(x) = \lim_{h \to 0} (2x + h) = 2x.$$

Example 5   Calculate $f'(x)$ if $f(x) = x^3$.

Solution   We look at the difference quotient

$$\frac{f(x+h) - f(x)}{h} = \frac{(x+h)^3 - x^3}{h}.$$

Multiplying out gives $(x+h)^3 = x^3 + 3x^2h + 3xh^2 + h^3$, so

$$f'(x) = \lim_{h \to 0} \frac{x^3 + 3x^2h + 3xh^2 + h^3 - x^3}{h} = \lim_{h \to 0} \frac{3x^2h + 3xh^2 + h^3}{h}.$$

Since in taking the limit as $h \to 0$, we consider values of h near, but not equal to, zero, we can cancel h giving

$$f'(x) = \lim_{h \to 0} \frac{3x^2h + 3xh^2 + h^3}{h} = \lim_{h \to 0} (3x^2 + 3xh + h^2).$$

As $h \to 0$, the value of $(3xh + h^2) \to 0$ so

$$f'(x) = \lim_{h \to 0} (3x^2 + 3xh + h^2) = 3x^2.$$

The previous two examples show how to compute the derivatives of power functions of the form $f(x) = x^n$, when n is 2 or 3. We can use the Binomial Theorem to show the *power rule* for a positive integer n:

$$\boxed{\text{If } f(x) = x^n \text{ then } f'(x) = nx^{n-1}.}$$

This result is in fact valid for any real value of n.

Exercises and Problems for Section 2.3

Exercises

For Exercises 1–9, sketch a graph of the derivative function of each of the given functions.

1.

2.

3.

4.

5.

6.

7.

8.

9.

10. For $f(x) = \ln x$, construct tables, rounded to four decimals, near $x = 1$, $x = 2$, $x = 5$, and $x = 10$. Use the tables to estimate $f'(1)$, $f'(2)$, $f'(5)$, and $f'(10)$. Then guess a general formula for $f'(x)$.

11. (a) Estimate $f'(2)$ using the values of f in the table.
(b) For what values of x does $f'(x)$ appear to be positive? Negative?

x	0	2	4	6	8	10	12
$f(x)$	10	18	24	21	20	18	15

12. Find approximate values for $f'(x)$ at each of the x-values given in the following table.

x	0	5	10	15	20
$f(x)$	100	70	55	46	40

In Exercises 13–14, find a formula for the derivative using the power rule. Confirm it using difference quotients.

13. $k(x) = 1/x$ **14.** $l(x) = 1/x^2$

Find a formula for the derivatives of the functions in Exercises 15–16 using difference quotients.

15. $g(x) = 2x^2 - 3$ **16.** $m(x) = 1/(x+1)$

For Exercises 17–20, sketch the graph of $f(x)$, and use this graph to sketch the graph of $f'(x)$.

17. $f(x) = x^2$ **18.** $f(x) = x(x-1)$

19. $f(x) = \cos x$ **20.** $f(x) = \log x$

Problems

21. Given the numerical values shown, find approximate values for the derivative of $f(x)$ at each of the x-values given. Where is the rate of change of $f(x)$ positive? Where is it negative? Where does the rate of change of $f(x)$ seem to be greatest?

x	0	1	2	3	4	5	6	7	8
$f(x)$	18	13	10	9	9	11	15	21	30

22. Values of x and $g(x)$ are given in the table. For what value of x is $g'(x)$ closest to 3?

x	2.7	3.2	3.7	4.2	4.7	5.2	5.7	6.2
$g(x)$	3.4	4.4	5.0	5.4	6.0	7.4	9.0	11.0

For Problems 23–32, sketch the graph of $f'(x)$.

23.

24.

25.

26.

27.

28.

29.

30.

31.

32.

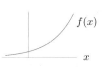

33. In the graph of f in Figure 2.30, at which of the labeled x-values is

(a) $f(x)$ greatest?   (b) $f(x)$ least?

(c) $f'(x)$ greatest?   (d) $f'(x)$ least?

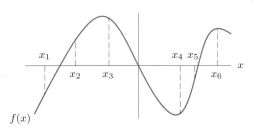

Figure 2.30

34. Figure 2.31 shows a graph of voltage across an electrical capacitor as a function of time. The current is proportional to the derivative of the voltage; the constant of proportionality is positive. Sketch a graph of the current as a function of time.

Figure 2.31

35. A vehicle moving along a straight road has distance $f(t)$ from its starting point at time t. Which of the graphs in Figure 2.32 could be $f'(t)$ for the following scenarios? (Assume the scales on the vertical axes are all the same.)

(a) A bus on a popular route, with no traffic
(b) A car with no traffic and all green lights
(c) A car in heavy traffic conditions

Figure 2.32

36. The derivative of f is the spike function in Figure 2.33. What can you say about the graph of f?

37. A child inflates a balloon, admires it for a while and then lets the air out at a constant rate. If $V(t)$ gives the volume of the balloon at time t, then Figure 2.34 shows $V'(t)$ as a function of t. At what time does the child:

(a) Begin to inflate the balloon?
(b) Finish inflating the balloon?
(c) Begin to let the air out?
(d) What would the graph of $V'(t)$ look like if the child had alternated between pinching and releasing the open end of the balloon, instead of letting the air out at a constant rate?

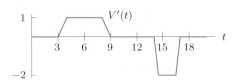

Figure 2.34

38. The population of a herd of deer is modeled by

$$P(t) = 4000 + 500 \sin\left(2\pi t - \frac{\pi}{2}\right)$$

where t is measured in years from January 1.

(a) How does this population vary with time? Sketch a graph of $P(t)$ for one year.
(b) Use the graph to decide when in the year the population is a maximum. What is that maximum? Is there a minimum? If so, when?
(c) Use the graph to decide when the population is growing fastest. When is it decreasing fastest?
(d) Estimate roughly how fast the population is changing on the first of July.

39. The graph in Figure 2.35 shows the accumulated federal debt since 1970. Sketch the derivative of this function. What does it represent?

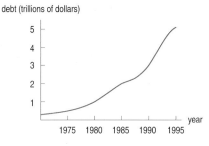

debt (trillions of dollars)

Figure 2.35

40. Draw the graph of a continuous function $y = f(x)$ that satisfies the following three conditions.

- $f'(x) > 0$ for $x < -2$,
- $f'(x) < 0$ for $-2 < x < 2$,
- $f'(x) = 0$ for $x > 2$.

41. Draw the graph of a continuous function $y = f(x)$ that satisfies the following three conditions.

- $f'(x) > 0$ for $-\frac{\pi}{2} < x < \frac{\pi}{2}$,
- $f'(x) < 0$ for $-\pi < x < -\frac{\pi}{2}$ and $\frac{\pi}{2} < x < \pi$,
- $f'(x) = 0$ at $x = -\frac{\pi}{2}$ and $x = \frac{\pi}{2}$.

42. Figure 2.36 is the graph of f', the derivative of a function f. On what interval(s) is the function f

(a) Increasing?   **(b)** Decreasing?

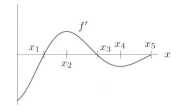

Figure 2.36: Graph of f', not f

43. Using a graph, explain why if $f(x)$ is an even function, then $f'(x)$ is odd.

44. Using a graph, explain why if $g(x)$ is an odd function, then $g'(x)$ is even.

2.4   INTERPRETATIONS OF THE DERIVATIVE

We have seen the derivative interpreted as a slope and as a rate of change. In this section, we see other interpretations. The purpose of these examples is not to make a catalog of interpretations but to illustrate the process of obtaining them.

An Alternative Notation for the Derivative

So far we have used the notation f' to stand for the derivative of the function f. An alternative notation for derivatives was introduced by the German mathematician Wilhelm Gottfried Leibniz

(1646–1716). If the variable y depends on the variable x, that is, if

$$y = f(x),$$

then he wrote dy/dx for the derivative, so

$$\frac{dy}{dx} = f'(x).$$

Leibniz's notation is quite suggestive if we think of the letter d in dy/dx as standing for "small difference in . . . ." The notation dy/dx reminds us that the derivative is a limit of ratios of the form

$$\frac{\text{Difference in } y\text{-values}}{\text{Difference in } x\text{-values}}.$$

The notation dy/dx suggests the units for the derivative: the units for y divided by the units for x. The separate entities dy and dx officially have no independent meaning: they are all part of one notation. In fact, a good way to view the notation dy/dx is to think of d/dx as a single symbol meaning "the derivative with respect to x of . . .". So dy/dx can be viewed as

$$\frac{d}{dx}(y), \quad \text{meaning "the derivative with respect to } x \text{ of } y."$$

On the other hand, many scientists and mathematicians think of dy and dx as separate entities representing "infinitesimally" small differences in y and x, even though it is difficult to say exactly how small "infinitesimal" is. Although not formally correct, it can be helpful to think of dy/dx as a small change in y divided by a small change in x.

For example, recall that if $s = f(t)$ is the position of a moving object at time t, then $v = f'(t)$ is the velocity of the object at time t. Writing

$$v = \frac{ds}{dt}$$

reminds us that v is a velocity, since the notation suggests a distance, ds, over a time, dt, and we know that distance over time is velocity. Similarly, we recognize

$$\frac{dy}{dx} = f'(x)$$

as the slope of the graph of $y = f(x)$ since slope is vertical rise, dy, over horizontal run, dx.

The disadvantage of Leibniz's notation is that it is awkward to specify the x-value at which we are evaluating the derivative. To specify $f'(2)$, for example, we have to write

$$\left.\frac{dy}{dx}\right|_{x=2}.$$

Using Units to Interpret the Derivative

The following examples illustrate how useful units can be in suggesting interpretations of the derivative. We use the fact that the units of the instantaneous and the average rate of change are the same.

For example, suppose $s = f(t)$ gives the distance, in meters, of a body from a fixed point as a function of time, t, in seconds. Then knowing that

$$\left.\frac{ds}{dt}\right|_{t=2} = f'(2) = 10 \text{ meters/sec}$$

tells us that when $t = 2$ seconds, the body is moving at an instantaneous velocity of 10 meters/sec. This means that if the body continued to move at this speed for a whole second, it would move 10 meters. In practice, however, the velocity of the body may not remain 10 meters/sec for long. Notice that the units of instantaneous velocity and of average velocity are the same.

Example 1   The cost C (in dollars) of building a house A square feet in area is given by the function $C = f(A)$. What is the practical interpretation of the function $f'(A)$?

Solution   In the alternative notation,

$$f'(A) = \frac{dC}{dA}.$$

This is a cost divided by an area, so it is measured in dollars per square foot. You can think of dC as the extra cost of building an extra dA square feet of house. Then you can think of dC/dA as the additional cost per square foot. So if you are planning to build a house roughly A square feet in area, $f'(A)$ is the cost per square foot of the *extra* area involved in building a slightly larger house, and is called the *marginal cost*. The marginal cost is probably smaller than the average cost per square foot for the entire house, since once you are already set up to build a large house, the cost of adding a few square feet is likely to be small.

Example 2   The cost of extracting T tons of ore from a copper mine is $C = f(T)$ dollars. What does it mean to say that $f'(2000) = 100$?

Solution   In the alternative notation,

$$f'(2000) = \left.\frac{dC}{dT}\right|_{T=2000} = 100.$$

Since C is measured in dollars and T is measured in tons, dC/dT must be measured in dollars per ton. So the statement $f'(2000) = 100$ says that when 2000 tons of ore have already been extracted from the mine, the cost of extracting the next ton is approximately \$100.

Example 3   If $q = f(p)$ gives the number of pounds of sugar produced when the price per pound is p dollars, then what are the units and the meaning of the statement $f'(3) = 50$?

Solution   Since $f'(3)$ is the limit as $h \to 0$ of the difference quotient

$$\frac{f(3+h) - f(3)}{h},$$

the units of $f'(3)$ and the difference quotient are the same. Since $f(3+h) - f(3)$ is in pounds and h is in dollars, the units of the difference quotient and $f'(3)$ are pounds/dollar. The statement

$$f'(3) = 50 \text{ pounds/dollar}$$

tells us that the instantaneous rate of change of q with respect to p is 50 when $p = 3$. In other words, when the price is \$3, the quantity produced is increasing at 50 pounds/dollar. Thus, if the price increased by a dollar, the quantity produced would increase by approximately 50 pounds.

Example 4   You are told that water is flowing through a pipe at a constant rate of 10 cubic feet per second. Interpret this rate as the derivative of some function.

Solution   You might think at first that the statement has something to do with the velocity of the water, but in fact a flow rate of 10 cubic feet per second could be achieved either with very slowly moving water through a large pipe, or with very rapidly moving water through a narrow pipe. If we look at the units—cubic feet per second—we realize that we are being given the rate of change of a quantity measured in cubic feet. But a cubic foot is a measure of volume, so we are being told the rate of change of a volume. One way to visualize this is to imagine all the water that is flowing through the pipe ending up in a tank somewhere. Let $V(t)$ be the volume of water in the tank at time t. Then we are being told that the rate of change of $V(t)$ is 10, or

$$V'(t) = \frac{dV}{dt} = 10.$$

Example 5   Suppose $P = f(t)$ is the population of Mexico in millions, where t is the number of years since 1980. Explain the meaning of the statements:

(a) $f'(6) = 2$ (b) $f^{-1}(95.5) = 16$ (c) $(f^{-1})'(95.5) = 0.46$

Solution

(a) The units of P are millions of people, the units of t are years, so the units of $f'(t)$ are millions of people per year. Therefore the statement $f'(6) = 2$ tells us that at $t = 6$ (that is, in 1986), the population of Mexico was increasing at 2 million people per year.

(b) The statement $f^{-1}(95.5) = 16$ tells us that the year when the population was 95.5 million was $t = 16$ (that is, in 1996).

(c) The units of $(f^{-1})'(P)$ are years per million of population. The statement $(f^{-1})'(95.5) = 0.46$ tells us that when the population was 95.5 million, it took about 0.46 years for the population to increase by 1 million.

Exercises and Problems for Section 2.4

Exercises

1. The temperature, H, in degrees Celsius, of a cup of coffee placed on the kitchen counter is given by $H = f(t)$, where t is in minutes since the coffee was put on the counter.

  (a) Is $f'(t)$ positive or negative? Give a reason for your answer.

  (b) What are the units of $f'(20)$? What is its practical meaning in terms of the temperature of the coffee?

2. The temperature, T, in degrees Fahrenheit, of a cold yam placed in a hot oven is given by $T = f(t)$, where t is the time in minutes since the yam was put in the oven.

  (a) What is the sign of $f'(t)$? Why?

  (b) What are the units of $f'(20)$? What is the practical meaning of the statement $f'(20) = 2$?

3. The cost, C (in dollars) to produce g gallons of ice cream can be expressed as $C = f(g)$. Using units, explain the meaning of the following statements in terms of ice cream.

  (a) $f(200) = 350$ **(b)** $f'(200) = 1.4$

4. The time for a chemical reaction, T (in minutes), is a function of the amount of catalyst present, a (in milliliters), so $T = f(a)$.

  (a) If $f(5) = 18$, what are the units of 5? What are the units of 18? What does this statement tell us about the reaction?

  (b) If $f'(5) = -3$, what are the units of 5? What are the units of -3? What does this statement tell us?

5. After investing $\$1000$ at an annual interest rate of 7% compounded continuously for t years, your balance is $\$B$, where $B = f(t)$. What are the units of dB/dt? What is the financial interpretation of dB/dt?

6. Suppose $C(r)$ is the total cost of paying off a car loan borrowed at an annual interest rate of $r\%$. What are the units of $C'(r)$? What is the practical meaning of $C'(r)$? What is its sign?

7. Suppose $P(t)$ is the monthly payment, in dollars, on a mortgage which will take t years to pay off. What are the units of $P'(t)$? What is the practical meaning of $P'(t)$? What is its sign?

8. Investing $\$1000$ at an annual interest rate of $r\%$, compounded continuously, for 10 years gives you a balance of $\$B$, where $B = g(r)$. Give a financial interpretation of the statements:

  (a) $g(5) \approx 1649$.

  (b) $g'(5) \approx 165$. What are the units of $g'(5)$?

9. Let $f(x)$ be the elevation in feet of the Mississippi river x miles from its source. What are the units of $f'(x)$? What can you say about the sign of $f'(x)$?

10. An economist is interested in how the price of a certain item affects its sales. At a price of $\$p$, a quantity, q, of the item is sold. If $q = f(p)$, explain the meaning of each of the following statements:

  (a) $f(150) = 2000$ **(b)** $f'(150) = -25$

Problems

11. A laboratory study investigating the relationship between diet and weight in adult humans found that the weight of a subject, W, in pounds, was a function, $W = f(c)$, of the average number of Calories per day, c, consumed by the subject.

  (a) Interpret the statements $f(1800) = 155$, $f'(2000) = 0$, and $f^{-1}(162) = 2200$ in terms of diet and weight.

  (b) What are the units of $f'(c) = dW/dc$?

12. A city grew in population throughout the 1980s. The population was at its largest in 1990, and then shrank throughout the 1990s. Let $P = f(t)$ represent the population of the city t years since 1980. Sketch graphs of $f(t)$ and $f'(t)$, labeling the units on the axes.

13. If t is the number of years since 1993, the population, P, of China, in billions, can be approximated by the function

$$P = f(t) = 1.15(1.014)^t.$$

Estimate $f(6)$ and $f'(6)$, giving units. What do these two numbers tell you about the population of China?

14. For some painkillers, the size of the dose, D, given depends on the weight of the patient, W. Thus, $D = f(W)$, where D is in milligrams and W is in pounds.

(a) Interpret the statements $f(140) = 120$ and $f'(140) = 3$ in terms of this painkiller.

(b) Use the information in the statements in part (a) to estimate $f(145)$.

15. Let $f(t)$ be the number of centimeters of rainfall that has fallen since midnight, where t is the time in hours. Interpret the following in practical terms, giving units.

(a) $f(10) = 3.1$ (b) $f^{-1}(5) = 16$

(c) $f'(10) = 0.4$ (d) $(f^{-1})'(5) = 2$

16. Let $p(h)$ be the pressure in dynes per cm^2 on a diver at a depth of h meters below the surface of the ocean. What do each of the following quantities mean to the diver? Give units for the quantities.

(a) $p(100)$ (b) h such that $p(h) = 1.2 \cdot 10^6$

(c) $p(h) + 20$ (d) $p(h + 20)$

(e) $p'(100)$ (f) h such that $p'(h) = 20$

17. If $g(v)$ is the fuel efficiency, in miles per gallon, of a car going at v miles per hour, what are the units of $g'(90)$? What is the practical meaning of the statement $g'(55) = -0.54$?

18. (a) If you jump out of an airplane without a parachute, you fall faster and faster until air resistance causes you to approach a steady velocity, called the *terminal* velocity. Sketch a graph of your velocity against time.

(b) Explain the concavity of your graph.

(c) Assuming air resistance to be negligible at $t = 0$, what natural phenomenon is represented by the slope of the graph at $t = 0$?

19. A company's revenue from car sales, C (in thousands of dollars), is a function of advertising expenditure, a, in thousands of dollars, so $C = f(a)$.

(a) What does the company hope is true about the sign of f'?

(b) What does the statement $f'(100) = 2$ mean in practical terms? How about $f'(100) = 0.5$?

(c) Suppose the company plans to spend about $100,000 on advertising. If $f'(100) = 2$, should the company spend more or less than $100,000 on advertising? What if $f'(100) = 0.5$?

20. Let $P(x)$ be the number of people of height $\leq x$ inches in the US. What is the meaning of $P'(66)$? What are its units? Estimate $P'(66)$ (using common sense). Is $P'(x)$ ever negative? [Hint: You may want to approximate $P'(66)$ by a difference quotient, using $h = 1$. Also, you may assume the US population is about 250 million, and note that 66 inches = 5 feet 6 inches.]

21. When you breathe, a muscle (called the diaphragm) reduces the pressure around your lungs and they expand to fill with air. The table shows the volume of a lung as a function of the reduction in pressure from the diaphragm. Pulmonologists (lung doctors) define the *compliance* of the lung as the derivative of this function.[2]

(a) What are the units of compliance?

(b) Estimate the maximum compliance of the lung.

(c) Explain why the compliance gets small when the lung is nearly full (around 1 liter).

Pressure reduction (cm of water)	Volume (liters)
0	0.20
5	0.29
10	0.49
15	0.70
20	0.86
25	0.95
30	1.00

22. The compressibility index, γ, of cold matter (in a neutron star or black hole) is given by

$$\gamma = \frac{\delta + (p/c^2)}{p} \frac{dp}{d\delta},$$

where p is the pressure (in dynes/cm^2), δ is the density (in g/cm^3), and $c \approx 3 \cdot 10^{10}$ is the speed of light (in cm/sec). Figure 2.37 shows the relationship between δ, γ, and p. Values of $\log p$ are marked along the graph.[3]

(a) Estimate $dp/d\delta$ for cold iron, which has a density of about 10 g/cm^3. What does the magnitude of your answer tell you about cold iron?

(b) Estimate $dp/d\delta$ for the matter inside a white dwarf star, which has a density of about 10^6 g/cm^3. What does your answer tell you about matter inside a white dwarf?

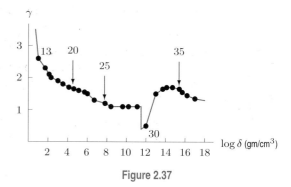

Figure 2.37

[2]Adapted from John B. West, *Respiratory Physiology* 4th Ed. (New York: Williams and Wilkins, 1990).

[3]From C. W. Misner, K. S. Thorne, and J. A. Wheeler, *Gravitation* (San Francisco: W. H. Freeman and Company, 1973).

2.5 THE SECOND DERIVATIVE

Since the derivative is itself a function, we can consider its derivative. For a function f, the derivative of its derivative is called the *second derivative*, and written f'' (read "f double-prime"). If $y = f(x)$, the second derivative can also be written as $\dfrac{d^2y}{dx^2}$, which means $\dfrac{d}{dx}\left(\dfrac{dy}{dx}\right)$, the derivative of $\dfrac{dy}{dx}$.

What Do Derivatives Tell Us?

Recall that the derivative of a function tells you whether a function is increasing or decreasing:

- If $f' > 0$ on an interval, then f is *increasing* over that interval.

- If $f' < 0$ on an interval, then f is *decreasing* over that interval.

Since f'' is the derivative of f',

- If $f'' > 0$ on an interval, then f' is *increasing* over that interval.

- If $f'' < 0$ on an interval, then f' is *decreasing* over that interval.

What does it mean for f' to be increasing or decreasing? An example in which f' is increasing is shown in Figure 2.38, where the curve is bending upward, or is *concave up*. In the example shown in Figure 2.39, in which f' is decreasing, the graph is bending downward, or is *concave down*. These figures suggest the following result:

If $f'' > 0$ on an interval, then f' is increasing, so the graph of f is concave up there.
If $f'' < 0$ on an interval, then f' is decreasing, so the graph of f is concave down there.

Figure 2.38: Meaning of f'': The slope increases from left to right, f'' is positive, and f is concave up

Figure 2.39: Meaning of f'': The slope decreases from left to right, f'' is negative, and f is concave down

Warning! The graph of a function f can be concave up everywhere and yet have $f'' = 0$ at some point. For instance, the graph of $f(x) = x^4$ in Figure 2.40 is concave up, but it can be shown that $f''(0) = 0$. If we are told that the graph of a function f is concave up, we can be sure that f'' is not negative, that is $f'' \geq 0$, but not that f'' is positive, $f'' > 0$.

If the graph of f is concave up on an interval, then $f'' \geq 0$ there.
If the graph of f is concave down on an interval, then $f'' \leq 0$ there.

Figure 2.40: Graph of $f(x) = x^4$

Example 1   For the functions graphed in Figure 2.41, what can be said about the sign of the second derivative?.

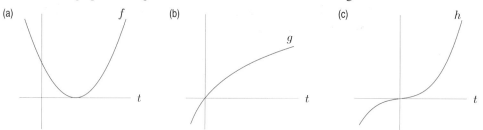

Figure 2.41: What signs do the second derivatives have?

Solution   (a) The graph of f is concave up everywhere, so $f'' \geq 0$ everywhere.
(b) The graph of g is concave down everywhere, so $g'' \leq 0$ everywhere.
(c) For $t < 0$, the graph of h is concave down, so $h'' \leq 0$ there. For $t > 0$, the graph of h is concave up, so $h'' \geq 0$ there.

Example 2   Sketch the second derivative f'' for the function f of Example 1 on page 81, graphed with its derivative, f', in Figure 2.42. Relate the resulting graph of f'' to the graphs of f and f'.

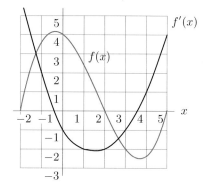

Figure 2.42: Function, f in color; derivative, f', in black

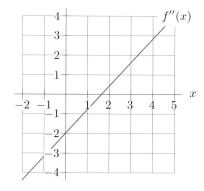

Figure 2.43: Graph of f''

Solution   We want to sketch the derivative of f'. We do this by estimating the slopes of f' and plotting them, obtaining Figure 2.43.

We observe that where $f'' > 0$, the graph of f is concave up and f' is increasing, and that where $f'' < 0$, the graph of f is concave down and f' is decreasing. Where $f''(x) = 0$, the graph of f changes from concave up to concave down, and f' changes from decreasing to increasing.

Interpretation of the Second Derivative as a Rate of Change

If we think of the derivative as a rate of change, then the second derivative is a rate of change of a rate of change. If the second derivative is positive, the rate of change of f is increasing; if the second derivative is negative, the rate of change of f is decreasing.

The second derivative can be a matter of practical concern. In 1985 a newspaper headline reported the Secretary of Defense as saying that Congress had cut the defense budget. As his opponents pointed out, however, Congress had merely cut the rate at which the defense budget was increasing.[4] In other words, the derivative of the defense budget was still positive (the budget was increasing), but the second derivative was negative (the budget's rate of increase had slowed).

---

[4]In the *Boston Globe*, March 13, 1985, Representative William Gray (D–Pa.) was reported as saying: "It's confusing to the American people to imply that Congress threatens national security with reductions when you're really talking about a reduction in the increase."

Example 3   A population, P, growing in a confined environment often follows a *logistic* growth curve, like that shown in Figure 2.44; Relate the sign of d^2P/dt^2 to how the rate of growth, dP/dt, changes over time. What are practical interpretations of t_0 and L?

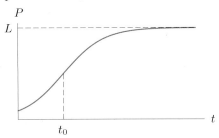

Figure 2.44: Logistic growth curve

Solution   For $t < t_0$, the rate of growth, dP/dt, is increasing and $d^2P/dt^2 \geq 0$. At t_0, the rate dP/dt is a maximum. In other words, at time t_0 the population is growing fastest. For $t > t_0$, the rate of growth, dP/dt, is decreasing and $dP^2/dt^2 \leq 0$. At t_0, the curve changes from concave up to concave down, and $d^2P/dt^2 = 0$ there.

 The quantity L represents the limiting value of the population as $t \to \infty$. Biologists call L the *carrying capacity* of the environment.

Example 4   Tests on the C5 Chevy Corvette sports car gave the results[5] in Table 2.10.

 (a) Estimate dv/dt for the time intervals shown.
 (b) What can you say about the sign of d^2v/dt^2 over the period shown?

Table 2.10   *Velocity of C5 Chevy Corvette*

Time, t (sec)	0	3	6	9	12
Velocity, v (meters/sec)	0	20	33	43	51

Solution   (a) For each time interval we can calculate the average rate of change of velocity. For example, from $t = 0$ to $t = 3$ we have

$$\frac{dv}{dt} \approx \text{Average rate of change of velocity} = \frac{20 - 0}{3 - 0} = 6.67\ \frac{\text{m/sec}}{\text{sec}}.$$

Estimated values of dv/dt are in Table 2.11.

 (b) Since the values of dv/dt are decreasing between the points shown, we expect $d^2v/dt^2 \leq 0$. The graph of v against t in Figure 2.45 supports this; it is concave down. The fact that $dv/dt > 0$ tells us that the car is speeding up; the fact that $d^2v/dt^2 \leq 0$ tells us that the rate of increase decreased (actually, did not increase) over this time period.

Table 2.11   *Estimates for dv/dt (meters/sec/sec)*

Time interval (sec)	$0 - 3$	$3 - 6$	$6 - 9$	$9 - 12$
Average rate of change (dv/dt)	6.67	4.33	3.33	2.67

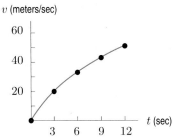

Figure 2.45: Velocity of C5 Chevy Corvette

[5]Adapted from the report in *Car and Driver*, February 1997.

Velocity and Acceleration

When a car is speeding up, we say that it is *accelerating*. We define *acceleration* as the rate of change of velocity with respect to time. If $v(t)$ is the velocity of an object at time t, we have

$$\text{Average acceleration from } t \text{ to } t+h = \frac{v(t+h) - v(t)}{h},$$

$$\text{Instantaneous acceleration} = v'(t) = \lim_{h \to 0} \frac{v(t+h) - v(t)}{h}.$$

If the term velocity or acceleration is used alone, it is assumed to be instantaneous. Since velocity is the derivative of position, acceleration is the second derivative of position. Summarizing:

If $y = s(t)$ is the position of an object at time t, then

- Velocity: $v(t) = \dfrac{dy}{dt} = s'(t)$.

- Acceleration: $a(t) = \dfrac{d^2y}{dt^2} = s''(t) = v'(t)$.

Example 5   A particle is moving along a straight line; its acceleration is zero only once. Its distance, s, to the right of a fixed point is given by Figure 2.46. Estimate:

(a) When the particle is moving to the right and when it is moving to the left.

(b) When the acceleration of the particle is zero, when it is negative, and when it is positive.

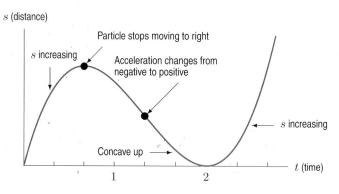

Figure 2.46: Distance of particle to right of a fixed point

Solution   (a) The particle is moving to the right whenever s is increasing. From the graph, this appears to be for $0 < t < \frac{2}{3}$ and for $t > 2$. For $\frac{2}{3} < t < 2$, the value of s is decreasing, so the particle is moving to the left.

(b) Since the acceleration is zero only once, this must be when the curve changes concavity, at about $t = \frac{4}{3}$. Then the acceleration is negative for $t < \frac{4}{3}$, since the graph is concave down there, and the acceleration is positive for $t > \frac{4}{3}$, since the graph is concave up there.

Exercises and Problems for Section 2.5

Exercises

1. For the function graphed in Figure 2.47, are the following nonzero quantities positive or negative?

(a) $f(2)$ ⁓   (b) $f'(2)$ ⁓   (c) $f''(2)$ +

Figure 2.47

2. At one of the labeled points on the graph in Figure 2.48 both dy/dx and d^2y/dx^2 are positive? Which is it?

2nd = Derivetive

Figure 2.48

3. At exactly two of the labeled points in Figure 2.49, the derivative f' is 0; the second derivative f'' is not zero at any of the labeled points. On a copy of the table, give the signs of f, f', f'' at each marked point.

Point	f	f'	f''
A			
B			
C			
D			

Figure 2.49

4. Sketch the graph of a function whose first and second derivatives are everywhere positive.

5. Sketch the graph of a function whose first derivative is everywhere negative and whose second derivative is positive for some x-values and negative for other x-values.

6. Sketch the graph of the height of a particle against time if velocity is positive and acceleration is negative.

For Exercises 7–12, give the signs of the first and second derivatives for each of the following functions. Each derivative is either positive everywhere, zero everywhere, or negative everywhere.

7.

8.

9.

10.

11.

12.

13. The position of a particle moving along the x-axis is given by $s(t) = 5t^2 + 3$. Use difference quotients to find the velocity $v(t)$ and acceleration $a(t)$.

Problems

14. The table gives the number of passenger cars, $C = f(t)$, in millions, in the US in the year t.

(a) Do $f'(t)$ and $f''(t)$ appear to be positive or negative during the period 1940–1980?

(b) Estimate $f'(1975)$. Using units, interpret your answer in terms of passenger cars.

t (year)	1940	1950	1960	1970	1980
C (cars, in millions)	27.5	40.3	61.7	89.3	121.6

15. An accelerating sports car goes from 0 mph to 60 mph in five seconds. Its velocity is given in the following table, converted from miles per hour to feet per second, so that all time measurements are in seconds. (Note: 1 mph is 22/15 ft/sec.) Find the average acceleration of the car over each of the first two seconds.

Time, t (sec)	0	1	2	3	4	5
Velocity, $v(t)$ (ft/sec)	0	30	52	68	80	88

16. Let $P(t)$ represent the price of a share of stock of a corporation at time t. What does each of the following statements tell us about the signs of the first and second derivatives of $P(t)$?

   (a) "The price of the stock is rising faster and faster."
   (b) "The price of the stock is close to bottoming out."

17. In economics, *total utility* refers to the total satisfaction from consuming some commodity. According to the economist Samuelson:[6]

> As you consume more of the same good, the total (psychological) utility increases. However, ...with successive new units of the good, your total utility will grow at a slower and slower rate because of a fundamental tendency for your psychological ability to appreciate more of the good to become less keen.

   (a) Sketch the total utility as a function of the number of units consumed.
   (b) In terms of derivatives, what is Samuelson saying?

18. An industry is being charged by the Environmental Protection Agency (EPA) with dumping unacceptable levels of toxic pollutants in a lake. Over a period of several months, an engineering firm makes daily measurements of the rate at which pollutants are being discharged into the lake. The engineers produce a graph similar to either Figure 2.50(a) or Figure 2.50(b). For each case, give an idea of what argument the EPA might make in court against the industry and of the industry's defense.

Figure 2.50

19. At which of the marked x-values in Figure 2.51 can the following statements be true?

   (a) $f(x) < 0$
   (b) $f'(x) < 0$
   (c) $f(x)$ is decreasing
   (d) $f'(x)$ is decreasing
   (e) Slope of $f(x)$ is positive
   (f) Slope of $f(x)$ is increasing

Figure 2.51

20. Figure 2.52 gives the position, $f(t)$, of a particle at time t. At which of the marked values of t can the following statements be true?

   (a) The position is positive
   (b) The velocity is positive
   (c) the acceleration is positive
   (d) The position is decreasing
   (e) The velocity is decreasing

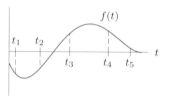

Figure 2.52

21. The graph of f' (not f) is given in Figure 2.53. At which of the marked values of x is

   (a) $f(x)$ greatest?   **(b)** $f(x)$ least?
   (c) $f'(x)$ greatest?   **(d)** $f'(x)$ least?
   (e) $f''(x)$ greatest?   **(f)** $f''(x)$ least?

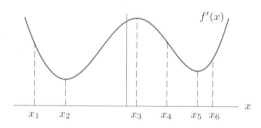

Figure 2.53: Graph of f', not f

22. A function f has $f(5) = 20$, $f'(5) = 2$, and $f''(x) < 0$, for $x \geq 5$. Which of the following are possible values for $f(7)$ and which are impossible?

   (a) 26   **(b)** 24   **(c)** 22

23. A continuous function defined for all x has the following properties:
   - f is increasing
   - $f(5) = 2$
   - f is concave down
   - $f'(5) = \frac{1}{2}$

   (a) Sketch a possible graph for f.
   (b) How many zeros does f have?
   (c) What can you say about the location of the zeros?
   (d) What is $\lim_{x \to -\infty} f(x)$?
   (e) Is it possible that $f'(1) = 1$?
   (f) Is it possible that $f'(1) = \frac{1}{4}$?

[6]From Paul A. Samuelson, *Economics*, 11th edition (New York: McGraw-Hill, 1981).

2.6   DIFFERENTIABILITY

What Does It Mean for a Function to be Differentiable?

A function is differentiable at a point if it has a derivative there. In other words:

> The function f is **differentiable** at x if
>
> $$\lim_{h \to 0} \frac{f(x+h) - f(x)}{h} \qquad \text{exists.,}$$
>
> Thus, the graph of f has a nonvertical tangent line at x. The value of the limit and the slope of the tangent line are the derivative of f at x.

Occasionally we meet a function which fails to have a derivative at a few points. A function fails to be differentiable at a point if:

- The function is not continuous at the point.
- The graph has a sharp corner at that point.
- The graph has a vertical tangent line.

Figure 2.54 shows a function which appears to be differentiable at all points except $x = a$ and $x = b$. There is no tangent at A because the graph has a corner there. As x approaches a from the left, the slope of the line joining P to A converges to some positive number. As x approaches a from the right, the slope of the line joining P to A converges to some negative number. Thus the slopes approach different numbers as we approach $x = a$ from different sides. Therefore the function is not differentiable at $x = a$. At B, the graph has a vertical tangent. As x approaches b, the slope of the line joining B to Q does not approach a limit; it just keeps growing larger and larger. Again, the limit defining the derivative does not exist and the function is not differentiable at $x = b$.

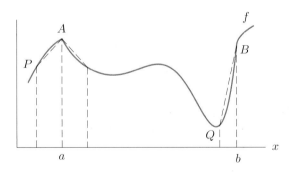

Figure 2.54: A function which is not differentiable at A or B

Figure 2.55: Graph of absolute value function, showing point of non-differentiability at $x = 0$

Examples of Nondifferentiable Functions

An example of a function whose graph has a corner is the *absolute value* function defined as follows:

$$f(x) = |x| = \begin{cases} x & \text{if } x \geq 0, \\ -x & \text{if } x < 0. \end{cases}$$

This function is called *piecewise linear* because each part of it is linear. Its graph is in Figure 2.55. Near $x = 0$, even close-up views of the graph of $f(x)$ look the same, so this is a corner which can't be straightened out by zooming in.

Example 1   Try to compute the derivative of the function $f(x) = |x|$ at $x = 0$. Is f differentiable there?

Solution   To find the derivative at $x = 0$, we want to look at

$$\lim_{h \to 0} \frac{f(h) - f(0)}{h} = \lim_{h \to 0} \frac{|h| - 0}{h} = \lim_{h \to 0} \frac{|h|}{h}.$$

As h approaches 0 from the right, h is positive, so $|h| = h$, and the ratio is always 1. As h approaches 0 from the left, h is negative, so $|h| = -h$, and the ratio is -1. Since the limits are different from each side, the limit of the difference quotient does not exist. Thus, the absolute value function is not differentiable at $x = 0$. The limits of 1 and -1 correspond to the fact that the slope of the right-hand part of the graph is 1, and the slope of the left-hand part is -1.

Example 2   Investigate the differentiability of $f(x) = x^{1/3}$ at $x = 0$.

Solution   This function is smooth at $x = 0$ (no sharp corners) but appears to have a vertical tangent there. (See Figure 2.56.) Looking at the difference quotient at $x = 0$, we see

$$\lim_{h \to 0} \frac{(0 + h)^{1/3} - 0^{1/3}}{h} = \lim_{h \to 0} \frac{h^{1/3}}{h} = \lim_{h \to 0} \frac{1}{h^{2/3}}.$$

As $h \to 0$ the denominator becomes small, so the fraction grows without bound. Hence, the function fails to have a derivative at $x = 0$.

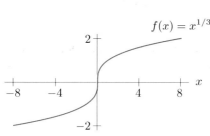

Figure 2.56: Continuous function not differentiable at $x = 0$: Vertical tangent

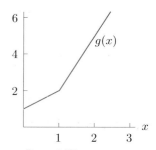

Figure 2.57: Continuous function not differentiable at $x = 1$

Example 3   Consider the function given by the formulas

$$g(x) = \begin{cases} x + 1 & \text{if } x \le 1 \\ 3x - 1 & \text{if } x > 1. \end{cases}$$

Draw the graph of g. Is g continuous? Is g differentiable at $x = 1$?

Solution   The graph in Figure 2.57 has no breaks in it, so the function is continuous. However, the graph has a corner at $x = 1$ which no amount of magnification will remove. To the left of $x = 1$, the slope is 1; to the right of $x = 1$, the slope is 3. Thus, the difference quotient at $x = 1$ has no limit, so the function g is not differentiable at $x = 1$.

The Relationship Between Differentiability and Continuity

The fact that a function which is differentiable at a point has a tangent line suggests that the function is continuous there, as the next theorem shows.

Theorem 2.1: A Differentiable Function Is Continuous

If $f(x)$ is differentiable at a point $x = a$, then $f(x)$ is continuous at $x = a$.

Proof We assume f is differentiable at $x = a$. Then we know that $f'(a)$ exists where

$$f'(a) = \lim_{x \to a} \frac{f(x) - f(a)}{x - a}.$$

To show that f is continuous at $x = a$, we want to show that $\lim_{x \to a} f(x) = f(a)$. We calculate $\lim_{x \to a} (f(x) - f(a))$, hoping to get 0. By algebra, we know that for $x \neq a$,

$$f(x) - f(a) = (x - a) \cdot \frac{f(x) - f(a)}{x - a}.$$

Taking the limits, we have

$$\lim_{x \to a} (f(x) - f(a)) = \lim_{x \to a} \left((x - a) \frac{f(x) - f(a)}{x - a} \right)$$

$$= \left(\lim_{x \to a} (x - a) \right) \cdot \left(\lim_{x \to a} \frac{f(x) - f(a)}{x - a} \right) \qquad \text{(By Theorem 1.2, Property 3)}$$

$$= 0 \cdot f'(a) = 0. \qquad \text{(Since } f'(a) \text{ exists)}$$

Thus we know that $\lim_{x \to a} f(x) = f(a)$, which means $f(x)$ is continuous at $x = a$.

Exercises and Problems for Section 2.6

Exercises

For the graphs in Exercises 1–2, list the x-values for which the function appears to be

(a) Not continuous.   (b) Not differentiable.

In Exercises 4–5, does the function appear to be differentiable on the interval of x-values shown?

1.

2.

4.

5.
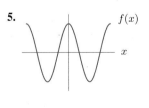

3. Are there any x-values at which the absolute value function, $f(x) = |x + 1|$, is

(a) Not continuous?   (b) Not differentiable?

Problems

6. Look at the graph of $f(x) = (x^2 + 0.0001)^{1/2}$ shown in Figure 2.58. The graph of f appears to have a sharp corner at $x = 0$. Do you think f has a derivative at $x = 0$?

Figure 2.58

Decide if the functions in Problems 7–9 are differentiable at $x = 0$. Try zooming in on a graphing calculator, or calculating the derivative $f'(0)$ from the definition.

7. $f(x) = (x + |x|)^2 + 1$

8. $f(x) = \begin{cases} x \sin(1/x) + x & \text{for } x \neq 0 \\ 0 & \text{for } x = 0 \end{cases}$

9. $f(x) = \begin{cases} x^2 \sin(1/x) & \text{for } x \neq 0 \\ 0 & \text{for } x = 0 \end{cases}$

10. In each of the following cases, sketch the graph of a continuous function $f(x)$ with the given properties.

   (a) $f''(x) > 0$ for $x < 2$ and for $x > 2$ and $f'(2)$ is undefined.

   (b) $f''(x) > 0$ for $x < 2$ and $f''(x) < 0$ for $x > 2$ and $f'(2)$ is undefined.

11. The acceleration due to gravity, g, varies with height above the surface of the earth, in a certain way. If you go down below the surface of the earth, g varies in a different way. It can be shown that g is given by

$$g = \begin{cases} \dfrac{GMr}{R^3} & \text{for } r < R \\[2mm] \dfrac{GM}{r^2} & \text{for } r \geq R \end{cases}$$

where R is the radius of the earth, M is the mass of the earth, G is the gravitational constant, and r is the distance to the center of the earth.

   (a) Sketch a graph of g against r.

   (b) Is g a continuous function of r? Explain your answer.

   (c) Is g a differentiable function of r? Explain your answer.

12. An electric charge, Q, in a circuit is given as a function of time, t, by

$$Q = \begin{cases} C & \text{for } t \leq 0 \\ Ce^{-t/RC} & \text{for } t > 0, \end{cases}$$

where C and R are positive constants. The electric current, I, is the rate of change of charge, so

$$I = \frac{dQ}{dt}.$$

   (a) Is the charge, Q, a continuous function of time?

   (b) Do you think the current, I, is defined for all times, t? [Hint: To graph this function, take, for example, $C = 1$ and $R = 1$.]

13. A magnetic field, B, is given as a function of the distance, r, from the center of a wire as follows:

$$B = \begin{cases} \dfrac{r}{r_0} B_0 & \text{for } r \leq r_0 \\[3mm] \dfrac{r_0}{r} B_0 & \text{for } r > r_0. \end{cases}$$

   (a) Sketch a graph of B against r. What is the meaning of the constant B_0?

   (b) Is B continuous at $r = r_0$? Give reasons.

   (c) Is B differentiable at $r = r_0$? Give reasons.

14. A cable is made of an insulating material in the shape of a long, thin cylinder of radius r_0. It has electric charge distributed evenly throughout it. The electric field, E, at a distance r from the center of the cable is given by

$$E = \begin{cases} kr & \text{for } r \leq r_0 \\[2mm] k\dfrac{r_0^2}{r} & \text{for } r > r_0. \end{cases}$$

   (a) Is E continuous at $r = r_0$?

   (b) Is E differentiable at $r = r_0$?

   (c) Sketch a graph of E as a function of r.

15. Graph the function defined by

$$g(r) = \begin{cases} 1 + \cos(\pi r/2) & \text{for } -2 \leq r \leq 2 \\ 0 & \text{for } r < -2 \quad \text{or} \quad r > 2. \end{cases}$$

   (a) Is g continuous at $r = 2$? Explain your answer.

   (b) Do you think g is differentiable at $r = 2$? Explain your answer.

16. The potential, ϕ, of a charge distribution at a point on the y-axis is given by

$$\phi = \begin{cases} 2\pi\sigma \left(\sqrt{y^2 + a^2} - y \right) & \text{for } y \geq 0 \\[2mm] 2\pi\sigma \left(\sqrt{y^2 + a^2} + y \right) & \text{for } y < 0 \end{cases}$$

where σ and a are positive constants. [Hint: To graph this function, take, for example, $2\pi\sigma = 1$ and $a = 1$.]

   (a) Is ϕ continuous at $y = 0$?

   (b) Do you think ϕ is differentiable at $y = 0$?

17. Sometimes, odd behavior can be hidden beneath the surface of a rather normal-looking function. Consider the following function:

$$f(x) = \begin{cases} 0 & \text{if } x < 0 \\ x^2 & \text{if } x \geq 0. \end{cases}$$

(a) Sketch a graph of this function. Does it have any vertical segments or corners? Is it differentiable everywhere? If so, sketch the derivative f' of this function. [Hint: You may want to use the result of Example 4 on page 84.]

(b) Is the derivative function, $f'(x)$, differentiable everywhere? If not, at what point(s) is it not differentiable? Draw the second derivative of $f(x)$ wherever it exists. Is the second derivative function, $f''(x)$, differentiable? Continuous?

CHAPTER SUMMARY (see also Ready Reference at the end of the book)

- **Rate of change**
  Average, instantaneous.

- **Definition of derivative**
  Difference quotient, limit.

- **Estimating and computing derivatives**
  Estimate derivatives from a graph, table of values, or formula. Use definition to find derivatives of simple functions algebraically. Know derivatives of constant, linear, and power functions.

- **Interpretation of derivatives**
  Rate of change, instantaneous velocity, slope, using units.

- **Second derivative**
  Concavity, acceleration.

- **Working with derivatives**
  Understand relation between sign of f' and whether f is increasing or decreasing. Sketch graph of f' from graph of f.

- **Differentiability**

REVIEW EXERCISES AND PROBLEMS FOR CHAPTER TWO

Exercises

Sketch the graphs of the derivatives of the functions shown in Exercises 1–6. Be sure your sketches are consistent with the important features of the graphs of the original functions.

Find a formula for the derivatives of the functions in Exercises 7–8 algebraically.

7. $f(x) = 5x^2 + x$

8. $n(x) = (1/x) + 1$

1.

2.

Use algebra to evaluate the limits in Exercises 9–13. Assume $a > 0$.

9. $\displaystyle\lim_{h \to 0} \frac{(a+h)^2 - a^2}{h}$

10. $\displaystyle\lim_{h \to 0} \frac{1/(a+h) - 1/a}{h}$

3.

4.
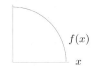

11. $\displaystyle\lim_{h \to 0} \frac{1/(a+h)^2 - 1/a^2}{h}$

12. $\displaystyle\lim_{h \to 0} \frac{\sqrt{a+h} - \sqrt{a}}{h}$ [Hint: Multiply by $\sqrt{a+h} + \sqrt{a}$ in numerator and denominator.]

5.

6.

13. $\displaystyle\lim_{h \to 0} \frac{1/\sqrt{a+h} - 1/\sqrt{a}}{h}$

Problems

14. (a) Give an example of a function with $\lim\limits_{x \to 2} f(x) = \infty$.

(b) Give an example of a function with $\lim\limits_{x \to 2} f(x) = -\infty$.

15. Suppose $f(2) = 3$ and $f'(2) = 1$. Find $f(-2)$ and $f'(-2)$, assuming that $f(x)$ is

(a) Even   **(b)** Odd.

16. Do the values for the function $y = k(x)$ in the table suggest that the graph of $k(x)$ is concave up or concave down for $1 \le x \le 3.3$? Write a sentence in support of your conclusion.

x	1.0	1.2	1.5	1.9	2.5	3.3
$k(x)$	4.0	3.8	3.6	3.4	3.2	3.0

17. Suppose that $f(x)$ is a function with $f(20) = 345$ and $f'(20) = 6$. Estimate $f(22)$.

18. Students were asked to evaluate $f'(4)$ from the following table which shows values of the function f:

x	1	2	3	4	5	6
$f(x)$	4.2	4.1	4.2	4.5	5.0	5.7

- Student A estimated the derivative as $f'(4) \approx \dfrac{f(5) - f(4)}{5 - 4} = 0.5$.
- Student B estimated the derivative as $f'(4) \approx \dfrac{f(4) - f(3)}{4 - 3} = 0.3$.
- Student C suggested that they should split the difference and estimate the average of these two results, that is, $f'(4) \approx \frac{1}{2}(0.5 + 0.3) = 0.4$.

(a) Sketch the graph of f, and indicate how the three estimates are represented on the graph.

(b) Explain which answer is likely to be best.

(c) Use Student C's method to find an algebraic formula to approximate $f'(x)$ using increments of size h.

19. Use Figure 2.59 to fill in the blanks in the following statements about the function f at point A.

(a) $f(\_\_) = \_\_$ **(b)** $f'(\_\_) = \_\_$

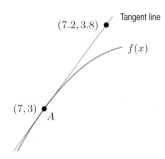

Figure 2.59

20. Use Figure 2.60. At point A, we are told that $x = 1$. In addition, $f(1) = 3$, $f'(1) = 2$, and $h = 0.1$. What are the values of $x_1, x_2, x_3, y_1, y_2, y_3$?

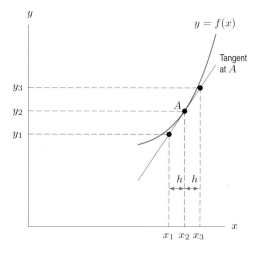

Figure 2.60

21. Given all of the following information about a function f, sketch its graph.

- $f(x) = 0$ at $x = -5$, $x = 0$, and $x = 5$
- $\lim\limits_{x \to -\infty} f(x) = \infty$
- $\lim\limits_{x \to \infty} f(x) = -3$
- $f'(x) = 0$ at $x = -3$, $x = 2.5$, and $x = 7$

22. A yam has just been taken out of the oven and is cooling off before being eaten. The temperature, T, of the yam (measured in degrees Fahrenheit) is a function of how long it has been out of the oven, t (measured in minutes). Thus, we have $T = f(t)$.

(a) Is $f'(t)$ positive or negative? Why?

(b) What are the units for $f'(t)$?

23. An economist is interested in how the price of a certain commodity affects its sales. Suppose that at a price of $\$p$, a quantity q of the commodity is sold. If $q = f(p)$, explain in economic terms the meaning of the statements $f(10) = 240,000$ and $f'(10) = -29,000$.

24. At time, t, in years, the US population is growing at 0.8% per year times its size, $P(t)$, at that moment. Using the derivative, write an equation representing this statement.

25. (a) Using the table, estimate $f'(0.6)$ and $f'(0.5)$.
   (b) Estimate $f''(0.6)$.
   (c) Where do you think the maximum and minimum values of f occur in the interval $0 \leq x \leq 1$?

x	0	0.2	0.4	0.6	0.8	1.0
$f(x)$	3.7	3.5	3.5	3.9	4.0	3.9

26. Let $g(x) = \sqrt{x}$ and $f(x) = kx^2$, where k is a constant.

   (a) Find the slope of the tangent line to the graph of g at the point $(4, 2)$.
   (b) Find the equation of this tangent line.
   (c) If the graph of f contains the point $(4, 2)$, find k.
   (d) Where does the graph of f intersect the tangent line found in part (b)?

27. A circle with center at the origin and radius of length $\sqrt{19}$ has equation $x^2 + y^2 = 19$. Graph the circle.

   (a) Just from looking at the graph, what can you say about the slope of the line tangent to the circle at the point $(0, \sqrt{19})$? What about the slope of the tangent at $(\sqrt{19}, 0)$?
   (b) Estimate the slope of the tangent to the circle at the point $(2, -\sqrt{15})$ by graphing the tangent carefully at that point.
   (c) Use the result of part (b) and the symmetry of the circle to find slopes of the tangents drawn to the circle at $(-2, \sqrt{15})$, $(-2, -\sqrt{15})$, and $(2, \sqrt{15})$.

28. Each of the graphs in Figure 2.61 shows the position of a particle moving along the x-axis as a function of time, $0 \leq t \leq 5$. The vertical scales of the graphs are the same. During this time interval, which particle has

   (a) Constant velocity?
   (b) The greatest initial velocity?
   (c) The greatest average velocity?
   (d) Zero average velocity?
   (e) Zero acceleration?
   (f) Positive acceleration throughout?

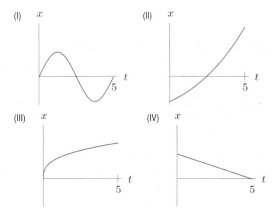

Figure 2.61

29. The population of a herd of deer is modeled by

$$P(t) = 4000 + 400 \sin\left(\frac{\pi}{6}t\right) + 180 \sin\left(\frac{\pi}{3}t\right)$$

where t is measured in months from the first of April.

   (a) Use a calculator or computer to sketch a graph showing how this population varies with time.

   Use the graph to answer the following questions.

   (b) When is the herd largest? How many deer are in it at that time?
   (c) When is the herd smallest? How many deer are in it then?
   (d) When is the herd growing the fastest? When is it shrinking the fastest?
   (e) How fast is the herd growing on April 1?

30. The number of hours, H, of daylight in Madrid is a function of t, the number of days since the start of the year. Figure 2.62 shows a one-month portion of the graph of H.

Figure 2.62

   (a) Comment on the shape of the graph. Why does it look like a straight line?
   (b) What month does this graph show? How do you know?
   (c) What is the approximate slope of this line? What does the slope represent in practical terms?

31. Suppose you put a yam in a hot oven, maintained at a constant temperature of $200°C$. As the yam picks up heat from the oven, its temperature rises.[7]

(a) Draw a possible graph of the temperature T of the yam against time t (minutes) since it is put into the oven. Explain any interesting features of the graph, and in particular explain its concavity.

(b) Suppose that, at $t = 30$, the temperature T of the yam is $120°$ and increasing at the (instantaneous) rate of $2°$/min. Using this information, plus what you know about the shape of the T graph, estimate the temperature at time $t = 40$.

(c) Suppose in addition you are told that at $t = 60$, the temperature of the yam is $165°$. Can you improve your estimate of the temperature at $t = 40$?

(d) Assuming all the data given so far, estimate the time at which the temperature of the yam is $150°$.

32. You are given the following values for the error function, $\mathrm{erf}(x)$.

$$\mathrm{erf}(0) = 0 \quad \mathrm{erf}(1) = 0.84270079$$
$$\mathrm{erf}(0.1) = 0.11246292 \quad \mathrm{erf}(0.01) = 0.01128342.$$

(a) Use all this information to determine your best estimate for $\mathrm{erf}'(0)$. (Give only those digits of which you feel reasonably certain.)

(b) Suppose you find that $\mathrm{erf}(0.001) = 0.00112838$. How does this extra information change your answer to part (a)?

33. (a) Use your calculator to approximate the derivative of the hyperbolic sine function (written $\sinh x$) at the points 0, 0.3, 0.7, and 1.

(b) Can you find a relation between the values of this derivative and the values of the hyperbolic cosine (written $\cosh x$)?

CAS Challenge Problems

34. Use a computer algebra system to find the derivative of $f(x) = \sin^2 x + \cos^2 x$ and simplify your answer. Explain your result.

35. (a) Use a computer algebra system to find the derivative of $f(x) = 2 \sin x \cos x$.

(b) Simplify $f(x)$ and $f'(x)$ using double angle formulas. Write down the derivative formula that you get after doing this simplification.

36. (a) Use a computer algebra system to find the second derivative of $g(x) = e^{-ax^2}$ with respect to x.

(b) Graph $g(x)$ and $g''(x)$ on the same axes for $a = 1, 2, 3$ and describe the relation between the two graphs.

(c) Explain your answer to part (b) in terms of concavity.

37. (a) Use a computer algebra system to find the derivative of $f(x) = \ln(x)$, $g(x) = \ln(2x)$, and $h(x) = \ln(3x)$. What is the relationship between the answers?

(b) Use the properties of logarithms to explain what you see in part (a).

38. (a) Use a computer algebra system to find the derivative of $(x^2 + 1)^2$, $(x^2 + 1)^3$, and $(x^2 + 1)^4$.

(b) Conjecture a formula for the derivative of $(x^2 + 1)^n$ that works for any integer n. Check your formula using the computer algebra system.

39. (a) Use a computer algebra system to find the derivatives of $\sin x$, $\cos x$ and $\sin x \cos x$.

(b) Is the derivative of a product of two functions always equal to the product of their derivatives?

CHECK YOUR UNDERSTANDING

Are the statements in Problems 1–19 true or false? Give an explanation for your answer.

1. If a car is going 50 miles per hour at 2 pm and 60 miles per hour at 3 pm then it travels between 50 and 60 miles during the hour between 2 pm and 3 pm.

2. If a car travels 80 miles between 2 and 4 pm, then its velocity is close to 40 mph at 2 pm.

3. If the time interval is short enough, then the average velocity of a car over the time interval and the instantaneous velocity at a time in the interval can be expected to be close.

4. If an object moves with the same average velocity over every time interval, then its average velocity equals its instantaneous velocity at any time.

5. The formula Distance traveled $=$ Average velocity $\times$ Time is valid for every moving object for every time interval.

6. By definition, the instantaneous velocity of an object equals a difference quotient.

7. If $f(x)$ is concave up, then $f'(a) < (f(b) - f(a))/(b - a)$ for $a < b$.

8. You cannot be sure of the exact value of a derivative of a function at a point using only the information in a table of values of the function. The best you can do is find an approximation.

9. If $f'(x)$ is increasing, then $f(x)$ is also increasing.

[7]From Peter D. Taylor, *Calculus: The Analysis of Functions* (Toronto: Wall & Emerson, Inc., 1992).

10. If $f(a) \neq g(a)$, then $f'(a) \neq g'(a)$.

11. The derivative of a linear function is constant.

12. If $g(x)$ is a vertical shift of $f(x)$, then $f'(x) = g'(x)$.

13. If $f(x)$ is defined for all x but $f'(0)$ is not defined, then the graph of $f(x)$ must have a corner at the point where $x = 0$.

14. If $y = f(x)$, then $\left. \dfrac{dy}{dx} \right|_{x=a} = f'(a)$.

15. If you zoom in (with your calculator) on the graph of $y = f(x)$ in a small interval around $x = 10$ and see a straight line, then the slope of that line equals the derivative $f'(10)$.

16. If $f''(x) > 0$ then $f'(x)$ is increasing.

17. The instantaneous acceleration of a moving particle at time t is the limit of difference quotients.

18. If $f(t)$ is the quantity in grams of a chemical produced after t minutes and $g(t)$ is the same quantity in kilograms, then $f'(t) = 1000g'(t)$.

19. If $f(t)$ is the quantity in kilograms of a chemical produced after t minutes and $g(t)$ is the quantity in kilograms produced after t seconds, then $f'(t) = 60g'(t)$.

Are the statements in Problems 20–24 true or false? If a statement is true, give an example illustrating it. If a statement is false, give a counterexample.

20. There is a function which is continuous on $[1, 5]$ but not differentiable at $x = 3$.

21. If a function is differentiable, then it is continuous.

22. If a function is continuous, then it is differentiable.

23. If a function is not continuous, then it is not differentiable.

24. If a function is not differentiable, then it is not continuous.

25. Which of the following would be a counterexample to the statement: "If f is differentiable at $x = a$ then f is continuous at $x = a$"?

(a) A function which is not differentiable at $x = a$ but is continuous at $x = a$.
(b) A function which is not continuous at $x = a$ but is differentiable at $x = a$.
(c) A function which is both continuous and differentiable at $x = a$.
(d) A function which is neither continuous nor differentiable at $x = a$.

PROJECTS FOR CHAPTER TWO

1. Hours of Daylight as a Function of Latitude

Let $S(x)$ be the number of sunlight hours on a cloudless June 21, as a function of latitude, x, measured in degrees.

(a) What is $S(0)$?

(b) Let x_0 be the latitude of the Arctic Circle ($x_0 \approx 66°30'$). In the northern hemisphere, $S(x)$ is given, for some constants a and b, by the formula:

$$S(x) = \begin{cases} a + b \arcsin \left(\dfrac{\tan x}{\tan x_0} \right) & \text{for } 0 \leq x < x_0 \\ 24 & \text{for } x_0 \leq x \leq 90. \end{cases}$$

Find a and b so that $S(x)$ is continuous.

(c) Calculate $S(x)$ for Tucson, Arizona ($x = 32°13'$) and Walla Walla, Washington ($46°4'$).

(d) Graph $S(x)$, for $0 \leq x \leq 90$.

(e) Does $S(x)$ appear to be differentiable?

2. US Population

Census figures for the US population (in millions) are listed in Table 2.12. Let f be the function such that $P = f(t)$ is the population (in millions) in year t.

Table 2.12   *US population (in millions), 1790–1990*

Year	Population	Year	Population	Year	Population	Year	Population
1790	3.9	1850	23.1	1910	92.0	1970	205.0
1800	5.3	1860	31.4	1920	105.7	1980	226.5
1810	7.2	1870	38.6	1930	122.8	1990	248.7
1820	9.6	1880	50.2	1940	131.7		
1830	12.9	1890	62.9	1950	150.7		
1840	17.1	1900	76.0	1960	179.0		

(a) (i) Estimate the rate of change of the population for the years 1900, 1945, and 1990.

(ii) When, approximately, was the rate of change of the population greatest?

(iii) Estimate the US population in 1956.

(iv) Based on the data from the table, what would you predict for the census in the year 2000?

(b) Assume that f is increasing (as the values in the table suggest). Then f is invertible.

(i) What is the meaning of $f^{-1}(100)$?

(ii) What does the derivative of $f^{-1}(P)$ at $P = 100$ represent? What are its units?

(iii) Estimate $f^{-1}(100)$.

(iv) Estimate the derivative of $f^{-1}(P)$ at $P = 100$.

(c) (i) Usually we think the US population $P = f(t)$ as a smooth function of time. To what extent is this justified? What happens if we zoom in at a point of the graph? What about events such as the Louisiana Purchase? Or the moment of your birth?

(ii) What do we in fact mean by the rate of change of the population at a particular time t?

(iii) Give another example of a real-world function which is not smooth but is usually treated as such.

Chapter Three

SHORT-CUTS TO DIFFERENTIATION

In Chapter 2, we defined the derivative function

$$f'(x) = \lim_{h \to 0} \frac{f(x+h) - f(x)}{h}$$

and saw how the derivative represents a slope and a rate of change. We learned how to approximate the derivative of a function given graphically (by estimating the slope of the tangent at each point) and numerically (by finding the average rate of change of the function between data values). We calculated the derivatives of x^2 and x^3 exactly using the definition.

In this chapter we make a systematic study of the derivatives of functions given by formulas, such as power, polynomial, exponential, logarithmic, trigonometric, and hyperbolic functions. The chapter also contains general rules, such as the product, quotient, and chain rules, which allow us to differentiate combinations of functions.

The chapter concludes with theorems about differentiable functions, including the Mean Value Theorem and its consequences.

3.1   POWERS AND POLYNOMIALS

Derivative of a Constant Times a Function

Figure 3.1 shows the graph of $y = f(x)$ and of three multiples: $y = 3f(x)$, $y = \frac{1}{2}f(x)$, and $y = -2f(x)$. What is the relationship between the derivatives of these functions? In other words, for a particular x-value, how are the slopes of these graphs related?

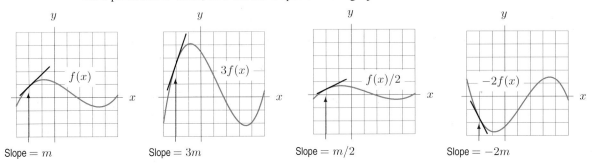

Figure 3.1: A function and its multiples: Derivative of multiple is multiple of derivative

Multiplying the value of a function by a constant stretches or shrinks the original graph (and reflects it in the x-axis if the constant is negative). This changes the slope of the curve at each point. If the graph has been stretched, the "rises" have all been increased by the same factor, whereas the "runs" remain the same. Thus, the slopes are all steeper by the same factor. If the graph has been shrunk, the slopes are all smaller by the same factor. If the graph has been reflected in the x-axis, the slopes will all have their signs reversed. In other words, if a function is multiplied by a constant, c, so is its derivative:

Theorem 3.1: Derivative of a Constant Multiple

If f is differentiable and c is a constant, then

$$\frac{d}{dx}\left[cf(x)\right] = cf'(x).$$

Proof   Although the graphical argument makes the theorem plausible, to prove it we must use the definition of the derivative:

$$\frac{d}{dx}\left[cf(x)\right] = \lim_{h \to 0} \frac{cf(x+h) - cf(x)}{h} = \lim_{h \to 0} c\,\frac{f(x+h) - f(x)}{h}$$

$$= c\lim_{h \to 0} \frac{f(x+h) - f(x)}{h} = cf'(x).$$

We can take c across the limit sign by the properties of limits (part 1 of Theorem 1.2 on page 51).

Derivatives of Sums and Differences

Suppose we have two functions, $f(x)$ and $g(x)$, with the values listed in Table 3.1. Values of the sum $f(x) + g(x)$ are given in the same table.

Table 3.1   *Sum of Functions*

x	$f(x)$	$g(x)$	$f(x) + g(x)$
0	100	0	100
1	110	0.2	110.2
2	130	0.4	130.4
3	160	0.6	160.6
4	200	0.8	200.8

We see that adding the increments of $f(x)$ and the increments of $g(x)$ gives the increments of $f(x) + g(x)$. For example, as x increases from 0 to 1, $f(x)$ increases by 10 and $g(x)$ increases by 0.2, while $f(x) + g(x)$ increases by $110.2 - 100 = 10.2$. Similarly, as x increases from 3 to 4, $f(x)$ increases by 40 and $g(x)$ by 0.2, while $f(x) + g(x)$ increases by $200.8 - 160.6 = 40.2$.

From this example, we see that the rate at which $f(x) + g(x)$ is increasing is the sum of the rates at which $f(x)$ and $g(x)$ are increasing. Similar reasoning applies to the difference, $f(x) - g(x)$. In terms of derivatives:

Theorem 3.2: Derivative of Sum and Difference

If f and g are differentiable, then

$$\frac{d}{dx}[f(x) + g(x)] = f'(x) + g'(x) \quad \text{and} \quad \frac{d}{dx}[f(x) - g(x)] = f'(x) - g'(x).$$

Proof Using the definition of the derivative:

$$\frac{d}{dx}[f(x) + g(x)] = \lim_{h \to 0} \frac{[f(x+h) + g(x+h)] - [f(x) + g(x)]}{h}$$

$$= \lim_{h \to 0} \left[\underbrace{\frac{f(x+h) - f(x)}{h}}_{\text{Limit of this is } f'(x)} + \underbrace{\frac{g(x+h) - g(x)}{h}}_{\text{Limit of this is } g'(x)} \right]$$

$$= f'(x) + g'(x).$$

We have used the fact that the limit of a sum is the sum of the limits, part 2 of Theorem 1.2 on page 51. The proof for $f(x) - g(x)$ is similar.

Powers of x

In Chapter 2 we showed that

$$f'(x) = \frac{d}{dx}(x^2) = 2x \quad \text{and} \quad g'(x) = \frac{d}{dx}(x^3) = 3x^2.$$

The graphs of $f(x) = x^2$ and $g(x) = x^3$ and their derivatives are shown in Figures 3.2 and 3.3. Notice $f'(x) = 2x$ has the behavior we expect. It is negative for $x < 0$ (when f is decreasing), zero for $x = 0$, and positive for $x > 0$ (when f is increasing). Similarly, $g'(x) = 3x^2$ is zero when $x = 0$, but positive everywhere else, as g is increasing everywhere else.

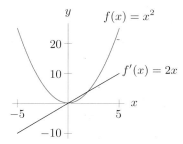

Figure 3.2: Graphs of $f(x) = x^2$ and its derivative $f'(x) = 2x$

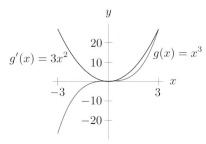

Figure 3.3: Graphs of $g(x) = x^3$ and its derivative $g'(x) = 3x^2$

These examples are special cases of the power rule which we justify for any positive integer n on page 113:

> ### The Power Rule
>
> For any constant real number n,
>
> $$\frac{d}{dx}(x^n) = nx^{n-1}.$$

Problem 71 asks you to show that this rule holds for negative integral powers; such powers can also be differentiated using the quotient rule (Section 3.3). In Section 3.6 we indicate how to justify the power rule for powers of the form $1/n$.

Example 1   Use the power rule to differentiate (a) $\dfrac{1}{x^3}$,   (b) $x^{1/2}$,   (c) $\dfrac{1}{\sqrt[3]{x}}$.

Solution

(a) For $n = -3$: $\dfrac{d}{dx}\left(\dfrac{1}{x^3}\right) = \dfrac{d}{dx}(x^{-3}) = -3x^{-3-1} = -3x^{-4} = -\dfrac{3}{x^4}.$

(b) For $n = 1/2$: $\dfrac{d}{dx}\left(x^{1/2}\right) = \dfrac{1}{2}x^{(1/2)-1} = \dfrac{1}{2}x^{-1/2} = \dfrac{1}{2\sqrt{x}}.$

(c) For $n = -1/3$: $\dfrac{d}{dx}\left(\dfrac{1}{\sqrt[3]{x}}\right) = \dfrac{d}{dx}\left(x^{-1/3}\right) = -\dfrac{1}{3}x^{(-1/3)-1} = -\dfrac{1}{3}x^{-4/3} = -\dfrac{1}{3x^{4/3}}.$

Example 2   Use the definition of the derivative to justify the power rule for $n = -2$: Show $\dfrac{d}{dx}(x^{-2}) = -2x^{-3}$.

Solution   Provided $x \neq 0$, we have

$$\frac{d}{dx}\left(x^{-2}\right) = \frac{d}{dx}\left(\frac{1}{x^2}\right) = \lim_{h\to 0}\left(\frac{\frac{1}{(x+h)^2} - \frac{1}{x^2}}{h}\right) = \lim_{h\to 0}\frac{1}{h}\left[\frac{x^2 - (x+h)^2}{(x+h)^2 x^2}\right] \quad \text{(Combining fractions over a common denominator)}$$

$$= \lim_{h\to 0}\frac{1}{h}\left[\frac{x^2 - (x^2 + 2xh + h^2)}{(x+h)^2 x^2}\right] \quad \text{(Multiplying out)}$$

$$= \lim_{h\to 0}\frac{-2xh - h^2}{h(x+h)^2 x^2} \quad \text{(Simplifying numerator)}$$

$$= \lim_{h\to 0}\frac{-2x - h}{(x+h)^2 x^2} \quad \text{(Dividing numerator and denominator by } h)$$

$$= \frac{-2x}{x^4} \quad \text{(Letting } h \to 0)$$

$$= -2x^{-3}.$$

The graphs of x^{-2} and its derivative, $-2x^{-3}$, are shown in Figure 3.4. Does the graph of the derivative have the features you expect?

Figure 3.4: Graphs of x^{-2} and its derivative, $-2x^{-3}$

Justification of $\frac{d}{dx}(x^n) = nx^{n-1}$, for n a Positive Integer

To calculate the derivatives of x^2 and x^3, we had to expand $(x + h)^2$ and $(x + h)^3$. To calculate the derivative of x^n, we must expand $(x + h)^n$. Let's look back at the previous expansions:

$$(x + h)^2 = x^2 + 2xh + h^2, \qquad (x + h)^3 = x^3 + 3x^2h + 3xh^2 + h^3,$$

and multiply out a few more examples:

$$(x + h)^4 = x^4 + 4x^3h + 6x^2h^2 + 4xh^3 + h^4,$$
$$(x + h)^5 = x^5 + 5x^4h + \underbrace{10x^3h^2 + 10x^2h^3 + 5xh^4 + h^5}.$$

Terms involving h^2 and higher powers of h

In general, we can say

$$(x + h)^n = x^n + nx^{n-1}h + \underbrace{\cdots\cdots + h^n}.$$

Terms involving h^2 and higher powers of h

We have just seen this is true for $n = 2, 3, 4, 5$. It can be proved in general using the Binomial Theorem (see the online theory supplement). Now to find the derivative,

$$\frac{d}{dx}(x^n) = \lim_{h \to 0} \frac{(x + h)^n - x^n}{h}$$
$$= \lim_{h \to 0} \frac{(x^n + nx^{n-1}h + \cdots + h^n) - x^n}{h}$$

Terms involving h^2 and higher powers of h

$$= \lim_{h \to 0} \frac{nx^{n-1}h + \overbrace{\cdots + h^n}}{h}.$$

When we factor out h from terms involving h^2 and higher powers of h, each term will still have an h in it. Factoring and dividing, we get:

Terms involving h and higher powers of h

$$\frac{d}{dx}(x^n) = \lim_{h \to 0} \frac{h(nx^{n-1} + \cdots + h^{n-1})}{h} = \lim_{h \to 0} (nx^{n-1} + \overbrace{\cdots + h^{n-1}}).$$

But as $h \to 0$, all terms involving an h will go to 0, so

$$\frac{d}{dx}(x^n) = \lim_{h \to 0} (nx^{n-1} + \underbrace{\cdots + h^{n-1}}) = nx^{n-1}.$$

These terms go to 0

Derivatives of Polynomials

Now that we know how to differentiate powers, constant multiples, and sums, we can differentiate any polynomial.

Example 3   Find the derivatives of   (a) $5x^2 + 3x + 2,$ (b) $\sqrt{3}x^7 - \dfrac{x^5}{5} + \pi.$

Solution

(a)

$$\frac{d}{dx}(5x^2 + 3x + 2) = 5\frac{d}{dx}(x^2) + 3\frac{d}{dx}(x) + \frac{d}{dx}(2)$$
$$= 5 \cdot 2x + 3 \cdot 1 + 0 \qquad \text{(Since the derivative of a constant, } \tfrac{d}{dx}(2), \text{ is zero.)}$$
$$= 10x + 3.$$

(b)

$$\frac{d}{dx}\left(\sqrt{3}x^7 - \frac{x^5}{5} + \pi\right) = \sqrt{3}\frac{d}{dx}(x^7) - \frac{1}{5}\frac{d}{dx}(x^5) + \frac{d}{dx}(\pi)$$

$$= \sqrt{3} \cdot 7x^6 - \frac{1}{5} \cdot 5x^4 + 0 \qquad \text{(Since } \pi \text{ is a constant, } d\pi/dx = 0.)$$

$$= 7\sqrt{3}x^6 - x^4.$$

We can also use the rules we have seen so far to differentiate expressions which are not polynomials.

Example 4   Differentiate (a) $5\sqrt{x} - \dfrac{10}{x^2} + \dfrac{1}{2\sqrt{x}}$ (b) $0.1x^3 + 2x^{\sqrt{2}}$

Solution   (a) $\dfrac{d}{dx}\left(5\sqrt{x} - \dfrac{10}{x^2} + \dfrac{1}{2\sqrt{x}}\right) = \dfrac{d}{dx}\left(5x^{1/2} - 10x^{-2} + \dfrac{1}{2}x^{-1/2}\right)$

$$= 5 \cdot \frac{1}{2}x^{-1/2} - 10(-2)x^{-3} + \frac{1}{2}\left(-\frac{1}{2}\right)x^{-3/2}$$

$$= \frac{5}{2\sqrt{x}} + \frac{20}{x^3} - \frac{1}{4x^{3/2}}.$$

(b) $\dfrac{d}{dx}(0.1x^3 + 2x^{\sqrt{2}}) = 0.1\dfrac{d}{dx}(x^3) + 2\dfrac{d}{dx}(x^{\sqrt{2}}) = 0.3x^2 + 2\sqrt{2}x^{\sqrt{2}-1}.$

Example 5   Find the second derivative and interpret its sign for
    (a) $f(x) = x^2,$ (b) $g(x) = x^3,$ (c) $k(x) = x^{1/2}.$

Solution   (a) If $f(x) = x^2$, then $f'(x) = 2x$, so $f''(x) = \dfrac{d}{dx}(2x) = 2$. Since f'' is always positive, f is concave up, as expected for a parabola opening upward. (See Figure 3.5.)

(b) If $g(x) = x^3$, then $g'(x) = 3x^2$, so $g''(x) = \dfrac{d}{dx}(3x^2) = 3\dfrac{d}{dx}(x^2) = 3 \cdot 2x = 6x$. This is positive for $x > 0$ and negative for $x < 0$, which means x^3 is concave up for $x > 0$ and concave down for $x < 0$. (See Figure 3.6.)

(c) If $k(x) = x^{1/2}$, then $k'(x) = \frac{1}{2}x^{(1/2)-1} = \frac{1}{2}x^{-1/2}$, so

$$k''(x) = \frac{d}{dx}\left(\frac{1}{2}x^{-1/2}\right) = \frac{1}{2} \cdot \left(-\frac{1}{2}\right)x^{-(1/2)-1} = -\frac{1}{4}x^{-3/2}.$$

Now k' and k'' are only defined on the domain of k, that is, $x \geq 0$. When $x > 0$, we see that $k''(x)$ is negative, so k is concave down. (See Figure 3.7.)

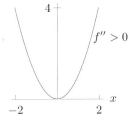

Figure 3.5: $f(x) = x^2$ and $f''(x) = 2$

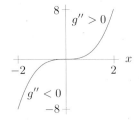

Figure 3.6: $g(x) = x^3$ and $g''(x) = 6x$

Figure 3.7: $k(x) = x^{1/2}$ and $k''(x) = -\frac{1}{4}x^{-3/2}$

Example 6   If the position of a body, in meters, is given as a function of time t, in seconds, by

$$s = -4.9t^2 + 5t + 6,$$

find the velocity and acceleration of the body at time t.

Solution   The velocity, v, is the derivative of the position:

$$v = \frac{ds}{dt} = \frac{d}{dt}(-4.9t^2 + 5t + 6) = -9.8t + 5,$$

and the acceleration, a, is the derivative of the velocity:

$$a = \frac{dv}{dt} = \frac{d}{dt}(-9.8t + 5) = -9.8.$$

Note that v is in meters/second and a is in meters/second^2.

Example 7   Figure 3.8 shows the graph of a cubic polynomial. Both graphically and algebraically, describe the behavior of the derivative of this cubic.

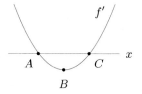

Figure 3.8: The cubic of Example 7   Figure 3.9: Derivative of the cubic of Example 7

Solution   Graphical approach: Suppose we move along the curve from left to right. To the left of A, the slope is positive; it starts very positive and decreases until the curve reaches A, where the slope is 0. Between A and C the slope is negative. Between A and B the slope is decreasing (getting more negative); it is most negative at B. Between B and C the slope is negative but increasing; at C the slope is zero. From C to the right, the slope is positive and increasing. The graph of the derivative function is shown in Figure 3.9.

     Algebraic approach: f is a cubic that goes to $+\infty$ as $x \to +\infty$, so

$$f(x) = ax^3 + bx^2 + cx + d$$

with $a > 0$. Hence,

$$f'(x) = 3ax^2 + 2bx + c,$$

whose graph is a parabola opening upward, as in Figure 3.9.

Exercises and Problems for Section 3.1

Exercises

1. Let $f(x) = 7$. Using the definition of the derivative, show that $f'(x) = 0$ for all values of x.

2. Let $f(x) = 17x + 11$. Use the definition of the derivative to calculate $f'(x)$.

For Exercises 3–44, find the derivatives of the given functions. Assume that a, b, c, and k are constants.

3. $y = x^{11}$

4. $y = x^{12}$

5. $y = x^{-12}$

6. $y = -x^{11}$

7. $y = x^{3.2}$

8. $y = x^{4/3}$

9. $y = x^{3/4}$

10. $y = x^{-3/4}$

11. $f(x) = \dfrac{1}{x^4}$

12. $g(t) = \dfrac{1}{t^5}$

13. $f(z) = -\dfrac{1}{z^{6.1}}$

14. $y = \dfrac{1}{r^{7/2}}$

15. $y = \sqrt{x}$

16. $f(x) = \sqrt[4]{x}$

17. $h(\theta) = \dfrac{1}{\sqrt[3]{\theta}}$

18. $f(x) = \sqrt{\dfrac{1}{x^3}}$

19. $f(x) = x^e$

20. $y = 4x^{3/2} - 5x^{1/2}$

21. $f(t) = 3t^2 - 4t + 1$

22. $y = 17x + 24x^{1/2}$

23. $y = z^2 + \dfrac{1}{2z}$

24. $f(x) = 5x^4 + \dfrac{1}{x^2}$

25. $h(w) = -2w^{-3} + 3\sqrt{w}$

26. $y = 6x^3 + 4x^2 - 2x$

27. $y = 3t^5 - 5\sqrt{t} + \dfrac{7}{t}$

28. $y = 3t^2 + \dfrac{12}{\sqrt{t}} - \dfrac{1}{t^2}$

29. $y = \sqrt{x}(x + 1)$

30. $y = t^{3/2}(2 + \sqrt{t})$

31. $h(t) = \dfrac{3}{t} + \dfrac{4}{t^2}$

32. $y = \sqrt{\theta}\left(\sqrt{\theta} + \dfrac{1}{\sqrt{\theta}}\right)$

33. $y = \dfrac{x^2 + 1}{x}$

34. $f(z) = \dfrac{z^2 + 1}{3z}$

35. $f(t) = \dfrac{t^2 + t^3 - 1}{t^4}$

36. $y = \dfrac{\theta - 1}{\sqrt{\theta}}$

37. $j(x) = \dfrac{x^3}{a} + \dfrac{a}{b}x^2 - cx$

38. $f(x) = \dfrac{ax + b}{x}$

39. $h(x) = \dfrac{ax + b}{c}$

40. $g(t) = \dfrac{\sqrt{t}(1 + t)}{t^2}$

41. $\dfrac{dV}{dr}$ if $V = \frac{4}{3}\pi r^2 b$

42. $\dfrac{dw}{dq}$ if $w = 3ab^2 q$

43. $\dfrac{dy}{dx}$ if $y = ax^2 + bx + c$

44. $\dfrac{dP}{dt}$ if $P = a + b\sqrt{t}$

Problems

For Problems 45–50, determine if the derivative rules from this section apply. If they do, find the derivative. If they don't apply, indicate why.

45. $y = (x + 3)^{1/2}$

46. $y = 3^x$

47. $g(x) = x^\pi - x^{-\pi}$

48. $y = 3x^2 + 4$

49. $y = \dfrac{1}{3x^2 + 4}$

50. $y = \dfrac{1}{3z^2} + \dfrac{1}{4}$

51. If $f(t) = 2t^3 - 4t^2 + 3t - 1$, find $f'(t)$ and $f''(t)$.

52. If $f(x) = 13 - 8x + \sqrt{2}x^2$ and $f'(r) = 4$, find r.

53. Find the equation of the line tangent to the graph of f at $(1, 1)$, where f is given by $f(x) = 2x^3 - 2x^2 + 1$.

54. (a) Find the equation of the tangent line to $f(x) = x^3$ at the point where $x = 2$.
(b) Graph the tangent line and the function on the same axes. If the tangent line is used to estimate values of the function, will the estimates be overestimates or underestimates?

55. If $f(x) = 4x^3 + 6x^2 - 23x + 7$, find the intervals on which $f'(x) \geq 1$.

56. Using a graph to help you, find the equations of all lines through the origin tangent to the parabola

$$y = x^2 - 2x + 4.$$

Sketch the lines on the graph.

57. On what intervals is the function $f(x) = x^4 - 4x^3$ both decreasing and concave up?

58. For what values of x is the graph of $y = x^5 - 5x$ both increasing and concave up?

59. If $f(x) = x^3 - 6x^2 - 15x + 20$, find analytically all values of x for which $f'(x) = 0$. Show your answers on a graph of f.

60. (a) Find the *eighth* derivative of $f(x) = x^7 + 5x^5 - 4x^3 + 6x - 7$. Think ahead!
(The nth derivative, $f^{(n)}(x)$, is the result of differentiating $f(x)$ n times.)
(b) Find the seventh derivative of $f(x)$.

61. The height of a sand dune (in centimeters) is represented by $f(t) = 700 - 3t^2$, where t is measured in years since 1995. Find $f(5)$ and $f'(5)$. Using units, explain what each means in terms of the sand dune.

62. A ball is dropped from the top of the Empire State building to the ground below. The height, y, of the ball above the ground (in feet) is given as a function of time, t, (in seconds) by

$$y = 1250 - 16t^2.$$

(a) Find the velocity of the ball at time t. What is the sign of the velocity? Why is this to be expected?
(b) Show that the acceleration of the ball is a constant. What are the value and sign of this constant?
(c) When does the ball hit the ground, and how fast is it going at that time? Give your answer in feet per second and in miles per hour (1 ft/sec = 15/22 mph).

63. At a time t seconds after it is thrown up in the air, a tomato is at a height of $f(t) = -4.9t^2 + 25t + 3$ meters.

(a) What is the average velocity of the tomato during the first 2 seconds? Give units.
(b) Find (exactly) the instantaneous velocity of the tomato at $t = 2$. Give units.
(c) What is the acceleration at $t = 2$?
(d) How high does the tomato go?
(e) How long is the tomato in the air?

64. The gravitational attraction, F, between the earth and a satellite of mass m at a distance r from the center of the earth is given by

$$F = \frac{GMm}{r^2},$$

where M is the mass of the earth, and G is a constant. Find the rate of change of force with respect to distance.

65. The period, T, of a pendulum is given in terms of its length, l, by

$$T = 2\pi\sqrt{\frac{l}{g}},$$

where g is the acceleration due to gravity (a constant).

(a) Find dT/dl.
(b) What is the sign of dT/dl? What does this tell you about the period of pendulums?

66. (a) Use the formula for the area of a circle of radius r, $A = \pi r^2$, to find dA/dr.

(b) The result from part (a) should look familiar. What does dA/dr represent geometrically?

(c) Use the difference quotient to explain the observation you made in part (b).

67. What is the formula for V, the volume of a sphere of radius r? Find dV/dr. What is the geometrical meaning of dV/dr?

68. Show that for any power function $f(x) = x^n$, we have $f'(1) = n$.

69. Given a power function of the form $f(x) = ax^n$, with $f'(2) = 3$ and $f'(4) = 24$, find n and a.

70. Is there a value of n which makes $y = x^n$ a solution to the equation $13x(dy/dx) = y$? If so, what value?

71. Using the definition of derivative, justify the formula $d(x^n)/dx = nx^{n-1}$.

(a) For $n = -1$; for $n = -3$.

(b) For any negative integer n.

3.2 THE EXPONENTIAL FUNCTION

What do we expect the graph of the derivative of the exponential function $f(x) = a^x$ to look like? The exponential function in Figure 3.10 increases slowly for $x < 0$ and more rapidly for $x > 0$, so the values of f' are small for $x < 0$ and larger for $x > 0$. Since the function is increasing for all values of x, the graph of the derivative must lie above the x-axis. It appears that the graph of f' may resemble the graph of f itself.

$f(x) = a^x$

Figure 3.10: $f(x) = a^x$, with $a > 1$

In this section we see that $f'(x) = k \cdot a^x$, so in fact $f'(x)$ is proportional to $f(x)$. This property of exponential functions makes them particularly useful in modeling because many quantities have rates of change which are proportional to themselves. For example, the simplest model of population growth has this property.

Derivatives of Exponential Functions and the Number e

We start by calculating the derivative of $g(x) = 2^x$, which is given by

$$g'(x) = \lim_{h \to 0} \left(\frac{2^{x+h} - 2^x}{h} \right) = \lim_{h \to 0} \left(\frac{2^x 2^h - 2^x}{h} \right) = \lim_{h \to 0} 2^x \left(\frac{2^h - 1}{h} \right)$$

$$= \lim_{h \to 0} \left(\frac{2^h - 1}{h} \right) \cdot 2^x. \qquad \text{(Since } x \text{ and } 2^x \text{ are fixed during this calculation).}$$

To find $\lim_{h \to 0}(2^h - 1)/h$, see Table 3.2 where we have substituted values of h near 0. The table suggests that the limit exists and has value 0.693. Let us call the limit k, so $k = 0.693$. Then

$$\frac{d}{dx}(2^x) = k \cdot 2^x = 0.693 \cdot 2^x.$$

So the derivative of 2^x is proportional to 2^x with constant of proportionality 0.693. A similar calculation shows that the derivative of $f(x) = a^x$ is

$$f'(x) = \lim_{h \to 0} \left(\frac{a^{x+h} - a^x}{h} \right) = \lim_{h \to 0} \left(\frac{a^h - 1}{h} \right) \cdot a^x.$$

Table 3.2

h	$(2^h - 1)/h$
-0.1	0.6697
-0.01	0.6908
-0.001	0.6929
0.001	0.6934
0.01	0.6956
0.1	0.7177

Table 3.3

a	$k = \lim_{h \to 0} \frac{a^h - 1}{h}$
2	0.693
3	1.099
4	1.396
5	1.609
6	1.792
7	1.946

Table 3.4

h	$(1 + h)^{1/h}$
-0.001	2.7196422
-0.0001	2.7184178
-0.00001	2.7182954
0.00001	2.7182682
0.0001	2.7181459
0.001	2.7169239

The quantity $\lim_{h \to 0}(a^h - 1)/h$ is also a constant, although the value of the constant depends on a. Writing $k = \lim_{h \to 0}(a^h - 1)/h$, we see that the derivative of $f(x) = a^x$ is proportional to a^x:

$$\frac{d}{dx}(a^x) = k \cdot a^x.$$

For particular values of a, we can estimate k by substituting values of h near 0 into the expression $(a^h - 1)/h$. Table 3.3 shows the results. Notice that for $a = 2$, the value of k is less than 1, while for $a = 3, 4, 5, \ldots$, the values of k are greater than 1. The values of k appear to be increasing, so we guess that there is a value of a between 2 and 3 for which $k = 1$. If so, we have found a value of a with the remarkable property that the function a^x is equal to its own derivative.

So let's look for such an a. This means we want to find a such that

$$\lim_{h \to 0} \frac{a^h - 1}{h} = 1, \qquad \text{or, for small } h, \qquad \frac{a^h - 1}{h} \approx 1.$$

Solving for a, we can estimate a as follows:

$$a^h - 1 \approx h, \qquad \text{or} \qquad a^h \approx 1 + h, \qquad \text{so} \qquad a \approx (1 + h)^{1/h}.$$

Taking small values of h, as in Table 3.4, we see $a \approx 2.718\ldots$. This is the number e introduced in Chapter 1. In fact, it can be shown that if

$$e = \lim_{h \to 0}(1 + h)^{1/h} = 2.718\ldots \qquad \text{then} \qquad \lim_{h \to 0} \frac{e^h - 1}{h} = 1.$$

This means that e^x is its own derivative:

$$\boxed{\frac{d}{dx}(e^x) = e^x.}$$

Figure 3.11 shows the graphs 2^x, 3^x, and e^x together with their derivatives. Notice that the derivative of 2^x is below the graph of 2^x, since $k < 1$ there, and the graph of the derivative of 3^x is above the graph of 3^x, since $k > 1$ there. With $e \approx 2.718$, the function e^x and its derivative are identical.

Note on Round-Off Error and Limits

If we try to evaluate $(1 + h)^{1/h}$ on a calculator by taking smaller and smaller values of h, the values of $(1+h)^{1/h}$ at first get closer to $2.718\ldots$. However, they will eventually move away again because of the *round-off error* (i.e., errors introduced by the fact that the calculator can only hold a certain number of digits).

As we try smaller and smaller values of h, how do we know when to stop? Unfortunately, there is no fixed rule. A calculator can only suggest the value of a limit, but can never confirm that this value is correct. In this case, it looks like the limit is $2.718\ldots$ because the values of $(e^h - 1)/h$ approach this number for a while. To be sure this is correct, we have to find the limit analytically.

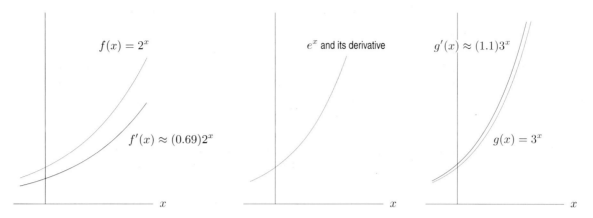

Figure 3.11: Graphs of the functions 2^x, e^x, and 3^x and their derivatives

A Formula for the Derivative of a^x

To get a formula for the derivative of a^x, we must calculate

$$f'(x) = \lim_{h \to 0} \frac{a^{x+h} - a^x}{h} = \underbrace{\left(\lim_{h \to 0} \frac{a^h - 1}{h} \right)}_{k} a^x.$$

However, without knowing the value of a, we can't use a calculator to estimate k. We take a different approach, rewriting $a = e^{\ln a}$, so

$$\lim_{h \to 0} \frac{a^h - 1}{h} = \lim_{h \to 0} \frac{\left(e^{\ln a} \right)^h - 1}{h} = \lim_{h \to 0} \frac{e^{(\ln a)h} - 1}{h}.$$

To evaluate this limit, we use a limit that we already know

$$\lim_{h \to 0} \frac{e^h - 1}{h} = 1.$$

In order to use this limit, we substitute $t = (\ln a)h$. Since t approaches 0 as h approaches 0, we have

$$\lim_{h \to 0} \frac{e^{(\ln a)h} - 1}{h} = \lim_{t \to 0} \frac{e^t - 1}{(t/\ln a)} = \lim_{t \to 0} \left(\ln a \cdot \frac{e^t - 1}{t} \right) = \ln a \left(\lim_{t \to 0} \frac{e^t - 1}{t} \right) = (\ln a) \cdot 1 = \ln a.$$

Thus, we have

$$f'(x) = \lim_{h \to 0} \frac{a^{x+h} - a^x}{h} = \left(\lim_{h \to 0} \frac{a^h - 1}{h} \right) a^x = (\ln a)a^x.$$

In Section 3.6 we obtain the same result by another method. We conclude that:

$$\boxed{\frac{d}{dx}(a^x) = (\ln a)a^x.}$$

Thus, for any a, the derivative of a^x is proportional to a^x. The constant of proportionality is $\ln a$. The derivative of a^x is equal to a^x if the constant of proportionality is 1, that is, if $\ln a = 1$, then $a = e$. The fact that the constant of proportionality is 1 when $a = e$ makes e a particularly convenient base for exponential functions.

Example 1   Differentiate $2 \cdot 3^x + 5e^x$.

Solution

$$\frac{d}{dx}(2 \cdot 3^x + 5e^x) = 2\frac{d}{dx}(3^x) + 5\frac{d}{dx}(e^x) = 2\ln 3 \cdot 3^x + 5e^x \approx (2.1972)3^x + 5e^x.$$

Exercises and Problems for Section 3.2

Exercises

Find the derivatives of the functions in Exercises 1–26. Assume that a, b, c, and k are constants.

1. $f(x) = 2e^x + x^2$

2. $y = 5t^2 + 4e^t$

3. $y = 5^x + 2$

4. $f(x) = 12e^x + 11^x$

5. $y = 5x^2 + 2^x + 3$

6. $f(x) = 2^x + 2 \cdot 3^x$

7. $y = 4 \cdot 10^x - x^3$

8. $y = 3x - 2 \cdot 4^x$

9. $y = 2^x + \frac{2}{x^3}$

10. $y = \frac{3^x}{3} + \frac{33}{\sqrt{x}}$

11. $z = (\ln 4)e^x$

12. $z = (\ln 4)4^x$

13. $f(t) = (\ln 3)^t$

14. $y = 5 \cdot 5^t + 6 \cdot 6^t$

15. $h(z) = (\ln 2)^z$

16. $f(x) = e^2 + x^e$

17. $f(x) = x^3 + 3^x$

18. $y = \pi^2 + \pi^x$

19. $f(x) = e^\pi + \pi^x$

20. $f(x) = \pi^x + x^\pi$

21. $f(x) = e^k + k^x$

22. $f(x) = e^{1+x}$

23. $f(t) = e^{t+2}$

24. $y = e^{\theta - 1}$

25. $y(x) = a^x + x^a$.

26. $f(x) = x^{\pi^2} + (\pi^2)^x$

Which of the functions in Exercises 27–35 can be differentiated using the rules we have developed so far? Differentiate if you can; otherwise, indicate why the rules discussed so far do not apply.

27. $y = x^2 + 2^x$

28. $y = \sqrt{x} - (\frac{1}{2})^x$

29. $y = x^2 \cdot 2^x$

30. $y = \frac{2^x}{x}$

31. $y = e^{x+5}$

32. $y = e^{5x}$

33. $y = 4^{(x^2)}$

34. $f(z) = (\sqrt{4})^z$

35. $f(\theta) = 4^{\sqrt{\theta}}$

Problems

36. With a yearly inflation rate of 5%, prices are given by

$$P = P_0(1.05)^t,$$

where P_0 is the price in dollars when $t = 0$ and t is time in years. Suppose $P_0 = 1$. How fast (in cents/year) are prices rising when $t = 10$?

37. Since January 1, 1960, the population of Slim Chance has been described by the formula

$$P = 35{,}000(0.98)^t,$$

where P is the population of the city t years after the start of 1960. At what rate was the population changing on January 1, 1983?

38. The population of the world in billions of people can be modeled by the function $f(t) = 5.3(1.018)^t$, where t is years since 1990. Find $f(0)$ and $f'(0)$. Find $f(30)$ and $f'(30)$. Using units, explain what each answer tells you about the population of the world.

39. Certain pieces of antique furniture increased very rapidly in price in the 1970s and 1980s. For example, the value of a particular rocking chair is well approximated by

$$V = 75(1.35)^t,$$

where V is in dollars and t is the number of years since 1975. Find the rate, in dollars per year, at which the price is increasing.

40. The value of a certain automobile purchased in 1997 can be approximated by the function $V(t) = 25(0.85)^t$, where t is the time, in years, from the date of purchase, and V is the value, in thousands of dollars.

(a) Evaluate and interpret $V(4)$.
(b) Find an expression for $V'(t)$, including units.
(c) Evaluate and interpret $V'(4)$.
(d) Use $V(t)$, $V'(t)$, and any other considerations you think are relevant to write a paragraph in support of or in opposition to the following statement: "From a monetary point of view, it is best to keep this vehicle as long as possible."

41. **(a)** Find the slope of the graph of $f(x) = 1 - e^x$ at the point where it crosses the x-axis.
    (b) Find the equation of the tangent line to the curve at this point.
    (c) Find the equation of the line perpendicular to the tangent line at this point. (This is the *normal* line.)

42. Find the value of c in Figure 3.12, where the line l tangent to the graph of $y = 2^x$ at $(0, 1)$ intersects the x-axis.

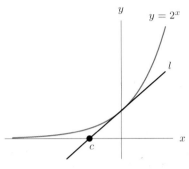

Figure 3.12

43. Find the quadratic polynomial $g(x) = ax^2 + bx + c$ which best fits the function $f(x) = e^x$ at $x = 0$, in the sense that

$$g(0) = f(0), \text{ and } g'(0) = f'(0), \text{ and } g''(0) = f''(0).$$

Using a computer or calculator, sketch graphs of f and g on the same axes. What do you notice?

44. Using the equation of the tangent line to the graph of e^x at $x = 0$, show that

$$e^x \geq 1 + x$$

for all values of x. A sketch may be helpful.

45. Find all solutions of the equation

$$2^x = 2x.$$

How do you know that you found all solutions?

46. For what value(s) of a are $y = a^x$ and $y = 1 + x$ tangent at $x = 0$? Explain.

3.3 THE PRODUCT AND QUOTIENT RULES

We now know how to find derivatives of powers and exponentials, and of sums and constant multiples of functions. This section shows how to find the derivatives of products and quotients.

Using △ Notation

To express the difference quotients of general functions, some additional notation is helpful. We write Δf, read "delta f," for a small change in the value of f at the point x,

$$\Delta f = f(x + h) - f(x).$$

In this notation, the derivative is the limit of the ratio $\Delta f / h$:

$$f'(x) = \lim_{h \to 0} \frac{\Delta f}{h}.$$

The Product Rule

Suppose we know the derivatives of $f(x)$ and $g(x)$ and want to calculate the derivative of the product, $f(x)g(x)$. The derivative of the product is calculated by taking the limit, namely,

$$\frac{d[f(x)g(x)]}{dx} = \lim_{h \to 0} \frac{f(x+h)g(x+h) - f(x)g(x)}{h}.$$

To picture the quantity $f(x + h)g(x + h) - f(x)g(x)$, imagine the rectangle with sides $f(x + h)$ and $g(x + h)$ in Figure 3.13, where $\Delta f = f(x + h) - f(x)$ and $\Delta g = g(x + h) - g(x)$. Then

$$f(x+h)g(x+h) - f(x)g(x) = (\text{Area of whole rectangle}) - (\text{Unshaded area})$$
$$= \text{Area of the three shaded rectangles}$$
$$= \Delta f \cdot g(x) + f(x) \cdot \Delta g + \Delta f \cdot \Delta g.$$

Now divide by h:

$$\frac{f(x+h)g(x+h) - f(x)g(x)}{h} = \frac{\Delta f}{h} \cdot g(x) + f(x) \cdot \frac{\Delta g}{h} + \frac{\Delta f \cdot \Delta g}{h}.$$

Figure 3.13: Illustration for the product rule (with Δf, Δg positive)

To evaluate the limit as $h \to 0$, we examine the three terms on the right separately. Notice that

$$\lim_{h \to 0} \frac{\Delta f}{h} \cdot g(x) = f'(x)g(x) \quad \text{and} \quad \lim_{h \to 0} f(x) \cdot \frac{\Delta g}{h} = f(x)g'(x).$$

In the third term we multiply the top and bottom by h to get $\dfrac{\Delta f}{h} \cdot \dfrac{\Delta g}{h} \cdot h$. Then,

$$\lim_{h \to 0} \frac{\Delta f \cdot \Delta g}{h} = \lim_{h \to 0} \frac{\Delta f}{h} \cdot \frac{\Delta g}{h} \cdot h = \lim_{h \to 0} \frac{\Delta f}{h} \cdot \lim_{h \to 0} \frac{\Delta g}{h} \cdot \lim_{h \to 0} h = f'(x) \cdot g'(x) \cdot 0 = 0.$$

Therefore, we conclude that

$$\lim_{h \to 0} \frac{f(x+h)g(x+h) - f(x)g(x)}{h} = \lim_{h \to 0} \left(\frac{\Delta f}{h} \cdot g(x) + f(x) \cdot \frac{\Delta g}{h} + \frac{\Delta f \cdot \Delta g}{h} \right)$$

$$= \lim_{h \to 0} \frac{\Delta f}{h} \cdot g(x) + \lim_{h \to 0} f(x) \cdot \frac{\Delta g}{h} + \lim_{h \to 0} \frac{\Delta f \cdot \Delta g}{h}$$

$$= f'(x)g(x) + f(x)g'(x).$$

Thus we have proved the following rule:

Theorem 3.3: The Product Rule

If $u = f(x)$ and $v = g(x)$ are differentiable, then

$$(fg)' = f'g + fg'.$$

The product rule can also be written

$$\frac{d(uv)}{dx} = \frac{du}{dx} \cdot v + u \cdot \frac{dv}{dx}.$$

In words:

> The derivative of a product is the derivative of the first times the second plus the first times the derivative of the second.

Another justification of the product rule is given in Problem 26 on page 155.

Example 1   Differentiate (a) $x^2 e^x$,   (b) $(3x^2 + 5x)e^x$,   (c) $\dfrac{e^x}{x^2}$.

Solution

(a)

$$\frac{d(x^2 e^x)}{dx} = \left(\frac{d(x^2)}{dx}\right)e^x + x^2\frac{d(e^x)}{dx} = 2xe^x + x^2 e^x = (2x + x^2)e^x.$$

(b)

$$\frac{d((3x^2 + 5x)e^x)}{dx} = \left(\frac{d(3x^2 + 5x)}{dx}\right)e^x + (3x^2 + 5x)\frac{d(e^x)}{dx}$$
$$= (6x + 5)e^x + (3x^2 + 5x)e^x = (3x^2 + 11x + 5)e^x.$$

(c) First we must write $\frac{e^x}{x^2}$ as the product $x^{-2}e^x$:

$$\frac{d}{dx}\left(\frac{e^x}{x^2}\right) = \frac{d(x^{-2}e^x)}{dx} = \left(\frac{d(x^{-2})}{dx}\right)e^x + x^{-2}\frac{d(e^x)}{dx}$$
$$= -2x^{-3}e^x + x^{-2}e^x = (-2x^{-3} + x^{-2})e^x.$$

The Quotient Rule

Suppose we want to differentiate a function of the form $Q(x) = f(x)/g(x)$. (Of course, we have to avoid points where $g(x) = 0$.) We want a formula for Q' in terms of f' and g'.

Assuming that $Q(x)$ is differentiable,[1] we can use the product rule on $f(x) = Q(x)g(x)$:

$$f'(x) = Q'(x)g(x) + Q(x)g'(x)$$
$$= Q'(x)g(x) + \frac{f(x)}{g(x)}g'(x).$$

Solving for $Q'(x)$ gives

$$Q'(x) = \frac{f'(x) - \dfrac{f(x)}{g(x)}g'(x)}{g(x)}.$$

Multiplying the top and bottom by $g(x)$ to simplify gives

$$\frac{d}{dx}\left(\frac{f(x)}{g(x)}\right) = \frac{f'(x)g(x) - f(x)g'(x)}{(g(x))^2}.$$

So we have the following rule:

Theorem 3.4: The Quotient Rule

If $u = f(x)$ and $v = g(x)$ are differentiable, then

$$\left(\frac{f}{g}\right)' = \frac{f'g - fg'}{g^2},$$

or equivalently,

$$\frac{d}{dx}\left(\frac{u}{v}\right) = \frac{\dfrac{du}{dx}\cdot v - u\cdot\dfrac{dv}{dx}}{v^2}.$$

In words:

  The derivative of a quotient is the derivative of the numerator times the denominator minus the numerator times the derivative of the denominator, all over the denominator squared.

---

[1] The method in Example 6 on page 130 can be used to explain why $Q(x)$ must be differentiable.

Example 2   Differentiate (a) $\dfrac{5x^2}{x^3+1}$,   (b) $\dfrac{1}{1+e^x}$,   (c) $\dfrac{e^x}{x^2}$.

Solution   (a)

$$\frac{d}{dx}\left(\frac{5x^2}{x^3+1}\right)=\frac{\left(\frac{d}{dx}(5x^2)\right)(x^3+1)-5x^2\frac{d}{dx}(x^3+1)}{(x^3+1)^2}=\frac{10x(x^3+1)-5x^2(3x^2)}{(x^3+1)^2}$$

$$=\frac{-5x^4+10x}{(x^3+1)^2}.$$

(b)

$$\frac{d}{dx}\left(\frac{1}{1+e^x}\right)=\frac{\left(\frac{d}{dx}(1)\right)(1+e^x)-1\frac{d}{dx}(1+e^x)}{(1+e^x)^2}=\frac{0(1+e^x)-1(0+e^x)}{(1+e^x)^2}$$

$$=\frac{-e^x}{(1+e^x)^2}.$$

(c) This is the same as part (c) of Example 1, but this time we do it by the quotient rule.

$$\frac{d}{dx}\left(\frac{e^x}{x^2}\right)=\frac{\left(\frac{d(e^x)}{dx}\right)x^2-e^x\left(\frac{d(x^2)}{dx}\right)}{(x^2)^2}=\frac{e^xx^2-e^x(2x)}{x^4}$$

$$=e^x\left(\frac{x^2-2x}{x^4}\right)=e^x\left(\frac{x-2}{x^3}\right).$$

This is, in fact, the same answer as before, although it looks different. Can you show that it is the same?

Exercises and Problems for Section 3.3

Exercises

1. If $f(x)=x^2(x^3+5)$, find $f'(x)$ two ways: by using the product rule and by multiplying out before taking the derivative. Do you get the same result? Should you?

2. If $f(x)=2^x\cdot3^x$, find $f'(x)$ two ways: by using the product rule and by using the fact that $2^x\cdot3^x=6^x$. Do you get the same result?

For Exercises 3–30, find the derivative. It may be to your advantage to simplify first. Assume that a, b, c, and k are constants.

3. $f(x)=xe^x$

4. $y=x\cdot2^x$

5. $y=\sqrt{x}\cdot2^x$

6. $y=(t^2+3)e^t$

7. $f(x)=(x^2-\sqrt{x})3^x$

8. $z=(s^2-\sqrt{s})(s^2+\sqrt{s})$

9. $f(y)=4^y(2-y^2)$

10. $y=(t^3-7t^2+1)e^t$

11. $f(x)=\dfrac{x}{e^x}$

12. $g(x)=\dfrac{25x^2}{e^x}$

13. $y=\dfrac{t+1}{2^t}$

14. $g(w)=\dfrac{w^{3.2}}{5^w}$

15. $q(r)=\dfrac{3r}{5r+2}$

16. $g(t)=\dfrac{t-4}{t+4}$

17. $z=\dfrac{3t+1}{5t+2}$

18. $z=\dfrac{t^2+5t+2}{t+3}$

19. $z=\dfrac{t^2+3t+1}{t+1}$

20. $f(x)=\dfrac{x^2+3}{x}$

21. $w=\dfrac{y^3-6y^2+7y}{y}$

22. $y=\dfrac{\sqrt{t}}{t^2+1}$

23. $f(z)=\dfrac{z^2+1}{\sqrt{z}}$

24. $g(t)=\dfrac{4}{3+\sqrt{t}}$

25. $h(r)=\dfrac{r^2}{2r+1}$

26. $f(z)=\dfrac{3z^2}{5z^2+7z}$

27. $w(x)=\dfrac{17e^x}{2^x}$

28. $h(p)=\dfrac{1+p^2}{3+2p^2}$

29. $f(x)=\dfrac{1+x}{2+3x+4x^2}$

30. $f(x)=\dfrac{ax+b}{cx+k}$

Problems

In Problems 31–33, use Figure 3.14 to estimate the derivative, or state that it does not exist. The graph of $f(x)$ has a sharp corner at $x = 2$.

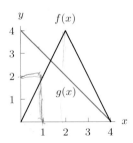

Figure 3.14

31. Let $h(x) = f(x) \cdot g(x)$. Find:

   (a) $h'(1)$ **(b)** $h'(2)$ **(c)** $h'(3)$

32. Let $k(x) = (f(x))/(g(x))$. Find:

   (a) $k'(1)$ **(b)** $k'(2)$ **(c)** $k'(3)$

33. Let $j(x) = (g(x))/(f(x))$. Find:

   (a) $j'(1)$ **(b)** $j'(2)$ **(c)** $j'(3)$

For Problems 34–39, let $h(x) = f(x) \cdot g(x)$, and $k(x) = f(x)/g(x)$, and $l(x) = g(x)/f(x)$. Use Figure 3.15 to estimate the derivatives.

 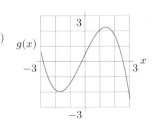

Figure 3.15

34. $h'(1)$ **35.** $k'(1)$ **36.** $h'(2)$

37. $k'(2)$ **38.** $l'(1)$ **39.** $l'(2)$

40. If $f(x) = (3x + 8)(2x - 5)$, find $f'(x)$ and $f''(x)$.

41. Differentiate $f(t) = e^{-t}$ by writing it as $f(t) = \dfrac{1}{e^t}$.

42. Differentiate $f(x) = e^{2x}$ by writing it as $f(x) = e^x \cdot e^x$.

43. Differentiate $f(x) = e^{3x}$ by writing it as $f(x) = e^x \cdot e^{2x}$ and using the result of Problem 42.

44. For what intervals is $f(x) = xe^x$ concave up?

45. For what intervals is $g(x) = \dfrac{1}{x^2 + 1}$ concave down?

46. Find the equation of the tangent line to the graph of $f(x) = \dfrac{2x - 5}{x + 1}$ at the point at which $x = 0$.

47. **(a)** Differentiate $y = \dfrac{e^x}{x}$, $y = \dfrac{e^x}{x^2}$, and $y = \dfrac{e^x}{x^3}$.

   (b) What do you anticipate the derivative of $y = \dfrac{e^x}{x^n}$ will be? Confirm your guess.

48. Using the product rule and the fact that $\dfrac{d(x)}{dx} = 1$, show that $\dfrac{d(x^2)}{dx} = 2x$ and $\dfrac{d(x^3)}{dx} = 3x^2$.

49. Use the product rule to show that $\dfrac{d}{dx}(x^{1/2}) = \dfrac{1}{2x^{1/2}}$.

   [Hint: Write $x = x^{1/2}x^{1/2}$.]

50. Suppose f and g are differentiable functions with the values shown in the following table. For each of the following functions h, find $h'(2)$.

   (a) $h(x) = f(x) + g(x)$ **(b)** $h(x) = f(x)g(x)$

   (c) $h(x) = \dfrac{f(x)}{g(x)}$

x	$f(x)$	$g(x)$	$f'(x)$	$g'(x)$
2	3	4	5	−2

51. If $H(3) = 1$, $H'(3) = 3$, $F(3) = 5$, $F'(3) = 4$, find:

   (a) $G'(3)$ if $G(z) = F(z) \cdot H(z)$
   (b) $G'(3)$ if $G(w) = F(w)/H(w)$

52. Find a possible formula for a function $y = f(x)$ such that $f'(x) = 10x^9 e^x + x^{10} e^x$.

53. The quantity, q, of a certain skateboard sold depends on the selling price, p, in dollars, so we write $q = f(p)$. You are given that $f(140) = 15{,}000$ and $f'(140) = -100$.

   (a) What do $f(140) = 15{,}000$ and $f'(140) = -100$ tell you about the sales of skateboards?

   (b) The total revenue, R, earned by the sale of skateboards is given by $R = pq$. Find $\left. \dfrac{dR}{dp} \right|_{p=140}$.

   (c) What is the sign of $\left. \dfrac{dR}{dp} \right|_{p=140}$? If the skateboards are currently selling for $140, what happens to revenue if the price is increased to $141?

54. When an electric current passes through two resistors with resistance r_1 and r_2, connected in parallel, the combined resistance, R, can be calculated from the equation

$$\frac{1}{R} = \frac{1}{r_1} + \frac{1}{r_2}.$$

Find the rate at which the combined resistance changes with respect to changes in r_1. Assume that r_2 is constant.

55. A museum has decided to sell one of its paintings and to invest the proceeds. If the picture is sold between the years 2000 and 2020 and the money from the sale is invested in a bank account earning 5% interest per year compounded annually, then $B(t)$, the balance in the year 2020, depends on the year, t, in which the painting is sold and the sale price $P(t)$. If t is measured from the year 2000 so that $0 < t < 20$ then

$$B(t) = P(t)(1.05)^{20-t}.$$

(a) Explain why $B(t)$ is given by this formula.
(b) Show that the formula for $B(t)$ is equivalent to

$$B(t) = (1.05)^{20}\frac{P(t)}{(1.05)^t}.$$

(c) Find $B'(10)$, given that $P(10) = 150{,}000$ and $P'(10) = 5000$.

56. Let $f(v)$ be the gas consumption (in liters/km) of a car going at velocity v (in km/hr). In other words, $f(v)$ tells you how many liters of gas the car uses to go one kilometer, if it is going at velocity v. You are told that

$$f(80) = 0.05 \text{ and } f'(80) = 0.0005.$$

(a) Let $g(v)$ be the distance the same car goes on one liter of gas at velocity v. What is the relationship between $f(v)$ and $g(v)$? Find $g(80)$ and $g'(80)$.
(b) Let $h(v)$ be the gas consumption in liters per hour. In other words, $h(v)$ tells you how many liters of gas the car uses in one hour if it is going at velocity v. What is the relationship between $h(v)$ and $f(v)$? Find $h(80)$ and $h'(80)$.
(c) How would you explain the practical meaning of the values of these functions and their derivatives to a driver who knows no calculus?

57. The function $f(x) = e^x$ has the properties

$$f'(x) = f(x) \text{ and } f(0) = 1.$$

Explain why $f(x)$ is the only function with both these properties. [Hint: Assume $g'(x) = g(x)$, and $g(0) = 1$, for some function $g(x)$. Define $h(x) = g(x)/e^x$, and compute $h'(x)$. Then use the fact that a function with a derivative of 0 must be a constant function.]

58. Find $f'(x)$ for the following functions with the product rule, rather than by multiplying out.

(a) $f(x) = (x-1)(x-2)$.
(b) $f(x) = (x-1)(x-2)(x-3)$.
(c) $f(x) = (x-1)(x-2)(x-3)(x-4)$.

59. Use the answer from Problem 58 to guess $f'(x)$ for the following function:

$$f(x) = (x-r_1)(x-r_2)(x-r_3)\cdots(x-r_n)$$

where $r_1, r_2, \ldots, r_n$ are any real numbers.

60. (a) Provide a three dimensional analogue for the geometrical demonstration of the formula for the derivative of a product, given in Figure 3.13 on page 122. In other words, find a formula for the derivative of $F(x) \cdot G(x) \cdot H(x)$ using Figure 3.16.
(b) Confirm your results by writing $F(x) \cdot G(x) \cdot H(x)$ as $[F(x) \cdot G(x)] \cdot H(x)$ and using the product rule twice.
(c) Generalize your result to n functions: what is the derivative of

$$f_1(x) \cdot f_2(x) \cdot f_3(x) \cdots f_n(x)?$$

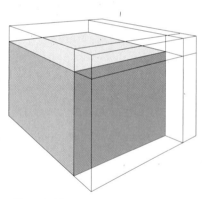

Figure 3.16: A graphical representation of the 3-dimensional product rule

61. If $P(x) = (x-a)^2 Q(x)$, where $Q(x)$ is a polynomial, we call $x = a$ a double zero of the polynomial $P(x)$.

(a) If $x = a$ is a double zero of a polynomial $P(x)$, show that $P(a) = P'(a) = 0$.
(b) If $P(x)$ is a polynomial and $P(a) = P'(a) = 0$, show that $x = a$ is a double zero of $P(x)$.

3.4 THE CHAIN RULE

Composite functions such as $\sin(3t)$ or e^{-x^2} occur frequently in practice. In this section we see how to differentiate such functions.

The Derivative of a Composition of Functions

Suppose $f(g(x))$ is a composite function, with f being the outside function and g being the inside. Let us write

$$z = g(x) \quad \text{and} \quad y = f(z), \quad \text{so} \quad y = f(g(x)).$$

Then a small change in x, called Δx, generates a small change in z, called Δz. In turn, Δz generates a small change in y called Δy. Provided Δx and Δz are not zero, we can say:

$$\frac{\Delta y}{\Delta x} = \frac{\Delta y}{\Delta z} \cdot \frac{\Delta z}{\Delta x}.$$

Since $\dfrac{dy}{dx} = \lim\limits_{\Delta x \to 0} \dfrac{\Delta y}{\Delta x}$, this suggests that in the limit as Δx, Δy, and Δz get smaller and smaller, we have:

> ### The Chain Rule
>
> $$\frac{dy}{dx} = \frac{dy}{dz} \cdot \frac{dz}{dx}.$$

Since $\dfrac{dy}{dz} = f'(z)$ and $\dfrac{dz}{dx} = g'(x)$, we can also write

$$\frac{d}{dx} f(g(x)) = f'(z) \cdot g'(x).$$

Substituting $z = g(x)$, we can rewrite this as follows:

> ### Theorem 3.5: The Chain Rule
>
> If f and g are differentiable functions, then
>
> $$\frac{d}{dx} f(g(x)) = f'(g(x)) \cdot g'(x).$$
>
> In words:
>   The derivative of a composite function is the product of the derivatives of the outside
>   and inside functions. The derivative of the outside function must be evaluated at the
>   inside function.

A justification of the chain rule is given in Problem 27 on page 155.

Example 1   The length, L, in cm, of a steel bar depends on the air temperature, $H°C$, and the temperature H depends on time, t, measured in hours. If the length increases by 2 cm for every degree increase in temperature, and the temperature is increasing at $3°C$ per hour, how fast is the length of the bar increasing? What are the units for your answer?

Solution   We expect the rate at which the length is increasing to be in cm/hr. We are told that

$$\begin{array}{cc} \text{Rate length increasing} \\ \text{with respect to temperature} \end{array} = \frac{dL}{dH} = 2 \text{ cm}/°C$$

$$\begin{array}{cc} \text{Rate temperature increasing} \\ \text{with respect to time} \end{array} = \frac{dH}{dt} = 3°C/\text{hr}.$$

We want to calculate the rate at which the length is increasing with respect to time, or dL/dt. We think of L as a function of H, and H as a function of t. By the chain rule we know that

$$\frac{dL}{dt} = \frac{dL}{dH} \cdot \frac{dH}{dt} = \left(2\frac{\text{cm}}{{}^\circ\text{C}}\right) \cdot \left(3\frac{{}^\circ\text{C}}{\text{hr}}\right) = 6 \text{ cm/hr}.$$

Thus, the length is increasing at 6 cm/hr.

Example 1 shows us how to interpret the chain rule in practical terms. The next examples show how the chain rule is used to compute derivatives of functions given by formulas.

Example 2   Find the derivatives of the following functions:

(a) $(x^2 + 1)^{100}$ (b) $\sqrt{3x^2 + 5x - 2}$ (c) $\dfrac{1}{x^2 + x^4}$ (d) e^{3x} (e) e^{x^2}

Solution   (a) Here $z = g(x) = x^2 + 1$ is the inside function; $f(z) = z^{100}$ is the outside function. Now $g'(x) = 2x$ and $f'(z) = 100z^{99}$, so

$$\frac{d}{dx}((x^2 + 1)^{100}) = 100z^{99} \cdot 2x = 100(x^2 + 1)^{99} \cdot 2x = 200x(x^2 + 1)^{99}.$$

(b) Here $z = g(x) = 3x^2 + 5x - 2$ and $f(z) = \sqrt{z}$, so $g'(x) = 6x + 5$ and $f'(z) = \dfrac{1}{2\sqrt{z}}$. Hence

$$\frac{d}{dx}(\sqrt{3x^2 + 5x - 2}) = \frac{1}{2\sqrt{z}} \cdot (6x + 5) = \frac{1}{2\sqrt{3x^2 + 5x - 2}} \cdot (6x + 5).$$

(c) Let $z = g(x) = x^2 + x^4$ and $f(z) = 1/z$, so $g'(x) = 2x + 4x^3$ and $f'(z) = -z^{-2} = -\dfrac{1}{z^2}$. Then

$$\frac{d}{dx}\left(\frac{1}{x^2 + x^4}\right) = -\frac{1}{z^2}(2x + 4x^3) = -\frac{2x + 4x^3}{(x^2 + x^4)^2}.$$

We could have done this problem using the quotient rule. Try it and see that you get the same answer!

(d) Let $z = g(x) = 3x$ and $f(z) = e^z$. Then $g'(x) = 3$ and $f'(z) = e^z$, so

$$\frac{d}{dx}\left(e^{3x}\right) = e^z \cdot 3 = 3e^{3x}.$$

(e) To figure out which is the inside function and which is the outside, notice that to evaluate e^{x^2} we first evaluate x^2 and then take e to that power. This tells us that the inside function is $z = g(x) = x^2$ and the outside function is $f(z) = e^z$. Therefore, $g'(x) = 2x$, and $f'(z) = e^z$, giving

$$\frac{d}{dx}(e^{x^2}) = e^z \cdot 2x = e^{x^2} \cdot 2x = 2xe^{x^2}.$$

To differentiate a complicated function, we may have to use the chain rule more than once, as in the following example.

Example 3   Differentiate: (a) $\sqrt{e^{-x/7} + 5}$ (b) $(1 - e^{2\sqrt{t}})^{19}$

Solution   (a) Let $z = g(x) = e^{-x/7} + 5$ be the inside function; let $f(z) = \sqrt{z}$ be the outside function. Now $f'(z) = \dfrac{1}{2\sqrt{z}}$, but we need the chain rule to find $g'(x)$.

We choose new inside and outside functions whose composition is $g(x)$. Let $u = h(x) = -x/7$ and $k(u) = e^u + 5$ so $g(x) = k(h(x)) = e^{-x/7} + 5$. Then $h'(x) = -1/7$ and $k'(u) = e^u$, so

$$g'(x) = e^u \cdot \left(-\frac{1}{7}\right) = -\frac{1}{7}e^{-x/7}.$$

Using the chain rule to combine the derivatives of $f(z)$ and $g(x)$, we have

$$\frac{d}{dx}\left(\sqrt{e^{-x/7} + 5}\right) = \frac{1}{2\sqrt{z}}\left(-\frac{1}{7}e^{-x/7}\right) = -\frac{e^{-x/7}}{14\sqrt{e^{-x/7} + 5}}.$$

(b) Let $z = g(t) = 1 - e^{2\sqrt{t}}$ be the inside function and $f(z) = z^{19}$ be the outside function. Then $f'(z) = 19z^{18}$ but we need the chain rule to differentiate $g(t)$.

Now we choose $u = h(t) = 2\sqrt{t}$ and $k(u) = 1 - e^u$, so $g(t) = k(h(t))$. Then $h'(t) = 2 \cdot \dfrac{1}{2}t^{-1/2} = \dfrac{1}{\sqrt{t}}$ and $k'(u) = -e^u$, so

$$g'(t) = -e^u \cdot \frac{1}{\sqrt{t}} = -\frac{e^{2\sqrt{t}}}{\sqrt{t}}.$$

Using the chain rule to combine the derivatives of $f(z)$ and $g(t)$, we have

$$\frac{d}{dx}(1 - e^{2\sqrt{t}})^{19} = 19z^{18}\left(-\frac{e^{2\sqrt{t}}}{\sqrt{t}}\right) = -19\frac{e^{2\sqrt{t}}}{\sqrt{t}}(1 - e^{2\sqrt{t}})^{18}.$$

It is often faster to use the chain rule without introducing new variables, as in the following examples.

Example 4   Differentiate $\sqrt{1 + e^{\sqrt{3+x^2}}}$.

Solution   The chain rule is needed four times:

$$\frac{d}{dx}\left(\sqrt{1 + e^{\sqrt{3+x^2}}}\right) = \frac{1}{2}\left(1 + e^{\sqrt{3+x^2}}\right)^{-1/2} \cdot \frac{d}{dx}\left(1 + e^{\sqrt{3+x^2}}\right)$$

$$= \frac{1}{2}\left(1 + e^{\sqrt{3+x^2}}\right)^{-1/2} \cdot e^{\sqrt{3+x^2}} \cdot \frac{d}{dx}\left(\sqrt{3 + x^2}\right)$$

$$= \frac{1}{2}\left(1 + e^{\sqrt{3+x^2}}\right)^{-1/2} \cdot e^{\sqrt{3+x^2}} \cdot \frac{1}{2}(3 + x^2)^{-1/2} \cdot \frac{d}{dx}(3 + x^2)$$

$$= \frac{1}{2}\left(1 + e^{\sqrt{3+x^2}}\right)^{-1/2} \cdot e^{\sqrt{3+x^2}} \cdot \frac{1}{2}(3 + x^2)^{-1/2} \cdot 2x.$$

Example 5   Find the derivative of e^{2x} by the chain rule and by the product rule.

Solution   Using the chain rule, we have

$$\frac{d}{dx}(e^{2x}) = e^{2x} \cdot \frac{d}{dx}(2x) = e^{2x} \cdot 2 = 2e^{2x}.$$

Using the product rule, we write $e^{2x} = e^x \cdot e^x$. Then

$$\frac{d}{dx}(e^{2x}) = \frac{d}{dx}(e^x e^x) = \left(\frac{d}{dx}(e^x)\right)e^x + e^x\left(\frac{d}{dx}(e^x)\right) = e^x \cdot e^x + e^x \cdot e^x = 2e^{2x}.$$

Using the Product and Chain Rules to Differentiate a Quotient

If you prefer, you can differentiate a quotient by the product and chain rules, instead of by the quotient rule. The resulting formulas may look different, but they will be equivalent.

Example 6   Find $k'(x)$ if $k(x) = \dfrac{x}{x^2 + 1}$.

Solution   One way is to use the quotient rule:

$$k'(x) = \frac{1 \cdot (x^2 + 1) - x \cdot (2x)}{(x^2 + 1)^2}$$
$$= \frac{1 - x^2}{(x^2 + 1)^2}.$$

Alternatively, we can write the original function as a product,

$$k(x) = x\frac{1}{x^2 + 1} = x \cdot (x^2 + 1)^{-1},$$

and use the product rule:

$$k'(x) = 1 \cdot (x^2 + 1)^{-1} + x \cdot \frac{d}{dx}\left[(x^2 + 1)^{-1}\right].$$

Now use the chain rule to differentiate $(x^2 + 1)^{-1}$, giving

$$\frac{d}{dx}\left[(x^2 + 1)^{-1}\right] = -(x^2 + 1)^{-2} \cdot 2x = \frac{-2x}{(x^2 + 1)^2}.$$

Therefore,

$$k'(x) = \frac{1}{x^2 + 1} + x \cdot \frac{-2x}{(x^2 + 1)^2} = \frac{1}{x^2 + 1} - \frac{2x^2}{(x^2 + 1)^2}.$$

Putting these two fractions over a common denominator gives the same answer as the quotient rule.

Exercises and Problems for Section 3.4

Exercises

Find the derivatives of the functions in Exercises 1–52. Assume that a, b, c, and k are constants.

1. $f(x) = (x + 1)^{99}$

2. $w = (t^2 + 1)^{100}$

3. $w = (t^3 + 1)^{100}$

4. $(4x^2 + 1)^7$

5. $f(x) = \sqrt{1 - x^2}$

6. $\sqrt{e^x + 1}$

7. $w = (\sqrt{t} + 1)^{100}$

8. $h(w) = (w^4 - 2w)^5$

9. $w(r) = \sqrt{r^4 + 1}$

10. $k(x) = (x^3 + e^x)^4$

11. $f(x) = e^{2x}\left(x^2 + 5^x\right)$

12. $f(t) = e^{3t}$

13. $g(x) = e^{\pi x}$

14. $f(\theta) = 2^{-\theta}$

15. $y = \pi^{(x+2)}$

16. $g(x) = 3^{(2x+7)}$

17. $f(t) = te^{5-2t}$

18. $p(t) = e^{4t+2}$

19. $v(t) = t^2 e^{-ct}$

20. $g(t) = e^{(1+3t)^2}$

21. $y = e^{3w/2}$

22. $y = e^{-4t}$

23. $y = \sqrt{s^3 + 1}$

24. $w = e^{\sqrt{s}}$

25. $y = te^{-t^2}$

26. $f(z) = \sqrt{z}e^{-z}$

27. $z(x) = \sqrt[3]{2^x + 5}$

28. $z = 2^{5t-3}$

29. $w = \sqrt{(x^2 \cdot 5^x)^3}$

30. $f(y) = \sqrt{10^{(5-y)}}$

31. $f(z) = \dfrac{\sqrt{z}}{e^z}$

32. $y = \dfrac{\sqrt{z}}{2^z}$

33. $y = \left(\dfrac{x^2 + 2}{3}\right)^2$

34. $h(x) = \sqrt{\dfrac{x^2 + 9}{x + 3}}$

35. $y = \dfrac{e^{2x}}{x^2 + 1}$

36. $y = \dfrac{1}{e^{3x} + x^2}$

37. $h(z) = \left(\dfrac{b}{a + z^2}\right)^4$

38. $f(z) = \dfrac{1}{(e^z + 1)^2}$

39. $w = (t^2 + 3t)(1 - e^{-2t})$

40. $h(x) = 2e^{e^{3x}}$

41. $f(x) = 6e^{5x} + e^{-x^2}$

42. $f(x) = e^{-(x-1)^2}$

43. $f(w) = (5w^2 + 3)e^{w^2}$

44. $f(\theta) = (e^\theta + e^{-\theta})^{-1}$

45. $y = \sqrt{e^{-3t^2} + 5}$

46. $z = (te^{3t} + e^{5t})^9$

47. $f(y) = e^{e^{(y^2)}}$

48. $f(t) = 2e^{-2e^{2t}}$

49. $f(x) = (ax^2 + b)^3$

50. $f(t) = ae^{bt}$

51. $f(x) = axe^{-bx}$

52. $g(\alpha) = e^{\alpha e^{-2\alpha}}$

Problems

In Problems 53–56, use Figure 3.14 to estimate the derivative, or state it does not exist. The graph of $f(x)$ has a sharp corner at $x = 2$.

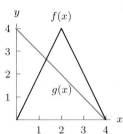

Figure 3.17

53. Let $h(x) = f(g(x))$. Find:
    (a) $h'(1)$ **(b)** $h'(2)$ **(c)** $h'(3)$

54. Let $u(x) = g(f(x))$. Find:
    (a) $u'(1)$ **(b)** $u'(2)$ **(c)** $u'(3)$

55. Let $v(x) = f(f(x))$. Find:
    (a) $v'(1)$ **(b)** $v'(2)$ **(c)** $v'(3)$

56. Let $w(x) = g(g(x))$. Find:
    (a) $w'(1)$ **(b)** $w'(2)$ **(c)** $w'(3)$

In Problems 57–60, use Figure 3.18 to evaluate the derivative.

 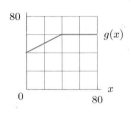

Figure 3.18

57. $\frac{d}{dx} f(g(x))|_{x=30}$

58. $\frac{d}{dx} f(g(x))|_{x=70}$

59. $\frac{d}{dx} g(f(x))|_{x=30}$

60. $\frac{d}{dx} g(f(x))|_{x=70}$

61. Find the equation of the tangent line to $f(x) = (x - 1)^3$ at the point where $x = 2$.

62. Find the equation of the line tangent to $y = f(x)$ at $x = 1$, where $f(x)$ is the function in Problem 41.

63. For what values of x is the graph of $y = e^{-x^2}$ concave down?

64. For what intervals is $f(x) = xe^{-x}$ concave down?

65. Suppose $f(x) = (2x + 1)^{10}(3x - 1)^7$. Find a formula for $f'(x)$. Decide on a reasonable way to simplify your result, and find a formula for $f''(x)$.

66. At any time, t, a population, $P(t)$, is growing at a rate proportional to the population at that moment.

    (a) Using derivatives, write an equation representing the growth of the population. Let k be the constant of proportionality.

    (b) Show that the function $P(t) = Ae^{kt}$ satisfies the equation in part (a) for any constant A.

67. Find the mean and variance of the normal distribution of statistics using the information in parts (a) and (b) with

$$m(t) = e^{\mu t + \sigma^2 t^2/2}.$$

    (a) Mean $= m'(0)$
    (b) Variance $= m''(0) - (m'(0))^2$

68. If the derivative of $y = k(x)$ equals 2 when $x = 1$, what is the derivative of

    (a) $k(2x)$ when $x = \dfrac{1}{2}$?
    (b) $k(x + 1)$ when $x = 0$?
    (c) $k\left(\dfrac{1}{4}x\right)$ when $x = 4$?

69. Is $x = \sqrt[3]{2t + 5}$ a solution to the equation $3x^2 \dfrac{dx}{dt} = 2$? Why or why not?

70. Find a possible formula for a function $m(x)$ such that $m'(x) = x^5 \cdot e^{(x^6)}$.

71. Given $F(2) = 1, F'(2) = 5, F(4) = 3, F'(4) = 7$ and $G(4) = 2, G'(4) = 6, G(3) = 4, G'(3) = 8$, find:

  (a) $H(4)$ if $H(x) = F(G(x))$
  (b) $H'(4)$ if $H(x) = F(G(x))$
  (c) $H(4)$ if $H(x) = G(F(x))$
  (d) $H'(4)$ if $H(x) = G(F(x))$
  (e) $H'(4)$ if $H(x) = F(x)/G(x)$

72. Given $y = f(x)$ with $f(1) = 4$ and $f'(1) = 3$, find

  (a) $g'(1)$ if $g(x) = \sqrt{f(x)}$.
  (b) $h'(1)$ if $h(x) = f(\sqrt{x})$.

In Problems 73–77, use Figures 3.19 and 3.20 and $h(x) = f(g(x))$.

Figure 3.19

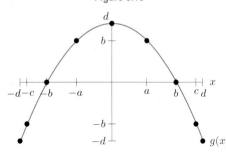

Figure 3.20

73. Evaluate $h(0)$ and $h'(0)$.

74. At $x = -c$, is h positive, negative, or zero? Increasing or decreasing?

75. At $x = a$, is h increasing or decreasing?

76. What are the signs of $h(d)$ and $h'(d)$?

77. How does the value of $h(x)$ change on the interval $-d < x < -b$?

78. The world's population is about $f(t) = 6e^{0.013t}$ billion, where t is time in years since 1999. Find $f(0), f'(0), f(10)$, and $f'(10)$. Using units, interpret your answers in terms of population.

79. If you invest P dollars in a bank account at an annual interest rate of $r\%$, then after t years you will have B dollars, where

$$B = P\left(1 + \frac{r}{100}\right)^t.$$

  (a) Find dB/dt, assuming P and r are constant. In terms of money, what does dB/dt represent?
  (b) Find dB/dr, assuming P and t are constant. In terms of money, what does dB/dr represent?

80. The theory of relativity predicts that an object whose mass is m_0 when it is at rest will appear heavier when moving at speeds near the speed of light. When the object is moving at speed v, its mass m is given by

$$m = \frac{m_0}{\sqrt{1 - (v^2/c^2)}}, \qquad \text{where } c \text{ is the speed of light.}$$

  (a) Find dm/dv.
  (b) In terms of physics, what does dm/dv tell you?

81. The charge, Q, on a capacitor which starts discharging at time $t = 0$ is given by

$$Q = \begin{cases} Q_0 & \text{for } t \leq 0 \\ Q_0 e^{-t/RC} & \text{for } t > 0, \end{cases}$$

where R and C are positive constants depending on the circuit and Q_0 is the charge at $t = 0$, where $Q_0 \neq 0$. The current, I, flowing in the circuit is given by $I = dQ/dt$.

  (a) Find the current I for $t < 0$ and for $t > 0$.
  (b) Is it possible to define I at $t = 0$?
  (c) Is the function Q differentiable at $t = 0$?

82. A particle is moving on the x-axis, where x is in centimeters. Its velocity, v, in cm/sec, when it is at the point with coordinate x is given by

$$v = x^2 + 3x - 2.$$

Find the acceleration of the particle when it is at the point $x = 2$. Give units in your answer.

83. A function f is said to have a *zero of multiplicity* m at $x = a$ if

$$f(x) = (x - a)^m h(x), \quad \text{with } h(a) \neq 0.$$

Explain why a function having a zero of multiplicity m at $x = a$ satisfies $f^{(p)}(a) = 0$, for $p = 1, 2, \ldots m - 1$. [Note: $f^{(p)}$ is the pth derivative.]

84. If $\dfrac{d}{dx} f(x^2 + 1) = \dfrac{2x}{x^2 + 1}$, what is $f'(x)$?

85. If $\dfrac{d}{dt} G(a - bt) = H(a - bt)$, what is $G'(t)$? (Give your answer in terms of $H(t)$.)

86. If $\dfrac{d}{dt}(tf(t)) = 1 + f(t)$, what is $f'(t)$?

87. If $\dfrac{d}{dx} f(e^x) = 2e^{2x}$, what is $f(x)$?

3.5 THE TRIGONOMETRIC FUNCTIONS

Derivatives of the Sine and Cosine

Since the sine and cosine functions are periodic, their derivatives must be periodic also. (Why?) Let's look at the graph of $f(x) = \sin x$ in Figure 3.21 and estimate the derivative function graphically.

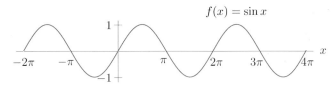

Figure 3.21: The sine function

First we might ask where the derivative is zero. (At $x = \pm\pi/2, \pm 3\pi/2, \pm 5\pi/2$, etc.) Then ask where the derivative is positive and where it is negative. (Positive for $-\pi/2 < x < \pi/2$; negative for $\pi/2 < x < 3\pi/2$, etc.) Since the largest positive slopes are at $x = 0, 2\pi$, and so on, and the largest negative slopes are at $x = \pi, 3\pi$, and so on, we get something like the graph in Figure 3.22.

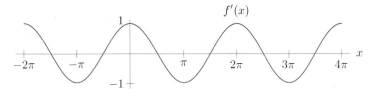

Figure 3.22: Derivative of $f(x) = \sin x$

The graph of the derivative in Figure 3.22 looks suspiciously like the graph of the cosine function. This might lead us to conjecture, quite correctly, that the derivative of the sine is the cosine.

Of course, we cannot be sure, just from the graphs, that the derivative of the sine really is the cosine. However, for now we'll assume that the derivative of the sine *is* the cosine and confirm the result at the end of the section.

One thing we can do now is to check that the derivative function in Figure 3.22 has amplitude 1 (as it ought to if it is the cosine). That means we have to convince ourselves that the derivative of $f(x) = \sin x$ is 1 when $x = 0$. The next example suggests that this is true when x is in radians.

Example 1   Using a calculator set in radians, estimate the derivative of $f(x) = \sin x$ at $x = 0$.

Solution   Since $f(x) = \sin x$,
$$f'(0) = \lim_{h \to 0} \frac{\sin(0 + h) - \sin 0}{h} = \lim_{h \to 0} \frac{\sin h}{h}.$$
Table 3.5 contains values of $(\sin h)/h$ which suggest that this limit is 1, so we estimate
$$f'(0) = \lim_{h \to 0} \frac{\sin h}{h} = 1.$$

Table 3.5

h (radians)	-0.1	-0.01	-0.001	-0.0001	0.0001	0.001	0.01	0.1
$(\sin h)/h$	0.99833	0.99998	1.0000	1.0000	1.0000	1.0000	0.99998	0.99833

Warning: It is important to notice that in the previous example h was in *radians*; any conclusions we have drawn about the derivative of $\sin x$ are valid *only* when x is in radians. If you find the derivative with h in degrees, you get a different result.

Example 2   Starting with the graph of the cosine function, sketch a graph of its derivative.

Solution The graph of $g(x) = \cos x$ is in Figure 3.23(a). Its derivative is 0 at $x = 0, \pm\pi, \pm2\pi$, and so on; it is positive for $-\pi < x < 0$, $\pi < x < 2\pi$, and so on; and it is negative for $0 < x < \pi$, $2\pi < x < 3\pi$, and so on. The derivative is in Figure 3.23(b).

(a)

(b)

Figure 3.23: $g(x) = \cos x$ and its derivative, $g'(x)$

As we did with the sine, we use the graphs to make a conjecture. The derivative of the cosine in Figure 3.23(b) looks exactly like the graph of sine, except reflected about the x-axis. But how can we be sure that the derivative is $-\sin x$?

Example 3 Use the relation $\dfrac{d}{dx}(\sin x) = \cos x$ to show that $\dfrac{d}{dx}(\cos x) = -\sin x$.

Solution Since the cosine function is the sine function shifted to the left by $\pi/2$ (that is, $\cos x = \sin(x + \pi/2)$), we expect the derivative of the cosine to be the derivative of the sine, shifted to the left by $\pi/2$. Differentiating using the chain rule:

$$\frac{d}{dx}(\cos x) = \frac{d}{dx}\left(\sin\left(x + \frac{\pi}{2}\right)\right) = \cos\left(x + \frac{\pi}{2}\right).$$

But $\cos(x + \pi/2)$ is the cosine shifted to the left by $\pi/2$, which gives a sine curve reflected about the x-axis. So we have

$$\frac{d}{dx}(\cos x) = \cos\left(x + \frac{\pi}{2}\right) = -\sin x.$$

At the end of this section and in Problems 51 and 52, we show that our conjectures for the derivatives of $\sin x$ and $\cos x$ are correct. Thus, we have:

$$\boxed{\text{For } x \text{ in radians,} \qquad \frac{d}{dx}(\sin x) = \cos x \quad \text{and} \quad \frac{d}{dx}(\cos x) = -\sin x.}$$

Example 4 Differentiate (a) $2\sin(3\theta)$,   (b) $\cos^2 x$,   (c) $\cos(x^2)$,   (d) $e^{-\sin t}$.

Solution Use the chain rule:

(a) $\dfrac{d}{d\theta}(2\sin(3\theta)) = 2\dfrac{d}{d\theta}(\sin(3\theta)) = 2(\cos(3\theta))\dfrac{d}{d\theta}(3\theta) = 2(\cos(3\theta))3 = 6\cos(3\theta).$

(b) $\dfrac{d}{dx}(\cos^2 x) = \dfrac{d}{dx}\left((\cos x)^2\right) = 2(\cos x) \cdot \dfrac{d}{dx}(\cos x) = 2(\cos x)(-\sin x) = -2\cos x \sin x.$

(c) $\dfrac{d}{dx}\left(\cos(x^2)\right) = -\sin(x^2) \cdot \dfrac{d}{dx}(x^2) = -2x\sin(x^2).$

(d) $\dfrac{d}{dt}(e^{-\sin t}) = e^{-\sin t}\dfrac{d}{dt}(-\sin t) = -(\cos t)e^{-\sin t}.$

Derivative of the Tangent Function

Since $\tan x = \sin x / \cos x$, we differentiate $\tan x$ using the quotient rule. Writing $(\sin x)'$ for $d(\sin x)/dx$, we have:

$$\frac{d}{dx}(\tan x) = \frac{d}{dx}\left(\frac{\sin x}{\cos x}\right) = \frac{(\sin x)'(\cos x) - (\sin x)(\cos x)'}{\cos^2 x} = \frac{\cos^2 x + \sin^2 x}{\cos^2 x} = \frac{1}{\cos^2 x}.$$

> For x in radians, $\qquad \dfrac{d}{dx}(\tan x) = \dfrac{1}{\cos^2 x}.$

The graphs of $f(x) = \tan x$ and $f'(x) = 1/\cos^2 x$ are in Figure 3.24. Is it reasonable that f' is always positive? Are the asymptotes of f' where we expect?

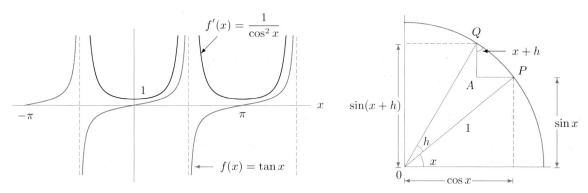

Figure 3.24: The function $\tan x$ and its derivative

Figure 3.25: Unit circle showing $\sin(x+h)$ and $\sin x$

Example 5   Differentiate   (a) $2\tan(3t)$,   (b) $\tan(1 - \theta)$,   (c) $\dfrac{1 + \tan t}{1 - \tan t}$.

Solution   (a) Use the chain rule:

$$\frac{d}{dt}(2\tan(3t)) = 2\frac{1}{\cos^2(3t)}\frac{d}{dt}(3t) = \frac{6}{\cos^2(3t)}.$$

(b) Use the chain rule:

$$\frac{d}{d\theta}(\tan(1-\theta)) = \frac{1}{\cos^2(1-\theta)} \cdot \frac{d}{d\theta}(1-\theta) = \frac{-1}{\cos^2(1-\theta)}.$$

(c) Use the quotient rule:

$$\frac{d}{dt}\left(\frac{1+\tan t}{1-\tan t}\right) = \frac{\left(\dfrac{d(1+\tan t)}{dt}\right)(1-\tan t) - (1+\tan t)\dfrac{d(1-\tan t)}{dt}}{(1-\tan t)^2}$$

$$= \frac{\dfrac{1}{\cos^2 t}(1-\tan t) - (1+\tan t)\left(-\dfrac{1}{\cos^2 t}\right)}{(1-\tan t)^2}$$

$$= \frac{2}{\cos^2 t \cdot (1-\tan t)^2}.$$

Informal Justification of $\frac{d}{dx}(\sin x) = \cos x$

Consider the unit circle in Figure 3.25. To find the derivative of $\sin x$, we need to estimate

$$\frac{\sin(x+h) - \sin x}{h}.$$

In Figure 3.25, the quantity $\sin(x+h) - \sin x$ is represented by the length QA. The arc QP is of length h, so

$$\frac{\sin(x+h) - \sin x}{h} = \frac{QA}{\text{Arc } QP}.$$

Now, if h is small, QAP is approximately a right triangle because the arc QP is almost a straight line. Furthermore, using geometry, we can show that angle $AQP = x + h$. For small h, we have

$$\frac{\sin(x+h) - \sin x}{h} = \frac{QA}{\text{Arc } QP} \approx \cos(x+h).$$

As $h \to 0$, the approximation gets better, so

$$\frac{d}{dx}(\sin x) = \lim_{h \to 0} \frac{\sin(x+h) - \sin x}{h} = \cos x.$$

Other derivations of this result are given in Problems 51 and 52 on page 137.

Exercises and Problems for Section 3.5

Exercises

1. Construct a table of values for $\cos x$, $x = 0, 0.1, 0.2, \ldots, 0.6$. Using the difference quotient, estimate the derivative at these points (use $h = 0.001$), and compare it with $(-\sin x)$.

Find the derivatives of the functions in Exercises 2–39. Assume a is a constant.

2. $r(\theta) = \sin\theta + \cos\theta$

3. $s(\theta) = \cos\theta\sin\theta$

4. $z = \cos(4\theta)$

5. $f(x) = \sin(3x)$

6. $g(x) = \sin(2 - 3x)$

7. $R(x) = 10 - 3\cos(\pi x)$

8. $g(\theta) = \sin^2(2\theta) - \pi\theta$

9. $f(x) = x^2\cos x$

10. $w = \sin(e^t)$

11. $f(x) = e^{\cos x}$

12. $f(y) = e^{\sin y}$

13. $z = \theta e^{\cos\theta}$

14. $R(\theta) = e^{\sin(3\theta)}$

15. $g(\theta) = \sin(\tan\theta)$

16. $w(x) = \tan(x^2)$

17. $f(x) = \sqrt{1 - \cos x}$

18. $f(x) = \cos(\sin x)$

19. $f(x) = \tan(\sin x)$

20. $k(x) = \sqrt{(\sin(2x))^3}$

21. $f(x) = 2x\sin(3x)$

22. $y = e^\theta\sin(2\theta)$

23. $f(x) = e^{-2x}\cdot\sin x$

24. $z = \sqrt{\sin t}$

25. $y = \sin^5\theta$

26. $g(z) = \tan(e^z)$

27. $z = \tan(e^{-3\theta})$

28. $w = e^{-\sin\theta}$

29. $h(t) = t\cos t + \tan t$

30. $f(\alpha) = \cos\alpha + 3\sin\alpha$

31. $k(\alpha) = \sin^5\alpha\cos^3\alpha$

32. $f(\theta) = \theta^3\cos\theta$

33. $y = \cos^2 w + \cos(w^2)$

34. $y = \sin(\sin x + \cos x)$

35. $y = \sin(2x)\cdot\sin(3x)$

36. $t(\theta) = \dfrac{\cos\theta}{\sin\theta}$

37. $f(x) = \sqrt{\dfrac{1 - \sin x}{1 - \cos x}}$

38. $r(y) = \dfrac{y}{\cos y + a}$

39. $G(x) = \dfrac{\sin^2 x + 1}{\cos^2 x + 1}$

Problems

40. Is the graph of $y = \sin(x^4)$ increasing or decreasing when $x = 10$? Is it concave up or concave down?

41. Find the fiftieth derivative of $y = \cos x$.

42. Find a possible formula for the function $q(x)$ such that

$$q'(x) = \frac{e^x\cdot\sin x - e^x\cdot\cos x}{(\sin x)^2}.$$

43. Find a function $F(x)$ satisfying $F'(x) = \sin(4x)$.

44. On page 32 the depth, y, in feet, of water in Boston harbor is given in terms of t, the number of hours since midnight, by

$$y = 5 + 4.9 \cos\left(\frac{\pi}{6}t\right).$$

    (a) Find dy/dt. What does dy/dt represent, in terms of water level?

    (b) For $0 \le t \le 24$, when is dy/dt zero? (Figure 1.45 on page 32 may be helpful.) Explain what it means (in terms of water level) for dy/dt to be zero.

45. A boat at anchor is bobbing up and down in the sea. The vertical distance, y, in feet, between the sea floor and the boat is given as a function of time, t, in minutes, by

$$y = 15 + \sin(2\pi t).$$

    (a) Find the vertical velocity, v, of the boat at time t.

    (b) Make rough sketches of y and v against t.

46. The voltage, V, in volts, in an electrical outlet is given as a function of time, t, in seconds, by the function $V = 156 \cos(120\pi t)$.

    (a) Give an expression for the rate of change of voltage with respect to time.

    (b) Is the rate of change ever zero? Explain.

    (c) What is the maximum value of the rate of change?

47. The function $y = A \sin\left(\left(\sqrt{\frac{k}{m}}\right)t\right)$ represents the oscillations of a mass m at the end of a spring. The constant k measures the stiffness of the spring.

    (a) Find a time at which the mass is farthest from its equilibrium position. Find a time at which the mass is moving fastest. Find a time at which the mass is accelerating fastest.

    (b) What is the period, T, of the oscillation?

    (c) Find dT/dm. What does the sign of dT/dm tell you?

48. Find the equations of the tangent lines to the graph of $f(x) = \sin x$ at $x = 0$ and at $x = \pi/3$. Use each tangent line to approximate $\sin(\pi/6)$. Would you expect these results to be equally accurate, since they are taken equally far away from $x = \pi/6$ but on opposite sides? If the accuracy is different, can you account for the difference?

49. If $k \ge 1$, the graphs of $y = \sin x$ and $y = ke^{-x}$ intersect for $x \ge 0$. Find the smallest value of k for which the graphs are tangent. What are the coordinates of the point of tangency?

50. Find d^2x/dt^2 as a function of x if $dx/dt = x \sin x$.

51. We will use the following identities to calculate the derivatives of $\sin x$ and $\cos x$:

$$\sin(a + b) = \sin a \cos b + \sin b \cos a$$
$$\cos(a + b) = \cos a \cos b - \sin a \sin b.$$

    (a) Use the definition of the derivative to show that if $f(x) = \sin x$,

$$f'(x) = \sin x \lim_{h \to 0} \frac{\cos h - 1}{h} + \cos x \lim_{h \to 0} \frac{\sin h}{h}.$$

    (b) Estimate the limits in part (a) with your calculator to explain why $f'(x) = \cos x$.

    (c) If $g(x) = \cos x$, use the definition of the derivative to show that $g'(x) = -\sin x$.

52. In this problem you will calculate the derivative of $\tan \theta$ rigorously (and without using the derivatives of $\sin \theta$ or $\cos \theta$). You will then use your result for $\tan \theta$ to calculate the derivatives of $\sin \theta$ and $\cos \theta$. Figure 3.26 shows $\tan \theta$ and $\Delta(\tan \theta)$, which is the change in $\tan \theta$, namely $\tan(\theta + \Delta\theta) - \tan \theta$.

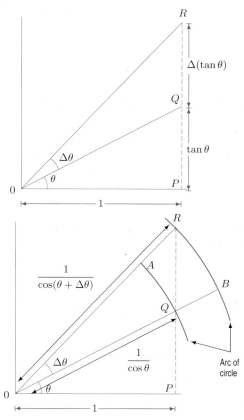

Figure 3.26: $\tan \theta$ and $\Delta(\tan \theta)$

    (a) By paying particular attention to how the two figures relate and using the fact that

Area of	$\le$	Area of	$\le$	Area of
Sector OAQ		Triangle OQR		Sector OBR

    explain why

$$\frac{\Delta\theta}{2\pi} \cdot \frac{\pi}{(\cos\theta)^2} \le \frac{\Delta(\tan\theta)}{2} \le \frac{\Delta\theta}{2\pi} \cdot \frac{\pi}{(\cos(\theta + \Delta\theta))^2}.$$

    [Hint: A sector of a circle with angle α at the center has area $\alpha/(2\pi)$ times the area of the whole circle.]

(b) Use part (a) to show as $\Delta\theta \to 0$ that

$$\frac{\Delta \tan\theta}{\Delta\theta} \to \left(\frac{1}{\cos\theta}\right)^2,$$

and hence that $\dfrac{d(\tan\theta)}{d\theta} = \left(\dfrac{1}{\cos\theta}\right)^2.$

(c) Derive the identity $(\tan\theta)^2 + 1 = \left(\dfrac{1}{\cos\theta}\right)^2$. Then

differentiate both sides of this identity with respect to θ, using the chain rule and the result of part (b) to show that $\dfrac{d}{d\theta}(\cos\theta) = -\sin\theta$.

(d) Differentiate both sides of the identity $(\sin\theta)^2 + (\cos\theta)^2 = 1$ and use the result of part (c) to show that $\dfrac{d}{d\theta}(\sin\theta) = \cos\theta$.

3.6   THE CHAIN RULE AND INVERSE FUNCTIONS

In this section we will use the chain rule to calculate the derivatives of fractional powers, logarithms, exponentials, and the inverse trigonometric functions.[2] The same method is used to obtain a formula for the derivative of a general inverse function.

Finding the Derivative of an Inverse Function: Derivative of $x^{1/2}$

Earlier we calculated the derivative of x^n with n an integer, but we have been using the result for non-integral values of n as well. We now confirm that the power rule holds for $n = 1/2$ by calculating the derivative of $f(x) = x^{1/2}$ using the chain rule. Since

$$[f(x)]^2 = x,$$

the derivative of $[f(x)]^2$ and the derivative of x must be equal, so

$$\frac{d}{dx}[f(x)]^2 = \frac{d}{dx}(x).$$

We can use the chain rule with $f(x)$ as the inside function to obtain:

$$\frac{d}{dx}[f(x)]^2 = 2f(x)\cdot f'(x) = 1.$$

Solving for $f'(x)$ gives

$$f'(x) = \frac{1}{2f(x)} = \frac{1}{2x^{1/2}},$$

or

$$\frac{d}{dx}(x^{1/2}) = \frac{1}{2x^{1/2}} = \frac{1}{2}x^{-1/2}.$$

A similar calculation can be used to obtain the derivative of $x^{1/n}$ where n is a positive integer.

Derivative of $\ln x$

We use the chain rule to differentiate an identity involving $\ln x$. Since $e^{\ln x} = x$, we have

$$\frac{d}{dx}(e^{\ln x}) = \frac{d}{dx}(x),$$

$$e^{\ln x}\cdot\frac{d}{dx}(\ln x) = 1. \qquad \text{(Since } e^x \text{ is outside function and } \ln x \text{ is inside function)}$$

Solving for $d(\ln x)/dx$ gives

$$\frac{d}{dx}(\ln x) = \frac{1}{e^{\ln x}} = \frac{1}{x},$$

so

$$\boxed{\frac{d}{dx}(\ln x) = \frac{1}{x}.}$$

---

[2]It requires a separate justification, not given here, that these functions are differentiable.

Example 1   Differentiate   (a) $\ln(x^2 + 1)$ (b) $t^2 \ln t$ (c) $\sqrt{1 + \ln(1 - y)}$.

Solution   (a) Using the chain rule:

$$\frac{d}{dx}\left(\ln(x^2 + 1)\right) = \frac{1}{x^2 + 1}\frac{d}{dx}(x^2 + 1) = \frac{2x}{x^2 + 1}.$$

(b) Using the product rule:

$$\frac{d}{dt}(t^2 \ln t) = \frac{d}{dt}(t^2)\cdot \ln t + t^2\frac{d}{dt}(\ln t) = 2t\ln t + t^2 \cdot \frac{1}{t} = 2t\ln t + t.$$

(c) Using the chain rule:

$$\frac{d}{dy}\left(\sqrt{1 + \ln(1 - y)}\right) = \frac{d}{dy}(1 + \ln(1 - y))^{1/2}$$

$$= \frac{1}{2}(1 + \ln(1 - y))^{-1/2}\cdot \frac{d}{dy}(1 + \ln(1 - y)) \qquad \text{(Using the chain rule)}$$

$$= \frac{1}{2\sqrt{1 + \ln(1 - y)}}\cdot \frac{1}{1 - y}\cdot \frac{d}{dy}(1 - y) \qquad \text{(Using the chain rule again)}$$

$$= \frac{-1}{2(1 - y)\sqrt{1 + \ln(1 - y)}}.$$

Derivative of a^x

In Section 3.2, we saw that the derivative of a^x is proportional to a^x. Now we see another way of calculating the constant of proportionality. We use the identity

$$\ln(a^x) = x\ln a.$$

Differentiating both sides, using $\dfrac{d}{dx}(\ln x) = \dfrac{1}{x}$ and the chain rule, and remembering that $\ln a$ is a constant, we obtain:

$$\frac{d}{dx}(\ln a^x) = \frac{1}{a^x}\cdot \frac{d}{dx}(a^x) = \ln a.$$

Solving gives the result we obtained earlier:

$$\boxed{\frac{d}{dx}(a^x) = (\ln a)a^x.}$$

Derivatives of Inverse Trigonometric Functions

In Section 1.5 we defined $\arcsin x$ as the angle between $-\pi/2$ and $\pi/2$ (inclusive) whose sine is x. Similarly, $\arctan x$ as the angle strictly between $-\pi/2$ and $\pi/2$ whose tangent is x. To find $\dfrac{d}{dx}(\arctan x)$ we use the identity $\tan(\arctan x) = x$. Differentiating using the chain rule gives

$$\frac{1}{\cos^2(\arctan x)}\cdot \frac{d}{dx}(\arctan x) = 1,$$

so

$$\frac{d}{dx}(\arctan x) = \cos^2(\arctan x).$$

Using the identity $1 + \tan^2\theta = \dfrac{1}{\cos^2\theta}$, and replacing θ by $\arctan x$, we have

$$\cos^2(\arctan x) = \frac{1}{1 + \tan^2(\arctan x)} = \frac{1}{1 + x^2}.$$

Thus we have

$$\frac{d}{dx}(\arctan x) = \frac{1}{1 + x^2}.$$

By a similar argument, we obtain the result:

$$\frac{d}{dx}(\arcsin x) = \frac{1}{\sqrt{1 - x^2}}.$$

Example 2 Differentiate (a) $\arctan(t^2)$ (b) $\arcsin(\tan \theta)$.

Solution Use the chain rule:

(a) $\dfrac{d}{dt}\left(\arctan(t^2)\right) = \dfrac{1}{1 + (t^2)^2} \cdot \dfrac{d}{dt}(t^2) = \dfrac{2t}{1 + t^4}.$

(b) $\dfrac{d}{dt}\left(\arcsin(\tan \theta)\right) = \dfrac{1}{\sqrt{1 - (\tan \theta)^2}} \cdot \dfrac{d}{d\theta}(\tan \theta) = \dfrac{1}{\sqrt{1 - \tan^2 \theta}} \cdot \dfrac{1}{\cos^2 \theta}.$

Derivative of a General Inverse Function

Each of the previous results gives the derivative of an inverse function. In general, if a function f has a differentiable inverse, f^{-1}, we find its derivative by differentiating $f(f^{-1}(x)) = x$ by the chain rule:

$$\frac{d}{dx}\left(f\left(f^{-1}(x)\right)\right) = 1$$

$$f'\left(f^{-1}(x)\right) \cdot \frac{d}{dx}\left(f^{-1}(x)\right) = 1$$

so

$$\frac{d}{dx}\left(f^{-1}(x)\right) = \frac{1}{f'(f^{-1}(x))}.$$

Thus, the derivative of the inverse is the reciprocal of the derivative of the original function, but evaluated at the point $f^{-1}(x)$ instead of the point x.

Example 3 Figure 3.27 shows $f(x)$ and $f^{-1}(x)$. Using Table 3.6, find

(a) (i) $f(2)$ (ii) $f^{-1}(2)$ (iii) $f'(2)$ (iv) $(f^{-1})'(2)$

(b) The equation of the tangent lines at the points P and Q.

(c) What is the relationship between the two tangent lines?

Table 3.6

x	$f(x)$	$f'(x)$
0	1	0.7
1	2	1.4
2	4	2.8
3	8	5.5

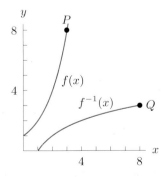

Figure 3.27

Solution   (a) Reading from the table, we have

(i) $f(2) = 4$.

(ii) $f^{-1}(2) = 1$.

(iii) $f'(2) = 2.8$.

(iv) To find the derivative of the inverse function, we use

$$(f^{-1})'(2) = \frac{1}{f'(f^{-1}(2))} = \frac{1}{f'(1)} = \frac{1}{1.4} = 0.714.$$

Notice that the derivative of f^{-1} is the reciprocal of the derivative of f. However, the derivative of f^{-1} is evaluated at 2, while the derivative of f is evaluated at 1, where $f^{-1}(2) = 1$ and $f(1) = 2$.

(b) At the point P, we have $f(3) = 8$ and $f'(3) = 5.5$, so the equation of the tangent line at P is

$$y - 8 = 5.5(x - 3).$$

At the point Q, we have $f^{-1}(8) = 3$, so the slope at Q is

$$(f^{-1})'(8) = \frac{1}{f'(f^{-1}(8))} = \frac{1}{f'(3)} = \frac{1}{5.5}.$$

Thus, the equation of the tangent line at Q is

$$y - 3 = \frac{1}{5.5}(x - 8).$$

(c) The two tangent lines have reciprocal slopes, and the points $(3, 8)$ and $(8, 3)$ are reflections of one another in the line $y = x$. Thus, the two tangent lines are reflections of one another in the line $y = x$.

Exercises and Problems for Section 3.6

Exercises

For Exercises 1–34, find the derivative. It may be to your advantage to simplify before differentiating. Assume a, b, c, and k are constants.

1. $f(t) = \ln(t^2 + 1)$

2. $f(x) = \ln(1 - x)$

3. $f(x) = \ln(e^{2x})$

4. $f(x) = e^{\ln(e^{2x^2 + 3})}$

5. $f(x) = \ln(1 - e^{-x})$

6. $f(\alpha) = \ln(\sin \alpha)$

7. $f(x) = \ln(e^x + 1)$

8. $y = x \ln x - x + 2$

9. $j(x) = \ln(e^{ax} + b)$

10. $h(w) = w^3 \ln(10w)$

11. $f(x) = \ln(e^{7x})$

12. $f(x) = e^{(\ln x) + 1}$

13. $f(w) = \ln(\cos(w - 1))$

14. $f(t) = \ln(e^{\ln t})$

15. $f(y) = \arcsin(y^2)$

16. $g(t) = \arctan(3t - 4)$

17. $g(\alpha) = \sin(\arcsin \alpha)$

18. $g(t) = e^{\arctan(3t^2)}$

19. $g(t) = \cos(\ln t)$

20. $h(z) = z^{\ln 2}$

21. $h(w) = w \arcsin w$

22. $f(x) = e^{\ln(kx)}$

23. $r(t) = \arcsin(2t)$

24. $j(x) = \cos\left(\sin^{-1} x\right)$

25. $f(x) = \cos(\arctan 3x)$

26. $g(x) = \arcsin(\sin \pi x)$

27. $f(z) = \dfrac{1}{\ln z}$

28. $f(x) = \dfrac{x}{1 + \ln x}$

29. $y = 2x(\ln x + \ln 2) - 2x + e$

30. $f(x) = \ln(\sin x + \cos x)$

31. $f(t) = \ln(\ln t) + \ln(\ln 2)$

32. $T(u) = \arctan\left(\dfrac{u}{1 + u}\right)$

33. $a(t) = \ln\left(\dfrac{1 - \cos t}{1 + \cos t}\right)^4$

34. $f(x) = \cos(\arcsin(x + 1))$

Problems

For Problems 35–38, let $h(x) = f(g(x))$ and $k(x) = g(f(x))$. Use Figure 3.28 to estimate the derivatives.

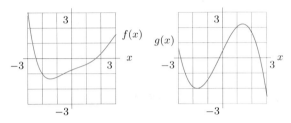

Figure 3.28

35. $h'(1)$ **36.** $k'(1)$ **37.** $h'(2)$ **38.** $k'(2)$

39. On what intervals is $\ln(x^2 + 1)$ concave up?

40. Use the chain rule to obtain the formula for $\frac{d}{dx}(\arcsin x)$.

41. Using the chain rule, find $\frac{d}{dx}(\log x)$.
(Recall $\log x = \log_{10} x$.)

42. To compare the acidity of different solutions, chemists use the pH (which is a single number, not the product of p and H). The pH is defined in terms of the concentration, x, of hydrogen ions in the solution as

$$\text{pH} = -\log x.$$

Find the rate of change of pH with respect to hydrogen ion concentration when the pH is 2. [Hint: Use the result of Problem 41.]

43. (a) Find the equation of the tangent line to $y = \ln x$ at $x = 1$.
(b) Use it to calculate approximate values for $\ln(1.1)$ and $\ln(2)$.
(c) Using a graph, explain whether the approximate values are smaller or larger than the true values. Would the same result have held if you had used the tangent line to estimate $\ln(0.9)$ and $\ln(0.5)$? Why?

44. (a) Find the equation of the best quadratic approximation to $y = \ln x$ at $x = 1$. The best quadratic approximation has the same first and second derivatives as $y = \ln x$ at $x = 1$.
(b) Use a computer or calculator to graph the approximation and $y = \ln x$ on the same set of axes. What do you notice?
(c) Use your quadratic approximation to calculate approximate values for $\ln(1.1)$ and $\ln(2)$.

45. (a) For $x > 0$, find and simplify the derivative of $f(x) = \arctan x + \arctan(1/x)$.
(b) What does your result tell you about f?

46. Imagine you are zooming in on the graph of each of the following functions near the origin:

$$y = x \qquad\qquad y = \sqrt{x}$$
$$y = x^2 \qquad\qquad y = \sin x$$
$$y = x \sin x \qquad\quad y = \tan x$$
$$y = \sqrt{x/(x+1)} \quad\ y = x^3$$
$$y = \ln(x+1) \qquad y = \tfrac{1}{2}\ln(x^2 + 1)$$
$$y = 1 - \cos x \qquad y = \sqrt{2x - x^2}$$

Which of them look the same? Group together those functions which become indistinguishable, and give the equations of the lines they look like.

In Problems 47–50, use Figure 3.29 to find a point x where $h(x) = n(m(x))$ has the given derivative.

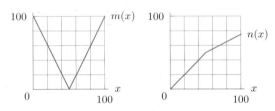

Figure 3.29

47. $h'(x) = -2$ **48.** $h'(x) = 2$

49. $h'(x) = 1$ **50.** $h'(x) = -1$

In Problems 51–53, use Figure 3.30 to calculate the derivative.

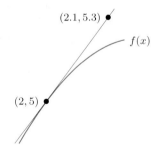

Figure 3.30

51. $h'(2)$ if $h(x) = (f(x))^3$

52. $k'(2)$ if $k(x) = (f(x))^{-1}$

53. $g'(5)$ if $g(x) = f^{-1}(x)$

54. (a) Given that $f(x) = x^3$, find $f'(2)$.
(b) Find $f^{-1}(x)$.
(c) Use your answer from part (b) to find $(f^{-1})'(8)$.
(d) How could you have used your answer from part (a) to find $(f^{-1})'(8)$?

55. (a) For $f(x) = 2x^5 + 3x^3 + x$, find $f'(x)$.
(b) How can you use your answer to part (a) to determine if $f(x)$ is invertible?
(c) Find $f(1)$.
(d) Find $f'(1)$.
(e) Find $(f^{-1})'(6)$.

56. Given that f and g are differentiable everywhere, g is the inverse of f, and that $f(3) = 4$, $f'(3) = 6$, $f'(4) = 7$, find $g'(4)$.

57. Use the table and the fact that $f(x)$ is invertible and differentiable everywhere to find $(f^{-1})'(3)$.

x	$f(x)$	$f'(x)$
3	1	7
6	2	10
9	3	5

58. Let $P = f(t)$ give the US population in millions in year t.

(a) What does the statement $f(2000) = 281$ tell you about the US population?
(b) Find and interpret $f^{-1}(281)$. Give units.
(c) What does the statement $f'(2000) = 3.476$ tell you about the population? Give units.
(d) Evaluate and interpret $(f^{-1})'(281)$. Give units.

59. Figure 3.31 shows the number of motor vehicles, $f(t)$, in millions, registered in the world t years after 1946. With units, estimate and interpret

(a) $f(21)$ **(b)** $f'(21)$
(c) $f^{-1}(200)$ **(d)** $(f^{-1})'(200)$

Figure 3.31

60. Using Figure 3.32, where $f'(2) = 2.1$, $f'(4) = 3.0$, $f'(6) = 3.7$, $f'(8) = 4.2$, find $(f^{-1})'(8)$.

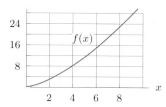

Figure 3.32

61. An increasing function $f(x)$ has the value $f(10) = 5$. Explain how you know that the calculations $f'(10) = 8$ and $(f^{-1})'(5) = 8$ cannot both be correct.

62. An invertible function $f(x)$ has values in the table. Evaluate

(a) $f'(a) \cdot (f^{-1})'(A)$ **(b)** $f'(b) \cdot (f^{-1})'(B)$
(c) $f'(c) \cdot (f^{-1})'(C)$

x	a	b	c	d
$f(x)$	A	B	C	D

63. If f is continuous, invertible, and defined for all x, why must at least one of the statements $(f^{-1})'(10) = 8$, $(f^{-1})'(20) = -6$ be wrong?

64. (a) Calculate $\lim_{h \to 0}(\ln(1 + h)/h)$ by identifying the limit as the derivative of $\ln(1 + x)$ at $x = 0$.
(b) Use the result of part (a) to show that $\lim_{h \to 0}(1 + h)^{1/h} = e$.
(c) Use the result of part (b) to calculate the related limit, $\lim_{n \to \infty}(1 + 1/n)^n$.

3.7 IMPLICIT FUNCTIONS

In earlier chapters, most functions were written in the form $y = f(x)$; here y is said to be an *explicit* function of x. An equation such as

$$x^2 + y^2 = 4$$

is said to give y as an *implicit* function of x. Its graph is the circle in Figure 3.33. Since there are x-values which correspond to two y-values, y is not a function of x on the whole circle. Solving gives

$$y = \pm\sqrt{4 - x^2},$$

where $y = \sqrt{4 - x^2}$ represents the top half of the circle and $y = -\sqrt{4 - x^2}$ represents the bottom half. So y is a function of x on the top half, and y is a different function of x on the bottom half.

But let's consider the circle as a whole. The equation does represent a curve which has a tangent line at each point. The slope of this tangent can be found by differentiating the equation of the circle with respect to x:

$$\frac{d}{dx}(x^2) + \frac{d}{dx}(y^2) = \frac{d}{dx}(4).$$

If we think of y as a function of x and use the chain rule, we get

$$2x + 2y\frac{dy}{dx} = 0.$$

Solving gives

$$\frac{dy}{dx} = -\frac{x}{y}.$$

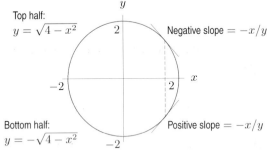

Top half:
$y = \sqrt{4 - x^2}$

Negative slope $= -x/y$

Bottom half:
$y = -\sqrt{4 - x^2}$

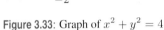

Positive slope $= -x/y$

Figure 3.33: Graph of $x^2 + y^2 = 4$

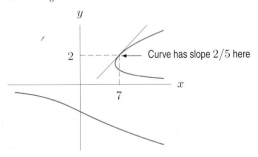

Curve has slope $2/5$ here

Figure 3.34: Graph of $y^3 - xy = -6$ and its tangent line at $(7, 2)$

The derivative here depends on both x and y (instead of just on x). This is because for many x-values there are two y-values, and the curve has a different slope at each one. Figure 3.33 shows that for x and y both positive, we are on the top right quarter of the curve and the slope is negative (as the formula predicts). For x positive and y negative, we are on the bottom right quarter of the curve and the slope is positive (as the formula predicts).

Differentiating the equation of the circle has given us the slope of the curve at all points except $(2, 0)$ and $(-2, 0)$, where the tangent is vertical. In general, this process of *implicit differentiation* leads to a derivative whenever the expression for the derivative does not have a zero in the denominator.

Example 1   Make a table of x and approximate y-values for the equation $y^3 - xy = -6$ near $x = 7, y = 2$. Your table should include the x-values $6.8, 6.9, 7.0, 7.1$, and 7.2.

Solution   We would like to solve for y in terms of x, but we cannot isolate y by factoring. There is a formula for solving cubics, somewhat like the quadratic formula, but it is too complicated to be useful here. Instead, first observe that $x = 7, y = 2$ does satisfy the equation. (Check this!) Now find dy/dx by implicit differentiation:

$$\frac{d}{dx}(y^3) - \frac{d}{dx}(xy) = \frac{d}{dx}(-6)$$

$$3y^2\frac{dy}{dx} - 1 \cdot y - x\frac{dy}{dx} = 0 \quad \text{(Differentiating with respect to } x\text{)}$$

$$3y^2\frac{dy}{dx} - x\frac{dy}{dx} = y$$

$$(3y^2 - x)\frac{dy}{dx} = y \quad \text{(Factoring out } \frac{dy}{dx}\text{)}$$

$$\frac{dy}{dx} = \frac{y}{3y^2 - x}.$$

When $x = 7$ and $y = 2$, we have

$$\frac{dy}{dx} = \frac{2}{12 - 7} = \frac{2}{5}.$$

(See Figure 3.34.) The equation of the tangent line at $(7, 2)$ is

$$y - 2 = \frac{2}{5}(x - 7)$$

or

$$y = 0.4x - 0.8.$$

Since the tangent lies very close to the curve near the point $(7, 2)$, we use the equation of the tangent line to calculate the following approximate y-values:

x	6.8	6.9	7.0	7.1	7.2
Approximate y	1.92	1.96	2.00	2.04	2.08

Notice that although the equation $y^3 - xy = -6$ leads to a curve which is difficult to deal with algebraically, it still looks like a straight line locally.

Example 2 Find all points where the tangent line to $y^3 - xy = -6$ is either horizontal or vertical.

Solution From the previous example, $\dfrac{dy}{dx} = \dfrac{y}{3y^2 - x}$. The tangent is horizontal when the numerator of dy/dx equals 0, so $y = 0$. Since we also must satisfy $y^3 - xy = -6$, we get $0^3 - x \cdot 0 = -6$, which is impossible. We conclude that there are no points on the curve where the tangent line is horizontal.

The tangent is vertical when the denominator of dy/dx is 0, giving $3y^2 - x = 0$. Thus, $x = 3y^2$ at any point with a vertical tangent line. Again, we must also satisfy $y^3 - xy = -6$, so

$$y^3 - (3y^2)y = -6,$$
$$-2y^3 = -6,$$
$$y = \sqrt[3]{3} \approx 1.442.$$

We can then find x by substituting $y = \sqrt[3]{3}$ in $y^3 - xy = -6$. We get $3 - x(\sqrt[3]{3}) = -6$, so $x = 9/(\sqrt[3]{3}) \approx 6.240$. So the tangent line is vertical at $(6.240, 1.442)$.

Using implicit differentiation and the expression for dy/dx to locate the points where the tangent is vertical or horizontal, as in the previous example, is a first step in obtaining an overall picture of the curve $y^3 - xy = -6$. However, filling in the rest of the graph, even roughly, by using the sign of dy/dx to tell us where the curve is increasing or decreasing can be difficult.

Exercises and Problems for Section 3.7

Exercises

For Exercises 1–18, find dy/dx. Assume a, b, c are constants.

1. $x^2 + y^2 = \sqrt{7}$

2. $x^2 + xy - y^3 = xy^2$

3. $xy + x + y = 5$

4. $x^2y - 2y + 5 = 0$

5. $\sqrt{x} = 5\sqrt{y}$

6. $\sqrt{x} + \sqrt{y} = 25$

7. $xy - x - 3y - 4 = 0$

8. $6x^2 + 4y^2 = 36$

9. $ax^2 - by^2 = c^2$

10. $\ln x + \ln(y^2) = 3$

11. $x \ln y + y^3 = \ln x$

12. $\sin(xy) = 2x + 5$

13. $\cos^2 y + \sin^2 y = y + 2$

14. $e^{\cos y} = x^3 \arctan y$

15. $\arctan(x^2 y) = xy^2$

16. $e^{x^2} + \ln y = 0$

17. $(x - a)^2 + y^2 = a^2$

18. $x^{2/3} + y^{2/3} = a^{2/3}$

In Exercises 19–22, find the slope of the tangent to the curve at the point specified.

19. $x^2 + y^2 = 1$ at $(0, 1)$

20. $\sin(xy) = x$ at $(1, \pi/2)$

21. $x^3 + 2xy + y^2 = 4$ at $(1, 1)$

22. $x^3 + 5x^2 y + 2y^2 = 4y + 11$ at $(1, 2)$

For Exercises 23–27, find the equations of the tangent lines to the following curves at the indicated points.

23. $xy^2 = 1$ at $(1, -1)$ **24.** $\ln(xy) = 2x$ at $(1, e^2)$

25. $y^2 = \dfrac{x^2}{xy - 4}$ at $(4, 2)$ **26.** $y = \dfrac{x}{y + a}$ at $(0, 0)$

27. $x^{2/3} + y^{2/3} = a^{2/3}$ at $(a, 0)$

Problems

28. (a) Find dy/dx given that $x^2 + y^2 - 4x + 7y = 15$.
   (b) Under what conditions on x and/or y is the tangent line to this curve horizontal? Vertical?

29. (a) Find the slope of the tangent line to the ellipse $\dfrac{x^2}{25} + \dfrac{y^2}{9} = 1$ at the point (x, y).
   (b) Are there any points where the slope is not defined?

30. (a) Find the equations of the tangent lines to the circle $x^2 + y^2 = 25$ at the points where $x = 4$.
   (b) Find the equations of the normal lines to this circle at the same points. (The normal line is perpendicular to the tangent line at that point.)
   (c) At what point do the two normal lines intersect?

31. (a) If $x^3 + y^3 - xy^2 = 5$, find dy/dx.
   (b) Using your answer to part (a), make a table of approximate y-values of points on the curve near $x = 1, y = 2$. Include $x = 0.96, 0.98, 1, 1.02, 1.04$.
   (c) Find the y-value for $x = 0.96$ by substituting $x = 0.96$ in the original equation and solving for y using a computer or calculator. Compare with your answer in part (b).
   (d) Find all points where the tangent line is horizontal

or vertical.

32. Find the equation of the tangent line to the curve $y = x^2$ at $x = 1$. Show that this line is also a tangent to a circle centered at $(8, 0)$ and find the equation of this circle.

33. At pressure P atmospheres, a certain fraction f of a gas decomposes. The quantities P and f are related, for some positive constant K, by the equation

$$\frac{4f^2 P}{1 - f^2} = K.$$

   (a) Find df/dP.
   (b) Show that $df/dP < 0$ always. What does this mean in practical terms?

34. Sketch the circles $y^2 + x^2 = 1$ and $y^2 + (x - 3)^2 = 4$. There is a line with positive slope that is tangent to both circles. Determine the points at which this tangent line touches each circle.

35. Show that the power rule for derivatives applies to rational powers of the form $y = x^{m/n}$ by raising both sides to the nth power and using implicit differentiation.

3.8 HYPERBOLIC FUNCTIONS

There are two combinations of e^x and e^{-x} which are used so often in engineering that they are given their own name. They are the *hyperbolic sine*, abbreviated sinh, and the *hyperbolic cosine*, abbreviated cosh. They are defined as follows:

Hyperbolic Functions

$$\cosh x = \frac{e^x + e^{-x}}{2} \qquad \sinh x = \frac{e^x - e^{-x}}{2}$$

Properties of Hyperbolic Functions

The graphs of $\cosh x$ and $\sinh x$ are given in Figures 3.35 and 3.36 together with the graphs of multiples of e^x and e^{-x}. The graph of $\cosh x$ is called a *catenary*; it is the shape of a hanging cable.

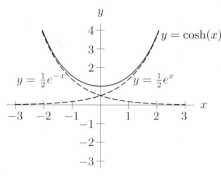

Figure 3.35: Graph of $y = \cosh x$

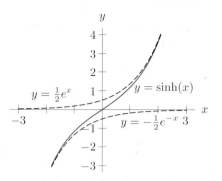

Figure 3.36: Graph of $y = \sinh x$

The graphs suggest that the following results hold:

$$\cosh 0 = 1 \qquad \sinh 0 = 0$$

$$\cosh(-x) = \cosh x \qquad \sinh(-x) = -\sinh x$$

To show that the hyperbolic functions really do have these properties, we use their formulas.

Example 1   Show that (a) $\cosh(0) = 1$ (b) $\cosh(-x) = \cosh x$

Solution   (a) Substituting $x = 0$ into the formula for $\cosh x$ gives

$$\cosh 0 = \frac{e^0 + e^{-0}}{2} = \frac{1 + 1}{2} = 1.$$

(b) Substituting $-x$ for x gives

$$\cosh(-x) = \frac{e^{-x} + e^{-(-x)}}{2} = \frac{e^{-x} + e^x}{2} = \cosh x.$$

Thus, we know that $\cosh x$ is an even function.

Example 2   Describe and explain the behavior of $\cosh x$ as $x \to \infty$ and $x \to -\infty$.

Solution   From Figure 3.35, it appears that as $x \to \infty$, the graph of $\cosh x$ resembles the graph of $\frac{1}{2}e^x$. Similarly, as $x \to -\infty$, the graph of $\cosh x$ resembles the graph of $\frac{1}{2}e^{-x}$. This behavior is explained by using the formula for $\cosh x$ and the facts that $e^{-x} \to 0$ as $x \to \infty$ and $e^x \to 0$ as $x \to -\infty$:

$$\text{As } x \to \infty, \qquad \cosh x = \frac{e^x + e^{-x}}{2} \to \frac{1}{2}e^x.$$

$$\text{As } x \to -\infty, \qquad \cosh x = \frac{e^x + e^{-x}}{2} \to \frac{1}{2}e^{-x}.$$

Identities Involving cosh x and sinh x

The reason the hyperbolic functions have names that remind us of the trigonometric functions is that they share similar properties. A familiar identity for trigonometric functions is

$$(\cos x)^2 + (\sin x)^2 = 1.$$

To discover an analogous identity relating $(\cosh x)^2$ and $(\sinh x)^2$, we first calculate

$$(\cosh x)^2 = \left(\frac{e^x + e^{-x}}{2}\right)^2 = \frac{e^{2x} + 2e^x e^{-x} + e^{-2x}}{4} = \frac{e^{2x} + 2 + e^{-2x}}{4}$$

$$(\sinh x)^2 = \left(\frac{e^x - e^{-x}}{2}\right)^2 = \frac{e^{2x} - 2e^x e^{-x} + e^{-2x}}{4} = \frac{e^{2x} - 2 + e^{-2x}}{4}.$$

If we add these expressions, the resulting right-hand side contains terms involving both e^{2x} and e^{-2x}. If, however, we subtract the expressions for $(\cosh x)^2$ and $(\sinh x)^2$, we obtain a simple result:

$$(\cosh x)^2 - (\sinh x)^2 = \frac{e^{2x} + 2 + e^{-2x}}{4} - \frac{e^{2x} - 2 + e^{-2x}}{4} = \frac{4}{4} = 1.$$

Thus, writing $\cosh^2 x$ for $(\cosh x)^2$ and $\sinh^2 x$ for $(\sinh x)^2$, we have the identity

$$\boxed{\cosh^2 x - \sinh^2 x = 1}$$

The Hyperbolic Tangent

Extending the analogy to the trigonometric functions, we define

$$\boxed{\tanh x = \frac{\sinh x}{\cosh x} = \frac{e^x - e^{-x}}{e^x + e^{-x}}}$$

Derivatives of Hyperbolic Functions

We calculate the derivatives using the fact that $\frac{d}{dx}(e^x) = e^x$. The results are again reminiscent of the trigonometric functions. For example,

$$\frac{d}{dx}(\cosh x) = \frac{d}{dx}\left(\frac{e^x + e^{-x}}{2}\right) = \frac{e^x - e^{-x}}{2} = \sinh x.$$

We find $\frac{d}{dx}(\sinh x)$ similarly, giving the following results:

$$\boxed{\frac{d}{dx}(\cosh x) = \sinh x \qquad \frac{d}{dx}(\sinh x) = \cosh x}$$

Example 3   Compute the derivative of $\tanh x$.

Solution   Using the quotient rule gives

$$\frac{d}{dx}(\tanh x) = \frac{d}{dx}\left(\frac{\sinh x}{\cosh x}\right) = \frac{(\cosh x)^2 - (\sinh x)^2}{(\cosh x)^2} = \frac{1}{\cosh^2 x}.$$

Exercises and Problems for Section 3.8

Exercises

Find the derivatives of the functions in Exercises 1–11.

1. $y = \cosh(2x)$

2. $y = \sinh(3z + 5)$

3. $f(t) = \cosh(\sinh t)$

4. $f(t) = t^3 \sinh t$

5. $g(t) = \cosh^2 t$

6. $y = \cosh(3t)\sinh(4t)$

7. $f(t) = \cosh(e^{t^2})$

8. $y = \tanh(3 + \sinh x)$

9. $f(y) = \sinh(\sinh(3y))$

10. $g(\theta) = \ln(\cosh(1 + \theta))$

11. $f(t) = \cosh^2 t - \sinh^2 t$

12. Show that $\sinh 0 = 0$.

13. Show that $\sinh(-x) = -\sinh(x)$.

14. Show that $d(\sinh x)/dx = \cosh x$.

Simplify the expressions in Exercises 15–16.

15. $\sinh(\ln t)$

16. $\cosh(\ln t)$

Problems

17. Describe and explain the behavior of $\sinh x$ as $x \to \infty$ and as $x \to -\infty$.

18. Is there an identity analogous to $\sin(2x) = 2\sin x \cos x$ for the hyperbolic functions? Explain.

19. Is there an identity analogous to $\cos(2x) = \cos^2 x - \sin^2 x$ for the hyperbolic functions? Explain.

Prove the identities in Problems 20–21.

20. $\sinh(A + B) = \sinh A \cosh B + \sinh B \cosh A$

21. $\cosh(A + B) = \cosh A \cosh B + \sinh B \sinh A$

In Problems 22–24, find the limit of the function as $x \to \infty$.

22. $\dfrac{\cosh(2x)}{\sinh(3x)}$

23. $\dfrac{e^{2x}}{\sinh(2x)}$

24. $\dfrac{\sinh(x^2)}{\cosh(x^2)}$

25. For what values of k is $\lim\limits_{x \to \infty} \dfrac{\sinh kx}{\cosh 2x}$ finite?

26. For what values of k is $\lim\limits_{x \to \infty} e^{-3x} \cosh kx$ finite?

27. **(a)** Using a calculator or computer, sketch the graph of $y = 2e^x + 5e^{-x}$ for $-3 \le x \le 3$, $0 \le y \le 20$. Observe that it looks like the graph of $y = \cosh x$. Approximately where is its minimum?

(b) Show algebraically that $y = 2e^x + 5e^{-x}$ can be written in the form $y = A\cosh(x - c)$. Calculate the values of A and c. Explain what this tells you about the graph in part (a).

28. The following problem is a generalization of Problem 27. Show that any function of the form

$$y = Ae^x + Be^{-x}, \quad A > 0, \; B > 0,$$

can be written, for some K and c, in the form

$$y = K\cosh(x - c).$$

What does this tell you about the graph of $y = Ae^x + Be^{-x}$?

29. The cable between the two towers of a suspension bridge hangs in the shape of the curve

$$y = \frac{T}{w} \cosh\left(\frac{wx}{T}\right),$$

where T is the tension in the cable at its lowest point and w is the weight of the cable per unit length. This curve is called a *catenary*.

(a) Suppose the cable stretches between the points $x = -T/w$ and $x = T/w$. Find an expression for the "sag" in the cable. (That is, find the difference between the height of the cable at the highest and lowest points.)

(b) Show that the shape of the cable satisfies the equation

$$\frac{d^2y}{dx^2} = \frac{w}{T}\sqrt{1 + \left(\frac{dy}{dx}\right)^2}.$$

30. The Saint Louis arch can be approximated by using a function of the form $y = b - a\cosh(x/a)$. Putting the origin on the ground in the center of the arch and the y-axis upward, find an approximate equation for the arch given the dimensions shown in Figure 3.37. (In other words, find a and b.)

615 ft

←530 ft→

Figure 3.37

31. **(a)** Find $\tanh 0$.

(b) For what values of x is $\tanh x$ positive? Negative? Explain your answer algebraically.

(c) On what intervals is $\tanh x$ increasing? Decreasing? Use derivatives to explain your answer.

(d) Find $\lim_{x \to \infty} \tanh x$ and $\lim_{x \to -\infty} \tanh x$. Show this information on a graph.

(e) Does $\tanh x$ have an inverse? Justify your answer using derivatives.

3.9   LINEAR APPROXIMATION AND THE DERIVATIVE

The Tangent Line Approximation

When we zoom in on the graph of a differentiable function, it looks like a straight line. In fact, the graph is not exactly a straight line when we zoom in; however, its deviation from straightness is so small that it can't be detected by the naked eye. Let's examine what this means. The straight line that we think we see when we zoom in on the graph of $f(x)$ at $x = a$ has slope equal to the derivative, $f'(a)$, so the equation is

$$y = f(a) + f'(a)(x - a).$$

The fact that the graph looks like a line means that y is a good approximation to $f(x)$. (See Figure 3.38.) This suggests the following definition:

The Tangent Line Approximation

Suppose f is differentiable at a. Then, for values of x near a, the tangent line approximation to $f(x)$ is

$$f(x) \approx f(a) + f'(a)(x - a).$$

The expression $f(a) + f'(a)(x - a)$ is called the *local linearization* of f near $x = a$. We are thinking of a as fixed, so that $f(a)$ and $f'(a)$ are constant.
The **error**, $E(x)$, in the approximation is defined by

$$E(x) = f(x) - f(a) - f'(a)(x - a).$$

It can be shown that the tangent line approximation is the best linear approximation to f near a. See Problem 28.

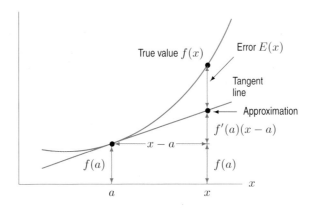

Figure 3.38: The tangent line approximation and its error

Example 1   What is the tangent line approximation for $f(x) = \sin x$ near $x = 0$?

Solution   The tangent line approximation of f near $x = 0$ is

$$f(x) \approx f(0) + f'(0)(x - 0).$$

If $f(x) = \sin x$, then $f'(x) = \cos x$, so $f(0) = \sin 0 = 0$ and $f'(0) = \cos 0 = 1$, and the approximation is

$$\sin x \approx x.$$

This means that, near $x = 0$, the function $f(x) = \sin x$ is well approximated by the function $y = x$. If we zoom in on the graphs of the functions $\sin x$ and x near the origin, we won't be able to tell them apart. (See Figure 3.39.)

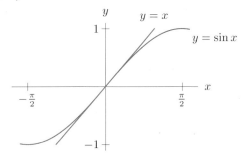

Figure 3.39: Tangent line approximation to $y = \sin x$

Example 2   What is the local linearization of e^{kx} near $x = 0$?

Solution   If $f(x) = e^{kx}$, then $f(0) = 1$ and, by the chain rule, $f'(x) = ke^{kx}$, so $f'(0) = ke^{k \cdot 0} = k$. Thus

$$f(x) \approx f(0) + f'(0)(x - 0)$$

becomes

$$e^{kx} \approx 1 + kx.$$

This is the tangent line approximation to e^{kx} near $x = 0$. In other words, if we zoom in on the functions $f(x) = e^{kx}$ and $y = 1 + kx$ near the origin, we won't be able to tell them apart.

Estimating the Error in the Approximation

Let us look at the error, $E(x)$, which is the difference between $f(x)$ and the local linearization. (Look back at Figure 3.38.) The fact that the graph of f looks like a line as we zoom in means that not only is $E(x)$ small for x near a, but also that $E(x)$ is small relative to $(x - a)$. To demonstrate this, we prove the following theorem about the ratio $E(x)/(x - a)$.

Theorem 3.6: Differentiability and Local Linearity

Suppose f is differentiable at $x = a$ and $E(x)$ is the error in the tangent line approximation, that is:

$$E(x) = f(x) - f(a) - f'(a)(x - a).$$

Then

$$\lim_{x \to a} \frac{E(x)}{x - a} = 0.$$

Proof Using the definition of $E(x)$, we have

$$\frac{E(x)}{x - a} = \frac{f(x) - f(a) - f'(a)(x - a)}{x - a} = \frac{f(x) - f(a)}{x - a} - f'(a).$$

Taking the limit as $x \to a$ and using the definition of the derivative, we see that

$$\lim_{x \to a} \frac{E(x)}{x - a} = \lim_{x \to a} \left(\frac{f(x) - f(a)}{x - a} - f'(a) \right) = f'(a) - f'(a) = 0.$$

Theorem 3.6 says that $E(x)$ approaches 0 faster than $(x - a)$. For the function in Example 3, we see that $E(x) \approx k(x - a)^2$ for constant k if x is near a.

Example 3   Let $E(x)$ be the error in the tangent line approximation to $f(x) = x^3 - 5x + 3$ for x near 2.

(a) What does a table of values for $E(x)/(x - 2)$ suggest about $\lim_{x \to 2} E(x)/(x - 2)$?
(b) Make another table to see that $E(x) \approx k(x-2)^2$. Estimate the value of k. Check that a possible value is $k = f''(2)/2$.

Solution   (a) Since $f(x) = x^3 - 5x + 3$, we have $f'(x) = 3x^2 - 5$, and $f''(x) = 6x$. Thus, $f(2) = 1$ and $f'(2) = 3 \cdot 2^2 - 5 = 7$, so the tangent line approximation for x near 2 is

$$f(x) \approx f(2) + f'(2)(x - 2)$$
$$f(x) \approx 1 + 7(x - 2).$$

Thus,

$$E(x) = \text{True value} - \text{Approximation} = (x^3 - 5x + 3) - (1 + 7(x - 2)).$$

The values of $E(x)/(x - 2)$ in Table 3.7 suggest that $E(x)/(x - 2)$ approaches 0 as $x \to 2$.
(b) Notice that if $E(x) \approx k(x - 2)^2$, then $E(x)/(x - 2)^2 \approx k$. Thus we make Table 3.8 showing values of $E(x)/(x - 2)^2$. Since the values are all approximately 6, we guess that $k = 6$ and $E(x) \approx 6(x - 2)^2$.
    Since $f''(2) = 12$, our value of k satisfies $k = f''(2)/2$.

Table 3.7

x	$E(x)/(x - 2)$
2.1	0.61
2.01	0.0601
2.001	0.006001
2.0001	0.00060001

Table 3.8

x	$E(x)/(x - 2)^2$
2.1	6.1
2.01	6.01
2.001	6.001
2.0001	6.0001

For all the functions in Problems 13–17, the error in the tangent line approximation behaves near $x = a$ as

$$E(x) \approx \frac{f''(a)}{2}(x - a)^2.$$

This is part of a general pattern for obtaining higher order approximations called Taylor polynomials, which are studied in Chapter 10.

Why Differentiability Makes A Graph Look Straight

We use the properties of the error $E(x)$ to understand why differentiability makes a graph look straight when we zoom in.

Example 4   Consider the graph of $f(x) = \sin x$ near $x = 0$, and its linear approximation computed in Example 1. Show that there is an interval around 0 with the property that the distance from $f(x) = \sin x$ to the linear approximation is less than $0.1|x|$ for all x in the interval.

Solution   The linear approximation of $f(x) = \sin x$ near 0 is $y = x$, so we write

$$\sin x = x + E(x).$$

Since $\sin x$ is differentiable at $x = 0$, Theorem 3.6 tells us that

$$\lim_{x \to 0} \frac{E(x)}{x} = 0.$$

If we take $\epsilon = 1/10$, then the definition of limit guarantees that there is a $\delta > 0$ such that

$$\left| \frac{E(x)}{x} \right| < 0.1 \quad \text{for all} \quad |x| < \delta.$$

In other words, for x in the interval $(-\delta, \delta)$, we have $|x| < \delta$, so

$$|E(x)| < 0.1|x|.$$

(See Figure 3.40.)

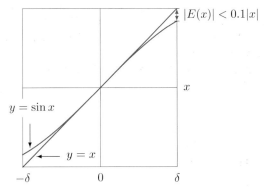

Figure 3.40: Graph of $y = \sin x$ and its linear approximation $y = x$, showing a window in which the magnitude of the error, $|E(x)|$, is less than $0.1|x|$ for all x in the window

We can generalize from this example to explain why differentiability makes the graph of f look straight when viewed over a small graphing window. Suppose f is differentiable at $x = a$. Then we know $\lim\limits_{x \to a} \left| \dfrac{E(x)}{x - a} \right| = 0$. So, for any $\epsilon > 0$, we can find a δ small enough so that

$$\left| \frac{E(x)}{x - a} \right| < \epsilon, \qquad \text{for} \quad a - \delta < x < a + \delta.$$

So, for any x in the interval $(a - \delta, a + \delta)$, we have

$$|E(x)| < \epsilon|x - a|.$$

Thus, the error, $E(x)$, is less than ϵ times $|x - a|$, the distance between x and a. So, as we zoom in on the graph by choosing smaller ϵ, the deviation, $|E(x)|$, of f from its tangent line shrinks, even relative to the scale on the x-axis. So, zooming makes a differentiable function look straight.

Exercises and Problems for Section 3.9

Exercises

1. Find the tangent line approximation for $\sqrt{1+x}$ near $x = 0$.

2. What is the tangent line approximation to e^x near $x = 0$?

3. Find the tangent line approximation to $1/x$ near $x = 1$.

4. Show that $1 - x/2$ is the tangent line approximation to $1/\sqrt{1+x}$ near $x = 0$.

5. Show that $e^{-x} \approx 1 - x$ near $x = 0$.

6. What is the local linearization of e^{x^2} near $x = 1$?

7. Using a graph like Figure 3.39, estimate to one decimal place the magnitude of the error in approximating $\sin x$ by x for $-1 \leq x \leq 1$. Is the approximation an over- or an underestimate?

8. For x near 0, local linearization gives

$$e^x \approx 1 + x.$$

Using a graph, decide if the approximation is an over- or underestimate, and estimate to one decimal place the magnitude of the error for $-1 \leq x \leq 1$.

Problems

9. (a) Find the tangent line approximation to $\cos x$ at $x = \pi/4$.
   (b) Use a graph to explain how you know whether the tangent line approximation is an under- or overestimate for $0 \leq x \leq \pi/2$.
   (c) To one decimal place, estimate the error in the approximation for $0 \leq x \leq \pi/2$.

10. (a) Show that $1 + kx$ is the local linearization of $(1+x)^k$ near $x = 0$.
    (b) Someone claims that the square root of 1.1 is about 1.05. Without using a calculator, do you think that this estimate is about right?
    (c) Is the actual number above or below 1.05?

11. Figure 3.41 shows $f(x)$ and its local linearization at $x = a$. What is the value of a? Of $f(a)$? Is the approximation an under- or overestimate? Use the linearization to approximate the value of $f(1.2)$.

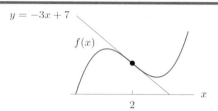

$y = -3x + 7$

$f(x)$

2

Figure 3.42

In Problems 13–17, find a formula for the error $E(x)$ in the tangent line approximation to the function near $x = a$. Using a table of values for $E(x)/(x - a)$ near $x = a$, find a value of k such that $E(x)/(x - a) \approx k(x - a)$. Check that, approximately, $k = f''(a)/2$ and that $E(x) \approx (f''(a)/2)(x - a)^2$.

13. $f(x) = x^4$, $a = 1$

14. $f(x) = \cos x$, $a = 0$

15. $f(x) = e^x$, $a = 0$

16. $f(x) = \sqrt{x}$, $a = 1$

17. $f(x) = \ln x$, $a = 1$

18. Multiply the local linearization of e^x near $x = 0$ by itself to obtain an approximation for e^{2x}. Compare this with the actual local linearization of e^{2x}. Explain why these two approximations are consistent, and discuss which one is more accurate.

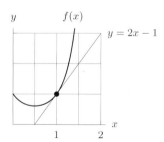

y $f(x)$

$y = 2x - 1$

1   2

x

Figure 3.41

12. Figure 3.42 shows the tangent line approximation to $f(x)$ near $x = a$.

    (a) Find a, $f(a)$, $f'(a)$.
    (b) Estimate $f(2.1)$ and $f(1.98)$. Are these under- or overestimates? Which estimate would you expect to be more accurate and why?

19. (a) Show that $1 - x$ is the local linearization of $\dfrac{1}{1+x}$ near $x = 0$.
    (b) From your answer to part (a), show that near $x = 0$,

$$\frac{1}{1+x^2} \approx 1 - x^2.$$

    (c) Without differentiating, what do you think the derivative of $\dfrac{1}{1+x^2}$ is at $x = 0$?

20. From the local linearizations of e^x and $\sin x$ near $x = 0$, write down the local linearization of the function $e^x \sin x$. From this result, write down the derivative of $e^x \sin x$ at $x = 0$. Using this technique, write down the derivative of $e^x \sin x / (1 + x)$ at $x = 0$.

21. The acceleration due to gravity, g, is given by

$$g = \frac{GM}{r^2},$$

where M is the mass of the earth, r is the distance from the center of the earth, and G is the universal gravitational constant.

   (a) Show that when r changes by Δr, the change in the acceleration due to gravity, Δg, is given by

$$\Delta g \approx -2g \frac{\Delta r}{r}.$$

   (b) What is the significance of the negative sign?

   (c) What is the percent change in g when moving from sea level to the top of Pike's Peak (4.315 km)? Assume the radius of the earth is 6400 km.

22. Writing g for the acceleration due to gravity, the period, T, of a pendulum of length l is given by

$$T = 2\pi \sqrt{\frac{l}{g}}.$$

   (a) Show that if the length of the pendulum changes by Δl, the change in the period, ΔT, is given by

$$\Delta T \approx \frac{T}{2l} \Delta l.$$

   (b) If the length of the pendulum increases by 2%, by what percent does the period change?

23. Suppose now the length of the pendulum in Problem 22 remains constant, but that the acceleration due to gravity changes.

   (a) Use the method of the preceding problem to relate ΔT approximately to Δg, the change in g.

   (b) If g increases by 1%, find the percent change in T.

24. Suppose f has a continuous positive second derivative for all x. Which is larger, $f(1+\Delta x)$ or $f(1)+f'(1)\Delta x$? Explain.

25. Suppose $f'(x)$ is a differentiable decreasing function for all x. In each of the following pairs, which number is the larger? Give a reason for your answer.

   (a) $f'(5)$ and $f'(6)$

   (b) $f''(5)$ and 0

   (c) $f(5 + \Delta x)$ and $f(5) + f'(5)\Delta x$

26. Use local linearization to derive the product rule,

$$[f(x)g(x)]' = f'(x)g(x) + f(x)g'(x).$$

[Hint: Use the definition of the derivative and the local linearizations $f(x+h) \approx f(x)+f'(x)h$ and $g(x+h) \approx g(x) + g'(x)h$.]

27. Derive the chain rule using local linearization. [Hint: In other words, differentiate $f(g(x))$, using $g(x + h) \approx g(x) + g'(x)h$ and $f(z + k) \approx f(z) + f'(z)k$.]

28. Consider a function f and a point a. Suppose there is a number L such that the linear function g

$$g(x) = f(a) + L(x - a)$$

is a good approximation to f. By good approximation, we mean that

$$\lim_{x \to a} \frac{E_L(x)}{x - a} = 0,$$

where $E_L(x)$ is the approximation error defined by

$$f(x) = g(x) + E_L(x) = f(a) + L(x - a) + E_L(x).$$

Show that f is differentiable at $x = a$ and that $f'(a) = L$. Thus the tangent line approximation is the only good linear approximation.

3.10   THEOREMS ABOUT DIFFERENTIABLE FUNCTIONS

A Relationship Between Local and Global: The Mean Value Theorem

We often want to infer a global conclusion (for example, f is increasing on an interval) from local information (for example, f' is positive at each point on an interval.) The following theorem relates the average rate of change of a function on an interval (global information) to the instantaneous rate of change at a point in the interval (local information at a point of the interval).

Theorem 3.7: The Mean Value Theorem

If f is continuous on $a \le x \le b$ and differentiable on $a < x < b$, then there exists a number c, with $a < c < b$, such that

$$f'(c) = \frac{f(b) - f(a)}{b - a}.$$

In other words, $f(b) - f(a) = f'(c)(b - a)$.

To understand what this theorem is saying geometrically, look at the graph in Figure 3.43. Join the points on the curve where $x = a$ and $x = b$ with a line and observe that the slope of this secant line AB is given by

$$m = \frac{f(b) - f(a)}{b - a}.$$

Now consider the tangent line drawn to the curve at each point between $x = a$ and $x = b$. In general, these lines have different slopes. For the curve shown in Figure 3.43, the tangent line at $x = a$ is flatter than the secant line from A to B. Similarly, the tangent line at $x = b$ is steeper than the secant line. However, there is at least one point between a and b where the slope of the tangent line to the curve is precisely the same as the slope of the secant line. Suppose this occurs at $x = c$. Then

$$f'(c) = m = \frac{f(b) - f(a)}{b - a}.$$

The Mean Value Theorem tells us that the point $x = c$ exists, but it does not tell us how to find c. Problems 41 and 42 in Section 4.3 show how the Mean Value Theorem can be derived.

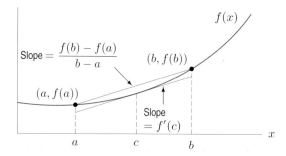

Figure 3.43: The point c with $f'(c) = \frac{f(b) - f(a)}{b - a}$

The Increasing Function Theorem

We say that a function f is *increasing* on an interval if, for any two numbers x_1 and x_2 in the interval such that $x_1 < x_2$, we have $f(x_1) < f(x_2)$. If instead we have $f(x_1) \leq f(x_2)$, we say f is *nondecreasing*.

Theorem 3.8: The Increasing Function Theorem

Suppose that f is continuous on $a \leq x \leq b$ and differentiable on $a < x < b$.
- If $f'(x) > 0$ on $a < x < b$, then f is increasing on $a \leq x \leq b$.
- If $f'(x) \geq 0$ on $a < x < b$, then f is nondecreasing on $a \leq x \leq b$.

Proof Suppose $a \leq x_1 < x_2 \leq b$. By the Mean Value Theorem, there is a number c, with $x_1 < c < x_2$, such that

$$f(x_2) - f(x_1) = f'(c)(x_2 - x_1).$$

If $f'(c) > 0$, this says $f(x_2) - f(x_1) > 0$, which means f is increasing. If $f'(c) \geq 0$, this says $f(x_2) - f(x_1) \geq 0$, which means f is nondecreasing.

It may seem that something as simple as the Increasing Function Theorem should follow immediately from the definition of the derivative, and you may be surprised that the Mean Value Theorem is needed.

The Constant Function Theorem

If f is constant on an interval, then we know that $f'(x) = 0$ on the interval. The following theorem is the converse.

> ### Theorem 3.9: The Constant Function Theorem
>
> Suppose that f is continuous on $a \leq x \leq b$ and differentiable on $a < x < b$. If $f'(x) = 0$ on $a < x < b$, then f is constant on $a \leq x \leq b$.

Proof   The proof is the same as for the Increasing Function Theorem, only in this case $f'(c) = 0$ so $f(x_2) - f(x_1) = 0$. Thus $f(x_2) = f(x_1)$ for $a \leq x_1 < x_2 \leq b$, so f is constant.

A proof of the Constant Function Theorem using the Increasing Function Theorem is given in Problems 13 and 20.

The Racetrack Principle

> ### Theorem 3.10: The Racetrack Principle[3]
>
> Suppose that g and h are continuous on $a \leq x \leq b$ and differentiable on $a < x < b$, and that $g'(x) \leq h'(x)$ for $a < x < b$.
> * If $g(a) = h(a)$, then $g(x) \leq h(x)$ for $a \leq x \leq b$.
> * If $g(b) = h(b)$, then $g(x) \geq h(x)$ for $a \leq x \leq b$.

The Racetrack Principle has the following interpretation. We can think of $g(x)$ and $h(x)$ as the positions of two racehorses at time x, with horse h always moving faster than horse g. If they start together, horse h is ahead during the whole race. If they finish together, horse g was ahead during the whole race.

Proof   Consider the function $f(x) = h(x) - g(x)$. Since $f'(x) = h'(x) - g'(x) \geq 0$, we know that f is nondecreasing by the Increasing Function Theorem. So $f(x) \geq f(a) = h(a) - g(a) = 0$. Thus $g(x) \leq h(x)$ for $a \leq x \leq b$. This proves the first part of the Racetrack Principle. Problem 19 asks for a proof of the second part.

Example 1   Explain graphically why $e^x \geq 1 + x$ for all values of x. Then use the Racetrack Principle to prove the inequality.

Solution   The graph of the function $y = e^x$ is concave up everywhere and the equation of its tangent line at the point $(0, 1)$ is $y = x + 1$. (See Figure 3.44.) Since the graph always lies above its tangent, we have the inequality
$$e^x \geq 1 + x.$$

[3] Based on the Racetrack Principle in *Calculus&Mathematica*, by William Davis, Horacio Porta, Jerry Uhl (Reading: Addison Wesley, 1994).

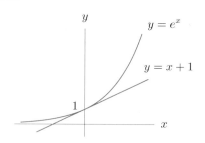

Figure 3.44: Graph showing that $e^x \geq 1 + x$

Now we prove the inequality using the Racetrack Principle. Let $g(x) = 1 + x$ and $h(x) = e^x$. Then $g(0) = h(0) = 1$. Furthermore, $g'(x) = 1$ and $h'(x) = e^x$. Hence $g'(x) \leq h'(x)$ for $x \geq 0$. So by the Racetrack Principle, with $a = 0$, we have $g(x) \leq h(x)$, that is, $1 + x \leq e^x$.

For $x \leq 0$ we have $h'(x) \leq g'(x)$. So by the Racetrack Principle, with $b = 0$, we have $g(x) \leq h(x)$, that is, $1 + x \leq e^x$.

Exercises and Problems for Section 3.10

Exercises

Decide if the statements in Exercises 1–5 are true or false. Give an explanation for your answer.

1. Let $f(x) = [x]$, the greatest integer less than or equal to x. Then $f'(x) = 0$, so $f(x)$ is constant by the Constant Function Theorem.

2. The Racetrack Principle can be used to justify the statement that if two horses start a race at the same time, the horse that wins must have been moving faster than the other throughout the race.

3. Two horses start a race at the same time and one runs slower than the other throughout the race. The Racetrack Principle can be used to justify the fact that the slower horse loses the race.

4. If $a < b$ and $f'(x)$ is positive on $[a, b]$ then $f(a) < f(b)$.

5. If $f(x)$ is increasing and differentiable on the interval $[a, b]$, then $f'(x) > 0$ on $[a, b]$.

Do the functions graphed in Exercises 6–9 appear to satisfy the hypotheses of the Mean Value Theorem on the interval $[a, b]$? Do they satisfy the conclusion?

6.

7.

8.

9.

Problems

10. Use the Racetrack Principle and the fact that $\sin 0 = 0$ to show that $\sin x \leq x$ for all $x \geq 0$.

11. Use the Racetrack Principle to show that $\ln x \leq x - 1$.

12. Use the fact that $\ln x$ and e^x are inverse functions to show that the inequalities $e^x \geq 1 + x$ and $\ln x \leq x - 1$ are equivalent for $x > 0$.

13. State a Decreasing Function Theorem, analogous to the Increasing Function Theorem. Deduce your theorem from the Increasing Function Theorem. [Hint: Apply the Increasing Function Theorem to $-f$.]

Use one of the theorems in this section to prove the statements in Problems 14–17.

14. If $f'(x) \leq 1$ for all x and $f(0) = 0$, then $f(x) \leq x$ for all $x \geq 0$.

15. If $f''(t) \leq 3$ for all t and $f(0) = f'(0) = 0$, then $f(t) \leq \frac{3}{2}t^2$ for all $t \geq 0$.

16. If $f'(x) = g'(x)$ for all x and $f(5) = g(5)$, then $f(x) = g(x)$ for all x.

17. If f is differentiable and $f(0) < f(1)$, then there is a number c, with $0 < c < 1$, such that $f'(c) > 0$.

18. The position of a particle on the x-axis is given by $s = f(t)$; its initial position and velocity are $f(0) = 3$ and $f'(0) = 4$. The acceleration is bounded by $5 \leq f''(t) \leq 7$ for $0 \leq t \leq 2$. What can we say about the position $f(2)$ of the particle at $t = 2$?

19. Suppose that g and h are continuous on $[a, b]$ and differentiable on (a, b). Prove that if $g'(x) \leq h'(x)$ for $a < x < b$ and $g(b) = h(b)$, then $h(x) \leq g(x)$ for $a \leq x \leq b$.

20. Deduce the Constant Function Theorem from the Increasing Function Theorem and the Decreasing Function Theorem. (See problem 13.)

21. Prove that if $f'(x) = g'(x)$ for all x in (a, b), then there is a constant C such that $f(x) = g(x) + C$ on (a, b). [Hint: Apply the Constant Function Theorem to $h(x) = f(x) - g(x)$.]

22. Suppose that $f'(x) = f(x)$ for all x. Prove that $f(x) = Ce^x$ for some constant C. [Hint: Consider $f(x)/e^x$.]

23. Suppose that f is continuous on $[a, b]$ and differentiable on (a, b) and that $m \leq f'(x) \leq M$ on (a, b). Use the Racetrack Principle to prove that $f(x) - f(a) \leq M(x - a)$ for all x in $[a, b]$, and that $m(x - a) \leq f(x) - f(a)$ for all x in $[a, b]$. Conclude that $m \leq (f(b) - f(a))/(b - a) \leq M$. This is called the Mean Value Inequality. In words: If the instantaneous rate of change of f is between m and M on an interval, so is the average rate of change of f over the interval.

24. Suppose that $f''(x) \geq 0$ for all x in (a, b). We will show the graph of f lies above the tangent line at $(c, f(c))$ for any c with $a < c < b$.

(a) Use the Increasing Function Theorem to prove that $f'(c) \leq f'(x)$ for $c \leq x < b$ and that $f'(x) \leq f'(c)$ for $a < x \leq c$.

(b) Use (a) and the Racetrack Principle to conclude that $f(c) + f'(c)(x - c) \leq f(x)$, for $a < x < b$.

CHAPTER SUMMARY (see also Ready Reference at the end of the book)

- **Derivatives of elementary functions**
  Power, polynomial, rational, exponential, logarithmic, trigonometric, inverse trigonometric, and hyperbolic functions.

- **Derivatives of sums, differences, and constant multiples**

- **Product and quotient rules**

- **Chain rule**

Differentiation of implicitly defined functions, inverse functions.

- **Tangent line approximation, local linearity**

- **Hyperbolic functions**

- **Theorems about differentiable functions**
  Mean value theorem, increasing function theorem, racetrack principle.

REVIEW EXERCISES AND PROBLEMS FOR CHAPTER THREE

Exercises

Find derivatives for the functions in Exercises 1–66. Assume a, b, c, and k are constants.

1. $f(t) = 2te^t - \dfrac{1}{\sqrt{t}}$

2. $w = \dfrac{5 - 3z}{5 + 3z}$

3. $f(x) = \dfrac{x^3}{9}(3\ln x - 1)$

4. $f(\theta) = \dfrac{1}{1 + e^{-\theta}}$

5. $h(\theta) = \theta(\theta^{-1/2} - \theta^{-2})$

6. $f(\theta) = \ln(\cos\theta)$

7. $f(y) = \ln\left(\ln(2y^3)\right)$

8. $g(x) = x^k + k^x$

9. $y = e^{-\pi} + \pi^{-e}$

10. $z = \sin^3\theta$

11. $f(t) = \cos^2(3t + 5)$

12. $M(\alpha) = \tan^2(2 + 3\alpha)$

13. $s(\theta) = \sin^2(3\theta - \pi)$

14. $h(t) = \ln\left(e^{-t} - t\right)$

15. $p(\theta) = \dfrac{\sin(5 - \theta)}{\theta^2}$

16. $w(\theta) = \dfrac{\theta}{\sin^2\theta}$

17. $g(x) = \dfrac{x^2 + \sqrt{x} + 1}{x^{3/2}}$

18. $g(w) = \dfrac{1}{2^w + e^w}$

19. $s(x) = \arctan(2 - x)$

20. $r(\theta) = e^{(e^\theta + e^{-\theta})}$

21. $m(n) = \sin(e^n)$

22. $k(\alpha) = e^{\tan(\sin\alpha)}$

23. $g(t) = t\cos(\sqrt{t}e^t)$

24. $f(r) = (\tan 2 + \tan r)^e$

25. $h(x) = xe^{\tan x}$

26. $y = e^{2x}\sin^2(3x)$

27. $g(x) = \tan^{-1}(3x^2 + 1)$

28. $y = 2^{\sin x}\cos x$

29. $h(x) = \ln e^{ax}$

30. $k(x) = \ln e^{ax} + \ln b$

31. $f(\theta) = e^{k\theta} - 1$

32. $f(t) = e^{-4kt}\sin t$

33. $H(t) = (at^2 + b)e^{-ct}$

34. $g(\theta) = \sqrt{a^2 - \sin^2\theta}$

35. $f(x) = a^{5x}$

36. $f(x) = \dfrac{a^2 - x^2}{a^2 + x^2}$

37. $w(r) = \dfrac{ar^2}{b + r^3}$

38. $f(s) = \dfrac{a^2 - s^2}{\sqrt{a^2 + s^2}}$

39. $y = \arctan\left(\dfrac{2}{x}\right)$

40. $r(t) = \ln\left(\sin\left(\dfrac{t}{k}\right)\right)$

41. $g(w) = \dfrac{5}{(a^2 - w^2)^2}$

42. $y = \dfrac{e^x - e^{-x}}{e^x + e^{-x}}$

43. $g(u) = \dfrac{e^{au}}{a^2 + b^2}$

44. $y = \dfrac{e^{ax} - e^{-ax}}{e^{ax} + e^{-ax}}$

45. $g(t) = \dfrac{\ln(kt) + t}{\ln(kt) - t}$

46. $z = \dfrac{e^{t^2} + t}{\sin(2t)}$

47. $f(t) = \sin\sqrt{e^t + 1}$

48. $g(y) = e^{2e^{(y^3)}}$

49. $g(x) = -\frac{1}{2}(x^5 + 2x - 9)$

50. $y = -3x^4 - 4x^3 - 6x + 2$

51. $g(z) = \dfrac{z^7 + 5z^6 - z^3}{z^2}$

52. $f(z) = (\ln 3)z^2 + (\ln 4)e^z$

53. $g(x) = 2x - \dfrac{1}{\sqrt[3]{x}} + 3^x - e$

54. $f(x) = (3x^2 + \pi)(e^x - 4)$

55. $f(\theta) = \theta^2 \sin\theta + 2\theta\cos\theta - 2\sin\theta$

56. $y = \sqrt{\cos(5\theta) + \sin^2(6\theta)}$

57. $r(\theta) = \sin\left((3\theta - \pi)^2\right)$

58. $y = \left(x^2 + 5\right)^3 \left(3x^3 - 2\right)^2$

59. $N(\theta) = \tan(\arctan(k\theta))$

60. $h(t) = e^{kt}(\sin at + \cos bt)$

61. $f(x) = (2 - 4x - 3x^2)(6x^e - 3\pi)$

62. $f(t) = (\sin(2t) - \cos(3t))^4$

63. $s(y) = \sqrt[3]{(\cos^2 y + 3 + \sin^2 y)}$

64. $f(x) = (4 - x^2 + 2x^3)(6 - 4x + x^7)$

65. $h(x) = \left(\dfrac{1}{x} - \dfrac{1}{x^2}\right)\left(2x^3 + 4\right)$

66. $f(z) = \sqrt{5z} + 5\sqrt{z} + \dfrac{5}{\sqrt{z}} - \sqrt{\dfrac{5}{z}} + \sqrt{5}$

For Exercises 67–68, assume that y is a differentiable function of x and find dy/dx.

67. $x^3 + y^3 - 4x^2y = 0$

68. $\sin(ay) + \cos(bx) = xy$

69. Find the slope of the curve $x^2 + 3y^2 = 7$ at $(2, -1)$.

70. Assume y is a differentiable function of x and that $y + \sin y + x^2 = 9$. Find dy/dx at the point $x = 3, y = 0$.

71. Find the equations for the lines tangent to the graph of $xy + y^2 = 4$ where $x = 3$.

Problems

For Problems 72–77, use Figure 3.45.

Figure 3.45

72. Let $h(x) = t(x)s(x)$ and $p(x) = t(x)/s(x)$. Estimate:

   (a) $h'(1)$ **(b)** $h'(0)$ **(c)** $p'(0)$

73. Let $r(x) = s(t(x))$. Estimate $r'(0)$.

74. Let $h(x) = s(s(x))$. Estimate:

   (a) $h'(1)$ **(b)** $h'(2)$

75. Estimate all values of x for which the tangent line to $y = s(s(x))$ is horizontal.

76. Let $h(x) = x^2 t(x)$ and $p(x) = t(x^2)$. Estimate:

   (a) $h'(-1)$ **(b)** $p'(-1)$

77. Find an approximate equation for the tangent line to $r(x) = s(t(x))$ at $x = 1$.

In Problems 78–80, use Figure 3.46 to evaluate the derivatives.

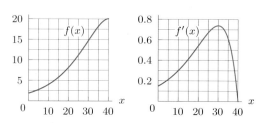

Figure 3.46

78. $(f^{-1})'(5)$ **79.** $(f^{-1})'(10)$ **80.** $(f^{-1})'(15)$

81. Suppose W is proportional to r^3. The derivative dW/dr is proportional to what power of r?

82. Using the information in the table about f and g, find:

    (a) $h(4)$ if $h(x) = f(g(x))$
    (b) $h'(4)$ if $h(x) = f(g(x))$
    (c) $h(4)$ if $h(x) = g(f(x))$
    (d) $h'(4)$ if $h(x) = g(f(x))$
    (e) $h'(4)$ if $h(x) = g(x)/f(x)$
    (f) $h'(4)$ if $h(x) = f(x)g(x)$

x	1	2	3	4
$f(x)$	3	2	1	4
$f'(x)$	1	4	2	3
$g(x)$	2	1	4	3
$g'(x)$	4	2	3	1

83. Given: $r(2) = 4$, $s(2) = 1$, $s(4) = 2$, $r'(2) = -1$, $s'(2) = 3$, $s'(4) = 3$. Compute the following derivatives, or state what additional information you would need to be able to do so.

    (a) $H'(2)$ if $H(x) = r(x) \cdot s(x)$
    (b) $H'(2)$ if $H(x) = \sqrt{r(x)}$
    (c) $H'(2)$ if $H(x) = r(s(x))$
    (d) $H'(2)$ if $H(x) = s(r(x))$

84. If $g(2) = 3$ and $g'(2) = -4$, find $f'(2)$ for the following:

    (a) $f(x) = x^2 - 4g(x)$ **(b)** $f(x) = \dfrac{x}{g(x)}$

    (c) $f(x) = x^2 g(x)$ **(d)** $f(x) = (g(x))^2$

    (e) $f(x) = x \sin(g(x))$ **(f)** $f(x) = x^2 \ln(g(x))$

85. For parts (a)–(f) of Problem 84, determine the equation of the line tangent to f at $x = 2$.

86. Imagine you are zooming in on the graphs of the following functions near the origin:

$$y = \arcsin x \qquad y = \sin x - \tan x \qquad y = x - \sin x$$

$$y = \arctan x \quad y = \frac{\sin x}{1 + \sin x} \qquad y = \frac{x^2}{x^2 + 1}$$

$$y = \frac{1 - \cos x}{\cos x} \quad y = \frac{x}{x^2 + 1} \qquad y = \frac{\sin x}{x} - 1$$

$$y = -x \ln x \quad y = e^x - 1 \qquad y = x^{10} + \sqrt[10]{x}$$

$$y = \frac{x}{x + 1}$$

Which of them look the same? Group together those functions which become indistinguishable, and give the equation of the line they look like. [Note: $(\sin x)/x - 1$ and $-x \ln x$ never quite make it to the origin.]

87. The graphs of $\sin x$ and $\cos x$ intersect once between 0 and $\pi/2$. What is the angle between the two curves at the point where they intersect? (You need to think about how the angle between two curves should be defined.)

In Problems 88–89, show that the curves meet at least once and determine whether the curves are perpendicular at the point of intersection.

88. $y = 1 + x - x^2$ and $y = 1 - x + x^2$

89. $y = 1 - x^3/3$ and $y = x - 1$

90. For some constant b and $x > 0$, let $y = x \ln x - bx$. Find the equation of the tangent line to this graph at the point at which the graph crosses the x-axis.

In Problems 91–93, find the limit of the function as $x \to -\infty$.

91. $\dfrac{\cosh(2x)}{\sinh(3x)}$

92. $\dfrac{e^{-2x}}{\sinh(2x)}$

93. $\dfrac{\sinh(x^2)}{\cosh(x^2)}$

94. The acceleration due to gravity, g, at a distance r from the center of the earth is given by

$$g = \frac{GM}{r^2},$$

where M is the mass of the earth and G is a constant.

    (a) Find dg/dr.
    (b) What is the practical interpretation (in terms of acceleration) of dg/dr? Why would you expect it to be negative?
    (c) You are told that $M = 6 \cdot 10^{24}$ and $G = 6.67 \cdot 10^{-20}$ where M is in kilograms and r in kilometers. What is the value of dg/dr at the surface of the earth ($r = 6400$ km)?
    (d) What does this tell you about whether or not it is reasonable to assume g is constant near the surface of the earth?

95. In 1990, the population of Mexico was about 84 million and growing at 2.6% annually, while the population of the US was about 250 million and growing at 0.7% annually. Which population was growing faster, if we measure growth rates in people/year? Explain your answer.

96. The distance, s, of a moving body from a fixed point is given as a function of time, t, by $s = 20e^{t/2}$.

    (a) Find the velocity, v, of the body as a function of t.
    (b) Find a relationship between v and s, then show that s satisfies the differential equation $s' = \frac{1}{2}s$.

97. Air pressure at sea level is 30 inches of mercury. At an altitude of h feet above sea level, the air pressure, P, in inches of mercury, is given by

$$P = 30e^{-3.23 \times 10^{-5}h}$$

(a) Sketch a graph of P against h.

(b) Find the equation of the tangent line at $h = 0$.

(c) A rule of thumb used by travelers is that air pressure drops about 1 inch for every 1000-foot increase in height above sea level. Write a formula for the air pressure given by this rule of thumb.

(d) What is the relation between your answers to parts (b) and (c)? Explain why the rule of thumb works.

(e) Are the predictions made by the rule of thumb too large or too small? Why?

98. The depth of the water, y, in meters, in the Bay of Fundy, Canada, is given as a function of time, t, in hours after midnight, by the function

$$y = 10 + 7.5 \cos(0.507t).$$

How quickly is the tide rising or falling (in meters/hour) at each of the following times?

(a) 6:00 am   **(b)** 9:00 am

(c) Noon   **(d)** 6:00 pm

99. A yam is put in a hot oven, maintained at a constant temperature 200°C. At time $t = 30$ minutes, the temperature T of the yam is 120° and is increasing at an (instantaneous) rate of 2°/min. Newton's law of cooling (or, in our case, warming) implies that the temperature at time t is given by

$$T(t) = 200 - ae^{-bt}.$$

Find a and b.

100. An object is oscillating at the end of a spring. Its position, in centimeters, relative to a fixed point, is given as a function of time, t, in seconds, by

$$y = y_0 \cos(2\pi\omega t), \quad \text{with } \omega \text{ a constant.}$$

(a) Find an expression for the velocity and acceleration of the object.

(b) How do the amplitudes of the position, velocity, and acceleration functions compare? How do the periods of these functions compare?

(c) Show that the function y satisfies the differential equation

$$\frac{d^2y}{dt^2} + 4\pi^2\omega^2 y = 0.$$

101. The total number of people, N, who have contracted a disease by a time t days after its outbreak is given by

$$N = \frac{1,000,000}{1 + 5,000e^{-0.1t}}.$$

(a) In the long run, how many people get the disease?

(b) Is there any day on which more than a million people fall sick? Half a million? Quarter of a million? (Note: You do not have to find on what days these things happen.)

102. At a particular location, $f(p)$ is the number of gallons of gas sold when the price is p dollars per gallon.

(a) What does the statement $f(2) = 4023$ tell you about gas sales?

(b) Find and interpret $f^{-1}(4023)$.

(c) What does the statement $f'(2) = -1250$ tell you about gas sales?

(d) Find and interpret $(f^{-1})'(4023)$

103. If f is increasing and $f(20) = 10$, which of the two options, (a) or (b), must be wrong?

(a) $f'(10)(f^{-1})'(20) = 1$.

(b) $f'(20)(f^{-1})'(10) = 2$.

104. If f is decreasing and $f(20) = 10$, which of the following must be incorrect?

(a) $(f^{-1})'(20) = -3$.   **(b)** $(f^{-1})'(10) = 12$.

105. Find the nth derivative of the following functions:

(a) $\ln x$ **(b)** xe^x **(c)** $e^x \cos x$

106. The derivative f' gives the (absolute) rate of change of a quantity f, and f'/f gives the relative rate of change of the quantity. In this problem, we show that the product rule is equivalent to an additive rule for relative rates of change. Assume $h = f \cdot g$ with $f \neq 0$ and $g \neq 0$.

(a) Show that the additive rule

$$\frac{f'}{f} + \frac{g'}{g} = \frac{h'}{h}$$

implies the product rule, by multiplying through by h and using the fact that $h = f \cdot g$.

(b) Show that the product rule implies the additive rule in part (a), by starting with the product rule and dividing through by $h = f \cdot g$.

107. The relative rate of change of a function f is defined to be f'/f. Find an expression for the relative rate of change of a quotient f/g in terms of the relative rates of change of the functions f and g.

CAS Challenge Problems

108. **(a)** Use a computer algebra system to differentiate $(x + 1)^x$ and $(\sin x)^x$.

    (b) Conjecture a rule for differentiating $(f(x))^x$, where f is any differentiable function.

    (c) Apply your rule to $g(x) = (\ln x)^x$. Does your answer agree with the answer given by the computer algebra system?

    (d) Prove your conjecture by rewriting $(f(x))^x$ in the form $e^{h(x)}$.

For Problems 109–111,

    (a) Use a computer algebra system to find and simplify the derivative of the given function.

    (b) Without a computer algebra system, use differentiation rules to calculate the derivative. Make sure that the answer simplifies to the same answer as in part (a).

    (c) Explain how you could have predicted the derivative by using algebra before taking the derivative.

109. $f(x) = \sin(\arcsin x)$

110. $g(r) = 2^{-2r} 4^r$

111. $h(t) = \ln(1 - 1/t) + \ln(t/(t - 1))$

CHECK YOUR UNDERSTANDING

Are the statements in Problems 1–14 true or false? Give an explanation for your answer.

1. The derivative of a polynomial is always a polynomial.

2. The derivative of π/x^2 is $-\pi/x$.

3. The derivative of $\tan \theta$ is periodic.

4. The graph of $\ln(x^2)$ is concave up for $x > 0$.

5. If $f'(x)$ is defined for all x, then $f(x)$ is defined for all x.

6. If $f'(2) = 3.1$ and $g'(2) = 7.3$, then the graph of $f(x) + g(x)$ has slope 10.4 at $x = 2$.

7. Let f and g be two functions whose second derivatives are defined. Then

$$(fg)'' = fg'' + f''g.$$

8. If a function is periodic, with period c, then so is its derivative.

9. If y satisfies the equation $y^2 + xy - 1 = 0$, then dy/dx exists everywhere.

10. The function $\tanh x$ is odd, that is, $\tanh(-x) = -\tanh x$.

11. The 100th derivative of $\sinh x$ is $\cosh x$.

12. $\sinh x + \cosh x = e^x$.

13. The function $\sinh x$ is periodic.

14. The function $\sinh^2 x$ is concave down everywhere.

Are the statements in Problems 15–21 true or false? If a statement is true, explain how you know. If a statement is false, give a counterexample.

15. If $f(x)$ is defined for all x, then $f'(x)$ is defined for all x.

16. If $f(x)$ is increasing, then $f'(x)$ is increasing.

17. The only functions whose fourth derivatives are equal to $\cos t$ are of the form $\cos t + C$, where C is any constant.

18. If $f(x)$ has an inverse function, $g(x)$, then the derivative of $g(x)$ is $1/f'(x)$.

19. $(fg)'(x)$ is never equal to $f'(x)g'(x)$.

20. If the function $f(x)/g(x)$ is defined but not differentiable at $x = 1$, then either $f(x)$ or $g(x)$ is not differentiable at $x = 1$.

21. If the derivative of $f(g(x))$ is equal to the derivative of $f(x)$ for all x, then $g(x) = x$ for all x.

Suppose that f'' and g'' exist and that f and g are concave up for all x. Are the statements in Problems 22–25 true or false for all such f and g? If a statement is true, explain how you know. If a statement is false, give a counterexample.

22. $f(x) + g(x)$ is concave up for all x.

23. $f(x)g(x)$ is concave up for all x.

24. $f(x) - g(x)$ cannot be concave up for all x.

25. $f(g(x))$ is concave up for all x.

26. Let f be a differentiable function and let L be the linear function $L(x) = f(a) + k(x - a)$ for some constant a. Decide whether the following statements are true or false for all constants k. Explain your answer.

    (a) L is the local linearization for f near $x = a$,

    (b) If $\lim_{x \to a} (f(x) - L(x)) = 0$, then L is the local linearization for f near $x = a$.

27. Which of the following would be a counterexample to the product rule?

    (a) Two differentiable functions f and g satisfying $(fg)' = f'g'$.

    (b) A differentiable function f such that $(xf(x))' = xf'(x) + f(x)$.

    (c) A differentiable function f such that $(f(x)^2)' = 2f(x)$.

    (d) Two differentiable functions f and g such that $f'(a) = 0$ and $g'(a) = 0$ and fg has positive slope at $x = a$.

Are the statements in Problems 28–31 true or false for a function f whose domain is all real numbers? If a statement is true, explain how you know. If a statement is false, give a counterexample.

28. If $f'(x) \geq 0$ for all x, then $f(a) \leq f(b)$ whenever $a \leq b$.

29. If $f'(x) \leq g'(x)$ for all x, then $f(x) \leq g(x)$ for all x.

30. If $f'(x) = g'(x)$ for all x, then $f(x) = g(x)$ for all x.

31. If $f'(x) \leq 1$ for all x and $f(0) = 0$, then $f(x) \leq x$ for all x.

In Problems 32–34, give an example of function(s) with the given properties.

32. A continuous function f on the interval $[-1, 1]$ that does not satisfy the conclusion of the Mean Value Theorem.

33. A function f that is differentiable on the interval $(0, 2)$, but does not satisfy the conclusion of the Mean Value Theorem on the interval $[0, 2]$.

34. A function that is differentiable on $(0, 1)$ and not continuous on $[0, 1]$, but which satisfies the conclusion of the Mean Value Theorem.

PROJECTS FOR CHAPTER THREE

1. Rule of 70

The "Rule of 70" is a rule of thumb to estimate how long it takes money in a bank to double. Suppose the money is in an account earning $i\%$ annual interest, compounded yearly. The Rule of 70 says that the time it takes the amount of money to double is approximately $70/i$ years, assuming i is small. Find the local linearization of $\ln(1 + x)$, and use it to explain why this rule works.

2. Newton's Method

Read about how to find roots using bisection and Newton's method in Appendices A and C.

(a) What is the smallest positive zero of the function $f(x) = \sin x$? Apply Newton's method, with initial guess $x_0 = 3$, to see how fast it converges to $\pi = 3.1415926536\ldots$.

   (i) Compute the first two approximations, x_1 and x_2; compare x_2 with π.

   (ii) Newton's method works very well here. Explain why. To do this, you will have to outline the basic idea behind Newton's method.

   (iii) Estimate the location of the zero using bisection, starting with the interval $[3, 4]$. How does bisection compare to Newton's method in terms of accuracy?

(b) Newton's method can be very sensitive to your initial estimate, x_0. For example, consider finding a zero of $f(x) = \sin x - \frac{2}{3}x$.

   (i) Use Newton's method with the following initial estimates to find a zero:

$$x_0 = 0.904, \quad x_0 = 0.905, \quad x_0 = 0.906.$$

   (ii) What happens?

Chapter Four

USING THE DERIVATIVE

In Chapter 2, we introduced the derivative and some of its interpretations. In Chapter 3, we saw how to differentiate all of the standard functions, including powers, exponentials, logarithms, and trigonometric functions. Now we use first and second derivatives to analyze the behavior of families of functions and to solve optimization problems.

4.1   USING FIRST AND SECOND DERIVATIVES

What Derivatives Tell Us About a Function and its Graph

As we saw in Chapter 2, the connection between derivatives of a function and the function itself is given by the following:

- If $f' > 0$ on an interval, then f is increasing on that interval.
- If $f' < 0$ on an interval, then f is decreasing on that interval.
- If $f'' > 0$ on an interval, then the graph of f is concave up on that interval.
- If $f'' < 0$ on an interval, then the graph of f is concave down on that interval.

We can do more with these principles now than we could in Chapter 2 because we now have formulas for the derivatives of the elementary functions.

When we graph a function on a computer or calculator, we often see only part of the picture, and we may miss some significant features. Information given by the first and second derivatives can help identify regions with interesting behavior.

Example 1   Use a computer or calculator to sketch a useful graph of the function $f(x) = x^3 - 9x^2 - 48x + 52$.

Solution   Since f is a cubic polynomial, we expect a graph that is roughly S-shaped. Graphing this function with $-10 \le x \le 10$, $-10 \le y \le 10$, gives the two nearly vertical lines in Figure 4.1. We know that there is more going on than this, but how do we know where to look?

Figure 4.1: Unhelpful graph of
$f(x) = x^3 - 9x^2 - 48x + 52$

We use the derivative to determine where the function is increasing and where it is decreasing. The derivative of f is

$$f'(x) = 3x^2 - 18x - 48.$$

To find where $f' > 0$ or $f' < 0$, we first find where $f' = 0$, that is, where $3x^2 - 18x - 48 = 0$. Factoring, we get $3(x - 8)(x + 2) = 0$, so $x = -2$ or $x = 8$. Since $f' = 0$ *only* at $x = -2$ and $x = 8$, and since f' is continuous, f' cannot change sign on any of the three intervals $x < -2$, or $-2 < x < 8$, or $8 < x$. How can we tell the sign of f' on each of these intervals? The easiest way is to pick a point and substitute into f'. For example, since $f'(-3) = 33 > 0$, we know f' is positive for $x < -2$, so f is increasing for $x < -2$. Similarly, since $f'(0) = -48$ and $f'(10) = 72$, we know that f decreases between $x = -2$ and $x = 8$ and increases for $x > 8$. Summarizing:

We find that $f(-2) = 104$ and $f(8) = -396$. Hence on the interval $-2 < x < 8$ the function decreases from a high of 104 to a low of -396. (Now we see why not much showed up in our first calculator graph.) One more point on the graph is easy to get: the y intercept, $f(0) = 52$. With just these three points we can get a much more helpful graph. By setting the plotting window to

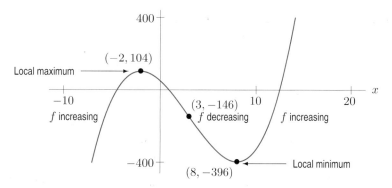

Figure 4.2: Useful graph of $f(x) = x^3 - 9x^2 - 48x + 52$

$-10 \leq x \leq 20$ and $-400 \leq y \leq 400$, we get Figure 4.2, which gives much more insight into the behavior of $f(x)$ than the graph in Figure 4.1.

In Figure 4.2, we see that part of the graph is concave up and part is concave down. We can use the second derivative to analyze concavity. We have

$$f''(x) = 6x - 18.$$

So, $f''(x) < 0$ when $x < 3$ and $f''(x) > 0$ when $x > 3$, and so the graph of f is concave down for $x < 3$ and concave up for $x > 3$. At $x = 3$, we have $f''(x) = 0$. Summarizing:

$$
\begin{array}{c|c}
 & x = 3 \\
f \text{ concave down } \cap & f \text{ concave up } \cup \\
\hline
f'' < 0 & f'' = 0 \quad f'' > 0
\end{array}
$$

Local Maxima and Minima

We are often interested in points such as those marked local maximum and local minimum in Figure 4.2. We have the following definition:

> Suppose p is a point in the domain of f:
> - f has a **local minimum** at p if $f(p)$ is less than or equal to the values of f for points near p.
> - f has a **local maximum** at p if $f(p)$ is greater than or equal to the values of f for points near p.

We use the adjective "local" because we are describing only what happens near p.

How Do We Detect a Local Maximum or Minimum?

In the preceding example, the points $x = -2$ and $x = 8$, where $f'(x) = 0$, played a key role in leading us to local maxima and minima. We give a name to such points:

> For any function f, a point p in the domain of f where $f'(p) = 0$ or $f'(p)$ is undefined is called a **critical point** of the function. In addition, the point $(p, f(p))$ on the graph of f is also called a critical point. A **critical value** of f is the value, $f(p)$, at a critical point, p.

Notice that "critical point of f" can refer either to points in the domain of f or to points on the graph of f. You will know which meaning is intended from the context.

Geometrically, at a critical point where $f'(p) = 0$, the line tangent to the graph of f at p is horizontal. At a critical point where $f'(p)$ is undefined, there is no horizontal tangent to the graph—

there's either a vertical tangent or no tangent at all. (For example, $x = 0$ is a critical point for the absolute value function $f(x) = |x|$.) However, most of the functions we work with are differentiable everywhere, and therefore most of our critical points are of the $f'(p) = 0$ variety.

The critical points divide the domain of f into intervals on which the sign of the derivative remains the same, either positive or negative. Therefore, if f is defined on the interval between two successive critical points, its graph cannot change direction on that interval; it is either increasing or decreasing. The following result, which is proved on page 186, tells us that all local maxima and minima which are not at endpoint occur at critical points.

Theorem 4.1: Local Extrema and Critical Points

Suppose f is defined on an interval and has a local maximum or minimum at the point $x = a$, which is not an endpoint of the interval. If f is differentiable at $x = a$, then $f'(a) = 0$. Thus, a is a critical point.

Warning! Not every critical point is a local maximum or local minimum. Consider $f(x) = x^3$, which has a critical point at $x = 0$. (See Figure 4.3.) The derivative, $f'(x) = 3x^2$, is positive on both sides of $x = 0$, so f increases on both sides of $x = 0$, and there is neither a local maximum nor a local minimum at $x = 0$.

Figure 4.3: Critical point which is not a local maximum or minimum

Testing For Local Maxima and Minima

If f' has different signs on either side of a critical point p, with $f'(p) = 0$, then the graph changes direction at p and looks like one of those in Figure 4.4. So we have the following criterion:

The First-Derivative Test for Local Maxima and Minima

Suppose p is a critical point of a continuous function f.
- If f' changes from negative to positive at p, then f has a local minimum at p.
- If f' changes from positive to negative at p, then f has a local maximum at p.

Figure 4.4: Changes in direction at a critical point, p: Local maxima and minima

---

Example 2   Use a graph of the function $f(x) = \dfrac{1}{x(x-1)}$ to observe its local maxima and minima. Confirm your observation analytically.

Solution   The graph in Figure 4.5 suggests that this function has no local minima but that there is a local maximum at about $x = \frac{1}{2}$. Confirming this analytically means using the formula for the derivative to show that what we expect is true. Since $f(x) = (x^2 - x)^{-1}$, we have

$$f'(x) = -1(x^2 - x)^{-2}(2x - 1) = -\frac{2x - 1}{(x^2 - x)^2}.$$

So $f'(x) = 0$ where $2x - 1 = 0$. Thus, the only critical point in the domain of f is $x = \frac{1}{2}$.

Furthermore, $f'(x) > 0$ where $0 < x < 1/2$, and $f'(x) < 0$ where $1/2 < x < 1$. Thus, f increases for $0 < x < 1/2$ and decreases for $1/2 < x < 1$. According to the first derivative test, the critical point $x = 1/2$ is a local maximum.

For $-\infty < x < 0$ or $1 < x < \infty$, there are no critical points and no local maxima or minima. Although $1/(x(x-1)) \to 0$ both as $x \to \infty$ and as $x \to -\infty$, we don't say 0 is a local minimum because $1/(x(x-1))$ never actually *equals* 0.

Notice that although $f' > 0$ everywhere that it is defined for $x < \frac{1}{2}$, the function f is not increasing throughout this interval. The problem is that f and f' are not defined at $x = 0$, so we cannot conclude that f is increasing when $x < 1/2$.

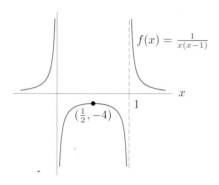

Figure 4.5: Find local maxima and minima

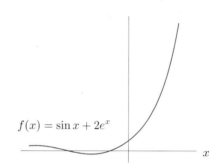

Figure 4.6: Explain the absence of local maxima and minima for $x \geq 0$

Example 3   The graph of $f(x) = \sin x + 2e^x$ is in Figure 4.6. Using the derivative, explain why there are no local maxima or minima for $x \geq 0$.

Solution   Local maxima and minima can occur only at critical points. Now $f'(x) = \cos x + 2e^x$, which is defined for all x. We know $\cos x$ is always between -1 and 1, and $2e^x \geq 2$ for $x \geq 0$, so $f'(x)$ cannot be 0 for any $x \geq 0$. Therefore there are no local maxima or minima there.

The Second-Derivative Test for Local Maxima and Minima

Knowing the concavity of a function can also be useful in testing if a critical point is a local maximum or a local minimum. Suppose p is a critical point of f, with $f'(p) = 0$, so that the graph of f has a horizontal tangent line at p. If the graph is concave up at p, then f has a local minimum at p. Likewise, if the graph is concave down, f has a local maximum. (See Figure 4.7.) This suggests:

Figure 4.7: Local maxima and minima and concavity

The Second-Derivative Test for Local Maxima and Minima

- If $f'(p) = 0$ and $f''(p) > 0$ then f has a local minimum at p.
- If $f'(p) = 0$ and $f''(p) < 0$ then f has a local maximum at p.
- If $f'(p) = 0$ and $f''(p) = 0$ then   the test tells us nothing.

Example 4   Classify as local maxima or local minima the critical points of $f(x) = x^3 - 9x^2 - 48x + 52$.

Solution   As we saw in Example 1 on page 166,

$$f'(x) = 3x^2 - 18x - 48$$

and the critical points of f are $x = -2$ and $x = 8$. We have

$$f''(x) = 6x - 18.$$

Thus $f''(8) = 30 > 0$, so f has a local minimum at $x = 8$. Since $f''(-2) = -30 < 0$, f has a local maximum at $x = -2$.

Warning! The second-derivative test does not tell us anything if both $f'(p) = 0$ and $f''(p) = 0$. For example, if $f(x) = x^3$ and $g(x) = x^4$, both $f'(0) = g'(0) = 0$ and $f''(0) = g''(0) = 0$. The point $x = 0$ is a minimum for g but is neither a maximum nor a minimum for f. However, the first-derivative test is still useful. For example, g' changes sign from negative to positive at $x = 0$, so we know g has a local minimum there.

Concavity and Inflection Points

We have studied points where the slope changes sign, which led us to critical points. Now we look at points where the concavity changes.

A point at which the graph of a function changes concavity is called an **inflection point** of f.

The words "inflection point of f" can refer either to a point in the domain of f or to a point on the graph of f. The context of the problem will tell you which is meant.

How Do We Locate an Inflection Point?

If f'' changes sign at a point, then the concavity of f changes there and we have an inflection point. Thus, points where f'' is zero or undefined are possible inflection points.

Example 5   Find the critical and inflection points for $g(x) = xe^{-x}$ and sketch the graph of g for $x \geq 0$.

Solution   Taking derivatives and simplifying, we have

$$g'(x) = (1 - x)e^{-x} \quad \text{and} \quad g''(x) = (x - 2)e^{-x}.$$

So $x = 1$ is a critical point, and $g' > 0$ for $x < 1$ and $g' < 0$ for $x > 1$. Hence g increases to a local maximum at $x = 1$ and then decreases. Also, $g(x) \to 0$ as $x \to \infty$. There is an inflection point at $x = 2$ since $g'' < 0$ for $x < 2$ and $g'' > 0$ for $x > 2$. The graph is sketched in Figure 4.8.

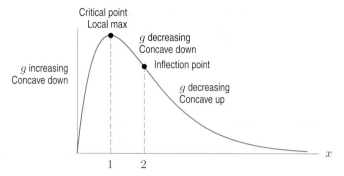

Figure 4.8: Graph of $g(x) = xe^{-x}$

Warning! Not every point x where $f''(x) = 0$ (or f'' is undefined) is an inflection point (just as not every point where $f' = 0$ is a local maximum or minimum). For instance $f(x) = x^4$ has $f''(x) = 12x^2$ so $f''(0) = 0$, but $f'' > 0$ when $x > 0$ and when $x < 0$, so there is *no* change in concavity at $x = 0$. See Figure 4.9.

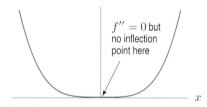

Figure 4.9: Graph of $f(x) = x^4$

Inflection Points and Local Maxima and Minima of the Derivative

Inflection points can also be interpreted in terms of first derivatives. Applying the First Derivative Test for local maxima and minima to f', we obtain the following result:

> Suppose a function f has a continuous derivative. If f'' changes sign at p, then f has an inflection point at p, and f' has a local minimum or a local maximum at p.

Figure 4.10 shows two inflection points. Notice that the curve crosses the tangent line at these points and that the slope of the curve is a maximum or a minimum there.

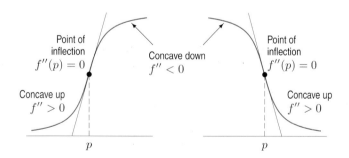

Figure 4.10: Change in concavity at p: Points of inflection

Example 6   Graph $f(x) = x + \sin x$, and determine where f is increasing most rapidly, and least rapidly.

Solution   The graph of $f(x) = x + \sin x$ is in Figure 4.11 and the graph of $f'(x) = 1 + \cos x$ is in Figure 4.12.
Where is f increasing most rapidly? At the points $x = \ldots, -2\pi, 0, 2\pi, 4\pi, 6\pi, \ldots$, because
these points are local maxima for f', and f' has the same value at each of them. Likewise f is
increasing least rapidly at the points $x = \ldots, -3\pi, -\pi, \pi, 3\pi, 5\pi, \ldots$, since these points are local
minima for f'. Notice that the points where f is increasing most rapidly and the points where it is
increasing least rapidly are inflection points of f.

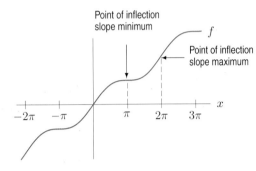

Figure 4.11: Graph of $f(x) = x + \sin x$

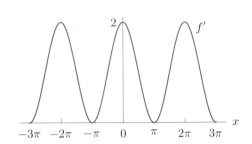

Figure 4.12: Graph of $f'(x) = 1 + \cos x$

Example 7   Water is being poured into the vase in Figure 4.13 at a constant rate, measured in volume per unit
time. Graph $y = f(t)$, the depth of the water against time, t. Explain the concavity and indicate the
inflection points.

Solution   At first the water level, y, rises slowly because the base of the vase is wide, and it takes a lot of
water to make the depth increase. However, as the vase narrows, the rate at which the water is
rising increases. Thus, y is increasing at an increasing rate and the graph is concave up. The rate of
increase in the water level is at a maximum when the water reaches the middle of the vase, where
the diameter is smallest; this is an inflection point. After that, the rate at which y increases decreases
again, so the graph is concave down. (See Figure 4.14.)

Figure 4.13: A vase

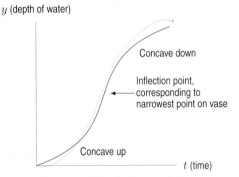

Figure 4.14: Graph of depth of water in the vase, y,
against time, t

Exercises and Problems for Section 4.1

Exercises

1. Graph a function which has exactly one critical point, at
$x = 2$, and exactly one inflection point, at $x = 4$.

2. Graph a function with exactly two critical points, one of
which is a local minimum and the other is neither a local
maximum nor a local minimum.

3. (a) Use a graph to estimate the x-values of any critical
points and inflection points of $f(x) = e^{-x^2}$.
   (b) Use derivatives to find the x-values of any critical
points and inflection points exactly.

4. Find all critical points of $f(x) = 10.2x^2 e^{-0.4x}$.

5. Find the inflection points of $f(x) = x^4 + x^3 - 3x^2 + 2$.

Classify the critical points of the functions in Exercises 6–7 as local maxima or local minima.

6. $g(x) = xe^{-x}$ **7.** $h(x) = x + 1/x$

Using a calculator or computer, graph the functions in Exercises 8–15. Describe briefly in words the interesting features of the graph including the location of the critical points and where the function is increasing/decreasing. Then use the derivative and algebra to explain the shape of the graph.

8. $f(x) = x^3 - 6x + 1$ **9.** $f(x) = x^3 + 6x + 1$

10. $f(x) = 3x^5 - 5x^3$ **11.** $f(x) = x + 2\sin x$

12. $f(x) = e^x - 10x$ **13.** $f(x) = e^x + \sin x$

14. $f(x) = xe^{-x^2}$ **15.** $f(x) = x\ln x, \quad x > 0$

16. Use the graphs you drew in Exercises 8–15 to describe in words the concavity of each graph, including approximate x-coordinates for all points of inflection. Then use f'' and algebra to explain what you see.

17. Indicate all critical points on the graph of f in Figure 4.15 and determine which correspond to local maxima of f, which to local minima, and which to neither.

Figure 4.15

18. Indicate on the graph of the derivative function f' in Figure 4.16 the x-values that are critical points of the function f itself. At which critical points does f have local maxima, local minima, or neither?

Figure 4.16

19. Indicate on the graph of the derivative f' in Figure 4.17 the x-values that are inflection points of the function f.

Figure 4.17

20. Indicate on the graph of the second derivative f'' in Figure 4.18 the x-values that are inflection points of the function f.

Figure 4.18

Problems

21. Find and classify the critical points of $f(x) = x^3(1-x)^4$ as local maxima and minima.

22. If $m, n \geq 2$ are integers, find and classify the critical points of $f(x) = x^m(1-x)^n$.

23. Give examples of graphs of functions with no, one, and infinitely many critical points.

24. Suppose f has a continuous derivative whose values are given in the following table.

(a) Estimate the x-coordinates of critical points of f for $0 \leq x \leq 10$.

(b) For each critical point, indicate if it is a local maximum of f, local minimum, or neither.

x	0	1	2	3	4	5	6	7	8	9	10
$f'(x)$	5	2	1	-2	-5	-3	-1	2	3	1	-1

25. (a) The following table gives values of the differentiable function $y = f(x)$. Estimate the x-values of critical points of $f(x)$ on the interval $0 < x < 10$. Classify each critical point as a local maximum, local minimum, or neither.

(b) Now assume that the table gives values of the continuous function $y = f'(x)$ (instead of $f(x)$). Estimate and classify critical points of the function $f(x)$.

x	0	1	2	3	4	5	6	7	8	9	10
y	1	2	1	-2	-5	-3	-1	2	3	1	-1

26. (a) Find and simplify the second derivative of

$$P = \frac{1}{1 + 10e^{-t}}.$$

(b) Sketch P and dP^2/dt^2 on separate axes. Label any asymptotes. On both graphs, label t_0, the t-coordinate of the inflection point on the graph of P.

27. Find values of a and b so that the function $f(x) = x^2 + ax + b$ has a local minimum at the point $(6, -5)$.

28. Find the value of a so that the function $f(x) = xe^{ax}$ has a critical point at $x = 3$.

29. Find constants a and b in the function $f(x) = axe^{bx}$ such that $f(\frac{1}{3}) = 1$ and the function has a local maximum at $x = \frac{1}{3}$.

30. You might think the graph of $f(x) = x^2 + \cos x$ should look like a parabola with some waves on it. Sketch the actual graph of $f(x)$ using a calculator or computer. Explain what you see using $f''(x)$.

31. The rabbit population on a small Pacific island is approximated by

$$P(t) = \frac{2000}{1 + e^{(5.3 - 0.4t)}}$$

with t measured in years since 1774, when Captain James Cook left 10 rabbits on the island.

  (a) Using a calculator, graph P. Does the population level off?
  (b) Estimate when the rabbit population grew most rapidly. How large was the population at that time?
  (c) What natural causes could lead to the shape of the graph of P?

32. (a) Water is flowing at a constant rate (i.e., constant volume per unit time) into a cylindrical container standing vertically. Sketch a graph showing the depth of water against time.
  (b) Water is flowing at a constant rate into a cone-shaped container standing on its point. Sketch a graph showing the depth of the water against time.

33. If water is flowing at a constant rate (i.e., constant volume per unit time) into the vase in Figure 4.19, sketch a graph of the depth of the water against time. Mark on the graph the time at which the water reaches the corner of the vase.

Figure 4.19

34. If water is flowing at a constant rate (i.e., constant volume per unit time) into the Grecian urn in Figure 4.20, sketch a graph of the depth of the water against time. Mark on the graph the time at which the water reaches the widest point of the urn.

Figure 4.20

35. Graph f given that:

  • $f'(x) = 0$ at $x = 2$, $f'(x) < 0$ for $x < 2$, $f'(x) > 0$ for $x > 2$,
  • $f''(x) = 0$ at $x = 4$, $f''(x) > 0$ for $x < 4$, $f''(x) < 0$ for $x > 4$.

36. Assume f is differentiable everywhere and has just one critical point, at $x = 3$. In parts (a)–(d), you are given additional conditions. In each case decide whether $x = 3$ is a local maximum, a local minimum, or neither. Explain your reasoning. Sketch possible graphs for all four cases.

  (a) $f'(1) = 3$ and $f'(5) = -1$
  (b) $\lim_{x \to \infty} f(x) = \infty$ and $\lim_{x \to -\infty} f(x) = \infty$
  (c) $f(1) = 1$, $f(2) = 2$, $f(4) = 4$, $f(5) = 5$
  (d) $f'(2) = -1$, $f(3) = 1$, $\lim_{x \to \infty} f(x) = 3$

37. Graph two continuous functions f and g, each of which has exactly five critical points, the points A–E in Figure 4.21, and which satisfy the following conditions:

  (a) $\lim_{x \to -\infty} f(x) = \infty$ and $\lim_{x \to \infty} f(x) = \infty$
  (b) $\lim_{x \to -\infty} g(x) = -\infty$ and $\lim_{x \to \infty} g(x) = 0$

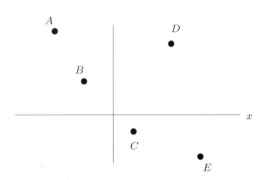

Figure 4.21

Problems 38–39 show graphs of the three functions f, f', f''. Identify which is which.

38.

39.

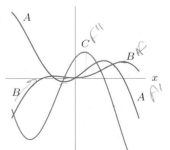

equato Polynomial (exponents) even =even

Problems 40–41 show graphs of f, f', f''. Each of these three functions is either odd or even. Decide which functions are odd and which are even. Use this information to identify which graph corresponds to f, which to f', and which to f''.

40.

41.

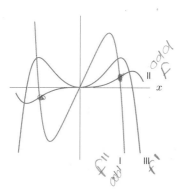

For Problems 42–45, sketch a possible graph of $y = f(x)$, using the given information about the derivatives $y' = f'(x)$ and $y'' = f''(x)$. Assume that the function is defined and continuous for all real x.

42.

43.

44.

45.

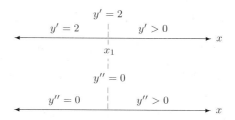

46. Assume that the polynomial f has exactly two local maxima and one local minimum, and that these are the only critical points of f.

  (a) Sketch a possible graph of f.
  (b) What is the largest number of zeros f could have?
  (c) What is the least number of zeros f could have?
  (d) What is the least number of inflection points f could have?
  (e) What is the smallest degree f could have?
  (f) Find a possible formula for $f(x)$.

47. Let f be a function with $f(x) > 0$ for all x. Set $g = 1/f$.

  (a) If f is increasing in an interval around x_0, what about g?
  (b) If f has a local maximum at x_1, what about g?
  (c) If f is concave down at x_2, what about g?

48. What happens to concavity when functions are added?

  (a) If $f(x)$ and $g(x)$ are concave up for all x, is $f(x) + g(x)$ concave up for all x?
  (b) If $f(x)$ is concave up for all x and $g(x)$ is concave down for all x, what can you say about the concavity of $f(x) + g(x)$? For example, what happens if $f(x)$ and $g(x)$ are both polynomials of degree 2?
  (c) If $f(x)$ is concave up for all x and $g(x)$ is concave down for all x, is it possible for $f(x) + g(x)$ to change concavity infinitely often?

4.2 FAMILIES OF CURVES

We saw in Chapter 1 that knowledge of one function can provide knowledge of the graphs of many others. The shape of the graph of $y = x^2$ also tells us, indirectly, about the graphs of $y = x^2 + 2$, $y = (x + 2)^2$, $y = 2x^2$, and countless other functions. We say that all functions of the form

$y = a(x + b)^2 + c$ form a *family of functions*; their graphs are like that of $y = x^2$, except for shifts and stretches determined by the values of a, b, and c. The constants a, b, c are called *parameters*. Different values of the parameters give different members of the family.

One reason for studying families of functions is their use in mathematical modeling. Confronted with the problem of modeling some phenomenon, a crucial first step involves recognizing families of functions which might fit the available data.

Motion Under Gravity: $y = -4.9t^2 + v_0t + y_0$

The position of an object moving vertically under the influence of gravity can be described by a function in the two-parameter family

$$y = -4.9t^2 + v_0t + y_0$$

where t is time in seconds and y is the distance in meters above the ground. Why do we need the parameters v_0 and y_0 to describe all such motions? Notice that at time $t = 0$ we have $y = y_0$. Thus the parameter y_0 gives the height above ground of the object at time $t = 0$. Since $dy/dt = -9.8t + v_0$, the parameter v_0 gives the velocity of the object at time $t = 0$.

Curves of the Form $y = A\sin(Bx)$

This family is used to model a wave. We saw in Section 1.5 that $|A|$ is the amplitude of the wave and that $2\pi/|B|$ is its period. Figures 4.22 and 4.23 illustrate these facts.

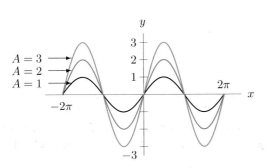

Figure 4.22: The family $y = A\sin x$
(with $B = 1$)

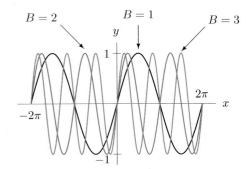

Figure 4.23: The family $y = \sin(Bx)$
(with $A = 1$)

Curves of the Form $y = e^{-(x-a)^2/b}$

This family is related to the *normal density* function, used in probability and statistics.[1] We assume that $b > 0$.

First we let $b = 1$; see Figure 4.24. The role of the parameter a is to shift the graph of $y = e^{-x^2}$ to the right or left. Notice that the value of y is always positive. Since $y \to 0$ as $x \to \pm\infty$, the x-axis is a horizontal asymptote. Thus $y = e^{-(x-a)^2}$ is the family of horizontal shifts of the bell-shaped curve.

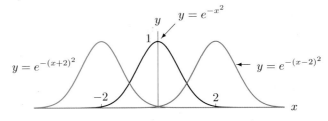

Figure 4.24: The family $y = e^{-(x-a)^2}$

---

[1]Probabilists divide our function by a constant, $\sqrt{\pi b}$, to get the normal density.

We now consider the role of the parameter b by studying the family with $a = 0$:

$$y = e^{-x^2/b}.$$

To investigate the critical points and points of inflection, we calculate

$$\frac{dy}{dx} = -\frac{2x}{b}e^{-x^2/b}$$

and, using the product rule, we get

$$\frac{d^2y}{dx^2} = -\frac{2}{b}e^{-x^2/b} - \frac{2x}{b}\left(-\frac{2x}{b}e^{-x^2/b}\right) = \frac{2}{b}\left(\frac{2x^2}{b} - 1\right)e^{-x^2/b}.$$

Critical points occur where $dy/dx = 0$, that is, where

$$\frac{dy}{dx} = -\frac{2x}{b}\,e^{-x^2/b} = 0.$$

Since $e^{-x^2/b}$ is never zero, the only critical point is $x = 0$. At that point, $y = 1$ and $d^2y/dx^2 < 0$. Hence, by the second derivative test, there is a local maximum at $x = 0$.

Inflection points occur where the second derivative changes sign; thus, we start by finding values of x for which $d^2y/dx^2 = 0$. Since $e^{-x^2/b}$ is never zero, $d^2y/dx^2 = 0$ when

$$\frac{2x^2}{b} - 1 = 0.$$

Solving for x gives

$$x = \pm\sqrt{\frac{b}{2}}.$$

Looking at the expression for d^2y/dx^2, we see that d^2y/dx^2 is negative for $x = 0$, and positive as $x \to \pm\infty$. Therefore the concavity changes at $x = -\sqrt{b/2}$ and at $x = \sqrt{b/2}$, so we have inflection points there.

Returning to the two-parameter family $y = e^{-(x-a)^2/b}$, we conclude that there is a maximum at $x = a$, obtained by horizontally shifting the maximum at $x = 0$ of $y = e^{-x^2/b}$ by a units. There are inflection points at $x = a \pm \sqrt{b/2}$ obtained by shifting the inflection points $x = \pm\sqrt{b/2}$ of $y = e^{-x^2/b}$ by a units. (See Figure 4.25.) At these points $y = e^{-1/2} \approx 0.6$.

With this information we can see the effect of the parameters. The parameter a determines the location of the center of the bell and the parameter b determines how narrow or wide the bell is. (See Figure 4.26.) If b is small, then the inflection points are close to a and the bell is sharply peaked near a; if b is large, the inflection points are farther away from a and the bell is spread out.

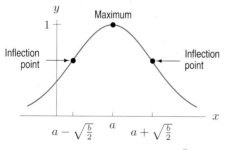

Figure 4.25: Graph of $y = e^{-(x-a)^2/b}$: bell-shaped curve with peak at $x = a$

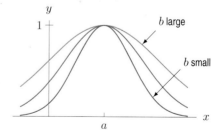

Figure 4.26: Graph of $y = e^{-(x-a)^2/b}$ for fixed a and various b

Curves of the Form $y = a(1 - e^{-bx})$

This is a two-parameter family. We consider $a, b > 0$. The graph of one member, with $a = 2$ and $b = 1$, is in Figure 4.27. Such a graph represents a quantity which is increasing but leveling off. For example, a body dropped in a thick fluid speeds up initially, but its velocity levels off as it approaches a terminal velocity. Similarly, if a pollutant pouring into a lake builds up toward a saturation level, its concentration may be described in this way. The graph might also represent the temperature of an object in an oven.

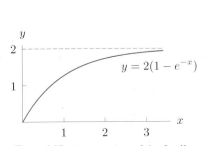

Figure 4.27: One member of the family $y = a(1 - e^{-bx})$, with $a = 2$

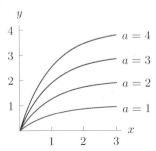

Figure 4.28: Fixing $b = 1$ gives $y = a(1 - e^{-x})$, graphed for various a

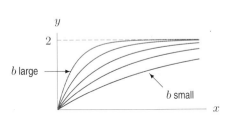

Figure 4.29: Fixing $a = 2$ gives $y = 2(1 - e^{-bx})$, graphed for various b

We examine the effect on the graph of varying a. Fix b at some positive number, say $b = 1$. Substitute different values for a and look at the graphs in Figure 4.28. We see that as x gets larger, y approaches a from below. The reason is $e^{-bx} \to 0$ as $x \to \infty$. Physically, the value of a represents the terminal velocity of a falling body or the saturation level of the pollutant in the lake.

Now examine the effect of varying b on the graph. Fix a at some positive number, say $a = 2$. Substitute different values for b and look at the graphs in Figure 4.29. The parameter b determines how sharply the curve rises and how quickly it gets close to the line $y = a$.

Let's confirm this last observation analytically. For $y = a(1 - e^{-bx})$, we have $dy/dx = abe^{-bx}$, so the slope of the tangent to the curve at $x = 0$ is ab. For larger b, the curve rises more rapidly at $x = 0$. How long does it take the curve to climb halfway up from $y = 0$ to $y = a$? When $y = a/2$, we have

$$a(1 - e^{-bx}) = \frac{a}{2}, \qquad \text{which leads to} \qquad x = \frac{\ln 2}{b}.$$

If b is large then $(\ln 2)/b$ is small, so in a short distance the curve is already half way up to a. If b is small, then $(\ln 2)/b$ is large and we have to go a long way out to get up to $a/2$. See Figure 4.30.

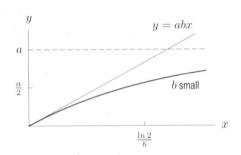

Figure 4.30: Tangent at $x = 0$ to $y = a(1 - e^{-bx})$, with fixed a, and large and small b

Exercises and Problems for Section 4.2

Exercises

Find formulas for the functions described in Exercises 1–15.

1. A line with slope 2 and x-intercept 5.

2. A parabola opening downward with its vertex at $(2, 5)$.

3. A parabola with x-intercepts ± 1 and y-intercept 3.

4. The top half of a circle centered at the origin and with radius 5.

5. The bottom half of a circle centered at the origin and with radius $\sqrt{2}$.

6. The top half of a circle with center $(-1, 2)$ and radius 3.

7. A function of the form $y = a(1 - e^{-bx})$ with $a, b > 0$ and a horizontal asymptote of $y = 5$.

8. A rational function of the form $y = ax/(x + b)$ with a vertical asymptote at $x = 2$ and a horizontal asymptote of $y - -5$.

9. A function of the form $y = A\sin(Bx) + C$ with a maximum at $(5, 2)$, a minimum at $(15, 1.5)$, and no critical points between these two points.

10. A function of the form $y = be^{-(x-a)^2/2}$ with its maximum at the point $(0, 3)$.

11. A curve of the form $y = e^{-(x-a)^2/b}$ for $b > 0$ with a local maximum at $x = 2$ and points of inflection at $x = 1$ and $x = 3$.

12. A function of the form $y = ax^b \ln x$, where a and b are nonzero constants, which has a local maximum at the point $(e^2, 6e^{-1})$.

13. A cubic polynomial having x-intercepts at $1, 5, 7$.

14. A cubic polynomial with a local maximum at $x = 1$, a local minimum at $x = 3$, a y-intercept of 5, and an x^3 term whose coefficient is 1.

15. A quartic polynomial whose graph is symmetric about the y-axis and has local maxima at $(-1, 4)$ and $(1, 4)$ and a y-intercept of 3.

Problems

16. Let $p(x) = x^3 - ax$, where a is constant.

   (a) If $a < 0$, show that $p(x)$ is always increasing.

   (b) If $a > 0$, show that $p(x)$ has a local maximum and a local minimum.

   (c) Sketch and label typical graphs for the cases when $a < 0$ and when $a > 0$.

17. Let $p(x) = x^3 - ax$, where a is constant and $a > 0$.

   (a) Find the local maxima and minima of p.

   (b) What effect does increasing the value of a have on the positions of the maxima and minima?

   (c) On the same axes, sketch and label the graphs of p for three positive values of a.

18. What effect does increasing the value of a have on the graph of $f(x) = x^2 + 2ax$? Consider roots, maxima and minima, and both positive and negative values of a.

19. (a) Graph $f(x) = x + a\sin x$ for $a = 0.5$ and $a = 3$.

   (b) For what values of a is $f(x)$ increasing for all x?

20. (a) Graph $f(x) = x^2 + a\sin x$ for $a = 1$ and $a = 20$.

   (b) For what values of a is $f(x)$ concave up for all x?

21. The number, N, of people who have heard a rumor spread by mass media at time, t, is given by

$$N(t) = a(1 - e^{-kt}).$$

There are 200,000 people in the population who hear the rumor eventually. If 10% of them heard it the first day, find a and k, assuming t is measured in days.

22. The temperature, T, in $^\circ$ C, of a yam put into a 200°C oven is given as a function of time, t, in minutes, by

$$T = a(1 - e^{-kt}) + b.$$

   (a) If the yam starts at 20°C, find a and b.

   (b) If the temperature of the yam is initially increasing at 2°C per minute, find k.

23. Consider the family of functions $y = f(x) = x - k\sqrt{x}$, with k a positive constant and $x \geq 0$. Show that the graph of $f(x)$ has a local minimum at a point whose x-coordinate is $1/4$ of the way between its x-intercepts.

24. Sketch graphs of $y = xe^{-bx}$ for $b = 1, 2, 3, 4$. Describe the graphical significance of b.

25. Find the coordinates of the critical point of $y = xe^{-bx}$ and use it to confirm your answer to Problem 24.

26. (a) Find all critical points of $f(x) = x^4 + ax^2 + b$.

   (b) Under what conditions on a and b does this function have exactly one critical point? What is the one critical point, and is it a local maximum, a local minimum, or neither?

   (c) Under what conditions on a and b does this function have exactly three critical points? What are they? Which are local maxima and which are local minima?

   (d) Is it ever possible for this function to have two critical points? No critical points? More than three critical points? Give an explanation in each case.

27. If $a > 0, b > 0$, show that $f(x) = a(1 - e^{-bx})$ is everywhere increasing and everywhere concave down.

28. A family of functions is given by

$$r(x) = \frac{1}{a + (x - b)^2}.$$

(a) For what values of a and b does the graph of r have a vertical asymptote? Where are the vertical asymptotes in this case?

(b) Find values of a and b so that the function r has a local maximum at the point $(3, 5)$.

29. For any constant a, let $f(x) = ax - x \ln x$ for $x > 0$.

(a) What is the x-intercept of the graph of $f(x)$?

(b) Graph $f(x)$ for $a = -1$ and $a = 1$.

(c) For what values of a does $f(x)$ have a critical point for $x > 0$? Find the coordinates of the critical point and decide if it is a local maximum, a local minimum, or neither.

30. Let $f(x) = x^2 + \cos(kx)$, for $k > 0$.

(a) Using a calculator or computer, graph f for $k = 0.5, 1, 3, 5$. Find the smallest number k at which you see points of inflection in the graph of f.

(b) Explain why the graph of f has no points of inflection if $k \leq \sqrt{2}$, and infinitely many points of inflection if $k > \sqrt{2}$.

(c) Explain why f has only a finite number of critical points, no matter what the value of k.

31. Let $f(x) = e^x - kx$, for $k > 0$.

(a) Using a calculator or computer, sketch the graph of f for $k = 1/4, 1/2, 1, 2, 4$. Describe what happens as k changes.

(b) Show that f has a local minimum at $x = \ln k$.

(c) Find the value of k for which the local minimum is the largest.

32. Let $y = Ae^{-Bx^2}$ for positive A, B. Analyze the effect of varying A and B on the shape of the curve. Illustrate your answer with sketches.

33. Consider the surge function $y = axe^{-bx}$ for $a, b > 0$.

(a) Find the local maxima, local minima, and points of inflection.

(b) How does varying a and b affect the shape of the graph?

(c) On one set of axes, graph this function for several values of a and b.

34. Sketch several members of the family $y = e^{-ax} \sin bx$ for $b = 1$, and describe the graphical significance of the parameter a.

35. Sketch several members of the family $e^{-ax} \sin bx$ for $a = 1$, and describe the graphical significance of the parameter b.

36. Consider the family

$$y = \frac{A}{x + B}.$$

(a) If $B = 0$, what is the effect of varying A on the graph?

(b) If $A = 1$, what is the effect of varying B?

(c) On one set of axes, graph the function for several values of A and B.

37. For positive a, b, the potential energy, U, of a particle is

$$U = b \left(\frac{a^2}{x^2} - \frac{a}{x} \right) \qquad \text{for } x > 0.$$

(a) Find the intercepts and asymptotes.

(b) Compute the local maxima and minima.

(c) Sketch the graph.

38. The force, F, on a particle with potential energy U is given by

$$F = -\frac{dU}{dx}.$$

Using the expression for U in Problem 37, graph F and U on the same axes, labeling intercepts and local maxima and minima.

39. For positive A, B, the force between two atoms is a function of the distance, r, between them:

$$f(r) = -\frac{A}{r^2} + \frac{B}{r^3} \qquad r > 0.$$

(a) What are the zeros and asymptotes of f?

(b) Find the coordinates of the critical points and inflection points of f.

(c) Sketch a graph of f.

(d) Illustrating your answers with a sketch, describe the effect on the graph of f of:

 (i) Increasing B, holding A fixed

 (ii) Increasing A, holding B fixed

40. Consider the family of functions $y = a \cosh(x/a)$ for $a > 0$. Sketch graphs for $a = 1, 2, 3$. Describe in words the effect of increasing a.

41. Let $y = Ae^x + Be^{-x}$ for any constants A, B.

(a) Sketch the function for

 (i) $A = 1, B = 1$ (ii) $A = 1, B = -1$

 (iii) $A = 2, B = 1$ (iv) $A = 2, B = -1$

 (v) $A = -2, B = -1$ (vi) $A = -2, B = 1$

(b) Describe in words the general shape of the graph if A and B have the same sign. What effect does the sign of A have on the graph?

(c) Describe in words the general shape of the graph if A and B have different signs. What effect does the sign of A have on the graph?

(d) For what values of A and B does the function have a local maximum? A local minimum? Justify your answer using derivatives.

42. The x-axis runs along the water surface in a straight, horizontal canal. At time t, the displacement, y, of the surface of the water at position x is given by

$$y = \sin(x - t).$$

(a) Graph and label y as a function of x for $t = 0, 0.5,$ $1, 1.5, 2$.

(b) What does the function $y = \sin(x - t)$ represent

(i) For fixed t? (ii) For fixed x?

(c) Assume t is constant. What does dy/dx represent?

(d) Assume x is constant. What does dy/dt represent?

43. An organism has size W at time t. For positive constants A, b, and c, the Gompertz growth function gives

$$W = Ae^{-e^{b-ct}}, \quad t \geq 0.$$

(a) Find the intercepts and asymptotes.

(b) Find the critical points and inflection points.

(c) Graph W for various values of A, b, and c.

(d) A certain organism grows fastest when it is about 1/3 of its final size. Would the Gompertz growth function be useful in modeling its growth? Explain.

4.3 OPTIMIZATION

It is often important to find the largest or smallest value of some quantity. For example, automobile engineers want to construct a car that uses the least amount of fuel, scientists want to calculate which wavelength carries the maximum radiation at a given temperature, and urban planners want to design traffic patterns to minimize delays. Such problems belong to the field of mathematics called *optimization*. The next three sections show how the derivative provides an efficient way of solving many optimization problems.

Global Maxima and Minima

The single greatest (or least) value of a function f over a specified domain is called the *global maximum* (or *minimum*) of f. Recall that the local maxima and minima tell us where a function is locally largest or smallest. Now we are interested in where the function is absolutely largest or smallest in a given domain. We define

- f has a **global minimum** at p if $f(p)$ is less than or equal to all values of f.
- f has a **global maximum** at p if $f(p)$ is greater than or equal to all values of f.

Global maxima and minima are sometimes called *extrema* or *optimal values*.

How Do We Find Global Maxima and Minima?

If f is a continuous function defined on a closed interval $a \leq x \leq b$ (that is, an interval containing its endpoints), then Theorem 4.2 on page 186 guarantees that global maxima and minima exist. Figure 4.31 illustrates that the global maximum or minimum of f occurs either at a critical point or at an endpoint of the interval, $x = a$ or $x = b$. These points are the candidates for global extrema.

Figure 4.31: Global maximum and minimum on a closed interval $a \leq x \leq b$

To find the global maximum and minimum of a continuous function on a closed interval: Compare values of the function at all the candidate points: the critical points in the interval and the endpoints.

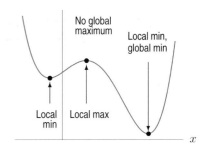

Figure 4.32: Global minimum on
$a < x < b$

Figure 4.33: Global minimum when the
domain is all real numbers

If the function is defined on an open interval $a < x < b$ (that is, an interval not including its endpoints) or on all real numbers, there may or may not be a global maximum or minimum. For example, there is no global maximum in Figure 4.32 because the function has no actual largest value. The global minimum in Figure 4.32 coincides with the local minimum. There is a global minimum but no global maximum in Figure 4.33.

To find the global maximum and minimum of a continuous function on an open interval or on all real numbers: Find the value of the function at all the critical points and sketch a graph. Look at the function values when x approaches the endpoints of the interval, or approaches $\pm\infty$, as appropriate.

Example 1   Find the global maxima and minima of $f(x) = x^3 - 9x^2 - 48x + 52$ on the following intervals:

(a) $-5 \leq x \leq 12$ (b) $-5 \leq x \leq 14$ (c) $-5 \leq x < \infty$.

Solution   (a) We have previously obtained the critical points $x = -2$ and $x = 8$ using

$$f'(x) = 3x^2 - 18x - 48 = 3(x + 2)(x - 8).$$

We evaluate f at the critical points and the endpoints of the interval:

$$f(-5) = (-5)^3 - 9(-5)^2 - 48(-5) + 52 = -58$$
$$f(-2) = 104$$
$$f(8) = -396$$
$$f(12) = -92.$$

Comparing these function values, we see that the global maximum on $[-5, 12]$ is 104 and occurs at $x = -2$, and the global minimum on $[-5, 12]$ is -396 and occurs at $x = 8$.

(b) For the interval $[-5, 14]$, we compare

$$f(-5) = -58, \quad f(-2) = 104, \quad f(8) = -396, \quad f(14) = 360.$$

The global maximum is now 360 and occurs at $x = 14$, and the global minimum is still -396 and occurs at $x = 8$. Since the function is increasing for $x > 8$, changing the right-hand end of the interval from $x = 12$ to $x = 14$ alters the global maximum but not the global minimum. See Figure 4.34.

(c) Figure 4.34 shows that for $-5 \leq x < \infty$ there is no global maximum, because we can make $f(x)$ as large as we please by choosing x large enough. The global minimum remains -396 at $x = 8$.

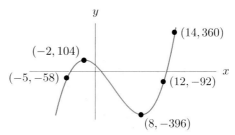

Figure 4.34: Graph of $f(x) = x^3 - 9x^2 - 48x + 52$

Example 2 When an arrow is shot into the air, its range, R, is defined as the horizontal distance from the archer to the point where the arrow hits the ground. If the ground is horizontal and we neglect air resistance, it can be shown that

$$R = \frac{{v_0}^2 \sin(2\theta)}{g},$$

where v_0 is the initial velocity of the arrow, g is the (constant) acceleration due to gravity, and θ is the angle above horizontal, so $0 \leq \theta \leq \pi/2$. (See Figure 4.35.) What initial angle maximizes R?

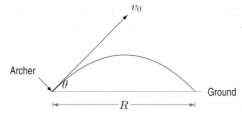

Figure 4.35: Arrow's path

Solution We can find the maximum of this function without using calculus. The maximum value of R occurs when $\sin(2\theta) = 1$, so $\theta = \arcsin(1)/2 = \pi/4$, giving $R = v_0^2/g$.

Let's see how we can do the same problem with calculus. We want to find the global maximum of R for $0 \leq \theta \leq \pi/2$. First we look for critical points:

$$\frac{dR}{d\theta} = 2\frac{v_0^2 \cos(2\theta)}{g}.$$

Setting $dR/d\theta$ equal to 0, we get

$$0 = \cos(2\theta), \quad \text{or} \quad 2\theta = \pm\frac{\pi}{2}, \pm\frac{3\pi}{2}, \pm\frac{5\pi}{2}, \ldots$$

so $\pi/4$ is the only critical point in the interval $0 \leq \theta \leq \pi/2$. The range at $\theta = \pi/4$ is $R = v_0^2/g$.

Now we must check the value of R at the endpoints $\theta = 0$ and $\theta = \pi/2$. Since $R = 0$ at each endpoint, the critical point $\theta = \pi/4$ gives both a local and a global maximum on $0 \leq \theta \leq \pi/2$. Therefore, the arrow goes farthest if shot at an angle of $\pi/4$, or $45°$.

A Graphical Example: Minimizing Gas Consumption

Next we look at an example in which a function is given graphically and the optimum values are read from a graph. You already know how to estimate the optimum values of $f(x)$ from a graph of $f(x)$—read off the highest and lowest values. In this example, we see how to estimate the optimum value of the quantity $f(x)/x$ from a graph of $f(x)$ against x.

The question we investigate is how to set driving speeds to maximize fuel efficiency.[2] We

[2]Adapted from Peter D. Taylor, *Calculus: The Analysis of Functions* (Toronto: Wall & Emerson, 1992).

Figure 4.36: Gas consumption versus velocity

assume that gas consumption, g (in gallons/hour), as a function of velocity, v (in mph) is as shown in Figure 4.36. We want to minimize the gas consumption per *mile*, not the gas consumption per hour. Let $G = g/v$ represent the average gas consumption per mile. (The units of G are gallons/mile.)

Example 3   Using Figure 4.36, estimate the velocity which minimizes G.

Solution   We want to find the minimum value of $G = g/v$ when g and v are related by the graph in Figure 4.36. We could use Figure 4.36 to sketch a graph of G against v and estimate a critical point. But there is an easier way.

Figure 4.37 shows that g/v is the slope of the line from the origin to the point P. Where on the curve should P be to make the slope a minimum? From the possible positions of the line shown in Figure 4.37, we see that the slope of the line is both a local and global minimum when the line is tangent to the curve. From Figure 4.38, we can see that the velocity at this point is about 50 mph. Thus to minimize gas consumption per mile, we should drive about 50 mph.

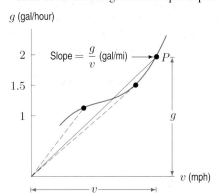

Figure 4.37: Graphical representation of gas consumption per mile, $G = g/v$

Figure 4.38: Velocity for maximum fuel efficiency

Finding Upper and Lower Bounds

A problem which is closely related to finding maxima and minima is finding the *bounds* of a function. In Example 1, the value of $f(x)$ on the interval $[-5, 12]$ ranges from -396 to 104. Thus

$$-396 \le f(x) \le 104,$$

and we say that -396 is a *lower bound* for f and 104 is an *upper bound* for f on $[-5, 12]$. (See Appendix A for more on bounds.) Of course, we could also say that

$$-400 \le f(x) \le 150,$$

so that f is also bounded below by -400 and above by 150 on $[-5, 12]$. However, we consider the -396 and 104 to be the *best possible bounds* because they describe more accurately how the function $f(x)$ behaves on $[-5, 12]$.

Example 4   An object on a spring oscillates about its equilibrium position at $y = 0$. Its distance from equilibrium is given as a function of time, t, by

$$y = e^{-t} \cos t.$$

Find the greatest distance the object goes above and below the equilibrium for $t \geq 0$.

Solution   We are looking for the bounds of the function. What does the graph of the function look like? We can think of it as a cosine curve with a decreasing amplitude of e^{-t}; in other words, it is a cosine curve squashed between the graphs of $y = e^{-t}$ and $y = -e^{-t}$, forming a wave with lower and lower crests and shallower and shallower troughs. (See Figure 4.39.)

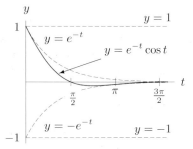

Figure 4.39: $f(t) = e^{-t} \cos t$ for $t \geq 0$

From the graph we can see that for $t \geq 0$, the graph lies between the horizontal lines $y = -1$ and $y = 1$. This means that -1 and 1 are bounds:

$$-1 \leq e^{-t} \cos t \leq 1.$$

The line $y = 1$ is the best possible upper bound because the graph does come up that high (at $t = 0$). However, we can find a better lower bound if we find the global minimum value of f for $t \geq 0$; this minimum occurs in the first trough between $t = \pi/2$ and $t = 3\pi/2$ because later troughs are squashed closer to the t-axis. At the minimum, $dy/dt = 0$. The product rule gives

$$\frac{dy}{dt} = (-e^{-t}) \cos t + e^{-t}(-\sin t) = -e^{-t}(\cos t + \sin t) = 0.$$

Since e^{-t} is never 0, we must have

$$\cos t + \sin t = 0, \quad \text{so} \quad \frac{\sin t}{\cos t} = -1.$$

Hence

$$\tan t = -1, \quad \text{giving} \quad t = \frac{3\pi}{4}.$$

Thus, the global minimum we see on the graph occurs at $t = 3\pi/4$. The value of y at that minimum is

$$y = e^{-3\pi/4} \cos\left(\frac{3\pi}{4}\right) \approx -0.067.$$

Rounding down so that the inequalities still hold for all $t \geq 0$ gives

$$-0.07 < e^{-t} \cos t \leq 1.$$

Notice how much smaller in magnitude the lower bound is than the upper. This is a reflection of how quickly the factor e^{-t} causes the oscillation to die out.

Existence and Location of Extrema

On page 181 we gave a method for finding global maxima and minima of a continuous function on a closed interval: Compare values of the function at all the critical points in the interval and at the endpoints. This method relies on the fact that global maxima and minima exist, and on knowing that they occur at critical points or endpoints. The existence of global extrema is guaranteed by the following theorem; their location is specified by Theorem 4.1 on page 168.

Theorem 4.2: The Extreme Value Theorem

If f is continuous on the closed interval $a \leq x \leq b$, then f has a global maximum and a global minimum on that interval.

For a proof of Theorem 4.2, see www.wiley.com/college/hugheshallett.

We now prove Theorem 4.1, which says that inside an interval, local maxima and minima can only occur at critical points. Suppose that f has a local maximum at $x = a$. Assuming that $f'(a)$ is defined, the definition of the derivative gives

$$f'(a) = \lim_{h \to 0} \frac{f(a+h) - f(a)}{h}.$$

Since this is a two-sided limit, we have

$$f'(a) = \lim_{h \to 0^-} \frac{f(a+h) - f(a)}{h} = \lim_{h \to 0^+} \frac{f(a+h) - f(a)}{h}.$$

By the definition of local maximum, $f(a+h) \leq f(a)$ for all sufficiently small h. Thus $f(a+h) - f(a) \leq 0$ for sufficiently small h. The denominator, h, is positive when we take the limit from the right and negative when we take the limit from the left. Thus

$$\lim_{h \to 0^-} \frac{f(a+h) - f(a)}{h} \geq 0 \quad \text{and} \quad \lim_{h \to 0^+} \frac{f(a+h) - f(a)}{h} \leq 0.$$

Since both these limits are equal to $f'(a)$, we have $f'(a) \geq 0$ and $f'(a) \leq 0$, so we must have $f'(a) = 0$. The proof for a local minimum at $x = a$ is similar.

Exercises and Problems for Section 4.3

Exercises

For Exercises 1–2, indicate all critical points on the given graphs. Determine which correspond to local minima, local maxima, global minima, global maxima, or none of these. (Note that the graphs are on closed intervals.)

1.

2.

3. (a) Find the critical points of $p(1-p)^4$.
  (b) Classify the critical points as local maxima, local minima, or neither.
  (c) What are the maximum and minimum values of $p(1-p)^4$ on $0 \leq x \leq 1$?

In Exercises 4–6, find the value(s) of x for which:

(a) $f(x)$ has a local maximum or local minimum. Indicate which ones are maxima and which are minima.

(b) $f(x)$ has a global maximum or global minimum.

4. $f(x) = x^{10} - 10x$, and $0 \leq x \leq 2$

5. $f(x) = x - \ln x$, and $0.1 \leq x \leq 2$

6. $f(x) = \sin^2 x - \cos x$, and $0 \leq x \leq \pi$

In Exercises 7–12, find the exact global maximum and minimum values of the function. The domain is all real numbers unless otherwise specified.

7. $g(x) = 4x - x^2 - 5$

8. $f(x) = x + 1/x$ for $x > 0$

9. $g(t) = te^{-t}$ for $t > 0$

10. $f(x) = x - \ln x$ for $x > 0$

11. $f(t) = \dfrac{t}{1 + t^2}$

12. $f(t) = (\sin^2 t + 2) \cos t$

Find the best possible bounds for the functions in Exercises 13–18.

13. $x^3 - 4x^2 + 4x$,   for $0 \leq x \leq 4$

14. e^{-x^2},   for $|x| \leq 0.3$

15. $x^3 e^{-x}$ for $x \geq 0$

16. $x + \sin x$,   for $0 \leq x \leq 2\pi$

17. $\ln(1 + x)$,   for $x \geq 0$

18. $\ln(1 + x^2)$,   for $-1 \leq x \leq 2$

Problems

19. A grapefruit is tossed straight up with an initial velocity of 50 ft/sec. The grapefruit is 5 feet above the ground when it is released. Its height at time t is given by

$$y = -16t^2 + 50t + 5.$$

How high does it go before returning to the ground?

20. When you cough, your windpipe contracts. The speed, v, with which air comes out depends on the radius, r, of your windpipe. If R is the normal (rest) radius of your windpipe, then for $r \leq R$, the speed is given by:

$$v = a(R - r)r^2 \quad \text{where } a \text{ is a positive constant.}$$

What value of r maximizes the speed?

21. For some positive constant C, a patient's temperature change, T, due to a dose, D, of a drug is given by

$$T = \left(\frac{C}{2} - \frac{D}{3}\right)D^2.$$

(a) What dosage maximizes the temperature change?

(b) The sensitivity of the body to the drug is defined as dT/dD. What dosage maximizes sensitivity?

22. A warehouse selling cement has to decide how often and in what quantities to reorder. It is cheaper, on average, to place large orders, because this reduces the ordering cost per unit. On the other hand, larger orders mean higher storage costs. The warehouse always reorders cement in the same quantity, q. The total weekly cost, C, of ordering and storage is given by

$$C = \frac{a}{q} + bq, \quad \text{where } a, b \text{ are positive constants.}$$

(a) Which of the terms, a/q and bq, represents the ordering cost and which represents the storage cost?

(b) What value of q gives the minimum total cost?

23. A chemical reaction converts substance A to substance Y; the presence of Y catalyzes the reaction. At the start of the reaction, the quantity of A present is a grams. At time t seconds later, the quantity of Y present is y grams. The rate of the reaction, in grams/sec, is given by

$$\text{Rate} = ky(a - y), \quad k \text{ is a positive constant.}$$

(a) For what values of y is the rate nonnegative? Graph the rate against y.

(b) For what values of y is the rate a maximum?

24. For positive constants A and B, the force between two atoms in a molecule is given by

$$f(r) = -\frac{A}{r^2} + \frac{B}{r^3},$$

where $r > 0$ is the distance between the atoms. What value of r minimizes the force between the atoms?

25. When an electric current passes through two resistors with resistance r_1 and r_2, connected in parallel, the combined resistance, R, can be calculated from the equation

$$\frac{1}{R} = \frac{1}{r_1} + \frac{1}{r_2},$$

where R, r_1, and r_2 are positive. Assume that r_2 is constant.

(a) Show that R is an increasing function of r_1.

(b) Where on the interval $a \leq r_1 \leq b$ does R take its maximum value?

26. As an epidemic spreads through a population, the number of infected people, I, is expressed as a function of the number of susceptible people, S, by

$$I = k \ln\left(\frac{S}{S_0}\right) - S + S_0 + I_0, \quad \text{for } k, S_0, I_0 > 0.$$

(a) Find the maximum number of infected people.

(b) The constant k is a characteristic of the particular disease; the constants S_0 and I_0 are the values of S and I when the disease starts. Which of the following affects the maximum possible value of I? Explain.

- The particular disease, but not how is starts.
- How the disease starts, but not the particular disease.
- Both the particular disease and how it starts.

27. The distance, s, traveled by a cyclist, who starts at 1 pm, is given in Figure 4.40. Time, t, is in hours since noon.

(a) Explain why the quantity, s/t, is represented by the slope of a line from the origin to the point (t, s) on the graph.

(b) Estimate the time at which the quantity s/t is a maximum.

(c) What is the relationship between the quantity s/t and the instantaneous speed of the cyclist at the time you found in part (b)?

Figure 4.40

28. A line goes through the origin and a point on the curve $y = x^2 e^{-3x}$, for $x \geq 0$. Find the maximum slope of such a line. At what x-value does it occur?

29. Two points on the curve $y = \dfrac{x^3}{1 + x^4}$ have opposite x-values, x and $-x$. Find the points making the slope of the line joining them greatest.

30. When birds lay eggs, they do so in clutches of several at a time. When the eggs hatch, each clutch gives rise to a brood of baby birds. We want to determine the clutch size which maximizes the number of birds surviving to adulthood per brood. If the clutch is small, there are few baby birds in the brood; if the clutch is large, there are so many baby birds to feed that most die of starvation. The number of surviving birds per brood as a function of clutch size is shown by the benefit curve in Figure 4.41.[3]

(a) Estimate the clutch size which maximizes the number of survivors per brood.

(b) Suppose also that there is a biological cost to having a larger clutch: the female survival rate is reduced by large clutches. This cost is represented by the dotted line in Figure 4.41. If we take cost into account by assuming that the optimal clutch size in fact maximizes the vertical distance between the curves, what is the new optimal clutch size?

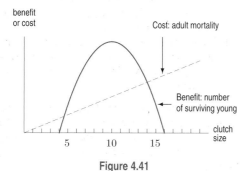

Figure 4.41

31. Let $f(v)$ be the amount of energy consumed by a flying bird, measured in joules per second (a joule is a unit of energy), as a function of its speed v (in meters/sec). Let $a(v)$ be the amount of energy consumed by the same bird, measured in joules per meter.

(a) Suggest a reason (in terms of the way birds fly) for the shape of the graph of $f(v)$ in Figure 4.42.

(b) What is the relationship between $f(v)$ and $a(v)$?

(c) Where is $a(v)$ a minimum?

(d) Should the bird try to minimize $f(v)$ or $a(v)$ when it is flying? Why?

Figure 4.42

32. The forward motion of an aircraft in level flight is reduced by two kinds of forces, known as *induced drag* and *parasite drag*. Induced drag is a consequence of the downward deflection of air as the wings produce lift. Parasite drag results from friction between the air and the entire surface of the aircraft. Induced drag is inversely proportional to the square of speed and parasite drag is directly proportional to the square of speed. The sum of induced drag and parasite drag is called total drag. The graph in Figure 4.43 shows a certain aircraft's induced drag and parasite drag functions.

(a) Sketch the total drag as a function of airspeed.

(b) Estimate two different airspeeds which each result in a total drag of 1000 pounds. Does the total drag function have an inverse? What about the induced and parasite drag functions?

(c) Fuel consumption (in gallons per hour) is roughly proportional to total drag. Suppose you are low on fuel and the control tower has instructed you to enter a circular holding pattern of indefinite duration to await the passage of a storm at your landing field. At what airspeed should you fly the holding pattern? Why?

Figure 4.43

[3]Data from C. M. Perrins and D. Lack reported by J. R. Krebs and N. B. Davies in *An Introduction to Behavioural Ecology* (Oxford: Blackwell, 1987).

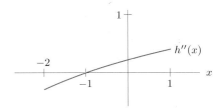

Figure 4.46

33. Let $f(v)$ be the fuel consumption, in gallons per hour, of a certain aircraft as a function of its airspeed, v, in miles per hour. A graph of $f(v)$ is given in Figure 4.44.

    (a) Let $g(v)$ be the fuel consumption of the same aircraft, but measured in gallons per mile instead of gallons per hour. What is the relationship between $f(v)$ and $g(v)$?
    (b) For what value of v is $f(v)$ minimized?
    (c) For what value of v is $g(v)$ minimized?
    (d) Should a pilot try to minimize $f(v)$ or $g(v)$?

Figure 4.44

34. The function $y - t(x)$ is positive and continuous with a global maximum at the point $(3, 3)$. Graph $t(x)$ if $t'(x)$ and $t''(x)$ have the same sign for $x < 3$, but opposite signs for $x > 3$.

35. Figure 4.45 gives the derivative of $g(x)$ on $-2 \leq x \leq 2$.

    (a) Write a few sentences describing the behavior of $g(x)$ on this interval.
    (b) Does the graph of $g(x)$ have any inflection points? If so, give the approximate x-coordinates of their locations. Explain your reasoning.
    (c) What are the global maxima and minima of g on $[-2, 2]$?
    (d) If $g(-2) = 5$, what do you know about $g(0)$ and $g(2)$? Explain.

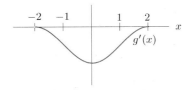

Figure 4.45

36. Figure 4.46 shows the second derivative of $h(x)$ for $-2 \leq x \leq 1$. If $h'(-1) = 0$ and $h(-1) = 2$,

    (a) Explain why $h'(x)$ is never negative on this interval.
    (b) Explain why $h(x)$ has a global maximum at $x = 1$.
    (c) Sketch a possible graph of $h(x)$ for $-2 \leq x \leq 1$.

Decide if the statements in Problems 37–39 are true or false. Give an explanation for your answer.

37. A critical point of f must be a local maximum or minimum of f.

38. Since the function $f(x) = 1/x$ is continuous for $x > 0$ and the interval $(0, 1)$ is bounded, f has a maximum on the interval $(0, 1)$.

39. The Extreme Value Theorem says that only continuous functions have global maxima and minima on every closed, bounded interval.

40. Show that $f''(x)$ is continuous and $f(x)$ has exactly two critical points, then $f'(x)$ has a local maximum or local minimum between the two critical points.

41. In this problem we prove a special case of the Mean Value Theorem where $f(a) = f(b) = 0$. This special case is called Rolle's Theorem: If f is continuous on $[a, b]$ and differentiable on (a, b), and if $f(a) - f(b) - 0$, then there is a number c, with $a < c < b$, such that

$$f'(c) = 0.$$

By the Extreme Value Theorem, f has a global maximum and a global minimum on $[a, b]$.

    (a) Prove Rolle's theorem in the case that both the global maximum and the global minimum are at endpoints of $[a, b]$. [Hint: $f(x)$ must be a very simple function in this case.]
    (b) Prove Rolle's theorem in the case that either the global maximum or the global minimum is not at an endpoint. [Hint: Think about local maxima and minima.]

42. Use Rolle's Theorem to prove the Mean Value Theorem. Suppose that $f(x)$ is continuous on $[a, b]$ and differentiable on (a, b). Let $g(x)$ be the difference between $f(x)$ and the y-value on the secant line joining $(a, f(a))$ to $(b, f(b))$, so

$$g(x) = f(x) - f(a) - \frac{f(b) - f(a)}{b - a}(x - a).$$

    (a) Show $g(x)$ on a sketch of $f(x)$.
    (b) Use Rolle's Theorem (Problem 41) to show that there must be a point c in (a, b) such that $g'(c) - 0$.
    (c) Show that if c is the point in part (b), then

$$f'(c) = \frac{f(b) - f(a)}{b - a}.$$

4.4   APPLICATIONS TO MARGINALITY

Management decisions within a particular business usually aim at maximizing profit for the company. In this section we will see how the derivative can be used to maximize profit. Profit depends on both production cost and revenue (or income) from sales. We begin by looking at the cost and revenue functions.

> The **cost function**, $C(q)$, gives the total cost of producing a quantity q of some good.

What sort of function do we expect C to be? The more goods that are made, the higher the total cost, so C is an increasing function. In fact, cost functions usually have the general shape shown in Figure 4.47. The intercept on the C-axis represents the *fixed costs*, which are incurred even if nothing is produced. (This includes, for instance, the machinery needed to begin production.) The cost function increases quickly at first and then more slowly because producing larger quantities of a good is usually more efficient than producing smaller quantities—this is called *economy of scale*. At still higher production levels, the cost function starts to increase faster again as resources become scarce, and sharp increases may occur when new factories have to be built. Thus, the graph of $C(q)$ may start out concave down and become concave up later on.

Figure 4.47: Cost as a function of quantity

> The **revenue function**, $R(q)$, gives the total revenue received by a firm from selling a quantity q of some good.

Revenue is income obtained from sales. If the price per item is p, and the quantity sold is q, then

$$\text{Revenue} = \text{Price} \times \text{Quantity}, \quad \text{so} \quad R = pq.$$

If the price per item does not depend on the quantity sold, then the graph of $R(q)$ is a straight line through the origin with slope equal to the price p. See Figure 4.48. In practice, for large values of q, the market may become glutted, causing the price to drop, giving $R(q)$ the shape in Figure 4.49.

Figure 4.48: Revenue: Constant price

Figure 4.49: Revenue: Decreasing price

The profit is usually written as π. (Economists use π to distinguish it from the price, p; this π has nothing to do with the area of a circle, and merely stands for the Greek equivalent of the letter "p.") The profit resulting from producing and selling q items is defined by

$$\text{Profit} = \text{Revenue} - \text{Cost}, \quad \text{so} \quad \pi(q) = R(q) - C(q).$$

Example 1   If cost, C, and revenue, R, are given by the graph in Figure 4.50, for what production quantities, q, does the firm make a profit? Approximately what production level maximizes profit?

Figure 4.50: Costs and revenues for Example 1

Solution   The firm makes a profit whenever revenues are greater than costs, that is, when $R > C$. The graph of R is above the graph of C approximately when $130 < q < 215$. Production between $q = 130$ units and $q = 215$ units generates a profit. The vertical distance between the cost and revenue curves is largest at q_0, so q_0 units gives maximum profit.

Marginal Analysis

Many economic decisions are based on an analysis of the costs and revenues "at the margin." Let's look at this idea through an example.

Suppose we are running an airline and we are trying to decide whether to offer an additional flight. How should we decide? We'll assume that the decision is to be made purely on financial grounds: if the flight will make money for the company, it should be added. Obviously we need to consider the costs and revenues involved. Since the choice is between adding this flight and leaving things the way they are, the crucial question is whether the *additional costs* incurred are greater or smaller than the *additional revenues* generated by the flight. These additional costs and revenues are called the *marginal costs* and *marginal revenues*.

Suppose $C(q)$ is the function giving the total cost of running q flights. If the airline had originally planned to run 100 flights, its costs would be $C(100)$. With the additional flight, its costs would be $C(101)$. Therefore,

$$\text{Marginal cost} = C(101) - C(100).$$

Now

$$C(101) - C(100) = \frac{C(101) - C(100)}{101 - 100},$$

and this quantity is the average rate of change of cost between 100 and 101 flights. In Figure 4.51 the average rate of change is the slope of the line joining the $C(100)$ and $C(101)$ points on the graph. If the graph of the cost function is not curving fast near the point, the slope of this line is close to the slope of the tangent line there. Therefore, the average rate of change is close to the instantaneous rate of change. Since these rates of change are not very different, many economists choose to define marginal cost, MC, as the instantaneous rate of change of cost with respect to quantity:

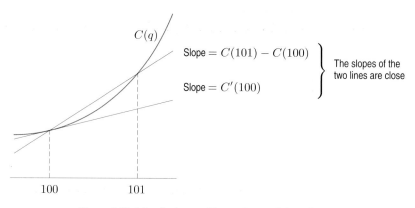

Figure 4.51: Marginal cost: Slope of one of these lines

$$\text{Marginal cost} = MC = C'(q).$$

Similarly if the revenue generated by q flights is $R(q)$, the additional revenue generated by increasing the number of flights from 100 to 101 is

$$\text{Marginal revenue} = R(101) - R(100).$$

Now $R(101) - R(100)$ is the average rate of change of revenue between 100 and 101 flights. As before, the average rate of change is usually almost equal to the instantaneous rate of change, so economists often define:

$$\text{Marginal revenue} = MR = R'(q).$$

We often refer to total cost and total revenue to distinguish them from marginal cost and marginal revenue. If the words cost and revenue are used alone, they are understood to mean total cost and total revenue.

Example 2   If $C(q)$ and $R(q)$ for the airline are given in Figure 4.52, should the company add the 101st flight?

Solution   The marginal revenue is the slope of the revenue curve, and the marginal cost is the slope of the cost curve at the point 100. From Figure 4.52, you can see that the slope at the point A is smaller than the slope at B, so $MC < MR$. This means that the airline will make more in extra revenue than it will spend in extra costs if it runs another flight, so it should go ahead and run the 101st flight.

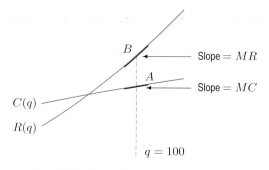

Figure 4.52: Cost and revenue for Example 2

Since MC and MR are derivative functions, they can be estimated from the graphs of total cost and total revenue.

Example 3   If R and C are given by the graphs in Figure 4.53, sketch graphs of $MR = R'(q)$ and $MC = C'(q)$.

Figure 4.53: Total revenue and total cost for Example 3

Solution   The revenue graph is a line through the origin, with equation

$$R = pq$$

where p is the price, which is a constant. The slope is p and

$$MR = R'(q) = p.$$

The total cost is increasing, so the marginal cost is always positive (above the q-axis). For small q values, the total cost curve is concave down, so the marginal cost is decreasing. For larger q, say $q > 100$, the total cost curve is concave up and the marginal cost is increasing. Thus the marginal cost has a minimum at about $q = 100$. (See Figure 4.54.)

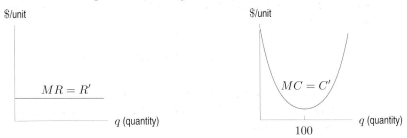

Figure 4.54: Marginal revenue and costs for Example 3

Maximizing Profit

Now let's look at how to maximize total profit, given functions for total revenue and total cost.

Example 4   Find the maximum profit if the total revenue and total cost are given, for $0 \le q \le 200$, by the curves R and C in Figure 4.55.

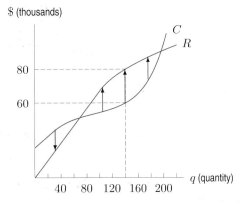

Figure 4.55: Maximum profit at $q = 140$

Solution   The profit is represented by the vertical difference between the curves and is marked by the vertical arrows on the graph. When revenue is below cost, the company is taking a loss; when revenue is above cost, the company is making a profit. We can see that the profit is largest at about $q = 140$, so this is the production level we're looking for. To be sure that the local maximum is a global maximum, we need to check the endpoints. At $q = 0$ and $q = 200$, the profit is negative, so the global maximum is indeed at $q = 140$.

To find the actual maximum profit, we estimate the vertical distance between the curves at $q = 140$. This gives a maximum profit of $\$80,000 - \$60,000 = \$20,000$.

Suppose we wanted to find the minimum profit. In this example, we must look at the endpoints, when $q = 0$ or $q = 200$. We see the minimum profit is negative (a loss), and it occurs at $q = 0$.

Maximum Profit Occurs Where $MR = MC$

In Example 4, observe that at $q = 140$ the slopes of the two curves in Figure 4.55 are equal. To the left of $q = 140$, the revenue curve has a larger slope than the cost curve, and the profit increases as q increases. The company will make more money by producing more units, so production should increase toward $q = 140$. To the right of $q = 140$, the slope of the revenue curve is less than the slope of the cost curve, and the profit is decreasing. The company will make more money by producing fewer units so production should decrease toward $q = 140$. At the point where the slopes are equal, the profit has a local maximum; otherwise the profit could be increased by moving toward that point. Since the slopes are equal at $q = 140$, we have $MR = MC$ there.

Now let's look at the general situation. To maximize or minimize profit over an interval, we optimize the profit, π, where

$$\pi(q) = R(q) - C(q).$$

We know that global maxima and minima can only occur at critical points or at endpoints of an interval. To find critical points of π, look for zeros of the derivative:

$$\pi'(q) = R'(q) - C'(q) = 0.$$

So

$$R'(q) = C'(q),$$

that is, the slopes of the revenue and cost curves are equal. This is the same observation that we made in the previous example. In economic language,

> The maximum (or minimum) profit can occur where
>
> Marginal cost = Marginal revenue.

Of course, maximal or minimal profit does not *have* to occur where $MR = MC$; there are also the endpoints to consider.

Example 5   Find the quantity q which maximizes profit if the total revenue, $R(q)$, and total cost, $C(q)$, are given in dollars by

$$R(q) = 5q - 0.003q^2$$
$$C(q) = 300 + 1.1q,$$

where $0 \le q \le 800$ units. What production level gives the minimum profit?

Solution   We look for production levels that give marginal revenue = marginal cost:

$$MR = R'(q) = 5 - 0.006q$$
$$MC = C'(q) = 1.1.$$

So $5 - 0.006q = 1.1$, giving

$$q = 3.9/0.006 = 650 \text{ units.}$$

Does this value of q represent a local maximum or minimum of π? We can tell by looking at production levels of 649 units and 651 units. When $q = 649$ we have $MR = \$1.106$, which is greater than the (constant) marginal cost of $\$1.1$. This means that producing one more unit will bring in more revenue than its cost, so profit will increase. When $q = 651$, $MR = \$1.094$, which is *less* than MC, so it is not profitable to produce the 651st unit. We conclude that $q = 650$ is a local maximum for the profit function π. The profit earned by producing and selling this quantity is $\pi(650) = R(650) - C(650) = \$967.50$.

To check for global maxima we need to look at the endpoints. If $q = 0$, the only cost is $\$300$ (the fixed costs) and there is no revenue, so $\pi(0) = -\$300$. At the upper limit of $q = 800$, we have $\pi(800) = \$900$. Therefore, the maximum profit is at the production level of 650 units, where $MR = MC$. The minimum profit (a loss) occurs when $q = 0$ and there is no production at all.

Exercises and Problems for Section 4.4

Exercises

1. Total cost and revenue are approximated by the functions $C = 5000 + 2.4q$ and $R = 4q$, both in dollars. Identify the fixed cost, marginal cost per item, and the price at which this commodity is sold.

2. **(a)** Fixed costs are $3 million; variable costs are $0.4 million per item. Write a formula for total cost as a function of quantity, q.
   (b) The item in part (a) is sold for $0.5 million each. Write a formula for revenue as a function of q.
   (c) Write a formula for the profit function for this item.

3. The revenue from selling q items is $R(q) = 500q - q^2$, and the total cost is $C(q) = 150 + 10q$. Write a function that gives the total profit earned, and find the quantity which maximizes the profit.

4. Revenue is given by $R(q) = 450q$ and cost is given by $C(q) = 10{,}000 + 3q^2$. At what quantity is profit maximized? What is the total profit at this production level?

5. Figure 4.56 shows cost and revenue. For what production levels is the profit function positive? Negative? Estimate the production at which profit is maximized.

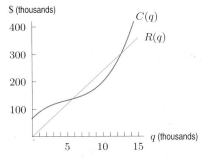

$ (thousands)

Figure 4.56

6. If $C'(500) = 75$ and $R'(500) = 100$, should the quantity produced be increased or decreased from $q = 500$ in order to increase profits?

7. When production is 2000, marginal revenue is $4 per unit and marginal cost is $3.25 per unit. Do you expect maximum profit to occur at a production level above or below 2000? Explain.

8. Figure 4.57 gives cost and revenue. What are fixed costs? What quantity maximizes profit, and what is the maximum profit earned?

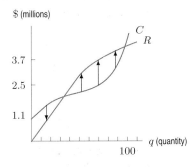

$ (millions)

Figure 4.57

9. Table 4.1 shows cost, $C(q)$, and revenue, $R(q)$,

   (a) At approximately what production level, q, is profit maximized? Explain your reasoning.
   (b) What is the price of the product?
   (c) What are the fixed costs?

Table 4.1

q	0	500	1000	1500	2000	2500	3000
$R(q)$	0	1500	3000	4500	6000	7500	9000
$C(q)$	3000	3800	4200	4500	4800	5500	7400

10. Table 4.2 shows marginal cost, MC, and marginal revenue, MR.

   (a) Use the marginal cost and marginal revenue at a production of $q = 5000$ to determine whether production should be increased or decreased from 5000.
   (b) Estimate the production level that maximizes profit.

Table 4.2

q	5000	6000	7000	8000	9000	10000
MR	60	58	56	55	54	53
MC	48	52	54	55	58	63

Problems

11. Using the cost and revenue graphs in Figure 4.58, sketch the following functions. Label the points q_1 and q_2.

   (a) Total profit   (b) Marginal cost
   (c) Marginal revenue

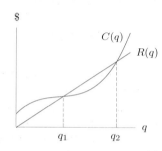

Figure 4.58

12. A manufacturer's cost of producing a product is given in Figure 4.58. The manufacturer can sell the product for a price p each (regardless of the quantity sold), so that the total revenue from selling a quantity q is $R(q) = pq$.

   (a) The difference $\pi(q) = R(q) - C(q)$ is the total profit. For which quantity q_0 is the profit a maximum? Mark your answer on a sketch of the graph.
   (b) What is the relationship between p and $C'(q_0)$? Explain your result both graphically and analytically. What does this mean in terms of economics? (Note that p is the slope of the line $R(q) = pq$. Note also that $\pi(q)$ has a maximum at $q = q_0$, so $\pi'(q_0) = 0$.)
   (c) Graph $C'(q)$ and p (as a horizontal line) on the same axes. Mark q_0 on the q-axis.

13. The marginal revenue and marginal cost for a certain item are graphed in Figure 4.59. Do the following quantities maximize profit for the company? Explain your answer.

   (a) $q = a$ (b) $q = b$

Figure 4.59

14. Let $C(q)$ be the total cost of producing a quantity q of a certain product. See Figure 4.60.

   (a) What is the meaning of $C(0)$?
   (b) Describe in words how the marginal cost changes as the quantity produced increases.
   (c) Explain the concavity of the graph (in terms of economics).
   (d) Explain the economic significance (in terms of marginal cost) of the point at which the concavity changes.
   (e) Do you expect the graph of $C(q)$ to look like this for all types of products?

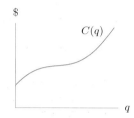

Figure 4.60

15. The total cost $C(q)$ of producing q goods is given by:

$$C(q) = 0.01q^3 - 0.6q^2 + 13q.$$

   (a) What is the fixed cost?
   (b) What is the maximum profit if each item is sold for $7? (Assume you sell everything you produce.)
   (c) Suppose exactly 34 goods are produced. They all sell when the price is $7 each, but for each $1 increase in price, 2 fewer goods are sold. Should the price be raised, and if so by how much?

16. (a) A cruise line offers a trip for $1000 per passenger. If at least 100 passengers sign up, the price is reduced for *all* the passengers by $5 for every additional passenger (beyond 100) who goes on the trip. The boat can accommodate 250 passengers. What number of passengers maximizes the cruise line's total revenue? What price does each passenger pay then?

   (b) The cost to the cruise line for q passengers is $40{,}000 + 200q$. What is the maximum profit that the cruise line can make on one trip? How many passengers must sign up for the maximum to be reached and what price will each pay?

17. A company manufactures only one product. The quantity, q, of this product produced per month depends on the amount of capital, K, invested (i.e., the number of machines the company owns, the size of its building, and so on) and the amount of labor, L, available each month. We assume that q can be expressed as a *Cobb-Douglas production function*:

$$q = cK^{\alpha}L^{\beta}$$

where c, α, β are positive constants, with $0 < \alpha < 1$ and $0 < \beta < 1$. In this problem we will see how the Russian government could use a Cobb-Douglas function to estimate how many people a newly privatized industry might employ. A company in such an industry has only a small amount of capital available to it and needs to use all of it, so K is fixed. Suppose L is measured in man-hours per month, and that each man-hour costs the company w rubles (a ruble is the unit of Russian currency). Suppose the company has no other costs besides labor, and that each unit of the good can be sold for a fixed price of p rubles. How many man-hours of labor per month should the company use in order to maximize its profit?

18. An agricultural worker in Uganda is planting clover to increase the number of bees making their home in the region. There are 100 bees in the region naturally, and for every acre put under clover, 20 more bees are found in the region.

  (a) Draw a graph of the total number, $N(x)$, of bees as a function of x, the number of acres devoted to clover.

  (b) Explain, both geometrically and algebraically, the shape of the graph of:

    (i) The marginal rate of increase of the number of bees with acres of clover, $N'(x)$.

    (ii) The average number of bees per acre of clover, $N(x)/x$.

19. If you invest x dollars in a certain project, your return is $R(x)$. You want to choose x to maximize your return per dollar invested,[4] which is

$$r(x) = \frac{R(x)}{x}.$$

  (a) The graph of $R(x)$ is in Figure 4.61, with $R(0) = 0$. Illustrate on the graph that the maximum value of $r(x)$ is reached at a point at which the line from the origin to the point is tangent to the graph of $R(x)$.

  (b) Also, the maximum of $r(x)$ occurs at a point at which the slope of the graph of $r(x)$ is zero. On the same axes as part (a), sketch $r(x)$. Illustrate that the maximum of $r(x)$ occurs where its slope is 0.

  (c) Show, by taking the derivative of the formula for $r(x)$, that the conditions in part (a) and (b) are equivalent: the x-value at which the line from the origin is tangent to the graph of R is the same as the x-value at which the graph of r has zero slope.

Figure 4.61

Problems 20–22 involve the *average cost* of manufacturing a quantity q of a good, which is defined to be

$$a(q) = \frac{C(q)}{q}.$$

20. Figure 4.62 shows the cost of production, $C(q)$, as a function of quantity produced, q.

  (a) For some q_0, sketch a line whose slope is the marginal cost, MC, at that point.

  (b) For the same q_0, explain why the average cost $a(q_0)$ can be represented by the slope of the line from that point on the curve to the origin.

  (c) Use the method of Example 3 on page 184 to explain why at the value of q which minimizes $a(q)$, the average and marginal costs are equal.

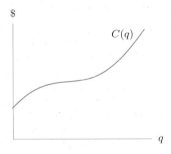

Figure 4.62

21. The average cost per item to produce q items is given by

$$a(q) = 0.01q^2 - 0.6q + 13, \quad \text{for} \quad q > 0.$$

  (a) What is the total cost, $C(q)$, of producing q goods?

  (b) What is the minimum marginal cost? What is the practical interpretation of this result?

  (c) At what production level is the average cost a minimum? What is the lowest average cost?

  (d) Compute the marginal cost at $q = 30$. How does this relate to your answer to part (c)? Explain this relationship both analytically and in words.

---

[4]From Peter D. Taylor, *Calculus: The Analysis of Functions* (Toronto: Wall & Emerson, 1992).

22. A reasonably realistic model of a firm's costs is given by the *short-run Cobb-Douglas cost curve*

$$C(q) = Kq^{1/a} + F,$$

where a is a positive constant, F is the fixed cost, and K measures the technology available to the firm.

(a) Show that C is concave down if $a > 1$.

(b) Assuming that $a < 1$ and that average cost is minimized when average cost equals marginal cost, find what value of q minimizes the average cost.

23. The production function $f(x)$ gives the number of units of an item that a manufacturing company can produce from x units of raw material. The company buys the raw material at price w dollars per unit and sells all it produces at a price of p dollars per unit. The quantity of raw material that maximizes profit is denoted by x^*.

(a) Do you expect the derivative $f'(x)$ to be positive or negative? Justify your answer.

(b) Explain why the formula $\pi(x) = pf(x) - wx$ gives the profit $\pi(x)$ that the company earns as a function of the quantity x of raw materials that it uses.

(c) Evaluate $f'(x^*)$.

(d) Assuming it is nonzero, is $f''(x^*)$ positive or negative?

(e) If the supplier of the raw materials is likely to change the price w, then it is appropriate to treat x^* as a function of w. Find a formula for the derivative dx^*/dw and decide whether it is positive or negative.

(f) If the price w goes up, should the manufacturing company buy more or less of the raw material?

4.5   OPTIMIZATION AND MODELING

Finding global maxima and minima is made much easier by having a formula for the function to be maximized or minimized. The process of translating a problem into a function whose formula we know is called *mathematical modeling*. By working through the examples that follow, you will get the flavor of some kinds of modeling.

Example 1   What are the dimensions of an aluminum can that holds 40 in^3 of juice and that uses the least material (i.e., aluminum)? Assume that the can is cylindrical, and is capped on both ends.

Solution   It is often a good idea to think about a problem in general terms before trying to solve it. Since we're trying to use as little material as possible, why not make the can very small, say, the size of a peanut? We can't, since the can must hold 40 in^3. If we make the can short, to try to use less material in the sides, we'll have to make it fat as well, so that it can hold 40 in^3. Saving aluminum in the sides might actually use more aluminum in the top and bottom in a short, fat can. See Figure 4.63(a).

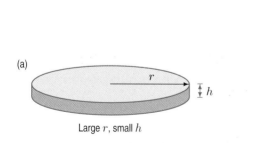

Figure 4.63: Various cylindrical-shaped cans

Table 4.3   *Material, M, used in can for various choices of radius, r, and height, h*

r (in)	h (in)	M (in^2)
0.2	318.31	400.25
1.0	12.73	86.27
2.0	3.18	65.09
3.0	1.41	83.13
4.0	0.80	120.64
10.0	0.13	636.49

If we try to save material by making the top and bottom small, the can has to be tall to accommodate the 40 in^3 of juice. So any savings we get by using a small top and bottom might be outweighed by the height of the sides. See Figure 4.63(b). This is, in fact, true, as Table 4.3 shows.

The table gives the amount of material used in the can for some choices of the radius, r, and the height, h. You can see that r and h change in opposite directions, and that more material is used at the extremes (very large or very small r and h) than in the middle. From the table it appears that the optimal radius for the can lies somewhere in $1.0 \leq r \leq 3.0$. If we consider the material used, M, as a function of the radius, r, a graph of this function looks like Figure 4.64. The graph shows that the global minimum we want is at a critical point.

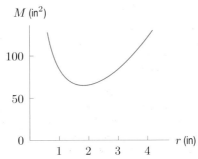

Figure 4.64: Total material used in can, M, as a function of radius, r

Both the table and the graph were obtained from a mathematical model, which in this case is a formula for the material used in making the can. Finding such a formula depends on knowing the geometry of a cylinder, in particular its area and volume. We have

$$M = \text{Material used in the can} = \text{Material in ends} + \text{Material in the side}$$

where

$$\text{Material in ends} = 2 \cdot \text{Area of a circle with radius } r = 2 \cdot \pi r^2,$$

$$\text{Material in the side} = \text{Surface area of cylinder with height } h \text{ and radius } r = 2\pi r h.$$

However, h is not independent of r: if r grows, h shrinks, and vice-versa. To find the relationship, we use the fact that the volume of the cylinder, $\pi r^2 h$, is equal to the constant 40 in^3:

$$\text{Volume of can} = \pi r^2 h = 40, \quad \text{giving} \quad h = \frac{40}{\pi r^2}.$$

This means

$$\text{Material in the side} = 2\pi r h = 2\pi r \frac{40}{\pi r^2} = \frac{80}{r}.$$

Thus we obtain the formula for the total material, M, used in a can of radius r:

$$M = 2\pi r^2 + \frac{80}{r}.$$

The domain of this function is all $r > 0$.

Now we use calculus to find the minimum of M. We look for critical points:

$$\frac{dM}{dr} = 4\pi r - \frac{80}{r^2} = 0 \quad \text{at a critical point,} \quad \text{so} \quad 4\pi r = \frac{80}{r^2}.$$

Therefore,

$$\pi r^3 = 20, \quad \text{so} \quad r = \left(\frac{20}{\pi}\right)^{1/3} \approx 1.85 \text{ inches,}$$

which agrees with our graph. We also have

$$h = \frac{40}{\pi r^2} \approx \frac{40}{\pi (1.85)^2} \approx 3.7 \text{ inches.}$$

Thus, the material used, $M = 2\pi(1.85)^2 + 80/1.85 = 64.7$ in^2.

Practical Tips for Modeling Optimization Problems

1. Make sure that you know what quantity or function is to be optimized.

2. If possible, make several sketches showing how the elements that vary are related. Label your sketches clearly by assigning variables to quantities which change.

3. Try to obtain a formula for the function to be optimized in terms of the variables that you identified in the previous step. If necessary, eliminate from this formula all but one variable. Identify the domain over which this variable varies.

4. Find the critical points and evaluate the function at these points and the endpoints (if relevant) to find the global maxima and minima.

The next example, another problem in geometry, illustrates this approach.

Example 2   Alaina wants to get to the bus stop as quickly as possible. The bus stop is across a grassy park, 2000 feet west and 600 feet north of her starting position. Alaina can walk west along the edge of the park on the sidewalk at a speed of 6 ft/sec. She can also travel through the grass in the park, but only at a rate of 4 ft/sec. What path will get her to the bus stop the fastest?

Solution

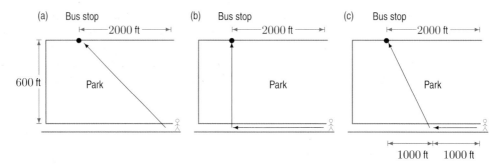

Figure 4.65: Three possible paths to the bus stop

We might first think that she should take a path that is the shortest distance. Unfortunately, the path that follows the shortest distance to the bus stop is entirely in the park, where her speed is slow. (See Figure 4.65(a).) That distance is $\sqrt{2000^2 + 600^2} \approx 2100$ feet, which takes her about 525 seconds to traverse. She could instead walk quickly the entire 2000 feet along the sidewalk, which leaves her just the 600-foot northward journey through the park. (See Figure 4.65(b).) This method would take $2000/6 + 600/4 \approx 483$ seconds total walking time.

But can she do even better? Perhaps another combination of sidewalk and park gives a shorter travel time. For example, what is the travel time if she walks 1000 feet west along the sidewalk and the rest of the way through the park? (See Figure 4.65(c).) The answer is about 458 seconds.

We make a model for this problem. We label the distance that Alaina walks west along the sidewalk x and the distance she walks through the park y, as in Figure 4.66. Then the total time, t, is

$$t = t_{\text{sidewalk}} + t_{\text{park}}.$$

Since

$$\text{Time} = \text{Distance}/\text{Speed},$$

and she can walk 6 ft/sec on the sidewalk and 4 ft/sec in the park, we have that

$$t = \frac{x}{6} + \frac{y}{4}.$$

Now, by the Pythagorean Theorem, $y = \sqrt{(2000 - x)^2 + 600^2}$. Therefore

$$t = \frac{x}{6} + \frac{\sqrt{(2000 - x)^2 + 600^2}}{4}.$$

We can find the critical points of this function analytically. (See Exercise 10 on page 203.) Alternatively, we can graph the function on a calculator and estimate the critical point, which is $x \approx 1463$ feet. This gives a minimum total time of about 445 seconds.

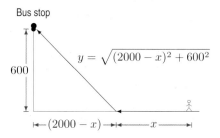

Figure 4.66: Modeling time to bus stop

Example 3   One hallway which is 4 feet wide meets another hallway which is 8 feet wide in a right angle. (See Figure 4.67.) What is the length of the longest ladder which can be carried horizontally around the corner?

Solution   We imagine the ladder being carried on its side and ignore its width. To allow the longest ladder possible we carry the ladder around the corner so that it just touches both walls (at A and C) and just touches the corner at B. Let's draw some lines that do this. (See Figure 4.67.) The length of the line $\overline{ABC}$ decreases as the corner is turned, then increases again. The minimum such length would be the length of the longest ladder that could make it around the corner. A smaller ladder would still work (it would not touch A, B, and C simultaneously), but a larger one would not fit.

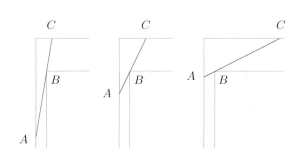

Figure 4.67: Various ladders that touch both walls and corner

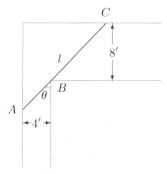

Figure 4.68: Ladder and hallway

We want the *smallest* length of the line $\overline{ABC}$. We express the length, l, as a function of θ, the angle between the line and the wall of the narrow hall. (See Figure 4.68.) We have

$$l = \overline{AB} + \overline{BC}.$$

Now $\overline{AB} = 4/\sin\theta$ and $\overline{BC} = 8/\cos\theta$, so

$$l = \frac{4}{\sin\theta} + \frac{8}{\cos\theta}.$$

Here the domain of θ is $0 < \theta < \pi/2$.

Next we differentiate:

$$\frac{dl}{d\theta} = -\frac{4}{(\sin\theta)^2}(\cos\theta) - \frac{8}{(\cos\theta)^2}(-\sin\theta).$$

To minimize l, we solve $dl/d\theta = 0$:

$$-4\frac{\cos\theta}{(\sin\theta)^2} + 8\frac{\sin\theta}{(\cos\theta)^2} = 0,$$

so

$$2(\sin \theta)^3 = (\cos \theta)^3, \quad \text{or} \quad \frac{(\sin \theta)^3}{(\cos \theta)^3} = \frac{1}{2}.$$

This means

$$\tan \theta = \sqrt[3]{0.5} \approx 0.79, \quad \text{so} \quad \theta \approx 0.67 \text{ radians.}$$

Thus $\theta \approx 0.67$ is a critical point, and an investigation of what happens for θ near 0 and θ near $\pi/2$ shows that l has a global minimum at $\theta = 0.67$. The minimum value of l is therefore

$$l = \frac{4}{\sin(0.67)} + \frac{8}{\cos(0.67)} \approx 16.65 \text{ feet,}$$

so the ladder can be at most 16.65 feet long and still make it around the corner.

Exercises and Problems for Section 4.5

Exercises

1. The bending moment M of a beam, supported at one end, at a distance x from the support is given by

$$M = \tfrac{1}{2}wLx - \tfrac{1}{2}wx^2,$$

where L is the length of the beam, and w is the uniform load per unit length. Find the point on the beam where the moment is greatest.

2. The potential energy, U, of a particle moving along the x-axis is given by

$$U = b\left(\frac{a^2}{x^2} - \frac{a}{x}\right),$$

where a and b are positive constants and $x > 0$. What value of x minimizes the potential energy?

3. An electric current, I, in amps, is given by

$$I = \cos(wt) + \sqrt{3}\sin(wt),$$

where $w \neq 0$ is a constant. What are the maximum and minimum values of I?

4. A smokestack deposits soot on the ground with a concentration inversely proportional to the square of the distance from the stack. With two smokestacks 20 miles apart, the concentration of the combined deposits on the line joining them, at a distance x from one stack, is given by

$$S = \frac{k_1}{x^2} + \frac{k_2}{(20 - x)^2}$$

where k_1 and k_2 are positive constants which depend on the quantity of smoke each stack is emitting. If $k_1 = 7k_2$, find the point on the line joining the stacks where the concentration of the deposit is a minimum.

5. In a chemical reaction, substance A combines with substance B to form substance Y. At the start of the reaction, the quantity of A present is a grams, and the quantity of B present is b grams. Assume $a < b$. At time t seconds

after the start of the reaction, the quantity of Y present is y grams. For certain types of reactions, the rate of the reaction, in grams/sec, is given by

$$\text{Rate} = k(a - y)(b - y), \quad k \text{ is a positive constant.}$$

(a) For what values of y is the rate nonnegative? Graph the rate against y.
(b) Use your graph to find the value of y at which the rate of the reaction is fastest.

6. A wave of wavelength λ traveling in deep water has speed, v, given by

$$v = k\sqrt{\frac{\lambda}{c} + \frac{c}{\lambda}},$$

where c and k are positive constants. As λ varies, does such a wave have a maximum or minimum velocity? If so, what is it? Explain.

7. The efficiency of a screw, E, is given by

$$E = \frac{(\theta - \mu\theta^2)}{\mu + \theta}, \quad \theta > 0,$$

where θ is the angle of pitch of the thread and μ is the coefficient of friction of the material, a (positive) constant. What value of θ maximizes E?

8. A woman pulls a sled which, together with its load, has a mass of m kg. If her arm makes an angle of θ with her body (assumed vertical) and the coefficient of friction (a positive constant) is μ, the least force, F, she must exert to move the sled is given by

$$F = \frac{mg\mu}{\sin \theta + \mu \cos \theta}.$$

If $\mu = 0.15$, find the maximum and minimum values of F for $0 \leq \theta \leq \pi/2$. Give answers as multiples of mg.

9. A circular ring of wire of radius r_0 lies in a plane perpendicular to the x-axis and is centered at the origin. The ring has a positive electric charge spread uniformly over it. The electric field in the x-direction, E, at the point x on the axis is given by

$$E = \frac{kx}{(x^2 + r_0^2)^{3/2}} \quad \text{for} \quad k > 0.$$

At what point on the x-axis is the field greatest? Least?

10. Find analytically the exact critical point of the function which represents the time, t, to walk to the bus stop in Example 2. Recall that t is given by

$$t = \frac{x}{6} + \frac{\sqrt{(2000 - x)^2 + 600^2}}{4}.$$

Problems

11. Figure 4.69 shows the curves $y = \sqrt{x}$, $x = 9$, $y = 0$, and a rectangle with its sides parallel to the axes and its left end at $x = a$. Find the dimensions of the rectangle having the maximum possible area.

$y = \sqrt{x}$

Figure 4.69

12. The hypotenuse of a right triangle has one end at the origin and one end on the curve $y = x^2 e^{-3x}$, with $x \geq 0$. One of the other two sides is on the x-axis, the other side is parallel to the y-axis. Find the maximum area of such a triangle. At what x-value does it occur?

13. A rectangle has one side on the x-axis and two vertices on the curve

$$y = \frac{1}{1 + x^2}.$$

Find the vertices of the rectangle with maximum area.

14. A rectangle has one side on the x-axis, one side on the y-axis, one vertex at the origin and one on the curve $y = e^{-2x}$ for $x \geq 0$. Find the

(a) Maximum area   **(b)** Minimum perimeter

15. A hemisphere of radius 1 sits on a horizontal plane. A cylinder stands with its axis vertical, the center of its base at the center of the sphere, and its top circular rim touching the hemisphere. Find the radius and height of the cylinder of maximum volume.

16. A closed box has a fixed surface area A and a square base with side x.

(a) Find a formula for its volume, V, as a function of x.
(b) Sketch a graph of V against x.
(c) Find the maximum value of V.

17. If you have 100 feet of fencing and want to enclose a rectangular area up against a long, straight wall, what is the largest area you can enclose?

18. A rectangular beam is cut from a cylindrical log of radius 30 cm. The strength of a beam of width w and height h is proportional to wh^2. (See Figure 4.70.) Find the width and height of the beam of maximum strength.

Figure 4.70

19. A landscape architect plans to enclose a 3000 square foot rectangular region in a botanical garden. She will use shrubs costing $25 per foot along three sides and fencing costing $10 per foot along the fourth side. Find the minimum total cost.

20. A rectangular swimming pool is to be built with an area of 1800 square feet. The owner wants 5-foot wide decks along either side and 10-foot wide decks at the two ends. Find the dimensions of the smallest piece of property on which the pool can be built satisfying these conditions.

21. A square-bottomed box with no top has a fixed volume, V. What dimensions minimize the surface area?

22. A light is suspended at a height h above the floor. (See Figure 4.71.) The illumination at the point P is inversely proportional to the square of the distance from the point P to the light and directly proportional to the cosine of the angle θ. How far from the floor should the light be to maximize the illumination at the point P?

Figure 4.71

23. Which point on the parabola $y = x^2$ is nearest to $(1, 0)$? Find the coordinates to two decimals. [Hint: Minimize the square of the distance—this avoids square roots.]

24. Find the coordinates of the point on the parabola $y = x^2$ which is closest to the point $(3,0)$.

25. The cross-section of a tunnel is a rectangle of height h surmounted by a semicircular roof section of radius r (See Figure 4.72 on page 204). If the cross-sectional area is A, determine the dimensions of the cross section which minimize the perimeter.

Figure 4.72

26. Of all rectangles with given area, A, which has the shortest diagonals?

27. You run a small furniture business. You sign a deal with a customer to deliver up to 400 chairs, the exact number to be determined by the customer later. The price will be $90 per chair up to 300 chairs, and above 300, the price will be reduced by $0.25 per chair (on the whole order) for every additional chair over 300 ordered. What are the largest and smallest revenues your company can make under this deal?

28. The cost of fuel to propel a boat through the water (in dollars per hour) is proportional to the cube of the speed. A certain ferry boat uses $100 worth of fuel per hour when cruising at 10 miles per hour. Apart from fuel, the cost of running this ferry (labor, maintenance, and so on) is $675 per hour. At what speed should it travel so as to minimize the cost *per mile* traveled?

29. (a) For which positive number x is $x^{1/x}$ largest? Justify your answer.

    [Hint: You may want to write $x^{1/x} = e^{\ln(x^{1/x})}$.]

    (b) For which positive integer n is $n^{1/n}$ largest? Justify your answer.

    (c) Use your answer to parts (a) and (b) to decide which is larger: $3^{1/3}$ or $\pi^{1/\pi}$.

30. The *arithmetic mean* of two numbers a and b is defined as $(a+b)/2$; the *geometric mean* of two positive numbers a and b is defined as $\sqrt{ab}$.

    (a) For two positive numbers, which of the two means is larger? Justify your answer.
    [Hint: Define $f(x) = (a+x)/2 - \sqrt{ax}$ for fixed a.]

    (b) For three positive numbers a, b, c, the arithmetic and geometric mean are $(a+b+c)/3$ and $\sqrt[3]{abc}$, respectively. Which of the two means of three numbers is larger? [Hint: Redefine $f(x)$ for fixed a *and* b.]

31. A bird such as a starling feeds worms to its young. To collect worms, the bird flies to a site where worms are to be found, picks up several in its beak, and flies back to its nest. The *loading curve* in Figure 4.73 shows how the number of worms (the load) a starling collects depends on the time it has been searching for them.[5] The curve is concave down because the bird can pick up worms more efficiently when its beak is empty; when its beak is partly full, the bird becomes much less efficient. The traveling time (from nest to site and back) is represented by the distance PO in Figure 4.73. The bird wants to maximize the rate at which it brings worms to the nest, where

$$\text{Rate worms arrive} = \frac{\text{Load}}{\text{Traveling time} + \text{Searching time}}$$

    (a) Draw a line in Figure 4.73 whose slope is this rate.
    (b) Using the graph, estimate the load which maximizes this rate.
    (c) If the traveling time is increased, does the optimal load increase or decrease? Why?

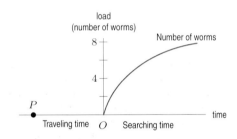

Figure 4.73

32. On the same side of a straight river are two towns, and the townspeople want to build a pumping station, S. See Figure 4.74. The pumping station is to be at the river's edge with pipes extending straight to the two towns. Where should the pumping station be located to minimize the total length of pipe?

Figure 4.74

[5]Alex Kacelnik (1984). Reported by J. R. Krebs and N. B. Davis, *An Introduction to Behavioural Ecology* (Oxford: Blackwell, 1987).

33. A pigeon is released from a boat (point B in Figure 4.75) floating on a lake. Because of falling air over the cool water, the energy required to fly one meter over the lake is twice the corresponding energy e required for flying over the bank ($e = 3$ joule/meter). To minimize the energy required to fly from B to the loft, L, the pigeon heads to a point P on the bank and then flies along the bank to L. The distance $\overline{AL}$ is 2000 m, and $\overline{AB}$ is 500 m.

   (a) Express the energy required to fly from B to L via P as a function of the angle θ (the angle BPA).

   (b) What is the optimal angle θ?

   (c) Does your answer change if $\overline{AL}$, $\overline{AB}$, and e have different numerical values?

Figure 4.75

34. To get the best view of the Statue of Liberty in Figure 4.76, you should be at the position where θ is a maximum. If the statue stands 92 meters high, including the pedestal, which is 46 meters high, how far from the base should you be? [Hint: Find a formula for θ in terms of your distance from the base. Use this function to maximize θ, noting that $0 \leq \theta \leq \pi/2$.]

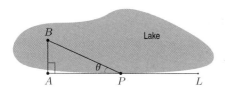

Figure 4.76

35. A light ray starts at the origin and is reflected off a mirror along the line $y = 1$ to the point $(2, 0)$. See Figure 4.77. Fermat's Principle[6] says that light's path minimizes the time of travel. The speed of light is a constant.

   (a) Using Fermat's principle, find the optimal position of P.

   (b) Using your answer to part (a), derive the Law of Reflection, that $\theta_1 = \theta_2$.

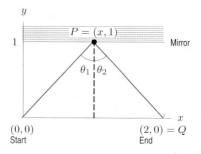

Figure 4.77

36. When a ray of light travels from one medium to another (for example, from air to water), it changes direction. This phenomenon is known as refraction. In Figure 4.78, light is traveling from A to B. The amount of refraction depends on the velocities, v_1 and v_2, of light in the two media and on Fermat's Principle which states that the light's time of travel, T, from A to B, is a minimum.

   (a) Find an expression for T in terms of x and the constants a, b, v_1, v_2, and c.

   (b) Show that if R is chosen so that the time of travel is minimized, then

$$\frac{\sin \theta_1}{\sin \theta_2} = \frac{v_1}{v_2}.$$

This result is known as Snell's Law and the ratio v_1/v_2 is called the index of refraction of the second medium with respect to the first medium.

Figure 4.78

37. Show that when the value of x in Figure 4.78 is chosen according to Snell's Law, the time taken by the light ray is a minimum.

In many applications, we want to maximize or minimize some quantity subject to a condition. Such constrained optimization problems are solved using Lagrange multipliers in multivariable calculus; Problems 38–40 show an alternate method.[7]

---

[6]See, for example, D. Halliday, R. Resnik, K. Kane, *Physics* Vol 2, 4th edn, p. 909 (New York: Wiley, 1992).

[7]Kelly Black "Putting Constraints in Optimization for First-Year Calculus Students", p. 310-312 *Siam Review* Vol. 39 No. 2, June 1997.

38. Minimize $x^2 + y^2$ while satisfying $x + y = 4$ using the following steps.

   (a) Graph $x + y = 4$. On the same axes, graph $x^2 + y^2 = 1$, $x^2 + y^2 = 4$, $x^2 + y^2 = 9$.

   (b) Explain why the minimum value of $x^2 + y^2$ on $x + y = 4$ occurs at the point at which a graph of $x^2 + y^2 = $ Constant is tangent to the line $x + y = 4$.

   (c) Using your answer to part (b) and implicit differentiation to find the slope of the circle, find the minimum value of $x^2 + y^2$ such that $x + y = 4$.

39. The quantity Q of an item which can be produced from quantities x and y of two raw materials is given by $Q = 10xy$ at a cost of $C = x + 2y$ thousand dollars. If there is a budget of $10 thousand for raw materials, find the maximum production using the following steps.

   (a) Graph $x + 2y = 10$ in the first quadrant. On the same axes, graph $Q = 10xy = 100$, $Q = 10xy = 200$, and $Q = 10xy = 300$.

   (b) Explain why the maximum production occurs at a point at which a production curve is tangent to the cost line $C = 10$.

   (c) Using your answer to part (b) and implicit differentiation to find the slope of the curve, find the maximum production under this budget.

40. With quantities x and y of two raw materials available, $Q = x^{1/2}y^{1/2}$ thousand items can be produced at a cost of $C = 2x + y$ thousand dollars. Using the following steps, find the minimum cost to produce 1 thousand items.

   (a) Graph $x^{1/2}y^{1/2} = 1$. On the same axes, graph $2x + y = 2$, $2x + y = 3$, and $2x + y = 4$.

   (b) Explain why the minimum cost occurs at a point at which a cost line is tangent to the production curve $Q = 1$.

   (c) Using your answer to part (b) and implicit differentiation to find the slope of the curve, find the minimum cost to meet this production level.

4.6 RATES AND RELATED RATES

Derivatives represent rates of change. In this section, we see how to calculate rates in a variety of situations.

Example 1   A spherical snowball is melting. Its radius decreases at a constant rate of 2 cm per minute from an initial value of 70 cm. How fast is the volume decreasing half an hour later?

Solution   The radius, r, starts at 70 cm and decreases at 2 cm/min. At time t minutes since the start,

$$r = 70 - 2t \text{ cm}.$$

The volume of the snowball is given by

$$V = \frac{4}{3}\pi r^3 = \frac{4}{3}\pi(70 - 2t)^3 \text{ cm}^3.$$

The rate at which the volume is changing at time t is

$$\frac{dV}{dt} = \frac{4}{3}\pi \cdot 3(70 - 2t)^2(-2) = -8\pi(70 - 2t)^2 \text{ cm}^3/\text{min}.$$

The volume is measured in cm^3, and time is in minutes, so after half an hour $t = 30$, and

$$\left.\frac{dV}{dt}\right|_{t=30} = -8\pi(70 - 2 \cdot 30)^2 = -800\pi \text{ cm}^3/\text{min}.$$

Thus, the rate at which the volume is increasing is $-800\pi \approx -2500 \text{ cm}^3/\text{min}$; the rate at which the volume is decreasing is about $2500 \text{ cm}^3/\text{min}$.

Example 2   A skydiver of mass m jumps from a plane at time $t = 0$. Under certain assumptions, the distance, $s(t)$, he has fallen in time t is given by

$$s(t) = \frac{m^2g}{k^2}\left(\frac{kt}{m} + e^{-kt/m} - 1\right) \quad \text{for some positive constant } k.$$

   (a) Find $s'(0)$ and $s''(0)$ and interpret in terms of the skydiver.

   (b) Relate the units of $s'(t)$ and $s''(t)$ to the units of t and $s(t)$.

Solution   (a) Differentiating using the chain rule gives

$$s'(t) = \frac{m^2 g}{k^2} \left(\frac{k}{m} + e^{-kt/m} \left(-\frac{k}{m} \right) \right) = \frac{mg}{k} \left(1 - e^{-kt/m} \right)$$

$$s''(t) = \frac{mg}{k} (-e^{kt/m}) \left(-\frac{k}{m} \right) = g e^{-kt/m}.$$

Since $e^{-k \cdot 0/m} = 1$, evaluating at $t = 0$ gives

$$s'(0) = \frac{mg}{k} (1 - 1) = 0 \quad \text{and} \quad s''(0) = g.$$

The first derivative of distance is velocity, so the fact that $s'(0) = 0$ tells us that the skydiver starts with zero velocity. The second derivative of distance is acceleration, so the fact that $s''(0) = g$ tells us that the skydiver's initial acceleration is g, the acceleration due to gravity.

(b) The units of velocity, $s'(t)$, and acceleration, $s''(t)$, are given by

Units of $s'(t)$ are $\dfrac{\text{Units of } s(t)}{\text{Units of } t} = \dfrac{\text{Units of distance}}{\text{Units of time}}$;   for example, meters/sec.

Units of $s''(t)$ are $\dfrac{\text{Units of } s'(t)}{\text{Units of } t} = \dfrac{\text{Units of distance}}{(\text{Units of time})^2}$;   for example, meters/sec^2.

Related Rates

In Example 1, the radius of the snowball decreased at a constant rate. A more realistic scenario is for the radius to decrease at different rates at different times. Then, we may not be able to write a formula for V as a function of t. However, we may still be able to calculate dV/dt, as in the following example.

Example 3   A spherical snowball melts in such a way that the instant at which its radius is 20 cm, its radius is decreasing at 3 cm/min. At what rate is the volume of the ball of snow changing at that instant?

Solution   Since the snowball is spherical, we again have that

$$V = \frac{4}{3} \pi r^3.$$

We can no longer write a formula for r in terms of t, but we know that

$$\frac{dr}{dt} = -3 \quad \text{when} \quad r = 20.$$

We want to know dV/dt when $r = 20$. Think of r as an (unknown) function of t and differentiate the expression for V with respect to t using the chain rule:

$$\frac{dV}{dt} = \frac{4}{3} \pi \cdot 3r^2 \frac{dr}{dt} = 4\pi r^2 \frac{dr}{dt}.$$

At the instant at which $r = 20$ and $dr/dt = -3$, we have

$$\frac{dV}{dt} = 4\pi \cdot 20^2 \cdot (-3) = -4800\pi \text{ cm}^3/\text{min}.$$

Notice that we have sidestepped the problem of not knowing r as a function of t by calculating the derivatives only at the moment we are interested in.

Example 4   Figure 4.79 shows the fuel consumption, g, in miles per gallon, of a car traveling at v mph. At one moment, the car was going 70 mph and its deceleration was 8000 miles/hour^2. How fast was the fuel consumption changing at that moment? Include units.

Figure 4.79: Fuel consumption versus velocity

Solution   Acceleration is rate of change of velocity, dv/dt, and we are told that the deceleration is 8000 miles/hour^2, so we know $dv/dt = -8000$ when $v = 70$. We want dg/dt. The chain rule gives

$$\frac{dg}{dt} = \frac{dg}{dv} \cdot \frac{dv}{dt}.$$

The value of dg/dv is the slope of the curve in Figure 4.79 at $v = 70$. Since the points $(30, 40)$ and $(100, 20)$ lie approximately on the tangent to the curve at $v = 70$, we can estimate the derivative

$$\frac{dg}{dv} \approx \frac{20 - 40}{100 - 30} = -\frac{2}{7}.$$

Thus,

$$\frac{dg}{dv} \approx \left(-\frac{2}{7}\right) \cdot (-8000) \approx 2300 \text{ mpg/hr}.$$

Since we approximated dg/dv, we can only get a rough estimate for dg/dt.

A famous problem involves the rate at which a ladder slips down a wall as the foot of the ladder moves.

Example 5   (a) A 3-meter ladder stands against a high wall. The foot of the ladder moves outward at a speed of 0.1 meter/sec when the foot is 1 meter from the wall. At that moment, how fast is the top of the ladder falling? What if the foot had been 2 meters from the wall?
    (b) If the foot of the ladder moves out at a constant speed, how does the speed at which the top falls change as the foot gets farther out?

Solution   (a) Let the foot be x meters from the base of the wall and let the top be y meters from the base. See Figure 4.80. Then, since the ladder is 3 meters long, by Pythagoras' Theorem,

$$x^2 + y^2 = 3^2 = 9.$$

Thinking of both x and y as functions of t, we differentiate this implicit relation, giving

$$2x\frac{dx}{dt} + 2y\frac{dy}{dt} = 0.$$

We are interested in the moment at which $dx/dt = 0.1$ and $x = 1$. We want to know dy/dt, so we solve, giving

$$\frac{dy}{dt} = -\frac{x}{y}\frac{dx}{dt}.$$

Figure 4.80: Side view of ladder standing against wall

When the foot of the ladder is 1 meter from the wall, $x = 1$ and $y = \sqrt{9 - 1^2} = \sqrt{8}$, so

$$\frac{dy}{dt} = -\frac{1}{\sqrt{8}}0.1 = -0.035 \text{ meter/sec.}$$

Thus, the top falls at 0.0035 meter/sec.

When the foot is 2 meters from the wall, $x = 2$ and $y = \sqrt{9 - 2^2} = \sqrt{5}$, so

$$\frac{dy}{dt} = -\frac{2}{\sqrt{5}}0.1 = -0.089 \text{ meter/sec.}$$

Thus, the top falls at 0.089 meter/sec. Notice that the top falls faster when the base of the ladder is farther from the wall.

(b) As the foot of the ladder moves out, x increases and y decreases. Looking at the expression

$$\frac{dy}{dt} = -\frac{x}{y}\frac{dx}{dt},$$

we see that if dx/dt is constant, the magnitude of dy/dt increases as the foot gets farther out.

Example 6   An airplane, flying at 450 km/hr at a constant altitude of 5 km, is approaching a camera mounted on the ground. Let θ be the angle of elevation above the ground at which the camera is pointed. See Figure 4.81. When $\theta = \pi/3$, how fast does the camera have to rotate in order to keep the plane in view?

Solution   Suppose the plane is vertically above the point B. Let x be the distance between B and C. The fact that the plane is moving toward C at 450 km/hr means that x is decreasing and $dx/dt = -450$ km/hr. From Figure 4.81, we see that $\tan\theta = 5/x$.

Differentiating $\tan\theta = 5/x$ with respect to t and using the chain rule gives

$$\frac{1}{\cos^2\theta}\frac{d\theta}{dt} = -5x^{-2}\frac{dx}{dt}.$$

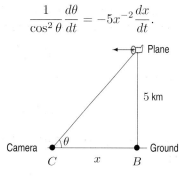

Figure 4.81: Plane approaching a camera at C (side view)

We want to calculate $d\theta/dt$ when $\theta = \pi/3$. At that moment, $\cos\theta = 1/2$ and $\tan\theta = \sqrt{3}$, so $x = 5/\sqrt{3}$. Substituting gives

$$\frac{1}{(1/2)^2}\frac{d\theta}{dt} = -5\left(\frac{5}{\sqrt{3}}\right)^{-2}\cdot(-450)$$

$$\frac{d\theta}{dt} = 67.500 \text{ radians/hour.}$$

This answer tells us that the camera must turn at roughly 1 radian per minute if it is to remain pointed at the plane.

Exercises and Problems for Section 4.6

Exercises

1. According to the US Census, the world population P, in billions, is approximately

   $$P = 6.342e^{0.011t},$$

   where t is in years since January 1, 2004. At what rate was the world's population increasing on that date? Give your answer in millions of people per year.

2. With time, t, in minutes, the temperature, H, in degrees Celsius, of a bottle of water put in the refrigerator at $t = 0$ is given by

   $$H = 4 + 16e^{-0.02t}.$$

   How fast is the water cooling initially? After 10 minutes? Give units.

3. The power, P, dissipated when a 9-volt battery is put across a resistance of R ohms is given by

   $$P = \frac{81}{R}.$$

   What is the rate of change of power with respect to resistance?

4. With length, l, in meters, the period T, in seconds, of a pendulum is given by

   $$T = 2\pi\sqrt{\frac{l}{9.8}}.$$

   (a) How fast does the period increase as l increases?
   (b) Does this rate of change increase or decrease as l increases?

5. At time t, in hours, a lake is covered with ice of thickness y cm, where $y = 0.2t^{1.5}$.

   (a) How fast is the ice forming when $t = 1$? When $t = 2$? Give units.
   (b) If ice forms for $0 \leq t \leq 3$, at what time in this interval is the ice thickest? At what time is the ice forming fastest?

6. A dose, D, of a drug causes a temperature change, T, in a patient. For C a positive constant, T is given by

   $$T = \left(\frac{C}{2} - \frac{D}{3}\right)D^3.$$

   (a) What is the rate of change of temperature change with respect to dose?
   (b) For what doses does the temperature change increase as the dose increases?

7. For positive constants k and g, the velocity, v, of a particle of mass m at time t is given by

   $$v = \frac{mg}{k}\left(1 - e^{-kt/m}\right).$$

   At what rate is the velocity is changing at time 0? At $t = 1$? What do your answers tell you about the motion?

8. The average cost per item, C, in dollars, of manufacturing a quantity q of cell phones is given by

   $$C = \frac{a}{q} + b \qquad \text{where } a, b \text{ are positive constants.}$$

   (a) Find the rate of change of C as q increases. What are its units?
   (b) If production increases at a rate of 100 cell phones per week, how fast is the average cost changing? Is the average cost increasing or decreasing?

9. For positive constants A and B, the force, F, between two atoms in a molecule at a distance r apart is given by

   $$F = -\frac{A}{r^2} + \frac{B}{r^3}.$$

   (a) How fast does force change as r increases? What type of units does it have?
   (b) If at some time t the distance is changing at a rate k, at what rate is the force changing with time? What type of units does this rate of change have?

10. An item costs $500 at time $t = 0$ and costs $P in year t. When inflation is $r\%$ per year, the price is given by

$$P = 500e^{rt/100}.$$

  (a) If r is a constant, at what rate is the price rising (in dollars per year)

   (i) Initially?   **(ii)** After 2 years?

  (b) Now suppose that r is increasing by 0.3 per year when $r = 4$ and $t = 2$. At what rate (dollars per year) is the price increasing at that time?

11. A voltage V across a resistance R generates a current $I = V/R$. If a constant voltage of 9 volts is put across a resistance that is increasing at a rate of 0.2 ohms per second when the resistance is 5 ohms, at what rate is the current changing?

12. The gravitational force, F, on a rocket at a distance, r, from the center of the earth is given by

$$F = \frac{k}{r^2},$$

where $k = 10^{13}$ newton $\cdot$ km^2. When the rocket is 10^4 km from the center of the earth, it is moving away at 0.2 km/sec. How fast is the gravitational force changing at that moment? Give units. (A newton is a unit of force.)

Problems

13. Point P moves around the unit circle.[8] (See Figure 4.82.) The angle θ, in radians, changes with time as shown in Figure 4.83.

  (a) Estimate the coordinates of P when $t = 2$.
  (b) When $t = 2$, approximately how fast is the point P moving in the x-direction? In the y-direction?

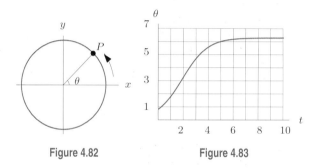

Figure 4.82   **Figure 4.83**

14. Figure 4.84 shows the number of gallons, G, of gasoline used on a trip of M miles.

  (a) The function f is linear on each of the intervals $0 < M < 70$ and $70 < M < 100$. What is the slope of these lines? What are the units of these slopes?
  (b) What is gas consumption (in miles per gallon) during the first 70 miles of this trip? During the next 30 miles?
  (c) Figure 4.85 shows distance traveled, M (in miles), as a function of time t, in hours since the start of the trip. Describe this trip in words. Give a possible explanation for what happens one hour into the trip. What do your answers to part (b) tell you about the trip?
  (d) If we let $G = k(t) = f(h(t))$, estimate $k(0.5)$ and interpret your answer in terms of the trip.
  (e) Find $k'(0.5)$ and $k'(1.5)$. Give units and interpret your answers.

[8] Based on an idea from Caspar Curjel.

Figure 4.84

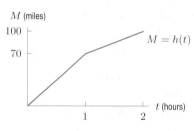

Figure 4.85:

15. Coroners estimate time of death using the rule of thumb that a body cools about 2°F during the first hour after death and about 1°F for each additional hour. Assuming an air temperature of 68°F and a living body temperature of 98.6°F, the temperature $T(t)$ in °F of a body at a time t hours since death is given by

$$T(t) = 68 + 30.6e^{-kt}.$$

  (a) For what value of k will the body cool by 2°F in the first hour?
  (b) Using the value of k found in part (a), after how many hours will the temperature of the body be decreasing at a rate of 1°F per hour?
  (c) Using the value of k found in part (a), show that, 24 hours after death, the coroner's rule of thumb gives approximately the same temperature as the formula.

16. A certain quantity of gas occupies a volume of $20\,\text{cm}^3$ at a pressure of 1 atmosphere. The gas expands without the addition of heat, so, for some constant k, its pressure, P, and volume, V, satisfy the relation

$$PV^{1.4} = k.$$

(a) Find the rate of change of pressure with volume. Give units.

(b) The volume is increasing at $2\,\text{cm}^3/\text{min}$ when the volume is $30\,\text{cm}^3$. At that moment, is the pressure increasing or decreasing? How fast? Give units.

17. (a) A hemispherical bowl of radius 10 cm contains water to a depth of h cm. Find the radius of the surface of the water as a function of h.

(b) The water level drops at a rate of 0.1 cm per hour. At what rate is the radius of the water decreasing when the depth is 5 cm?

18. A cone-shaped coffee filter of radius 6 cm and depth 10 cm contains water, which drips out through a hole at the bottom at a constant rate of $1.5\,\text{cm}^3$ per second.

(a) If the filter starts out full, how long does it take to empty?

(b) Find the volume of water in the filter when the depth of the water is h cm.

(c) How fast is the water level falling when the depth is 8 cm?

19. A spherical snowball is melting. Its radius is decreasing at 0.2 cm per hour when the radius is 15 cm. How fast is its volume decreasing at that time?

20. A ruptured oil tanker causes a circular oil slick on the surface of the ocean. When its radius is 150 meters, the radius of the slick is expanding by 0.1 meter/minute and its thickness is 0.02 meter. At that moment:

(a) How fast is the area of the slick expanding?

(b) The circular slick has the same thickness everywhere, and the volume of oil spilled remains fixed. How fast is the thickness of the slick decreasing?

21. A potter forms a piece of clay into a cylinder. As he rolls it, the length, L, of the cylinder increases and the radius, r, decreases. If the length of the cylinder is increasing at 0.1 cm per second, find the rate at which the radius is changing when the radius is 1 cm and the length is 5 cm.

22. The London Eye is a large ferris wheel that has diameter 135 meters and revolves continuously. Passengers enter the cabins at the bottom of the wheel and complete one revolution in 20 minutes. One minute into the ride a passenger is rising at 0.1 meters per second. How fast is the horizontal motion of the passenger at that moment?

23. A gas station stands at the intersection of a north-south road and an east-west road. A police car is traveling toward the gas station from the east, chasing a stolen truck which is traveling north away from the gas station. The speed of the police car is 100 mph at the moment it is 3

miles from the gas station. At the same time, the truck is 4 miles from the gas station going 80 mph. At this moment:

(a) Is the distance between the car and truck increasing or decreasing? How fast? (Distance is measured along a straight line joining the car and the truck.)

(b) How does your answer change if the truck is going 70 mph instead of 80 mph?

24. A train is traveling at 0.8 km/min along a long straight track, moving in the direction shown in Figure 4.86. A movie camera, 0.5 km away from the track, is focused on the train.

(a) Express z, the distance between the camera and the train, as a function of x.

(b) How fast is the distance from the camera to the train changing when the train is 1 km from the camera? Give units.

(c) How fast is the camera rotating (in radians/min) at the moment when the train is 1 km from the camera?

Figure 4.86

25. A lighthouse is 2 km from the long, straight coastline shown in Figure 4.87. Find the rate of change of the distance of the spot of light from the point O with respect to the angle θ.

Figure 4.87

26. A train is heading due west from St. Louis. At noon, a plane flying horizontally due north at a fixed altitude of 4 miles passes directly over the train. When the train has traveled another mile, it is going 80 mph, and the plane has traveled another 5 miles and is going 500 mph. At that moment, how fast is the distance between the train and the plane increasing?

27. The radius of a spherical balloon is increasing by 2 cm/sec. At what rate is air being blown into the balloon at the moment when the radius is 10 cm? Give units in your answer.

28. A spherical cell is growing at a constant rate of $400 \ \mu\text{m}^3/\text{day}$ (1 $\mu\text{m} = 10^{-6}$ m). At what rate is its radius increasing when the radius is 10 μm?

29. A raindrop is a perfect sphere with radius r cm and surface area S cm^2. Condensation accumulates on the raindrop at a rate equal to kS, where $k = 2$ cm/sec. Show that the radius of the raindrop increases at a constant rate and find that rate.

30. Sand falls from a hopper at a rate of 0.1 cubic meters per hour and forms a conical pile beneath. If the side of the cone makes an angle of $\pi/6$ radians with the vertical, find the rate at which the height of the cone increases. At what rate does the radius of the base increase? Give both answers in terms of h, the height of the pile in meters.

31. A circular region is irrigated by a 20 meter long pipe, fixed at one end and rotating horizontally, spraying water. One rotation takes 5 minutes. A road passes 30 meters from the edge of the circular area. See Figure 4.88.

    (a) How fast is the end of the pipe, P, moving?
    (b) How fast is the distance PQ changing when θ is $\pi/2$? When θ is 0?

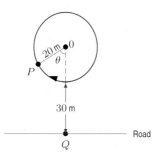

Figure 4.88

32. A water tank is in the shape of an inverted cone with depth 10 meters and top radius 8 meters. Water is flowing into the tank at 0.1 cubic meters/min but leaking out at a rate of $0.001h^2$ cubic meters/min, where h is the depth of the water in the tank in meters. Can the tank ever overflow?

33. For the amusement of the guests, some hotels have elevators on the outside of the building. One such hotel is 300 feet high. You are standing by a window 100 feet above the ground and 150 feet away from the hotel, and the elevator descends at a constant speed of 30 ft/sec, starting at time $t = 0$, where t is time in seconds. Let θ be the angle between the line of your horizon and your line of sight to the elevator. (See Figure 4.89.)

    (a) Find a formula for $h(t)$, the elevator's height above the ground as it descends from the top of the hotel.
    (b) Using your answer to part (a), express θ as a function of time t and find the rate of change of θ with respect to t.
    (c) The rate of change of θ is a measure of how fast the elevator appears to you to be moving. At what height is the elevator when it appears to be moving fastest?

Figure 4.89

34. In a romantic relationship between Angela and Brian, who are unsuited for each other, $a(t)$ represents the affection Angela has for Brian at time t days after they meet, while $b(t)$ represents the affection Brian has for Angela at time t. If $a(t) > 0$ then Angela likes Brian; if $a(t) < 0$ then Angela dislikes Brian; if $a(t) = 0$ then Angela neither likes nor dislikes Brian. Their affection for each other is given by the relation $a^2(t) + b^2(t) = c$, where c is a constant.

    (a) Show that $a(t) \cdot a'(t) = -b(t) \cdot b'(t)$.
    (b) At any time during their relationship, the rate per day at which Brian's affection for Angela changes is $b'(t) = -a(t)$. Explain what this means if Angela
        (i) Likes Brian,   (ii) Dislikes Brian.
    (c) Use parts (a) and (b) to show that $a'(t) = b(t)$. Explain what this means if Brian
        (i) Likes Angela,   (ii) Dislikes Angela.
    (d) If $a(0) = 1$ and $b(0) = 1$ who first dislikes the other?

35. In a 19th century sea-battle, the number of ships on each side remaining t hours after the start are given by $x(t)$ and $y(t)$. If the ships are equally equipped, the relation between them is $x^2(t) - y^2(t) = c$, where c is a positive constant. The battle ends when one side has no ships remaining.

    (a) If, at the start of the battle, 50 ships on one side oppose 40 ships on the other, what is the value of c?
    (b) If $y(3) = 16$, what is $x(3)$? What does this represent in terms of the battle?
    (c) There is a time T when $y(T) = 0$. What does this T represent in terms of the battle?
    (d) At the end of the battle, how many ships remain on the victorious side?
    (e) At any time during the battle, the rate per hour at which y loses ships is directly proportional to the number of x ships, with constant of proportionality k. Write an equation that represents this. Is k positive or negative?
    (f) Show that the rate per hour at which x loses ships is directly proportional to the number of y ships, with constant of proportionality k.
    (g) Three hours after the start of the battle, x is losing ships at the rate of 32 ships per hour. What is k? At what rate is y losing ships at this time?

4.7   L'HOPITAL'S RULE, GROWTH, AND DOMINANCE

Suppose we want to calculate the exact value of the limit

$$\lim_{x \to 0} \frac{e^{2x} - 1}{x}.$$

Substituting $x = 0$ gives us $0/0$, which is undefined:

$$\frac{e^{2(0)} - 1}{0} = \frac{1 - 1}{0} = \frac{0}{0}.$$

Substituting values of x near 0 gives us an approximate value for the limit.

However, the limit can be calculated exactly using local linearity. Suppose we let $f(x)$ be the numerator, so $f(x) = e^{2x} - 1$, and $g(x)$ be the denominator, so $g(x) = x$. Then $f(0) = 0$ and $f'(x) = 2e^{2x}$, so $f'(0) = 2$. When we zoom in on the graph of $f(x) = e^{2x} - 1$ near the origin, we see its tangent line $y = 2x$ shown in Figure 4.90. We are interested in the ratio $f(x)/g(x)$, which is approximately the ratio of the y-values in Figure 4.90. So, for x near 0,

$$\frac{f(x)}{g(x)} = \frac{e^{2x} - 1}{x} \approx \frac{2x}{x} = \frac{2}{1} = \frac{f'(0)}{g'(0)}.$$

As $x \to 0$, this approximation gets better, and we have

$$\lim_{x \to 0} \frac{e^{2x} - 1}{x} = 2.$$

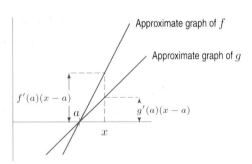

Figure 4.90: Ratio $(e^{2x} - 1)/x$ is approximated by ratio of slopes as we zoom in near the origin

Figure 4.91: Ratio $f(x)/g(x)$ is approximated by ratio of slopes, $f'(a)/g'(a)$, as we zoom in at a

L'Hopital's Rule

If $f(a) = g(a) = 0$, we can use the same method to investigate limits of the form

$$\lim_{x \to a} \frac{f(x)}{g(x)}.$$

As in the previous case, we zoom in on the graphs of $f(x)$ and $g(x)$. Figure 4.91 shows that both graphs cross the x-axis at $x = a$. This suggests that the limit of $f(x)/g(x)$ as $x \to a$ is the ratio of slopes, giving the following result:

L'Hopital's rule: If f and g are differentiable, $f(a) = g(a) = 0$, and $g'(a) \neq 0$, then

$$\lim_{x \to a} \frac{f(x)}{g(x)} = \frac{f'(a)}{g'(a)}.$$

To justify this result, let us assume $g'(a) \neq 0$ and consider the quantity $f'(a)/g'(a)$. Using the definition of the derivative and the fact that $f(a) = g(a) = 0$, we have

$$\frac{f'(a)}{g'(a)} = \frac{\displaystyle\lim_{h \to 0} \frac{f(a+h) - f(a)}{h}}{\displaystyle\lim_{h \to 0} \frac{g(a+h) - g(a)}{h}} = \frac{\displaystyle\lim_{h \to 0} \frac{f(a+h)}{h}}{\displaystyle\lim_{h \to 0} \frac{g(a+h)}{h}}$$

$$= \lim_{h \to 0} \frac{f(a+h)}{g(a+h)} = \lim_{x \to a} \frac{f(x)}{g(x)}.$$

Note that if $f'(a) \neq 0$ and $g'(a) = 0$, the limit of $f(x)/g(x)$ does not exist.

Example 1   Use l'Hopital's rule to confirm that $\displaystyle\lim_{x \to 0} \frac{\sin x}{x} = 1$.

Solution   Let $f(x) = \sin x$ and $g(x) = x$. Then $f(0) = g(0) = 0$ and $f'(x) = \cos x$ and $g'(x) = 1$. Thus,

$$\lim_{x \to 0} \frac{\sin x}{x} = \frac{\cos 0}{1} = 1.$$

If we also have $f'(a) = g'(a) = 0$, then we can use the following result:

More general form of l'Hopital's rule: If f and g are differentiable and $f(a) = g(a) = 0$, then

$$\lim_{x \to a} \frac{f(x)}{g(x)} = \lim_{x \to a} \frac{f'(x)}{g'(x)},$$

provided the limit on the right exists.

Example 2   Calculate $\displaystyle\lim_{t \to 0} \frac{e^t - 1 - t}{t^2}$.

Solution   Let $f(t) = e^t - 1 - t$ and $g(t) = t^2$. Then $f(0) = e^0 - 1 - 0 = 0$ and $g(0) = 0$, and $f'(t) = e^t - 1$ and $g'(t) = 2t$. So

$$\lim_{t \to 0} \frac{e^t - 1 - t}{t^2} = \lim_{t \to 0} \frac{e^t - 1}{2t}.$$

Since $f'(0) = g'(0) = 0$, the ratio $f'(0)/g'(0)$ is not defined. So we use l'Hopital's rule again:

$$\lim_{t \to 0} \frac{e^t - 1 - t}{t^2} = \lim_{t \to 0} \frac{e^t - 1}{2t} = \lim_{t \to 0} \frac{e^t}{2} = \frac{1}{2}.$$

We can also use L'Hopital's rule in the following cases.

L'Hopital's rule applies to limits involving infinity, provided f and g are differentiable:
- When $\lim_{x \to a} f(x) = \pm\infty$ and $\lim_{x \to a} g(x) = \pm\infty$,

or

- When $a = \infty$ (or $a = -\infty$) and $\displaystyle\lim_{x \to \infty} f(x) = \lim_{x \to \infty} g(x) = 0$ or $\displaystyle\lim_{x \to \infty} f(x) = \pm\infty$ and $\displaystyle\lim_{x \to \infty} g(x) = \pm\infty$.

It can be shown that under these circumstances:

$$\lim_{x \to a} \frac{f(x)}{g(x)} = \lim_{x \to a} \frac{f'(x)}{g'(x)}$$

(where a may be $\pm\infty$), provided the limit on the right-hand side exists.

Notice that we cannot evaluate $f'(x)/g'(x)$ directly when $a = \pm\infty$. The next example shows how this version of l'Hopital's rule is used.

Example 3   Calculate $\displaystyle\lim_{x\to\infty} \frac{5x + e^{-x}}{7x}$.

Solution   Let $f(x) = 5x + e^{-x}$ and $g(x) = 7x$. Then $\displaystyle\lim_{x\to\infty} f(x) = \lim_{x\to\infty} g(x) = \infty$, and $f'(x) = 5 - e^{-x}$ and $g'(x) = 7$, so

$$\lim_{x\to\infty} \frac{5x + e^{-x}}{7x} = \lim_{x\to\infty} \frac{(5 - e^{-x})}{7} = \frac{5}{7}.$$

We can also use l'Hopital's rule to calculate some limits of the form $\displaystyle\lim_{x\to\infty} f(x)g(x)$, providing we rewrite them appropriately.

Example 4   Calculate $\displaystyle\lim_{x\to\infty} xe^{-x}$.

Solution   Since $\displaystyle\lim_{x\to\infty} x = \infty$ and $\displaystyle\lim_{x\to\infty} e^{-x} = 0$, we see that

$$xe^{-x} \to 0 \cdot \infty \qquad \text{as } x \to \infty.$$

Since $0 \cdot \infty$ is undefined, we rewrite the function xe^{-x} as

$$xe^{-x} = \frac{x}{e^x}.$$

Now we use l'Hopital's rule, since

$$xe^{-x} = \frac{x}{e^x} \to \frac{\infty}{\infty} \qquad \text{as } x \to \infty.$$

Taking $f(x) = x$ and $g(x) = e^x$ gives $f'(x) = 1$ and $g'(x) = e^x$, so

$$\lim_{x\to\infty} xe^{-x} = \lim_{x\to\infty} \frac{x}{e^x} = \lim_{x\to\infty} \frac{1}{e^x} = 0.$$

A Famous Limit

In the following example, l'Hopital's rule is applied to calculate a limit that can be used to define e.

Example 5   Evaluate $\displaystyle\lim_{x\to\infty} \left(1 + \frac{1}{x}\right)^x$.

Solution   Since $\left(1 + \dfrac{1}{x}\right)^x \to 1^\infty$ as $x \to \infty$, and 1^∞ is undefined, we write

$$y = \left(1 + \frac{1}{x}\right)^x$$

and find the limit of $\ln y$:

$$\ln y = \ln \left(1 + \frac{1}{x}\right)^x = x \ln \left(1 + \frac{1}{x}\right).$$

As in the previous example, we rewrite the product as fraction, giving

$$\ln y = \frac{\ln(1 + 1/x)}{1/x}.$$

Since $\lim_{x \to \infty} \ln(1 + 1/x) = 0$ and $\lim_{x \to \infty}(1/x) = 0$, we can use l'Hopital's rule with $f(x) = \ln(1 + 1/x)$ and $g(x) = 1/x$. We have

$$f'(x) = \frac{1}{1 + 1/x}\left(-\frac{1}{x^2}\right) \quad \text{and} \quad g'(x) = -\frac{1}{x^2},$$

so

$$\lim_{x \to \infty} \ln y = \lim_{x \to \infty} \frac{\ln(1 + 1/x)}{1/x} = \lim_{x \to \infty} \frac{1}{1 + 1/x}\left(-\frac{1}{x^2}\right) \bigg/ \left(-\frac{1}{x^2}\right)$$

$$= \lim_{x \to \infty} \frac{1}{1 + 1/x} = 1.$$

Since $\lim_{x \to \infty} \ln y = 1$, we have

$$\lim_{x \to \infty} y = e^1 = e.$$

Dominance: Powers, Polynomials, Exponentials, and Logarithms

In Chapter 1, we saw that some functions were much larger than others as $x \to \infty$. We say that g *dominates* f as $x \to \infty$ if $\lim_{x \to \infty} \frac{f(x)}{g(x)} = 0$. L'Hopital's rule gives us an easy way of checking this.

Example 6   Check that $x^{1/2}$ dominates $\ln x$ as $x \to \infty$.

Solution   We apply l'Hopital's rule to $(\ln x)/x^{1/2}$:

$$\lim_{x \to \infty} \frac{\ln x}{x^{1/2}} = \lim_{x \to \infty} \frac{1/x}{\frac{1}{2}x^{-1/2}}.$$

To evaluate this limit, we simplify and get

$$\lim_{x \to \infty} \frac{1/x}{\frac{1}{2}x^{-1/2}} = \lim_{x \to \infty} \frac{2x^{1/2}}{x} = \lim_{x \to \infty} \frac{2}{x^{1/2}} = 0.$$

Therefore we have

$$\lim_{x \to \infty} \frac{\ln x}{x^{1/2}} = 0,$$

which tells us that $x^{1/2}$ dominates $\ln x$ as $x \to \infty$.

Example 7   Check any exponential function of the form e^{kx} (with $k > 0$) dominates any power function of the form Ax^p (with A and p positive) as $x \to \infty$.

Solution   We apply l'Hopital's rule repeatedly to Ax^p/e^{kx}:

$$\lim_{x \to \infty} \frac{Ax^p}{e^{kx}} = \lim_{x \to \infty} \frac{Apx^{p-1}}{ke^{kx}} = \lim_{x \to \infty} \frac{Ap(p-1)x^{p-2}}{k^2 e^{kx}} = \cdots$$

Keep applying l'Hopital's rule until the power of x is no longer positive. Then the limit of the numerator must be a finite number, while the limit of the denominator must be ∞. Therefore we have

$$\lim_{x \to \infty} \frac{Ax^p}{e^{kx}} = 0,$$

so e^{kx} dominates Ax^p.

Exercises and Problems for Section 4.7

Exercises

For Exercises 1–4, find the sign of $\lim\limits_{x \to a} \dfrac{f(x)}{g(x)}$ from the figure.

1. **2.**

3. **4.**

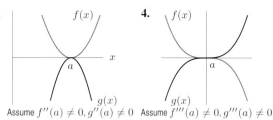

Assume $f''(a) \neq 0, g''(a) \neq 0$ Assume $f'''(a) \neq 0, g'''(a) \neq 0$

Based on your knowledge of the behavior of the numerator and denominator, predict the value of the limits in Exercises 5–8. Then find each limit using l'Hopital's rule.

5. $\lim\limits_{x \to 0} \dfrac{x^2}{\sin x}$ **6.** $\lim\limits_{x \to 0} \dfrac{\sin^2 x}{x}$

7. $\lim\limits_{x \to 0} \dfrac{\sin x}{x^{1/3}}$ **8.** $\lim\limits_{x \to 0} \dfrac{x}{(\sin x)^{1/3}}$

In Exercises 9–12, which function dominates as $x \to \infty$?

9. x^5 and $0.1x^7$ **10.** $0.01x^3$ and $50x^2$

11. $\ln(x + 3)$ and $x^{0.2}$ **12.** x^{10} and $e^{0.1x}$

13. Evaluate $\lim\limits_{x \to 0^+} x \ln x$. [Hint: Write $x \ln x = \dfrac{\ln x}{1/x}$.]

Problems

14. Find the horizontal asymptote of $f(x) = \dfrac{2x^3 + 5x^2}{3x^3 - 1}$.

15. The functions f and g and their tangent lines at $(4, 0)$ are shown in Figure 4.92. Find $\lim\limits_{x \to 4} \dfrac{f(x)}{g(x)}$.

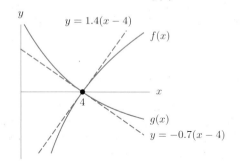

Figure 4.92

16. **(a)** What is the slope of $f(x) = \sin(3x)$ at $x = 0$?
  (b) What is the slope of $g(x) = 5x$ at $x = 0$?
  (c) Use the results of parts (a) and (b) to calculate $\lim\limits_{x \to 0} \dfrac{\sin(3x)}{5x}$.

In Problems 17–25 determine whether the limit exists, and where possible evaluate it.

17. $\lim\limits_{x \to 1} \dfrac{\ln x}{x^2 - 1}$ **18.** $\lim\limits_{t \to \pi} \dfrac{\sin^2 t}{t - \pi}$

19. $\lim\limits_{x \to 0} \dfrac{\sinh(2x)}{x}$ **20.** $\lim\limits_{x \to 0} \dfrac{1 - \cosh(3x)}{x}$

21. $\lim\limits_{x \to 0^+} x^a \ln x$, where a is a positive constant.

22. $\lim\limits_{x \to 1^-} \dfrac{\cos^{-1} x}{x - 1}$ **23.** $\lim\limits_{t \to 0^+} \dfrac{3 \sin t - \sin 3t}{3 \tan t - \tan 3t}$

24. $\lim\limits_{x \to 0} \left(\dfrac{1}{x} - \dfrac{1}{\sin x} \right)$ **25.** $\lim\limits_{x \to \infty} \left(1 + \sin \left(\dfrac{3}{x} \right) \right)^x$

Explain why l'Hopital's rule cannot be used to calculate the limits in Problems 26–28. Then evaluate the limit if it exists.

26. $\lim\limits_{x \to 1} \dfrac{\sin(2x)}{x}$ **27.** $\lim\limits_{x \to 0} \dfrac{\cos x}{x}$ **28.** $\lim\limits_{x \to \infty} \dfrac{e^{-x}}{\sin x}$

29. If a, b, x, and y are all positive and $a + b = 1$, evaluate

  (a) $\lim\limits_{p \to 0} \dfrac{\ln(ax^p + by^p)}{p}$ **(b)** $\lim\limits_{p \to 0} (ax^p + by^p)^{1/p}$

30. **(a)** Explain why l'Hopital's rule cannot be used to evaluate $\lim\limits_{x \to \infty} \dfrac{\pi - 2 \tan^{-1} x}{e^{-x}}$.
  (b) Use a graph to estimate the value of this limit if it exists.

In Problems 31–33, evaluate the limit using the fact that

$$\lim_{n \to \infty} \left(1 + \frac{1}{n}\right)^n = e.$$

31. $\displaystyle\lim_{x \to 0^+} (1 + x)^{1/x}$ **32.** $\displaystyle\lim_{n \to \infty} \left(1 + \frac{2}{n}\right)^n$

33. $\displaystyle\lim_{x \to 0^+} (1 + kx)^{t/x};\ k > 0$

34. Show that $\displaystyle\lim_{n \to \infty} \left(1 - \frac{1}{n}\right)^n = e^{-1}$.

35. Use the result of Problem 34 to evaluate

$$\lim_{n \to \infty} \left(1 - \frac{\lambda}{n}\right)^n.$$

Evaluate the limits in Problems 36–38 where

$$f(t) = \left(\frac{3^t + 5^t}{2}\right)^{1/t} \qquad \text{for } t \ne 0.$$

36. $\displaystyle\lim_{t \to -\infty} f(t)$ **37.** $\displaystyle\lim_{t \to +\infty} f(t)$ **38.** $\displaystyle\lim_{t \to 0} f(t)$

39. (a) Graph the functions $\dfrac{1}{1 - x}$ and $\dfrac{\ln((x - a)/(x - 1))}{a - 1}$ for $a = 1.1$, $a = 1.01$, and $a = 1.001$. Describe what you see.

   (b) Evaluate $\displaystyle\lim_{a \to 1} \frac{\ln((x - a)/(x - 1))}{a - 1}$.

   (c) Explain the connection between parts (a) and (b).

40. (a) Graph the functions $\dfrac{1}{x}$ and $\dfrac{\ln((x + a)/(x - a))}{2a}$ for $a = 0.1$, $a = 0.01$, and $a = 0.001$. Describe what you see.

   (b) Evaluate $\displaystyle\lim_{a \to 0} \frac{\ln((x + a)/(x - a))}{2a}$.

   (c) Explain the connection between parts (a) and (b).

41. (a) For $x > 0$ and $r \ne 0$, let $h_r(x) = (x^r - 1)/r$. Compute $h_0(x)$, which is defined by $h_0(x) = \lim_{r \to 0}(x^r - 1)/r$.

   (b) On the same set of axes, graph $h_r(x)$ for $r = -5, -2, 0, 1$, and 2.

4.8 PARAMETRIC EQUATIONS

Representing Motion in the Plane

To represent the motion of a particle in the xy-plane we use two equations, one for the x-coordinate of the particle, $x = f(t)$, and another for the y-coordinate, $y = g(t)$. Thus at time t the particle is at the point $(f(t), g(t))$. The equation for x describes the right-left motion; the equation for y describes the up-down motion. The two equations for x and y are called *parametric equations* with *parameter* t.

Example 1   Describe the motion of the particle whose coordinates at time t are $x = \cos t$ and $y = \sin t$.

Solution   Since $(\cos t)^2 + (\sin t)^2 = 1$, we have $x^2 + y^2 = 1$. That is, at any time t the particle is at a point (x, y) on the unit circle $x^2 + y^2 = 1$. We plot points at different times to see how the particle moves on the circle. (See Figure 4.93 and Table 4.4.) The particle moves at a uniform speed, completing one full trip counterclockwise around the circle every 2π units of time. Notice how the x-coordinate goes repeatedly back and forth from -1 to 1 while the y-coordinate goes repeatedly up and down from -1 to 1. The two motions combine to trace out a circle.

Figure 4.93: The circle parameterized by $x = \cos t$, $y = \sin t$

Table 4.4   *Points on the circle with $x = \cos t$, $y = \sin t$*

t	x	y
0	1	0
$\pi/2$	0	1
π	-1	0
$3\pi/2$	0	-1
2π	1	0

Example 2   Figure 4.94 shows the graphs of two functions, $f(t)$ and $g(t)$. Describe the motion of the particle whose coordinates at time t are $x = f(t)$ and $y = g(t)$.

Figure 4.94: Graphs of $x = f(t)$ and $y = g(t)$ used to trace out the path $(f(t), g(t))$ in Figure 4.95

Solution   Between times $t = 0$ and $t = 1$, the x-coordinate goes from 0 to 1, while the y-coordinate stays fixed at 0. So the particle moves along the x-axis from $(0, 0)$ to $(1, 0)$. Then, between times $t = 1$ and $t = 2$, the x-coordinate stays fixed at $x = 1$, while the y-coordinate goes from 0 to 1. Thus, the particle moves along the vertical line from $(1, 0)$ to $(1, 1)$. Similarly, between times $t = 2$ and $t = 3$, it moves horizontally back to $(0, 1)$, and between times $t = 3$ and $t = 4$ it moves down the y-axis to $(0, 0)$. Thus, it traces out the square in Figure 4.95.

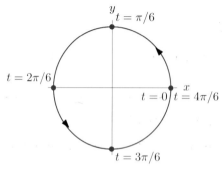

Figure 4.95: The square parameterized by $(f(t), g(t))$

Different Motions Along the Same Path

Example 3   Describe the motion of the particle whose x and y coordinates at time t are given by the equations

$$x = \cos(3t), \quad y = \sin(3t).$$

Solution   Since $(\cos(3t))^2 + (\sin(3t))^2 = 1$, we have $x^2 + y^2 = 1$, giving motion around the unit circle. But from Table 4.5, we see that the particle in this example is moving three times as fast as the particle in Example 1. (See Figure 4.96.)

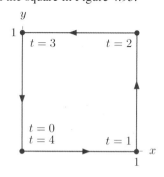

Figure 4.96: The circle parameterized by $x = \cos(3t), y = \sin(3t)$

Table 4.5   *Points on circle with* $x = \cos(3t), y = \sin(3t)$

t	x	y
0	1	0
$\pi/6$	0	1
$2\pi/6$	-1	0
$3\pi/6$	0	-1
$4\pi/6$	1	0

Example 3 is obtained from Example 1 by replacing t by $3t$; this is called a *change in parameter*. If we make a change in parameter, the particle traces out the same curve (or a part of it) but at a different speed or in a different direction.

Example 4   Describe the motion of the particle whose x and y coordinates at time t are given by

$$x = \cos(e^{-t^2}), \quad y = \sin(e^{-t^2}).$$

Solution   As in Examples 1 and 3, we have $x^2 + y^2 = 1$ so the motion lies on the unit circle. As time t goes from $-\infty$ (way back in the past) to 0 (the present) to ∞ (way off in the future), e^{-t^2} goes from near 0 to 1 back to near 0. So $(x, y) = (\cos(e^{-t^2}), \sin(e^{-t^2}))$ goes from near $(1, 0)$ to $(\cos 1, \sin 1)$ and back to near $(1, 0)$. The particle does not actually reach the point $(1, 0)$. (See Figure 4.97 and Table 4.6.)

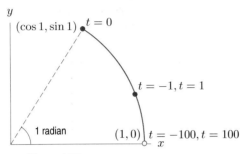

Figure 4.97: The circle parameterized by $x = \cos(e^{-t^2})$, $y = \sin(e^{-t^2})$

Table 4.6   *Points on circle with $x = \cos(e^{-t^2})$, $y = \sin(e^{-t^2})$*

t	x	y
-100	~ 1	~ 0
-1	0.93	0.36
0	0.54	0.84
1	0.93	0.36
100	~ 1	~ 0

Motion in a Straight Line

An object moves with constant speed along a straight line through the point (x_0, y_0). Both the x- and y-coordinates have a constant rate of change. Let $a = dx/dt$ and $b = dy/dt$. Then at time t the object has coordinates $x = x_0 + at$, $y = y_0 + bt$. (See Figure 4.98.) Notice that a represents the change in x in one unit of time, and b represents the change in y. Thus the line has slope $m = b/a$.

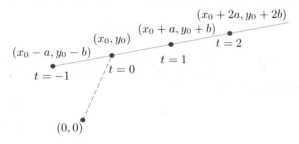

Figure 4.98: The line $x = x_0 + at$, $y = y_0 + bt$

This yields the following:

Parametric Equations for a Straight Line

An object moving along a line through the point (x_0, y_0), with $dx/dt = a$ and $dy/dt = b$, has parametric equations

$$x = x_0 + at, \quad y = y_0 + bt.$$

The slope of the line is $m = b/a$.

Example 5   Find parametric equations for:

(a) The line passing through the points $(2, -1)$ and $(-1, 5)$.
(b) The line segment from $(2, -1)$ to $(-1, 5)$.

Solution   (a) Imagine an object moving with constant speed along a straight line from $(2, -1)$ to $(-1, 5)$, making the journey from the first point to the second in one unit of time. Then $dx/dt = ((-1) - 2)/1 = -3$ and $dy/dt = (5 - (-1))/1 = 6$. Thus the parametric equation are

$$x = 2 - 3t, \quad y = -1 + 6t.$$

(b) In the parameterization in part (a), $t = 0$ corresponds to the point $(2, -1)$ and $t = 1$ corresponds to the point $(-1, 5)$. So the parameterization of the segment is

$$x = 2 - 3t, \quad y = -1 + 6t, \quad 0 \leq t \leq 1.$$

There are many other possible parametric equations for this line.

Speed and Velocity

An object moves along a straight line at a constant speed, with $dx/dt = a$ and $dy/dt = b$. In one unit of time, the object moves a units horizontally and b units vertically. Thus, by the Pythagorean Theorem, it travels a distance $\sqrt{a^2 + b^2}$. So its speed is

$$\text{Speed} = \frac{\text{Distance traveled}}{\text{Time taken}} = \frac{\sqrt{a^2 + b^2}}{1} = \sqrt{a^2 + b^2}.$$

For general motion along a curve with varying speed, we make the following definition:

> The **instantaneous speed** of a moving object is defined to be
>
> $$v = \sqrt{\left(\frac{dx}{dt}\right)^2 + \left(\frac{dy}{dt}\right)^2}.$$
>
> The quantity $v_x = dx/dt$ is the **instantaneous velocity** in the x-direction; $v_y = dy/dt$ is the **instantaneous velocity** in the y-direction.

The quantities v_x and v_y are called the *components* of the velocity in the x- and y-directions.

Example 6   A particle moves in the xy-plane with $x = 2t^3 - 9t^2 + 12t$ and $y = 3t^4 - 16t^3 + 18t^2$, where t is time.

(a) At what times is the particle
   (i) Stopped   (ii)   Moving parallel to the x- or y- axis?
(b) Find the speed of the particle at time t.

Solution   (a) Differentiating gives

$$\frac{dx}{dt} = 6t^2 - 18t + 12 \quad \frac{dy}{dt} = 12t^3 - 48t^2 + 36t.$$

We are interested in the points at which $dx/dt = 0$ or $dy/dt = 0$. Solving gives

$$\frac{dx}{dt} = 6(t^2 - 3t + 2) = 6(t - 1)(t - 2) \quad \text{so} \quad \frac{dx}{dt} = 0 \quad \text{if } t = 1 \text{ or } t = 2.$$

$$\frac{dy}{dt} = 12t(t^2 - 4t + 3) = 12t(t - 1)(t - 3) \quad \text{so} \quad \frac{dy}{dt} = 0 \quad \text{if } t = 0, t = 1, \text{ or } t = 3.$$

(i) The particle is stopped if both dx/dt and dy/dt are 0, which occurs at $t = 1$.

(ii) The particle is moving parallel to the x-axis if $dy/dt = 0$ but $dx/dt \neq 0$. This occurs at $t = 0$ and $t = 3$. The particle is moving parallel to the y-axis if $dx/dt = 0$ but $dy/dt \neq 0$. This occurs at $t = 2$.

(b) We have

$$\text{Speed} = \sqrt{\left(\frac{dx}{dt}\right)^2 + \left(\frac{dy}{dt}\right)^2} = \sqrt{(6t^2 - 18t + 12)^2 + (12t^3 - 48t^2 + 36t)^2}$$

$$= 6\sqrt{4t^6 - 32t^5 + 89t^4 - 102t^3 + 49t^2 - 12t + 4}.$$

Example 7   A child is sitting on a ferris wheel of diameter 10 meters, making one revolution every 2 minutes. Find the speed of the child

(a) Using geometry.   (b) Using a parameterization of the motion.

Solution   (a) The child moves at a constant speed around a circle of radius 5 meters, completing one revolution every 2 minutes. One revolution around a circle of radius 5 is a distance of 10π, so the child's speed is $10\pi/2 = 5\pi \approx 15.7$ m/min. See Figure 4.99.

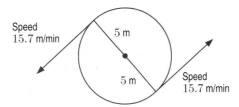

Figure 4.99: Motion of a child on a ferris wheel at two different times is represented by the arrows. The direction of each arrow is the direction of motion at that time.

(b) The ferris wheel has radius 5 meters and completes 1 revolution counterclockwise every 2 minutes. If the origin is at the center of the circle and we measure x and y in meters, the motion is parameterized by equations of the form

$$x = 5\cos(\omega t), \quad y = 5\sin(\omega t),$$

where ω is chosen to make the period 2 minutes. Since the period of $\cos(\omega t)$ and $\sin(\omega t)$ is $2\pi/\omega$, we must have

$$\frac{2\pi}{\omega} = 2, \quad \text{so} \quad \omega = \pi.$$

Thus, for t in minutes, the motion is described by the equations

$$x = 5\cos(\pi t), \quad y = 5\sin(\pi t).$$

So the speed is given by

$$v = \sqrt{\left(\frac{dx}{dt}\right)^2 + \left(\frac{dy}{dt}\right)^2}$$

$$= \sqrt{(-5\pi)^2\sin^2(\pi t) + (5\pi)^2\cos^2(\pi t)} = 5\pi\sqrt{\sin^2(\pi t) + \cos^2(\pi t)} = 5\pi \approx 15.7 \text{ m/min},$$

which agrees with the speed we calculated in part (a).

Tangent Lines

To find the tangent line at a point (x_0, y_0) to a curve given parametrically, we find the straight line motion through (x_0, y_0) with the same velocity in the x and y directions as the curve.

Example 8   Find the tangent line at the point $(1, 2)$ to the curve defined by the parametric equation

$$x = t^3, \quad y = 2t.$$

Solution   At time $t = 1$ the particle is at the point $(1, 2)$. The velocity in the x-direction at time t is $v_x = dx/dt = 3t^2$, and the velocity in the y-direction is $v_y = dy/dt = 2$. So at $t = 1$ the velocity in the x-direction is 3 and the velocity in the y-direction is 2. Thus the tangent line has parametric equations

$$x = 1 + 3t, \quad y = 2 + 2t.$$

Parametric Representations of Curves in the Plane

Sometimes we are more interested in the curve traced out by the particle than we are in the motion itself. In that case we will call the parametric equations a *parameterization* of the curve. As we can see by comparing Examples 1 and 3, two different parameterizations can describe the same curve in the xy-plane. Though the parameter, which we usually denote by t, may not have physical meaning, it is often helpful to think of it as time.

Example 9   Give a parameterization of the semicircle of radius 1 shown in Figure 4.100.

Figure 4.100: Parameterization of semicircle for Example refex:paramex8:chap4:sec8:07MAY2004

Figure 4.101: Parameterization of the ellipse $4x^2 + y^2 = 1$ for Example 10

Solution   We can use the equations $x = \cos t$ and $y = \sin t$ for counterclockwise motion in a circle, from Example 1 on page 219. The particle passes $(0, 1)$ at $t = \pi/2$, moves counterclockwise around the circle, and reaches $(0, -1)$ at $t = 3\pi/2$. So a parameterization is

$$x = \cos t, \; y = \sin t, \quad \frac{\pi}{2} \le t \le \frac{3\pi}{2}.$$

To find the xy-equation of a curve given parametrically, we eliminate the parameter t in the parametric equations. In the previous example, we use the Pythagorean identity, so

$$\cos^2 t + \sin^2 t = 1 \quad \text{gives} \quad x^2 + y^2 = 1.$$

Example 10   Give a parameterization of the ellipse $4x^2 + y^2 = 1$ shown in Figure 4.101.

Solution   Since $(2x)^2 + y^2 = 1$, we adapt the parameterization of the circle in Example 1. Replacing x by $2x$ gives the equations $2x = \cos t, y = \sin t$. A parameterization of the ellipse is thus

$$x = \tfrac{1}{2} \cos t, \quad y = \sin t, \quad 0 \le t \le 2\pi.$$

We usually require that the parameterization of a curve go from one end of the curve to the other without retracing any portion of the curve. This is different from parameterizing the motion of a particle, where, for example, a particle may move around the same circle many times.

Parameterizing the Graph of a Function

The graph of any function $y = f(x)$ can be parameterized by letting the parameter t be x:

$$x = t, \quad y = f(t).$$

Example 11   Give parametric equations for the curve $y = x^3 - x$. In which direction does this parameterization trace out the curve?

Solution   Let $x = t$, $y = t^3 - t$. Thus, $y = t^3 - t = x^3 - x$. Since $x = t$, as time increases the x-coordinate moves from left to right, so the particle traces out the curve $y = x^3 - x$ from left to right.

Curves Given Parametrically

Some complicated curves can be graphed more easily using parametric equations; the next example shows such a curve.

Example 12   Assume t is time in seconds. Sketch the curve traced out by the particle whose motion is given by

$$x = \cos(3t), \quad y = \sin(5t).$$

Solution   The x-coordinate oscillates back and forth between 1 and -1, completing 3 oscillations every 2π seconds. The y-coordinate oscillates up and down between 1 and -1, completing 5 oscillations every 2π seconds. Since both the x- and y-coordinates return to their original values every 2π seconds, the curve is retraced every 2π seconds. The result is a pattern called a Lissajous figure. (See Figure 4.102.) Problems 45–48 concern Lissajous figures $x = \cos(at)$, $y = \sin(bt)$ for other values of a and b.

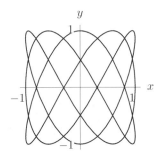

Figure 4.102: A Lissajous figure: $x = \cos(3t), y = \sin(5t)$

Slope and Concavity of Parametric Curves

Suppose we have a curve traced out by the parametric equations $x = f(t)$, $y = g(t)$. To find the slope at a point on the curve, we could, in theory, eliminate the parameter t and then differentiate the function we obtain. However, the chain rule gives us an easier way.

Suppose the curve traced out by the parametric equations is represented by $y = h(x)$. (It may be represented by an implicit function.) Thinking of x and y as functions of t, the chain rule gives

$$\frac{dy}{dt} = \frac{dy}{dx} \cdot \frac{dx}{dt},$$

so we obtain the slope of the curve as a function of t:

$$\boxed{\text{Slope of curve} = \frac{dy}{dx} = \frac{dy/dt}{dx/dt}.}$$

We can find the second derivative, d^2y/dx^2, by a similar method and use it to investigate the concavity of the curve. The chain rule tells us that if w is any differentiable function of x, then

$$\frac{dw}{dx} = \frac{dw/dt}{dx/dt}.$$

For $w = dy/dx$, we have

$$\frac{dw}{dx} = \frac{d}{dx}\left(\frac{dy}{dx}\right) = \frac{d^2y}{dx^2},$$

so the chain rule gives the second derivative at any point on a parametric curve:

$$\frac{d^2y}{dx^2} = \frac{d}{dt}\left(\frac{dy}{dx}\right)\Big/ \frac{dx}{dt}.$$

Example 13 If $x = \cos t$, $y = \sin t$, find the point corresponding to $t = \pi/4$, the slope of the curve at the point, and d^2y/dx^2 at the point.

Solution The point corresponding to $t = \pi/4$ is $(\cos(\pi/4), \sin(\pi/4)) = (1/\sqrt{2}, 1/\sqrt{2})$.
   To find the slope, we use

$$\frac{dy}{dx} = \frac{dy/dt}{dx/dt} = \frac{\cos t}{-\sin t},$$

so when $t = \pi/4$,

$$\text{Slope} = \frac{\cos(\pi/4)}{-\sin(\pi/4)} = -1.$$

Thus, the curve has slope -1 at the point $(1/\sqrt{2}, 1/\sqrt{2})$. This is as we would expect, since the curve traced out is the circle of Example 8.
   To find d^2y/dx^2, we use $w = dy/dx = -(\cos t)/(\sin t)$, so

$$\frac{d^2y}{dx^2} = \frac{d}{dt}\left(-\frac{\cos t}{\sin t}\right)\Big/(-\sin t) = -\frac{(-\sin t)(\sin t) - (\cos t)(\cos t)}{\sin^2 t} \cdot \left(-\frac{1}{\sin t}\right) = -\frac{1}{\sin^3 t}.$$

Thus, at $t = \pi/4$

$$\frac{d^2y}{dx^2}\bigg|_{t=\pi/4} = -\frac{1}{(\sin(\pi/4))^3} = -2\sqrt{2}.$$

The concavity is negative since the point is on the top half of the circle where the graph is bending downward.

Exercises and Problems for Section 4.8

Exercises

For Exercises 1–4, use the graphs of f and g to describe the motion of a particle whose position at time t is given by $x = f(t)$, $y = g(t)$.

1.

2.

3.

4.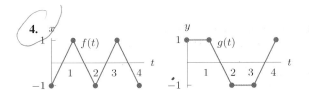

Exercises 5–10 give parameterizations of the unit circle or a part of it. Describe in words how the circle is traced out, including when and where the particle is moving clockwise and when and where the particle is moving counterclockwise.

5. $x = \sin t, \quad y = \cos t$

6. $x = \cos t, \quad y = -\sin t$

7. $x = \cos(t^2), \quad y = \sin(t^2)$

8. $x = \cos(t^3 - t), \quad y = \sin(t^3 - t)$

9. $x = \cos(\ln t), \quad y = \sin(\ln t)$

10. $x = \cos(\cos t), \quad y = \sin(\cos t)$

For Exercises 11–14, find the speed for the given motion of a particle. Find any times when the particle comes to a stop.

11. $x = t^2, \quad y = t^3$

12. $x = \cos(t^2), \quad y = \sin(t^2)$

13. $x = \cos 2t, \quad y = \sin t$

14. $x = t^2 - 2t, y = t^3 - 3t$

15. Find parametric equations for the tangent line at $t = 2$ for Problem 11.

In Exercises 16–22, write a parameterization for the curves in the xy-plane.

16. A circle of radius 3 centered at the origin and traced out clockwise.

17. A vertical line through the point $(-2, -3)$.

18. A circle of radius 5 centered at the point $(2, 1)$ and traced out counterclockwise.

19. A circle of radius 2 centered at the origin traced clockwise starting from $(-2, 0)$ when $t = 0$.

20. The line through the points $(2, -1)$ and $(1, 3)$.

21. An ellipse centered at the origin and crossing the x-axis at ± 5 and the y-axis at ± 7.

22. An ellipse centered at the origin, crossing the x-axis at ± 3 and the y-axis at ± 7. Start at the point $(-3, 0)$ and trace out the ellipse counterclockwise.

In Exercises 23–25, find an equation of the tangent line to the curve for the given value of t.

23. $x = t^3 - t, \quad y = t^2 \quad$ when $t = 2$

24. $x = t^2 - 2t, \quad y = t^2 + 2t \quad$ when $t = 1$

25. $x = \sin(3t), \quad y = \sin(4t) \quad$ when $t = \pi$

Problems

26. A line is parameterized by $x = 10 + t$ and $y = 2t$.

    (a) What part of the line do we get by restricting t to $t < 0$?

    (b) What part of the line do we get by restricting t to $0 \le t \le 1$?

27. A line is parameterized by $x = 2 + 3t$ and $y = 4 + 7t$.

    (a) What part of the line is obtained by restricting t to nonnegative numbers?

    (b) What part of the line is obtained if t is restricted to $-1 \le t \le 0$?

    (c) How should t be restricted to give the part of the line to the left of the y-axis?

28. **(a)** Explain how you know that the following two pairs of equations parameterize the same line:

$$x = 2 + t, y = 4 + 3t \text{ and } x = 1 - 2t, y = 1 - 6t.$$

    (b) What are the slope and y intercept of this line?

29. Describe the similarities and differences among the motions in the plane given by the following three pairs of parametric equations:

    (a) $x = t, \quad y = t^2$ **(b)** $x = t^2, \quad y = t^4$

    (c) $x = t^3, \quad y = t^6.$

30. What can you say about the values of a, b and k if the equations

$$x = a + k \cos t, \quad y = b + k \sin t, \qquad 0 \le t \le 2\pi,$$

trace out the following circles in Figure 4.103?

    (a) C_1 **(b)** C_2 **(c)** C_3

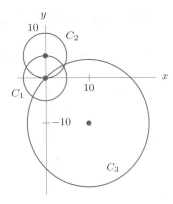

Figure 4.103

31. Suppose $a, b, c, d, m, n, p, q > 0$. Match each pair of parametric equations with one of the lines l_1, l_2, l_3, l_4 in Figure 4.104.

I. $\begin{cases} x = a + ct, \\ y = -b + dt. \end{cases}$ II. $\begin{cases} x = m + pt, \\ y = n - qt. \end{cases}$

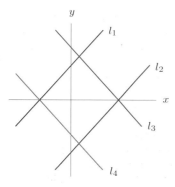

Figure 4.104

32. Describe in words the curve represented by the parametric equations

$$x = 3 + t^3, \quad y = 5 - t^3.$$

33. (a) Sketch the parameterized curve $x = t \cos t$, $y = t \sin t$ for $0 \le t \le 4\pi$.
   (b) By calculating the position at $t = 2$ and $t = 2.01$, estimate the speed at $t = 2$.
   (c) Use derivatives to calculate the speed at $t = 2$ and compare your answer to part (b).

34. The position of a particle at time t is given by $x = e^t$ and $y = 2e^{2t}$.
   (a) Find dy/dx in terms of t.
   (b) Eliminate the parameter and write y in terms of x.
   (c) Using your answer to part (b), find dy/dx in terms of x.

35. For x and y in meters, the motion of the particle given by

$$x = t^3 - 3t, \quad y = t^2 - 2t,$$

where the y-axis is vertical and the x-axis is horizontal.
   (a) Does the particle ever come to a stop? If so, when and where?
   (b) Is the particle ever moving straight up or down? If so, when and where?
   (c) Is the particle ever moving straight horizontally right or left? If so, when and where?

36. At time t, the position of a particle moving on a curve is given by $x = e^{2t} - e^{-2t}$ and $y = 3e^{2t} + e^{-2t}$.
   (a) Find all values of t at which the curve has
      (i) A horizontal tangent.
      (ii) A vertical tangent.
   (b) Find dy/dx in terms of t.
   (c) Find $\lim_{t \to \infty} dy/dx$.

37. At tme t, the position of a particle is $x(t) = 5\sin(2t)$ and $y(t) = 4\cos(2t)$, with $0 \le t < 2\pi$.
   (a) Graph the path of the particle for $0 \le t < 2\pi$, indicating the direction of motion.
   (b) Find the position and velocity of the particle when $t = \pi/4$.
   (c) How many times does the particle pass through the point found in part (b)?
   (d) What does your answer to part (b) tell you about the direction of the motion relative to the coordinate axes when $t = \pi/4$?
   (e) What is the speed of the particle at time $t = \pi$?

38. At time t, a projectile launched with angle of elevation α and initial velocity v_0 has position $x(t) = (v_0 \cos \alpha)t$ and $y(t) = (v_0 \sin \alpha)t - \frac{1}{2}gt^2$, where g is the acceleration due to gravity.
   (a) A football player kicks a ball at an angle of $36°$ above the ground with an initial velocity of 60 feet per second. Write the parametric equations for the position of the football at time t seconds. Use $g = 32\text{ft/sec}^2$.
   (b) Graph the path that the football follows.
   (c) How long does it take for the football to hit the ground? How far is it from the spot where the football player kicked it?
   (d) What is the maximum height the football reaches during its flight?
   (e) At what speed is the football traveling 1 second after it was kicked?

39. Two particles move in the xy-plane. At time t, the position of particle A is given by $x(t) = 4t - 4$ and $y(t) = 2t - k$, and the position of particle B is given by $x(t) = 3t$ and $y(t) = t^2 - 2t - 1$.
   (a) If $k = 5$, do the particles ever collide? Explain.
   (b) Find k so that the two particles do collide.
   (c) At the time that the particles collide in part (b), which particle is moving faster?

40. (a) Find d^2y/dx^2 for $x = t^3 + t, y = t^2$.
   (b) Is the curve concave up or down at $t = 1$?

41. (a) An object moves along the path $x = 3t$ and $y = \cos(2t)$, where t is time. Write the equation for the line tangent to this path at $t = \pi/3$.
   (b) Find the smallest positive value of t for which the y-coordinate is a local maximum.
   (c) Find d^2y/dx^2 when $t = 2$. What does this tell you about the concavity of the graph at $t = 2$?

42. The position of a particle at time t is given by $x = e^t + 3$ and $y = e^{2t} + 6e^t + 9$.

(a) Find dy/dx in terms of t.

(b) Find d^2y/dx^2. What does this tell you about the concavity of the graph?

(c) Eliminate the parameter and write y in terms of x.

(d) Using your answer from part (c), find dy/dx and d^2y/dx^2 in terms of x. Show that these answers are the same as the answers to parts (a) and (b).

43. A particle moves in the xy-plane so that its position at time t is given by $x = \sin t$ and $y = \cos(2t)$ for $0 \leq t < 2\pi$.

(a) At what time does the particle first touch the x-axis? What is the speed of the particle at that time?

(b) Is the particle ever at rest?

(c) Discuss the concavity of the graph.

44. Derive the general formula for the second derivative d^2y/dx^2 of a parametrically defined curve:

$$\frac{d^2y}{dx^2} = \frac{(dx/dt)(d^2y/dt^2) - (dy/dt)(d^2x/dt^2)}{(dx/dt)^3}.$$

Graph the Lissajous figures in Problems 45–48 using a calculator or computer.

45. $x = \cos 2t, \quad y = \sin 5t$

46. $x = \cos 3t, \quad y = \sin 7t$

47. $x = \cos 2t, \quad y = \sin 4t$

48. $x = \cos 2t, \quad y = \sin \sqrt{3}t$

49. A hypothetical moon orbits a planet which in turn orbits a star. Suppose that the orbits are circular and that the moon orbits the planet 12 times in the time it takes for the planet to orbit the star once. In this problem we will investigate whether the moon could come to a stop at some instant. (See Figure 4.105.)

(a) Suppose the radius of the moon's orbit around the planet is 1 unit and the radius of the planet's orbit around the star is R units. Explain why the motion of the moon relative to the star can be described by the parametric equations

$$x = R\cos t + \cos(12t), \quad y = R\sin t + \sin(12t).$$

(b) Find values for R and t such that the moon stops relative to the star at time t.

(c) On a graphing calculator, plot the path of the moon for the value of R you obtained in part (b). Experiment with other values for R.

Figure 4.105

CHAPTER SUMMARY (see also Ready Reference at the end of the book)

- **Local extrema**
  Maximum, minimum, critical point, tests for local maxima/minima.

- **Using the second derivative**
  Concavity, inflection point.

- **Families of curves**
  Role of parameters

- **Optimization**
  Global extremum, modeling problems, graphical optimization, upper and lower bounds, extreme value theorem.

- **Marginality**
  Cost/revenue functions, marginal cost/marginal revenue functions.

- **L'Hopital's Rule**
  Limits, growth, dominance

- **Rates and Related Rates**

- **Parametric equations**
  Motion of a particle in the plane, parametric equations for lines and other curves, velocity, tangent lines, slope, concavity.

REVIEW EXERCISES AND PROBLEMS FOR CHAPTER FOUR

Exercises

For Exercises 1–2, indicate all critical points on the given graphs. Which correspond to local minima, local maxima, global maxima, global minima, or none of these? (Note that the graphs are on closed intervals.)

1.

2.

In Exercises 3–6, do the following:

(a) Find f' and f''.

(b) Find the critical points of f.

(c) Find any inflection points.

(d) Evaluate f at the critical points and the endpoints. Identify the global maxima and minima of f.

(e) Sketch f. Indicate clearly where f is increasing or decreasing, and its concavity.

3. $f(x) = x^3 - 3x^2 \quad (-1 \le x \le 3)$

4. $f(x) = x + \sin x \quad (0 \le x \le 2\pi)$

5. $f(x) = e^{-x} \sin x \quad (0 \le x \le 2\pi)$

6. $f(x) = x^{-2/3} + x^{1/3} \quad (1.2 \le x \le 3.5)$

In Exercises 7–9, find the limits as x tends to $+\infty$ and $-\infty$, and then proceed as in Exercises 3–6. (That is, find f', etc.).

7. $f(x) = 2x^3 - 9x^2 + 12x + 1$

8. $f(x) = \dfrac{4x^2}{x^2 + 1}$

9. $f(x) = xe^{-x}$

In Exercises 10–12, find the exact global maximum and minimum values of the function.

10. $h(z) = \dfrac{1}{z} + 4z^2$ for $z > 0$

11. $g(t) = \dfrac{1}{t^3 + 1}$ for $t \ge 0$

12. $f(x) = \dfrac{1}{(x - 1)^2 + 2}$

In Exercises 13–18, use derivatives to identify local maxima and minima and points of inflection. Graph the function. Confirm your answers using a calculator or computer.

13. $f(x) = x^3 + 3x^2 - 9x - 15$

14. $f(x) = x^5 - 15x^3 + 10$

15. $f(x) = x - 2\ln x \quad$ for $x > 0$

16. $f(x) = x^2 e^{5x}$

17. $f(x) = e^{-x^2}$

18. $f(x) = \dfrac{x^2}{x^2 + 1}$

Find the best possible bounds for the functions in Exercises 19–21.

19. $e^{-x} \sin x, \quad$ for $x \ge 0$

20. $x \sin x, \quad$ for $0 \le x \le 2\pi$

21. $x^3 - 6x^2 + 9x + 5 \quad$ for $0 \le x \le 5$

Problems

22. For the function, f, graphed in Figure 4.106:

(a) Sketch $f'(x)$.

(b) Where does $f'(x)$ change its sign?

(c) Where does $f'(x)$ have local maxima or minima?

Figure 4.106

23. Using your answer to Problem 22 as a guide, write a short paragraph (using complete sentences) which describes the relationships between the following features of a function f:

- The local maxima and minima of f.
- The points at which the graph of f changes concavity.
- The sign changes of f'.
- The local maxima and minima of f'.

24. Figure 4.107 is a graph of f'. For what values of x does f have a local maximum? A local minimum?

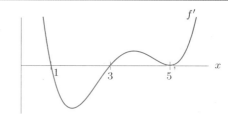

Figure 4.107: Graph of f' (not f)

25. On the graph of f' in Figure 4.108, indicate the x-values that are critical points of the function f itself. Are they local maxima, local minima, or neither?

Figure 4.108: Graph of f', not f

26. Figure 4.109 is the graph of f', the derivative of a function f. At which of the points $0, x_1, x_2, x_3, x_4, x_5$, is the function f:

(a) At a local maximum value?
(b) At a local minimum value?
(c) Climbing fastest?
(d) Falling most steeply?

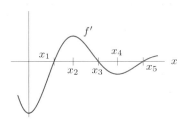

Figure 4.109: Graph of f' not f

For the graphs of f' in Problems 27–30 decide:
(a) Over what intervals is f increasing? Decreasing?
(b) Does f have maxima or minima? If so, which, and where?

27.

28.

29.

30.

31. Find values of a and b so that the function $y = axe^{-bx}$ has a local maximum at the point $(2, 10)$.

32. A drug is injected into a patient at a rate given by $r(t) = ate^{-bt}$ ml/sec, where t is in seconds since the injection started, $0 \le t \le 5$, and a and b are constants. The maximum rate of 0.3 ml/sec occurs half a second after the injection starts. Find a formula for a and b.

33. Any body radiates energy at various wavelengths. Figure 4.110 shows the intensity of the radiation of a black body at a temperature $T = 3000°$ kelvin as a function of the wavelength. The intensity of the radiation is highest in the infrared range, that is, at wavelengths longer than

that of visible light (0.4–0.7μm). Max Planck's radiation law, announced to the Berlin Physical Society on October 19, 1900, states that

$$r(\lambda) = \frac{a}{\lambda^5(e^{b/\lambda} - 1)}.$$

Find constants a and b so that the formula fits the graph. (Later in 1900 Planck showed from theory that $a = 2\pi c^2 h$ and $b = \frac{hc}{Tk}$ where c = speed of light, h = Planck's constant, and k = Boltzmann's constant.)

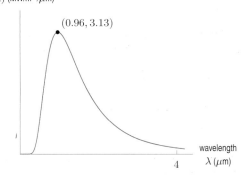

Figure 4.110

34. Sketch several members of the family $y = x^3 - ax^2$ on the same axes. Show that the critical points lie on the curve $y = -\frac{1}{2}x^3$.

35. A right triangle has one vertex at the origin and one vertex on the curve $y = e^{-x/3}$ for $1 \le x \le 5$. One of the two perpendicular sides is along the x-axis; the other is parallel to the y-axis. Find the maximum and minimum areas for such a triangle.

36. A rectangle has one side on the x-axis and two corners on the top half of the circle of radius 1 centered at the origin. Find the maximum area of such a rectangle. What are the coordinates of its vertices?

37. A square-bottomed box with a top has a fixed volume, V. What dimensions minimize the surface area?

38. A business sells an item at a constant rate of r units per month. It reorders in batches of q units, at a cost of $a + bq$ dollars per order. Storage costs are k dollars per item per month, and, on average, $q/2$ items are in storage, waiting to be sold. [Assume r, a, b, k are positive constants.]

(a) How often does the business reorder?
(b) What is the average monthly cost of reordering?
(c) What is the total monthly cost, C of ordering and storage?
(d) Obtain Wilson's lot size formula, the optimal batch size which minimizes cost.

39. A ship is steaming due north at 12 knots (1 knot = 1.85 kilometers/hour) and sights a large tanker 3 kilometers away northwest steaming at 15 knots due east. For reasons of safety, the ships want to maintain a distance of at least 100 meters between them. Use a calculator or computer to determine the shortest distance between them if they remain on their current headings, and hence decide if they need to change course.

40. Boise, Idaho, is about 300 miles inland from the nearest point on the Pacific coast; San Diego is about 1000 miles south of that point down the coast. (See Figure 4.111.) Assuming the coast is a straight line going north-south, C is the point along the coast directly west of Boise. It costs 2 cents per mile to transport a ton of potatoes by truck and 1 cent per mile to transport them by ship. The Idaho Potato Company wants to find the point, P, on the Pacific coast to which it should truck its potatoes before loading them aboard a cargo ship in order to minimize the total cost of transporting them from Boise to San Diego.

  (a) Set up the function which The Idaho Potato Company must minimize.
  (b) Find the position of P which minimizes cost.

Figure 4.111

41. A polystyrene cup is in the shape of a frustum (the part of a cone between two parallel planes cutting the cone), has top radius $2r$, base radius r and height h. The surface area S of such a cup is given by $S = 3\pi r\sqrt{r^2 + h^2}$ and its volume V by $V = 7\pi r^2 h/3$. If the cup is to hold 200 ml, use a calculator or a computer to estimate the value of r that minimizes its surface area.

42. You are given the n numbers $a_1, a_2, a_3, \cdots, a_n$. For a variable x, consider the expression

$$D = (x-a_1)^2 + (x-a_2)^2 + (x-a_3)^2 + \cdots + (x-a_n)^2.$$

Show that D is a minimum when x is the average of $a_1, a_2, a_3, \cdots, a_n$.

43. Suppose $g(t) = (\ln t)/t$ for $t > 0$.

  (a) Does g have either a global maximum or a global minimum on $0 < t < \infty$? If so, where, and what are their values?

  (b) What does your answer to part (a) tell you about the number of solutions to the equation

$$\frac{\ln x}{x} = \frac{\ln 5}{5}?$$

  (Note: There are many ways to investigate the number of solutions to this equation. We are asking you to draw a conclusion from your answer to part (a).)
  (c) Estimate the solution(s).

44. For $a > 0$, the following line forms a triangle in the first quadrant with the x- and y-axes:

$$a(a^2 + 1)y = a - x.$$

  (a) In terms of a, find the x- and y-intercepts of the line.
  (b) Find the area of the triangle, as a function of a.
  (c) Find the value of a making the area a maximum.
  (d) What is this greatest area?
  (e) If you want the triangle to have area $1/5$, what choices do you have for a?

45. A piece of wire of length L cm is cut into two pieces. One piece, of length x cm, is made into a circle; the rest is made into a square.

  (a) Find the value of x that makes the sum of the areas of the circle and square a minimum. Find the value of x giving a maximum.
  (b) For the values of x found in part (a), show that the ratio of the length of wire in the square to the length of wire in the circle is equal to the ratio of the area of the square to the area of the circle.[9]
  (c) Are the values of x found in part (a) the only values of x for which the ratios in part (b) are equal?

46. The vase in Figure 4.112 is filled with water at a constant rate (i.e., constant volume per unit time).

  (a) Graph $y = f(t)$, the depth of the water, against time, t. Show on your graph the points at which the concavity changes.
  (b) At what depth is $y = f(t)$ growing most quickly? Most slowly? Estimate the ratio between the growth rates at these two depths.

Figure 4.112

---

[9]From Sally Thomas.

47. (a) Graph the functions $-\dfrac{1}{x}$ and $\dfrac{1}{a}\left(\arctan\left(\dfrac{x}{a}\right) - \dfrac{\pi}{2}\right)$ when $x > 0$ for $a = 0.1$, $a = 0.01$, and $a = 0.001$. Describe what you see.

(b) Evaluate $\displaystyle\lim_{a \to 0^+} \dfrac{1}{a}\left(\arctan\left(\dfrac{x}{a}\right) - \dfrac{\pi}{2}\right)$.

(c) Explain the connection between parts (a) and (b).

In Problems 48–49 determine whether the limit exists, and where possible evaluate it.

48. $\displaystyle\lim_{x \to 0} \dfrac{1 - \cosh(5x)}{x^2}$ **49.** $\displaystyle\lim_{x \to 0} \dfrac{x - \sinh(x)}{x^3}$

50. The rate of change of a population depends on the current population, P, and is given by

$$\frac{dP}{dt} = kP(L - P) \qquad \text{for positive constants } k, L.$$

(a) For what nonnegative values of P is the population increasing? Decreasing? For what values of P does the population remain constant?

(b) Find d^2P/dt^2 as a function of P.

51. A spherical balloon is inflated so that its radius is increasing at a constant rate of 1 cm per second. At what rate is air being blown into the balloon when its radius is 5 cm?

52. When the growth of a spherical cell depends on the flow of nutrients through the surface, it is reasonable to assume that the growth rate, dV/dt, is proportional to the surface area, S. Assume that for a particular cell $dV/dt = \frac{1}{3} \cdot S$. At what rate is its radius r increasing?

53. A horizontal disk of radius a centered at the origin in the xy-plane is rotating about a vertical axis through the center. The angle between the positive x-axis and a radial line painted on the disk is θ radians.

(a) What does $d\theta/dt$ represent?

(b) What is the relationship between $d\theta/dt$ and the speed v of a point on the rim?

54. A chemical storage tank is in the shape of an inverted cone with depth 12 meters and top radius 5 meters. When the depth of the chemical in the tank is 1 meter, the level is falling at 0.1 meters per minute. How fast is the volume of chemical changing?

55. A voltage, V volts, applied to a resistor of R ohms produces an electric current of I amps where $V = IR$. As the current flows the resistor heats up and its resistance falls. If 100 volts is applied to a resistor of 1000 ohms the current is initially 0.1 amps but rises by 0.001 amps/minute. At what rate is the resistance falling if the voltage remains constant?

56. A radio navigation system used by aircraft gives a cockpit readout of the distance, s, in miles, between a fixed ground station and the aircraft. The system also gives a readout of the instantaneous rate of change, ds/dt, of this distance in miles/hour. An aircraft on a straight flight path at a constant altitude of 10,560 feet (2 miles) has passed directly over the ground station and is now flying away from it. What is the speed of this aircraft along its constant altitude flight path when the cockpit readouts are $s = 4.6$ miles and $ds/dt = 210$ miles/hour?

Problems 57–58 involve Boyle's Law which states that for a fixed quantity of gas at a constant temperature, the pressure, P, and the volume, V, are inversely related. Thus, for some constant k

$$PV = k.$$

57. A fixed quantity of gas is allowed to expand at constant temperature. Find the rate of change of pressure with respect to volume.

58. A certain quantity of gas occupies 10 cm^3 at a pressure of 2 atmospheres. The pressure is increased, while keeping the temperature constant.

(a) Does the volume increase or decrease?

(b) If the pressure is increasing at a rate of 0.05 atmospheres/minute when the pressure is 2 atmospheres, find the rate at which the volume is changing at that moment. What are the units of your answer?

CAS Challenge Problems

59. A population, P, in a restricted environment may grow with time, t, according to the *logistic function*

$$P = \frac{L}{1 + Ce^{-kt}}$$

where L is called the carrying capacity and L, C and k are positive constants.

(a) Find $\displaystyle\lim_{t \to \infty} P$. Explain why L is called the carrying capacity.

(b) Using a computer algebra system, show that the graph of P has an inflection point at $P = L/2$.

60. For positive a, consider the family of functions

$$y = \arctan\left(\frac{\sqrt{x} + \sqrt{a}}{1 - \sqrt{ax}}\right), \qquad x > 0.$$

(a) Graph several curves in the family and describe how the graph changes as a varies.

(b) Use a computer algebra system to find dy/dx, and graph the derivative for several values of a. What do you notice?

(c) Do your observations in part (b) agree with the answer to part (a)? Explain. [Hint: Use the fact that $\sqrt{ax} = \sqrt{a}\sqrt{x}$ for $a > 0$, $x > 0$.]

61. The function $\mathrm{arcsinh}\,x$ is the inverse function of $\sinh x$.

(a) Use a computer algebra system to find a formula for the derivative of $\mathrm{arcsinh}\,x$.

(b) Derive the formula by hand by differentiating both sides of the equation

$$\sinh(\mathrm{arcsinh}\,x) = x.$$

[Hint: Use the identity $\cosh^2 x - \sinh^2 x = 1$.]

62. The function $\mathrm{arccosh}\,x$, for $x \geq 0$, is the inverse function of $\cosh x$, for $x \geq 0$.

(a) Use a computer algebra system to find the derivative of $\mathrm{arccosh}\,x$.

(b) Derive the formula by hand by differentiating both sides of the equation

$$\cosh(\mathrm{arccosh}\,x) = x, \quad x \geq 1.$$

[Hint: Use the identity $\cosh^2 x - \sinh^2 x = 1$.]

63. Consider the family of functions

$$f(x) = \frac{\sqrt{a+x}}{\sqrt{a}+\sqrt{x}}, \quad x \geq 0, \quad \text{for positive } a.$$

(a) Using a computer algebra system, find the local maxima and minima of f.

(b) On one set of axes, graph this function for several values of a. How does varying a affect the shape of the graph? Explain your answer in terms of the answer to part (a).

(c) Use your computer algebra system to find the inflection points of f when $a = 2$.

64. (a) Use a computer algebra system to find the derivative of

$$y = \arctan\left(\sqrt{\frac{1-\cos x}{1+\cos x}}\right).$$

(b) Graph the derivative. Does the graph agree with the answer you got in part (a)? Explain using the identity $\cos(x) = \cos^2(x/2) - \sin^2(x/2)$.

65. In 1696, the first calculus textbook was published by the Marquis de l'Hopital. The following problem is a simplified version of a problem from this text.

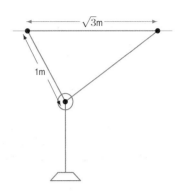

Figure 4.113

In Figure 4.113, two ropes are attached to the ceiling at points $\sqrt{3}$ meters apart. The rope on the left is 1 meter long and has a pulley attached at its end. The rope on the right is 3 meters long; it passes through the pulley and has a weight tied to its end. When the weight is released, the ropes and pulley arrange themselves so that the distance from the weight to the ceiling is maximized.

(a) Show that the maximum distance occurs when the weight is exactly halfway between the the points where the ropes are attached to the ceiling. [Hint: Write the vertical distance from the weight to the ceiling in terms of its horizontal distance to the point at which the left rope is tied to the ceiling. A computer algebra system will be useful.]

(b) Does the weight always end up halfway between the ceiling anchor points no matter how long the left-hand rope is? Explain.

CHECK YOUR UNDERSTANDING

Are the statements in Problems 1–8 true or false for a function f whose domain is all real numbers? If a statement is true, explain how you know. If a statement is false, give a counterexample.

1. A local minimum of f occurs at a critical point of f.

2. If $x = p$ is not a critical point of f, then $x = p$ is not a local maximum of f.

3. A local maximum of f occurs at a point where

$$f'(x) = 0.$$

4. If $x = p$ is not a local maximum of f, then $x = p$ is not a critical point of f.

5. If $f'(p) = 0$, then $f(x)$ has a local minimum or local maximum at $x = p$.

6. If $f'(x)$ is continuous and $f(x)$ has no critical points, then f is everywhere increasing or everywhere decreasing.

7. If $f''(p) = 0$, then the graph of f has an inflection point at $x = p$.

8. If $f''(x)$ is continuous and the graph of f has an inflection point at $x = p$, then $f''(p) = 0$.

In Problems 9–10, give an example of function(s) with the given properties.

9. A family of functions, $f(x)$, depending on a parameter a, such that each member of the family has exactly one critical point.

10. A family of functions, $g(x)$, depending on two parameters, a and b, such that each member of the family has exactly two critical points and one inflection point. You may want to restrict a and b.

11. Let $f(x) = x^2$. Decide if the following statements are true or false. Explain your answer.

(a) f has an upper bound on the interval $(0, 2)$.
(b) f has a global maximum on the interval $(0, 2)$.
(c) f does not have a global minimum on the interval $(0, 2)$.
(d) f does not have a global minimum on any interval (a, b).
(e) f has a global minimum on any interval $[a, b]$.

12. Which of the following statements is implied by the statement "If f is continuous on $[a, b]$ then f has a global maximum on $[a, b]$"?

(a) If f has a global maximum on $[a, b]$ then f must be continuous on $[a, b]$.
(b) If f is not continuous on $[a, b]$ then f does not have a global maximum on $[a, b]$.
(c) If f does not have a global maximum on $[a, b]$ then f is not continuous on $[a, b]$.

Are the statements in Problems 13–20 true of false? Give an explanation for your answer.

13. The global maximum of $f(x) = x^2$ on every closed interval is at one of the endpoints of the interval.

14. A function can have two different upper bounds.

15. If a differentiable function $f(x)$ has a global maximum on the interval $0 \le x \le 10$ at $x = 0$, then $f'(0) \le 0$.

16. If the radius of a circle is increasing at a constant rate, then so is the circumference.

17. If the radius of a circle is increasing at a constant rate, then so is the area.

18. The curve given parametrically by $x = f(t)$ and $y = g(t)$ has no sharp corners if f and g are differentiable.

19. If a curve is given parametrically by $x = \cos(t^2)$, $y = \sin(t^2)$, then its slope is $\tan(t^2)$.

20. If $g'(a) \ne 0$, then $\lim_{x \to a} \dfrac{f(x)}{g(x)} = \dfrac{f'(a)}{g'(a)}$.

In Problems 21–26, give an example of a function f that makes the statement true, or say why such an example is impossible. Assume that f'' exists everywhere.

21. f is concave up and $f(x)$ is positive for all x.

22. f is concave down and $f(x)$ is positive for all x.

23. f is concave down and $f(x)$ is negative for all x.

24. f is concave up and $f(x)$ is negative for all x.

25. $f(x)f''(x) < 0$ for all x.

26. $f(x)f'(x)f''(x)f'''(x) < 0$ for all x.[10]

PROJECTS FOR CHAPTER FOUR

1. Building a Greenhouse

Your parents are going to knock out the bottom of the entire length of the south wall of their house and turn it into a greenhouse by replacing the bottom portion of the wall with a huge sloped piece of glass (which is expensive). They have already decided they are going to spend a certain fixed amount. The triangular ends of the greenhouse will be made of various materials they already have lying around.[11]

The floor space in the greenhouse is only considered usable if they can both stand up in it, so part of it will be unusable. They want to choose the dimensions of the greenhouse to get the most usable floor space. What should the dimensions of the greenhouse be and how much usable space will your parents get?

2. Fitting a Line to Data

(a) The line which best fits the data points $(x_1, y_1), (x_2, y_2) \ldots (x_n, y_n)$ is the one which minimizes the sum of the squares of the vertical distances from the points to the line. These are the distances marked in Figure 4.114. Find the best fitting line of the form $y = mx$ for the points $(2, 3.5), (3, 6.8), (5, 9.1)$.

---

[10]From the 1998 William Lowell Putnam Mathematical Competition, by permission of the Mathematical Association of America.

[11]Adapted from M. Cohen, E. Gaughan, A. Knoebel, D. Kurtz, D. Pengelley, *Student Research Projects in Calculus* (Washington DC: Mathematical Association of America, 1992).

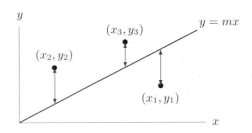

Figure 4.114

(b) A cone with height and radius both equal to r has volume, V, proportional to r^3; that is, $V = kr^3$ for some constant k. A lab experiment is done to measure the volume of several cones; the results are in the following table. Using the method of part (a), determine the best value of k. [Note: Since the volumes were determined experimentally, the values may not be accurate. Assume that the radii were measured accurately.]

Radius (cm)	2	5	7	8
Volume (cm^3)	8.7	140.3	355.8	539.2

(c) Using the method of part (a), show that the best fitting line of the form $y = mx$ for the points $(x_1, y_1), (x_2, y_2) \ldots (x_n, y_n)$ has

$$m = \frac{x_1 y_1 + x_2 y_2 + \cdots + x_n y_n}{x_1^2 + x_2^2 + \cdots + x_n^2}.$$

3. Interstate Trucking

In this project we investigate the effect of the price of diesel fuel and drivers' wages on a trucker's optimal driving speed. We analyze the relationship between fuel cost and wages and the 55 mph speed limit. [12]

After an extensive study, the Interstate Commerce Commission has concluded that the primary variable expenses for any over-the-road freight hauler are the wages for the driver and the cost of fuel. The study concluded that maintenance and replacement costs for vehicles, although considerable, did not vary significantly from carrier to carrier. On the other hand, the fuel costs and wages did vary. Given this result, your project is to determine how these two variables affect the cost of transporting freight. In particular, you are to determine if there is an optimal wage to fuel cost relationship for which most haulers would encourage their drivers to abide by a 55 mile per hour speed limit.

Reading the study tells you that under ideal conditions, an interstate freight hauling vehicle (18 wheeler) gets 6 miles per gallon of fuel. This mileage is affected by the speed of the vehicle and its weight. In general, the miles per gallon of fuel decreases by 0.2 for each increase of 10,000 pounds in the weight of the truck and freight over 25,000 pounds. Also, the miles per gallon of fuel decreases by 0.1 mile per gallon for each mile per hour the truck averages above 45 miles per hour.

(a) Using this information, create an expression for the cost per mile of driving, taking into account only the drivers' wages and fuel cost.

(b) Suppose that the national average for diesel fuel is $1.25 per gallon, that drivers on the average earn $15.00 per hour, and that the average weight of a loaded truck is 75,000 pounds. What is the optimal average speed under these conditions?

(c) Compute the cost per mile for 55 and 60 miles per hour. Suppose you hauled produce from California to Maine. What average speed would you choose?

[12] Adapted from L. Carl Leinbach, *Calculus Laboratories Using Derive*, (Belmont, CA: Wadsworth, 1991). Reprinted by permission of Brooks/Cole Publishing Company, a division of International Thompson Publishing, Inc.

(d) In solving parts (a)-(c), you found an equation relating cost, wages, fuel cost, weight of the truck, and the average speed of the truck. This equation is a mathematical model of interstate freight hauling costs. Using this model, determine the fuel cost that makes 55 miles per hour the optimum average speed for trucking companies.

(e) Each year drivers' wages change as a result of contract renewals and inflation. The cost of fuel also fluctuates, but can be adjusted by surcharges and fuel taxes. Thus, the relationship between wages and the cost of fuel can be altered by various government agencies. Assuming that 75,000 pounds remains the national average for the weight of a truck involved in interstate freight hauling, what should be the relationship between fuel cost and wages to maintain 55 miles per hour as the optimal speed?

(f) There have been some suggestions of changing the road use tax to lower the average weight of over-the-road freight haulers. Assuming the relationship of wages to fuel cost found in part (e) is maintained, how would lowering the average weight affect the optimal average speed?

4. Firebreaks

The summer of 2000 was devastating for forests in the western US: over 3.5 million acres of trees were lost to fires, making this the worst fire season in 30 years. This project studies a fire management technique called *firebreaks*, which reduce the damage done by forest fires. A firebreak is a strip where trees have been removed in a forest so that a fire started on one side of the strip will not spread to the other side. Having many firebreaks helps confine a fire to a small area. On the other hand, having too many firebreaks involves removing large swaths of trees.[13]

(a) A forest in the shape of a 50 km by 50 km square has firebreaks in rectangular strips 50 km by 0.01 km. The trees between two firebreaks are called a stand of trees. All firebreaks in this forest are parallel to each other and to one edge of the forest, with the first firebreak at the edge of the forest. The firebreaks are evenly spaced throughout the forest. (For example, Figure 4.115 shows four firebreaks.) The total area lost in the case of a fire is the area of the stand of trees in which the fire started plus the area of all the firebreaks.

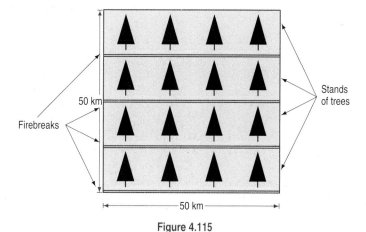

Figure 4.115

(i) Find the number of firebreaks that minimizes the total area lost to the forest in the case of a fire.

(ii) If a firebreak is 50 km by b km, find the optimal number of firebreaks as a function of b. If the width, b, of a firebreak is quadrupled, how does the optimal number of firebreaks change?

---

[13] Adapted from D. Quinney and R. Harding, *Calculus Connections* (New York: John Wiley & Sons, 1996).

(b) Now suppose firebreaks are arranged in two equally spaced sets of parallel lines, as shown in Figure 4.116. The forest is a 50 km by 50 km square, and each firebreak is a rectangular strip 50 km by 0.01 km. Find the number of firebreaks in each direction that minimizes the total area lost to the forest in the case of a fire.

Figure 4.116

Chapter Five

KEY CONCEPT: THE DEFINITE INTEGRAL

We started Chapter 2 by calculating the velocity from the distance traveled. This led us to the notion of the derivative, or rate of change, of a function. Now we consider the reverse problem: given the velocity of a moving object, how can we calculate the distance traveled? This leads to our second key concept, the *definite integral*, which computes the total change in a function from its rate of change. We then see that the definite integral can be used to calculate not only distance but also other quantities, such as the area under a curve and the average value of a function.

The central result in this chapter is the Fundamental Theorem of Calculus, which tells us that we can use a definite integral to get information about a function from its derivative. Calculating derivatives and calculating definite integrals are, in a sense, reverse processes.

5.1 HOW DO WE MEASURE DISTANCE TRAVELED?

For positive constant velocities, we can find the distance a moving object travels using the formula

$$\text{Distance} = \text{Velocity} \times \text{Time.}$$

In this section we see how to estimate the distance when the velocity is not a constant.

A Thought Experiment: How Far Did the Car Go?

Velocity Data Every Two Seconds

A car is moving with increasing velocity. Table 5.1 shows the velocity every two seconds:

Table 5.1 *Velocity of car every two seconds*

Time (sec)	0	2	4	6	8	10
Velocity (ft/sec)	20	30	38	44	48	50

How far has the car traveled? Since we don't know how fast the car is moving at every moment, we can't calculate the distance exactly, but we can make an estimate. The velocity is increasing, so the car is going at least 20 ft/sec for the first two seconds. Since Distance = Velocity × Time, the car goes at least $20 \cdot 2 = 40$ feet during the first two seconds. Likewise, it goes at least $30 \cdot 2 = 60$ feet during the next two seconds, and so on. During the ten-second period it goes at least

$$20 \cdot 2 + 30 \cdot 2 + 38 \cdot 2 + 44 \cdot 2 + 48 \cdot 2 = 360 \text{ feet.}$$

Thus, 360 feet is an underestimate of the total distance traveled during the ten seconds.

To get an overestimate, we can reason this way: During the first two seconds, the car's velocity is at most 30 ft/sec, so it moved at most $30 \cdot 2 = 60$ feet. In the next two seconds it moved at most $38 \cdot 2 = 76$ feet, and so on. Therefore, over the ten-second period it moved at most

$$30 \cdot 2 + 38 \cdot 2 + 44 \cdot 2 + 48 \cdot 2 + 50 \cdot 2 = 420 \text{ feet.}$$

Therefore,

$$360 \text{ feet} \leq \text{Total distance traveled} \leq 420 \text{ feet.}$$

There is a difference of 60 feet between the upper and lower estimates.

Velocity Data Every One Second

What if we want a more accurate estimate? Then we make more frequent velocity measurements, say every second, as in Table 5.2.

As before, we get a lower estimate for each second by using the velocity at the beginning of that second. During the first second the velocity is at least 20 ft/sec, so the car travels at least $20 \cdot 1 = 20$ feet. During the next second the car moves at least 26 feet, and so on. We have

$$\text{New lower estimate} = 20 \cdot 1 + 26 \cdot 1 + 30 \cdot 1 + 34 \cdot 1 + 38 \cdot 1$$
$$+ 41 \cdot 1 + 44 \cdot 1 + 46 \cdot 1 + 48 \cdot 1 + 49 \cdot 1$$
$$= 376 \text{ feet.}$$

Table 5.2 *Velocity of car every second*

Time (sec)	0	1	2	3	4	5	6	7	8	9	10
Velocity (ft/sec)	20	26	30	34	38	41	44	46	48	49	50

Notice that this lower estimate is greater than the old lower estimate of 360 feet.

We get a new upper estimate by considering the velocity at the end of each second. During the first second the velocity is at most 26 ft/sec, so the car moves at most $26 \cdot 1 = 26$ feet; in the next second it moves at most 30 feet, and so on.

$$\text{New upper estimate} = 26 \cdot 1 + 30 \cdot 1 + 34 \cdot 1 + 38 \cdot 1 + 41 \cdot 1$$
$$+ \, 44 \cdot 1 + 46 \cdot 1 + 48 \cdot 1 + 49 \cdot 1 + 50 \cdot 1$$
$$= 406 \text{ feet.}$$

This is less than the old upper estimate of 420 feet. Now we know that

$$376 \text{ feet} \leq \text{Total distance traveled} \leq 406 \text{ feet.}$$

The difference between upper and lower estimates is now 30 feet, half of what it was before. By halving the interval of measurement, we have halved the difference between the upper and lower estimates.

Visualizing Distance on the Velocity Graph: Two Second Data

We can represent both upper and lower estimates on a graph of the velocity. The graph also shows how changing the time interval between velocity measurements changes the accuracy of our estimates.

The velocity can be graphed by plotting the two-second data in Table 5.1 and drawing a curve through the data points. (See Figure 5.1.) The area of the first dark rectangle is $20 \cdot 2 = 40$, the lower estimate of the distance moved during the first two seconds. The area of the second dark rectangle is $30 \cdot 2 = 60$, the lower estimate for the distance moved in the next two seconds. The total area of the dark rectangles represents the lower estimate for the total distance moved during the ten seconds.

If the dark and light rectangles are considered together, the first area is $30 \cdot 2 = 60$, the upper estimate for the distance moved in the first two seconds. The second area is $38 \cdot 2 = 76$, the upper estimate for the next two seconds. The upper estimate for the total distance is represented by the sum of the areas of the dark and light rectangles. Therefore, the area of the light rectangles alone represents the difference between the two estimates.

To visualize the difference between the two estimates, look at Figure 5.1 and imagine the light rectangles all pushed to the right and stacked on top of each other. This gives a rectangle of width 2 and height 30. The height, 30, is the difference between the initial and final values of the velocity: $30 = 50 - 20$. The width, 2, is the time interval between velocity measurements.

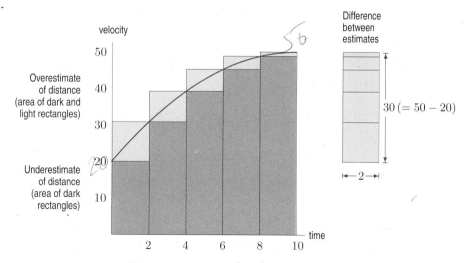

Figure 5.1: Velocity measured every 2 seconds

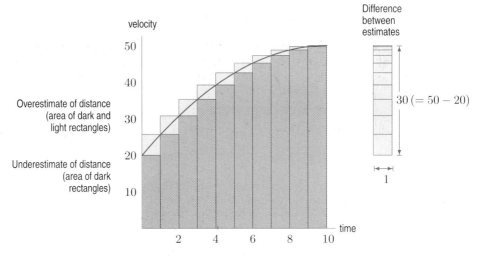

Figure 5.2: Velocity measured every second

Visualizing Distance on the Velocity Graph: One Second Data

Figure 5.2 shows the velocities measured every second. The area of the dark rectangles again represents the lower estimate, and the area of the dark and light rectangles together represent the upper estimate. As before, the difference between the two estimates is represented by the area of the light rectangles. This difference can be calculated by stacking the light rectangles vertically, giving a rectangle of the same height as before but of half the width. Its area is therefore half what it was before. Again, the height of this stack is $50 - 20 = 30$, but its width is now 1.

Example 1   What would be the difference between the upper and lower estimates if the velocity were given every tenth of a second? Every hundredth of a second? Every thousandth of a second?

Solution   Every tenth of a second: Difference between estimates $= (50 - 20)(1/10) = 3$ feet.
Every hundredth of a second: Difference between estimates $= (50 - 20)(1/100) = 0.3$ feet.
Every thousandth of a second: Difference between estimates $= (50 - 20)(1/1000) = 0.03$ feet.

Example 2   How frequently must the velocity be recorded in order to estimate the total distance traveled to within 0.1 feet?

Solution   The difference between the velocity at the beginning and end of the observation period is $50 - 20 = 30$. If the time between the measurements is h, then the difference between the upper and lower estimates is $(30)h$. We want

$$(30)h < 0.1,$$

or

$$h < \frac{0.1}{30} = 0.0033.$$

So if the measurements are made less than 0.0033 seconds apart, the distance estimate is accurate to within 0.1 feet.

Visualizing Distance on the Velocity Graph: Area Under Curve

As we make more frequent velocity measurements, the rectangles used to estimate the distance traveled fit the curve more closely. See Figures 5.3 and 5.4. In the limit, as the number of subdivisions increases, we see that the distance traveled is given by the area between the velocity curve and the horizontal axis. See Figure 5.5. In general:

> If the velocity is positive, the total distance traveled is the area under the velocity curve.

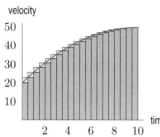

Figure 5.3: Velocity measured every 1/2 second

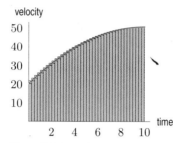

Figure 5.4: Velocity measured every 1/4 second

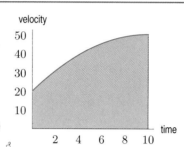

Figure 5.5: Distance traveled is area under curve

Example 3 With time t in seconds, the velocity of a bicycle, in feet per second, is given by $v(t) = 5t$. How far does the bicycle travel in 3 seconds?

Solution The velocity is linear. See Figure 5.6. The distance traveled is the area between the line $v(t) = 5t$ and the t-axis. Since this region is a triangle of height 15 and base 3

$$\text{Distance traveled } = \text{ Area of triangle } = \frac{1}{2} \cdot 15 \cdot 3 = 22.5 \text{ feet.}$$

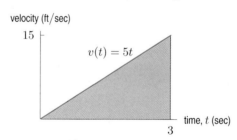

Figure 5.6: Shaded area represents distance traveled

Negative Velocity and Change in Position

In the thought experiment, the velocity is positive and our sums represent distance traveled. What if the velocity is sometimes negative?

Example 4 A particle moves along the y-axis with velocity 30 cm/sec for 5 seconds and velocity -10 cm/sec for the next 5 seconds. Positive velocity indicates upward motion; negative velocity represents downward motion. What is represented by the sum

$$30 \cdot 5 + (-10) \cdot 5?$$

Solution The first term in the sum represents an upward motion of $30 \cdot 5 = 150$ centimeters. The second term represents a motion of $(-10) \cdot 5 = -50$ centimeters, that is, 50 centimeters downward. Thus, the sum represents a change in position of $150 - 50 = 100$ centimeters upward.

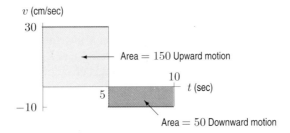

Figure 5.7: Difference in areas gives change in position

Figure 5.7 shows velocity versus time. The area of the rectangle above the t-axis represents upward distance, while the area of the rectangle below the t-axis represents downward distance.

In general, if the velocity can be negative as well as positive, the limit of the sums represents change in position, rather than distance.

Left and Right Sums

We now write the estimates for the distance traveled by the car in new notation. Let $v = f(t)$ denote any nonnegative velocity function. We want to find the distance traveled between times $t = a$ and $t = b$. We take measurements of $f(t)$ at equally spaced times $t_0, t_1, t_2, \ldots, t_n$, with time $t_0 = a$ and time $t_n = b$. The time interval between any two consecutive measurements is

$$\Delta t = \frac{b - a}{n},$$

where Δt means the change, or increment, in t.

During the first time interval, from t_0 and t, the velocity can be approximated by $f(t_0)$, so the distance traveled is approximately

$$f(t_0)\Delta t.$$

During the second time interval, the velocity is about $f(t_1)$, so the distance traveled is about

$$f(t_1)\Delta t.$$

Continuing in this way and adding all the estimates, we get an estimate for the total distance traveled. In the last interval, the velocity is approximately $f(t_{n-1})$, so the last term is $f(t_{n-1})\Delta t$:

Total distance traveled
between $t = a$ and $t = b$ $\approx f(t_0)\Delta t + f(t_1)\Delta t + f(t_2)\Delta t + \cdots + f(t_{n-1})\Delta t.$

This is called a *left-hand sum* because we used the value of velocity from the left end of each time interval. It is represented by the sum of the areas of the rectangles in Figure 5.8.

We can also calculate a *right-hand sum* by using the value of the velocity at the right end of each time interval. In that case the estimate for the first interval is $f(t_1)\Delta t$, for the second interval it is $f(t_2)\Delta t$, and so on. The estimate for the last interval is now $f(t_n)\Delta t$, so

Total distance traveled
between $t = a$ and $t = b$ $\approx f(t_1)\Delta t + f(t_2)\Delta t + f(t_3)\Delta t + \cdots + f(t_n)\Delta t.$

The right-hand sum is represented by the area of the rectangles in Figure 5.9.

If f is an increasing function, as in Figures 5.8 and 5.9, the left-hand sum is an underestimate and the right-hand sum is an overestimate of the total distance traveled. If f is decreasing, as in Figure 5.10, then the roles of the two sums are reversed.

Figure 5.8: Left-hand sums

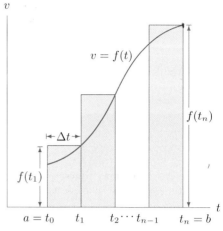

Figure 5.9: Right-hand sums

Accuracy of Estimates

For either increasing or decreasing velocity functions, the exact value of the distance traveled lies somewhere between the two estimates. Thus, the accuracy of our estimate depends on how close these two sums are. For a function which is increasing throughout or decreasing throughout the interval $[a, b]$:

$$\left| \begin{array}{c} \text{Difference between} \\ \text{upper and lower estimates} \end{array} \right| = \left| \begin{array}{c} \text{Difference between} \\ f(a) \text{ and } f(b) \end{array} \right| \cdot \Delta t = |f(b) - f(a)| \cdot \Delta t.$$

(Absolute values make the differences nonnegative.) In Figure 5.10, the area of the light rectangles is the difference between estimates. By making the time interval, Δt, between measurements small enough, we can make this difference between lower and upper estimates as small as we like.

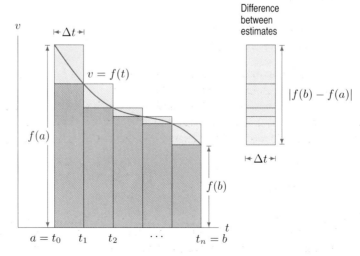

Figure 5.10: Left and right sums if f is decreasing

Exercises and Problems for Section 5.1

Exercises

1. A car comes to a stop six seconds after the driver applies the brakes. While the brakes are on, the following velocities are recorded:

Time since brakes applied (sec)	0	2	4	6
Velocity (ft/sec)	88	45	16	0

    (a) Give lower and upper estimates for the distance the car traveled after the brakes were applied.

    (b) On a sketch of velocity against time, show the lower and upper estimates of part (a).

2. The velocity $v(t)$ in Table 5.3 is decreasing, $2 \leq t \leq 12$. Using $n = 5$ subdivisions to approximate the total distance traveled, find

    (a) An upper estimate   **(b)** A lower estimate

Table 5.3

t	2	4	6	8	10	12
$v(t)$	44	42	41	40	37	35

3. The velocity $v(t)$ in Table 5.4 is increasing, $0 \leq t \leq 12$.

    (a) Find an upper estimate for the total distance traveled using

        (i) $n = 4$ (ii) $n = 2$

    (b) Which of the two answers in part (a) is more accurate? Why?

    (c) Find a lower estimate of the total distance traveled using $n = 4$.

Table 5.4

t	0	3	6	9	12
$v(t)$	34	37	38	40	45

4. Use the expressions for left and right sums on page 244 and Table 5.5.

    (a) If $n = 4$, what is Δt? What are t_0, t_1, t_2, t_3, t_4? What are $f(t_0), f(t_1), f(t_2), f(t_3), f(t_4)$?

    (b) Find the left and right sums using $n = 4$.

    (c) If $n = 2$, what is Δt? What are t_0, t_1, t_2? What are $f(t_0), f(t_1), f(t_2)$?

    (d) Find the left and right sums using $n = 2$.

Table 5.5

t	15	17	19	21	23
$f(t)$	10	13	18	20	30

5. Use the expressions for left and right sums on page 244 and Table 5.6.

    (a) If $n = 4$, what is Δt? What are t_0, t_1, t_2, t_3, t_4? What are $f(t_0), f(t_1), f(t_2), f(t_3), f(t_4)$?

    (b) Find the left and right sums using $n = 4$.

    (c) If $n = 2$, what is Δt? What are t_0, t_1, t_2? What are $f(t_0), f(t_1), f(t_2)$?

    (d) Find the left and right sums using $n = 2$.

Table 5.6

t	0	4	8	12	16
$f(t)$	25	23	22	20	17

6. At time, t, in seconds, your velocity, v, in meters/second, is given by

$$v(t) = 1 + t^2 \quad \text{for} \quad 0 \leq t \leq 6.$$

Use $\Delta t = 2$ to estimate the distance traveled during this time. Find the upper and lower estimates, and then average the two.

7. For time, t, in hours, $0 \leq t \leq 1$, a bug is crawling at a velocity, v, in meters/hour given by

$$v = \frac{1}{1 + t}.$$

Use $\Delta t = 0.2$ to estimate the distance that the bug crawls during this hour. Find an overestimate and an underestimate. Then average the two to get a new estimate.

8. Figure 5.11 shows the velocity, v, of an object (in meters/sec). Estimate the total distance the object traveled between $t = 0$ and $t = 6$.

Figure 5.11

9. The velocity of a car is $f(t) = 5t$ meters/sec. Use a graph of $f(t)$ to find the exact distance traveled by the car, in meters, from $t = 0$ to $t = 10$ seconds.

In Exercises 10–13, find the difference between the upper and lower estimates of the distance traveled at velocity $f(t)$ on the interval $a \leq t \leq b$ for n subdivisions.

10. $f(t) = 5t + 8, a = 1, b = 3, n = 100$

11. $f(t) = 25 - t^2, a = 1, b = 4, n = 500$

12. $f(t) = \sin t, a = 0, b = \pi/2, n = 100$

13. $f(t) = e^{-t^2/2}, a = 0, b = 2, n = 20$

14. In Figure 5.12, use the grid to get upper and lower estimates of the area of the region bounded by the curve, the horizontal axis and the vertical lines $x = 3$ and $x = -3$.

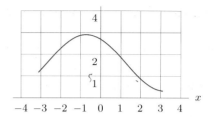

Figure 5.12

Problems

15. A student is speeding down Route 11 in his fancy red Porsche when his radar system warns him of an obstacle 400 feet ahead. He immediately applies the brakes, starts to slow down, and spots a skunk in the road directly ahead of him. The "black box" in the Porsche records the car's speed every two seconds, producing the following table. The speed decreases throughout the 10 seconds it takes to stop, although not necessarily at a uniform rate.

Time since brakes applied (sec)	0	2	4	6	8	10
Speed (ft/sec)	100	80	50	25	10	0

(a) What is your best estimate of the total distance the student's car traveled before coming to rest?

(b) Which one of the following statements can you justify from the information given?

(i) The car stopped before getting to the skunk.

(ii) The "black box" data is inconclusive. The skunk may or may not have been hit.

(iii) The skunk was hit by the car.

16. Roger runs a marathon. His friend Jeff rides behind him on a bicycle and clocks his speed every 15 minutes. Roger starts out strong, but after an hour and a half he is so exhausted that he has to stop. Jeff's data follow:

Time since start (min)	0	15	30	45	60	75	90
Speed (mph)	12	11	10	10	8	7	0

(a) Assuming that Roger's speed is never increasing, give upper and lower estimates for the distance Roger ran during the first half hour.

(b) Give upper and lower estimates for the distance Roger ran in total during the entire hour and a half.

(c) How often would Jeff have needed to measure Roger's speed in order to find lower and upper estimates within 0.1 mile of the actual distance he ran?

Problems 17–20 show the velocity, in cm/sec, of a particle moving along the x-axis. Compute the particle's change in position, left (negative) or right (positive), between times $t = 0$ and $t = 5$ seconds.

17.

18.

19.

20.

21. The velocity of a particle moving along the x-axis is given by $f(t) = 6 - 2t$ cm/sec. Use a graph of $f(t)$ to find the exact change in position of the particle from time $t = 0$ to $t = 4$ seconds.

22. A baseball thrown directly upward at 96 ft/sec has velocity $v(t) = 96 - 32t$ ft/sec at time t seconds.

(a) Graph the velocity from $t = 0$ to $t = 6$.

(b) When does the baseball reach the peak of its flight? How high does it go?

(c) How high is the baseball at time $t = 5$?

23. A car initially going 50 ft/sec brakes at a constant rate (constant negative acceleration), coming to a stop in 5 seconds.

(a) Graph the velocity from $t = 0$ to $t = 5$.

(b) How far does the car travel?

(c) How far does the car travel if its initial velocity is doubled, but it brakes at the same constant rate?

24. A woman drives 10 miles, accelerating uniformly from rest to 60 mph. Graph her velocity versus time. How long does it take for her to reach 60 mph?

25. Two cars start at the same time and travel in the same direction along a straight road. Figure 5.13 gives the velocity, v, of each car as a function of time, t. Which car:

 (a) Attains the larger maximum velocity?
 (b) Stops first?
 (c) Travels farther?

Figure 5.13

26. Two cars travel in the same direction along a straight road. Figure 5.14 shows the velocity, v, of each car at time t. Car B starts 2 hours after car A and car B reaches a maximum velocity of 50 km/hr.

 (a) For approximately how long does each car travel?
 (b) Estimate car A's maximum velocity.
 (c) Approximately how far does each car travel?

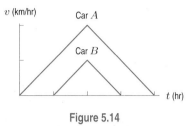

Figure 5.14

5.2   THE DEFINITE INTEGRAL

In Section 5.1, we saw how distance traveled can be approximated by a sum of areas of rectangles. We also saw how the approximation improves as the width of the rectangles gets smaller. In this section, we construct these sums for any function f, whether or not it represents a velocity.

Sigma Notation

Suppose $f(t)$ is a continuous function for $a \leq t \leq b$. We divide the interval from a to b into n equal subdivisions, and we call the width of an individual subdivision Δt, so

$$\Delta t = \frac{b-a}{n}.$$

Let $t_0, t_1, t_2, \ldots, t_n$ be endpoints of the subdivisions. Both the left-hand and right-hand sums can be written more compactly using *sigma*, or summation, notation. The symbol $\sum$ is a capital sigma, or Greek letter "S." We write

$$\text{Right-hand sum} = f(t_1)\Delta t + f(t_2)\Delta t + \cdots + f(t_n)\Delta t = \sum_{i=1}^{n} f(t_i)\Delta t.$$

The $\sum$ tells us to add terms of the form $f(t_i)\Delta t$. The "$i = 1$" at the base of the sigma sign tells us to start at $i = 1$, and the "n" at the top tells us to stop at $i = n$.

In the left-hand sum we start at $i = 0$ and stop at $i = n - 1$, so we write

$$\text{Left-hand sum} = f(t_0)\Delta t + f(t_1)\Delta t + \cdots + f(t_{n-1})\Delta t = \sum_{i=0}^{n-1} f(t_i)\Delta t.$$

Taking the Limit to Obtain the Definite Integral

Now we take the limit of these sums as n goes to infinity. If f is continuous for $a \leq t \leq b$, the limits of the left- and right-hand sums exist and are equal. The *definite integral* is the limit of these sums. A formal definition of the definite integral is given in the online supplement to the text at www.wiley.com/college/hugheshallett.

Suppose f is continuous for $a \leq t \leq b$. The **definite integral** of f from a to b, written

$$\int_a^b f(t)\, dt,$$

is the limit of the left-hand or right-hand sums with n subdivisions of $a \leq t \leq b$ as n gets arbitrarily large. In other words,

$$\int_a^b f(t)\, dt = \lim_{n \to \infty} (\text{Left-hand sum}) = \lim_{n \to \infty} \left(\sum_{i=0}^{n-1} f(t_i) \Delta t \right)$$

and

$$\int_a^b f(t)\, dt = \lim_{n \to \infty} (\text{Right-hand sum}) = \lim_{n \to \infty} \left(\sum_{i=1}^{n} f(t_i) \Delta t \right).$$

Each of these sums is called a *Riemann sum*, f is called the *integrand*, and a and b are called the *limits of integration*.

The "$\int$" notation comes from an old-fashioned "S," which stands for "sum" in the same way that $\sum$ does. The "dt" in the integral comes from the factor Δt. Notice that the limits on the $\sum$ symbol are 0 and $n-1$ for the left-hand sum, and 1 and n for the right-hand sum, whereas the limits on the $\int$ sign are a and b.

Computing a Definite Integral

In practice, we often approximate definite integrals numerically using a calculator or computer. They use programs which compute sums for larger and larger values of n, and eventually give a value for the integral. Some (but not all) definite integrals can be computed exactly. However, any definite integral can be approximated numerically.

In the next example, we see how numerical approximation works. For each value of n, we show an over- and an under-estimate for the integral $\int_1^2 (1/t)\, dt$. As we increase the value of n, the over- and under-estimates get closer together, trapping the value of the integral between them. By increasing the value of n sufficiently, we can calculate the integral to any desired accuracy.

Example 1   Calculate the left-hand and right-hand sums with $n = 2$ for $\int_1^2 \dfrac{1}{t}\, dt$. What is the relation between the left- and right-hand sums for $n = 10$ and $n = 250$ and the integral?

Solution   Here $a = 1$ and $b = 2$, so for $n = 2$, $\Delta t = (2 - 1)/2 = 0.5$. Therefore, $t_0 = 1$, $t_1 = 1.5$ and $t_2 = 2$. (See Figure 5.15.) We have

$$\text{Left-hand sum} = f(1)\Delta t + f(1.5)\Delta t = 1(0.5) + \frac{1}{1.5}(0.5) = 0.8333,$$

$$\text{Right-hand sum} = f(1.5)\Delta t + f(2)\Delta t = \frac{1}{1.5}(0.5) + \frac{1}{2}(0.5) = 0.5833.$$

In Figure 5.15 we see that the left-hand sum is bigger than the area under the curve and the right-hand sum is smaller. So the area under the curve $f(t) = 1/t$ from $t = 1$ to $t = 2$ is between them:

$$0.5833 < \int_1^2 \frac{1}{t}\, dt < 0.8333.$$

Since $1/t$ is decreasing, when $n = 10$ in Figure 5.16 we again see that the left-hand sum is larger than the area under the curve, and the right-hand sum smaller. A calculator or computer gives

$$0.6688 < \int_1^2 \frac{1}{t}\, dt < 0.7188.$$

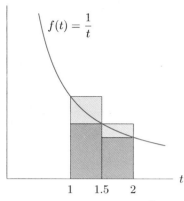

Figure 5.15: Approximating $\int_1^2 \frac{1}{t}\, dt$ with $n = 2$

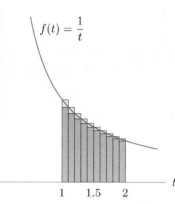

Figure 5.16: Approximating $\int_1^2 \frac{1}{t}\, dt$ with $n = 10$

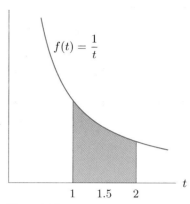

Figure 5.17: Shaded area is exact value of $\int_1^2 \frac{1}{t}\, dt$

The left- and right-hand sums trap the exact value of the integral between them. As the subdivisions become finer, the left- and right-hand sums get closer together.

When $n = 250$, a calculator or computer gives

$$0.6921 < \int_1^2 \frac{1}{t}\, dt < 0.6941.$$

So, to two decimal places, we can say that

$$\int_1^2 \frac{1}{t}\, dt \approx 0.69.$$

The exact value is known to be $\int_1^2 \frac{1}{t}\, dt = \ln 2 = 0.693147\ldots$. See Figure 5.17.

The Definite Integral as an Area

If $f(x)$ is positive we can interpret each term $f(x_0)\Delta x, f(x_1)\Delta x, \ldots$ in a left- or right-hand Riemann sum as the area of a rectangle. See Figure 5.18. As the width Δx of the rectangles approaches zero, the rectangles fit the curve of the graph more exactly, and the sum of their areas gets closer and closer to the area under the curve shaded in Figure 5.19. This suggests that:

When $f(x) \geq 0$ and $a < b$:

$$\text{Area under graph of } f \text{ and above } x\text{-axis between } a \text{ and } b = \int_a^b f(x)\, dx.$$

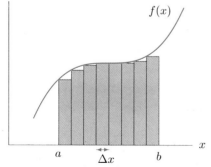

Figure 5.18: Area of rectangles approximating the area under the curve

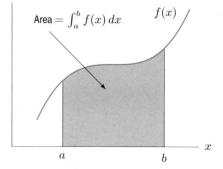

Figure 5.19: The definite integral $\int_a^b f(x)\, dx$

Example 2 Consider the integral $\int_{-1}^{1} \sqrt{1 - x^2} \, dx$.

(a) Interpret the integral as an area, and find its exact value.

(b) Estimate the integral using a calculator or computer. Compare your answer to the exact value.

Solution (a) The integral is the area under the graph of $y = \sqrt{1 - x^2}$ between -1 and 1. See Figure 5.20. Rewriting this equation as $x^2 + y^2 = 1$, we see that the graph is a semicircle of radius 1 and area $\pi/2$.

(b) A calculator gives the value of the integral as $1.5707963\ldots$.

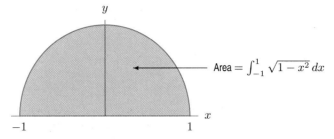

Figure 5.20: Area interpretation of $\int_{-1}^{1} \sqrt{1 - x^2} \, dx$

When $f(x)$ is Not Positive

We have assumed in drawing Figure 5.19 that the graph of $f(x)$ lies above the x-axis. If the graph lies below the x-axis, then each value of $f(x)$ is negative, so each $f(x)\Delta x$ is negative, and the area gets counted negatively. In that case, the definite integral is the negative of the area.

When $f(x)$ is positive for some x values and negative for others, and $a < b$:

$\int_{a}^{b} f(x) \, dx$ is the sum of areas above the x-axis, counted positively, and areas below the x-axis, counted negatively.

Example 3 How does the definite integral $\int_{-1}^{1} (x^2 - 1) \, dx$ relate to the area between the parabola $y = x^2 - 1$ and the x-axis?

Solution A calculator gives $\int_{-1}^{1} (x^2 - 1) \, dx = -1.33$. The parabola lies below the axis between $x = -1$ and $x = 1$. (See Figure 5.21.) So the area between the parabola and the x-axis is approximately 1.33.

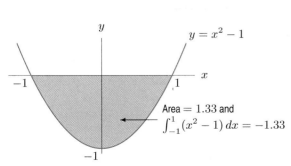

Figure 5.21: Integral $\int_{-1}^{1} (x^2 - 1) \, dx$ is negative of shaded area

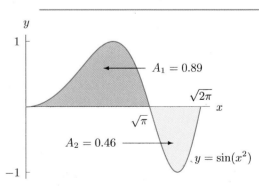

Figure 5.22: Integral $\int_{0}^{\sqrt{2\pi}} \sin(x^2) \, dx = A_1 - A_2$

Example 4   Interpret the definite integral $\int_0^{\sqrt{2\pi}} \sin(x^2)\, dx$ in terms of areas.

Solution   The integral is the area above the x-axis, A_1, minus the area below the x-axis, A_2. See Figure 5.22. Estimating the integral with a calculator gives

$$\int_0^{\sqrt{2\pi}} \sin(x^2)\, dx = 0.43.$$

The graph of $y = \sin(x^2)$ crosses the x-axis where $x^2 = \pi$, that is, at $x = \sqrt{\pi}$. The next crossing is at $x = \sqrt{2\pi}$. Breaking the integral into two parts and calculating each one separately gives

$$\int_0^{\sqrt{\pi}} \sin(x^2)\, dx = 0.89 \quad \text{and} \quad \int_{\sqrt{\pi}}^{\sqrt{2\pi}} \sin(x^2)\, dx = -0.46.$$

So $A_1 = 0.89$ and $A_2 = 0.46$. Then, as we would expect,

$$\int_0^{\sqrt{2\pi}} \sin(x^2)\, dx = A_1 - A_2 = 0.89 - 0.46 = 0.43.$$

More General Riemann Sums

Left- and right-hand sums are special cases of Riemann sums. For a general Riemann sum we allow subdivisions to have different lengths. Also, instead of evaluating f only at the left or right endpoint of each subdivision, we allow it to be evaluated anywhere in the subdivision. Thus, a general Riemann sum has the form

$$\sum_{i=1}^{n} \text{Value of } f(t) \text{ at some point in } i^{\text{th}} \text{ subdivision} \times \text{Length of } i^{\text{th}} \text{ subdivision.}$$

(See Figure 5.23.) As before, we let $t_0, t_1, \ldots, t_n$ be the endpoints of the subdivisions, so the length of the i-th subdivision is $\Delta t_i = t_i - t_{i-1}$. For each i we choose a point c_i in the i-th subinterval at which to evaluate f, leading to the following definition:

A general Riemann sum for f on the interval $[a, b]$ is a sum of the form

$$\sum_{i=1}^{n} f(c_i)\Delta t_i,$$

where $a = t_0 < t_1 < \cdots < t_n = b$, and, for $i = 1, \ldots, n$, $\Delta t_i = t_i - t_{i-1}$, and $t_{i-1} \leq c_i \leq t_i$.

If f is continuous, we can make a general Riemann sum as close as we like to the value of the definite integral by making the interval lengths small enough. Thus, in approximating definite integrals or in proving theorems about them, we can use general Riemann sums rather than left- or right-hand sums. The online theory supplement at www.wiley.com/college/hugheshallett shows how general Riemann sums are used in proving theorems.

Figure 5.23: A general Riemann sum approximating $\int_a^b f(t)\, dt$

Exercises and Problems for Section 5.2

Exercises

1. Using Figure 5.24, draw rectangles representing each of the following Riemann sums for the function f on the interval $0 \le t \le 8$. Calculate the value of each sum.

    (a) Left-hand sum with $\Delta t = 4$
    (b) Right-hand sum with $\Delta t = 4$
    (c) Left-hand sum with $\Delta t = 2$
    (d) Right-hand sum with $\Delta t = 2$

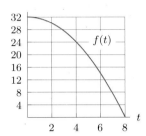

Figure 5.24

2. Use Figure 5.25 to estimate $\int_0^3 f(x)\,dx$.

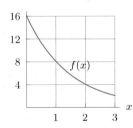

Figure 5.25

3. Use Figure 5.26 to estimate $\int_{-10}^{15} f(x)dx$.

Figure 5.26

4. Use Figure 5.27 to estimate $\int_{-15}^{20} f(x)\,dx$.

Figure 5.27

5. The graph of a function $f(t)$ is given in Figure 5.28. Which of the following four numbers could be an estimate of $\int_0^1 f(t)dt$ accurate to two decimal places? Explain how you chose your answer.

    (a) -98.35 **(b)** 71.84
    (c) 100.12 **(d)** 93.47

Figure 5.28

6. Use the table to estimate $\int_0^{40} f(x)dx$. What values of n and Δx did you use?

x	0	10	20	30	40
$f(x)$	350	410	435	450	460

7. Use the table to estimate $\int_0^{15} f(x)\,dx$.

x	0	3	6	9	12	15
$f(x)$	50	48	44	36	24	8

8. Use the table to estimate $\int_0^{12} f(x)\,dx$.

x	0	3	6	9	12
$f(x)$	32	22	15	11	9

In Exercises 9–11, use a calculator or a computer to find the value of the definite integral to four decimal places.

9. $\displaystyle\int_0^3 2^x\,dx$ **10.** $\displaystyle\int_0^1 \sin(t^2)dt$ **11.** $\displaystyle\int_{-1}^1 e^{-x^2}\,dx$

In Exercises 12–18, find the area of the regions between the curve and the horizontal axis

12. Under the curve $y = \cos t$ for $0 \le t \le \pi/2$.

13. Under $y = 6x^3 - 2$ for $5 \le x \le 10$.

14. Under $y = \ln x$ for $1 \le x \le 4$.

15. Under the curve $y = \cos\sqrt{x}$ for $0 \le x \le 2$.

16. Under $y = 2\cos(t/10)$ for $1 \le t \le 2$.

17. Under the curve $y = 7 - x^2$ and above the x-axis.

18. Above the curve $y = x^4 - 8$ and below the x-axis.

Problems

19. (a) On a sketch of $y = \ln x$, represent the left Riemann sum with $n = 2$ approximating $\int_1^2 \ln x \, dx$. Write out the terms in the sum, but do not evaluate it.

(b) On another sketch, represent the right Riemann sum with $n = 2$ approximating $\int_1^2 \ln x \, dx$. Write out the terms in the sum, but do not evaluate it.

(c) Which sum is an overestimate? Which sum is an underestimate?

20. (a) Use a calculator or computer to find $\int_0^6 (x^2 + 1) \, dx$. Represent this value as the area under a curve.

(b) Estimate $\int_0^6 (x^2 + 1) \, dx$ using a left-hand sum with $n = 3$. Represent this sum graphically on a sketch of $f(x) = x^2 + 1$. Is this sum an overestimate or underestimate of the true value found in part (a)?

(c) Estimate $\int_0^6 (x^2 + 1) \, dx$ using a right-hand sum with $n = 3$. Represent this sum on your sketch. Is this sum an overestimate or underestimate?

21. Estimate $\int_1^2 x^2 \, dx$ using left- and right-hand sums with four subdivisions. How far from the true value of the integral could your estimate be?

22. Without computing the sums, find the difference between the right- and left-hand Riemann sums if we use $n = 500$ subintervals to approximate $\int_{-1}^1 (2x^3 + 4) \, dx$.

23. (a) Graph $f(x) = x(x + 2)(x - 1)$.

(b) Find the total area between the graph and the x-axis between $x = -2$ and $x = 1$.

(c) Find $\int_{-2}^1 f(x) \, dx$ and interpret it in terms of areas.

24. Compute the definite integral $\int_0^4 \cos \sqrt{x} \, dx$ and interpret the result in terms of areas.

25. Without computation, decide if $\int_0^{2\pi} e^{-x} \sin x \, dx$ is positive or negative. [Hint: Sketch $e^{-x} \sin x$.]

26. (a) Graph $f(x) = \begin{cases} 1 - x & 0 \le x \le 1 \\ x - 2 & 1 < x \le 2. \end{cases}$

(b) Find $\displaystyle\int_0^2 f(x) \, dx$.

(c) Calculate the 4-term left Riemann sum approximation to the definite integral. How does the approximation compare to the exact value?

27. Use Figure 5.29 to find the values of

(a) $\int_a^b f(x) \, dx$ **(b)** $\int_b^c f(x) \, dx$

(c) $\int_a^c f(x) \, dx$ **(d)** $\int_a^c |f(x)| \, dx$

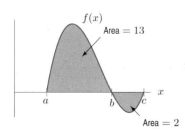

Figure 5.29

28. Given $\int_{-2}^0 f(x) \, dx = 4$ and Figure 5.30, estimate:

(a) $\int_0^2 f(x) \, dx$ **(b)** $\int_{-2}^2 f(x) \, dx$

(c) The total shaded area.

Figure 5.30

29. (a) Using Figure 5.31, find $\int_{-3}^0 f(x) \, dx$.

(b) If the area of the shaded region is A, estimate $\int_{-3}^4 f(x) \, dx$.

Figure 5.31

30. Write a few sentences in support of or in opposition to the following statement:

"If a left-hand sum underestimates a definite integral by a certain amount, then the corresponding right-hand sum will overestimate the integral by the same amount."

31. Using the graph of $2 + \cos x$, for $0 \le x \le 4\pi$, list the following quantities in increasing order: the value of the integral $\int_0^{4\pi} (2 + \cos x) \, dx$, the left sum with $n = 2$ subdivisions, and the right sum with $n = 2$ subdivisions.

32. Sketch the graph of a function f (you do not need to give a formula for f) on an interval $[a, b]$ with the property that with $n = 2$ subdivisions,

$$\int_a^b f(x) \, dx < \text{Left-hand sum} < \text{Right-hand sum}.$$

33. Three terms of a left-hand sum used to approximate a definite integral $\int_a^b f(x) \, dx$ are as follows.

$$\left(2 + 0 \cdot \frac{4}{3}\right)^2 \cdot \frac{4}{3} + \left(2 + 1 \cdot \frac{4}{3}\right)^2 \cdot \frac{4}{3} + \left(2 + 2 \cdot \frac{4}{3}\right)^2 \cdot \frac{4}{3}.$$

Find possible values for a and b and a possible formula for $f(x)$.

34. Consider the integral $\int_1^2 (1/t)\, dt$ in Example 1. By dividing the interval $1 \leq t \leq 2$ into 10 equal parts, we can show that

$$0.1 \left[\frac{1}{1.1} + \frac{1}{1.2} + \ldots + \frac{1}{2} \right] \leq \int_1^2 \frac{1}{t}\, dt$$

and

$$\int_1^2 \frac{1}{t}\, dt \leq 0.1 \left[\frac{1}{1} + \frac{1}{1.1} + \ldots + \frac{1}{1.9} \right].$$

(a) Now divide the interval $1 \leq t \leq 2$ into n equal parts to show that

$$\sum_{r=1}^{n} \frac{1}{n+r} \;<\; \int_1^2 \frac{1}{t}\, dt \;<\; \sum_{r=0}^{n-1} \frac{1}{n+r}.$$

(b) Show that the difference between the upper and lower sums in part (a) is $1/(2n)$.

(c) The exact value of $\int_1^2 (1/t)\, dt$ is $\ln 2$. How large should n be to approximate $\ln 2$ with an error of at most $5 \cdot 10^{-6}$, using one of the sums in part (a)?

5.3 THE FUNDAMENTAL THEOREM AND INTERPRETATIONS

The Notation and Units for the Definite Integral

Just as the Leibniz notation dy/dx for the derivative reminds us that the derivative is the limit of a ratio of differences, the notation for the definite integral helps us recall the meaning of the integral. The symbol

$$\int_a^b f(x)\, dx$$

reminds us that an integral is a limit of sums of terms of the form "$f(x)$ times a small difference of x." Officially, dx is not a separate entity, but a part of the whole integral symbol. Just as one thinks of d/dx as a single symbol meaning "the derivative with respect to x of ...," one can think of $\int_a^b \ldots dx$ as a single symbol meaning "the integral of ... with respect to x."

However, many scientists and mathematicians informally think of dx as an "infinitesimally" small bit of x multiplied by $f(x)$. This viewpoint is often the key to interpreting the meaning of a definite integral. For example, if $f(t)$ is the velocity of a moving particle at time t, then $f(t)\, dt$ may by thought of informally as velocity $\times$ time, giving the distance traveled by the particle during a small bit of time dt. The integral $\int_a^b f(t)\, dt$ may then be thought of as the sum of all these small distances, giving us the net change in position of the particle between $t = a$ and $t = b$. The notation for the integral suggests units for the value of the integral. Since the terms being added up are products of the form "$f(x)$ times a difference in x," the unit of measurement for $\int_a^b f(x)\, dx$ is the product of the units for $f(x)$ and the units for x. For example, if $f(t)$ is velocity measured in meters/second and t is time measured in seconds, then

$$\int_a^b f(t)\, dt$$

has units of (meters/sec)$\times$(sec) = meters. This is what we expect, since the value of this integral represents change in position.

As another example, graph $y = f(x)$ with the same units of measurement of length along the x- and y-axes, say cm. Then $f(x)$ and x are measured in the same units, so

$$\int_a^b f(x)\, dx$$

is measured in square units of cm $\times$ cm = cm^2. Again, this is what we would expect since in this context the integral represents an area.

The Fundamental Theorem of Calculus

We have seen that change in position can be calculated as the limit of Riemann sums of the velocity function $v = f(t)$. Thus, change in position is given by the definite integral $\int_a^b f(t)\, dt$. If we let

$F(t)$ denote the position function, then the change in position can also be written as $F(b) - F(a)$. Thus we have:

$$\int_a^b f(t)\, dt = \begin{matrix} \text{Change in position from} \\ t = a \text{ to } t = b \end{matrix} = F(b) - F(a)$$

We also know that the position F and velocity f are related using derivatives: $F'(t) = f(t)$. Thus, we have uncovered a connection between the integral and derivative, which is so important it is called the Fundamental Theorem of Calculus. It applies to any function F with a continuous derivative $f = F'$.

Theorem 5.1: The Fundamental Theorem of Calculus[1]

If f is continuous on the interval $[a, b]$ and $f(t) = F'(t)$, then

$$\int_a^b f(t)\, dt = F(b) - F(a).$$

To understand the Fundamental Theorem of Calculus, think of $f(t) = F'(t)$ as the rate of change of the quantity $F(t)$. To calculate the total change in $F(t)$ between times $t = a$ and $t = b$, we divide the interval $a \leq t \leq b$ into n subintervals, each of length Δt. For each small interval, we estimate the change in $F(t)$, written ΔF, and add these. In each subinterval we assume the rate of change of $F(t)$ is approximately constant, so that we can say

$$\Delta F \approx \text{Rate of change of } F \times \text{Time elapsed}.$$

For the first subinterval, from t_0 to t_1, the rate of change of $F(t)$ is approximately $F'(t_0)$, so

$$\Delta F \approx F'(t_0)\, \Delta t.$$

Similarly, for the second interval

$$\Delta F \approx F'(t_1)\, \Delta t.$$

Summing over all the subintervals, we get

$$\begin{matrix} \text{Total change in } F(t) \\ \text{between } t = a \text{ and } t = b \end{matrix} = \sum_{i=0}^{n-1} \Delta F \approx \sum_{i=0}^{n-1} F'(t_i)\, \Delta t.$$

We have approximated the change in $F(t)$ as a left-hand sum.

However, the total change in $F(t)$ between the times $t = a$ and $t = b$ is simply $F(b) - F(a)$. Taking the limit as n goes to infinity converts the Riemann sum to a definite integral and suggests the following interpretation of the Fundamental Theorem of Calculus:[2]

$$F(b) - F(a) = \begin{matrix} \text{Total change in } F(t) \\ \text{between } t = a \text{ and } t = b \end{matrix} = \int_a^b F'(t)\, dt.$$

In words, the definite integral of a rate of change gives the total change.

This argument does not, however, constitute a proof of the Fundamental Theorem. The errors in the various approximations must be investigated using the definition of limit. For a proof using the Mean Value Theorem, see the online supplement at www.wiley.com/college/hugheshallett.

---

[1]This result is sometimes called the First Fundamental Theorem of Calculus, to distinguish it from the Second Fundamental Theorem of Section 6.4.

[2]We could equally well have used a right-hand sum, since the definite integral is their common limit.

Example 1   If $F'(t) = f(t)$ and $f(t)$ is velocity in miles/hour, with t in hours, what are the units of $\int_a^b f(t)\, dt$ and $F(b) - F(a)$?

Solution   Since the units of $f(t)$ are miles/hour and the units of t are hours, the units of $\int_a^b f(t)\, dt$ are (miles/hour) × hours = miles. Since F measures change in position, the units of $F(b) - F(a)$ are miles. As expected, the units of $\int_a^b f(t)\, dt$ and $F(b) - F(a)$ are the same.

The Definite Integral of a Rate of Change: Applications of the Fundamental Theorem

Many applications are based on the Fundamental Theorem, which tells us that the definite integral of a rate of change gives the total change.

Example 2   Let $F(t)$ represent a bacteria population which is 5 million at time $t = 0$. After t hours, the population is growing at an instantaneous rate of 2^t million bacteria per hour. Estimate the total increase in the bacteria population during the first hour, and the population at $t = 1$.

Solution   Since the rate at which the population is growing is $F'(t) = 2^t$, we have

$$\text{Change in population} = F(1) - F(0) = \int_0^1 2^t\, dt.$$

Using a calculator to evaluate the integral,

$$\text{Change in population} = \int_0^1 2^t\, dt = 1.44 \text{ million bacteria.}$$

Since $F(0) = 5$, the population at $t = 1$ is given by

$$\text{Population} = F(1) = F(0) + \int_0^1 2^t\, dt = 5 + 1.44 = 6.44 \text{ million.}$$

The following example shows how representing a quantity as a definite integral, and thereby as an area, can be helpful even if we don't evaluate the integral.

Example 3   Two cars start from rest at a traffic light and accelerate for several minutes. Figure 5.32 shows their velocities as a function of time.
(a)   Which car is ahead after one minute?   (b)   Which car is ahead after two minutes?

Figure 5.32: Velocities of two cars in Example 3.
Which is ahead when?

Solution   (a) For the first minute car 1 goes faster than car 2, and therefore car 1 must be ahead at the end of one minute.

(b) At the end of two minutes the situation is less clear, since car 1 was going faster for the first minute and car 2 for the second. However, if $v = f(t)$ is the velocity of a car after t minutes, then we know that

$$\text{Distance traveled in two minutes} = \int_0^2 f(t)\, dt,$$

since the integral of velocity is distance traveled. This definite integral may also be interpreted as the area under the graph of f between 0 and 2. Since the area representing the distance traveled by car 2 is clearly larger than the area for car 1 (see Figure 5.32), we know that car 2 has traveled farther than car 1.

Example 4   Biological activity in water is reflected in the rate at which carbon dioxide, CO_2, is added or removed. Plants take CO_2 out of the water during the day for photosynthesis and put CO_2 into the water at night. Animals put CO_2 into the water all the time as they breathe. Figure 5.33 shows the rate of change of the CO_2 level in a pond.[3] At dawn, there were 2.600 mmol of CO_2 per liter of water.

(a) At what time was the CO_2 level lowest? Highest?
(b) Estimate how much CO_2 enters the pond during the night ($t = 12$ to $t = 24$).
(c) Estimate the CO_2 level at dusk (twelve hours after dawn).
(d) Does the CO_2 level appear to be approximately in equilibrium?

rate
(mmol/liter per hour)

Figure 5.33: Rate at which CO_2 enters a pond over a 24-hour period

Solution   Let $f(t)$ be the rate at which CO_2 is entering the water at time t and let $F(t)$ be the concentration of CO_2 in the water at time t, so $F'(t) = f(t)$.

(a) From Figure 5.33, we see $f(t)$ is negative for $0 \le t \le 12$, so the CO_2 level is decreasing during this interval (daytime). Since $f(t)$ is positive for $12 < t < 24$, the CO_2 level is increasing during this interval (night). The CO_2 is lowest at $t = 12$ (dusk) and highest at $t = 0$ and $t = 24$ (dawn).

(b) We want to calculate the total change in the CO_2 level in the pond, $F(24) - F(12)$. By the Fundamental Theorem of Calculus,

$$F(24) - F(12) = \int_{12}^{24} f(t)\, dt.$$

We use values of $f(t)$ from the graph (displayed in Table 5.7) to construct a left Riemann sum approximation to this integral with $n = 6$, $\Delta t = 2$:

$$\int_{12}^{24} f(t)\, dt \approx f(12) \cdot 2 + f(14) \cdot 2 + f(16) \cdot 2 + \cdots + f(22) \cdot 2$$

$$\approx (0.000)2 + (0.045)2 + (0.035)2 + \cdots + (0.012)2 = 0.278.$$

---

[3]Data from R. J. Beyers, *The Pattern of Photosynthesis and Respiration in Laboratory Microsystems* (Mem. 1st. Ital. Idrobiol., 1965).

Thus, between $t = 12$ and $t = 24$,

$$\text{Change in } CO_2 \text{ level} = F(24) - F(12) = \int_{12}^{24} f(t)\, dt \approx 0.278 \text{ mmol/liter.}$$

(c) To find the CO_2 level at $t = 12$, we use the Fundamental Theorem to estimate the change in CO_2 level during the day:

$$F(12) - F(0) = \int_0^{12} f(t)\, dt$$

Using a left Riemann sum as in part (c), we have

$$F(12) - F(0) = \int_0^{12} f(t)\, dt \approx -0.328.$$

Since initially there were $F(0) = 2.600$ mmol/liter, we have

$$F(12) = F(0) - 0.328 = 2.272 \text{ mmol/liter.}$$

(d) The amount of CO_2 removed during the day is represented by the area of the region below the t-axis; the amount of CO_2 added during the night is represented by the area above the t-axis. These areas look approximately equal, so the CO_2 level is approximately in equilibrium.

Using Riemann sums to estimate these areas, we find that about 0.278 mmol/l of CO_2 was released into the pond during the night and about 0.328 mmol/l of CO_2 was absorbed from the pond during the day. These quantities are sufficiently close that the difference could be due to measurement error, or to errors from the Riemann sum approximation.

Table 5.7   *Rate, $f(t)$, at which CO_2 is entering or leaving water (read from Figure 5.33)*

t	$f(t)$	t	$f(t)$	t	$f(t)$	t	$f(t)$	t	$f(t)$	t	$f(t)$
0	0.000	4	−0.039	8	−0.026	12	0.000	16	0.035	20	0.020
2	−0.044	6	−0.035	10	−0.020	14	0.045	18	0.027	22	0.012

The Definite Integral as an Average

We know how to find the average of n numbers: Add them and divide by n. But how do we find the average value of a continuously varying function? Let us consider an example. Suppose $f(t)$ is the temperature at time t, measured in hours since midnight, and that we want to calculate the average temperature over a 24-hour period. One way to start is to average the temperatures at n equally spaced times, $t_1, t_2, \ldots, t_n$, during the day.

$$\text{Average temperature} \approx \frac{f(t_1) + f(t_2) + \cdots + f(t_n)}{n}.$$

The larger we make n, the better the approximation. We can rewrite this expression as a Riemann sum over the interval $0 \le t \le 24$ if we use the fact that $\Delta t = 24/n$, so $n = 24/\Delta t$:

$$\begin{aligned} \text{Average temperature} &\approx \frac{f(t_1) + f(t_2) + \cdots + f(t_n)}{24/\Delta t} \\ &= \frac{f(t_1)\Delta t + f(t_2)\Delta t + \cdots + f(t_n)\Delta t}{24} \\ &= \frac{1}{24}\sum_{i=1}^{n} f(t_i)\Delta t. \end{aligned}$$

As $n \to \infty$, the Riemann sum tends toward an integral, and $1/24$ of the sum also approximates the average temperature better. It makes sense, then, to write

$$\text{Average temperature} = \lim_{n \to \infty} \frac{1}{24} \sum_{i=1}^{n} f(t_i)\Delta t = \frac{1}{24} \int_{0}^{24} f(t)\, dt.$$

We have found a way of expressing the average temperature over an interval in terms of an integral. Generalizing for any function f, if $a < b$, we define

$$\begin{array}{c} \text{Average value of } f \\ \text{from } a \text{ to } b \end{array} = \frac{1}{b-a} \int_{a}^{b} f(x)\, dx.$$

How to Visualize the Average on a Graph

The definition of average value tells us that

$$(\text{Average value of } f) \cdot (b - a) = \int_{a}^{b} f(x)\, dx.$$

Let's interpret the integral as the area under the graph of f. Then the average value of f is the height of a rectangle whose base is $(b - a)$ and whose area is the same as the area under the graph of f. (See Figure 5.34.)

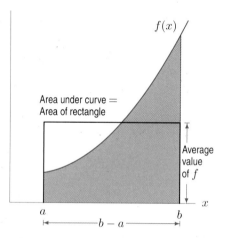

Figure 5.34: Area and average value

Example 5   Suppose that $C(t)$ represents the daily cost of heating your house, measured in dollars per day, where t is time measured in days and $t = 0$ corresponds to January 1, 2005. Interpret $\int_{0}^{90} C(t)\, dt$ and $\frac{1}{90 - 0} \int_{0}^{90} C(t)\, dt$.

Solution   The units for the integral $\int_{0}^{90} C(t)\, dt$ are (dollars/day) × (days) = dollars. The integral represents the total cost in dollars to heat your house for the first 90 days of 2005, namely the months of January, February, and March. The second expression is measured in (1/days)(dollars) or dollars per day, the same units as $C(t)$. It represents the average cost per day to heat your house during the first 90 days of 2005.

Example 6   On page 10, we saw that the population of Mexico could be modeled by the function

$$P = f(t) = 67.38(1.026)^t,$$

where P is in millions of people and t is in years since 1980. Use this function to predict the average population of Mexico between the years 2000 and 2020.

Solution   We want the average value of $f(t)$ between $t = 20$ and $t = 40$. This is given by

$$\text{Average population} = \frac{1}{40 - 20} \int_{20}^{40} f(t)\, dt = \frac{1}{20}(2942.66) = 147.1.$$

We used a calculator to evaluate the integral. We see that the average population of Mexico between 2000 and 2020 is predicted to be about 147 million people.

Exercises and Problems for Section 5.3

Exercises

1. If $f(t)$ is measured in dollars per year and t is measured in years, what are the units of $\int_a^b f(t)\, dt$?

2. If $f(t)$ is measured in meters/second^2 and t is measured in seconds, what are the units of $\int_a^b f(t)\, dt$?

3. If $f(x)$ is measured in pounds and x is measured in feet, what are the units of $\int_a^b f(x)\, dx$?

In Exercises 4–7, explain in words what the integral represents and give units.

4. $\int_1^3 v(t)\, dt$, where $v(t)$ is velocity in meters/sec and t is time in seconds.

5. $\int_0^6 a(t)\, dt$, where $a(t)$ is acceleration in km/hr^2 and t is time in hours.

6. $\int_{2000}^{2004} f(t)\, dt$, where $f(t)$ is the rate at which the world's population is growing in year t, in billion people per year.

7. $\int_0^5 s(x)\, dx$, where $s(x)$ is rate of change of salinity (salt concentration) in gm/liter per cm in sea water, where x is depth below the surface of the water in cm.

8. For the two cars in Example 3, page 257, estimate:

(a) The distances moved by car 1 and car 2 during the first minute.

(b) The time at which the two cars have gone the same distance.

In Exercises 9–11, find the average value of the function over the given interval.

9. $g(t) = 1 + t$ over $[0, 2]$ **10.** $g(t) = e^t$ over $[0, 10]$

11. $f(x) = 4x + 7$ over $[1, 3]$.

12. How do the units for the average value relate to the units for $f(x)$ and the units for x?

13. Oil leaks out of a tanker at a rate of $r = f(t)$ gallons per minute, where t is in minutes. Write a definite integral expressing the total quantity of oil which leaks out of the tanker in the first hour.

14. Water is leaking out of a tank at a rate of $R(t)$ gallons/hour, where t is measured in hours.

(a) Write a definite integral that expresses the total amount of water that leaks out in the first two hours.

(b) In Figure 5.35, shade the region whose area represents the total amount of water that leaks out in the first two hours.

(c) Give an upper and lower estimate of the total amount of water that leaks out in the first two hours.

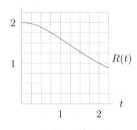

Figure 5.35

Problems

15. Coal gas is produced at a gasworks. Pollutants in the gas are removed by scrubbers, which become less and less efficient as time goes on. The following measurements, made at the start of each month, show the rate at which pollutants are escaping (in tons/month) in the gas:

Time (months)	0	1	2	3	4	5	6
Rate pollutants escape	5	7	8	10	13	16	20

(a) Make an overestimate and an underestimate of the total quantity of pollutants that escape during the first month.

(b) Make an overestimate and an underestimate of the total quantity of pollutants that escape during the six months.

(c) How often would measurements have to be made to find overestimates and underestimates which differ by less than 1 ton from the exact quantity of pollutants that escaped during the first six months?

16. The rate at which the world's oil is being consumed is continuously increasing. Suppose the rate of oil consumption (in billions of barrels per year) is given by the function $r = f(t)$, where t is measured in years and $t = 0$ is the start of 1990.

(a) Write a definite integral which represents the total quantity of oil used between the start of 1990 and the start of 1995.

(b) Suppose $r = 32e^{0.05t}$. Using a left-hand sum with five subdivisions, find an approximate value for the total quantity of oil used between the start of 1990 and the start of 1995.

(c) Interpret each of the five terms in the sum from part (b) in terms of oil consumption.

17. The following table gives the emissions, E, of nitrogen oxides in millions of metric tons per year in the US.[4] Let t be the number of years since 1940 and $E = f(t)$.

(a) What are the units and meaning of $\int_0^{50} f(t)dt$?

(b) Estimate $\int_0^{50} f(t)dt$.

Year	1940	1950	1960	1970	1980	1990
E	6.9	9.4	13.0	18.5	20.9	19.6

18. A two-day environmental clean up started at 9 am on the first day. The number of workers fluctuated as shown in Figure 5.36. If the workers were paid $10 per hour, how much was the total personnel cost of the clean up?

[4]*Statistical Abstract of the US*, 1992.

Figure 5.36

19. Suppose in Problem 18 that the workers were paid $10 per hour for work during the time period 9 am to 5 pm and were paid $15 per hour for work during the rest of the day. What would the total personnel costs of the clean up have been under these conditions?

20. A warehouse charges its customers $5 per day for every 10 cubic feet of space used for storage. Figure 5.37 records the storage used by one company over a month. How much will the company have to pay?

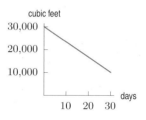

Figure 5.37

21. The amount of waste a company produces, W, in metric tons per week, is approximated by $W = 3.75e^{-0.008t}$, where t is in weeks since January 1, 2000. Waste removal for the company costs $15/ton. How much does the company pay for waste removal during the year 2000?

22. A force F parallel to the x-axis is given by the graph in Figure 5.38. Estimate the work, W, done by the force, where $W = \int_0^{16} F(x)\, dx$.

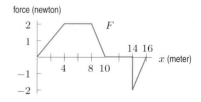

Figure 5.38

23. (a) Using Figures 5.39 and 5.40, find the average value on $0 \leq x \leq 2$ of

    (i) $f(x)$ (ii) $g(x)$ (iii) $f(x) \cdot g(x)$

(b) Is the following statement true? Explain your answer.

$$\text{Average}(f) \cdot \text{Average}(g) = \text{Average}(f \cdot g)$$

Figure 5.39   **Figure 5.40**

24. (a) Using Figure 5.41, find $\int_1^6 f(x)\,dx$.
(b) What is the average value of f on $[1, 6]$?

Figure 5.41

25. (a) Using Figure 5.42, estimate $\int_{-3}^{3} f(x)\,dx$.
(b) Which of the following average values of $f(x)$ is larger?

    (i) Between $x = -3$ and $x = 3$
    (ii) Between $x = 0$ and $x = 3$

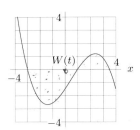

Figure 5.42

26. For the even function f in Figure 5.43, write an expression involving one or more definite integrals that denotes:

(a) The average value of f for $0 \leq x \leq 5$.
(b) The average value of $|f|$ for $0 \leq x \leq 5$.

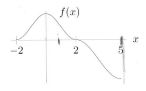

Figure 5.43

27. For the even function f in Figure 5.43, consider the average value of f over the following intervals:

  I. $0 \leq x \leq 1$ II. $0 \leq x \leq 2$
  III. $0 \leq x \leq 5$ IV. $-2 \leq x \leq 2$

(a) For which interval is the average value of f least?
(b) For which interval is the average value of f greatest?
(c) For which pair of intervals are the average values equal?

28. (a) Without computing any integrals, explain why the average value of $f(x) = \sin x$ on $[0, \pi]$ must be between 0.5 and 1.
(b) Compute this average.

29. (a) What is the average value of $f(x) = \sqrt{1 - x^2}$ over the interval $0 \leq x \leq 1$?
(b) How can you tell whether this average value is more or less than 0.5 without doing any calculations?

30. The value, V, of a Tiffany lamp, worth \$225 in 1965, increases at 15% per year. Its value in dollars t years after 1965 is given by

$$V = 225(1.15)^t.$$

Find the average value of the lamp over the period 1965–2000.

31. A news broadcast in early 1993 said the average American's annual income is changing at a rate of $r(t) = 40(1.002)^t$ dollars per month, where t is in months from January 1, 1993. How much did the average American's income change during 1993?

32. The number of hours, H, of daylight in Madrid as a function of date is approximated by the formula

$$H = 12 + 2.4 \sin[0.0172(t - 80)],$$

where t is the number of days since the start of the year. Find the average number of hours of daylight in Madrid:

(a) in January   **(b)** in June   **(c)** over a year

(d) Explain why the relative magnitudes of your answers to parts (a), (b), and (c) are reasonable.

33. Figure 5.44 shows the rate, $f(x)$, in thousands of algae per hour, at which a population of algae is growing, where x is in hours.

(a) Estimate the average value of the rate over the interval $x = -1$ to $x = 3$.
(b) Estimate the total change in the population over the interval $x = -3$ to $x = 3$.

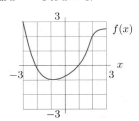

Figure 5.44

34. A bicyclist pedals along a straight road with velocity, v, given in Figure 5.45. She starts 5 miles from a lake; positive velocities take her away from the lake and negative velocities take her toward the lake. When is the cyclist farthest from the lake, and how far away is she then?

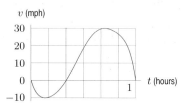

Figure 5.45

35. A car speeds up at a constant rate from 10 to 70 mph over a period of half an hour. Its fuel efficiency (in miles per gallon) at various speeds is shown in the table. Make lower and upper estimates of the quantity of fuel used during the half hour.

Speed (mph)	10	20	30	40	50	60	70
Fuel efficiency (mpg)	15	18	21	23	24	25	26

36. Height velocity graphs are used by endocrinologists to follow the progress of children with growth deficiencies. Figure 5.46 shows the height velocity curves of an average boy and an average girl between age 3 and 18.

    (a) Which curve is for girls and which is for boys? Explain how you can tell.

(b) About how much does the average boy grow between ages 3 and 10?

(c) The growth spurt associated with adolescence and the onset of puberty occurs between ages 12 and 15 for the average boy and between ages 10 and 12.5 for the average girl. Estimate the height gained by each average child during this growth spurt.

(d) When fully grown, about how much taller is the average man than the average woman? (The average boy and girl are about the same height at age 3.)

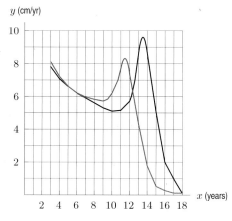

Figure 5.46

37. In Chapter 2, the average velocity over the time interval $a \leq t \leq b$ was defined to be $(s(b) - s(a))/(b - a)$, where $s(t)$ is the position function. Use the Fundamental Theorem of Calculus to show that the average value of the velocity function $v(t)$, on the interval $a \leq t \leq b$ is also $(s(b) - s(a))/(b - a)$.

5.4   THEOREMS ABOUT DEFINITE INTEGRALS

Properties of the Definite Integral

For the definite integral $\int_a^b f(x)\, dx$, we have so far only considered the case $a < b$. We now allow $a \geq b$. We still set $x_0 = a$, $x_n = b$, and $\Delta x = (b - a)/n$. As before, we have $\int_a^b f(x)dx = \lim_{n \to \infty} \sum_{i=1}^{n} f(x_i)\Delta x$.

Theorem 5.2: Properties of Limits of Integration

If a, b, and c are any numbers and f is a continuous function, then

1. $\displaystyle\int_b^a f(x)\, dx = -\int_a^b f(x)\, dx.$

2. $\displaystyle\int_a^c f(x)\, dx + \int_c^b f(x)\, dx = \int_a^b f(x)\, dx.$

In words:

    1. The integral from b to a is the negative of the integral from a to b.

    2. The integral from a to c plus the integral from c to b is the integral from a to b.

By interpreting the integrals as areas, we can justify these results for $f \geq 0$. In fact, they are true for all functions for which the integrals make sense. For a formal proof, see the online supplement at www.wiley.com/college/hugheshallett.

Why is $\int_b^a f(x)\, dx = -\int_a^b f(x)\, dx$?

By definition, both integrals are approximated by sums of the form $\sum f(x_i)\Delta x$. The only difference in the sums for $\int_b^a f(x)\, dx$ and $\int_a^b f(x)\, dx$ is that in the first $\Delta x = (a - b)/n = -(b - a)/n$ and in the second $\Delta x = (b - a)/n$. Since everything else about the sums is the same, we must have $\int_b^a f(x)\, dx = -\int_a^b f(x)\, dx$.

Why is $\int_a^c f(x)\, dx + \int_c^b f(x)\, dx = \int_a^b f(x)\, dx$?

Suppose $a < c < b$. Figure 5.47 suggests that $\int_a^c f(x)\, dx + \int_c^b f(x)\, dx = \int_a^b f(x)\, dx$ since the area under f from a to c plus the area under f from c to b together make up the whole area under f from a to b.

This property holds for all numbers a, b, and c, not just those satisfying $a < c < b$. (See Figure 5.48.) For example, the area under f from 3 to 6 is equal to the area from 3 to 8 *minus* the area from 6 to 8, so

$$\int_3^6 f(x)\, dx = \int_3^8 f(x)\, dx - \int_6^8 f(x)\, dx = \int_3^8 f(x)\, dx + \int_8^6 f(x)\, dx.$$

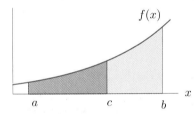

Figure 5.47: Additivity of the definite integral $(a < c < b)$

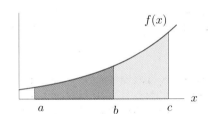

Figure 5.48: Additivity of the definite integral $(a < b < c)$

Example 1   Suppose $\int_0^{1.25} \cos(x^2)\, dx = 0.98$ and $\int_0^1 \cos(x^2)\, dx = 0.90$. (See Figure 5.49.) What are the values of the following integrals?

(a) $\displaystyle\int_1^{1.25} \cos(x^2)\, dx$ (b) $\displaystyle\int_{-1}^1 \cos(x^2)\, dx$ (c) $\displaystyle\int_{1.25}^{-1} \cos(x^2)\, dx$

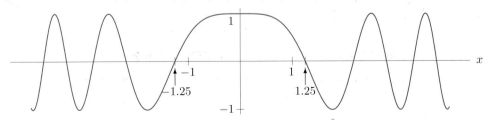

Figure 5.49: Graph of $f(x) = \cos(x^2)$

Solution   (a) Since, by the additivity property,

$$\int_0^{1.25} \cos(x^2)\, dx = \int_0^1 \cos(x^2)\, dx + \int_1^{1.25} \cos(x^2)\, dx,$$

we get

$$0.98 = 0.90 + \int_1^{1.25} \cos(x^2)\, dx,$$

so

$$\int_1^{1.25} \cos(x^2)\, dx = 0.08.$$

(b) By the additivity property, we have

$$\int_{-1}^1 \cos(x^2)\, dx = \int_{-1}^0 \cos(x^2)\, dx + \int_0^1 \cos(x^2)\, dx.$$

By the symmetry of $\cos(x^2)$ about the y-axis,

$$\int_{-1}^0 \cos(x^2)\, dx = \int_0^1 \cos(x^2)\, dx = 0.90.$$

So

$$\int_{-1}^1 \cos(x^2)\, dx = 0.90 + 0.90 = 1.80.$$

(c) Using both properties in Theorem 5.2, we have

$$\int_{1.25}^{-1} \cos(x^2)\, dx = -\int_{-1}^{1.25} \cos(x^2)\, dx = -\left(\int_{-1}^0 \cos(x^2)\, dx + \int_0^{1.25} \cos(x^2)\, dx \right)$$

$$= -(0.90 + 0.98) = -1.88.$$

Theorem 5.3: Properties of Sums and Constant Multiples of the Integrand

Let f and g be continuous functions and let c be a constant.

1. $\displaystyle \int_a^b (f(x) \pm g(x))\, dx = \int_a^b f(x)\, dx \pm \int_a^b g(x)\, dx.$

2. $\displaystyle \int_a^b cf(x)\, dx = c \int_a^b f(x)\, dx.$

In words:

1. The integral of the sum (or difference) of two functions is the sum (or difference) of their integrals.

2. The integral of a constant times a function is that constant times the integral of the function.

Why Do these Properties Hold?

Both can be visualized by thinking of the definite integral as the limit of a sum of areas of rectangles.

For property 1, suppose that f and g are positive on the interval $[a, b]$ so that the area under $f(x) + g(x)$ is approximated by the sum of the areas of rectangles like the one shaded in Figure 5.50. The area of this rectangle is

$$(f(x_i) + g(x_i))\Delta x = f(x_i)\Delta x + g(x_i)\Delta x.$$

Since $f(x_i)\Delta x$ is the area of a rectangle under the graph of f, and $g(x_i)\Delta x$ is the area of a rectangle under the graph of g, the area under $f(x) + g(x)$ is the sum of the areas under $f(x)$ and $g(x)$.

For property 2, notice that multiplying a function by c stretches or shrinks the graph in the vertical direction by a factor of c. Thus, it stretches or shrinks the height of each approximating rectangle by c, and hence multiplies the area by c.

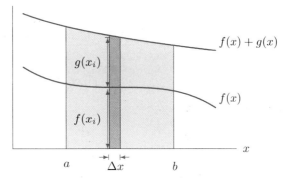

Figure 5.50: Area $= \int_a^b [f(x) + g(x)]\, dx = \int_a^b f(x)\, dx + \int_a^b g(x)\, dx$

Example 2   Evaluate the definite integral $\int_0^2 (1 + 3x)\, dx$ exactly.

Solution   We can break this integral up as follows:

$$\int_0^2 (1 + 3x)\, dx = \int_0^2 1\, dx + \int_0^2 3x\, dx = \int_0^2 1\, dx + 3 \int_0^2 x\, dx.$$

From Figures 5.51 and 5.52 and the area interpretation of the integral, we see that

$$\int_0^2 1\, dx = \begin{array}{c} \text{Area of} \\ \text{rectangle} \end{array} = 2 \quad \text{and} \quad \int_0^2 x\, dx = \begin{array}{c} \text{Area of} \\ \text{triangle} \end{array} = \frac{1}{2} \cdot 2 \cdot 2 = 2.$$

Therefore,

$$\int_0^2 (1 + 3x)\, dx = \int_0^2 1\, dx + 3 \int_0^2 x\, dx = 2 + 3 \cdot 2 = 8.$$

Figure 5.51: Area representing $\int_0^2 1\, dx$

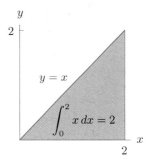

Figure 5.52: Area representing $\int_0^2 x\, dx$

Area Between Curves

Theorem 5.3 enables us to find the area of a region between curves. We have the following result:

> If the graph of $f(x)$ lies above the graph of $g(x)$ for $a \le x \le b$, then
>
> $$\begin{array}{c} \text{Area between } f \text{ and } g \\ \text{for } a \le x \le b \end{array} = \int_a^b (f(x) - g(x))\, dx.$$

Example 3   Find the area of the shaded region in Figure 5.53.

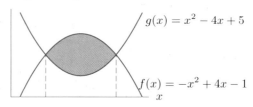

Figure 5.53: Area between two parabolas

Solution   The curves cross where

$$x^2 - 4x + 5 = -x^2 + 4x - 1$$
$$2x^2 - 8x + 6 = 0$$
$$2(x - 1)(x - 3) = 0$$
$$x = 1, 3.$$

Since $f(x) = -x^2 + 4x - 1$ is above $g(x) = x^2 - 4x + 5$ for x between 1 and 3, we find the shaded area by subtraction:

$$\text{Area} = \int_1^3 f(x)\,dx - \int_1^3 g(x)\,dx = \int_1^3 (f(x) - g(x))\,dx$$

$$= \int_1^3 ((-x^2 + 4x - 1) - (x^2 - 4x + 5))\,dx$$

$$= \int_1^3 (-2x^2 + 8x - 6)\,dx = 2.667.$$

Using Symmetry to Evaluate Integrals

Symmetry can be useful in evaluating definite integrals. An even function is symmetric about the y-axis. An odd function is symmetric about the origin. Figures 5.54 and 5.55 suggest the following results:

If f is even, then $\int_{-a}^{a} f(x)\,dx = 2\int_0^a f(x)\,dx$.   If g is odd, then $\int_{-a}^{a} g(x)\,dx = 0$.

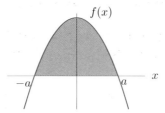

Figure 5.54: For an even function,
$\int_{-a}^{a} f(x)\,dx = 2\int_0^a f(x)\,dx$

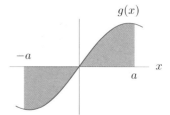

Figure 5.55: For an odd function,
$\int_{-a}^{a} g(x)\,dx = 0$

Example 4   Given that $\int_0^\pi \sin t\,dt = 2$, find (a) $\int_{-\pi}^{\pi} \sin t\,dt$ (b) $\int_{-\pi}^{\pi} |\sin t|\,dt$

Solution   Graphs of $\sin t$ and $|\sin t|$ are in Figures 5.56 and 5.57.

(a) Since $\sin t$ is an odd function

$$\int_{-\pi}^{\pi} \sin t \, dt = 0.$$

(b) Since $|\sin t|$ is an even function

$$\int_{-\pi}^{\pi} |\sin t| \, dt = 2 \int_{0}^{\pi} |\sin t| \, dt = 4.$$

Figure 5.56

Figure 5.57

Using the Fundamental Theorem to Compute Integrals

The Fundamental Theorem provides an exact way of computing certain definite integrals.

Example 5   Compute $\displaystyle\int_{1}^{3} 2x \, dx$ by two different methods.

Solution   Using left- and right-hand sums, we can approximate this integral as accurately as we want. With $n = 100$, for example, the left-sum is 7.96 and the right sum is 8.04. Using $n = 500$ we learn

$$7.992 < \int_{1}^{3} 2x \, dx < 8.008.$$

The Fundamental Theorem, on the other hand, allows us to compute the integral exactly. We take $f(x) = 2x$. We know that if $F(x) = x^2$, then $F'(x) = 2x$. So we use $f(x) = 2x$ and $F(x) = x^2$ and obtain

$$\int_{1}^{3} 2x \, dx = F(3) - F(1) = 3^2 - 1^2 = 8.$$

Comparing Integrals

Suppose we have constants m and M such that $m \leq f(x) \leq M$ for $a \leq x \leq b$. We say f is *bounded above* by M and *bounded below* by m. Then the graph of f lies between the horizontal lines $y = m$ and $y = M$. So the definite integral lies between $m(b - a)$ and $M(b - a)$. See Figure 5.58.

Suppose $f(x) \leq g(x)$ for $a \leq x \leq b$, as in Figure 5.59. Then the definite integral of f is less than or equal to the definite integral of g. This leads us to the following results:

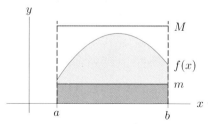

Figure 5.58: The area under the graph of f lies between the areas of the rectangles

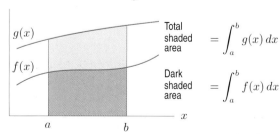

Figure 5.59: If $f(x) \leq g(x)$ then $\int_{a}^{b} f(x) \, dx \leq \int_{a}^{b} g(x) \, dx$

Theorem 5.4: Comparison of Definite Integrals

Let f and g be continuous functions.

1. If $m \leq f(x) \leq M$ for $a \leq x \leq b$, then $\quad m(b - a) \leq \int_a^b f(x) \, dx \leq M(b - a)$.

2. If $f(x) \leq g(x)$ for $a \leq x \leq b$, then $\quad \int_a^b f(x) \, dx \leq \int_a^b g(x) \, dx$.

Example 6   Explain why $\int_0^{\sqrt{\pi}} \sin(x^2) \, dx \leq \sqrt{\pi}$.

Solution   Since $\sin(x^2) \leq 1$ for all x (see Figure 5.60), part 2 of Theorem 5.4 gives

$$\int_0^{\sqrt{\pi}} \sin(x^2) \, dx \leq \int_0^{\sqrt{\pi}} 1 \, dx = \sqrt{\pi}.$$

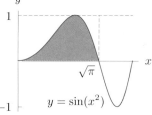

Figure 5.60: Graph showing that $\int_0^{\sqrt{\pi}} \sin(x^2) \, dx < \sqrt{\pi}$

Exercises and Problems for Section 5.4

Exercises

1. (a) Suppose $f'(x) = \sin(x^2)$ and $f(0) = 2$. Use a graph of $f'(x)$ to decide which is larger:

     (i) $f(0)$ or $f(1)$ (ii) $f(2)$ or $f(2.5)$

   (b) Estimate $f(b)$ for $b = 0, 1, 2, 3$.

2. The graph of a derivative $f'(x)$ is shown in Figure 5.61. Fill in the table of values for $f(x)$ given that $f(0) = 2$.

x	0	1	2	3	4	5	6
$f(x)$	2						

Figure 5.61: Graph of f', not f

3. Figure 5.62 shows f. If $F' = f$ and $F(0) = 0$, find $F(b)$ for $b = 1, 2, 3, 4, 5, 6$.

Figure 5.62

Find the area of the regions in Exercises 4–9.

4. Under $y = e^x$ and above $y = 1$ for $0 \leq x \leq 2$.

5. Under $y = 5 \ln(2x)$ and above $y = 3$ for $3 \leq x \leq 5$.

6. Between $y = x^2$ and $y = x^3$ for $0 \leq x \leq 1$.

7. Between $y = x^{1/2}$ and $y = x^{1/3}$ for $0 \leq x \leq 1$.

8. Between $y = \sin x + 2$ and $y = 0.5$ for $6 \leq x \leq 10$.

9. Under $y = \cos t$ and above $y = \sin t$ for $0 \leq t \leq \pi$.

In Exercises 10–13, let $f(t) = F'(t)$. Write the integral $\int_a^b f(t) \, dt$ and evaluate it using the Fundamental Theorem of Calculus.

10. $F(t) = t^2$; $a = 1, b = 3$

11. $F(t) = \ln t$; $a = 1, b = 5$

12. $F(t) = \sin t$; $a = 0, b = \pi/2$

13. $F(t) = t^4$, $a = 1, b = 1$

Problems

Let $\int_a^b f(x)\,dx = 8$, $\int_a^b (f(x))^2\,dx = 12$, $\int_a^b g(t)\,dt = 2$, and $\int_a^b (g(t))^2\,dt = 3$. Find the integrals in Problems 14–19.

14. $\int_a^b (f(x) + g(x))\,dx$ **15.** $\int_a^b cf(z)\,dz$

16. $\int_a^b \left((f(x))^2 - (g(x))^2\right)\,dx$

17. $\int_a^b (f(x))^2\,dx - \left(\int_a^b f(x)\,dx\right)^2$

18. $\int_a^b \left(c_1 g(x) + (c_2 f(x))^2\right)\,dx$

19. $\int_{a+5}^{b+5} f(x - 5)\,dx$

20. In Exercises 6 and 7 we calculated the areas between $y = x^2$ and $y = x^3$ and between $y = x^{1/2}$ and $y = x^{1/3}$ on $0 \le x \le 1$. Explain why you would expect these two areas to be equal.

In Problems 21–24, find $\int_2^5 f(x)\,dx$.

21. $f(x)$ is odd and $\int_{-2}^5 f(x)\,dx = 8$

22. $f(x)$ is even, $\int_{-2}^2 f(x)\,dx = 6$, and $\int_{-5}^5 f(x)\,dx = 14$

23. $\int_2^5 (3f(x) + 4)\,dx = 18$

24. $\int_2^4 2f(x)\,dx = 8$ and $\int_5^4 f(x) = 1$

25. (a) Using a graph, decide if the area under $y = e^{-x^2/2}$ between 0 and 1 is more or less than 1.
(b) Find the area.

26. Without any computation, find the values of

(a) $\displaystyle\int_{-2}^2 \sin x\,dx,$ **(b)** $\displaystyle\int_{-\pi}^{\pi} x^{113}\,dx.$

27. (a) Is $\int_{-1}^1 e^{x^2}\,dx$ positive, negative, or zero? Explain.
(b) Explain why $0 < \int_0^1 e^{x^2}\,dx < 3$.

28. Without computation, show that $2 \le \displaystyle\int_0^2 \sqrt{1 + x^3}\,dx \le 6$.

29. Without calculating the integral, explain why the following statements are false.

(a) $\displaystyle\int_{-2}^{-1} e^{x^2}\,dx = -3$

(b) $\displaystyle\int_{-1}^1 \left|\frac{\cos(x + 2)}{1 + \tan^2 x}\right|\,dx = 0$

30. Using Figure 5.63, write $\int_0^3 f(x)\,dx$ in terms of $\int_{-1}^1 f(x)\,dx$ and $\int_1^3 f(x)\,dx$.

Figure 5.63

31. (a) Let $\int_0^3 f(x)\,dx = 6$. What is the average value of $f(x)$ on the interval $x = 0$ to $x = 3$?
(b) If $f(x)$ is even, what is $\int_{-3}^3 f(x)\,dx$? What is the average value of $f(x)$ on the interval $x = -3$ to $x = 3$?
(c) If $f(x)$ is odd, what is $\int_{-3}^3 f(x)\,dx$? What is the average value of $f(x)$ on the interval $x = -3$ to $x = 3$?

32. (a) Use Figure 5.64 to explain why $\int_{-3}^3 xe^{-x^2}\,dx = 0$.
(b) Find the left-hand sum approximation with $n = 3$ to $\int_0^3 xe^{-x^2}\,dx$. Give your answer to four decimal places.
(c) Repeat part (b) for $\int_{-3}^0 xe^{-x^2}\,dx$.
(d) Do your answers to parts (b) and (c) add to 0? Explain.

Figure 5.64

33. (a) For any continuous function f, is $\int_1^2 f(x)\,dx + \int_2^3 f(x)\,dx = \int_1^3 f(x)\,dx$?
(b) For any function f, add the left-hand sum approximation with 10 subdivisions to $\int_1^2 f(x)\,dx$ to the left-hand sum approximation with 10 subdivisions to $\int_2^3 f(x)\,dx$. Do you get the left sum approximations with 10 subdivisions to $\int_1^3 f(x)\,dx$? If not, interpret the result as a different Riemann Sum.

Problems 34–35 concern the graph of f' in Figure 5.65.

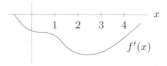

Figure 5.65: Graph of f', not f

34. Which is greater, $f(0)$ or $f(1)$?

35. List the following in increasing order:
$$\frac{f(4) - f(2)}{2}, \quad f(3) - f(2), \quad f(4) - f(3).$$

For Problems 36–39, mark the following quantities on a copy of the graph of f in Figure 5.66.

Figure 5.66

36. A length representing $f(b) - f(a)$.

37. A slope representing $\dfrac{f(b) - f(a)}{b - a}$.

38. An area representing $F(b) - F(a)$, where $F' = f$.

39. A length roughly approximating

$$\frac{F(b) - F(a)}{b - a}, \text{ where } F' = f.$$

40. Figure 5.67 shows the *standard normal distribution* from statistics, which is given by

$$\frac{1}{\sqrt{2\pi}} e^{-x^2/2}.$$

Statistics books often contain tables such as the following, which show the area under the curve from 0 to b for various values of b.

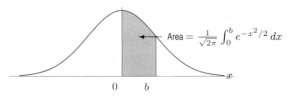

Figure 5.67

b	$\frac{1}{\sqrt{2\pi}} \int_0^b e^{-x^2/2}\, dx$
1	0.3413
2	0.4772
3	0.4987
4	0.5000

Use the information given in the table and the symmetry of the curve about the y-axis to find:

(a) $\dfrac{1}{\sqrt{2\pi}} \displaystyle\int_1^3 e^{-x^2/2}\, dx$ (b) $\dfrac{1}{\sqrt{2\pi}} \displaystyle\int_{-2}^3 e^{-x^2/2}\, dx$

41. Use the property $\int_b^a f(x)\, dx = -\int_a^b f(x)\, dx$ to show that $\int_a^a f(x)\, dx = 0$.

42. The average value of $y = v(x)$ equals 4 for $1 \leq x \leq 6$, and equals 5 for $6 \leq x \leq 8$. What is the average value of $v(x)$ for $1 \leq x \leq 8$?

CHAPTER SUMMARY (see also Ready Reference at the end of the book)

- **Definite integral as limit of right or left sums**
- **Fundamental Theorem of Calculus**
- **Interpretations of the definite integral**
  Area, total change from rate of change, change in position given velocity, $(b - a)\cdot$ average value.
- **Properties of the definite integral**

Properties involving integrand, properties involving limits, comparison between integrals.

- **Working with the definite integral**
  Estimate definite integral from graph, table of values, or formula.
- **Theorems about definite integrals**

REVIEW EXERCISES AND PROBLEMS FOR CHAPTER FIVE

Exercises

1. A village wishes to measure the quantity of water that is piped to a factory during a typical morning. A gauge on the water line gives the flowrate (in cubic meters per hour) at any instant. The flowrate is about 100 m³/hr at 6 am and increases steadily to about 280 m³/hr at 9 am.

   (a) Using only this information, give your best estimate of the total volume of water used by the factory between 6 am and 9 am.
   (b) How often should the flowrate gauge be read to obtain an estimate of this volume to within 6 m³?

2. You jump out of an airplane. Before your parachute opens you fall faster and faster, but your acceleration decreases as you fall because of air resistance. The table

gives your acceleration, a (in m/sec²), after t seconds.

t	0	1	2	3	4	5
a	9.81	8.03	6.53	5.38	4.41	3.61

   (a) Give upper and lower estimates of your speed at $t = 5$.
   (b) Get a new estimate by taking the average of your upper and lower estimates. What does the concavity of the graph of acceleration tell you about your new estimate?

3. Use Figure 5.68 to estimate $\int_0^{20} f(x)\,dx$.

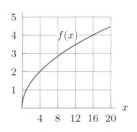

Figure 5.68

4. Using Figure 5.69, estimate $\int_{-3}^{5} f(x)dx$.

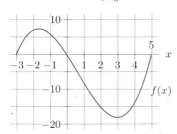

Figure 5.69

5. Using the table, estimate $\int_0^{100} f(t)\,dt$.

t	0	20	40	60	80	100
$f(t)$	1.2	2.8	4.0	4.7	5.1	5.2

6. (a) What is the area between the graph of $f(x)$ in Figure 5.70 and the x-axis, between $x = 0$ and $x = 5$?
(b) What is $\int_0^5 f(x)\,dx$?

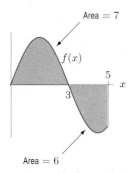

Area = 7

$f(x)$

Area = 6

Figure 5.70

Find the area of the regions in Exercises 7–13.

7. Between the parabola $y = 4 - x^2$ and the x-axis.

8. Between $y = x^2 - 9$ and the x-axis.

9. Under one arch of $y = \sin x$ and above the x-axis.

10. Between the line $y = 1$ and one arch of $y = \sin \theta$.

11. Between $y = -x^2 + 5x - 4$ and the x-axis, $0 \le x \le 3$.

12. Between $y = \cos x + 7$ and $y = \ln(x - 3)$, $5 \le x \le 7$.

13. Above the curve $y = -e^x + e^{2(x-1)}$ and below the x-axis, for $x \ge 0$.

14. A car going 80 ft/sec (about 55 mph) brakes to a stop in 8 seconds. Its velocity is recorded every 2 seconds and is given in the following table.

(a) Give your best estimate of the distance traveled by the car during the 8 seconds.
(b) To estimate the distance traveled accurate to within 20 feet, how often should you record the velocity?

t (seconds)	0	2	4	6	8
$v(t)$ (ft/sec)	80	52	28	10	0

15. A car accelerates smoothly from 0 to 60 mph in 10 seconds with the velocity given in Figure 5.71. Estimate how far the car travels during the 10-second period.

Figure 5.71

16. Your velocity is $v(t) = \ln(t^2 + 1)$ ft/sec for $0 \le t \le 3$. Estimate the distance traveled during this time.

17. Your velocity is $v(t) = \sin(t^2)$ mph for $0 \le t \le 1.1$. Estimate the distance traveled during this time.

Problems

18. If $\int_2^5 (2f(x) + 3)\,dx = 17$, find $\int_2^5 f(x)\,dx$.

19. If $f(x)$ is odd and $\int_{-2}^3 f(x)\,dx = 30$, find $\int_2^3 f(x)\,dx$.

20. If $f(x)$ is even and $\int_{-2}^2 (f(x) - 3)\,dx = 8$, find $\int_0^2 f(x)\,dx$.

21. If the average value of f on the interval $2 \le x \le 5$ is 4, find $\int_2^5 (3f(x) + 2)\,dx$.

22. Find $\int_{-1}^1 |x|\,dx$ geometrically.

23. Without any computation, find $\int_{-\pi/4}^{\pi/4} x^3 \cos x^2\,dx$.

24. (a) Sketch a graph of $f(x) = \sin(x^2)$ and mark on it the points $x = \sqrt{\pi}, \sqrt{2\pi}, \sqrt{3\pi}, \sqrt{4\pi}$.

(b) Use your graph to decide which of the four numbers

$$\int_0^{\sqrt{n\pi}} \sin(x^2)\,dx \quad n = 1, 2, 3, 4$$

is largest. Which is smallest? How many of the numbers are positive?

25. Two trains travel along parallel tracks. The velocity, v, of the trains as functions of time t are shown in Figure 5.72.

(a) Describe in words the trips taken by each train.

(b) Estimate the ratio of the following quantities for Train A to Train B:

   (i)   Maximum velocity (ii)   Time traveled

  (iii)   Distance traveled

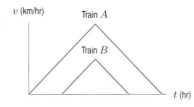

Figure 5.72

26. Annual coal production in the US (in quadrillion BTU per year) is given in the table.[5] Estimate the total amount of coal produced in the US between 1960 and 1990. If $r = f(t)$ is the rate of coal production t years since 1960, write an integral to represent the 1960–1990 coal production.

Year	1960	1965	1970	1975	1980	1985	1990
Rate	10.82	13.06	14.61	14.99	18.60	19.33	22.46

27. An old rowboat has sprung a leak. Water is flowing into the boat at a rate, $r(t)$, given in the following table.

t minutes	0	5	10	15
$r(t)$ liters/min	12	20	24	16

(a) Compute upper and lower estimates for the volume of water that has flowed into the boat during the 15 minutes.

(b) Draw a graph to illustrate the lower estimate.

28. Figure 5.73 gives your velocity during a trip starting from home. Positive velocities take you away from home and negative velocities take you toward home. Where are you at the end of the 5 hours? When are you farthest from home? How far away are you at that time?

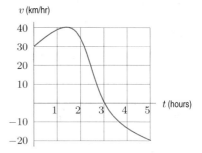

Figure 5.73

29. A bicyclist is pedaling along a straight road for one hour with a velocity v shown in Figure 5.74. She starts out five kilometers from the lake and positive velocities take her toward the lake. [Note: The vertical lines on the graph are at 10 minute (1/6 hour) intervals.]

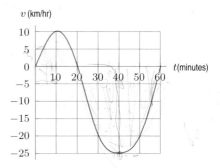

Figure 5.74

(a) Does the cyclist ever turn around? If so, at what time(s)?

(b) When is she going the fastest? How fast is she going then? Toward the lake or away?

(c) When is she closest to the lake? Approximately how close to the lake does she get?

(d) When is she farthest from the lake? Approximately how far from the lake is she then?

30. Figure 5.75 shows the rate of change of the quantity of water in a water tower, in liters per day, during the month of April. If the tower had 12,000 liters of water in it on April 1, estimate the quantity of water in the tower on April 30.

Figure 5.75

---

[5] *World Almanac*, 1995.

31. A bar of metal is cooling from $1000°C$ to room temperature, $20°C$. The temperature, H, of the bar t minutes after it starts cooling is given, in $°C$, by

$$H = 20 + 980e^{-0.1t}.$$

(a) Find the temperature of the bar at the end of one hour.

(b) Find the average value of the temperature over the first hour.

(c) Is your answer to part (b) greater or smaller than the average of the temperatures at the beginning and the end of the hour? Explain this in terms of the concavity of the graph of H.

32. A cup of coffee at $90°C$ is put into a $20°C$ room when $t = 0$. The coffee's temperature is changing at a rate of $r(t) = -7e^{-0.1t}$ $°C$ per minute, with t in minutes. Estimate the coffee's temperature when $t = 10$.

33. Water is pumped out of a holding tank at a rate of $5 - 5e^{-0.12t}$ liters/minute, where t is in minutes since the pump is started. If the holding tank contains 1000 liters of water when the pump is started, how much water does it hold one hour later?

34. The graph of a continuous function f is given in Figure 5.76. Rank the following integrals in ascending numerical order. Explain your reasons.

(i) $\int_0^2 f(x)\,dx$ (ii) $\int_0^1 f(x)\,dx$

(iii) $\int_0^2 (f(x))^{1/2}\,dx$ (iv) $\int_0^2 (f(x))^2\,dx$.

Figure 5.76

35. Using Figure 5.77, list the following integrals in increasing order (from smallest to largest). Which integrals are negative, which are positive? Give reasons.

I. $\int_a^b f(x)\,dx$ II. $\int_a^c f(x)\,dx$ III. $\int_a^e f(x)\,dx$

IV. $\int_b^e f(x)\,dx$ V. $\int_b^c f(x)\,dx$

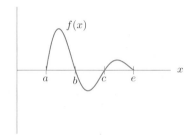

Figure 5.77

36. For the even function f graphed in Figure 5.78:

(a) Suppose you know $\int_0^2 f(x)\,dx$. What is $\int_{-2}^2 f(x)\,dx$?

(b) Suppose you know $\int_0^5 f(x)\,dx$ and $\int_2^5 f(x)\,dx$. What is $\int_0^2 f(x)\,dx$?

(c) Suppose you know $\int_{-2}^5 f(x)\,dx$ and $\int_{-2}^2 f(x)\,dx$. What is $\int_0^5 f(x)\,dx$?

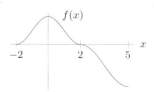

Figure 5.78

37. For the even function f graphed in Figure 5.78:

(a) Suppose you know $\int_{-2}^2 f(x)\,dx$ and $\int_0^5 f(x)\,dx$. What is $\int_2^5 f(x)\,dx$?

(b) Suppose you know $\int_{-2}^5 f(x)\,dx$ and $\int_{-2}^0 f(x)\,dx$. What is $\int_2^5 f(x)\,dx$?

(c) Suppose you know $\int_2^5 f(x)\,dx$ and $\int_{-2}^5 f(x)\,dx$. What is $\int_0^2 f(x)\,dx$?

38. The graphs in Figure 5.79 represent the velocity, v, of a particle moving along the x-axis for time $0 \le t \le 5$. The vertical scales of all graphs are the same. Identify the graph showing which particle:

(a) Has a constant acceleration.

(b) Ends up farthest to the left of where it started.

(c) Ends up the farthest from its starting point.

(d) Experiences the greatest initial acceleration.

(e) Has the greatest average velocity.

(f) Has the greatest average acceleration.

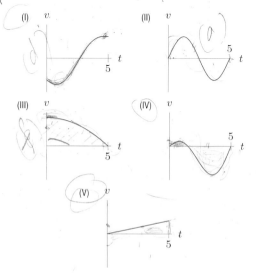

Figure 5.79

39. Assume w, b_1, and b_2 are positive constants, with $b_2 > b_1$. Compute the following integral using its geometric interpretation.

$$\int_0^w \left(b_1 + \frac{b_2 - b_1}{w}x\right) dx$$

40. Water is run into a large tank through a hose at a constant rate. After 5 minutes a hole is opened in the bottom of the tank, and water starts to flow out. Initially the flow rate through the hole is twice as great as the rate through the hose, but as the water level in the tank goes down, the flow rate through the hole decreases; after another 10 minutes the water level in the tank appears to be constant. Plot graphs of the flow rates through the hose and through the hole against time on the same pair of axes. Show how the volume of water in the tank at any time can be interpreted as an area (or the difference between two areas) on the graph. In particular, interpret the steady-state volume of water in the tank.[6]

41. Figure 5.80 shows thrust-time curves for two model rockets. The thrust or force, F, of the engine (in newtons) is plotted against time, t, (in seconds). The total impulse of the rocket's engine is defined as the definite integral of F with respect to t. The total impulse is a measure of the strength of the engine.

   (a) For approximately how many seconds is the thrust of rocket B greater than 10 newtons?
   (b) Estimate the total impulse for model rocket A.
   (c) What are the units for the total impulse calculated in part (b)?
   (d) Which rocket has the largest total impulse?
   (e) Which rocket has the largest maximum thrust?

Figure 5.80

42. The Glen Canyon Dam at the top of the Grand Canyon prevents natural flooding. In 1996, scientists decided an artificial flood was necessary to restore the environmental balance. Water was released through the dam at a controlled rate[7] shown in Figure 5.81. The figure also shows the rate of flow of the last natural flood in 1957.

   (a) At what rate was water passing through the dam in 1996 before the artificial flood?

   (b) At what rate was water passing down the river in the pre-flood season in 1957?
   (c) Estimate the maximum rates of discharge for the 1996 and 1957 floods.
   (d) Approximately how long did the 1996 flood last? How long did the 1957 flood last?
   (e) Estimate how much additional water passed down the river in 1996 as a result of the artificial flood.
   (f) Estimate how much additional water passed down the river in 1957 as a result of the flood.

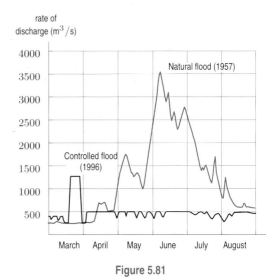

Figure 5.81

43. The Montgolfier brothers (Joseph and Etienne) were eighteenth-century pioneers in the field of hot-air ballooning. Had they had the appropriate instruments, they might have left us a record, like that shown in Figure 5.82, of one of their early experiments. The graph shows their vertical velocity, v, with upward as positive.

   (a) Over what intervals was the acceleration positive? Negative?
   (b) What was the greatest altitude achieved, and at what time?
   (c) At what time was the upward acceleration greatest?
   (d) At what time was the deceleration greatest?
   (e) What might have happened during this flight to explain the answer to part (d)?
   (f) This particular flight ended on top of a hill. How do you know that it did, and what was the height of the hill above the starting point?

---

[6]From *Calculus: The Analysis of Functions*, by Peter D. Taylor (Toronto: Wall & Emerson, Inc., 1992)

[7]Adapted from M. Collier, R. Webb, E. Andrews "Experimental Flooding in Grand Canyon" in *Scientific American* (January 1997).

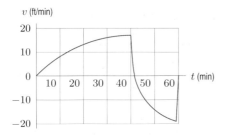

Figure 5.82

44. A mouse moves back and forth in a straight tunnel, attracted to bits of cheddar cheese alternately introduced to and removed from the ends (right and left) of the tunnel. The graph of the mouse's velocity, v, is given in Figure 5.83, with positive velocity corresponding to motion toward the right end. Assuming that the mouse starts $(t = 0)$ at the center of the tunnel, use the graph to estimate the time(s) at which:

(a) The mouse changes direction.
(b) The mouse is moving most rapidly to the right; to the left.
(c) The mouse is farthest to the right of center; farthest to the left.
(d) The mouse's speed (i.e., the magnitude of its velocity) is decreasing.
(e) The mouse is at the center of the tunnel.

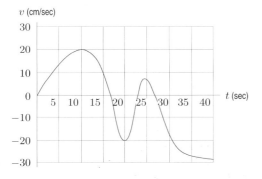

Figure 5.83

45. When an aircraft attempts to climb as rapidly as possible, its climb rate decreases with altitude. (This occurs because the air is less dense at higher altitudes.) The table shows performance data for a single-engine aircraft.

Altitude (1000 ft)	0	1	2	3	4	5
Climb rate (ft/min)	925	875	830	780	730	685
Altitude (1000 ft)	6	7	8	9	10	
Climb rate (ft/min)	635	585	535	490	440	

(a) Calculate upper and lower estimates for the time required for this aircraft to climb from sea level to 10,000 ft.

(b) If climb rate data were available in increments of 500 ft, what would be the difference between a lower and upper estimate of climb time based on 20 subdivisions?

46. Pollution is being dumped into a lake at a rate which is increasing at a constant rate from 10 kg/year to 50 kg/year until a total of 270 kg has been dumped. Sketch a graph of the rate at which pollution is being dumped in the lake against time. How long does it take until 270 kg has been dumped?

47. Using Figure 5.84, list from least to greatest,

(a) $f'(1)$.
(b) The average value of $f(x)$ on $0 \le x \le a$.
(c) The average value of the rate of change of $f(x)$, for $0 \le x \le a$.
(d) $\int_0^a f(x)\,dx$.

Figure 5.84

48. The number of days a cold-blooded organism, such as an insect, takes to mature depends on the surrounding temperature, H. Each organism has a minimum temperature $H_{\min}$ below which no development takes place.[8] For an interval of time, Δt, on which the temperature is constant, the increase in maturity of the organism can be measured by the number of degree-days, ΔS, where t is in days and

$$\Delta S = (H - H_{\min})\Delta t.$$

(a) If H varies with time, so $H = f(t)$, write an integral that represents the total number of degree-days, S, required if development to maturity takes T days.
(b) An organism, which has $H_{\min} = 15°$C, requires 125 degree-days to develop to maturity. Estimate the development time if the temperature, $H°$C, at time t days is in Table 5.8.

Table 5.8

t	1	2	3	4	5	6	7	8	9	10	11	12
H	20	22	27	28	27	31	29	30	28	25	24	26

---

[8]Information from http://www.ento.vt.edu/~sharov/PopEcol/popecol.html (Accessed Nov. 18, 2003).

CAS Challenge Problems

49. Consider the definite integral $\int_0^1 x^4\,dx$.

   (a) Write an expression for a right-hand Riemann sum approximation for this integral using n subdivisions. Express each x_i, $i = 1,2,\ldots,n$, in terms of i.

   (b) Use a computer algebra system to obtain a formula for the sum you wrote in part (a) in terms of n.

   (c) Take the limit of this expression for the sum as $n \to \infty$, thereby finding the exact value of this integral.

50. Repeat Problem 49, using the definite integral $\int_0^1 x^5\,dx$.

For Problems 51–53, you will write a Riemann sum approximating a definite integral and use a computer algebra system to find a formula for the Riemann sum. By evaluating the limit of this sum as the number of subdivisions approaches infinity, you will obtain the definite integral.

51. **(a)** Using summation notation, write the left-hand Riemann sum with n subdivisions for $\int_1^2 t\,dt$.

   (b) Use a computer algebra system to find a formula for the Riemann sum.

   (c) Evaluate the limit of the sum as n approaches infinity.

   (d) Calculate directly the area under the graph of $y = t$ between $t = 1$ and $t = 2$, and compare it with your answer to part (c).

52. **(a)** Using summation notation, write the left-hand Riemann sum with n subdivisions for $\int_1^2 t^2\,dt$.

   (b) Use a computer algebra system to find a formula for the Riemann sum.

   (c) Evaluate the limit of the sum as n approaches infinity.

   (d) What is the area under the graph of $y = t^2$ between $t = 1$ and $t = 2$?

53. **(a)** Using summation notation, write the right-hand Riemann sum with n subdivisions for $\int_0^\pi \sin x\,dx$.

   (b) Use a computer algebra system to find a formula for the Riemann sum. [Note: Not all computer algebra systems can evaluate this sum.]

   (c) Use a computer algebra system to evaluate the limit of the sum as n approaches infinity.

   (d) Confirm your answer to part (c) by calculating the definite integral with the computer algebra system.

In Problems 54–55:

   (a) Use a computer algebra system to compute the given definite integral.

   (b) From your answer to part (a) and the Fundamental Theorem of Calculus, guess a function whose derivative is the integrand. Check your guess using the computer algebra system.

[Hint: Make sure that the constants a, b, and c do not have previously assigned values in your computer algebra system.]

54. $\displaystyle\int_a^b \sin(cx)\,dx$

55. $\displaystyle\int_a^c \frac{x}{1 + bx^2}\,dx, \quad b > 0$

CHECK YOUR UNDERSTANDING

Are the statements in Problems 1–5 true of false? Give an explanation for your answer.

1. The units for an integral of a function $f(x)$ are the same as the units for $f(x)$.

2. For an increasing function, the left-hand sum on a given interval with a given number of subdivisions is always less than the right-hand sum.

3. For a decreasing function, the difference between the left-hand sum and right-hand sum is halved when the number of subdivisions is doubled.

4. For a given function on a given interval, the difference between the left-hand sum and right-hand sums gets smaller as the number of subdivisions gets larger.

5. On the interval $a \le t \le b$, the integral of the velocity is the total distance traveled from $t = a$ to $t = b$.

In Problems 6–22, are the statements true for all continuous functions $f(x)$ and $g(x)$? Give an explanation for your answer.

6. If $\int_0^2 (f(x) + g(x))dx = 10$ and $\int_0^2 f(x)dx = 3$, then $\int_0^2 g(x)dx = 7$.

7. If $\int_0^2 (f(x) + g(x))dx = 10$, then $\int_0^2 f(x)dx = 3$ and $\int_0^2 g(x)dx = 7$.

8. If $\int_0^2 f(x)dx = 6$, then $\int_0^4 f(x)dx = 12$.

9. If $\int_0^2 f(x)dx = 6$ and $g(x) = 2f(x)$, then $\int_0^2 g(x)dx = 12$.

10. If $\int_0^2 f(x)dx = 6$ and $h(x) = f(5x)$, then $\int_0^2 h(x)dx = 30$.

11. If $a = b$, then $\int_a^b f(x)\,dx = 0$.

12. If $a \neq b$, then $\int_a^b f(x)\,dx \neq 0$.

13. $\int_1^2 f(x)\,dx + \int_2^3 g(x)\,dx = \int_1^3 (f(x) + g(x))\,dx$.

14. $\int_{-1}^1 f(x)\,dx = 2\int_0^1 f(x)\,dx$.

15. $\int_0^2 f(x)dx \le \int_0^3 f(x)dx$.

16. $\int_0^2 f(x)dx = \int_0^2 f(t)dt.$

17. If $\int_2^6 f(x)\,dx \leq \int_2^6 g(x)\,dx$, then $f(x) \leq g(x)$ for $2 \leq x \leq 6$.

18. If $f(x) \leq g(x)$ on the interval $[a, b]$, then the average value of f is less than or equal to the average value of g on the interval $[a, b]$.

19. The average value of f on the interval $[0, 10]$ is the average of the average value of f on $[0, 5]$ and the average value of f on $[5, 10]$.

20. If $a < c < d < b$, then the average value of f on the interval $[c, d]$ is less than the average value of f on the interval $[a, b]$.

21. Suppose that A is the average value of f on the interval $[1, 4]$ and B is the average value of f on the interval $[4, 9]$. Then the average value of f on $[1, 9]$ is the weighted average $(3/8)A + (5/8)B$.

22. On the interval $[a, b]$, the average value of $f(x) + g(x)$ is the average value of $f(x)$ plus the average value of $g(x)$.

In Problems 23–25 decide whether the statement is true or false. Justify your answer.

23. The average value of the product, $f(x)g(x)$, of two functions on an interval equals the product of the average values of $f(x)$ and $g(x)$ on the interval.

24. A 4-term left-hand Riemann sum approximation cannot give the exact value of a definite integral.

25. If $f(x)$ is decreasing and $g(x)$ is increasing, then $\int_a^b f(x)\,dx \neq \int_a^b g(x)\,dx$.

In Problems 26–27, graph a continuous function $f(x) \geq 0$ on $[0, 10]$ with the given properties.

26. The maximum value taken on by $f(x)$ for $0 \leq x \leq 10$ is 1. In addition $\int_0^{10} f(x)\,dx = 5$.

27. The maximum value taken on by $f(x)$ for $0 \leq x \leq 10$ is 5. In addition $\int_0^{10} f(x)\,dx = 1$.

28. Which of the following statements follow directly from the rule

$$\int_a^b (f(x) + g(x))\,dx = \int_a^b f(x)\,dx + \int_a^b g(x)\,dx?$$

(a) If $\int_a^b (f(x) + g(x))\,dx = 5 + 7$, then $\int_a^b f(x)\,dx = 5$ and $\int_a^b g(x)\,dx = 7$.

(b) If $\int_a^b f(x)\,dx = \int_a^b g(x)\,dx = 7$, then $\int_a^b (f(x) + g(x))\,dx = 14$.

(c) If $h(x) = f(x) + g(x)$, then $\int_a^b (h(x) - g(x))\,dx = \int_a^b h(x)\,dx - \int_a^b g(x)\,dx$.

PROJECTS FOR CHAPTER FIVE

1. An Orbiting Satellite

A NASA satellite orbits the earth every 90 minutes. During an orbit, the satellite's electric power comes either from solar array wings, when these are illuminated by the sun, or from batteries. The batteries discharge whenever the satellite uses more electricity than the solar array can provide or whenever the satellite is in the shadow of the earth (where the solar array cannot be used). If the batteries are overused, however, they can be damaged. [9]

You are to determine whether the batteries could be damaged in either of the following operations. You are told that the battery capacity is 50 ampere-hours. If the total battery discharge does not exceed 40% of battery capacity, the batteries will not be damaged.

(a) Operation 1 is performed by the satellite while orbiting the earth. At the beginning of a given 90-minute orbit, the satellite performs a 15-minute maneuver which requires more current than the solar array can deliver, causing the batteries to discharge. The maneuver causes a sinusoidally varying battery discharge of period 30 minutes with a maximum discharge of ten amperes at 7.5 minutes. For the next 45 minutes the solar array meets the total satellite current demand, and the batteries do not discharge. During the last 30 minutes, the satellite is in the shadow of the earth and the batteries supply the total current demand of 30 amperes.

    (i) The battery current in amperes is a function of time. Plot the function, showing the current in amperes as a function of time for the 90-minute orbit. Write a formula (or formulas) for the battery current function.

    (ii) Calculate the total battery discharge (in units of ampere-hours) for the 90-minute orbit for Operation 1.

---

[9] Adapted from Amy C. R. Gerson, "Electrical Engineering: Space Systems," in *She Does Math! Real Life Problems from Women on the Job*, ed. Marla Parker, p. 61 (Washington, DC: Mathematical Association of America, 1995).

(iii) What is your recommendation regarding the advisability of Operation 1?

(b) Operation 2 is simulated at NASA's laboratory in Houston. The following graph was produced by the laboratory simulation of the current demands on the battery during the 90-minute orbit required for Operation 2.

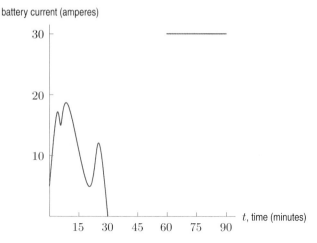

Figure 5.85: Battery discharge simulation graph for Operation 2

(i) Calculate the total battery discharge (in units of ampere-hours) for the 90-minute orbit for Operation 2.

(ii) What is your recommendation regarding the advisability of Operation 2?

2. Gas Tanks

Gasoline is stored at a gas station in an underground tank which is a circular cylinder lying on its side. You may have seen a gas station attendant dropping a long measuring stick into the tank to measure the amount of gasoline remaining. Suppose the remaining fuel comes to a height of h inches and the diameter of the tank is d inches. It might be tempting to guess that the tank is h/d full. The purpose of this project is to see how far off such a guess is.

Assume a cylindrical gas tank has diameter d and length l, as in Figure 5.86.

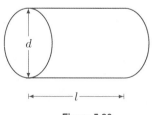

Figure 5.86

(a) Suppose another gas tank has the shape of a w by d by l rectangular box. What is w if this tank holds the same amount as the cylindrical tank? Does h/d give an exact measure of how full the rectangular tank is?

(b) When computing the accuracy of the stick method for determining what fraction of the tank is full, why can we ignore the length of the tank?

(c) Draw cross-sections of both tanks on one set of axes, with the x-axis measuring height and with both tanks centered at the origin.

(d) Let $E(h)$ be the error, as a function of h, when using h/d to measure how full the cylindrical tank is. Use the picture you drew in part (c) to write $E(h)$ as an integral.

(e) For what value of h is the error maximal?

Chapter Six

CONSTRUCTING ANTIDERIVATIVES

In Chapter 2, we saw how to calculate velocity given position, and in Chapter 5, we saw how to reconstruct distance from velocity.

In this chapter we look in more detail at how to reconstruct a function from its derivative using the Fundamental Theorem of Calculus. We introduce antiderivatives and show how to construct them numerically, graphically, and analytically. We introduce differential equations and the Second Fundamental Theorem of Calculus.

6.1   ANTIDERIVATIVES GRAPHICALLY AND NUMERICALLY

The Family of Antiderivatives

If the derivative of F is f, we call F an *antiderivative* of f. For example, since the derivative of x^2 is $2x$, we say that

$$x^2 \text{ is an antiderivative of } 2x.$$

Notice that $2x$ has many antiderivatives, since $x^2 + 1$, $x^2 + 2$, and $x^2 + 3$, all have derivative $2x$. In fact, if C is any constant, we have

$$\frac{d}{dx}(x^2 + C) = 2x + 0 = 2x,$$

so any function of the form $x^2 + C$ is an antiderivative of $2x$. The function $f(x) = 2x$ has a *family of antiderivatives*.

Let us look at another example. If v is the velocity of a car and s is its position, then $v = ds/dt$ and s is an antiderivative of v. As before, $s + C$ is an antiderivative of v for any constant C. In terms of the car, adding C to s is equivalent to adding C to the odometer reading. Adding a constant to the odometer reading simply means measuring distance from a different point, which does not alter the car's velocity.

Visualizing Antiderivatives Using Slopes

Suppose we have the graph of f', and we want to sketch an approximate graph of f. We are looking for the graph of f whose slope at any point is equal to the value of f' there. Where f' is above the x-axis, f is increasing; where f' is below the x-axis, f is decreasing. If f' is increasing, f is concave up; if f' is decreasing, f is concave down.

Example 1   The graph of f' is given in Figure 6.1. Sketch a graph of f in the cases when $f(0) = 0$ and $f(0) = 1$.

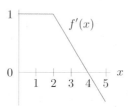

Figure 6.1: Graph of f'

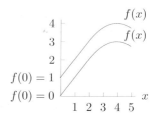

Figure 6.2: Two different f's which have the same derivative f'

Solution   For $0 \leq x \leq 2$, the function f has a constant slope of 1, so the graph of f is a straight line. For $2 \leq x \leq 4$, the function f is increasing but more and more slowly; it has a maximum at $x = 4$ and decreases thereafter. (See Figure 6.2.) The solutions with $f(0) = 0$ and $f(0) = 1$ start at different points on the vertical axis but have the same shape.

Example 2   Sketch a graph of the antiderivative F of $f(x) = e^{-x^2}$ satisfying $F(0) = 0$.

Solution   The graph of $f(x) = e^{-x^2}$ is shown in Figure 6.3. The slope of the antiderivative $F(x)$ is given by $f(x)$. Since $f(x)$ is always positive, the antiderivative $F(x)$ is always increasing. Since $f(x)$ is increasing for negative x, we know that $F(x)$ is concave up for negative x. Since $f(x)$ is decreasing

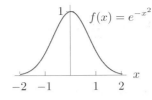

Figure 6.3: Graph of $f(x) = e^{-x^2}$

Figure 6.4: An antiderivative $F(x)$ of
$f(x) = e^{-x^2}$

for positive x, we know that $F(x)$ is concave down for positive x. Since $f(x) \to 0$ as $x \to \pm\infty$, the graph of $F(x)$ levels off at both ends. See Figure 6.4.

Example 3   For the function f' given in Figure 6.5, sketch a graph of three antiderivative functions f, one with $f(0) = 0$, one with $f(0) = 1$, and one with $f(0) = 2$.

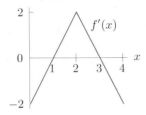

Figure 6.5: Slope function, f'

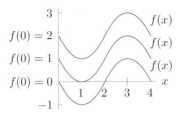

Figure 6.6: Antiderivatives f

Solution   To graph f, start at the point on the vertical axis specified by the initial condition and move with slope given by the value of f' in Figure 6.5. Different initial conditions lead to different graphs for f, but for a given x-value they all have the same slope (because the value of f' is the same for each). Thus, the different f curves are obtained from one another by a vertical shift. See Figure 6.6.

- Where f' is positive ($1 < x < 3$), we see f is increasing; where f' is negative ($0 < x < 1$ or $3 < x < 4$), we see f is decreasing.
- Where f' is increasing ($0 < x < 2$), we see f is concave up; where f' is decreasing ($2 < x < 4$), we see f is concave down.
- Where $f' = 0$, we see f has a local maximum at $x = 3$ and a local minimum at $x = 1$.
- Where f' has a maximum ($x = 2$), we see f has a point of inflection.

Computing Values of an Antiderivative Using the Fundamental Theorem

A graph of f' shows where f is increasing and where f is decreasing. We can calculate the actual value of the function f by using the Fundamental Theorem of Calculus (Theorem 5.1 on page 256): If f' is continuous, then

$$\int_a^b f'(x)\,dx = f(b) - f(a).$$

Example 4   Figure 6.7 is the graph of the derivative $f'(x)$ of a function $f(x)$. It is given that $f(0) = 100$. Sketch the graph of $f(x)$, showing all critical points and inflection points of f and giving their coordinates.

Figure 6.7: Graph of derivative

Solution   The critical points of f occur at $x = 0$, $x = 20$, and $x = 30$, where $f'(x) = 0$. The inflection points of f occur at $x = 10$ and $x = 25$, where $f'(x)$ has a maximum or minimum. To find the coordinates of the critical points and inflection points of f, we evaluate $f(x)$ for $x = 0, 10, 20, 25, 30$. Using the Fundamental Theorem, we can express the values of $f(x)$ in terms of definite integrals. We evaluate the definite integrals using the areas of triangular regions under the graph of $f'(x)$, remembering that areas below the x-axis are subtracted. (See Figure 6.8.)

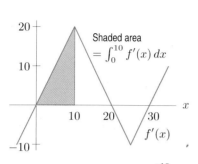

Figure 6.8: Finding $f(10) = f(0) + \int_0^{10} f'(x)\, dx$

Figure 6.9: Graph of $f(x)$

Since $f(0) = 100$, the Fundamental Theorem gives us the following values of f, which are marked in Figure 6.9.

$$f(10) = f(0) + \int_0^{10} f'(x)\, dx = 100 + (\text{Shaded area in Figure 6.8}) = 100 + \frac{1}{2}(10)(20) = 200,$$

$$f(20) = f(10) + \int_{10}^{20} f'(x)\, dx = 200 + \frac{1}{2}(10)(20) = 300,$$

$$f(25) = f(20) + \int_{20}^{25} f'(x)\, dx = 300 - \frac{1}{2}(5)(10) = 275,$$

$$f(30) = f(25) + \int_{25}^{30} f'(x)\, dx = 275 - \frac{1}{2}(5)(10) = 250.$$

Example 5   Suppose $F'(t) = t \cos t$ and $F(0) = 2$. Find $F(b)$ at the points $b = 0, 0.1, 0.2, \ldots, 1.0$.

Solution   We apply the Fundamental Theorem with $f(t) = t \cos t$ and $a = 0$ to get values for $F(b)$:

$$F(b) - F(0) = \int_0^b F'(t)\, dt = \int_0^b t \cos t\, dt.$$

Since $F(0) = 2$, we have

$$F(b) = 2 + \int_0^b t \cos t\, dt.$$

Calculating the definite integral $\int_0^b t \cos t\, dt$ numerically for $b = 0, 0.1, 0.2, \ldots, 1.0$ gives the values for F in Table 6.1:

Table 6.1   *Approximate values for F*

b	0	0.1	0.2	0.3	0.4	0.5	0.6	0.7	0.8	0.9	1.0
$F(b)$	2.000	2.005	2.020	2.044	2.077	2.117	2.164	2.216	2.271	2.327	2.382

Notice that $F(b)$ appears to be increasing between $b = 0$ and $b = 1$. This could have been predicted from the fact that $t \cos t$, the derivative of $F(t)$, is positive for t between 0 and 1.

Exercises and Problems for Section 6.1

Exercises

In Exercises 1–4, sketch two functions F such that $F' = f$. In one case let $F(0) = 0$ and in the other, let $F(0) = 1$.

1.

2.

3.

4.

7. (a) Using Figure 6.10, estimate $\int_0^7 f(x)\,dx$.
(b) If F is an antiderivative of the same function f and $F(0) = 25$, estimate $F(7)$.

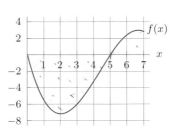

Figure 6.10

5. Given the values of the derivative $f'(x)$ in the table and that $f(0) = 100$, estimate $f(x)$ for $x = 2, 4, 6$.

x	0	2	4	6
$f'(x)$	10	18	23	25

6. Estimate $f(x)$ for $x = 2,\ 4,\ 6$, using the given values of $f'(x)$ and the fact that $f(0) = 50$.

x	0	2	4	6
$f'(x)$	17	15	10	2

8. Use Figure 6.11 and the fact that $P = 2$ when $t = 0$ to find values of P when $t = 1,\ 2,\ 3,\ 4$ and 5.

Figure 6.11

Problems

In Problems 9–10, a graph of f is given. Let $F'(x) = f(x)$.
(a) What are the critical points of $F(x)$?
(b) Which critical points are local maxima, which are local minima, and which are neither?
(c) Sketch a possible graph of $F(x)$.

9.
10.

11.
12.
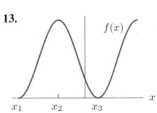

13.

In Problems 11–13, sketch two functions F with $F'(x) = f(x)$. In one, let $F(0) = 0$; in the other, let $F(0) = 1$. Mark x_1, x_2, and x_3 on the x-axis of your graph. Identify local maxima, minima, and inflection points of $F(x)$.

14. A particle moves back and forth along the x-axis. Figure 6.12 approximates the velocity of the particle as a function of time. Positive velocities represent movement to the right and negative velocities represent movement to the left. The particle starts at the point $x = 5$. Graph the distance of the particle from the origin, with distance measured in kilometers and time in hours.

Figure 6.12

15. Assume f' is given by the graph in Figure 6.13. Suppose f is continuous and that $f(3) = 0$.

(a) Sketch a graph of f.

(b) Find $f(0)$ and $f(7)$.

(c) Find $\int_0^7 f'(x)\,dx$ in two different ways.

Figure 6.13

16. Use Figure 6.14 and the fact that $F(2) = 3$ to sketch the graph of $F(x)$. Label the values of at least four points.

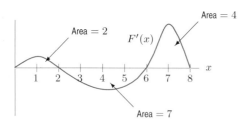

Figure 6.14

17. Using Figure 6.15, sketch a graph of an antiderivative $G(t)$ of $g(t)$ satisfying $G(0) = 5$. Label each critical point of $G(t)$ with its coordinates.

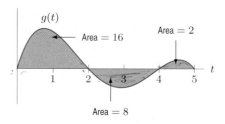

Figure 6.15

18. Using the graph of g' in Figure 6.16 and the fact that $g(0) = 50$, sketch the graph of $g(x)$. Give the coordinates of all critical points and inflection points of g.

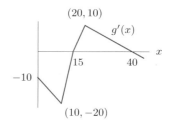

Figure 6.16

19. The vertical velocity of a cork bobbing up and down on the waves in the sea is given by Figure 6.17. Upward is considered positive. Describe the motion of the cork at each of the labeled points. At which point(s), if any, is the acceleration zero? Sketch a graph of the height of the cork above the sea floor as a function of time.

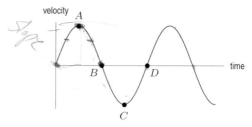

Figure 6.17

20. Figure 6.18 shows the rate of change of the concentration of adrenaline, in micrograms per milliliter per minute, in a person's body. Sketch a graph of the concentration of adrenaline, in micrograms per milliliter, in the body as a function of time, in minutes.

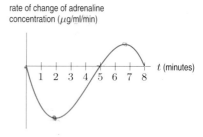

Figure 6.18

21. Urologists are physicians who specialize in the health of the bladder. In a common diagnostic test, urologists monitor the emptying of the bladder using a device that produces two graphs. In one of the graphs the flow rate (in milliliters per second) is measured as a function of time (in seconds). In the other graph, the volume emptied from the bladder is measured (in milliliters) as a function of time (in seconds). See Figure 6.19.

(a) Which graph is the flow rate and which is the volume?

(b) Which one of these graphs is an antiderivative of the other?

(I)

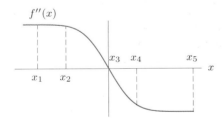
(II)

Figure 6.19

22. Use a graph of $f(x) = 2\sin(x^2)$ to determine where an antiderivative, F, of this function reaches its maximum on $0 \leq x \leq 3$. If $F(1) = 5$, find the maximum value attained by F.

23. The graph of f'' is given in Figure 6.20. Draw graphs of f and f', assuming both go through the origin, and use them to decide at which of the labeled x-values:

(a) $f(x)$ is greatest.
(b) $f(x)$ is least.
(c) $f'(x)$ is greatest.
(d) $f'(x)$ is least.
(e) $f''(x)$ is greatest.
(f) $f''(x)$ is least.

Figure 6.20: Graph of f''

24. Two functions, $f(x)$ and $g(x)$, are shown in Figure 6.21. Let F and G be antiderivatives of f and g, respectively. On the same axes, sketch graphs of the antiderivatives $F(x)$ and $G(x)$ satisfying $F(0) = 0$ and $G(0) = 0$. Compare F and G, including a discussion of zeros and x- and y-coordinates of critical points.

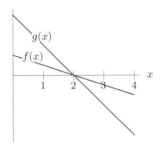

Figure 6.21

25. The Quabbin Reservoir in the western part of Massachusetts provides most of Boston's water. The graph in Figure 6.22 represents the flow of water in and out of the Quabbin Reservoir throughout 1993.

(a) Sketch a possible graph for the quantity of water in the reservoir, as a function of time.

(b) When, in the course of 1993, was the quantity of water in the reservoir largest? Smallest? Mark and label these points on the graph you drew in part (a).

(c) When was the quantity of water increasing most rapidly? Decreasing most rapidly? Mark and label these times on both graphs.

(d) By July 1994 the quantity of water in the reservoir was about the same as in January 1993. Draw plausible graphs for the flow into and the flow out of the reservoir for the first half of 1994. Explain your graphs.

Figure 6.22

6.2 CONSTRUCTING ANTIDERIVATIVES ANALYTICALLY

What Is an Antiderivative of $f(x) = 0$?

A function whose derivative is zero everywhere on an interval must have a horizontal tangent line at every point of its graph, and the only way this can happen is if the function is constant. Alternatively, if we think of the derivative as a velocity, and if the velocity is always zero, then the object is standing still; the position function is constant. A rigorous proof of this result using the definition of the derivative is surprisingly subtle. (See the Constant Function Theorem on page 157.)

> If $F'(x) = 0$ on an interval, then $F(x) = C$ on this interval, for some constant C.

What Is the Most General Antiderivative of f ?

We know that if a function f has an antiderivative F, then it has a family of antiderivatives of the form $F(x) + C$, where C is any constant. You might wonder if there are any others. To decide, suppose that we have two functions F and G with $F' = f$ and $G' = f$: that is, F and G are both antiderivatives of the same function f. Since $F' = G'$ we have $(F - G)' = 0$. But this means that we must have $F - G = C$, so $F(x) = G(x) + C$, where C is a constant. Thus, any two antiderivatives of the same function differ only by a constant.

> If F and G are both antiderivatives of f on an interval, then $F(x) = G(x) + C$.

The Indefinite Integral

All antiderivatives of $f(x)$ are of the form $F(x) + C$. We introduce a notation for the general antiderivative that looks like the definite integral without the limits and is called the *indefinite integral*:

$$\int f(x)\, dx = F(x) + C.$$

It is important to understand the difference between

$$\int_a^b f(x)\, dx \qquad \text{and} \qquad \int f(x)\, dx.$$

The first is a number and the second is a family of *functions*. The word "integration" is frequently used for the process of finding the antiderivative as well as of finding the definite integral. The context usually makes clear which is intended.

What is an Antiderivative of $f(x) = k$?

If k is a constant, the derivative of kx is k, so we have

An antiderivative of k is kx.

Using the indefinite integral notation, we have

> If k is constant,
>
> $$\int k\, dx = kx + C.$$

Finding Antiderivatives

Finding antiderivatives of functions is like taking square roots of numbers: if we pick a number at random, such as 7 or 493, we may have trouble finding its square root without a calculator. But if we happen to pick a number such as 25 or 64, which we know is a perfect square, then we can find its square root exactly. Similarly, if we pick a function which we recognize as a derivative, then we can find its antiderivative easily.

For example, to find an antiderivative of $f(x) = x$, notice that $2x$ is the derivative of x^2; this tells us that x^2 is an antiderivative of $2x$. If we divide by 2, then we guess that

$$\text{An antiderivative of } x \text{ is } \frac{x^2}{2}.$$

To check this statement, take the derivative of $x^2/2$:

$$\frac{d}{dx}\left(\frac{x^2}{2}\right) = \frac{1}{2} \cdot \frac{d}{dx} x^2 = \frac{1}{2} \cdot 2x = x.$$

What about an antiderivative of x^2? The derivative of x^3 is $3x^2$, so the derivative of $x^3/3$ is $3x^2/3 = x^2$. Thus,

$$\text{An antiderivative of } x^2 \text{ is } \frac{x^3}{3}.$$

The pattern looks like

$$\text{An antiderivative of } x^n \text{ is } \frac{x^{n+1}}{n+1}.$$

(We assume $n \neq -1$, or we would have $x^0/0$, which does not make sense.) It is easy to check this formula by differentiation:

$$\frac{d}{dx}\left(\frac{x^{n+1}}{n+1}\right) = \frac{(n+1)x^n}{n+1} = x^n.$$

In indefinite integral notation, we have shown that

$$\boxed{\int x^n \, dx = \frac{x^{n+1}}{n+1} + C, \quad n \neq -1.}$$

What about when $n = -1$? In other words, what is an antiderivative of $1/x$? Fortunately, we know a function whose derivative is $1/x$, namely, the natural logarithm. Thus, since

$$\frac{d}{dx}(\ln x) = \frac{1}{x},$$

we know that

$$\int \frac{1}{x} \, dx = \ln x + C, \quad \text{for } x > 0.$$

If $x < 0$, then $\ln x$ is not defined, so it can't be an antiderivative of $1/x$. In this case, we can try $\ln(-x)$:

$$\frac{d}{dx}\ln(-x) = (-1)\frac{1}{-x} = \frac{1}{x}$$

so

$$\int \frac{1}{x} \, dx = \ln(-x) + C, \quad \text{for } x < 0.$$

This means $\ln x$ is an antiderivative of $1/x$ if $x > 0$, and $\ln(-x)$ is an antiderivative of $1/x$ if $x < 0$. Since $|x| = x$ when $x > 0$ and $|x| = -x$ when $x < 0$ we can collapse these two formulas into:

An antiderivative of $\dfrac{1}{x}$ is $\ln|x|$.

Therefore

$$\int \frac{1}{x}\, dx = \ln|x| + C.$$

Since the exponential function is its own derivative, it is also its own antiderivative; thus

$$\int e^x\, dx = e^x + C.$$

Also, antiderivatives of the sine and cosine are easy to guess. Since

$$\frac{d}{dx}\sin x = \cos x \qquad \text{and} \qquad \frac{d}{dx}\cos x = -\sin x,$$

we get

$$\int \cos x\, dx = \sin x + C \qquad \text{and} \qquad \int \sin x\, dx = -\cos x + C.$$

Example 1   Find $\displaystyle\int (3x + x^2)\, dx$.

Solution   We know that $x^2/2$ is an antiderivative of x and that $x^3/3$ is an antiderivative of x^2, so we expect

$$\int (3x + x^2)\, dx = 3\left(\frac{x^2}{2}\right) + \frac{x^3}{3} + C.$$

You should always check your antiderivatives by differentiation—it's easy to do. Here

$$\frac{d}{dx}\left(\frac{3}{2}x^2 + \frac{x^3}{3} + C\right) = \frac{3}{2}\cdot 2x + \frac{3x^2}{3} = 3x + x^2.$$

The preceding example illustrates that the sum and constant multiplication rules of differentiation work in reverse:

Theorem 6.1: Properties of Antiderivatives: Sums and Constant Multiples

In indefinite integral notation,

1. $\displaystyle\int [f(x) \pm g(x)]\, dx = \int f(x)\, dx \pm \int g(x)\, dx$

2. $\displaystyle\int cf(x)\, dx = c\int f(x)\, dx.$

In words,

1. An antiderivative of the sum (or difference) of two functions is the sum (or difference) of their antiderivatives.

2. An antiderivative of a constant times a function is the constant times an antiderivative of the function.

Example 2   Find $\int (\sin x + 3 \cos x)\, dx$.

Solution   We break the antiderivative into two terms:

$$\int (\sin x + 3 \cos x)\, dx = \int \sin x\, dx + 3 \int \cos x\, dx = -\cos x + 3 \sin x + C.$$

Check by differentiating:

$$\frac{d}{dx}(-\cos x + 3 \sin x + C) = \sin x + 3 \cos x.$$

Using Antiderivatives to Compute Definite Integrals

The Fundamental Theorem of Calculus gives us a new way of calculating definite integrals. The theorem says that if $F' = f$, then

$$\int_a^b f(x)\, dx = F(b) - F(a),$$

To find $\int_a^b f(x)\, dx$, we first try to find F, and then calculate $F(b) - F(a)$. This method of computing definite integrals gives an exact answer. However, the method only works in situations where we can find the antiderivative $F(x)$. This is not always easy; for example, none of the functions we have encountered so far is an antiderivative of $\sin(x^2)$.

Example 3   Compute $\int_1^2 3x^2\, dx$ using the Fundamental Theorem.

Solution   Since $F(x) = x^3$ is an antiderivative of $f(x) = 3x^2$, we have

$$\int_1^2 3x^2\, dx = F(2) - F(1) = 2^3 - 1^3 = 7.$$

Notice in this example we used the antiderivative x^3, but $x^3 + C$ works just as well because the constant cancels out:

$$\int_1^2 3x^2\, dx = F(2) - F(1) = (2^3 + C) - (1^3 + C) = 7.$$

It is helpful to introduce a shorthand notation for $F(b) - F(a)$: we write it as $F(x)\Big|_a^b$. For example:

$$\int_1^2 3x^2\, dx = x^3 \Big|_1^2 = 2^3 - 1^3 = 7.$$

Exercises and Problems for Section 6.2

Exercises

In Exercises 1–21, find an antiderivative.

1. $f(x) = 5$

2. $f(x) = 5x$

3. $f(x) = x^2$

4. $g(t) = t^2 + t$

5. $h(t) = \cos t$

6. $g(z) = \sqrt{z}$

7. $h(z) = \dfrac{1}{z}$

8. $r(t) = \dfrac{1}{t^2}$

9. $g(z) = \dfrac{1}{z^3}$

10. $f(z) = e^z$

11. $g(t) = \sin t$

12. $f(t) = 2t^2 + 3t^3 + 4t^4$

13. $p(t) = t^3 - \dfrac{t^2}{2} - t$

14. $q(y) = y^4 + \dfrac{1}{y}$

15. $p(t) = \cos t + \dfrac{1}{\cos^2 t}$

16. $f(t) = \dfrac{t^2 + 1}{t}$

17. $p(\theta) = 2\sin(2\theta)$

18. $r(t) = e^t + 5e^{5t}$

19. $q(t) = (t + 1)^2$

20. $f(x) = 5^x$

21. $f(x) = 5x - \sqrt{x}$

In Exercises 22–32, find the general antiderivative.

22. $f(t) = 6t$

23. $h(x) = x^3 - x$

24. $f(x) = x^2 - 4x + 7$

25. $r(t) = t^3 + 5t - 1$

26. $f(z) = z + e^z$

27. $g(t) = \sqrt{t}$

28. $g(x) = \sin x + \cos x$

29. $h(x) = 4x^3 - 7$

30. $p(t) = 2 + \sin t$

31. $p(t) = \dfrac{1}{\sqrt{t}}$

32. $g(x) = \dfrac{5}{x^3}$

In Exercises 33–40, find an antiderivative $F(x)$ with $F'(x) = f(x)$ and $F(0) = 0$. Is there only one possible solution?

33. $f(x) = 3$

34. $f(x) = 2x$

35. $f(x) = -7x$

36. $f(x) = \dfrac{1}{4}x$

37. $f(x) = x^2$

38. $f(x) = \sqrt{x}$

39. $f(x) = 2 + 4x + 5x^2$

40. $f(x) = \sin x$

Find the indefinite integrals in Exercises 41–59.

41. $\displaystyle\int 5x\,dx$

42. $\displaystyle\int x^3\,dx$

43. $\displaystyle\int \sin\theta\,d\theta$

44. $\displaystyle\int (x^3 - 2)\,dx$

45. $\displaystyle\int \left(t^2 + \dfrac{1}{t^2}\right) dt$

46. $\displaystyle\int 4\sqrt{w}\,dw$

47. $\displaystyle\int (x^2 + 5x + 8)\,dx$

48. $\displaystyle\int \dfrac{4}{t^2}\,dt$

49. $\displaystyle\int (4t + 7)\,dt$

50. $\displaystyle\int \cos\theta\,d\theta$

51. $\displaystyle\int 5e^z\,dz$

52. $\displaystyle\int \left(x + \dfrac{1}{\sqrt{x}}\right) dx$

53. $\displaystyle\int \sin t\,dt$

54. $\displaystyle\int (\pi + x^{11})\,dx$

55. $\displaystyle\int \left(t\sqrt{t} + \dfrac{1}{t\sqrt{t}}\right) dt$

56. $\displaystyle\int \cos(x + 1)\,dx$

57. $\displaystyle\int e^{2r}\,dr$

58. $\displaystyle\int \dfrac{1}{e^z}\,dz$

59. $\displaystyle\int \left(\dfrac{y^2 - 1}{y}\right)^2 dy$

In Exercises 60–71, evaluate the definite integrals exactly [as in $\ln(3\pi)$], using the Fundamental Theorem, and numerically [$\ln(3\pi) \approx 2.243$]:

60. $\displaystyle\int_0^3 (x^2 + 4x + 3)\,dx$

61. $\displaystyle\int_1^3 \dfrac{1}{t}\,dt$

62. $\displaystyle\int_0^{\pi/4} \sin x\,dx$

63. $\displaystyle\int_0^2 3e^x\,dx$

64. $\displaystyle\int_2^5 (x^3 - \pi x^2)\,dx$

65. $\displaystyle\int_0^1 \sin\theta\,d\theta$

66. $\displaystyle\int_1^2 \dfrac{1 + y^2}{y}\,dy$

67. $\displaystyle\int_0^2 \left(\dfrac{x^3}{3} + 2x\right) dx$

68. $\displaystyle\int_0^{\pi/4} (\sin t + \cos t)\,dt$

69. $\displaystyle\int_0^1 2e^x\,dx$

70. $\displaystyle\int_{-3}^{-1} \dfrac{2}{r^3}\,dr$

71. $\displaystyle\int_0^{\pi/4} \dfrac{1}{\cos^2 x}\,dx$

Problems

72. Use the Fundamental Theorem to find the area under $f(x) = x^2$ between $x = 1$ and $x = 4$.

73. Calculate the exact area between the x-axis and the graph of $y = 7 - 8x + x^2$.

74. Find the exact area enclosed by the curve $y = x^2(1-x)^2$ and the x-axis.

75. Find the exact area below the curve $y = x^3(1-x)$ and above the x-axis.

76. Calculate the exact area above the graph of $y = \sin \theta$ and below the graph of $y = \cos \theta$ for $0 \le \theta \le \pi/4$.

77. Find the exact value of the area between the graphs of $y = \cos x$ and $y = e^x$ for $0 \le x \le 1$.

78. Find the exact value of the area between the graphs of $y = \sinh x$, $y = \cosh x$, for $-1 \le x \le 1$.

79. Use the Fundamental Theorem to determine the value of b if the area under the graph of $f(x) = 8x$ between $x = 1$ and $x = b$ is equal to 192. Assume $b > 1$.

80. Find the exact positive value of c if the area between the graph of $y = x^2 - c^2$ and the x-axis is 36.

81. Use the Fundamental Theorem to find the average value of $f(x) = x^2 + 1$ on the interval $x = 0$ to $x = 10$. Illustrate your answer on a graph of $f(x)$.

82. The average value of the function $v(x) = 6/x^2$ on the interval $[1, c]$ is equal to 1. Find the value of c.

83. **(a)** What is the average value of $f(t) = \sin t$ over $0 \le t \le 2\pi$? Why is this a reasonable answer?
   (b) Find the average of $f(t) = \sin t$ over $0 \le t \le \pi$.

84. The origin and the point (a, a) are at opposite corners of a square. Calculate the ratio of the areas of the two parts into which the curve $\sqrt{x} + \sqrt{y} = \sqrt{a}$ divides the square.

85. If A_n is the area between the curves $y = x$ and $y = x^n$, show that $A_n \to \frac{1}{2}$ as $n \to \infty$ and explain this result graphically.

86. In drilling an oil well, the total cost, C, consists of fixed costs (independent of the depth of the well) and marginal costs, which depend on depth; drilling becomes more expensive, per meter, deeper into the earth. Suppose the fixed costs are 1,000,000 riyals (the riyal is the unit of currency of Saudi Arabia), and the marginal costs are

$$C'(x) = 4000 + 10x \text{ riyals/meter},$$

where x is the depth in meters. Find the total cost of drilling a well x meters deep.

87. One of the earliest pollution problems brought to the attention of the Environmental Protection Agency (EPA) was the case of the Sioux Lake in eastern South Dakota. For years a small paper plant located nearby had been discharging waste containing carbon tetrachloride (CCl_4) into the waters of the lake. At the time the EPA learned of the situation, the chemical was entering at a rate of 16 cubic yards/year.

The agency ordered the installation of filters designed to slow (and eventually stop) the flow of CCl_4 from the mill. Implementation of this program took exactly three years, during which the flow of pollutant was steady at 16 cubic yards/year. Once the filters were installed, the flow declined. If t is time measured in years since the EPA learned of the situation, between the time the filters were installed and the time the flow stopped, the rate of flow was well approximated by

$$\text{Rate (in cubic yards/year)} = t^2 - 14t + 49.$$

   (a) Graph the rate of CCl_4 flow into the lake as a function of time, beginning at the time the EPA first learned of the situation.
   (b) How many years elapsed between the time the EPA learned of the situation and the time the pollution flow stopped entirely?
   (c) How much CCl_4 entered the waters during the time shown in the graph in part (a)?

6.3 DIFFERENTIAL EQUATIONS

In Chapter 2 we saw that velocity is the derivative of distance and that acceleration is the derivative of velocity. In this section we analyze the motion of an object falling freely under the influence of gravity. This involves going "backward" from acceleration to velocity to position.

Motion With Constant Velocity

Let's briefly consider a familiar problem: An object moving in a straight line with constant velocity. Imagine a car moving at 50 mph. How far does it go in a given time? The answer is given by

$$\text{Distance} = \text{Rate} \times \text{Time}$$

or

$$s = 50t$$

where s is the distance of the car (in miles) from a fixed reference point and t is the time in hours. Alternatively, we can describe the motion by writing the equation

$$\frac{ds}{dt} = 50.$$

This is called a *differential equation* for the *function s*. The solution to this equation is the antiderivative

$$s = 50t + C.$$

The equation $s = 50t + C$ tells us that $s = C$ when $t = 0$. Thus, the constant C represents the initial distance, s_0, of the car from the reference point.

Uniformly Accelerated Motion

Now we consider an object moving with *constant acceleration* along a straight line, or *uniformly accelerated motion*. It has been known since Galileo's time that an object moving under the influence of gravity (ignoring air resistance) has constant acceleration, g. In the most frequently used units, its value is approximately

$$g = 9.8 \text{ m/sec}^2, \quad \text{or} \quad g = 32 \text{ ft/sec}^2.$$

Thus, if v is the upward velocity and t is the time,

$$\frac{dv}{dt} = -g.$$

The negative sign represents the fact that velocity is measured upward, whereas gravity acts downward.

Example 1   A stone is dropped from a 100-foot-high building. Find, as a function of time, its position and velocity. When does it hit the ground, and how fast is it going at that time?

Solution   Suppose t is measured in seconds from when the stone was dropped. If we measure distance, s, in feet above the ground, then the velocity, v, is in ft/sec upward, and the acceleration due to gravity is 32 ft/sec^2 downward, so

$$\frac{dv}{dt} = -32.$$

From what we know about antiderivatives, we must have

$$v = -32t + C$$

where C is some constant. Since $v = C$ when $t = 0$, the constant C represents the initial velocity, v_0. The fact that the stone is dropped rather than thrown off the top of the building tells us that the initial velocity is zero, so $v_0 = 0$. Substituting gives

$$0 = -32(0) + C \quad \text{so} \quad C = 0.$$

Thus,

$$v = -32t.$$

But now we can write

$$v = \frac{ds}{dt} = -32t.$$

The general antiderivative of $-32t$ is

$$s = -16t^2 + K,$$

where K is another constant.

Since the stone starts at the top of the building, $s = 100$ when $t = 0$. Substituting gives

$$100 = -16(0^2) + K, \quad \text{so} \quad K = 100,$$

and therefore

$$s = -16t^2 + 100.$$

Thus, we have found both v and s as functions of time.

The stone hits the ground when $s = 0$, so we must solve

$$0 = -16t^2 + 100$$

giving $t^2 = 100/16$ or $t = \pm 10/4 = \pm 2.5$ sec. Since t must be positive, $t = 2.5$ sec. At that time, $v = -32(2.5) = -80$ ft/sec. (The velocity is negative because we are considering up as positive and down as negative.)

Example 2   An object is thrown vertically upward with a speed of 10 m/sec from a height of 2 meters. Find the highest point it reaches and when it hits the ground.

Solution   We must find the position as a function of time. In this example, the velocity is in m/sec, so we use $g = 9.8$ m/sec^2. Measuring distance in meters upward from the ground, we have

$$\frac{dv}{dt} = -9.8.$$

As before, v is a function whose derivative is constant, so

$$v = -9.8t + C.$$

Since the initial velocity is 10 m/sec upward, we know that $v = 10$ when $t = 0$. Substituting gives

$$10 = -9.8(0) + C \quad \text{so} \quad C = 10.$$

Thus,

$$v = -9.8t + 10.$$

To find s, we use

$$v = \frac{ds}{dt} = -9.8t + 10$$

and look for a function that has $-9.8t + 10$ as its derivative. The general antiderivative of $-9.8t + 10$ is

$$s = -4.9t^2 + 10t + K,$$

where K is any constant. To find K, we use the fact that the object starts at a height of 2 meters, so $s = 2$ when $t = 0$. Substituting gives

$$2 = -4.9(0)^2 + 10(0) + K, \quad \text{so} \quad K = 2,$$

and therefore

$$s = -4.9t^2 + 10t + 2.$$

The object reaches its highest point when the velocity is 0, so at that time

$$v = -9.8t + 10 = 0.$$

This occurs when

$$t = \frac{10}{9.8} \approx 1.02 \text{ sec.}$$

When $t = 1.02$ seconds,

$$s = -4.9(1.02)^2 + 10(1.02) + 2 \approx 7.10 \text{ m.}$$

So the maximum height reached is 7.10 meters. The object reaches the ground when $s = 0$:

$$0 = -4.9t^2 + 10t + 2.$$

Solving this using the quadratic formula gives

$$t \approx -0.18 \quad \text{and} \quad t \approx 2.22 \text{ sec.}$$

Since the time at which the object hits the ground must be positive, $t \approx 2.22$ seconds.

Antiderivatives and Differential Equations

We solved the problem of uniformly accelerated motion by working backward from the derivative of a function to the function itself. If f is a known function, finding the *general solution* to the differential equation

$$\frac{dy}{dx} = f(x)$$

means finding the general antiderivative $y = F(x) + C$ with $F'(x) = f(x)$.

Example 3   Find and graph the general solution of the differential equation

$$\frac{dy}{dx} = \sin x + 2.$$

Solution   We are asking for a function whose derivative is $\sin x + 2$. One antiderivative of $\sin x + 2$ is

$$y = -\cos x + 2x.$$

The general solution is therefore

$$y = -\cos x + 2x + C,$$

where C is any constant. Figure 6.23 shows several curves in this family.

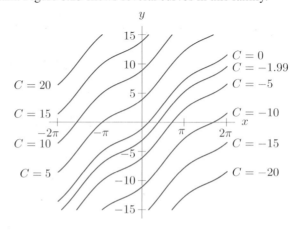

Figure 6.23: Solution curves of $dy/dx = \sin x + 2$

How Can We Pick One Solution to the Differential Equation $\dfrac{dy}{dx} = f(x)$?

Picking one antiderivative is equivalent to selecting a value of C. This requires knowing an extra piece of information, such as the initial velocity or initial position. In general, we pick a particular curve in the family of solutions by specifying that the curve should pass through a given point (x_0, y_0). The differential equation plus an extra condition

$$\frac{dy}{dx} = f(x), \qquad y(x_0) = y_0$$

is called an *initial value problem*. An initial value problem usually has a unique solution. Note that $y(x_0) = y_0$ is shorthand for $y = y_0$ when $x = x_0$.

Example 4 Find the solution of the initial value problem

$$\frac{dy}{dx} = \sin x + 2, \quad y(3) = 5.$$

Solution We have already seen that the general solution to the differential equation is $y = -\cos x + 2x + C$. The initial condition allows us to determine the constant C. Substituting $y(3) = 5$ gives

$$5 = y(3) = -\cos 3 + 2 \cdot 3 + C,$$

so C is given by

$$C = 5 + \cos 3 - 6 \approx -1.99.$$

Thus, the (unique) solution is

$$y = -\cos x + 2x - 1.99.$$

Figure 6.23 shows this particular solution, marked $C = -1.99$.

Exercises and Problems for Section 6.3

Exercises

In Exercises 1–4, find the general solution of the differential equations.

1. $\dfrac{dy}{dx} = x^3 + 5$

2. $\dfrac{dy}{dx} = 8x + \dfrac{1}{x}$

3. $\dfrac{dW}{dt} = 4\sqrt{t}$

4. $\dfrac{dr}{dp} = 3\sin p$

5. Show that $y = x + \sin x - \pi$ satisfies the initial value problem

$$\frac{dy}{dx} = 1 + \cos x, \quad y(\pi) = 0.$$

In Exercises 6–9, find the solution of the initial value problem.

6. $\dfrac{dy}{dx} = 6x^2 + 4x, \quad y(2) = 10$

7. $\dfrac{dP}{dt} = 10e^t, \quad P(0) = 25$

8. $\dfrac{ds}{dt} = -32t + 100, \quad s = 50$ when $t = 0$

9. $\dfrac{dq}{dz} = 2 + \sin z, \quad q = 5$ when $z = 0.$

10. Show that $y = xe^{-x} + 2$ is a solution of the initial value problem

$$\frac{dy}{dx} = (1 - x)e^{-x}, \quad y(0) = 2.$$

Problems

11. A tomato is thrown upward from a bridge 25 m above the ground at 40 m/sec.

    (a) Give formulas for the acceleration, velocity, and height of the tomato at time t.

    (b) How high does the tomato go, and when does it reach its highest point?

    (c) How long is it in the air?

12. (a) Find the general solution of the differential equation $dy/dx = 2x + 1$.

    (b) Sketch a graph of at least three solutions.

    (c) Find the solution satisfying $y(1) = 5$. Graph this solution with the others from part (b).

13. Ice is forming on a pond at a rate given by

$$\frac{dy}{dt} = k\sqrt{t},$$

where y is the thickness of the ice in inches at time t measured in hours since the ice started forming, and k is a positive constant. Find y as a function of t.

14. A water balloon launched from the roof of a building at time $t = 0$ has vertical velocity $v(t) = -32t + 40$ feet/sec at time t seconds, with $v > 0$ corresponding to upward motion.

    (a) If the roof of the building is 30 feet above the ground, find an expression for the height of the water balloon above the ground at time t.

    (b) What is the average velocity of the balloon between $t = 1.5$ and $t = 3$ seconds?

    (c) A 6-foot person is standing on the ground. How fast is the water balloon falling when it strikes the person on the top of the head?

15. If a car goes from 0 to 80 mph in six seconds with constant acceleration, what is that acceleration?

16. A car starts from rest at time $t = 0$ and accelerates at $-0.6t + 4$ meters/sec^2 for $0 \le t \le 12$. How long does it take for the car to go 100 meters?

17. A car going 80 ft/sec (about 55 mph) brakes to a stop in five seconds. Assume the deceleration is constant.

    (a) Graph the velocity against time, t, for $0 \le t \le 5$ seconds.

    (b) Represent, as an area on the graph, the total distance traveled from the time the brakes are applied until the car comes to a stop.

    (c) Find this area and hence the distance traveled.

    (d) Now find the total distance traveled using antidifferentiation.

18. A 727 jet needs to attain a speed of 200 mph to take off. If it can accelerate from 0 to 200 mph in 30 seconds, how long must the runway be? (Assume constant acceleration.)

19. A car going at 30 ft/sec decelerates at a constant 5 ft/sec^2.

    (a) Draw up a table showing the velocity of the car every half second. When does the car come to rest?

    (b) Using your table, find left and right sums which estimate the total distance traveled before the car comes to rest. Which is an overestimate, and which is an underestimate?

    (c) Sketch a graph of velocity against time. On the graph, show an area representing the distance traveled before the car comes to rest. Use the graph to calculate this distance.

    (d) Now find a formula for the velocity of the car as a function of time, and then find the total distance traveled by antidifferentiation. What is the relationship between your answer to parts (c) and (d) and your estimates in part (b)?

20. An object is shot vertically upward from the ground with a velocity of 160 ft/sec.

    (a) Sketch a graph of the velocity of the object (with upward as positive) against time.

    (b) Mark on the graph the points when the object reaches its highest point and when it lands.

    (c) Find the maximum height reached by the object by considering an area on the graph.

    (d) Now express velocity as a function of time, and find the greatest height by antidifferentiation.

21. A stone thrown upward from the top of a 320-foot cliff at 128 ft/sec eventually falls to the beach below.

    (a) How long does the stone take to reach its highest point?

    (b) What is its maximum height?

    (c) How long before the stone hits the beach?

    (d) What is the velocity of the stone on impact?

22. On the moon, the acceleration due to gravity is about 1.6 m/sec^2 (compared to $g \approx 9.8$ m/sec^2 on earth). If you drop a rock on the moon (with initial velocity 0), find formulas for:

    (a) Its velocity, $v(t)$, at time t.

    (b) The distance, $s(t)$, it falls in time t.

23. (a) Imagine throwing a rock straight up in the air. What is its initial velocity if the rock reaches a maximum height of 100 feet above its starting point?

    (b) Now imagine being transplanted to the moon and throwing a moon rock vertically upward with the same velocity as in part (a). How high will it go? (On the moon, $g = 5$ ft/sec^2.)

24. A cat, walking along the window ledge of a New York apartment, knocks off a flower pot, which falls to the street 200 feet below. How fast is the flower pot traveling when it hits the street? (Give your answer in ft/sec and in mph, given that 1 ft/sec = 15/22 mph.)

25. An Acura NSX going at 70 mph stops in 157 feet. Find the acceleration, assuming it is constant.

6.4 SECOND FUNDAMENTAL THEOREM OF CALCULUS

Suppose f is an elementary function, that is, a combination of constants, powers of x, $\sin x$, $\cos x$, e^x, and $\ln x$. Then we have to be lucky to find an antiderivative F which is also an elementary function. But if we can't find F as an elementary function, how can we be sure that F exists at all? In this section we learn to use the definite integral to construct antiderivatives.

Construction of Antiderivatives Using the Definite Integral

Consider the function $f(x) = e^{-x^2}$. We would like to find a way of calculating values of its antiderivative, F, which is not an elementary function. However, we know from the Fundamental Theorem of Calculus that

$$F(b) - F(a) = \int_a^b e^{-t^2}\, dt.$$

Setting $a = 0$ and replacing b by x, we have

$$F(x) - F(0) = \int_0^x e^{-t^2}\, dt.$$

Suppose we want the antiderivative that satisfies $F(0) = 0$. Then we get

$$F(x) = \int_0^x e^{-t^2}\, dt.$$

This is a formula for F. For any value of x, there is a unique value for $F(x)$, so F is a function. For any fixed x, we can calculate $F(x)$ numerically. For example,

$$F(2) = \int_0^2 e^{-t^2}\, dt = 0.88208\ldots.$$

Notice that our expression for F is not an elementary function; we have *created* a new function using the definite integral. The next theorem says that this method of constructing antiderivatives works in general. This means that if we define F by

$$F(x) = \int_a^x f(t)\, dt$$

then F must be an antiderivative of f.

Theorem 6.2: Construction Theorem for Antiderivatives

(Second Fundamental Theorem of Calculus) If f is a continuous function on an interval, and if a is any number in that interval, then the function F defined as follows is an antiderivative of f:

$$F(x) = \int_a^x f(t)\, dt.$$

Proof   Our task is to show that F, defined by this integral, is an antiderivative of f. We want to show that $F'(x) = f(x)$. By the definition of the derivative,

$$F'(x) = \lim_{h \to 0} \frac{F(x+h) - F(x)}{h}.$$

To gain some geometric insight, let's suppose f is positive and h is positive. Then we can visualize

$$F(x) = \int_a^x f(t)\,dt \quad \text{and} \quad F(x+h) = \int_a^{x+h} f(t)\,dt$$

as areas, which leads to representing

$$F(x+h) - F(x) = \int_x^{x+h} f(t)\,dt$$

as a difference of two areas. From Figure 6.24, we see that $F(x+h) - F(x)$ is roughly the area of a rectangle of height $f(x)$ and width h (shaded darker in Figure 6.24), so we have

$$F(x+h) - F(x) \approx f(x)h,$$

hence

$$\frac{F(x+h) - F(x)}{h} \approx f(x).$$

More precisely, we can use Theorem 5.4 on comparing integrals on page 270 to conclude that

$$mh \le \int_x^{x+h} f(t)\,dt \le Mh,$$

where m is the greatest lower bound for f on the interval from x to $x+h$ and M is the least upper bound on that interval. (See Figure 6.25.) Hence

$$mh \le F(x+h) - F(x) \le Mh,$$

so

$$m \le \frac{F(x+h) - F(x)}{h} \le M.$$

Since f is continuous, both m and M approach $f(x)$ as h approaches zero. Thus

$$f(x) \le \lim_{h \to 0} \frac{F(x+h) - F(x)}{h} \le f(x).$$

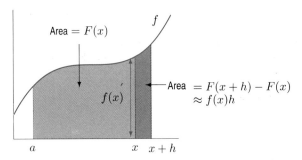

Figure 6.24: $F(x+h) - F(x)$ is area of roughly rectangular region

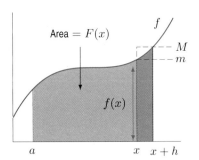

Figure 6.25: Upper and lower bounds for $F(x+h) - F(x)$

Thus both inequalities must actually be equalities, so we have the result we want:

$$f(x) = \lim_{h \to 0} \frac{F(x+h) - F(x)}{h} = F'(x).$$

Relationship between the Construction Theorem and the Fundamental Theorem of Calculus

If F is constructed as in Theorem 6.2, then we have just shown that $F' = f$. Suppose G is any other antiderivative of f, so $G' = f$, and therefore $F' = G'$. Since the derivative of $F - G$ is zero, the Constant Function Theorem on page 157 tells us that $F - G$ is a constant, so $F(x) = G(x) + C$.
Since we know $F(a) = 0$ (by the definition of F), we can write

$$\int_a^b f(t)\, dt = F(b) = F(b) - F(a) = (G(b) + C) - (G(a) + C) = G(b) - G(a).$$

This result, that the definite integral $\int_a^b f(t)\, dt$ can be evaluated using any antiderivative of f, is the (First) Fundamental Theorem of Calculus. Thus, we have shown that the First Fundamental Theorem of Calculus can be obtained from the Construction Theorem (the Second Fundamental Theorem of Calculus).

Using the Construction Theorem for Antiderivatives

The construction theorem enables us to write down antiderivatives of functions that do not have elementary antiderivatives. For example, an antiderivative of $(\sin x)/x$ is

$$F(x) = \int_0^x \frac{\sin t}{t}\, dt.$$

Notice that F is a function; we can calculate its values to any degree of accuracy. This function already has a name: it is called the *sine-integral*, and it is denoted $\text{Si}(x)$.

Example 1   Construct a table of values of $\text{Si}(x)$ for $x = 0, 1, 2, 3$.

Solution   Using numerical methods, we calculate the values of $\text{Si}(x) = \int_0^x \sin t/t\, dt$ given in Table 6.2. Since the integrand is undefined at $t = 0$, we took the lower limit as 0.00001 instead of 0.

Table 6.2   *A table of values of Si(x)*

x	0	1	2	3
$\text{Si}(x)$	0	0.95	1.61	1.85

The reason the sine-integral has a name is that some scientists and engineers use it all the time (for example, in optics). For them, it is just another common function like sine or cosine. Its derivative is given by

$$\frac{d}{dx}\, \text{Si}(x) = \frac{\sin x}{x}.$$

Example 2   Find the derivative of $x\, \text{Si}(x)$.

Solution   Using the product rule,

$$\frac{d}{dx}\,(x\, \text{Si}(x)) = \left(\frac{d}{dx}\, x\right) \text{Si}(x) + x\left(\frac{d}{dx}\, \text{Si}(x)\right)$$
$$= 1 \cdot \text{Si}(x) + x\frac{\sin x}{x}$$
$$= \text{Si}(x) + \sin x.$$

Exercises and Problems for Section 6.4

Exercises

In Exercises 1–4, let $F(x) = \int_0^x f(t)\, dt$. Graph $F(x)$ as a function of x.

1.

2.

3.

4.

5. For $x = 0$, 0.5, 1.0, 1.5, and 2.0, make a table of values for $I(x) = \int_0^x \sqrt{t^4 + 1}\, dt$.

6. Assume that $F'(t) = \sin t \cos t$ and $F(0) = 1$. Find $F(b)$ for $b = 0$, 0.5, 1, 1.5, 2, 2.5, and 3.

7. (a) Continue the table of values for $\mathrm{Si}(x) = \int_0^x (\sin t/t)\, dt$ on page 301 for $x = 4$ and $x = 5$.
   (b) Why is $\mathrm{Si}(x)$ decreasing between $x = 4$ and $x = 5$?

In Exercises 8–10, write an expression for the function, $f(x)$, with the given properties.

8. $f'(x) = \sin(x^2)$ and $f(0) = 7$

9. $f'(x) = (\sin x)/x$ and $f(1) = 5$

10. $f'(x) = \mathrm{Si}(x)$ and $f(0) = 2$

Problems

11. Use Figure 6.26 to sketch a graph of $F(x) = \int_0^x f(t)\, dt$. Label the points x_1, x_2, x_3.

Figure 6.26

12. The graph of the derivative F' of some function F is given in Figure 6.27. If $F(20) = 150$, estimate the maximum value attained by F.

Figure 6.27

In Problems 13–14, find the value of the function with the given properties.

13. $F(1)$, where $F'(x) = e^{-x^2}$ and $F(0) = 2$
14. $G(-1)$, where $G'(x) = \cos(x^2)$ and $G(0) = -3$

Find the derivatives in Problems 15–20.

15. $\dfrac{d}{dx} \displaystyle\int_0^x \cos(t^2)\, dt$

16. $\dfrac{d}{dx} \displaystyle\int_1^x (1+t)^{200}\, dt$

17. $\dfrac{d}{dx} \displaystyle\int_{0.5}^x \arctan(t^2)\, dt$

18. $\dfrac{d}{dt} \displaystyle\int_t^\pi \cos(z^3)\, dz$

19. $\dfrac{d}{dx} \displaystyle\int_x^1 \ln t\, dt$

20. $\dfrac{d}{dx} \left[\mathrm{Si}(x^2) \right]$

21. Let $g(x) = \int_0^x f(t)\, dt$. Using Figure 6.28, find
   (a) $g(0)$ (b) $g'(1)$
   (c) The interval where g is concave up.
   (d) The value of x where g takes its maximum on the interval $0 \le x \le 8$.

Figure 6.28

22. Let $F(x) = \int_0^x \sin(2t)\, dt$.
   (a) Evaluate $F(\pi)$.
   (b) Draw a sketch to explain geometrically why the answer to part (a) is correct.
   (c) For what values of x is $F(x)$ positive? negative?

23. Let $F(x) = \int_2^x (1/\ln t)\, dt$ for $x \ge 2$.
   (a) Find $F'(x)$.
   (b) Is F increasing or decreasing? What can you say about the concavity of its graph?
   (c) Sketch a graph of $F(x)$.

Calculate the derivatives in Problems 24–27.

24. $\dfrac{d}{dx} \displaystyle\int_2^{x^3} \sin(t^2)\, dt$

25. $\dfrac{d}{dt} \displaystyle\int_1^{\sin t} \cos(x^2)\, dx$

26. $\dfrac{d}{dx} \displaystyle\int_{\cos x}^3 e^{t^2}\, dt$

27. $\dfrac{d}{dt} \displaystyle\int_{e^t}^{t^3} \sqrt{1 + x^2}\, dx$

28. If $\int v(t-t_0)\, dt = s(t)+C$, where t_0 is a constant, what is $s'(t)$?

29. If $\int_{t_0}^{t} v(u)\, du = s(t)$, what is $\int_{a}^{b} v(t)\, dt$?

30. If $\int v(t)\, dt = s(t-t_0)+C$, where t_0 is a constant, what is $s'(t)$?

31. If $\int f(2x)\, dx = g(2x) + C$, what is $g'(x)$?

$f\left(x^2\right) =$

32. If $\int af(x)\, dx = g(ax) + C$, where a is a nonzero constant, what is $g'(x)$?

In Problems 33–36, find the given quantities. The *error function*, $\mathrm{erf}(x)$, is defined by

$$\mathrm{erf}(x) = \frac{2}{\sqrt{\pi}} \int_0^x e^{-t^2}\, dt.$$

33. $\dfrac{d}{dx}\left(x\,\mathrm{erf}(x)\right)$

34. $\dfrac{d}{dx}\left(\mathrm{erf}(\sqrt{x})\right)$

35. $\dfrac{d}{dx}\left(\int_0^{x^3} e^{-t^2}\, dt\right)$

36. $\dfrac{d}{dx}\left(\int_x^{x^3} e^{-t^2}\, dt\right)$

6.5 THE EQUATIONS OF MOTION

The problem of a body moving freely under the influence of gravity near the surface of the earth intrigued mathematicians and philosophers from Greek times onward and was finally solved by Galileo and Newton. The question to be answered was: How do the velocity and the position of the body vary with time? We define s to be the position, or height, of the body above a fixed point (often ground level); v then is the velocity of the body measured upward. We assume that the acceleration of the body is a constant, $-g$ (the negative sign means that the acceleration is downward), so

$$\text{Acceleration} = \frac{dv}{dt} = -g.$$

Thus, velocity is the antiderivative of $-g$:

$$v = -gt + C.$$

If the initial velocity is v_0, then $C = v_0$, so

$$v = -gt + v_0.$$

How about the position? We know that

$$\frac{ds}{dt} = v = -gt + v_0.$$

Therefore, we can find s by antidifferentiating again, giving:

$$s = -\frac{gt^2}{2} + v_0 t + C.$$

If the initial position is s_0, then we must have

$$s = -\frac{gt^2}{2} + v_0 t + s_0.$$

Our derivation of the formulas for the velocity and the position of the body took little effort. It hides an almost 2000-year struggle to understand the mechanics of falling bodies, from Aristotle's *Physics* to Galileo's *Dialogues Concerning Two New Sciences*.

Though it is an oversimplification of his ideas, we can say that Aristotle's conception of motion was primarily in terms of *change of position*. This seems entirely reasonable; it is what we commonly observe, and this view dominated discussions of motion for centuries. But it misses a subtlety that was brought to light by Descartes, Galileo, and, with a different emphasis, by Newton. That subtlety is now usually referred to as the *principle of inertia*.

This principle holds that a body traveling undisturbed at constant velocity in a straight line will continue in this motion indefinitely. Stated another way, it says that one cannot distinguish in any absolute sense (that is, by performing an experiment), between being at rest and moving with constant velocity in a straight line. If you are reading this book in a closed room and have no external reference points, there is no experiment that will tell you, one way or the other, whether you are at rest or whether you, the room, and everything in it are moving with constant velocity in a straight line. Therefore, as Newton saw, an understanding of motion should be based on *change of velocity* rather than change of position. Since acceleration is the rate of change of velocity, it is acceleration that must play a central role in the description of motion.

How does acceleration come about? How does the velocity change? Through the action of *forces*. Newton placed a new emphasis on the importance of forces. Newton's laws of motion do not say what a force *is*, they say how it *acts*. His first law is the principle of inertia, which says what happens in the *absence* of a force—there is no change in velocity. His second law says that a force acts to produce a change in velocity, that is, an acceleration. It states that $F = ma$, where m is the mass of the object, F is the net force, and a is the acceleration produced by this force.

Let's return to Galileo. He demonstrated that a body falling under the influence of gravity does so with constant acceleration. Furthermore, assuming we can neglect air resistance, this constant acceleration is independent of the mass of the body. This last fact was the outcome of Galileo's famous observation that a heavy ball and a light ball dropped off the Leaning Tower of Pisa hit the ground at the same time. Whether or not he actually performed this experiment, Galileo presented a very clear thought experiment in the *Dialogues* to prove the same point. (This point was counter to Aristotle's more common sense notion that the heavier ball would reach the ground first.) Galileo showed that the mass of the object did not appear as a variable in the equation of motion. Thus, the same constant acceleration equation applies to all bodies falling under the influence of gravity.

Nearly a hundred years after Galileo's experiment, Newton formulated his laws of motion and gravity, which led to a differential equation describing the motion of a falling body. According to Newton, acceleration is caused by force, and in the case of falling bodies, that force is the force of gravity. Newton's law of gravity says that the gravitational force between two bodies is attractive and given by

$$F = \frac{GMm}{r^2},$$

where G is the gravitational constant, m and M are the masses of the two bodies, and r is the distance between them. This is the famous *inverse square law*. For a falling body, we take M to be the mass of the earth and r to be the distance from the body to the center of the earth. So, actually, r changes as the body falls, but for anything we can easily observe (say, a ball dropped from the Tower of Pisa), it won't change significantly over the course of the motion. Hence, as an approximation, it is reasonable to assume that the force is constant. According to Newton's second law,

$$\text{Force} = \text{Mass} \times \text{Acceleration}.$$

Since the gravitational force is acting downward

$$-\frac{GMm}{r^2} = m\frac{d^2s}{dt^2}.$$

Hence,

$$\frac{d^2s}{dt^2} = -\frac{GM}{r^2} = \text{Constant}.$$

If we define $g = GM/r^2$, then

$$\frac{d^2s}{dt^2} = -g.$$

The fact that the mass cancels out of Newton's equations of motion reflects Galileo's experimental observation that the acceleration due to gravity is independent of the mass of the body.

Exercises and Problems for Section 6.5

Exercises

1. An object is thrown upward at time $t = 0$. After t seconds, its height is $y = -4.9t^2 + 7t + 1.5$ meters above the ground.

   $-9.8t + 7$

   (a) From what height was the object thrown? 1.5
   (b) What is the initial velocity of the object? 7

   (c) What is the acceleration due to gravity?

2. At time $t = 0$, a stone is thrown off a 250-meter cliff with velocity 20 meters/sec downward. Express its height, $h(t)$, in meters above the ground as a function of time, t, in seconds.

Problems

3. An object is dropped from a 400-foot tower. When does it hit the ground and how fast is it going at the time of the impact?

4. The object in Problem 3 falls off the same 400-foot tower. What would the acceleration due to gravity have to be to make it reach the ground in half the time?

5. A ball that is dropped from a window hits the ground in five seconds. How high is the window? (Give your answer in feet.)

6. On the moon the acceleration due to gravity is 5 ft/sec². An astronaut jumps into the air with an initial upward velocity of 10 ft/sec. How high does he go? How long is the astronaut off the ground?

7. Galileo was the first person to show that the distance traveled by a body falling from rest is proportional to the square of the time it has traveled, and independent of the mass of the body. Derive this result from the fact that the acceleration due to gravity is a constant.

8. While attempting to understand the motion of bodies under gravity, Galileo stated that:

   > The time in which any space is traversed by a body starting at rest and uniformly accelerated is equal to the time in which that same space would be traversed by the same body moving at a uniform speed whose value is the mean of the highest velocity and the velocity just before acceleration began.

(a) Write Galileo's statement in symbols, defining all the symbols you use.
(b) Check Galileo's statement for a body dropped off a 100-foot building accelerating from rest under gravity until it hits the ground.
(c) Show why Galileo's statement is true in general.

9. In his *Dialogues Concerning Two New Sciences*, Galileo wrote:

   > The distances traversed during equal intervals of time by a body falling from rest stand to one another in the same ratio as the odd numbers beginning with unity.

   Assume, as is now believed, that $s = -(gt^2)/2$, where s is the total distance traveled in time t, and g is the acceleration due to gravity.

   (a) How far does a falling body travel in the first second (between $t = 0$ and $t = 1$)? During the second second (between $t = 1$ and $t = 2$)? The third second? The fourth second?
   (b) What do your answers tell you about the truth of Galileo's statement?

10. The acceleration due to gravity 2 meters from the ground is 9.8 m/sec². What is the acceleration due to gravity 100 meters from the ground? At 100,000 meters? (The radius of the earth is $6.4 \cdot 10^6$ meters.)

$5 \ell \, sec$

$v = 5t$

$p = -\dfrac{5t^2}{2} + 10t$

CHAPTER SUMMARY (see also Ready Reference at the end of the book)

- **Constructing antiderivatives**
  Graphically, numerically, analytically.
- **The family of antiderivatives**
  The indefinite integral.
- **Differential equations**

Initial value problems, uniform motion.

- **Construction theorem (Second Fundamental Theorem of Calculus)**
  Constructing antiderivatives using definite integrals.

REVIEW EXERCISES AND PROBLEMS FOR CHAPTER SIX

Exercises

In Exercises 1–2, sketch two functions F such that $F' = f$. In one case let $F(0) = 0$ and in the other, let $F(0) = 1$.

1.

2.

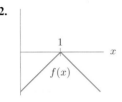

Find the indefinite integrals in Exercises 3–23.

3. $\int (5x + 7)\, dx$

4. $\int \left(4t + \dfrac{1}{t}\right) dt$

5. $\int (2 + \cos t)\, dt$

6. $\int 7e^x\, dx$

7. $\int (3e^x + 2\sin x)\, dx$

8. $\int (x + 3)^2\, dx$

9. $\int \dfrac{8}{\sqrt{x}}\, dx$

10. $\int \left(\dfrac{3}{t} - \dfrac{2}{t^2}\right) dt$

11. $\int (e^x + 5)\, dx$

12. $\int \left(\sqrt{x^3} - \dfrac{2}{x}\right) dx$

13. $\int \dfrac{1}{\cos^2 x}\, dx$

14. $\int 2^x\, dx$

15. $\int (x + 1)^2\, dx$

16. $\int (x + 1)^3\, dx$

17. $\int (x + 1)^9\, dx$

18. $\int \left(\dfrac{x+1}{x}\right) dx$

19. $\int \left(\dfrac{x^2 + x + 1}{x}\right) dx$

20. $\int \left(3\cos t + 3\sqrt{t}\right) dt$

21. $\int (3\cos x - 7\sin x)\, dx$

22. $\int \left(\dfrac{2}{x} + \pi \sin x\right) dx$

23. $\int (2e^x - 8\cos x)\, dx$

Find antiderivatives for the functions in Exercises 24–31. Check by differentiation.

24. $p(t) = \dfrac{1}{t}$

25. $f(x) = \dfrac{1}{x^2}$

26. $f(x) = \cos x$

27. $g(x) = \sin x$

28. $f(x) = e^x - 1$

29. $f(x) = 5e^x$

30. $h(t) = \dfrac{5}{t}$

31. $f(t) = t + \dfrac{1}{t}$

For Exercises 32–37, find an antiderivative $F(x)$ with $F'(x) = f(x)$ and $F(0) = 4$.

32. $f(x) = x^2$

33. $f(x) = x^3 + 6x^2 - 4$

34. $f(x) = \sqrt{x}$

35. $f(x) = e^x$

36. $f(x) = \sin x$

37. $f(x) = \cos x$

38. Use the Fundamental Theorem of Calculus to evaluate $\int_1^3 (6x^2 + 8x + 5)\,dx$.

Problems

39. Use the Fundamental Theorem to find the area under $f(x) = x^2$ between $x = 0$ and $x = 3$.

40. Find the exact area of the region bounded by the x-axis and the graph of $y = x^3 - x$.

41. Calculate the exact area above the graph of $y = \dfrac{1}{2}\left(\dfrac{3}{\pi}x\right)^2$ and below the graph of $y = \cos x$. The curves intersect at $x = \pm\pi/3$.

42. Find the exact area of the shaded region in Figure 6.29 between $y = 3x^2 - 3$ and the x-axis.

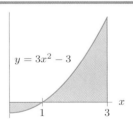

Figure 6.29

43. **(a)** Find the exact area between $f(x) = x^3 - 7x^2 + 10x$, the x-axis, $x = 0$, and $x = 5$.
(b) Find $\int_0^5 (x^3 - 7x^2 + 10x)\,dx$ exactly and interpret this integral in terms of areas.

44. The area under $1/\sqrt{x}$ on the interval $1 \le x \le b$ is equal to 6. Find the value of b using the Fundamental Theorem.

45. Find the exact positive value of c which makes the area under the graph of $y = c(1 - x^2)$ and above the x-axis equal to 1.

46. Sketch the parabola $y = x(x - \pi)$ and the curve $y = \sin x$, showing their points of intersection. Find the exact area between the two graphs.

47. Find the exact average value of $f(x) = \sqrt{x}$ on the interval $0 \le x \le 9$. Illustrate your answer on a graph of $f(x) = \sqrt{x}$.

48. For t in years, $0 \le t \le 3$, the rate of discharge of a pollutant is estimated to be $t^2 - 14t + 49$ cubic meters per year. What is the total amount discharged during the three years?

49. For a function f, you are given the graph of the derivative f' in Figure 6.30 and that $f(0) = 50$.

    (a) On the interval $0 \le t \le 5$, at what value of t does f appear to reach its maximum value? Its minimum value?

    (b) Estimate these maximum and minimum values.

    (c) Estimate $f(5) - f(0)$.

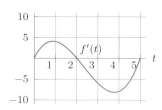

Figure 6.30

50. Assume f' is given by the graph in Figure 6.31. Suppose f is continuous and that $f(0) = 0$.

    (a) Find $f(3)$ and $f(7)$.

    (b) Find all x with $f(x) = 0$.

    (c) Sketch a graph of f over the interval $0 \le x \le 7$.

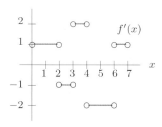

Figure 6.31

For Problems 51–52 the graph of $f'(x)$ is given. Sketch a possible graph for $f(x)$. Mark the points $x_1 \ldots x_4$ on your graph and label local maxima, local minima and points of inflection.

51.   **52.**

Calculate the derivatives in Problems 53–58.

53. $\dfrac{d}{dx} \displaystyle\int_2^x \arccos(t^7)\, dt$ **54.** $\dfrac{d}{dt} \displaystyle\int_t^7 \log(x^6)\, dx$

55. $\dfrac{d}{dx} \displaystyle\int_5^{e^x} \cos(t^3)\, dt$ **56.** $\dfrac{d}{dx} \displaystyle\int_{\sin x}^{17} \tan^3 t\, dt$

57. $\dfrac{d}{dt} \displaystyle\int_{t^5}^{\cos t} 4^{7x}\, dx$ **58.** $\dfrac{d}{dt} \displaystyle\int_{e^t}^{4 \sin t} \dfrac{1+x}{1+x^2}\, dx$

59. Let $F(x) = \int_{\pi/2}^x (\sin t / t)\, dt$. Find the value(s) of x between $\pi/2$ and $3\pi/2$ for which $F(x)$ has a global maximum and global minimum.

60. The graphs of three functions are given in Figure 6.32. Determine which is f, which is f', and which is $\int_0^x f(t)\, dt$. Explain your answer.

Figure 6.32

61. Write an expression for the function which is an antiderivative of e^{x^2} and which passes through the point

    (a) $(0, 3)$ **(b)** $(-1, 5)$

62. An object thrown in the air on a planet in a distant galaxy is at height $s = -25t^2 + 72t + 40$ feet at time t seconds after it is thrown. What is the acceleration due to gravity on this planet? With what velocity was the object thrown? From what height?

63. An object is attached to a coiled spring which is suspended from the ceiling of a room. The function $h(t)$ gives the height of the object above the floor of the room at time, t. The graph of $h'(t)$ is given in Figure 6.33. Sketch a possible graph of $h(t)$.

Figure 6.33

64. The acceleration, a, of a particle as a function of time is shown in Figure 6.34. Sketch graphs of velocity and position against time. The particle starts at rest at the origin.

Figure 6.34

65. The angular speed of a car engine increases from 1100 revs/min to 2500 revs/min in 6 sec.

 (a) Assuming that it is constant, find the angular acceleration in revs/min^2.
 (b) How many revolutions does the engine make in this time?

66. A helicopter rotor slows down at a constant rate from 350 revs/min to 260 revs/min in 1.5 minutes.

 (a) Find the angular acceleration during this time interval. What are the units of this acceleration?
 (b) Assuming the angular acceleration remains constant, how long does it take for the rotor to stop? (Measure time from the moment when speed was 350 revs/min.)
 (c) How many revolutions does the rotor make between the time the angular speed was 350 revs/min and stopping?

67. An object is thrown vertically upward with a velocity of 80 ft/sec.

 (a) Make a table showing its velocity every second.
 (b) When does it reach its highest point? When does it hit the ground?
 (c) Using your table, write left and right sums which under- and overestimate the height the object attains.
 (d) Use antidifferentiation to find the greatest height it reaches.

68. A car, initially moving at 60 mph, has a constant deceleration and stops in a distance of 200 feet. What is its deceleration? (Give your answer in ft/sec^2. Note that 1 mph = 22/15 ft/sec.)

69. The birth rate, B, in births per hour, of a bacteria population is given in Figure 6.35. The curve marked D gives the death rate, in deaths per hour, of the same population.

 (a) Explain what the shape of each of these graphs tells you about the population.
 (b) Use the graphs to find the time at which the net rate of increase of the population is at a maximum.
 (c) At time $t = 0$ the population has size N. Sketch the graph of the total number born by time t. Also sketch the graph of the number alive at time t. Estimate the time at which the population is a maximum.

Figure 6.35

70. Water flows at a constant rate into the left side of the W-shaped container in Figure 6.36. Sketch a graph of the height, H, of the water in the left side of the container as a function of time, t. The container starts empty.

Figure 6.36

71. A particle of mass, m, acted on by a force, F, moves in a straight line. Its acceleration, a, is given by Newton's Law:

$$F = ma.$$

The work, W, done by a constant force when the particle moves through a displacement, d, is

$$W = Fd.$$

The velocity, v, of the particle as a function of time, t, is given in Figure 6.37. What is the sign of the work done during each of the time intervals: $[0, t_1]$, $[t_1, t_2]$, $[t_2, t_3]$, $[t_3, t_4]$,$[t_2, t_4]$?

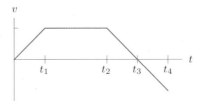

Figure 6.37

CAS Challenge Problems

72. (a) Set up a right-hand Riemann sum for $\int_a^b x^3 \, dx$ using n subdivisions. What is Δx? Express each x_i, for $i = 1, 2, \ldots, n$, in terms of i.

(b) Use a computer algebra system to find an expression for the Riemann sum in part (a); then find the limit of this expression as $n \to \infty$.

(c) Simplify the final expression and compare the result to that obtained using the Fundamental Theorem of Calculus.

73. (a) Use a computer algebra system to find $\int e^{2x} \, dx$, $\int e^{3x} \, dx$, and $\int e^{3x+5} \, dx$.

(b) Using your answers to part (a), conjecture a formula for $\int e^{ax+b} \, dx$, where a and b are constants.

(c) Check your formula by differentiation. Explain which differentiation rules you are using.

74. (a) Use a computer algebra system to find $\int \sin(3x) \, dx$, $\int \sin(4x) \, dx$, and $\int \sin(3x - 2) \, dx$.

(b) Using your answers to part (a), conjecture a formula for $\int \sin(ax + b) \, dx$, where a and b are constants.

(c) Check your formula by differentiation. Explain which differentiation rules you are using.

75. (a) Use a computer algebra system to find

$$\int \frac{x-2}{x-1} \, dx, \quad \int \frac{x-3}{x-1} \, dx, \quad \text{and} \quad \int \frac{x-1}{x-2} \, dx.$$

(b) If a and b are constants, use your answers to part (a) to conjecture a formula for

$$\int \frac{x-a}{x-b} \, dx.$$

(c) Check your formula by differentiation. Explain which rules of differentiation you are using.

76. (a) Use a computer algebra system to find

$$\int \frac{1}{(x-1)(x-3)} \, dx, \quad \int \frac{1}{(x-1)(x-4)} \, dx$$

and

$$\int \frac{1}{(x-1)(x+3)} \, dx.$$

(b) If a and b are constants, use your answers to part (a) to conjecture a formula for

$$\int \frac{1}{(x-a)(x-b)} \, dx.$$

(c) Check your formula by differentiation. Explain which rules of differentiation you are using.

CHECK YOUR UNDERSTANDING

Are the statements in Problems 1–26 true or false? Give an explanation for your answer.

1. A function $f(x)$ has at most one derivative.

2. An antiderivative of $3\sqrt{x+1}$ is $2(x+1)^{3/2}$.

3. An antiderivative of $3x^2$ is $x^3 + \pi$.

4. An antiderivative of $1/x$ is $\ln|x| + \ln 2$.

5. An antiderivative of e^{-x^2} is $-e^{-x^2}/2x$.

6. $\int f(x)dx = (1/x) \int x f(x)dx$.

7. If $F(x)$ is an antiderivative of $f(x)$ and $G(x) = F(x) + 2$, then $G(x)$ is an antiderivative of $f(x)$.

8. If $F(x)$ is an antiderivative of $f(x)$, then $y = F(x)$ is a solution to the differential equation $dy/dx = f(x)$.

9. If $y = F(x)$ is a solution to the differential equation $dy/dx = f(x)$, then $F(x)$ is an antiderivative of $f(x)$.

10. If an object has constant nonzero acceleration, then the position of the object as a function of time is a quadratic polynomial.

11. If $F(x)$ and $G(x)$ are two antiderivatives of $f(x)$ for $-\infty < x < \infty$ and $F(5) > G(5)$, then $F(10) > G(10)$.

12. In an initial value problem for the differential equation $dy/dx = f(x)$, the value of y at $x = 0$ is always specified.

13. If $f(x)$ is positive for all x, then there is a solution of the differential equation $dy/dx = f(x)$ where $y(x)$ is positive for all x.

14. If $f(x) > 0$ for all x then every solution of the differential equation $dy/dx = f(x)$ is an increasing function.

15. If two solutions of a differential equation $dy/dx = f(x)$ have different values at $x = 3$ then they have different values at every x.

16. If the function $y = f(x)$ is a solution of the differential equation $dy/dx = \sin x/x$, then the function $y = f(x) + 5$ is also a solution.

17. There is only one solution $y(t)$ to the initial value problem $dy/dt = 3t^2$, $y(1) = \pi$.

18. If $F(x)$ is an antiderivative of $f(x)$ and $G(x)$ is an antiderivative of $g(x)$, then $F(x) \cdot G(x)$ is an antiderivative of $f(x) \cdot g(x)$.

19. If $F(x)$ and $G(x)$ are both antiderivatives of $f(x)$ on an interval then $F(x) - G(x)$ is a constant function.

20. Every continuous function has an antiderivative.

21. A ball thrown downward at 10 feet per second from the top of a 100 foot building hits the ground in less than 3 seconds.

22. $\int_0^x \sin(t^2)dt$ is an antiderivative of $\sin(x^2)$.

23. If $F(x) = \int_0^x f(t)dt$, then $F(5) - F(3) = \int_3^5 f(t)dt$.

24. If $F(x) = \int_0^x f(t)dt$, then $F(x)$ must be increasing.

25. If $F(x) = \int_0^x f(t)dt$ and $G(x) = \int_2^x f(t)dt$, then $F(x) = G(x) + C$.

26. If $F(x) = \int_0^x f(t)dt$ and $G(x) = \int_0^x g(t)dt$, then $F(x) + G(x) = \int_0^x (f(t) + g(t))dt$.

PROJECTS FOR CHAPTER SIX

1. Distribution of Resources

Whether a resource is distributed evenly among members of a population is often an important political or economic question. How can we measure this? How can we decide if the distribution of wealth in this country is becoming more or less equitable over time? How can we measure which country has the most equitable income distribution? This problem describes a way of making such measurements. Suppose the resource is distributed evenly. Then any 20% of the population will have 20% of the resource. Similarly, any 30% will have 30% of the resource and so on. If, however, the resource is not distributed evenly, the poorest $p\%$ of the population (in terms of this resource) will not have $p\%$ of the goods. Suppose $F(x)$ represents the fraction of the resources owned by the poorest fraction x of the population. Thus $F(0.4) = 0.1$ means that the poorest 40% of the population owns 10% of the resource.

(a) What would F be if the resource were distributed evenly?

(b) What must be true of any such F? What must $F(0)$ and $F(1)$ equal? Is F increasing or decreasing? Is the graph of F concave up or concave down?

(c) Gini's index of inequality, G, is one way to measure how evenly the resource is distributed. It is defined by

$$G = 2 \int_0^1 [x - F(x)] \, dx.$$

Show graphically what G represents.

2. Yield from an Apple Orchard

Figure 6.38 is a graph of the annual yield, $y(t)$, in bushels per year, from an orchard t years after planting. The trees take about 10 years to get established, but for the next 20 years they give a substantial yield. After about 30 years, however, age and disease start to take their toll, and the annual yield falls off.[1]

(a) Represent on a sketch of Figure 6.38 the total yield, $F(M)$, up to M years, with $0 \le M \le 60$. Write an expression for $F(M)$ in terms of $y(t)$.

(b) Sketch a graph of $F(M)$ against M for $0 \le M \le 60$.

(c) Write an expression for the average annual yield, $a(M)$, up to M years.

(d) When should the orchard be cut down and replanted? Assume that we want to maximize average revenue per year, and that fruit prices remain constant, so that this is achieved by maximizing average annual yield. Use the graph of $y(t)$ to estimate the time at which the average annual yield is a maximum. Explain your answer geometrically and symbolically.

Figure 6.38

---

[1] From Peter D. Taylor, *Calculus: The Analysis of Functions*, (Toronto: Wall & Emerson, Inc., 1992).

Chapter Seven

INTEGRATION

In Chapter 5, we introduced the definite integral, and in Chapter 6, we introduced antiderivatives. In this chapter, we focus on methods of finding antiderivatives and definite integrals. We first see additional ways to evaluate antiderivatives and definite integrals exactly, and we then develop more efficient numerical methods for approximating definite integrals. Finally, we introduce improper integrals and discuss methods of evaluating them.

7.1   INTEGRATION BY SUBSTITUTION

In Chapter 3, we learned rules to differentiate any function obtained by combining constants, powers of x, $\sin x$, $\cos x$, e^x, and $\ln x$, using addition, multiplication, division, or composition of functions. Such functions are called *elementary*.

In the next few sections, we introduce two methods of antidifferentiation: substitution and integration by parts, which reverse the chain and product rules, respectively. However, there is a great difference between looking for derivatives and looking for antiderivatives. Every elementary function has elementary derivatives, but many elementary functions do not have elementary antiderivatives. Some examples are $\sqrt{x^3 + 1}$, $(\sin x)/x$, and e^{-x^2}. These are not exotic functions, but ordinary functions that arise naturally.

The Guess-and-Check Method

A good strategy for finding simple antiderivatives is to *guess* an answer (using knowledge of differentiation rules) and then *check* the answer by differentiating it. If we get the expected result, then we're done; otherwise, we revise the guess and check again.

The method of guess-and-check is useful in reversing the chain rule. According to the chain rule,

$$\frac{d}{dx}(f(g(x))) = \underbrace{f'}_{\text{Derivative of outside}} \overbrace{(g(x))}^{\text{Inside}} \cdot \underbrace{g'(x)}_{\text{Derivative of inside}}.$$

Thus, any function which is the result of applying the chain rule is the product of two factors: the "derivative of the outside" and the "derivative of the inside." If a function has this form, its antiderivative is $f(g(x))$.

Example 1   Find $\displaystyle\int 3x^2 \cos(x^3)\, dx$.

Solution   The function $3x^2 \cos(x^3)$ looks like the result of applying the chain rule: there is an "inside" function x^3 and its derivative $3x^2$ appears as a factor. Since the outside function is a cosine which has a sine as an antiderivative, we guess $\sin(x^3)$ for the antiderivative. Differentiating to check gives

$$\frac{d}{dx}(\sin(x^3)) = \cos(x^3) \cdot (3x^2).$$

Since this is what we began with, we know that

$$\int 3x^2 \cos(x^3)\, dx = \sin(x^3) + C.$$

The basic idea of this method is to try to find an inside function whose derivative appears as a factor. This works even when the derivative is missing a constant factor, as in the next example.

Example 2   Find $\displaystyle\int t e^{(t^2+1)}\, dt$.

Solution   It looks like $t^2 + 1$ is an inside function. So we guess $e^{(t^2+1)}$ for the antiderivative, since taking the derivative of an exponential results in the reappearance of the exponential together with other terms from the chain rule. Now we check:

$$\frac{d}{dt}\left(e^{(t^2+1)}\right) = \left(e^{(t^2+1)}\right) \cdot 2t.$$

The original guess was too large by a factor of 2. We change the guess to $\frac{1}{2}e^{(t^2+1)}$ and check again:

$$\frac{d}{dt}\left(\frac{1}{2}e^{(t^2+1)}\right) = \frac{1}{2}e^{(t^2+1)} \cdot 2t = e^{(t^2+1)} \cdot t.$$

Thus, we know that

$$\int te^{(t^2+1)}\,dt = \frac{1}{2}e^{(t^2+1)} + C.$$

Example 3   Find $\int x^3\sqrt{x^4+5}\,dx$.

Solution   Here the inside function is $x^4 + 5$, and its derivative appears as a factor, with the exception of a missing 4. Thus, the integrand we have is more or less of the form

$$g'(x)\sqrt{g(x)},$$

with $g(x) = x^4 + 5$. Since $x^{3/2}/(3/2)$ is an antiderivative of the outside function $\sqrt{x}$, we might guess that an antiderivative is

$$\frac{(g(x))^{3/2}}{3/2} = \frac{(x^4+5)^{3/2}}{3/2}.$$

Let's check and see:

$$\frac{d}{dx}\left(\frac{(x^4+5)^{3/2}}{3/2}\right) = \frac{3}{2}\frac{(x^4+5)^{1/2}}{3/2} \cdot 4x^3 = 4x^3(x^4+5)^{1/2},$$

so $\dfrac{(x^4+5)^{3/2}}{3/2}$ is too big by a factor of 4. The correct antiderivative is

$$\frac{1}{4}\frac{(x^4+5)^{3/2}}{3/2} = \frac{1}{6}(x^4+5)^{3/2}.$$

Thus

$$\int x^3\sqrt{x^4+5}\,dx = \frac{1}{6}(x^4+5)^{3/2} + C.$$

As a final check:

$$\frac{d}{dx}\left(\frac{1}{6}(x^4+5)^{3/2}\right) = \frac{1}{6}\cdot\frac{3}{2}(x^4+5)^{1/2}\cdot 4x^3 = x^3(x^4+5)^{1/2}.$$

As we have seen in the preceding examples, antidifferentiating a function often involves "correcting for" constant factors: if differentiation produces an extra factor of 4, antidifferentiation will require a factor of $\frac{1}{4}$.

The Method of Substitution

When the integrand is complicated, it helps to formalize this guess-and-check method as follows:

> **To Make a Substitution**
>
> Let w be the "inside function" and $dw = w'(x)\,dx = \dfrac{dw}{dx}\,dx$.

Let's redo the first example using a substitution.

Example 4   Find $\int 3x^2\cos(x^3)\,dx$.

Solution   As before, we look for an inside function whose derivative appears—in this case x^3. We let $w = x^3$. Then $dw = w'(x)\,dx = 3x^2\,dx$. The original integrand can now be completely rewritten in terms of the new variable w:

$$\int 3x^2 \cos(x^3)\,dx = \int \cos\underbrace{(x^3)}_{w} \cdot \underbrace{3x^2\,dx}_{dw} = \int \cos w\,dw = \sin w + C = \sin(x^3) + C.$$

By changing the variable to w, we can simplify the integrand. We now have $\cos w$, which can be antidifferentiated more easily. The final step, after antidifferentiating, is to convert back to the original variable, x.

Why Does Substitution Work?

The substitution method makes it look as if we can treat dw and dx as separate entities, even canceling them in the equation $dw = (dw/dx)dx$. Let's see why this works. Suppose we have an integral of the form $\int f(g(x))g'(x)\,dx$, where $g(x)$ is the inside function and $f(x)$ is the outside function. If F is an antiderivative of f, then $F' = f$, and by the chain rule $\frac{d}{dx}(F(g(x))) = f(g(x))g'(x)$. Therefore,

$$\int f(g(x))g'(x)\,dx = F(g(x)) + C.$$

Now write $w = g(x)$ and $dw/dx = g'(x)$ on both sides of this equation:

$$\int f(w)\frac{dw}{dx}\,dx = F(w) + C.$$

On the other hand, knowing that $F' = f$ tells us that

$$\int f(w)\,dw = F(w) + C.$$

Thus, the following two integrals are equal:

$$\int f(w)\frac{dw}{dx}\,dx = \int f(w)\,dw.$$

Substituting w for the inside function and writing $dw = w'(x)dx$ leaves the indefinite integral unchanged.

Let's revisit the second example that we did by guess-and-check.

Example 5   Find $\displaystyle\int te^{(t^2+1)}\,dt$.

Solution   Here the inside function is $t^2 + 1$, with derivative $2t$. Since there is a factor of t in the integrand, we try

$$w = t^2 + 1.$$

Then

$$dw = w'(t)\,dt = 2t\,dt.$$

Notice, however, that the original integrand has only $t\,dt$, not $2t\,dt$. We therefore write

$$\frac{1}{2}\,dw = t\,dt$$

and then substitute:

$$\int te^{(t^2+1)}\,dt = \int e^{\overbrace{(t^2+1)}^{w}} \cdot \underbrace{t\,dt}_{\frac{1}{2}\,dw} = \int e^w \frac{1}{2}\,dw = \frac{1}{2}\int e^w\,dw = \frac{1}{2}e^w + C = \frac{1}{2}e^{(t^2+1)} + C.$$

This gives the same answer as we found using guess-and-check.

Why didn't we put $\frac{1}{2}\int e^w\,dw = \frac{1}{2}e^w + \frac{1}{2}C$ in the preceding example? Since the constant C is arbitrary, it doesn't really matter whether we add C or $\frac{1}{2}C$. The convention is always to add C to whatever antiderivative we have calculated.

Now let's redo the third example that we solved previously by guess-and-check.

Example 6 Find $\displaystyle\int x^3\sqrt{x^4+5}\,dx$.

Solution   The inside function is x^4+5, with derivative $4x^3$. The integrand has a factor of x^3, and since the only thing missing is a constant factor, we try

$$w = x^4 + 5.$$

Then

$$dw = w'(x)\,dx = 4x^3\,dx,$$

giving

$$\frac{1}{4}dw = x^3\,dx.$$

Thus,

$$\int x^3\sqrt{x^4+5}\,dx = \int \sqrt{w}\,\frac{1}{4}dw = \frac{1}{4}\int w^{1/2}\,dw = \frac{1}{4}\cdot\frac{w^{3/2}}{3/2} + C = \frac{1}{6}(x^4+5)^{3/2} + C.$$

Once again, we get the same result as with guess-and-check.

Warning

We saw in the preceding examples that we can apply the substitution method when a *constant* factor is missing from the derivative of the inside function. However, we may not be able to use substitution if anything other than a constant factor is missing. For example, setting $w = x^4 + 5$ to find

$$\int x^2\sqrt{x^4+5}\,dx$$

does us no good because $x^2\,dx$ is not a constant multiple of $dw = 4x^3\,dx$. Substitution works if the integrand contains the derivative of the inside function, *to within a constant factor.*

Some people prefer the substitution method over guess-and-check since it is more systematic, but both methods achieve the same result. For simple problems, guess-and-check can be faster.

Example 7 Find $\displaystyle\int e^{\cos\theta}\sin\theta\,d\theta$.

Solution   We let $w = \cos\theta$ since its derivative is $-\sin\theta$ and there is a factor of $\sin\theta$ in the integrand. This gives

$$dw = w'(\theta)\,d\theta = -\sin\theta\,d\theta,$$

so

$$-dw = \sin\theta\,d\theta.$$

Thus

$$\int e^{\cos\theta}\sin\theta\,d\theta = \int e^w\,(-dw) = (-1)\int e^w\,dw = -e^w + C = -e^{\cos\theta} + C.$$

Example 8   Find $\displaystyle\int \frac{e^t}{1+e^t}\, dt$.

Solution   Observing that the derivative of $1+e^t$ is e^t, we see $w = 1 + e^t$ is a good choice. Then $dw = e^t\, dt$, so that

$$\int \frac{e^t}{1+e^t}\, dt = \int \frac{1}{1+e^t}\, e^t\, dt = \int \frac{1}{w}\, dw = \ln|w| + C$$
$$= \ln|1+e^t| + C$$
$$= \ln(1+e^t) + C. \qquad \text{(Since $(1+e^t)$ is always positive.)}$$

Since the numerator is $e^t\, dt$, we might also have tried $w = e^t$. This substitution leads to the integral $\int (1/(1+w))\,dw$, which is better than the original integral but requires another substitution, $u = 1 + w$, to finish. There are often several different ways of doing an integral by substitution.

Notice the pattern in the previous example: having a function in the denominator and its derivative in the numerator leads to a natural logarithm. The next example follows the same pattern.

Example 9   Find $\displaystyle\int \tan\theta\, d\theta$.

Solution   Recall that $\tan\theta = (\sin\theta)/(\cos\theta)$. If $w = \cos\theta$, then $dw = -\sin\theta\, d\theta$, so

$$\int \tan\theta\, d\theta = \int \frac{\sin\theta}{\cos\theta}\, d\theta = \int \frac{-dw}{w} = -\ln|w| + C = -\ln|\cos\theta| + C.$$

Definite Integrals by Substitution

Example 10   Compute $\displaystyle\int_0^2 xe^{x^2}\, dx$.

Solution   To evaluate this definite integral using the Fundamental Theorem of Calculus, we first need to find an antiderivative of $f(x) = xe^{x^2}$. The inside function is x^2, so we let $w = x^2$. Then $dw = 2x\, dx$, so $\frac{1}{2}\, dw = x\, dx$. Thus,

$$\int xe^{x^2}\, dx = \int e^w \frac{1}{2}\, dw = \frac{1}{2}e^w + C = \frac{1}{2}e^{x^2} + C.$$

Now we find the definite integral

$$\int_0^2 xe^{x^2}\, dx = \frac{1}{2}e^{x^2}\Big|_0^2 = \frac{1}{2}(e^4 - e^0) = \frac{1}{2}(e^4 - 1).$$

There is another way to look at the same problem. After we established that

$$\int xe^{x^2}\, dx = \frac{1}{2}e^w + C,$$

our next two steps were to replace w by x^2, and then x by 2 and 0. We could have directly replaced the original limits of integration, $x = 0$ and $x = 2$, by the corresponding w limits. Since $w = x^2$, the w limits are $w = 0^2 = 0$ (when $x = 0$) and $w = 2^2 = 4$ (when $x = 2$), so we get

$$\int_{x=0}^{x=2} xe^{x^2}\, dx = \frac{1}{2}\int_{w=0}^{w=4} e^w\, dw = \frac{1}{2}e^w\Big|_0^4 = \frac{1}{2}(e^4 - e^0) = \frac{1}{2}(e^4 - 1).$$

As we would expect, both methods give the same answer.

To Use Substitution to Find Definite Integrals

Either
- Compute the indefinite integral, expressing an antiderivative in terms of the original variable, and then evaluate the result at the original limits,

or
- Convert the original limits to new limits in terms of the new variable and do not convert the antiderivative back to the original variable.

Example 11 Evaluate $\int_0^{\pi/4} \dfrac{\tan^3 \theta}{\cos^2 \theta}\, d\theta$.

Solution To use substitution, we must decide what w should be. There are two possible inside functions, $\tan \theta$ and $\cos \theta$. Now

$$\frac{d}{d\theta}(\tan \theta) = \frac{1}{\cos^2 \theta} \quad \text{and} \quad \frac{d}{d\theta}(\cos \theta) = -\sin \theta,$$

and since the integral contains a factor of $1/\cos^2 \theta$ but not of $\sin \theta$, we try $w = \tan \theta$. Then $dw = (1/\cos^2 \theta)d\theta$. When $\theta = 0$, $w = \tan 0 = 0$, and when $\theta = \pi/4$, $w = \tan(\pi/4) = 1$, so

$$\int_0^{\pi/4} \frac{\tan^3 \theta}{\cos^2 \theta}\, d\theta = \int_0^{\pi/4} (\tan \theta)^3 \cdot \frac{1}{\cos^2 \theta}\, d\theta = \int_0^1 w^3 \, dw = \frac{1}{4}w^4 \Big|_0^1 = \frac{1}{4}.$$

Example 12 Evaluate $\int_1^3 \dfrac{dx}{5 - x}$.

Solution Let $w = 5 - x$, so $dw = -dx$. When $x = 1$, $w = 4$, and when $x = 3$, $w = 2$, so

$$\int_1^3 \frac{dx}{5 - x} = \int_4^2 \frac{-dw}{w} = -\ln|w| \Big|_4^2 = -(\ln 2 - \ln 4) = \ln\left(\frac{4}{2}\right) = \ln 2 \approx 0.69.$$

Notice that we write the limit $w = 4$ at the bottom, even though it is larger than $w = 2$, because $w = 4$ corresponds to the lower limit $x = 1$.

More Complex Substitutions

In the examples of substitution presented so far, we guessed an expression for w and hoped to find dw (or some constant multiple of it) in the integrand. What if we are not so lucky? It turns out that it often works to let w be some messy expression contained inside, say, a cosine or under a root, even if we cannot see immediately how such a substitution helps.

Example 13 Find $\int \sqrt{1 + \sqrt{x}}\, dx$.

Solution This time, the derivative of the inside function is nowhere to be seen. Nevertheless, we try $w = 1 + \sqrt{x}$. Then $w - 1 = \sqrt{x}$, so $(w - 1)^2 = x$. Therefore $2(w - 1)\, dw = dx$. We have

$$\int \sqrt{1 + \sqrt{x}}\, dx = \int \sqrt{w}\, 2(w - 1)\, dw = 2 \int w^{1/2}(w - 1)\, dw$$

$$= 2 \int (w^{3/2} - w^{1/2})\, dw = 2\left(\frac{2}{5}w^{5/2} - \frac{2}{3}w^{3/2}\right) + C$$

$$= 2\left(\frac{2}{5}(1 + \sqrt{x})^{5/2} - \frac{2}{3}(1 + \sqrt{x})^{3/2}\right) + C.$$

Notice that the substitution in the preceding example again converts the inside of the messiest function into something simple. In addition, since the derivative of the inside function is not waiting for us, we have to solve for x so that we can get dx entirely in terms of w and dw.

Example 14   Find $\displaystyle\int (x+7)\sqrt[3]{3-2x}\,dx$.

Solution   Here, instead of the derivative of the inside function (which is -2), we have the factor $(x+7)$. However, substituting $w = 3-2x$ turns out to help anyway. Then $dw = -2\,dx$, so $(-1/2)\,dw = dx$. Now we must convert everything to w, including $x+7$. If $w = 3 - 2x$, then $2x = 3 - w$, so $x = 3/2 - w/2$, and therefore we can write $x+7$ in terms of w. Thus

$$
\int (x+7)\sqrt[3]{3-2x}\,dx = \int \left(\frac{3}{2} - \frac{w}{2} + 7\right) \sqrt[3]{w} \left(-\frac{1}{2}\right) dw
$$

$$
= -\frac{1}{2} \int \left(\frac{17}{2} - \frac{w}{2}\right) w^{1/3}\,dw
$$

$$
= -\frac{1}{4} \int (17 - w)w^{1/3}\,dw
$$

$$
= -\frac{1}{4} \int (17w^{1/3} - w^{4/3})\,dw
$$

$$
= -\frac{1}{4} \left(17\frac{w^{4/3}}{4/3} - \frac{w^{7/3}}{7/3}\right) + C
$$

$$
= -\frac{1}{4} \left(\frac{51}{4}(3 - 2x)^{4/3} - \frac{3}{7}(3 - 2x)^{7/3}\right) + C.
$$

Looking back over the solution, the reason this substitution works is that it converts $\sqrt[3]{3-2x}$, the messiest part of the integrand, to $\sqrt[3]{w}$, which can be combined with the other term and then integrated.

Exercises and Problems for Section 7.1

Exercises

1. Use substitution to express each of the following integrals as a multiple of $\int_a^b (1/w)\,dw$ for some a and b. Then evaluate the integrals.

   (a) $\displaystyle\int_0^1 \frac{x}{1+x^2}\,dx$ (b) $\displaystyle\int_0^{\pi/4} \frac{\sin x}{\cos x}\,dx$

2. (a) Find the derivatives of $\sin(x^2+1)$ and $\sin(x^3+1)$.
   (b) Use your answer to part (a) to find antiderivatives of:

   (i) $x\cos(x^2+1)$ (ii) $x^2\cos(x^3+1)$

   (c) Find the general antiderivatives of:

   (i) $x\sin(x^2+1)$ (ii) $x^2\sin(x^3+1)$

Find the integrals in Exercises 3–44. Check your answers by differentiation.

3. $\displaystyle\int e^{3x}\,dx$

4. $\displaystyle\int e^{-x}\,dx$

5. $\displaystyle\int 25e^{-0.2t}\,dt$

6. $\displaystyle\int t\cos(t^2)\,dt$

7. $\displaystyle\int \sin(2x)\,dx$

8. $\displaystyle\int \sin(3-t)\,dt$

9. $\displaystyle\int xe^{-x^2}\,dx$

10. $\displaystyle\int y(y^2+5)^8\,dy$

11. $\displaystyle\int t^2(t^3-3)^{10}\,dt$

12. $\displaystyle\int x^2(1+2x^3)^2\,dx$

13. $\displaystyle\int x(x^2+3)^2\,dx$

14. $\displaystyle\int x(x^2-4)^{7/2}\,dx$

15. $\displaystyle\int y^2(1+y)^2\,dy$

16. $\displaystyle\int (2t-7)^{73}\,dt$

17. $\displaystyle\int \frac{dy}{y+5}$

18. $\displaystyle\int \frac{1}{\sqrt{4-x}}\,dx$

19. $\displaystyle\int (x^2+3)^2\,dx$

20. $\displaystyle\int x^2 e^{x^3+1}\,dx$

21. $\displaystyle\int \sin\theta(\cos\theta+5)^7\,d\theta$

22. $\displaystyle\int \sqrt{\cos 3t}\sin 3t\,dt$

23. $\displaystyle\int \sin^6\theta\cos\theta\,d\theta$

24. $\displaystyle\int \sin^3\alpha\cos\alpha\,d\alpha$

25. $\displaystyle\int \sin^6(5\theta)\cos(5\theta)\,d\theta$

26. $\displaystyle\int \tan(2x)\,dx$

27. $\int \dfrac{(\ln z)^2}{z}\, dz$

28. $\int \dfrac{e^t + 1}{e^t + t}\, dt$

29. $\int \dfrac{y}{y^2 + 4}\, dy$

30. $\int \dfrac{\cos \sqrt{x}}{\sqrt{x}}\, dx$

31. $\int \dfrac{e^{\sqrt{y}}}{\sqrt{y}}\, dy$

32. $\int \dfrac{1 + e^x}{\sqrt{x + e^x}}\, dx$

33. $\int \dfrac{e^x}{2 + e^x}\, dx$

34. $\int \dfrac{x + 1}{x^2 + 2x + 19}\, dx$

35. $\int \dfrac{t}{1 + 3t^2}\, dt$

36. $\int \dfrac{e^x - e^{-x}}{e^x + e^{-x}}\, dx$

37. $\int \dfrac{(t + 1)^2}{t^2}\, dt$

38. $\int \dfrac{x \cos(x^2)}{\sqrt{\sin(x^2)}}\, dx$

39. $\int \cosh x\, dx$

40. $\int \sinh 3t\, dt$

41. $\int (\sinh z) e^{\cosh z}\, dz$

42. $\int \cosh(2w + 1)\, dw$

43. $\int x \cosh x^2\, dx$

44. $\int \cosh^2 x \sinh x\, dx$

For the functions in Exercises 45–52, find the general antiderivative. Check your answers by differentiation.

45. $p(t) = \pi t^3 + 4t$

46. $f(x) = \sin 3x$

47. $f(x) = 2x \cos(x^2)$

48. $r(t) = 12t^2 \cos(t^3)$

49. $f(x) = \sin(2 - 5x)$

50. $f(x) = e^{\sin x} \cos x$

51. $f(x) = \dfrac{x}{x^2 + 1}$

52. $f(x) = \dfrac{1}{3\cos^2(2x)}$

For Exercises 53–60, use the Fundamental Theorem to calculate the definite integrals.

53. $\int_0^\pi \cos(x + \pi)\, dx$

54. $\int_0^{1/2} \cos(\pi x)\, dx$

55. $\int_0^{\pi/2} e^{-\cos \theta} \sin \theta\, d\theta$

56. $\int_1^2 2x e^{x^2}\, dx$

57. $\int_1^8 \dfrac{e^{\sqrt[3]{x}}}{\sqrt[3]{x^2}}\, dx$

58. $\int_{-1}^{e-2} \dfrac{1}{t + 2}\, dt$

59. $\int_1^4 \dfrac{\cos \sqrt{x}}{\sqrt{x}}\, dx$

60. $\int_0^2 \dfrac{x}{(1 + x^2)^2}\, dx$

For Exercises 61–66, evaluate the definite integrals. Whenever possible, use the Fundamental Theorem of Calculus, perhaps after a substitution. Otherwise, use numerical methods.

61. $\int_{-1}^3 (x^3 + 5x)\, dx$

62. $\int_{-1}^1 \dfrac{1}{1 + y^2}\, dy$

63. $\int_1^3 \dfrac{1}{x}\, dx$

64. $\int_1^3 \dfrac{dt}{(t + 7)^2}$

65. $\int_{-1}^2 \sqrt{x + 2}\, dx$

66. $\int_1^2 \dfrac{\sin t}{t}\, dt$

Find the integrals in Exercises 67–74.

67. $\int y \sqrt{y + 1}\, dy$

68. $\int z(z + 1)^{1/3}\, dz$

69. $\int \dfrac{t^2 + t}{\sqrt{t + 1}}\, dt$

70. $\int \dfrac{dx}{2 + 2\sqrt{x}}$

71. $\int x^2 \sqrt{x - 2}\, dx$

72. $\int (z + 2)\sqrt{1 - z}\, dz$

73. $\int \dfrac{t}{\sqrt{t + 1}}\, dt$

74. $\int \dfrac{3x - 2}{\sqrt{2x + 1}}\, dx$

Problems

75. If appropriate, evaluate the following integrals by substitution. If substitution is not appropriate, say so, and do not evaluate.

(a) $\int x \sin(x^2)\, dx$

(b) $\int x^2 \sin x\, dx$

(c) $\int \dfrac{x^2}{1 + x^2}\, dx$

(d) $\int \dfrac{x}{(1 + x^2)^2}\, dx$

(e) $\int x^3 e^{x^2}\, dx$

(f) $\int \dfrac{\sin x}{2 + \cos x}\, dx$

76. Suppose $\int_0^2 g(t)\, dt = 5$. Calculate the following:

(a) $\int_0^4 g(t/2)\, dt$

(b) $\int_0^2 g(2 - t)\, dt$

77. Suppose $\int_0^1 f(t)\, dt = 3$. Calculate the following:

(a) $\int_0^{0.5} f(2t)\, dt$

(b) $\int_0^1 f(1 - t)\, dt$

(c) $\int_1^{1.5} f(3 - 2t)\, dt$

78. (a) Calculate exactly: $\int_{-\pi}^{\pi} \cos^2 \theta \sin \theta\, d\theta$.

(b) Calculate the exact area under the curve $y = \cos^2 \theta \sin \theta$ between $\theta = 0$ and $\theta = \pi$.

79. Use the Fundamental Theorem to find the area under the graph of $f(x) = 1/(x + 1)$ between $x = 0$ and $x = 2$.

80. Find the exact area under the graph of $f(x) = x e^{x^2}$ between $x = 0$ and $x = 2$.

81. Find the exact value of the area under one arch of the curve $V(t) = V_0 \sin(\omega t)$, where $V_0 > 0$ and $\omega > 0$.

82. Find the exact average value of $f(x) = 1/(x + 1)$ on the interval $x = 0$ to $x = 2$. Sketch a graph showing the function and the average value.

83. Find $\int 4x(x^2 + 1)\, dx$ using two methods:

(a) Do the multiplication first, and then antidifferentiate.
(b) Use the substitution $w = x^2 + 1$.
(c) Explain how the expressions from parts (a) and (b) are different. Are they both correct?

84. (a) Find $\int \sin\theta \cos\theta\, d\theta$.

(b) You probably solved part (a) by making the substitution $w = \sin\theta$ or $w = \cos\theta$. (If not, go back and do it that way.) Now find $\int \sin\theta \cos\theta\, d\theta$ by making the *other* substitution.

(c) There is yet another way of finding this integral which involves the trigonometric identities

$$\sin(2\theta) = 2\sin\theta\cos\theta$$
$$\cos(2\theta) = \cos^2\theta - \sin^2\theta.$$

Find $\int \sin\theta \cos\theta\, d\theta$ using one of these identities and then the substitution $w = 2\theta$.

(d) You should now have three different expressions for the indefinite integral $\int \sin\theta \cos\theta\, d\theta$. Are they really different? Are they all correct? Explain.

85. Let $I_{m,n} = \int_0^1 x^m(1-x)^n\, dx$ for constant m, n. Show that $I_{m,n} = I_{n,m}$.

86. If t is in years since 1990, the population, P, of the world in billions can be modeled by $P = 5.3e^{0.014t}$.

(a) What does this model give for the world population in 1990? In 2000?
(b) Use the Fundamental Theorem to find the average population of the world during the 1990s.

87. Oil is leaking out of a ruptured tanker at the rate of $r(t) = 50e^{-0.02t}$ thousand liters per minute.

(a) At what rate, in liters per minute, is oil leaking out at $t = 0$? At $t = 60$?
(b) How many liters leak out during the first hour?

88. Throughout much of the 20th century, the yearly consumption of electricity in the US increased exponentially at a continuous rate of 7% per year. Assume this trend continues and that the electrical energy consumed in 1900 was 1.4 million megawatt-hours.

(a) Write an expression for yearly electricity consumption as a function of time, t, in years since 1900.
(b) Find the average yearly electrical consumption throughout the 20th century.

(c) During what year was electrical consumption closest to the average for the century?
(d) Without doing the calculation for part (c), how could you have predicted which half of the century the answer would be in?

89. If we assume that wind resistance is proportional to velocity, then the downward velocity, v, of a body of mass m falling vertically is given by

$$v = \frac{mg}{k}\left(1 - e^{-kt/m}\right),$$

where g is the acceleration due to gravity and k is a constant. Find the height, h, above the surface of the earth as a function of time. Assume the body starts at height h_0.

90. If we assume that wind resistance is proportional to the square of velocity, then the downward velocity, v, of a falling body is given by

$$v = \sqrt{\frac{g}{k}}\left(\frac{e^{t\sqrt{gk}} - e^{-t\sqrt{gk}}}{e^{t\sqrt{gk}} + e^{-t\sqrt{gk}}}\right).$$

Use the substitution $w = e^{t\sqrt{gk}} + e^{-t\sqrt{gk}}$ to find the height, h, of the body above the surface of the earth as a function of time. Assume the body starts at a height h_0.

91. (a) During the 1970s, ACME Widgets sold at a continuous rate of $R = R_0e^{0.15t}$ widgets per year, where t is time in years since January 1, 1970. Suppose they were selling widgets at a rate of 1000 per year on the first day of the decade. How many widgets did they sell during the decade? How many did they sell if the rate on January 1, 1970 was 150,000,000 widgets per year?

(b) In the first case above (1000 widgets per year on January 1, 1970), how long did it take for half the widgets in the 1970s to be sold? In the second case (150,000,000 widgets per year on January 1, 1970), when had half the widgets in the 1970s been sold?

(c) In 1980 ACME began an advertising campaign claiming that half the widgets it had sold in the previous ten years were still in use. Based on your answer to part (b), about how long must a widget last in order to justify this claim?

7.2 INTEGRATION BY PARTS

The method of substitution reverses the chain rule. Now we introduce *integration by parts*, which is based on the product rule.

Example 1 Find $\int xe^x\, dx$.

Solution We are looking for a function whose derivative is xe^x. The product rule might lead us to guess xe^x, because we know that the derivative has two terms, one of which is xe^x:

$$\frac{d}{dx}(xe^x) = \frac{d}{dx}(x)e^x + x\frac{d}{dx}(e^x) = e^x + xe^x.$$

Of course, our guess is wrong because of the extra e^x. But we can adjust our guess by subtracting e^x; this leads us to try $xe^x - e^x$. Let's check it:

$$\frac{d}{dx}(xe^x - e^x) = \frac{d}{dx}(xe^x) - \frac{d}{dx}(e^x) = e^x + xe^x - e^x = xe^x.$$

It works, so $\int xe^x \, dx = xe^x - e^x + C$.

Example 2   Find $\int \theta \cos \theta \, d\theta$.

Solution   We guess the antiderivative is $\theta \sin \theta$ and use the product rule to check:

$$\frac{d}{d\theta}(\theta \sin \theta) = \frac{d(\theta)}{d\theta} \sin \theta + \theta \frac{d}{d\theta}(\sin \theta) = \sin \theta + \theta \cos \theta.$$

To correct for the extra $\sin \theta$ term, we must subtract from our original guess something whose derivative is $\sin \theta$. Since $\frac{d}{d\theta}(\cos \theta) = -\sin \theta$, we try:

$$\frac{d}{d\theta}(\theta \sin \theta + \cos \theta) = \frac{d}{d\theta}(\theta \sin \theta) + \frac{d}{d\theta}(\cos \theta) = \sin \theta + \theta \cos \theta - \sin \theta = \theta \cos \theta.$$

Thus, $\int \theta \cos \theta \, d\theta = \theta \sin \theta + \cos \theta + C$.

The General Formula for Integration by Parts

We can formalize the process illustrated in the last two examples in the following way. We begin with the product rule:

$$\frac{d}{dx}(uv) = u'v + uv'$$

where u and v are functions of x with derivatives u' and v', respectively. We rewrite this as:

$$uv' = \frac{d}{dx}(uv) - u'v$$

and then integrate both sides:

$$\int uv' \, dx = \int \frac{d}{dx}(uv) \, dx - \int u'v \, dx.$$

Since an antiderivative of $\frac{d}{dx}(uv)$ is just uv, we get the following formula:

Integration by Parts

$$\int uv' \, dx = uv - \int u'v \, dx.$$

This formula is useful when the integrand can be viewed as a product and when the integral on the right-hand side is simpler than that on the left. In effect, we were using integration by parts in the previous two examples. In Example 1, we let $xe^x = (x) \cdot (e^x) = uv'$, and choose $u = x$ and $v' = e^x$. Thus, $u' = 1$ and $v = e^x$, so

$$\int \underbrace{(x)}_{u} \underbrace{(e^x)}_{v'} \, dx = \underbrace{(x)}_{u} \underbrace{(e^x)}_{v} - \int \underbrace{(1)}_{u'} \underbrace{(e^x)}_{v} \, dx = xe^x - e^x + C.$$

So uv represents our first guess, and $\int u'v \, dx$ represents the correction to our guess.

Notice what would have happened if we took $v = e^x + C_1$. Then

$$\int xe^x \, dx = x(e^x + C_1) - \int (e^x + C_1) \, dx$$
$$= xe^x + C_1 x - e^x - C_1 x + C$$
$$= xe^x - e^x + C,$$

as before. Thus, it is not necessary to include an arbitrary constant in the antiderivative for v; any antiderivative will do.

What would have happened if we had picked u and v' the other way around? If $u = e^x$ and $v' = x$, then $u' = e^x$ and $v = x^2/2$. The formula for integration by parts then gives

$$\int xe^x \, dx = \frac{x^2}{2} e^x - \int \frac{x^2}{2} \cdot e^x \, dx,$$

which is true but not helpful since the integral on the right is worse than the one on the left. To use this method, we must choose u and v' to make the integral on the right easier to find than the integral on the left.

How to Choose u and v'

- Whatever you let v' be, you need to be able to find v.
- It helps if u' is simpler than u (or at least no more complicated than u).
- It helps if v is simpler than v' (or at least no more complicated than v').

If we pick $v' = x$ in Example 1, then $v = x^2/2$, which is certainly "worse" than v'.

There are some examples which don't look like good candidates for integration by parts because they don't appear to involve products, but for which the method works well. Such examples often involve $\ln x$ or the inverse trigonometric functions. Here is one:

Example 3   Find $\displaystyle\int_2^3 \ln x \, dx$.

Solution   This does not look like a product unless we write $\ln x = (1)(\ln x)$. Then we might say $u = 1$ so $u' = 0$, which certainly makes things simpler. But if $v' = \ln x$, what is v? If we knew, we would not need integration by parts. Let's try the other way: if $u = \ln x$, $u' = 1/x$ and if $v' = 1$, $v = x$, so

$$\int_2^3 \underbrace{(\ln x)}_{u} \underbrace{(1)}_{v'} \, dx = \underbrace{(\ln x)}_{u} \underbrace{(x)}_{v} \Big|_2^3 - \int_2^3 \underbrace{\left(\frac{1}{x}\right)}_{u'} \cdot \underbrace{(x)}_{v} \, dx$$

$$= x \ln x \Big|_2^3 - \int_2^3 1 \, dx = (x \ln x - x) \Big|_2^3$$

$$= 3 \ln 3 - 3 - 2 \ln 2 + 2 = 3 \ln 3 - 2 \ln 2 - 1.$$

Notice that when doing a definite integral by parts, we must remember to put the limits of integration (here 2 and 3) on the uv term (in this case $x \ln x$) as well as on the integral $\int u'v \, dx$.

Example 4   Find $\displaystyle\int x^6 \ln x \, dx$.

Solution   View $x^6 \ln x$ as uv' where $u = \ln x$ and $v' = x^6$. Then $v = \frac{1}{7}x^7$ and $u' = 1/x$, so integration by parts gives us:

$$\int x^6 \ln x \, dx = \int (\ln x)x^6 \, dx = (\ln x)\left(\frac{1}{7}x^7\right) - \int \frac{1}{7}x^7 \cdot \frac{1}{x} \, dx$$

$$= \frac{1}{7}x^7 \ln x - \frac{1}{7}\int x^6 \, dx$$

$$= \frac{1}{7}x^7 \ln x - \frac{1}{49}x^7 + C.$$

In Example 4 we did not choose $v' = \ln x$, because it is not immediately clear what v would be. In fact, we used integration by parts in Example 3 to find the antiderivative of $\ln x$. Also, using $u = \ln x$, as we have done, gives $u' = 1/x$, which can be considered simpler than $u = \ln x$. This shows that u does not have to be the first factor in the integrand (here x^6).

Example 5   Find $\int x^2 \sin 4x \, dx$.

Solution   If we let $v' = \sin 4x$, then $v = -\frac{1}{4}\cos 4x$, which is no worse than v'. Also letting $u = x^2$, we get $u' = 2x$, which is simpler than $u = x^2$. Using integration by parts:

$$\int x^2 \sin 4x \, dx = x^2\left(-\frac{1}{4}\cos 4x\right) - \int 2x\left(-\frac{1}{4}\cos 4x\right) dx$$

$$= -\frac{1}{4}x^2 \cos 4x + \frac{1}{2}\int x \cos 4x \, dx.$$

The trouble is we still have to grapple with $\int x \cos 4x \, dx$. This can be done by using integration by parts again with a new u and v, namely $u = x$ and $v' = \cos 4x$:

$$\int x \cos 4x \, dx = x\left(\frac{1}{4}\sin 4x\right) - \int 1 \cdot \frac{1}{4}\sin 4x \, dx$$

$$= \frac{1}{4}x \sin 4x - \frac{1}{4} \cdot \left(-\frac{1}{4}\cos 4x\right) + C$$

$$= \frac{1}{4}x \sin 4x + \frac{1}{16}\cos 4x + C.$$

Thus,

$$\int x^2 \sin 4x \, dx = -\frac{1}{4}x^2 \cos 4x + \frac{1}{2}\int x \cos 4x \, dx$$

$$= -\frac{1}{4}x^2 \cos 4x + \frac{1}{2}\left(\frac{1}{4}x \sin 4x + \frac{1}{16}\cos 4x + C\right)$$

$$= -\frac{1}{4}x^2 \cos 4x + \frac{1}{8}x \sin 4x + \frac{1}{32}\cos 4x + C.$$

Notice that, in this example, each time we used integration by parts, the exponent of x went down by 1. In addition, when the arbitrary constant C is multiplied by $\frac{1}{2}$, it is still represented by C.

Example 6   Find $\int \cos^2 \theta \, d\theta$.

Solution   Using integration by parts with $u = \cos \theta$, $v' = \cos \theta$ gives $u' = -\sin \theta$, $v = \sin \theta$, so we get

$$\int \cos^2 \theta \, d\theta = \cos \theta \sin \theta + \int \sin^2 \theta \, d\theta.$$

Substituting $\sin^2 \theta = 1 - \cos^2 \theta$ leads to

$$\int \cos^2 \theta \, d\theta = \cos \theta \sin \theta + \int (1 - \cos^2 \theta) \, d\theta$$

$$= \cos \theta \sin \theta + \int 1 \, d\theta - \int \cos^2 \theta \, d\theta.$$

Looking at the right side, we see that the original integral has reappeared. If we move it to the left, we get

$$2 \int \cos^2 \theta \, d\theta = \cos \theta \sin \theta + \int 1 \, d\theta = \cos \theta \sin \theta + \theta + C.$$

Dividing by 2 gives

$$\int \cos^2 \theta \, d\theta = \frac{1}{2} \cos \theta \sin \theta + \frac{1}{2} \theta + C.$$

Problem 41 asks you to do this integral by another method.

The previous example illustrates a useful technique : Use integration by parts to transform the integral into an expression containing another copy of the same integral, possibly multiplied by a coefficient, then solve for the original integral. It may be necessary to apply integration by parts twice, as the next example shows.

Example 7   Use integration by parts twice to find $\int e^{2x} \sin(3x) \, dx$.

Solution   Using integration by parts with $u = e^{2x}$ and $v' = \sin(3x)$ gives $u' = 2e^{2x}$, $v = -\frac{1}{3} \cos(3x)$, so we get

$$\int e^{2x} \sin(3x) \, dx = -\frac{1}{3} e^{2x} \cos(3x) + \frac{2}{3} \int e^{2x} \cos(3x) \, dx.$$

On the right side we have an integral similar to the original one, with the sine replaced by a cosine. Using integration by parts on that integral in the same way gives

$$\int e^{2x} \cos(3x) \, dx = \frac{1}{3} e^{2x} \sin(3x) - \frac{2}{3} \int e^{2x} \sin(3x) \, dx.$$

Substituting this into the expression we obtained for the original integral gives

$$\int e^{2x} \sin(3x) \, dx = -\frac{1}{3} e^{2x} \cos(3x) + \frac{2}{3} \left(\frac{1}{3} e^{2x} \sin(3x) - \frac{2}{3} \int e^{2x} \sin(3x) \, dx \right)$$

$$= -\frac{1}{3} e^{2x} \cos(3x) + \frac{2}{9} e^{2x} \sin(3x) - \frac{4}{9} \int e^{2x} \sin(3x) \, dx.$$

The right side now has a copy of the original integral, multiplied by $-4/9$. Moving it to the left, we get

$$\left(1 + \frac{4}{9}\right) \int e^{2x} \sin(3x) \, dx = -\frac{1}{3} e^{2x} \cos(3x) + \frac{2}{9} e^{2x} \sin(3x).$$

Dividing through by the coefficient on the left, $(1 + 4/9) = 13/9$, we get

$$\int e^{2x} \sin(3x) \, dx = \frac{9}{13} \left(-\frac{1}{3} e^{2x} \cos(3x) + \frac{2}{9} e^{2x} \sin(3x) \right)$$

$$= \frac{1}{13} e^{2x} \left(2 \sin(3x) - 3 \cos(3x) \right) + C.$$

Exercises and Problems for Section 7.2

Exercises

1. Write $\arctan x = 1 \cdot \arctan x$ to find $\int \arctan x \, dx$.

Find the integrals in Exercises 2–29.

2. $\int t \sin t \, dt$

3. $\int t^2 \sin t \, dt$

4. $\int te^{5t} \, dt$

5. $\int t^2 e^{5t} \, dt$

6. $\int pe^{-0.1p} \, dp$

7. $\int (z+1)e^{2z} \, dz$

8. $\int y \ln y \, dy$

9. $\int x^3 \ln x \, dx$

10. $\int q^5 \ln 5q \, dq$

11. $\int \theta^2 \cos 3\theta \, d\theta$

12. $\int \sin^2 \theta \, d\theta$

13. $\int \cos^2(3\alpha + 1) \, d\alpha$

14. $\int (\ln t)^2 \, dt$

15. $\int y\sqrt{y+3} \, dy$

16. $\int (t+2)\sqrt{2+3t} \, dt$

17. $\int (\theta+1)\sin(\theta+1) \, d\theta$

18. $\int \dfrac{z}{e^z} \, dz$

19. $\int \dfrac{\ln x}{x^2} \, dx$

20. $\int \dfrac{y}{\sqrt{5-y}} \, dy$

21. $\int \dfrac{t+7}{\sqrt{5-t}} \, dt$

22. $\int x(\ln x)^4 \, dx$

23. $\int \arcsin w \, dw$

24. $\int \arctan 7z \, dz$

25. $\int x \arctan x^2 \, dx$

26. $\int x^3 e^{x^2} \, dx$

27. $\int x^5 \cos x^3 \, dx$

28. $\int x \sinh x \, dx$

29. $\int (x-1)\cosh x \, dx$

Evaluate the integrals in Exercises 30–37 both exactly [e.g. $\ln(3\pi)$] and numerically [e.g. $\ln(3\pi) \approx 2.243$]:

30. $\int_1^5 \ln t \, dt$

31. $\int_3^5 x \cos x \, dx$

32. $\int_0^{10} ze^{-z} \, dz$

33. $\int_1^3 t \ln t \, dt$

34. $\int_0^1 \arctan y \, dy$

35. $\int_0^5 \ln(1+t) \, dt$

36. $\int_0^1 \arcsin z \, dz$

37. $\int_0^1 u \arcsin u^2 \, du$

Problems

38. For each of the following integrals, indicate whether integration by substitution or integration by parts is more appropriate. Do not evaluate the integrals.

(a) $\int x \sin x \, dx$

(b) $\int \dfrac{x^2}{1+x^3} \, dx$

(c) $\int xe^{x^2} \, dx$

(d) $\int x^2 \cos(x^3) \, dx$

(e) $\int \dfrac{1}{\sqrt{3x+1}} \, dx$

(f) $\int x^2 \sin x \, dx$

(g) $\int \ln x \, dx$

39. Find the exact value of the area under the first arch of $f(x) = x \sin x$.

40. In Exercise 12, you evaluated $\int \sin^2 \theta \, d\theta$ using integration by parts. (If you did not do it by parts, do so now!) Redo this integral using the identity $\sin^2 \theta = (1 - \cos 2\theta)/2$. Explain any differences in the form of the answer obtained by the two methods.

41. Compute $\int \cos^2 \theta \, d\theta$ in two different ways and explain any differences in the form of your answers. (The identity $\cos^2 \theta = (1 + \cos 2\theta)/2$ may be useful.)

42. Use integration by parts twice to find $\int e^x \sin x \, dx$.

43. Use integration by parts twice to find $\int e^\theta \cos \theta \, d\theta$.

44. Use the results from Problems 42 and 43 and integration by parts to find $\int xe^x \sin x \, dx$.

45. Use the results from Problems 42 and 43 and integration by parts to find $\int \theta e^\theta \cos \theta \, d\theta$.

In Problems 46–49, derive the given formulas.

46. $\int x^n e^x \, dx = x^n e^x - n \int x^{n-1} e^x \, dx$

47. $\int x^n \cos ax \, dx = \dfrac{1}{a}x^n \sin ax - \dfrac{n}{a} \int x^{n-1} \sin ax \, dx$

48. $\int x^n \sin ax \, dx = -\dfrac{1}{a}x^n \cos ax + \dfrac{n}{a} \int x^{n-1} \cos ax \, dx$

49. $\int \cos^n x \, dx = \dfrac{1}{n}\cos^{n-1} x \sin x + \dfrac{n-1}{n} \int \cos^{n-2} x \, dx$

50. Integrating $e^{ax} \sin bx$ by parts twice yields a result of the form

$$\int e^{ax} \sin bx \, dx = e^{ax} (A \sin bx + B \cos bx) + C.$$

(a) Find the constants A and B in terms of a and b. [Hint: Don't actually perform the integration by parts.]

(b) Evaluate $\int e^{ax} \cos bx \, dx$ by modifying the result in part (a). [Again, it is not necessary to perform integration by parts, as the result is of the same form as that in part (a).]

51. Estimate $\int_0^{10} f(x) g'(x) \, dx$ if $f(x) = x^2$ and g has the values in the following table.

x	0	2	4	6	8	10
$g(x)$	2.3	3.1	4.1	5.5	5.9	6.1

52. Let f be twice differentiable with $f(0) = 6$, $f(1) = 5$, and $f'(1) = 2$. Evaluate the integral $\int_0^1 x f''(x) dx$.

53. Let $F(a)$ be the area under the graph of $y = x^2 e^{-x}$ between $x = 0$ and $x = a$, for $a > 0$.

(a) Find a formula for $F(a)$.
(b) Is F an increasing or decreasing function?
(c) Is F concave up or concave down for $0 < a < 2$?

54. The concentration, C, in ng/ml, of a drug in the blood as a function of the time, t, in hours since the drug was administered is given by $C = 15te^{-0.2t}$. The area under the concentration curve is a measure of the overall effect of the drug on the body, called the bioavailability. Find the bioavailability of the drug between $t = 0$ and $t = 3$.

55. The voltage, V, in an electric circuit is given as a function of time, t, by

$$V = V_0 \cos(\omega t + \phi).$$

Suppose each of the positive constants, V_0, ω, ϕ is increased (while the other two are held constant). What is the effect of each increase on the following quantities:

(a) The maximum value of V?
(b) The maximum value of dV/dt?
(c) The average value of V^2 over one period?

56. During a surge in the demand for electricity, the rate, r, at which energy is used can be approximated by

$$r = te^{-at},$$

where t is the time in hours and a is a positive constant.

(a) Find the total energy, E, used in the first T hours. Give your answer as a function of a.
(b) What happens to E as $T \to \infty$?

57. In describing the behavior of an electron, we use wave functions $\Psi_1, \Psi_2, \Psi_3, \ldots$ of the form

$$\Psi_n(x) = C_n \sin(n\pi x) \qquad \text{for } n = 1, 2, 3, \ldots$$

where x is the distance from a fixed point and C_n is a positive constant.

(a) Find C_1 so that Ψ_1 satisfies

$$\int_0^1 (\Psi_1(x))^2 \, dx = 1.$$

This is called normalizing the wave function Ψ_1.
(b) For any integer n, find C_n so that Ψ_n is normalized.

7.3 TABLES OF INTEGRALS

Since so few functions have elementary antiderivatives, they have been compiled in a list called a table of integrals.[1] A short table of indefinite integrals is given inside the back cover of this book. The key to using these tables is being able to recognize the general class of function that you are trying to integrate, so you can know in what section of the table to look.

Warning: This section involves long division of polynomials and completing the square. You may want to review these topics!

Using the Table of Integrals

Part I of the table inside the back cover gives the antiderivatives of the basic functions x^n, a^x, $\ln x$, $\sin x$, $\cos x$, and $\tan x$. (The antiderivative for $\ln x$ is found using integration by parts and is a special case of the more general formula III-13.) Most of these are already familiar.

Part II of the table contains antiderivatives of functions involving products of e^x, $\sin x$, and $\cos x$. All of these antiderivatives were obtained using integration by parts.

---

[1] See, for example, *CRC Standard Mathematical Tables* (Boca Raton, Fl: CRC Press). Many computer programs and calculators can compute antiderivatives as well.

Example 1   Find $\int \sin 7z \sin 3z \, dz$.

Solution   Since the integrand is the product of two sines, we should use II-10 in the table,

$$\int \sin 7z \sin 3z \, dz = -\frac{1}{40}(7 \cos 7z \sin 3z - 3 \cos 3z \sin 7z) + C.$$

Part III of the table contains antiderivatives for products of a polynomial and e^x, $\sin x$, or $\cos x$. It also has an antiderivative for $x^n \ln x$, which can easily be used to find the antiderivatives of the product of a general polynomial and $\ln x$. Each *reduction formula* is used repeatedly to reduce the degree of the polynomial until a zero degree polynomial is obtained.

Example 2   Find $\int (x^5 + 2x^3 - 8)e^{3x} \, dx$.

Solution   Since $p(x) = x^5 + 2x^3 - 8$ is a polynomial multiplied by e^{3x}, this is of the form in III-14. Now $p'(x) = 5x^4 + 6x^2$ and $p''(x) = 20x^3 + 12x$, and so on, giving

$$\int (x^5 + 2x^3 - 8)e^{3x} \, dx = e^{3x} \left[\frac{1}{3}(x^5 + 2x^3 - 8) - \frac{1}{9}(5x^4 + 6x^2) + \frac{1}{27}(20x^3 + 12x) \right.$$
$$\left. -\frac{1}{81}(60x^2 + 12) + \frac{1}{243}(120x) - \frac{1}{729} \cdot 120 \right] + C.$$

Here we have the successive derivatives of the original polynomial $x^5 + 2x^3 - 8$, occurring with alternating signs and multiplied by successive powers of $1/3$.

Part IV of the table contains reduction formulas for the antiderivatives of $\cos^n x$ and $\sin^n x$, which can be obtained by integration by parts. When n is a positive integer, formulas IV-17 and IV-18 can be used repeatedly to reduce the power n until it is 0 or 1.

Example 3   Find $\int \sin^6 \theta \, d\theta$.

Solution   Use IV-17 repeatedly:

$$\int \sin^6 \theta \, d\theta = -\frac{1}{6}\sin^5 \theta \cos \theta + \frac{5}{6}\int \sin^4 \theta \, d\theta$$
$$\int \sin^4 \theta \, d\theta = -\frac{1}{4}\sin^3 \theta \cos \theta + \frac{3}{4}\int \sin^2 \theta \, d\theta$$
$$\int \sin^2 \theta \, d\theta = -\frac{1}{2}\sin \theta \cos \theta + \frac{1}{2}\int 1 \, d\theta.$$

Calculate $\int \sin^2 \theta \, d\theta$ first, and use this to find $\int \sin^4 \theta \, d\theta$; then calculate $\int \sin^6 \theta \, d\theta$. Putting this all together, we get

$$\int \sin^6 \theta \, d\theta = -\frac{1}{6}\sin^5 \theta \cos \theta - \frac{5}{24}\sin^3 \theta \cos \theta - \frac{15}{48}\sin \theta \cos \theta + \frac{15}{48}\theta + C.$$

The last item in **Part IV** of the table is not a formula: it is advice on how to antidifferentiate products of integer powers of $\sin x$ and $\cos x$. There are various techniques to choose from, depending on the nature (odd or even, positive or negative) of the exponents.

Example 4   Find $\int \cos^3 t \sin^4 t\, dt$.

Solution   Here the exponent of $\cos t$ is odd, so IV-23 recommends making the substitution $w = \sin t$. Then $dw = \cos t\, dt$. To make this work, we'll have to separate off one of the cosines to be part of dw. Also, the remaining even power of $\cos t$ can be rewritten in terms of $\sin t$ by using $\cos^2 t = 1 - \sin^2 t = 1 - w^2$, so that

$$\int \cos^3 t \sin^4 t\, dt = \int \cos^2 t \sin^4 t \cos t\, dt$$

$$= \int (1 - w^2) w^4\, dw = \int (w^4 - w^6)\, dw$$

$$= \frac{1}{5} w^5 - \frac{1}{7} w^7 + C = \frac{1}{5} \sin^5 t - \frac{1}{7} \sin^7 t + C.$$

Example 5   Find $\int \cos^2 x \sin^4 x\, dx$.

Solution   In this example, both exponents are even. The advice given in IV-23 is to convert to all sines or all cosines. We'll convert to all sines by substituting $\cos^2 x = 1 - \sin^2 x$, and then we'll multiply out the integrand:

$$\int \cos^2 x \sin^4 x\, dx = \int (1 - \sin^2 x) \sin^4 x\, dx = \int \sin^4 x\, dx - \int \sin^6 x\, dx.$$

In Example 3 we found $\int \sin^4 x\, dx$ and $\int \sin^6 x\, dx$. Put them together to get

$$\int \cos^2 x \sin^4 x\, dx = -\frac{1}{4} \sin^3 x \cos x - \frac{3}{8} \sin x \cos x + \frac{3}{8} x$$

$$- \left(-\frac{1}{6} \sin^5 x \cos x - \frac{5}{24} \sin^3 x \cos x - \frac{15}{48} \sin x \cos x + \frac{15}{48} x \right) + C$$

$$= \frac{1}{6} \sin^5 x \cos x - \frac{1}{24} \sin^3 x \cos x - \frac{3}{48} \sin x \cos x + \frac{3}{48} x + C.$$

The last two parts of the table are concerned with quadratic functions: **Part V** has expressions with quadratic denominators; **Part VI** contains square roots of quadratics. The quadratics that appear in these formulas are of the form $x^2 \pm a^2$ or $a^2 - x^2$, or in factored form $(x - a)(x - b)$, where a and b are different constants. Quadratics can be converted to these forms by factoring or completing the square.

Preparing to Use the Table: Transforming the Integrand

To use the integral table, we often need to manipulate or reshape integrands to fit entries in the table. The manipulations that tend to be useful are factoring, long division, completing the square, and substitution.

Using Factoring

Example 6   Find $\int \dfrac{3x + 7}{x^2 + 6x + 8}\, dx$.

Solution   In this case we factor the denominator to get it into a form in the table:

$$x^2 + 6x + 8 = (x + 2)(x + 4).$$

Now in V-27 we let $a = -2$, $b = -4$, $c = 3$, and $d = 7$, to obtain

$$\int \frac{3x + 7}{x^2 + 6x + 8}\, dx = \frac{1}{2} (\ln |x + 2| - (-5) \ln |x + 4|) + C.$$

Long Division

Example 7   Find $\int \dfrac{x^2}{x^2 + 4}\, dx$.

Solution   A good rule of thumb when integrating a rational function whose numerator has a degree greater than or equal to that of the denominator is to start by doing *long division*. This results in a polynomial plus a simpler rational function as a remainder. Performing long division here, we obtain:

$$\frac{x^2}{x^2 + 4} = 1 - \frac{4}{x^2 + 4}.$$

Then, by V-24 with $a = 2$, we obtain:

$$\int \frac{x^2}{x^2 + 4}\, dx = \int 1\, dx - 4 \int \frac{1}{x^2 + 4}\, dx = x - 4 \cdot \frac{1}{2} \arctan \frac{x}{2} + C = x - 2 \arctan \frac{x}{2} + C.$$

Completing the Square to Rewrite the Quadratic in the Form $w^2 + a^2$

Example 8   Find $\int \dfrac{1}{x^2 + 6x + 14}\, dx$.

Solution   By completing the square, we can get this integrand into a form in the table:

$$x^2 + 6x + 14 = (x^2 + 6x + 9) - 9 + 14$$
$$= (x + 3)^2 + 5.$$

Let $w = x + 3$. Then $dw = dx$ and so the substitution gives

$$\int \frac{1}{x^2 + 6x + 14}\, dx = \int \frac{1}{w^2 + 5}\, dw = \frac{1}{\sqrt{5}} \arctan \frac{w}{\sqrt{5}} + C = \frac{1}{\sqrt{5}} \arctan \frac{x + 3}{\sqrt{5}} + C,$$

where the antidifferentiation uses V-24 with $a^2 = 5$.

Substitution

Example 9   Find $\int e^t \sin(5t + 7)\, dt$.

Solution   This looks similar to II-8. To make the correspondence more complete, let's try the substitution $w = 5t + 7$. Then $dw = 5\, dt$, so $dt = \frac{1}{5}\, dw$. Also, $t = (w - 7)/5$. Then the integral becomes

$$\int e^t \sin(5t + 7)\, dt = \int e^{(w-7)/5} \sin w\, \frac{dw}{5}$$

$$= \frac{e^{-7/5}}{5} \int e^{w/5} \sin w\, dw. \qquad \text{\footnotesize (Since } e^{(w-7)/5} = e^{w/5} e^{-7/5} \text{ and } e^{-7/5} \text{ is a constant)}$$

Now we can use II-8 with $a = \frac{1}{5}$ and $b = 1$ to write

$$\int e^{w/5} \sin w\, dw = \frac{1}{(\frac{1}{5})^2 + 1^2} e^{w/5} \left(\frac{\sin w}{5} - \cos w \right) + C,$$

so

$$\int e^t \sin(5t + 7)\, dt = \frac{e^{-7/5}}{5} \left(\frac{25}{26} e^{(5t+7)/5} \left(\frac{\sin(5t + 7)}{5} - \cos(5t + 7) \right) \right) + C$$

$$= \frac{5 e^t}{26} \left(\frac{\sin(5t + 7)}{5} - \cos(5t + 7) \right) + C.$$

Exercises and Problems for Section 7.3

Exercises

For Exercises 1–40, antidifferentiate using the table of integrals. You may need to transform the integrand first.

1. $\int e^{-3\theta} \cos\theta \, d\theta$

2. $\int x^5 \ln x \, dx$

3. $\int x^3 \sin 5x \, dx.$

4. $\int (x^2 + 3) \ln x \, dx.$

5. $\int (x^3 + 5)^2 \, dx.$

6. $\int \sin w \cos^4 w \, dw$

7. $\int \sin^4 x \, dx$

8. $\int \frac{1}{3 + y^2} \, dy$

9. $\int x^3 e^{2x} \, dx$

10. $\int \frac{dx}{9x^2 + 16}$

11. $\int \frac{dx}{\sqrt{25 - 16x^2}}$

12. $\int \frac{dx}{\sqrt{9x^2 + 25}}$

13. $\int \sin 3\theta \cos 5\theta \, d\theta$

14. $\int \sin 3\theta \sin 5\theta \, d\theta$

15. $\int x^2 e^{3x} \, dx$

16. $\int x^2 e^{x^3} \, dx$

17. $\int x^4 e^{3x} \, dx$

18. $\int u^5 \ln(5u) \, du$

19. $\int \frac{1}{\cos^3 x} \, dx$

20. $\int \frac{t^2 + 1}{t^2 - 1} \, dt$

21. $\int x^3 \sin x^2 \, dx$

22. $\int \cos 2y \cos 7y \, dy$

23. $\int y^2 \sin 2y \, dy$

24. $\int e^{5x} \sin 3x \, dx$

25. $\int \frac{1}{\cos^5 x} \, dx.$

26. $\int \frac{1}{\sin^2 2\theta} \, d\theta$

27. $\int \frac{1}{\sin^3 3\theta} \, d\theta$

28. $\int \frac{1}{\cos^4 7x} \, dx$

29. $\int \frac{1}{x^2 + 4x + 3} \, dx$

30. $\int \tan^4 x \, dx$

31. $\int \frac{dz}{z(z - 3)}$

32. $\int \frac{dy}{4 - y^2}$

33. $\int \frac{1}{1 + (z + 2)^2} \, dz$

34. $\int \frac{1}{y^2 + 4y + 5} \, dy$

35. $\int \frac{1}{x^2 + 4x + 4} \, dx$

36. $\int \sin^3 x \, dx$

37. $\int \sin^3 3\theta \cos^2 3\theta \, d\theta$

38. $\int z e^{2z^2} \cos(2z^2) \, dz$

39. $\int \sinh^3 x \cosh^2 x \, dx$

40. $\int \sinh^2 x \cosh^3 x \, dx$

For Problems 41–44, evaluate the definite integrals. Whenever possible, use the Fundamental Theorem of Calculus, perhaps after a substitution. Otherwise, use numerical methods.

41. $\int_0^{\pi/12} \sin(3\alpha) \, d\alpha$

42. $\int_0^1 \frac{1}{x^2 + 2x + 1} \, dx$

43. $\int_0^1 \frac{(x + 2)}{(x + 2)^2 + 1} \, dx$

44. $\int_0^{1/\sqrt{2}} \frac{x}{\sqrt{1 - x^4}} \, dx$

Problems

45. Show that for all integers m and n, with $m \neq \pm n$, $\int_{-\pi}^{\pi} \sin m\theta \sin n\theta \, d\theta = 0$.

46. Show that for all integers m and n, with $m \neq \pm n$, $\int_{-\pi}^{\pi} \cos m\theta \cos n\theta \, d\theta = 0$.

47. The voltage, V, in an electrical outlet is given as a function of time, t, by the function $V = V_0 \cos(120\pi t)$, where V is in volts and t is in seconds, and V_0 is a positive constant representing the maximum voltage.

  (a) What is the average value of the voltage over 1 second?

  (b) Engineers do not use the average voltage. They use the root mean square voltage defined by $\overline{V} = \sqrt{\text{average of } (V^2)}$. Find $\overline{V}$ in terms of V_0. (Take the average over 1 second.)

  (c) The standard voltage in an American house is 110 volts, meaning that $\overline{V} = 110$. What is V_0?

48. For some constants A and B, the rate of production, $R(t)$, of oil in a new oil well is modeled by:
$$R(t) = A + Be^{-t} \sin(2\pi t)$$

where t is the time in years, A is the equilibrium rate, and B is the "variability" coefficient.

  (a) Find the total amount of oil produced in the first N years of operation. (Take N to be an integer.)

  (b) Find the average amount of oil produced per year over the first N years (where N is an integer).

  (c) From your answer to part (b), find the average amount of oil produced per year as $N \to \infty$.

  (d) Looking at the function $R(t)$, explain how you might have predicted your answer to part (c) without doing any calculations.

  (e) Do you think it is reasonable to expect this model to hold over a very long period? Why or why not?

49. For a positive integer n, let $\Psi_n(x) = C_n \sin(n\pi x)$ be the wave function used in describing the behavior of an electron. If n and m are different positive integers, find
$$\int_0^1 \Psi_n(x) \cdot \Psi_m(x) \, dx.$$

7.4   ALGEBRAIC IDENTITIES AND TRIGONOMETRIC SUBSTITUTIONS

Although not all functions have elementary antiderivatives, many do. In this section we introduce two powerful methods of integration which show that large classes of functions have elementary antiderivatives. The first is the method of partial fractions, which depends on an algebraic identity, and allows us to integrate rational functions. The second is the method of trigonometric substitutions, which allows us to handle expressions involving the square root of a quadratic polynomial. Some of the formulas in the table of integrals can be derived using the techniques of this section.

Method of Partial Fractions

The integral of some rational functions can be obtained by splitting the integrand into *partial fractions*. For example, to find

$$\int \frac{1}{(x-2)(x-5)}\,dx,$$

the integrand is split into partial fractions with denominators $(x-2)$ and $(x-5)$. We write

$$\frac{1}{(x-2)(x-5)} = \frac{A}{x-2} + \frac{B}{x-5},$$

where A and B are constants that need to be found. Multiplying by $(x-2)(x-5)$ gives the identity

$$1 = A(x-5) + B(x-2)$$

so

$$1 = (A+B)x - 5A - 2B.$$

Since this equation holds for all x, the constant terms on both sides must be equal. Similarly, the coefficients of x on both sides must be equal. So

$$\begin{aligned} -5A - 2B &= 1 \\ A + B &= 0. \end{aligned}$$

Solving these equations gives $A = -1/3$, $B = 1/3$. Thus,

$$\frac{1}{(x-2)(x-5)} = \frac{-1/3}{x-2} + \frac{1/3}{x-5}.$$

Example 1   Use partial fractions to integrate $\int \dfrac{1}{(x-2)(x-5)}\,dx$.

Solution   We split the integrand into partial fractions, each of which can be integrated:

$$\int \frac{1}{(x-2)(x-5)}\,dx = \int \left(\frac{-1/3}{x-2} + \frac{1/3}{x-5} \right) dx = -\frac{1}{3}\ln|x-2| + \frac{1}{3}\ln|x-5| + C.$$

You can check that using formula V-26 in the integral table gives the same result.

This method can be used to derive formulas V-26 and V-27 in the integral table. A similar method works whenever the denominator of the integrand factors into distinct linear factors and the numerator has degree less than the denominator.

Example 2   Find $\displaystyle\int \frac{x+2}{x^2+x}\,dx$.

Solution   We factor the denominator and split the integrand into partial fractions:

$$\frac{x+2}{x^2+x} = \frac{x+2}{x(x+1)} = \frac{A}{x} + \frac{B}{x+1}.$$

Multiplying by $x(x+1)$ gives the identity

$$x+2 = A(x+1) + Bx$$
$$= (A+B)x + A.$$

Equating constant terms and coefficients of x gives $A = 2$ and $A + B = 1$, so $B = -1$. Then we split the integrand into two parts and integrate:

$$\int \frac{x+2}{x^2+x}\,dx = \int \left(\frac{2}{x} - \frac{1}{x+1}\right)\,dx = 2\ln|x| - \ln|x+1| + C.$$

The next example illustrates what to do if there is a repeated factor in the denominator.

Example 3   Calculate $\displaystyle\int \frac{10x - 2x^2}{(x-1)^2(x+3)}\,dx$ using partial fractions of the form $\dfrac{A}{x-1}, \dfrac{B}{(x-1)^2}, \dfrac{C}{x+3}$.

Solution   We are given that the squared factor, $(x-1)^2$, leads to partial fractions of the form:

$$\frac{10x - 2x^2}{(x-1)^2(x+3)} = \frac{A}{x-1} + \frac{B}{(x-1)^2} + \frac{C}{x+3}$$

Multiplying through by $(x-1)^2(x+3)$ gives

$$10x - 2x^2 = A(x-1)(x+3) + B(x+3) + C(x-1)^2$$
$$= (A+C)x^2 + (2A + B - 2C)x - 3A + 3B + C.$$

Equating the coefficients of x^2 and x and the constant terms, we get the simultaneous equations:

$$A + C = -2$$
$$2A + B - 2C = 10$$
$$-3A + 3B + C = 0$$

Solving gives $A = 1, B = 2, C = -3$. Thus, we obtain three integrals which can be evaluated:

$$\int \frac{10x - 2x^2}{(x-1)^2(x+3)}\,dx = \int \left(\frac{1}{x-1} + \frac{2}{(x-1)^2} - \frac{3}{x+3}\right)\,dx$$
$$= \ln|x-1| - \frac{2}{(x-1)} - 3\ln|x+3| + K.$$

For the second integral, we use the fact that $\int 2/(x-1)^2\,dx = 2\int(x-1)^{-2}\,dx = -2(x-1)^{-1} + K$.

If there is a quadratic in the denominator which cannot be factored, we must allow a numerator of the form $Ax + B$ in the numerator, as the next example shows.

Example 4 Find $\int \dfrac{2x^2 - x - 1}{(x^2 + 1)(x - 2)}\, dx$ using partial fractions of the form $\dfrac{Ax + B}{x^2 + 1}$ and $\dfrac{C}{x - 2}$.

Solution We are given that the quadratic denominator, $(x^2 + 1)$, which cannot be factored further, has a numerator of the form $Ax + B$, so we have

$$\frac{2x^2 - x - 1}{(x^2 + 1)(x - 2)} = \frac{Ax + B}{x^2 + 1} + \frac{C}{x - 2}.$$

Multiplying by $(x^2 + 1)(x - 2)$ gives

$$2x^2 - x - 1 = (Ax + B)(x - 2) + C(x^2 + 1)$$
$$= (A + C)x^2 + (B - 2A)x + C - 2B.$$

Equating the coefficients of x^2 and x and the constant terms gives the simultaneous equations

$$A + C = 2$$
$$B - 2A = -1$$
$$C - 2B = -1.$$

Solving gives $A = B = C = 1$, so we rewrite the integral as follows:

$$\int \frac{2x^2 - x - 1}{(x^2 + 1)(x - 2)}\, dx = \int \left(\frac{x + 1}{x^2 + 1} + \frac{1}{x - 2} \right) dx.$$

This identity is useful provided we can perform the integration on the right-hand side. The first integral can be done if it is split into two; the second integral is similar to those in the previous examples. We have

$$\int \frac{2x^2 - x - 1}{(x^2 + 1)(x - 2)}\, dx = \int \frac{x}{x^2 + 1}\, dx + \int \frac{1}{x^2 + 1}\, dx + \int \frac{1}{x - 2}\, dx.$$

To calculate $\int (x/(x^2 + 1))\, dx$, substitute $w = x^2 + 1$, or guess and check. The final result is

$$\int \frac{2x^2 - x - 1}{(x^2 + 1)(x - 2)}\, dx = \frac{1}{2}\ln|x^2 + 1| + \arctan x + \ln|x - 2| + K.$$

Finally, the next example shows what to do if the numerator has degree larger than the denominator.

Example 5 Calculate $\int \dfrac{x^3 - 7x^2 + 10x + 1}{x^2 - 7x + 10}\, dx$ using long division.

Solution The degree of the numerator is greater than the degree of the denominator, so we divide first:

$$\frac{x^3 - 7x^2 + 10x + 1}{x^2 - 7x + 10} = \frac{x(x^2 - 7x + 10) + 1}{x^2 - 7x + 10} = x + \frac{1}{x^2 - 7x + 10}.$$

The remainder, in this case $1/(x^2 - 7x + 10)$, is a rational function on which we try to use partial fractions. We have

$$\frac{1}{x^2 - 7x + 10} = \frac{1}{(x - 2)(x - 5)}$$

so in this case we use the result of Example 1 to obtain

$$\int \frac{x^3 - 7x^2 + 10x + 1}{x^2 - 7x + 10}\, dx = \int \left(x + \frac{1}{(x - 2)(x - 5)} \right) dx = \frac{x^2}{2} - \frac{1}{3}\ln|x - 2| + \frac{1}{3}\ln|x - 5| + C.$$

Many, though not all, rational functions can be integrated by the strategy suggested by the previous examples.

Strategy for Integrating a Rational Function, $\frac{P(x)}{Q(x)}$

- If degree of $P(x) \geq$ degree of $Q(x)$, try long division and the method of partial fractions on the remainder.
- If $Q(x)$ is the product of distinct linear factors, use partial fractions of the form

$$\frac{A}{(x-c)}.$$

- If $Q(x)$ contains a repeated linear factor, $(x-c)^n$, use partial fractions of the form

$$\frac{A_1}{(x-c)} + \frac{A_2}{(x-c)^2} + \cdots + \frac{A_n}{(x-c)^n}.$$

- If $Q(x)$ contains an unfactorable quadratic $q(x)$, try a partial fraction of the form

$$\frac{Ax+B}{q(x)}.$$

To use this method, we need to be able to integrate each partial fraction. We know how to integrate terms of the form $A/(x-c)^n$ using the power rule (if $n > 1$) and logarithms (if $n = 1$). Next we see how to integrate terms of the form $(Ax+B)/q(x)$, where $q(x)$ is an unfactorable quadratic.

Trigonometric Substitutions

Section 7.1 showed how substitutions could be used to transform complex integrands. Now we see how substitution of $\sin\theta$ or $\tan\theta$ can be used for integrands involving square roots of quadratics or unfactorable quadratics.

Sine Substitutions

Substitutions involving $\sin\theta$ make use of the Pythagorean identity, $\cos^2\theta + \sin^2\theta = 1$, to simplify an integrand involving $\sqrt{a^2 - x^2}$.

Example 6   Find $\displaystyle\int \frac{1}{\sqrt{1-x^2}}\,dx$ using the substitution $x = \sin\theta$.

Solution   If $x = \sin\theta$, then $dx = \cos\theta\,d\theta$, and substitution gives

$$\int \frac{1}{\sqrt{1-x^2}}\,dx = \int \frac{1}{\sqrt{1-\sin^2\theta}}\cos\theta\,d\theta = \int \frac{\cos\theta}{\sqrt{\cos^2\theta}}\,d\theta.$$

Now either $\sqrt{\cos^2\theta} = \cos\theta$ or $\sqrt{\cos^2\theta} = -\cos\theta$ depending on the values taken by θ. If we choose $-\pi/2 \leq \theta \leq \pi/2$, then $\cos\theta \geq 0$, so $\sqrt{\cos^2\theta} = \cos\theta$. Then

$$\int \frac{\cos\theta}{\sqrt{\cos^2\theta}}\,d\theta = \int \frac{\cos\theta}{\cos\theta}\,d\theta = \int 1\,d\theta = \theta + C = \arcsin x + C.$$

The last step uses the fact that $\theta = \arcsin x$ if $x = \sin\theta$ and $-\pi/2 \leq \theta \leq \pi/2$.

From now on, when we substitute $\sin\theta$, we assume that the interval $-\pi/2 \leq \theta \leq \pi/2$ has been chosen. Notice that the previous example is the case $a = 1$ of VI-28 in the table of integrals. The next example illustrates how to choose the substitution when $a \neq 1$.

Example 7   Use a trigonometric substitution to find $\displaystyle\int \frac{1}{\sqrt{4-x^2}}\,dx$.

Solution   This time we choose $x = 2\sin\theta$, with $-\pi/2 \le \theta \le \pi/2$, so that $4 - x^2$ becomes a perfect square:

$$\sqrt{4 - x^2} = \sqrt{4 - 4\sin^2\theta} = 2\sqrt{1 - \sin^2\theta} = 2\sqrt{\cos^2\theta} = 2\cos\theta.$$

Then $dx = 2\cos\theta\,d\theta$, so substitution gives

$$\int \frac{1}{\sqrt{4 - x^2}}\,dx = \int \frac{1}{2\cos\theta} 2\cos\theta\,d\theta = \int 1\,d\theta = \theta + C = \arcsin\left(\frac{x}{2}\right) + C.$$

The general rule for choosing a sine substitution is:

> To simplify $\sqrt{a^2 - x^2}$, for constant a, try $x = a\sin\theta$, with $-\pi/2 \le \theta \le \pi/2$.

Notice $\sqrt{a^2 - x^2}$ is only defined on the interval $[-a, a]$. Assuming that the domain of the integrand is $[-a, a]$, the substitution $x = a\sin\theta$, with $-\pi/2 \le \theta \le \pi/2$, is valid for all x in the domain, because its range is $[-a, a]$ and it has an inverse $\theta = \arcsin(x/a)$ on $[-a, a]$.

Example 8   Find the area of the ellipse $4x^2 + y^2 = 9$.

Solution   Solving for y shows that $y = \sqrt{9 - 4x^2}$ gives the upper half of the ellipse. From Figure 7.1, we see that

$$\text{Area} = 4\int_0^{3/2} \sqrt{9 - 4x^2}\,dx.$$

To decide which trigonometric substitution to use, we write the integrand as

$$\sqrt{9 - 4x^2} = 2\sqrt{\frac{9}{4} - x^2} = 2\sqrt{\left(\frac{3}{2}\right)^2 - x^2}.$$

This suggests that we should choose $x = (3/2)\sin\theta$, so that $dx = (3/2)\cos\theta\,d\theta$ and

$$\sqrt{9 - 4x^2} = 2\sqrt{\left(\frac{3}{2}\right)^2 - \left(\frac{3}{2}\right)^2\sin^2\theta} = 2\left(\frac{3}{2}\right)\sqrt{1 - \sin^2\theta} = 3\cos\theta.$$

When $x = 0$, $\theta = 0$, and when $x = 3/2$, $\theta = \pi/2$, so

$$4\int_0^{3/2} \sqrt{9 - 4x^2}\,dx = 4\int_0^{\pi/2} 3\cos\theta\left(\frac{3}{2}\right)\cos\theta\,d\theta = 18\int_0^{\pi/2} \cos^2\theta\,d\theta.$$

Using Example 6 on page 323 or table of integrals IV-18, we find

$$\int \cos^2\theta\,d\theta = \frac{1}{2}\cos\theta\sin\theta + \frac{1}{2}\theta + C.$$

So we have

$$\text{Area} = 4\int_0^{3/2} \sqrt{9 - 4x^2}\,dx = \frac{18}{2}\left(\cos\theta\sin\theta + \theta\right)\Big|_0^{\pi/2} = 9\left(0 + \frac{\pi}{2}\right) = \frac{9\pi}{2}.$$

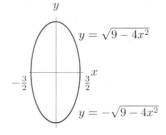

Figure 7.1: The ellipse $4x^2 + y^2 = 9$

Tangent Substitutions

Integrals involving $a^2 + x^2$ may be simplified by a substitution involving $\tan\theta$ and the trigonometric identities $\tan\theta = \sin\theta/\cos\theta$ and $\cos^2\theta + \sin^2\theta = 1$.

Example 9   Find $\displaystyle\int \frac{1}{x^2 + 9}\, dx$ using the substitution $x = 3\tan\theta$.

Solution   If $x = 3\tan\theta$, then $dx = (3/\cos^2\theta)\, d\theta$, so

$$\int \frac{1}{x^2+9}\, dx = \int \left(\frac{1}{9\tan^2\theta + 9}\right)\left(\frac{3}{\cos^2\theta}\right) d\theta = \frac{1}{3}\int \frac{1}{\left(\frac{\sin^2\theta}{\cos^2\theta} + 1\right)\cos^2\theta}\, d\theta$$

$$= \frac{1}{3}\int \frac{1}{\sin^2\theta + \cos^2\theta}\, d\theta = \frac{1}{3}\int 1\, d\theta = \frac{1}{3}\theta + C = \frac{1}{3}\arctan\left(\frac{x}{3}\right) + C.$$

> To simplify $a^2 + x^2$ or $\sqrt{a^2 + x^2}$, for constant a, try $x = a\tan\theta$, with $-\pi/2 < \theta < \pi/2$.

Note that $a^2 + x^2$ and $\sqrt{a^2 + x^2}$ are defined on $(-\infty, \infty)$. Assuming that the domain of the integrand is $(-\infty, \infty)$, the substitution $x = a\tan\theta$, with $-\pi/2 < \theta < \pi/2$, is valid for all x in the domain, because its range is $(-\infty, \infty)$ and it has an inverse $\theta = \arctan(x/a)$ on $(-\infty, \infty)$.

Example 10   Use a tangent substitution to show that the following two integrals are equal:

$$\int_0^1 \sqrt{1+x^2}\, dx = \int_0^{\pi/4} \frac{1}{\cos^3\theta}\, d\theta.$$

What area do these integrals represent?

Solution   We put $x = \tan\theta$, with $-\pi/2 < \theta < \pi/2$, so that $dx = (1/\cos^2\theta)\, d\theta$, and

$$\sqrt{1+x^2} = \sqrt{1 + \frac{\sin^2\theta}{\cos^2\theta}} = \sqrt{\frac{\cos^2\theta + \sin^2\theta}{\cos^2\theta}} = \frac{1}{\cos\theta}.$$

When $x = 0$, $\theta = 0$, and when $x = 1$, $\theta = \pi/4$, so

$$\int_0^1 \sqrt{1+x^2}\, dx = \int_0^{\pi/4}\left(\frac{1}{\cos\theta}\right)\left(\frac{1}{\cos^2\theta}\right) d\theta = \int_0^{\pi/4}\frac{1}{\cos^3\theta}\, d\theta.$$

The left-hand integral represents the area under the hyperbola $y^2 - x^2 = 1$ in Figure 7.2.

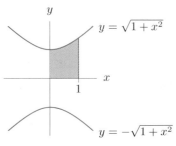

Figure 7.2: The hyperbola $y^2 - x^2 = 1$

Completing the Square to Use a Trigonometric Substitution

To make a trigonometric substitution, we may first need to complete the square.

Example 11   Find $\int \dfrac{3}{\sqrt{2x - x^2}}\, dx$.

Solution   To use a sine or tangent substitution, the expression under the square root sign should be in the form $a^2 + x^2$ or $a^2 - x^2$. Completing the square, we get

$$2x - x^2 = 1 - (x - 1)^2.$$

This suggests we substitute $x - 1 = \sin\theta$, or $x = \sin\theta + 1$. Then $dx = \cos\theta\, d\theta$, and

$$\int \frac{3}{\sqrt{2x - x^2}}\, dx = \int \frac{3}{\sqrt{1 - (x - 1)^2}}\, dx = \int \frac{3}{\sqrt{1 - \sin^2\theta}} \cos\theta\, d\theta$$

$$= \int \frac{3}{\cos\theta} \cos\theta\, d\theta = \int 3\, d\theta = 3\theta + C.$$

Since $x - 1 = \sin\theta$, we have $\theta = \arcsin(x - 1)$, so

$$\int \frac{3}{\sqrt{2x - x^2}}\, dx = 3\arcsin(x - 1) + C.$$

Example 12   Find $\int \dfrac{1}{x^2 + x + 1}\, dx$.

Solution   Completing the square, we get

$$x^2 + x + 1 = \left(x + \frac{1}{2}\right)^2 + \frac{3}{4} = \left(x + \frac{1}{2}\right)^2 + \left(\frac{\sqrt{3}}{2}\right)^2.$$

This suggests we substitute $x + 1/2 = (\sqrt{3}/2)\tan\theta$, or $x = -1/2 + (\sqrt{3}/2)\tan\theta$. Then $dx = (\sqrt{3}/2)(1/\cos^2\theta)\, d\theta$, so

$$\int \frac{1}{x^2 + x + 1}\, dx = \int \left(\frac{1}{(x + \frac{1}{2})^2 + \frac{3}{4}}\right)\left(\frac{\sqrt{3}}{2}\frac{1}{\cos^2\theta}\right) d\theta$$

$$= \frac{\sqrt{3}}{2} \int \left(\frac{1}{\frac{3}{4}\tan^2\theta + \frac{3}{4}}\right)\left(\frac{1}{\cos^2\theta}\right) d\theta = \frac{2}{\sqrt{3}} \int \frac{1}{(\tan^2\theta + 1)\cos^2\theta}\, d\theta$$

$$= \frac{2}{\sqrt{3}} \int \frac{1}{\sin^2\theta + \cos^2\theta}\, d\theta = \frac{2}{\sqrt{3}} \int 1\, d\theta = \frac{2}{\sqrt{3}}\theta + C.$$

Since $x + 1/2 = (\sqrt{3}/2)\tan\theta$, we have $\theta = \arctan((2/\sqrt{3})x + 1/\sqrt{3})$, so

$$\int \frac{1}{x^2 + x + 1}\, dx = \frac{2}{\sqrt{3}}\arctan\left(\frac{2}{\sqrt{3}}x + \frac{1}{\sqrt{3}}\right) + C.$$

Exercises and Problems for Section 7.4

Exercises

Split the functions in Exercises 1–7 into partial fractions.

1. $\dfrac{20}{25 - x^2}$

2. $\dfrac{x + 1}{6x + x^2}$

3. $\dfrac{8}{y^3 - 4y}$

4. $\dfrac{2(1 + s)}{s(s^2 + 3s + 2)}$

5. $\dfrac{2}{s^4 - 1}$

6. $\dfrac{2y}{y^3 - y^2 + y - 1}$

7. $\dfrac{1}{w^4 - w^3}$

8. Integrate the function in Exercise 1 with respect to x.

9. Integrate the function in Exercise 2 with respect to x.

10. Integrate the function in Exercise 3 with respect to y.

11. Integrate the function in Exercise 4 with respect to s.

12. Integrate the function in Exercise 5 with respect to s.

13. Integrate the function in Exercise 6 with respect to y.

14. Integrate the function in Exercise 7 with respect to w.

In Exercises 15–19, evaluate the integral.

15. $\int \dfrac{3x^2 - 8x + 1}{x^3 - 4x^2 + x + 6}\,dx$; use $\dfrac{A}{x-2} + \dfrac{B}{x+1} + \dfrac{C}{x-3}$.

16. $\int \dfrac{dx}{x^3 - x^2}$; use $\dfrac{A}{x} + \dfrac{B}{x^2} + \dfrac{C}{x-1}$.

17. $\int \dfrac{10x + 2}{x^3 - 5x^2 + x - 5}\,dx$; use $\dfrac{A}{x-5} + \dfrac{Bx + C}{x^2 + 1}$.

18. $\int \dfrac{x^4 + 12x^3 + 15x^2 + 25x + 11}{x^3 + 12x^2 + 11x}\,dx$;

use division and $\dfrac{A}{x} + \dfrac{B}{x+1} + \dfrac{C}{x+11}$.

19. $\int \dfrac{x^4 + 3x^3 + 2x^2 + 1}{x^2 + 3x + 2}\,dx$; use division.

Use the given substitution in Exercises 20–22.

20. $\int \dfrac{1}{\sqrt{9 - 4x^2}}\,dx$, $\quad x = \dfrac{3}{2}\sin t$.

21. $\int \dfrac{1}{x^2 + 4x + 5}\,dx$, $\quad x = \tan t - 2$.

22. $\int \dfrac{1}{\sqrt{4x - 3 - x^2}}\,dx$, $\quad x = \sin t + 2$.

23. Give a substitution (not necessarily trigonometric) which could be used to compute the following integrals:

(a) $\int \dfrac{x}{\sqrt{x^2 + 10}}\,dx$ (b) $\int \dfrac{1}{\sqrt{x^2 + 10}}\,dx$

Problems

Complete the square and give a substitution (not necessarily trigonometric) which could be used to compute the integrals in Problems 24–31.

24. $\int \dfrac{1}{x^2 + 2x + 2}\,dx$

25. $\int \dfrac{1}{x^2 + 6x + 25}\,dx$

26. $\int \dfrac{dy}{y^2 + 3y + 3}$

27. $\int \dfrac{x + 1}{x^2 + 2x + 2}\,dx$

28. $\int \dfrac{4}{\sqrt{2z - z^2}}\,dz$

29. $\int \dfrac{z - 1}{\sqrt{2z - z^2}}\,dz$

30. $\int (t + 2)\sin(t^2 + 4t + 7)\,dt$

31. $\int (2 - \theta)\cos(\theta^2 - 4\theta)\,d\theta$

Calculate the integrals in Problems 32–49.

32. $\int \dfrac{1}{(x - 5)(x - 3)}\,dx$

33. $\int \dfrac{1}{(x + 2)(x + 3)}\,dx$

34. $\int \dfrac{1}{(x + 7)(x - 2)}\,dx$

35. $\int \dfrac{x}{x^2 - 3x + 2}\,dx$

36. $\int \dfrac{dz}{z^2 + z}$

37. $\int \dfrac{dx}{x^2 + 5x + 4}$

38. $\int \dfrac{dP}{3P - 3P^2}$

39. $\int \dfrac{3x + 1}{x^2 - 3x + 2}\,dx$

40. $\int \dfrac{y + 2}{2y^2 + 3y + 1}\,dy$

41. $\int \dfrac{x + 1}{x^3 + x}\,dx$

42. $\int \dfrac{x - 2}{x^2 + x^4}\,dx$

43. $\int \dfrac{x^2}{\sqrt{9 - x^2}}\,dx$

44. $\int \dfrac{y^2}{25 + y^2}\,dy$

45. $\int \dfrac{dt}{t^2\sqrt{1 + t^2}}$

46. $\int \dfrac{dz}{(4 - z^2)^{3/2}}$

47. $\int \dfrac{10}{(s + 2)(s^2 + 1)}\,ds$

48. $\int \dfrac{1}{x^2 + 4x + 13}\,dx$

49. $\int \dfrac{e^x\,dx}{(e^x - 1)(e^x + 2)}$

Find the exact area of the regions in Problems 50–55.

50. Bounded by $3x/((x - 1)(x - 4))$, $y = 0$, $x = 2$, $x = 3$.

51. Bounded by $y = (3x^2 + x)/((x^2 + 1)(x + 1))$, $y = 0$, $x = 0$, $x = 1$.

52. Bounded by $y = x^2/\sqrt{1 - x^2}$, $y = 0$, $x = 0$, $x = 1/2$.

53. Bounded by $y = x^3/\sqrt{4 - x^2}$, $y = 0$, $x = 0$, $x = \sqrt{2}$.

54. Bounded by $y = 1/\sqrt{x^2 + 9}$, $y = 0$, $x = 0$, $x = 3$.

55. Bounded by $y = 1/(x\sqrt{x^2 + 9})$, $y = 0$, $x = \sqrt{3}$, $x = 3$.

Calculate the integrals in Problems 56–58 by partial fractions and using the indicated substitution. Show that the results you get are the same.

56. $\int \dfrac{dx}{1 - x^2}$; substitution $x = \sin\theta$.

57. $\int \dfrac{2x}{x^2 - 1}\,dx$; substitution $w = x^2 - 1$.

58. $\int \dfrac{3x^2 + 1}{x^3 + x}\,dx$; substitution $w = x^3 + x$.

59. (a) Show $\int \dfrac{1}{\sin^2\theta}\,d\theta = -\dfrac{1}{\tan\theta} + C$.

(b) Calculate $\int \dfrac{dy}{y^2\sqrt{5 - y^2}}$.

60. Calculate the integral $\int \dfrac{1}{(x - a)(x - b)}\,dx$ for

(a) $a \neq b$ (b) $a = b$

61. Calculate the integral $\int \dfrac{x}{(x - a)(x - b)}\,dx$ for

(a) $a \neq b$ (b) $a = b$

62. Calculate the integral $\int \dfrac{1}{x^2 - a}\, dx$ for

    (a) $a > 0$ **(b)** $a = 0$ **(c)** $a < 0$

63. A rumor is spread in a school. For $0 < a < 1$ and $b > 0$, the time t at which a fraction p of the school population has heard the rumor is given by

$$t(p) = \int_a^p \frac{b}{x(1 - x)}\, dx.$$

    (a) Evaluate the integral to find an explicit formula for $t(p)$. Write your answer so it has only one ln term.
    (b) At time $t = 0$ one percent of the school population ($p = 0.01$) has heard the rumor. What is a?
    (c) At time $t = 1$ half the school population ($p = 0.5$) has heard the rumor. What is b?

    (d) At what time has 90% of the school population ($p = 0.9$) heard the rumor?

64. The Law of Mass Action tells us that the time, T, taken by a chemical to create a quantity x_0 of the product (in molecules) is given by

$$T = \int_0^{x_0} \frac{k\, dx}{(a - x)(b - x)}$$

where a and b are initial quantities of the two ingredients used to make the product, and k is a positive constant. Suppose $0 < a < b$.

    (a) Find the time taken to make a quantity $x_0 = a/2$ of the product.
    (b) What happens to T as $x_0 \to a$?

7.5 APPROXIMATING DEFINITE INTEGRALS

The methods of the last few sections allow us to get exact answers for definite integrals in a variety of cases. However, many functions do not have elementary antiderivatives. To evaluate the definite integrals of such functions, we cannot use the Fundamental Theorem; we must use numerical methods.

We already know how to approximate a definite integral numerically using left- and right-hand Riemann sums. In the next two sections we introduce better methods for approximating definite integrals—better in the sense that they give more accurate results with less work than that required to find the left- and right-hand sums.

The Midpoint Rule

In the left- and right-hand Riemann sums, the heights of the rectangles are found using the left-hand or right-hand endpoints, respectively, of the subintervals. For the *midpoint rule*, we use the midpoint of each of the subintervals.

For example, in approximating $\int_1^2 f(x)\, dx$ by a Riemann sum with two subdivisions, we first divide the interval $1 \leq x \leq 2$ into two pieces. The midpoint of the first subinterval is 1.25 and the midpoint of the second is 1.75. The heights of the two rectangles are $f(1.25)$ and $f(1.75)$, respectively. (See Figure 7.3.) The Riemann sum is

$$f(1.25)0.5 + f(1.75)0.5.$$

Figure 7.3 shows that evaluating f at the midpoint of each subdivision usually gives a better approximation to the area under the curve than evaluating f at either end. For this particular f, it appears that each rectangle is partly above and partly below the graph on each subinterval. Furthermore, the area under the curve which is not under the rectangle appears to be nearly equal to the area under the rectangle which is above the curve. In fact, this new midpoint Riemann sum is generally a better approximation to the definite integral than the left- or right-hand sum with the same number of subdivisions, n.

Figure 7.3: Midpoint rule with two subdivisions

So far, we have three ways of estimating an integral using a Riemann sum:

1. The **left rule** uses the left endpoint of each subinterval.
2. The **right rule** uses the right endpoint of each subinterval.
3. The **midpoint rule** uses the midpoint of each subinterval.

We write $\text{LEFT}(n)$, $\text{RIGHT}(n)$, and $\text{MID}(n)$ to denote the results obtained by using these rules with n subdivisions.

Example 1   For $\int_1^2 \frac{1}{x}\,dx$, compute $\text{LEFT}(2)$, $\text{RIGHT}(2)$ and $\text{MID}(2)$, and compare your answers with the exact value of the integral.

Solution   For $n = 2$ subdivisions of the interval $[1, 2]$, we use $\Delta x = 0.5$. Then

$$\text{LEFT}(2) = f(1)(0.5) + f(1.5)(0.5) = \frac{1}{1}(0.5) + \frac{1}{1.5}(0.5) = 0.8333\ldots$$

$$\text{RIGHT}(2) = f(1.5)(0.5) + f(2)(0.5) = \frac{1}{1.5}(0.5) + \frac{1}{2}(0.5) = 0.5833\ldots$$

$$\text{MID}(2) = f(1.25)(0.5) + f(1.75)(0.5) = \frac{1}{1.25}(0.5) + \frac{1}{1.75}(0.5) = 0.6857\ldots$$

All three Riemann sums in this example are approximating

$$\int_1^2 \frac{1}{x}\,dx = \ln x \Big|_1^2 = \ln 2 - \ln 1 = \ln 2 = 0.6931\ldots.$$

With only two subdivisions, the left and right rules give quite poor approximations but the midpoint rule is already fairly close to the exact answer.

Figure 7.4(a) illustrates why the left and right rules are so inaccurate. Since $f(x) = 1/x$ is decreasing from 1 to 2, the left rule overestimates on each subdivision while the right rule underestimates. However, the midpoint rule approximates with rectangles on each subdivision that are each partly above and partly below the graph, so the errors tend to balance out. (See Figure 7.4(b).)

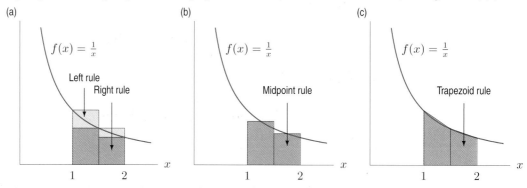

Figure 7.4: Left, right, midpoint, and trapezoid approximations to $\int_1^2 \frac{1}{x}\,dx$

The Trapezoid Rule

We have just seen how the midpoint rule can have the effect of balancing out the errors of the left and right rules. There is another way of balancing these errors: we average the results from the left and right rules. This approximation is called the *trapezoid rule*:

$$\text{TRAP}(n) = \frac{\text{LEFT}(n) + \text{RIGHT}(n)}{2}.$$

The trapezoid rule averages the values of f at the left and right endpoints of each subinterval and multiplies by Δx. This is the same as approximating the area under the graph of f in each subinterval by a trapezoid (see Figure 7.5).

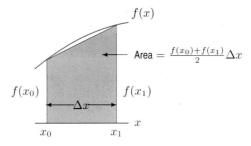

Figure 7.5: Area used in the trapezoid rule

Example 2   For $\displaystyle\int_1^2 \frac{1}{x}\, dx$, compare the trapezoid rule with two subdivisions with the left, right, and midpoint rules.

Solution   In the previous example we got $\mathrm{LEFT}(2) = 0.8333\ldots$ and $\mathrm{RIGHT}(2) = 0.5833\ldots$. The trapezoid rule is the average of these, so $\mathrm{TRAP}(2) = 0.7083\ldots$. (See Figure 7.4(c).) The exact value of the integral is $0.6931\ldots$, so the trapezoid rule is better than the left or right rules. The midpoint rule is still the best, however, since $\mathrm{MID}(2) = 0.6857\ldots$.

Is the Approximation an Over- or Underestimate?

It is useful to know when a rule is producing an overestimate and when it is producing an underestimate. In Chapter 5 we saw that the following relationship holds.

If f is increasing on $[a, b]$, then

$$\mathrm{LEFT}(n) \le \int_a^b f(x)\, dx \le \mathrm{RIGHT}(n).$$

If f is decreasing on $[a, b]$, then

$$\mathrm{RIGHT}(n) \le \int_a^b f(x)\, dx \le \mathrm{LEFT}(n).$$

The Trapezoid Rule

If the graph of the function is concave down on $[a, b]$, then each trapezoid lies below the graph and the trapezoid rule underestimates. If the graph is concave up on $[a, b]$, the trapezoid rule overestimates. (See Figure 7.6.)

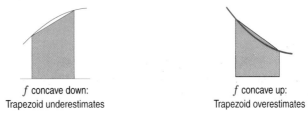

f concave down:
Trapezoid underestimates

f concave up:
Trapezoid overestimates

Figure 7.6: Error in the trapezoid rule

The Midpoint Rule

To understand the relationship between the midpoint rule and concavity, take a rectangle whose top intersects the curve at the midpoint of a subinterval. Draw a tangent to the curve at the midpoint; this gives a trapezoid. See Figure 7.7. (This is *not* the same trapezoid as in the trapezoid rule.) The midpoint rectangle and the new trapezoid have the same area, because the shaded triangles in Figure 7.7 are congruent. Hence, if the graph of the function is concave down, the midpoint rule overestimates; if the graph is concave up, the midpoint rule underestimates. (See Figure 7.8.)

If the graph of f is concave down on $[a, b]$, then

$$\text{TRAP}(n) \leq \int_a^b f(x)\, dx \leq \text{MID}(n).$$

If the graph of f is concave up on $[a, b]$, then

$$\text{MID}(n) \leq \int_a^b f(x)\, dx \leq \text{TRAP}(n).$$

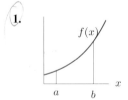

Figure 7.7: Midpoint rectangle and trapezoid with same area

f concave down: f concave up:
Midpoint overestimates Midpoint underestimates

Figure 7.8: Error in the midpoint rule

Exercises and Problems for Section 7.5

Exercises

In Exercises 1–6, sketch the area given by the following approximations to $\int_a^b f(x)\,dx$. Identify each approximation as an overestimate or an underestimate.

(a) LEFT(2)   (b) RIGHT(2)
(c) TRAP(2)   (d) MID(2)

1.

2.

3.

4.

5.

6.

7. Calculate the following approximations to $\int_0^6 x^2 dx$.

   (a) LEFT(2)   (b) RIGHT(2)
   (c) TRAP(2)   (d) MID(2)

8. (a) Find LEFT(2) and RIGHT(2) for $\int_0^4 (x^2 + 1)\, dx$.
   (b) Illustrate your answers to part (a) graphically. Is each approximation an underestimate or overestimate?

9. (a) Find MID(2) and TRAP(2) for $\int_0^4 (x^2 + 1)\, dx$.
   (b) Illustrate your answers to part (a) graphically. Is each approximation an underestimate or overestimate?

10. Calculate the following approximations to $\int_0^\pi \sin\theta\, d\theta$.

   (a) LEFT(2)   (b) RIGHT(2)
   (c) TRAP(2)   (d) MID(2)

Problems

11. (a) Estimate $\int_0^1 1/(1+x^2)\,dx$ by subdividing the interval into eight parts using:

(i) The left Riemann sum.

(ii) The right Riemann sum.

(iii) The trapezoidal rule.

(b) Since the exact value of the integral is $\pi/4$, you can estimate the value of π using part(a). Explain why your first estimate is too large and your second estimate too small.

12. Using the table, estimate the total distance traveled from time $t = 0$ to time $t = 6$ using LEFT, RIGHT, and TRAP.

Time, t	0	1	2	3	4	5	6
Velocity, v	3	4	5	4	7	8	11

13. Using Figure 7.9, order the following approximations to the integral $\int_0^3 f(x)dx$ and its exact value from smallest to largest:
LEFT(n), RIGHT(n), MID(n), TRAP(n), Exact value.

Figure 7.9

14. The results from the left, right, trapezoid, and midpoint rules used to approximate $\int_0^1 g(t)\,dt$, with the same number of subdivisions for each rule, are as follows:
$0.601, 0.632, 0.633, 0.664$.

(a) Using Figure 7.10, match each rule with its approximation.

(b) Between which two consecutive approximations does the true value of the integral lie?

Figure 7.10

For the functions in Problems 15–18, pick which approximation — left, right, trapezoid, or midpoint — is guaranteed to give an overestimate for $\int_0^5 f(x)\,dx$, and which is guaranteed to give an underestimate. (There may be more than one.)

15.

16.

17.

18.

19. Using a fixed number of subdivisions, we approximate the integrals of f and g on the interval in Figure 7.11.

(a) For which function, f or g, is LEFT more accurate? RIGHT? Explain.

(b) For which function, f or g, is TRAP more accurate? MID? Explain.

Figure 7.11

20. (a) Values for $f(x)$ are in the table. Which of the four approximation methods in this section is most likely to give the best estimate of $\int_0^{12} f(x)\,dx$? Estimate the integral using this method.

(b) Assume $f(x)$ is continuous with no critical points or points of inflection on the interval $0 \le x \le 12$. Is the estimate found in part (a) an over- or underestimate? Explain.

x	0	3	6	9	12
$f(x)$	100	97	90	78	55

21. (a) Find the exact value of $\int_0^{2\pi} \sin\theta\,d\theta$.

(b) Explain, using pictures, why the MID(1) and MID(2) approximations to this integral give the exact value.

(c) Does MID(3) give the exact value of this integral? How about MID(n)? Explain.

22. (a) Show geometrically why $\int_0^1 \sqrt{2-x^2}\,dx = \frac{\pi}{4} + \frac{1}{2}$. [Hint: Break up the area under $y = \sqrt{2-x^2}$ from $x = 0$ to $x = 1$ into two pieces: a sector of a circle and a right triangle.]

(b) Approximate $\int_0^1 \sqrt{2-x^2}\,dx$ for $n = 5$ using the left, right, trapezoid, and midpoint rules. Compute the error in each case using the answer to part (a), and compare the errors.

23. The width, in feet, at various points along the fairway of a hole on a golf course is given in Figure 7.12. If one pound of fertilizer covers 200 square feet, estimate the amount of fertilizer needed to fertilize the fairway.

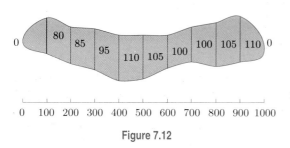

Figure 7.12

Problems 24–28 involve approximating $\int_a^b f(x)\,dx$.

24. Show $\text{RIGHT}(n) = \text{LEFT}(n) + f(b)\Delta x - f(a)\Delta x$.

25. Show $\text{TRAP}(n) = \text{LEFT}(n) + \frac{1}{2}\left(f(b) - f(a)\right)\Delta x$.

26. Show $\text{LEFT}(2n) = \frac{1}{2}\left(\text{LEFT}(n) + \text{MID}(n)\right)$.

27. Check that the equations in Problems 24 and 25 hold for $\int_1^2 (1/x)\,dx$ when $n = 10$.

28. Suppose that $a = 2$, $b = 5$, $f(2) = 13$, $f(5) = 21$ and that $\text{LEFT}(10) = 3.156$ and $\text{MID}(10) = 3.242$. Use Problems 24–26 to compute $\text{RIGHT}(10)$, $\text{TRAP}(10)$, $\text{LEFT}(20)$, $\text{RIGHT}(20)$, and $\text{TRAP}(20)$.

7.6 APPROXIMATION ERRORS AND SIMPSON'S RULE

When we compute an approximation, we are always concerned about the error, namely the difference between the exact answer and the approximation. We usually do not know the exact error; if we did, we would also know the exact answer. Often the best we can get is an upper bound on the error and some idea of how much work is involved in making the error smaller. The study of numerical approximations is really the study of errors. The errors for some methods are much smaller than those for others. The errors for the midpoint and trapezoid rules are related to each other in a way that suggests an even better method, called Simpson's rule. We work with the example $\int_1^2 (1/x)\,dx$ because we know the exact value of this integral ($\ln 2$) and we can investigate the behavior of the errors.

Error in Left and Right Rules

For any approximation, we take

$$\text{Error} = \text{Actual value} - \text{Approximate value}.$$

Let us see what happens to the error in the left and right rules as we increase n. We increase n each time by a factor of 5 starting at $n = 2$. The results are in Table 7.1. A positive error indicates that the Riemann sum is less than the exact value, $\ln 2$. Notice that the errors for the left and right rules have opposite signs but are approximately equal in magnitude. (See Figure 7.13.) The best way to try to get the errors to cancel is to average the left and right rules; this average is the trapezoid rule. If we had not already thought of the trapezoid rule, we might have been led to invent it by this observation.

There is another pattern to the errors in Table 7.1. If we compute the *ratio* of the errors in Table 7.2, we see that the error[2] in both the left and right rules decreases by a factor of about 5 as n increases by a factor of 5.

Table 7.1 *Errors for the left and right rule approximation to* $\int_1^2 \frac{1}{x}\,dx = \ln 2 \approx 0.6931471806$

n	Error in left rule	Error in right rule
2	−0.1402	0.1098
10	−0.0256	0.0244
50	−0.0050	0.0050
250	−0.0010	0.0010

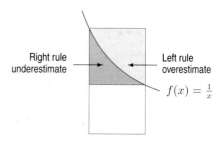

Figure 7.13: Errors in left and right sums

[2]The values in Table 7.1 are rounded to 4 decimal places; those in Table 7.2 were computed using more decimal places and then rounded.

Table 7.2   *Ratio of the errors as n increases for $\int_1^2 \frac{1}{x}\,dx$*

	Ratio of errors in left rule	Ratio of errors in right rule
Error(2)/Error(10)	5.47	4.51
Error(10)/Error(50)	5.10	4.90
Error(50)/Error(250)	5.02	4.98

There is nothing special about the number 5; the same holds for any factor. To get one extra digit of accuracy in any calculation, we must make the error $1/10$ as big, so we must increase n by a factor of 10. In fact, *for the left or right rules, each extra digit of accuracy requires about 10 times the work.* The calculator used to produce these tables took about half a second to compute the left rule approximation for $n = 50$, and this yields $\ln 2$ to two digits. To get three correct digits, n would need to be around 500 and the time would be about 5 seconds. Four digits requires $n = 5000$ and 50 seconds. Ten digits requires $n = 5 \cdot 10^9$ and $5 \cdot 10^7$ seconds, which is more than a year! Clearly, the errors for the left and right rules do not decrease fast enough as n increases for practical use.

Error in Trapezoid and Midpoint Rules

Table 7.3 shows that the trapezoid and midpoint rules produce much better approximations to $\int_1^2 (1/x)\,dx$ than the left and right rules.

Again there is a pattern to the errors. For each n, the midpoint rule is noticeably better than the trapezoid rule; the error for the midpoint rule, in absolute value, seems to be about half the error of the trapezoid rule. To see why, compare the shaded areas in Figure 7.14. Also, notice in Table 7.3 that the errors for the two rules have opposite signs; this is due to concavity.

We are interested in how the errors behave as n increases. Table 7.4 gives the ratios of the errors for each rule. For each rule, we see that as n increases by a factor of 5, the error decreases by a factor of about $25 = 5^2$. In fact, it can be shown that this squaring relationship holds for any factor, so increasing n by a factor of 10 will decrease the error by a factor of about $100 = 10^2$. Reducing the error by a factor of 100 is equivalent to adding two more decimal places of accuracy to the result.

Table 7.3   *The errors for the trapezoid and midpoint rules for $\int_1^2 \frac{1}{x}\,dx$*

n	Error in trapezoid rule	Error in midpoint rule
2	−0.0152	0.0074
10	−0.00062	0.00031
50	−0.0000250	0.0000125
250	−0.0000010	0.0000005

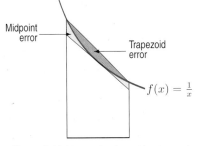

Figure 7.14: Errors in the midpoint and trapezoid rules

In other words: *In the trapezoid or midpoint rules, each extra 2 digits of accuracy requires about 10 times the work.*

This result shows the advantage of the midpoint and trapezoid rules over the left and right rules: less additional work needs to be done to get another decimal place of accuracy. The calculator used to produce these tables again took about half a second to compute the midpoint rule for $\int_1^2 \frac{1}{x}\,dx$ with $n = 50$, and this gets 4 digits correct. Thus to get 6 digits would take $n = 500$ and 5 seconds, to get 8 digits would take 50 seconds, and to get 10 digits would take 500 seconds, or about 10 minutes. That is still not great, but it is certainly better than the 1 year required by the left or right rule.

Table 7.4   *Ratios of the errors as n increases for $\int_1^2 \frac{1}{x}\,dx$*

	Ratio of errors in trapezoid rule	Ratio of errors in midpoint rule
Error(2)$\big/$Error(10)	24.33	23.84
Error(10)$\big/$Error(50)	24.97	24.95
Error(50)$\big/$Error(250)	25.00	25.00

Simpson's Rule

Still more improvement is possible. Observing that the trapezoid error has the opposite sign and about twice the magnitude of the midpoint error, we may guess that a weighted average of the two rules, with the midpoint rule weighted twice the trapezoid rule, will have a much smaller error. This approximation is called *Simpson's rule*[3]:

$$\text{SIMP}(n) = \frac{2 \cdot \text{MID}(n) + \text{TRAP}(n)}{3}.$$

Table 7.5 gives the errors for Simpson's rule. Notice how much smaller the errors are than the previous errors. Of course, it is a little unfair to compare Simpson's rule at $n = 50$, say, with the previous rules, because Simpson's rule must compute the value of f at both the midpoint and the endpoints of each subinterval and hence involves evaluating the function at twice as many points. We know by our previous analysis, however, that even if we did compute the other rules at $n = 100$ to compare with Simpson's rule at $n = 50$, the other errors would only decrease by a factor of 2 for the left and right rules and by a factor of 4 for the trapezoid and midpoint rules.

We see in Table 7.5 that as n increases by a factor of 5, the errors decrease by a factor of about 600, or about 5^4. Again this behavior holds for any factor, so increasing n by a factor of 10 decreases the error by a factor of about 10^4. In other words: *In Simpson's rule, each extra 4 digits of accuracy requires about 10 times the work.*

Table 7.5   *The errors for Simpson's rule and the ratios of the errors*

n	Error	Ratio
2	-0.0001067877	
		550.15
10	-0.0000001940	
		632.27
50	-0.0000000003	

This is a great improvement over either the midpoint or trapezoid rules, which only give two extra digits of accuracy when n is increased by a factor of 10. Simpson's rule is so efficient that we get 9 digits correct with $n = 50$ in about 1 second on our calculator. Doubling n will decrease the error by a factor of about $2^4 = 16$ and hence will give the tenth digit. The total time is 2 seconds, which is pretty good.

In general, Simpson's rule achieves a reasonable degree of accuracy when using relatively small values of n, and is a good choice for an all-purpose method for estimating definite integrals.

Analytical View of the Trapezoid and Simpson's Rules

Our approach to approximating $\int_a^b f(x)\,dx$ numerically has been empirical: try a method, see how the error behaves, and then try to improve it. We can also develop the various rules for numerical integration by making better and better approximations to the integrand, f. The left, right, and midpoint rules are all examples of approximating f by a constant (flat) function on each subinterval. The trapezoid rule is obtained by approximating f by a linear function on each subinterval. Simpson's rule can, in the same spirit, be obtained by approximating f by quadratic functions. The details are given in Problems 8 and 9 on page 348.

---

[3]Some books and computer programs use slightly different terminology for Simpson's rule; what we call $n = 50$, they call $n = 100$.

How the Error Depends on the Integrand

Other factors besides the size of n affect the size of the error in each of the rules. Instead of looking at how the error behaves as we increase n, let's leave n fixed and imagine trying our approximation methods on different functions. We observe that the error in the left or right rule depends on how steeply the graph of f rises or falls. A steep curve makes the triangular regions missed by the left or right rectangles tall and hence large in area. This observation suggests that the error in the left or right rules depends on the size of the derivative of f (see Figure 7.15).

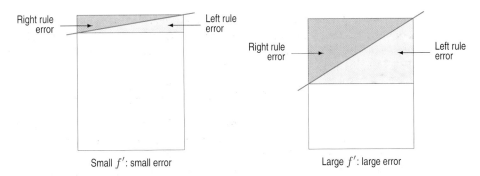

Figure 7.15: The error in the left and right rules depends on the steepness of the curve

From Figure 7.16 it appears that the errors in the trapezoid and midpoint rules depend on how much the curve is bent up or down. In other words, the concavity, and hence the size of the second derivative of f, has an effect on the errors of these two rules. Finally, it can be shown[4] that the error in Simpson's rule depends on the size of the *fourth* derivative of f, written $f^{(4)}$.

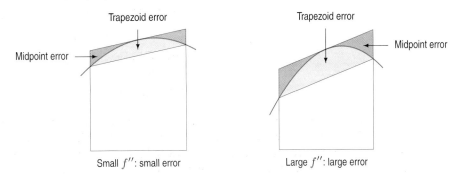

Figure 7.16: The error in the trapezoid and midpoint rules depends on how bent the curve is

Exercises and Problems for Section 7.6

Exercises

1. Estimate $\int_0^6 x^2 dx$ using SIMP(2).

2. (a) Using the result of Problem 9 on page 342, compute SIMP(2) for $\int_0^4 (x^2 + 1)\, dx$.

(b) Use the Fundamental Theorem of Calculus to find $\int_0^4 (x^2 + 1)\, dx$ exactly.

(c) What is the error in SIMP(2) for this integral?

---

[4]See Kendall E. Atkinson, *An Introduction to Numerical Analysis* (New York: John Wiley and Sons, 1978).

Problems

3. In this problem you will investigate the behavior of the errors in the approximation of the integral

$$\int_1^2 \frac{1}{x}\, dx \approx 0.6931471806\ldots .$$

(a) For $n = 2$, 4, 8, 16, 32, 64, 128 subdivisions, find the left and right approximations and the errors in each.

(b) What are the signs of the errors in the left and right approximations? How do the errors change if n is doubled?

(c) For the values of n in part (a), compute the midpoint and trapezoid approximations and the errors in each.

(d) What are the signs of the errors in the midpoint and trapezoid approximations? How do the errors change if n is doubled?

(e) For $n = 2$, 4, 8, 16, 32, compute Simpson's rule approximation and the error in each. How do these errors change as n doubles?

4. (a) What is the exact value of $\int_0^2 (x^3 + 3x^2)\, dx$?
(b) Find SIMP(n) for $n = 2, 4, 100$. What do you notice?

5. (a) What is the exact value of $\int_0^4 e^x\, dx$?
(b) Find LEFT(2), RIGHT(2), TRAP(2), MID(2), and SIMP(2). Compute the error for each.
(c) Repeat part (b) with $n = 4$ (instead of $n = 2$).
(d) For each rule in part (b), as n goes from $n = 2$ to $n = 4$, does the error go down approximately as you would expect? Explain.

6. The approximation to a definite integral using $n = 10$ is 2.346; the exact value is 4.0. If the approximation was found using each of the following rules, use the same rule to estimate the integral with $n = 30$.

(a) LEFT   **(b)** TRAP   **(c)** SIMP

7. (a) A certain computer takes two seconds to compute a certain definite integral accurate to 4 digits to the right of the decimal point, using the left rectangle

rule. How long (in years) will it take to get 8 digits correct using the left rectangle rule? How about 12 digits? 20 digits?

(b) Repeat part (a) but this time assume that the trapezoidal rule is being used throughout.

Problems 8–9 show how Simpson's rule can be obtained by approximating the integrand, f, by quadratic functions.

8. Suppose that $a < b$ and that m is the midpoint $m = (a+b)/2$. Let $h = b - a$. The purpose of this problem is to show that if f is a quadratic function, then

$$\int_a^b f(x)\, dx = \frac{h}{3}\left(\frac{f(a)}{2} + 2f(m) + \frac{f(b)}{2}\right).$$

(a) Show that this equation holds for the functions $f(x) = 1$, $f(x) = x$, and $f(x) = x^2$.

(b) Use part (a) and the properties of the integral on page 266 to show that the equation holds for any quadratic function, $f(x) = Ax^2 + Bx + C$.

9. Consider the following method for approximating $\int_a^b f(x)\, dx$. Divide the interval $[a, b]$ into n equal subintervals. On each subinterval approximate f by a quadratic function that agrees with f at both endpoints and at the midpoint of the subinterval.

(a) Explain why the integral of f on the subinterval $[x_i, x_{i+1}]$ is approximately equal to the expression

$$\frac{h}{3}\left(\frac{f(x_i)}{2} + 2f(m_i) + \frac{f(x_{i+1})}{2}\right),$$

where m_i is the midpoint of the subinterval, $m_i = (x_i + x_{i+1})/2$. (See Problem 8.)

(b) Show that if we add up these approximations for each subinterval, we get Simpson's rule:

$$\int_a^b f(x)\, dx \approx \frac{2 \cdot \text{MID}(n) + \text{TRAP}(n)}{3}.$$

7.7 IMPROPER INTEGRALS

Our original discussion of the definite integral $\int_a^b f(x)\, dx$ assumed that the interval $a \le x \le b$ was of finite length and that f was continuous. Integrals that arise in applications don't necessarily have these nice properties. In this section we investigate a class of integrals, called *improper* integrals, in which one limit of integration is infinite or the integrand is unbounded. As an example, to estimate the mass of the earth's atmosphere, we might calculate an integral which sums the mass of the air up to different heights. In order to represent the fact that the atmosphere does not end at a specific height, we let the upper limit of integration get larger and larger, or tend to infinity.

We will usually consider only improper integrals with positive integrands since they are the most common.

One Type of Improper Integral: When the Limit of Integration Is Infinite

Here is an example of an improper integral:

$$\int_1^\infty \frac{1}{x^2}\,dx.$$

To evaluate this integral, we first compute the definite integral $\int_1^b (1/x^2)\,dx$:

$$\int_1^b \frac{1}{x^2}\,dx = -x^{-1}\Big|_1^b = -\frac{1}{b} + \frac{1}{1}.$$

Now take the limit as $b \to \infty$. Since

$$\lim_{b\to\infty} \int_1^b \frac{1}{x^2}\,dx = \lim_{b\to\infty}\left(-\frac{1}{b} + 1\right) = 1,$$

we say that the improper integral $\int_1^\infty (1/x^2)\,dx$ *converges* to 1.

If we think in terms of areas, the integral $\int_1^\infty (1/x^2)\,dx$ represents the area under $f(x) = 1/x^2$ from $x = 1$ extending infinitely far to the right. (See Figure 7.17(a).) It may seem strange that this region has finite area. What our limit computations are saying is that

$$\text{When } b = 10:\quad \int_1^{10} \frac{1}{x^2}\,dx = -\frac{1}{x}\Big|_1^{10} = -\frac{1}{10} + 1 = 0.9$$

$$\text{When } b = 100:\quad \int_1^{100} \frac{1}{x^2}\,dx = -\frac{1}{100} + 1 = 0.99$$

$$\text{When } b = 1000:\quad \int_1^{1000} \frac{1}{x^2}\,dx = -\frac{1}{1000} + 1 = 0.999$$

and so on. In other words, as b gets larger and larger, the area between $x = 1$ and $x = b$ tends to 1. See Figure 7.17(b). Thus, it does make sense to declare that $\int_1^\infty (1/x^2)\,dx = 1$.

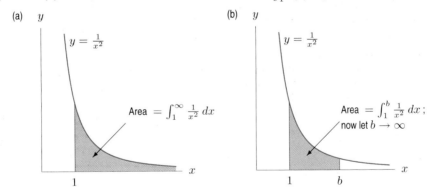

Figure 7.17: Area representation of improper integral

Of course, in another example, we might not get a finite limit as b gets larger and larger. In that case we say the improper integral *diverges*.

Suppose $f(x)$ is positive for $x \geq a$.

If $\displaystyle\lim_{b\to\infty} \int_a^b f(x)\,dx$ is a finite number, we say that $\displaystyle\int_a^\infty f(x)\,dx$ **converges** and define

$$\int_a^\infty f(x)\,dx = \lim_{b\to\infty} \int_a^b f(x)\,dx.$$

Otherwise, we say that $\displaystyle\int_a^\infty f(x)\,dx$ **diverges**. We define $\displaystyle\int_{-\infty}^b f(x)\,dx$ similarly.

Example 1   Does the improper integral $\int_1^\infty \frac{1}{\sqrt{x}}\,dx$ converge or diverge?

Solution   We consider

$$\int_1^b \frac{1}{\sqrt{x}}\,dx = \int_1^b x^{-1/2}\,dx = 2x^{1/2}\Big|_1^b = 2b^{1/2} - 2.$$

We see that $\int_1^b (1/\sqrt{x})\,dx$ grows without bound as $b \to \infty$. We have shown that the area under the curve in Figure 7.18 is not finite. Thus we say the integral $\int_1^\infty (1/\sqrt{x})\,dx$ *diverges*. We could also say $\int_1^\infty (1/\sqrt{x})\,dx = \infty$.

Notice that $f(x) \to 0$ as $x \to \infty$ does not guarantee convergence of $\int_a^\infty f(x)\,dx$.

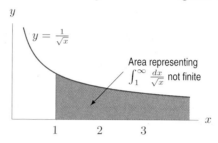

Figure 7.18: $\int_1^\infty \frac{1}{\sqrt{x}}\,dx$ diverges

What is the difference between the functions $1/x^2$ and $1/\sqrt{x}$ that makes the area under the graph of $1/x^2$ approach 1 as $x \to \infty$, whereas the area under $1/\sqrt{x}$ grows very large? Both functions approach 0 as x grows, so as b grows larger, smaller bits of area are being added to the definite integral. The difference between the functions is subtle: the values of the function $1/\sqrt{x}$ *don't shrink fast enough* for the integral to have a finite value. Of the two functions, $1/x^2$ drops to 0 much faster than $1/\sqrt{x}$, and this feature keeps the area under $1/x^2$ from growing beyond 1.

Example 2   Find $\int_0^\infty e^{-5x}\,dx$.

Solution   First we consider $\int_0^b e^{-5x}\,dx$:

$$\int_0^b e^{-5x}\,dx = -\frac{1}{5}e^{-5x}\Big|_0^b = -\frac{1}{5}e^{-5b} + \frac{1}{5}.$$

Since $e^{-5b} = \frac{1}{e^{5b}}$, this term tends to 0 as b approaches infinity, so $\int_0^\infty e^{-5x}\,dx$ converges. Its value is

$$\int_0^\infty e^{-5x}\,dx = \lim_{b\to\infty} \int_0^b e^{-5x}\,dx = \lim_{b\to\infty}\left(-\frac{1}{5}e^{-5b} + \frac{1}{5}\right) = 0 + \frac{1}{5} = \frac{1}{5}.$$

Since e^{5x} grows very rapidly, we expect that e^{-5x} will approach 0 rapidly. The fact that the area approaches $1/5$ instead of growing without bound is a consequence of the speed with which the integrand e^{-5x} approaches 0.

Example 3   Determine for which values of the exponent, p, the improper integral $\int_1^\infty \frac{1}{x^p}\,dx$ diverges.

Solution   For $p \neq 1$,

$$\int_1^b x^{-p}\,dx = \frac{1}{-p+1}x^{-p+1}\Big|_1^b = \left(\frac{1}{-p+1}b^{-p+1} - \frac{1}{-p+1}\right).$$

The important question is whether the exponent of b is positive or negative. If it is negative, then as b approaches infinity, b^{-p+1} approaches 0. If the exponent is positive, then b^{-p+1} grows without bound as b approaches infinity.

What happens if $p = 1$? In this case we get

$$\int_1^\infty \frac{1}{x}\, dx = \lim_{b \to \infty} \ln x \Big|_1^b = \lim_{b \to \infty} \ln b - \ln 1.$$

Since $\ln b$ becomes arbitrarily large as b approaches infinity, the integral grows without bound. We conclude that $\int_1^\infty (1/x^p)\, dx$ diverges precisely when $p \le 1$. For $p > 1$ the integral has the value

$$\int_1^\infty \frac{1}{x^p}\, dx = \lim_{b \to \infty} \int_1^b \frac{1}{x^p}\, dx = \lim_{b \to \infty} \left(\frac{1}{-p+1} b^{-p+1} - \frac{1}{-p+1} \right) = - \left(\frac{1}{-p+1} \right) = \frac{1}{p-1}.$$

Application of Improper Integrals to Energy

The energy, E, required to separate two charged particles, originally a distance a apart, to a distance b, is given by the integral

$$E = \int_a^b \frac{k q_1 q_2}{r^2}\, dr$$

where q_1 and q_2 are the magnitudes of the charges and k is a constant. If q_1 and q_2 are in coulombs, a and b are in meters, and E is in joules, the value of the constant k is $9 \cdot 10^9$.

Example 4   A hydrogen atom consists of a proton and an electron, with opposite charges of magnitude $1.6 \cdot 10^{-19}$ coulombs. Find the energy required to take a hydrogen atom apart (that is, to move the electron from its orbit to an infinite distance from the proton). Assume that the initial distance between the electron and the proton is the Bohr radius, $R_B = 5.3 \cdot 10^{-11}$ meter.

Solution   Since we are moving from an initial distance of R_B to a final distance of ∞, the energy is represented by the improper integral

$$E = \int_{R_B}^\infty k \frac{q_1 q_2}{r^2}\, dr = k q_1 q_2 \lim_{b \to \infty} \int_{R_B}^b \frac{1}{r^2}\, dr$$

$$= k q_1 q_2 \lim_{b \to \infty} -\frac{1}{r} \Big|_{R_B}^b = k q_1 q_2 \lim_{b \to \infty} \left(-\frac{1}{b} + \frac{1}{R_B} \right) = \frac{k q_1 q_2}{R_B}.$$

Substituting numerical values, we get

$$E = \frac{(9 \cdot 10^9)(1.6 \cdot 10^{-19})^2}{5.3 \cdot 10^{-11}} \approx 4.35 \cdot 10^{-18} \text{ joules.}$$

This is about the amount of energy needed to lift a speck of dust 0.000000025 inch off the ground. (In other words, not much!)

What happens if the limits of integration are $-\infty$ and ∞? In this case, we break the integral at any point and write the original integral as a sum of two new improper integrals.

We can use any (finite) number c to define

$$\int_{-\infty}^\infty f(x)\, dx = \int_{-\infty}^c f(x)\, dx + \int_c^\infty f(x)\, dx.$$

If *either* of the two new improper integrals diverges, we say the original integral diverges. Only if both of the new integrals have a finite value do we add the values to get a finite value for the original integral.

It is not hard to show that the preceding definition does not depend on the choice for c.

Another Type of Improper Integral: When the Integrand Becomes Infinite

There is another way for an integral to be improper. The interval may be finite but the function may be unbounded near some points in the interval. For example, consider $\int_0^1 (1/\sqrt{x})\,dx$. Since the graph of $y = 1/\sqrt{x}$ has a vertical asymptote at $x = 0$, the region between the graph, the x-axis, and the lines $x = 0$ and $x = 1$ is unbounded. Instead of extending to infinity in the horizontal direction as in the previous improper integrals, this region extends to infinity in the vertical direction. See Figure 7.19(a). We handle this improper integral in a similar way as before: we compute $\int_a^1 (1/\sqrt{x})\,dx$ for values of a slightly larger than 0 and look at what happens as a approaches 0 from the positive side. (This is written as $a \to 0^+$.)

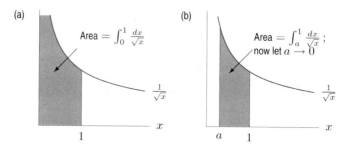

Figure 7.19: Area representation of improper integral

First we compute the integral:

$$\int_a^1 \frac{1}{\sqrt{x}}\,dx = 2x^{1/2}\Big|_a^1 = 2 - 2a^{1/2}.$$

Now we take the limit:

$$\lim_{a \to 0^+} \int_a^1 \frac{1}{\sqrt{x}}\,dx = \lim_{a \to 0^+} (2 - 2a^{1/2}) = 2.$$

Since the limit is finite, we say the improper integral converges, and that

$$\int_0^1 \frac{1}{\sqrt{x}}\,dx = 2.$$

Geometrically, what we have done is to calculate the finite area between $x = a$ and $x = 1$ and take the limit as a tends to 0 from the right. See Figure 7.19(b). Since the limit exists, the integral converges to 2. If the limit did not exist, we would say the improper integral diverges.

Example 5   Investigate the convergence of $\displaystyle\int_0^2 \frac{1}{(x-2)^2}\,dx$.

Solution   This is an improper integral since the integrand tends to infinity as x approaches 2, and is undefined at $x = 2$. Since the trouble is at the right endpoint, we replace the upper limit by b, and let b tend to 2 from the left. This is written $b \to 2^-$, with the "$-$" signifying that 2 is approached from below. See Figure 7.20.

$$\int_0^2 \frac{1}{(x-2)^2}\,dx = \lim_{b \to 2^-} \int_0^b \frac{1}{(x-2)^2}\,dx = \lim_{b \to 2^-} (-1)(x-2)^{-1}\Big|_0^b = \lim_{b \to 2^-} \left(-\frac{1}{(b-2)} - \frac{1}{2}\right).$$

Therefore, since $\displaystyle\lim_{b \to 2^-}\left(-\frac{1}{b-2}\right)$ does not exist, the integral diverges.

Figure 7.20: Shaded area represents $\int_0^2 \frac{1}{(x-2)^2}\,dx$

Figure 7.21: Shaded area represents $\int_{-1}^2 \frac{1}{x^4}\,dx$

Suppose $f(x)$ is positive and continuous on $a \le x < b$ and tends to infinity as $x \to b$.

If $\displaystyle\lim_{c \to b^-} \int_a^c f(x)\,dx$ is a finite number, we say that $\displaystyle\int_a^b f(x)\,dx$ **converges** and define

$$\int_a^b f(x)\,dx = \lim_{c \to b^-} \int_a^c f(x)\,dx.$$

Otherwise, we say that $\displaystyle\int_a^b f(x)\,dx$ **diverges**.

When $f(x)$ tends to infinity as x approaches a, we define convergence in a similar way. In addition, an integral can be improper because the integrand tends to infinity *inside* the interval of integration rather than at an endpoint. In this case, we break the given integral into two (or more) improper integrals so that the integrand tends to infinity only at endpoints.

Suppose that $f(x)$ is positive and continuous on $[a, b]$ except at the point c. If $f(x)$ tends to infinity as $x \to c$, then we define

$$\int_a^b f(x)\,dx = \int_a^c f(x)\,dx + \int_c^b f(x)\,dx.$$

If *either* of the two new improper integrals diverges, we say the original integral diverges. Only if *both* of the new integrals have a finite value do we add the values to get a finite value for the original integral.

Example 6   Investigate the convergence of $\displaystyle\int_{-1}^2 \frac{1}{x^4}\,dx$.

Solution   See the graph in Figure 7.21. The trouble spot is $x = 0$, rather than $x = -1$ or $x = 2$. To handle this situation, we break the given improper integral into two other improper integrals each of which have $x = 0$ as one of the endpoints:

$$\int_{-1}^2 \frac{1}{x^4}\,dx = \int_{-1}^0 \frac{1}{x^4}\,dx + \int_0^2 \frac{1}{x^4}\,dx.$$

We can now use the previous technique to evaluate the new integrals, if they converge. Since

$$\int_0^2 \frac{1}{x^4}\,dx = \lim_{a \to 0^+} -\frac{1}{3}x^{-3}\Big|_a^2 = \lim_{a \to 0^+} \left(-\frac{1}{3}\right)\left(\frac{1}{8} - \frac{1}{a^3}\right)$$

the integral $\int_0^2 (1/x^4)\, dx$ diverges. Thus, the original integral diverges. A similar computation shows that $\int_{-1}^0 (1/x^4)\, dx$ also diverges.

It is easy to miss an improper integral when the integrand tends to infinity inside the interval. For example, it is fundamentally incorrect to say that $\int_{-1}^2 (1/x^4)\, dx = -\frac{1}{3}x^{-3}\Big|_{-1}^2 = -\frac{1}{24} - \frac{1}{3} = -\frac{3}{8}$.

Example 7   Find $\displaystyle\int_0^6 \frac{1}{(x-4)^{2/3}}\, dx$.

Solution   Figure 7.22 shows that the trouble spot is at $x = 4$, so we break the integral at $x = 4$ and consider the separate parts.

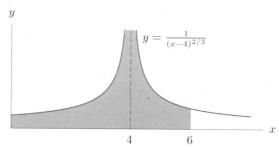

Figure 7.22: Shaded area represents $\int_0^6 \frac{1}{(x-4)^{2/3}}\, dx$

We have

$$\int_0^4 \frac{1}{(x-4)^{2/3}}\, dx = \lim_{b \to 4^-} 3(x-4)^{1/3}\Big|_0^b = \lim_{b \to 4^-}\left(3(b-4)^{1/3} - 3(-4)^{1/3}\right) = 3(4)^{1/3}.$$

Similarly,

$$\int_4^6 \frac{1}{(x-4)^{2/3}}\, dx = \lim_{a \to 4^+} 3(x-4)^{1/3}\Big|_a^6 = \lim_{a \to 4^+}\left(3 \cdot 2^{1/3} - 3(a-4)^{1/3}\right) = 3(2)^{1/3}.$$

Since both of these integrals converge, the original integral converges:

$$\int_0^6 \frac{1}{(x-4)^{2/3}}\, dx = 3(4)^{1/3} + 3(2)^{1/3} \approx 8.54.$$

Finally, there is a question of what to do when an integral is improper at both endpoints. In this case, we just break the integral at any interior point of the interval. The original integral diverges if either or both of the new integrals diverge.

Example 8   Investigate the convergence of $\displaystyle\int_0^\infty \frac{1}{x^2}\, dx$.

Solution   This integral is improper both because the upper limit is ∞ and because the function is undefined at $x = 0$. We break the integral into two parts at, say, $x = 1$. We know by Example 3 that $\int_1^\infty (1/x^2)\, dx$ has a finite value. However, the other part, $\int_0^1 (1/x^2)\, dx$, diverges since:

$$\int_0^1 \frac{1}{x^2}\, dx = \lim_{a \to 0^+} -x^{-1}\Big|_a^1 = \lim_{a \to 0^+}\left(\frac{1}{a} - 1\right).$$

Therefore $\displaystyle\int_0^\infty \frac{1}{x^2}\, dx$ diverges as well.

Exercises and Problems for Section 7.7

Exercises

1. Shade the area represented by:
   (a) $\int_1^\infty (1/x^2)\, dx$ (b) $\int_0^1 (1/\sqrt{x})\, dx$

2. Evaluate the improper integral $\int_0^\infty e^{-0.4x}\, dx$ and sketch the area it represents.

3. (a) Use a calculator or computer to estimate $\int_0^b xe^{-x}dx$ for $b = 5, 10, 20$.
   (b) Use your answers to part (a) to estimate the value of $\int_0^\infty xe^{-x}dx$, assuming it is finite.

4. (a) Sketch the the area represented by the improper integral $\int_{-\infty}^\infty e^{-x^2}\, dx$.
   (b) Use a calculator or computer to estimate $\int_{-a}^a e^{-x^2}\, dx$ for $a = 1,2,3,4,5$.
   (c) Use the answers to part (b) to estimate the value of $\int_{-\infty}^\infty e^{-x^2}\, dx$, assuming it is finite.

Calculate the integrals in Exercises 5–32, if they converge.

5. $\int_1^\infty \dfrac{1}{5x + 2}\, dx$

6. $\int_1^\infty \dfrac{1}{(x + 2)^2}\, dx$

7. $\int_0^\infty xe^{-x^2}\, dx$

8. $\int_1^\infty e^{-2x}\, dx$

9. $\int_0^\infty \dfrac{x}{e^x}\, dx$

10. $\int_1^\infty \dfrac{x}{4 + x^2}\, dx$

11. $\int_{-\infty}^0 \dfrac{e^x}{1 + e^x}\, dx$

12. $\int_{-\infty}^\infty \dfrac{dz}{z^2 + 25}$

13. $\int_0^4 \dfrac{dx}{\sqrt{16 - x^2}}$

14. $\int_{\pi/4}^{\pi/2} \dfrac{\sin x}{\sqrt{\cos x}}\, dx$

15. $\int_{-1}^1 \dfrac{1}{v}\, dv$

16. $\int_0^1 \dfrac{x^4 + 1}{x}\, dx$

17. $\int_1^\infty \dfrac{1}{x^2 + 1}\, dx$

18. $\int_1^\infty \dfrac{1}{\sqrt{x^2 + 1}}\, dx$

19. $\int_0^4 \dfrac{1}{u^2 - 16}\, du$

20. $\int_1^\infty \dfrac{y}{y^4 + 1}\, dy$

21. $\int_2^\infty \dfrac{dx}{x \ln x}$

22. $\int_0^1 \dfrac{\ln x}{x}\, dx$

23. $\int_{16}^{20} \dfrac{1}{y^2 - 16}\, dy$

24. $\int_1^2 \dfrac{dx}{x \ln x}$

25. $\int_0^\pi \dfrac{1}{\sqrt{x}} e^{-\sqrt{x}}\, dx$

26. $\int_3^\infty \dfrac{dx}{x(\ln x)^2}$

27. $\int_0^2 \dfrac{1}{\sqrt{4 - x^2}}\, dx$

28. $\int_4^\infty \dfrac{dx}{(x - 1)^2}$

29. $\int_4^\infty \dfrac{dx}{x^2 - 1}$

30. $\int_7^\infty \dfrac{dy}{\sqrt{y - 5}}$

31. $\int_{-3}^3 \dfrac{y\, dy}{\sqrt{9 - y^2}}$

32. $\int_3^6 \dfrac{d\theta}{(4 - \theta)^2}$

Problems

33. Find the area under the curve $y = xe^{-x}$ for $x \geq 0$.

34. Find the area under the curve $y = 1/\cos^2 t$ between $t = 0$ and $t = \pi/2$.

For what values of p do the integrals in Problems 35–36 converge or diverge? What is the value of the integral when it converges?

35. $\int_e^\infty x^p \ln x\, dx$

36. $\int_0^e x^p \ln x\, dx$

37. For $\alpha > 0$, calculate
    (a) $\int_0^\infty \dfrac{e^{-y/\alpha}}{\alpha}\, dy$ (b) $\int_0^\infty \dfrac{ye^{-y/\alpha}}{\alpha}\, dy$
    (c) $\int_0^\infty \dfrac{y^2 e^{-y/\alpha}}{\alpha}\, dy$

38. Given that $\int_{-\infty}^\infty e^{-x^2}\, dx = \sqrt{\pi}$, calculate the exact value of
    $$\int_{-\infty}^\infty e^{-(x-a)^2/b}\, dx.$$

The kth moment, m_k of the normal distribution is defined by
$$m_k = \frac{1}{\sqrt{2\pi}} \int_{-\infty}^\infty x^k e^{-x^2/2}dx.$$

In Problems 39–42, use the fact that $\int_{-\infty}^\infty e^{-x^2/2}dx = \sqrt{2\pi}$ to calculate the moments. Assume all the integrals converge.

39. m_1 40. m_2 41. m_3 42. m_4

43. The gamma function is defined for all $x > 0$ by the rule
    $$\Gamma(x) = \int_0^\infty t^{x-1}e^{-t}\, dt.$$
    (a) Find $\Gamma(1)$ and $\Gamma(2)$.
    (b) Integrate by parts with respect to t to show that, for positive n,
    $$\Gamma(n + 1) = n\Gamma(n).$$
    (c) Find a simple expression for $\Gamma(n)$ for positive integers n.

44. The rate, r, at which people get sick during an epidemic of the flu can be approximated by $r = 1000te^{-0.5t}$, where r is measured in people/day and t is measured in days since the start of the epidemic.

    (a) Sketch a graph of r as a function of t.

    (b) When are people getting sick fastest?

    (c) How many people get sick altogether?

45. Find the energy required to separate opposite electric charges of magnitude 1 coulomb. The charges are initially 1 meter apart and one is moved infinitely far from the other. (The definition of energy is on page 351.)

7.8 COMPARISON OF IMPROPER INTEGRALS

Making Comparisons

Sometimes it is difficult to find the exact value of an improper integral by antidifferentiation, but it may be possible to determine whether an integral converges or diverges. The key is to *compare* the given integral to one whose behavior we already know. Let's look at an example.

Example 1   Determine whether $\displaystyle\int_1^\infty \frac{1}{\sqrt{x^3 + 5}}\, dx$ converges.

Solution   First, let's see what this integrand does as $x \to \infty$. For large x, the 5 becomes insignificant compared with the x^3, so

$$\frac{1}{\sqrt{x^3 + 5}} \approx \frac{1}{\sqrt{x^3}} = \frac{1}{x^{3/2}}.$$

Since

$$\int_1^\infty \frac{1}{\sqrt{x^3}}\, dx = \int_1^\infty \frac{1}{x^{3/2}}\, dx = \lim_{b\to\infty} \int_1^b \frac{1}{x^{3/2}}\, dx = \lim_{b\to\infty} -2x^{-1/2}\Big|_1^b = \lim_{b\to\infty}\left(2 - 2b^{-1/2}\right) = 2,$$

the integral $\int_1^\infty (1/x^{3/2})\, dx$ converges. So we expect our integral to converge as well.

In order to confirm this, we observe that for $0 \le x^3 \le x^3 + 5$, we have

$$\frac{1}{\sqrt{x^3 + 5}} \le \frac{1}{\sqrt{x^3}}.$$

and so for $b \ge 1$,

$$\int_1^b \frac{1}{\sqrt{x^3 + 5}}\, dx \le \int_1^b \frac{1}{\sqrt{x^3}}\, dx.$$

(See Figure 7.23.) Since $\int_1^b (1/\sqrt{x^3 + 5})\, dx$ increases as b approaches infinity but is always smaller than $\int_1^b (1/x^{3/2})\, dx < \int_1^\infty (1/x^{3/2})\, dx = 2$, we know $\int_1^\infty (1/\sqrt{x^3 + 5})\, dx$ must have a finite value less than 2. Thus,

$$\int_1^\infty \frac{dx}{\sqrt{x^3 + 5}} \text{ converges to a value less than 2.}$$

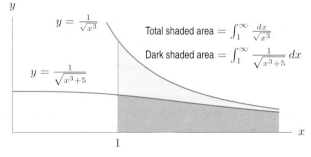

Figure 7.23: Graph showing $\int_1^\infty \frac{1}{\sqrt{x^3+5}}\, dx \le \int_1^\infty \frac{dx}{\sqrt{x^3}}$

A little more work is required to estimate the value of a convergent improper integral.

Example 2   Estimate the value of $\int_1^\infty \dfrac{1}{\sqrt{x^3+5}}\,dx$ with an error of less than 0.01 using the approximation

$$\int_1^\infty \frac{1}{\sqrt{x^3+5}}\,dx \approx \int_1^b \frac{1}{\sqrt{x^3+5}}\,dx.$$

Solution   We must figure out how large a value of b to take. Since

$$\int_1^\infty \frac{1}{\sqrt{x^3+5}}\,dx = \int_1^b \frac{1}{\sqrt{x^3+5}}\,dx + \int_b^\infty \frac{1}{\sqrt{x^3+5}}\,dx,$$

we find b such that the tail of the integral satisfies the inequality

$$\left| \int_b^\infty \frac{1}{\sqrt{x^3+5}}\,dx \right| < 0.01.$$

From the solution of Example 1, we have

$$0 < \int_b^\infty \frac{1}{\sqrt{x^3+5}}\,dx < \int_b^\infty \frac{1}{\sqrt{x^3}}\,dx = \frac{2}{\sqrt{b}}.$$

We choose b such that $2/\sqrt{b} < 0.01$, which means that $b > 40{,}000$. With an error of less than 0.01, we have

$$\int_1^\infty \frac{1}{\sqrt{x^3+5}}\,dx \approx \int_1^{50{,}000} \frac{1}{\sqrt{x^3+5}}\,dx = 1.699.$$

Notice that we first looked at the behavior of the integrand as $x \to \infty$. This is useful because the convergence or divergence of the integral is determined by what happens as $x \to \infty$.

The Comparison Test for $\displaystyle\int_a^\infty f(x)\,dx$

Assume $f(x)$ is positive. Making a comparison involves two stages:
1. Guess, by looking at the behavior of the integrand for large x, whether the integral converges or not. (This is the "behaves like" principle.)
2. Confirm the guess by comparison:
   - If $0 \le f(x) \le g(x)$ and $\int_a^\infty g(x)\,dx$ converges, then $\int_a^\infty f(x)\,dx$ converges.
   - If $0 \le g(x) \le f(x)$ and $\int_a^\infty g(x)\,dx$ diverges, then $\int_a^\infty f(x)\,dx$ diverges.

Example 3   Decide whether $\int_4^\infty \dfrac{dt}{(\ln t) - 1}$ converges or diverges.

Solution   Since $\ln t$ grows without bound as $t \to \infty$, the -1 is eventually going to be insignificant in comparison to $\ln t$. Thus, as far as convergence is concerned,

$$\int_4^\infty \frac{1}{(\ln t) - 1}\,dt \quad \text{behaves like} \quad \int_4^\infty \frac{1}{\ln t}\,dt.$$

Does $\int_4^\infty (1/\ln t)\,dt$ converge or diverge? Since $\ln t$ grows very slowly, $1/\ln t$ goes to zero very slowly, and so the integral probably does not converge. We know that $(\ln t) - 1 < \ln t < t$ for all positive t. So, provided $t > e$, we take reciprocals:

$$\frac{1}{(\ln t) - 1} > \frac{1}{\ln t} > \frac{1}{t}.$$

Since $\int_4^\infty (1/t)\,dt$ diverges, we conclude that

$$\int_4^\infty \frac{1}{(\ln t) - 1}\,dt \text{ diverges.}$$

How Do We Know What To Compare With?

In Examples 1 and 3, we investigated the convergence of an integral by comparing it with an easier integral. How did we pick the easier integral? This is a matter of trial and error, guided by any information we get by looking at the original integrand as $x \to \infty$. We want the comparison integrand to be easy and, in particular, to have a simple antiderivative.

Useful Integrals for Comparison

- $\displaystyle\int_1^\infty \frac{1}{x^p}\,dx$ converges for $p > 1$ and diverges for $p \le 1$.

- $\displaystyle\int_0^1 \frac{1}{x^p}\,dx$ converges for $p < 1$ and diverges for $p \ge 1$.

- $\displaystyle\int_0^\infty e^{-ax}\,dx$ converges for $a > 0$.

Of course, we can use any function for comparison, provided we can determine its behavior.

Example 4   Investigate the convergence of $\displaystyle\int_1^\infty \frac{(\sin x) + 3}{\sqrt{x}}\,dx$.

Solution   Since it looks difficult to find an antiderivative of this function, we try comparison. What happens to this integrand as $x \to \infty$? Since $\sin x$ oscillates between -1 and 1,

$$\frac{2}{\sqrt{x}} = \frac{-1+3}{\sqrt{x}} \le \frac{(\sin x)+3}{\sqrt{x}} \le \frac{1+3}{\sqrt{x}} = \frac{4}{\sqrt{x}},$$

the integrand oscillates between $2/\sqrt{x}$ and $4/\sqrt{x}$. (See Figure 7.24.)

What do $\int_1^\infty (2/\sqrt{x})\,dx$ and $\int_1^\infty (4/\sqrt{x})\,dx$ do? As far as convergence is concerned, they certainly do the same thing, and whatever that is, the original integral does it too. It is important to notice that $\sqrt{x}$ grows very slowly. This means that $1/\sqrt{x}$ gets small slowly, which means that convergence is unlikely. Since $\sqrt{x} = x^{1/2}$, the result in the preceding box (with $p = \frac{1}{2}$) tells us that $\int_1^\infty (1/\sqrt{x})\,dx$ diverges. So the comparison test tells us that the original integral diverges.

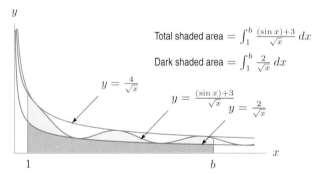

Total shaded area $= \int_1^b \frac{(\sin x)+3}{\sqrt{x}}\,dx$

Dark shaded area $= \int_1^b \frac{2}{\sqrt{x}}\,dx$

Figure 7.24: Graph showing $\int_1^b \frac{2}{\sqrt{x}}\,dx \le \int_1^b \frac{(\sin x)+3}{\sqrt{x}}\,dx$, for $b \ge 1$

Notice that there are two possible comparisons we could have made in Example 4:

$$\frac{2}{\sqrt{x}} \le \frac{(\sin x)+3}{\sqrt{x}} \qquad \text{or} \qquad \frac{(\sin x)+3}{\sqrt{x}} \le \frac{4}{\sqrt{x}}.$$

Since both $\int_1^\infty (2/\sqrt{x})\,dx$ and $\int_1^\infty (4/\sqrt{x})\,dx$ diverge, only the first comparison is useful. Knowing that an integral is *smaller* than a divergent integral is of no help whatsoever!

The next example shows what to do if the comparison does not hold throughout the interval of integration.

Example 5   Show $\int_{1}^{\infty} e^{-x^2/2}\, dx$ converges to a finite value.

Solution   We know that $e^{-x^2/2}$ goes very rapidly to zero as $x \to \infty$, so we expect this integral to converge. Hence we look for some larger integrand which has a convergent integral. One possibility is $\int_{1}^{\infty} e^{-x}\, dx$, because e^{-x} has an elementary antiderivative and $\int_{1}^{\infty} e^{-x}\, dx$ converges. What is the relationship between $e^{-x^2/2}$ and e^{-x}? We know that for $x \geq 2$,

$$x \leq \frac{x^2}{2} \quad \text{so} \quad -\frac{x^2}{2} \leq -x,$$

and so, for $x \geq 2$

$$e^{-x^2/2} \leq e^{-x}.$$

Since this inequality holds only for $x \geq 2$, we split the interval of integration into two pieces:

$$\int_{1}^{\infty} e^{-x^2/2}\, dx = \int_{1}^{2} e^{-x^2/2}\, dx + \int_{2}^{\infty} e^{-x^2/2}\, dx.$$

Now $\int_{1}^{2} e^{-x^2/2}\, dx$ is finite (it is not improper) and $\int_{2}^{\infty} e^{-x^2/2}\, dx$ is finite by comparison with $\int_{2}^{\infty} e^{-x}\, dx$. Therefore, $\int_{1}^{\infty} e^{-x^2/2}\, dx$ is the sum of two finite pieces and therefore must be finite.

Exercises and Problems for Section 7.8

Exercises

In Exercises 1–9, use the box on page 358 and the behavior of rational and exponential functions as $x \to \infty$ to predict whether the integrals converge or diverge.

1. $\int_{1}^{\infty} \frac{x^2}{x^4 + 1}\, dx$

2. $\int_{2}^{\infty} \frac{x^3}{x^4 - 1}\, dx$

3. $\int_{1}^{\infty} \frac{x^2 + 1}{x^3 + 3x + 2}\, dx$

4. $\int_{1}^{\infty} \frac{1}{x^2 + 5x + 1}\, dx$

5. $\int_{1}^{\infty} \frac{x}{x^2 + 2x + 4}\, dx$

6. $\int_{1}^{\infty} \frac{x^2 - 6x + 1}{x^2 + 4}\, dx$

7. $\int_{1}^{\infty} \frac{5x + 2}{x^4 + 8x^2 + 4}\, dx$

8. $\int_{1}^{\infty} \frac{1}{e^{5t} + 2}\, dt$

9. $\int_{1}^{\infty} \frac{x^2 + 4}{x^4 + 3x^2 + 11}\, dx$

In Exercises 10–25 decide if the improper integral converges or diverges. Explain your reasoning.

10. $\int_{50}^{\infty} \frac{dz}{z^3}$

11. $\int_{1}^{\infty} \frac{dx}{1 + x}$

12. $\int_{1}^{\infty} \frac{dx}{x^3 + 1}$

13. $\int_{5}^{8} \frac{6}{\sqrt{t - 5}}\, dt$

14. $\int_{0}^{1} \frac{1}{x^{19/20}}\, dx$

15. $\int_{-1}^{5} \frac{dt}{(t + 1)^2}$

16. $\int_{-\infty}^{\infty} \frac{du}{1 + u^2}$

17. $\int_{1}^{\infty} \frac{du}{u + u^2}$

18. $\int_{1}^{\infty} \frac{d\theta}{\sqrt{\theta^2 + 1}}$

19. $\int_{2}^{\infty} \frac{d\theta}{\sqrt{\theta^3 + 1}}$

20. $\int_{0}^{1} \frac{d\theta}{\sqrt{\theta^3 + \theta}}$

21. $\int_{0}^{\infty} \frac{dy}{1 + e^y}$

22. $\int_{1}^{\infty} \frac{2 + \cos \phi}{\phi^2}\, d\phi$

23. $\int_{0}^{\infty} \frac{dz}{e^z + 2^z}$

24. $\int_{0}^{\pi} \frac{2 - \sin \phi}{\phi^2}\, d\phi$

25. $\int_{4}^{\infty} \frac{3 + \sin \alpha}{\alpha}\, d\alpha$

Estimate the values of the integrals in Exercises 26–27 to two decimal places by integrating the functions on your calculator or computer for large values of the upper limit of integration.

26. $\int_{1}^{\infty} e^{-x^2}\, dx$

27. $\int_{0}^{\infty} e^{-x^2} \cos^2 x\, dx$

Problems

28. The graphs of $y = 1/x, y = 1/x^2$ and the functions $f(x)$, $g(x)$, $h(x)$, and $k(x)$ are shown in Figure 7.25.

  (a) Is the area between $y = 1/x$ and $y = 1/x^2$ on the interval from $x = 1$ to ∞ finite or infinite? Explain.

  (b) Using the graph, decide whether the integral of each of the functions $f(x)$, $g(x)$, $h(x)$ and $k(x)$ on the interval from $x = 1$ to ∞ converges, diverges, or whether it is impossible to tell.

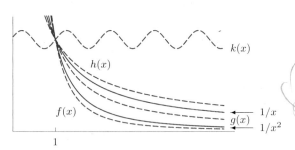

Figure 7.25

29. Suppose $\int_a^\infty f(x)\,dx$ converges. What does Figure 7.26 suggest about the convergence of $\int_a^\infty g(x)\,dx$?

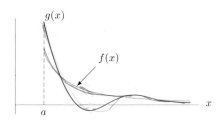

Figure 7.26

For what values of p do the integrals in Problems 30–31 converge or diverge?

30. $\displaystyle\int_2^\infty \frac{dx}{x(\ln x)^p}$ **31.** $\displaystyle\int_1^2 \frac{dx}{x(\ln x)^p}$

32. Find the value of a (to three decimal places) that makes

$$\int_{-\infty}^\infty a e^{-x^2/2}\,dx = 1.$$

33. (a) The function $g(x) = a e^{-(x-k)^2/2}$ is used by statisticians. To three decimal places, what value of a should be chosen to ensure that

$$\int_{-\infty}^\infty g(x)\,dx = 1?$$

  (b) Is your answer the same as or different from your answer to Problem 32? Why?

34. (a) Find an upper bound for

$$\int_3^\infty e^{-x^2}\,dx.$$

  [Hint: $e^{-x^2} \le e^{-3x}$ for $x \ge 3$.]

  (b) For any positive n, generalize the result of part (a) to find an upper bound for

$$\int_n^\infty e^{-x^2}\,dx$$

  by noting that $nx \le x^2$ for $x \ge n$.

35. In Planck's Radiation Law, we encounter the integral

$$\int_1^\infty \frac{dx}{x^5(e^{1/x} - 1)}.$$

  (a) Explain why a graph of the tangent line to e^t at $t = 0$ tells us that for all t

$$1 + t \le e^t.$$

  (b) Substituting $t = 1/x$, show that for all x

$$e^{1/x} - 1 > \frac{1}{x}.$$

  (c) Use the comparison test to show that the original integral converges.

CHAPTER SUMMARY (see also Ready Reference at the end of the book)

- **Integration techniques**
  Substitution, parts, partial fractions, trigonometric substitution, using tables.
- **Numerical approximations**

  Riemann sums (left, right, midpoint), trapezoid rule, Simpson's rule, approximation errors.
- **Improper integrals**
  Convergence/divergence, comparison test for integrals.

REVIEW EXERCISES AND PROBLEMS FOR CHAPTER SEVEN

Exercises

For Exercises 1–106, evaluate the following integrals. Assume $a, b, c,$ and k are constants. Exercises 1– 66 can be done without an integral table, as can some of the later problems.

1. $\int \sin t \, dt$

2. $\int \cos 2t \, dt$

3. $\int e^{5z} \, dz$

4. $\int (3w + 7) \, dw$

5. $\int \sin 2\theta \, d\theta$

6. $\int (x^3 - 1)^4 x^2 \, dx$

7. $\int \left(x^{3/2} + x^{2/3} \right) dx$

8. $\int (e^x + 3^x) \, dx$

9. $\int (r + 1)^3 \, dr$

10. $\int \left(\dfrac{4}{x^2} - \dfrac{3}{x^3} \right) dx$

11. $\int \dfrac{x^3 + x + 1}{x^2} dx$

12. $\int \dfrac{(1 + \ln x)^2}{x} dx$

13. $\int t e^{t^2} \, dt$

14. $\int x \cos x \, dx$

15. $\int x^2 e^{2x} \, dx$

16. $\int x\sqrt{1 - x} \, dx$

17. $\int x \ln x \, dx$

18. $\int y \sin y \, dy$

19. $\int (\ln x)^2 \, dx$

20. $\int \ln(x^2) \, dx$

21. $\int e^{0.5 - 0.3t} \, dt$

22. $\int \sin^2 \theta \cos \theta \, d\theta$

23. $\int x\sqrt{4 - x^2} \, dx$

24. $\int \dfrac{(u + 1)^3}{u^2} \, du$

25. $\int \dfrac{\cos \sqrt{y}}{\sqrt{y}} \, dy$

26. $\int \dfrac{1}{\cos^2 z} \, dz$

27. $\int \cos^2 \theta \, d\theta$

28. $\int t^{10}(t - 10) \, dt$

29. $\int \tan(2x - 6) \, dx$

30. $\int \dfrac{(\ln x)^2}{x} \, dx$

31. $\int \dfrac{(t + 2)^2}{t^3} \, dt$

32. $\int \left(x^2 + 2x + \dfrac{1}{x} \right) dx$

33. $\int \dfrac{t + 1}{t^2} \, dt$

34. $\int t e^{t^2 + 1} \, dt$

35. $\int \tan \theta \, d\theta$

36. $\int \sin(5\theta) \cos(5\theta) \, d\theta$

37. $\int \dfrac{x}{x^2 + 1} \, dx$

38. $\int \dfrac{dz}{1 + z^2}$

39. $\int \dfrac{dz}{1 + 4z^2}$

40. $\int \cos^3 2\theta \sin 2\theta \, d\theta$

41. $\int \sin 5\theta \cos^3 5\theta \, d\theta$

42. $\int \sin^3 z \cos^3 z \, dz$

43. $\int t(t - 10)^{10} \, dt$

44. $\int \cos \theta \sqrt{1 + \sin \theta} \, d\theta$

45. $\int x e^x \, dx$

46. $\int t^3 e^t \, dt$

47. $\int_1^3 x(x^2 + 1)^{70} \, dx$

48. $\int (3z + 5)^3 \, dz$

49. $\int \dfrac{du}{9 + u^2}$

50. $\int \dfrac{\cos w}{1 + \sin^2 w} \, dw$

51. $\int \dfrac{1}{x} \tan(\ln x) \, dx$

52. $\int \dfrac{1}{x} \sin(\ln x) \, dx$

53. $\int \dfrac{dx}{\sqrt{1 - 4x^2}}$

54. $\int \dfrac{w \, dw}{\sqrt{16 - w^2}}$

55. $\int \dfrac{e^{2y} + 1}{e^{2y}} \, dy$

56. $\int \dfrac{\sin w \, dw}{\sqrt{1 - \cos w}}$

57. $\int \dfrac{dx}{x \ln x}$

58. $\int \dfrac{du}{3u + 8}$

59. $\int \dfrac{x \cos \sqrt{x^2 + 1}}{\sqrt{x^2 + 1}} \, dx$

60. $\int \dfrac{t^3}{\sqrt{1 + t^2}} \, dt$

61. $\int u e^{ku} \, du$

62. $\int (w + 5)^4 w \, dw$

63. $\int e^{\sqrt{2x + 3}} \, dx$

64. $\int r(\ln r)^2 \, dr$

65. $\int (e^x + x)^2 \, dx$

66. $\int u^2 \ln u \, du$

67. $\int \dfrac{5x + 6}{x^2 + 4} \, dx$

68. $\int \dfrac{1}{\sin^3(2x)} \, dx$

69. $\int \dfrac{dr}{r^2 - 100}$

70. $\int y^2 \sin(cy) \, dy$

71. $\int e^{-ct} \sin kt \, dt$

72. $\int e^{5x} \cos(3x) \, dx$

73. $\displaystyle\int \left(x^{\sqrt{k}} + \sqrt{k}^{x}\right) dx$

74. $\displaystyle\int \sqrt{3 + 12x^2}\, dx.$

75. $\displaystyle\int \frac{1}{\sqrt{x^2 - 3x + 2}}\, dx$

76. $\displaystyle\int \frac{x^3}{x^2 + 3x + 2}\, dx$

77. $\displaystyle\int \frac{x^2 + 1}{x^2 - 3x + 2}\, dx$

78. $\displaystyle\int \frac{dx}{ax^2 + bx}$

79. $\displaystyle\int \frac{ax + b}{ax^2 + 2bx + c}\, dx$

80. $\displaystyle\int \left(\frac{x}{3} + \frac{3}{x}\right)^2 dx$

81. $\displaystyle\int \frac{2^t}{2^t + 1}\, dt$

82. $\displaystyle\int 10^{1-x}\, dx$

83. $\displaystyle\int (x^2 + 5)^3\, dx$

84. $\displaystyle\int v \arcsin v\, dv$

85. $\displaystyle\int \sqrt{4 - x^2}\, dx$

86. $\displaystyle\int \frac{z^3}{z - 5}\, dz$

87. $\displaystyle\int \frac{\sin w \cos w}{1 + \cos^2 w}\, dw$

88. $\displaystyle\int \frac{1}{\tan(3\theta)}\, d\theta$

89. $\displaystyle\int \frac{x}{\cos^2 x}\, dx$

90. $\displaystyle\int \frac{x + 1}{\sqrt{x}}\, dx$

91. $\displaystyle\int \frac{x}{\sqrt{x + 1}}\, dx$

92. $\displaystyle\int \frac{\sqrt{\sqrt{x} + 1}}{\sqrt{x}}\, dx$

93. $\displaystyle\int \frac{e^{2y}}{e^{2y} + 1}\, dy$

94. $\displaystyle\int \frac{z}{(z^2 - 5)^3}\, dz$

95. $\displaystyle\int \frac{z}{(z - 5)^3}\, dz$

96. $\displaystyle\int \frac{(1 + \tan x)^3}{\cos^2 x}\, dx$

97. $\displaystyle\int \frac{(2x - 1)e^{x^2}}{e^x}\, dx$

98. $\displaystyle\int (2x + 1)e^{x^2} e^x\, dx$

99. $\displaystyle\int \sin x (\sqrt{2 + 3\cos x})\, dx$

100. $\displaystyle\int (x^2 - 3x + 2)e^{-4x}\, dx$

101. $\displaystyle\int \sin^2(2\theta) \cos^3(2\theta)\, d\theta$

102. $\displaystyle\int \cos(2 \sin x) \cos x\, dx$

103. $\displaystyle\int (x + \sin x)^3 (1 + \cos x)\, dx$

104. $\displaystyle\int \left(2x^3 + 3x + 4\right) \cos(2x)\, dx$

105. $\displaystyle\int \sinh^2 x \cosh x\, dx$

106. $\displaystyle\int (x + 1) \sinh(x^2 + 2x)\, dx$

For Exercises 107–120, evaluate the definite integrals using the Fundamental Theorem of Calculus and check your answers numerically.

107. $\displaystyle\int_0^1 x(1 + x^2)^{20}\, dx$

108. $\displaystyle\int_4^1 x\sqrt{x^2 + 4}\, dx$

109. $\displaystyle\int_0^{\pi} \sin\theta(\cos\theta + 5)^7\, d\theta$

110. $\displaystyle\int_0^1 \frac{x}{1 + 5x^2}\, dx$

111. $\displaystyle\int_1^2 \frac{x^2 + 1}{x}\, dx$

112. $\displaystyle\int_1^3 \ln(x^3)\, dx$

113. $\displaystyle\int_1^e (\ln x)^2\, dx$

114. $\displaystyle\int_{-\pi}^{\pi} e^{2x} \sin 2x\, dx$

115. $\displaystyle\int_0^{10} z e^{-z}\, dz$

116. $\displaystyle\int_{-\pi/3}^{\pi/4} \sin^3\theta \cos\theta\, d\theta$

117. $\displaystyle\int_1^4 \frac{e^{\sqrt{x}}}{\sqrt{x}}\, dx$

118. $\displaystyle\int_0^1 \frac{dx}{x^2 + 1}$

119. $\displaystyle\int_{-\pi/4}^{\pi/4} \cos^2\theta \sin^5\theta\, d\theta$

120. $\displaystyle\int_{-2}^{0} \frac{2x + 4}{x^2 + 4x + 5}\, dx$

For a, b positive constants, evaluate the integrals in Exercises 121–132 using partial fractions or a trigonometric substitution.

121. $\displaystyle\int \frac{1}{(x - 2)(x + 2)}\, dx$

122. $\displaystyle\int \frac{1}{\sqrt{25 - x^2}}\, dx,$

123. $\displaystyle\int \frac{1}{x(x + 5)}\, dx$

124. $\displaystyle\int \frac{1}{\sqrt{1 - 9x^2}}\, dx$

125. $\displaystyle\int \frac{2x + 3}{x(x + 2)(x - 1)}\, dx$

126. $\displaystyle\int \frac{3x + 1}{x(x^2 - 1)}\, dx$

127. $\displaystyle\int \frac{1 + x^2}{x(1 + x)^2}\, dx$

128. $\displaystyle\int \frac{1}{x^2 + 2x + 2}\, dx$

129. $\displaystyle\int \frac{1}{x^2 + 4x + 5}\, dx$

130. $\displaystyle\int \frac{1}{\sqrt{a^2 - (bx)^2}}\, dx$

131. $\displaystyle\int \frac{\cos x}{\sin^3 x + \sin x}\, dx$

132. $\displaystyle\int \frac{e^x}{e^{2x} - 1}\, dx$

For Exercises 133–146 decide if the integral converges or diverges. If the integral converges, find its value or give a bound on its value.

133. $\displaystyle\int_4^{\infty} \frac{dt}{t^{3/2}}$

134. $\displaystyle\int_{10}^{\infty} \frac{dx}{x \ln x}$

135. $\displaystyle\int_0^\infty we^{-w}\,dw$

136. $\displaystyle\int_{-1}^1 \frac{1}{x^4}\,dx$

137. $\displaystyle\int_{-\pi/4}^{\pi/4} \tan\theta\,d\theta$

138. $\displaystyle\int_2^\infty \frac{1}{4+z^2}\,dz$

139. $\displaystyle\int_{10}^\infty \frac{1}{z^2-4}\,dz$

140. $\displaystyle\int_{-5}^{10} \frac{dt}{\sqrt{t+5}}$

141. $\displaystyle\int_0^{\pi/2} \frac{1}{\sin\phi}\,d\phi$

142. $\displaystyle\int_0^{\pi/4} \tan 2\theta\,d\theta$

143. $\displaystyle\int_1^\infty \frac{x}{x+1}\,dx$

144. $\displaystyle\int_0^\infty \frac{\sin^2\theta}{\theta^2+1}\,d\theta$

145. $\displaystyle\int_0^\pi \tan^2\theta\,d\theta$

146. $\displaystyle\int_0^1 (\sin x)^{-3/2}\,dx$

Problems

147. A function is defined by $f(t) = t^2$ for $0 \le t \le 1$ and $f(t) = 2 - t$ for $1 < t \le 2$. Compute $\int_0^2 f(t)\,dt$.

148. (a) Find $\int (x+5)^2\,dx$ in two ways:
  (i) By multiplying out
  (ii) By substituting $w = x + 5$
  (b) Are the results the same? Explain.

149. Suppose $\int_{-1}^1 h(z)\,dz = 7$, and that $h(z)$ is even. Calculate the following:
  (a) $\displaystyle\int_0^1 h(z)\,dz$ (b) $\displaystyle\int_{-4}^{-2} 5h(z+3)\,dz$

150. The curves $y = \sin x$ and $y = \cos x$ cross each other infinitely often. What is the area of the region bounded by these two curves between two consecutive crossings?

151. Find the average (vertical) height of the shaded area in Figure 7.27.

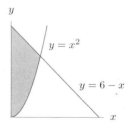

Figure 7.27

152. Find the average (horizontal) width of the shaded area in Figure 7.27.

153. (a) Find the average value of the following functions over one cycle:
  (i) $f(t) = \cos t$
  (ii) $g(t) = |\cos t|$
  (iii) $k(t) = (\cos t)^2$
  (b) Write the averages you have just found in ascending order. Using words and graphs, explain why the averages come out in the order they do.

154. What, if anything, is wrong with the following calculation?

$$\int_{-2}^2 \frac{1}{x^2}\,dx = -\frac{1}{x}\Big|_{-2}^2 = -\frac{1}{2} - \left(-\frac{1}{-2}\right) = -1.$$

155. Let $E(x) = \int \frac{e^x}{e^x+e^{-x}}\,dx$ and $F(x) = \int \frac{e^{-x}}{e^x+e^{-x}}\,dx$.
  (a) Calculate $E(x) + F(x)$.
  (b) Calculate $E(x) - F(x)$.
  (c) Use your results from parts (a) and (b) to calculate $E(x)$ and $F(x)$.

156. Using Figure 7.28, put the following approximations to the integral $\int_a^b f(x)\,dx$ and its exact value in order from smallest to largest: LEFT(5), LEFT(10), RIGHT(5), RIGHT(10), MID(10), TRAP(10), Exact value

Figure 7.28

157. Given that $\displaystyle\int_0^\infty e^{-x^2}\,dx = \frac{\sqrt{\pi}}{2}$, find $\displaystyle\int_0^\infty x^2 e^{-x^2}\,dx$.

158. The moment generating function, $m(t)$, which gives useful information about the normal distribution of statistics, is defined by

$$m(t) = \int_{-\infty}^\infty e^{tx}\frac{e^{-x^2/2}}{\sqrt{2\pi}}\,dx.$$

Find a formula for $m(t)$. [Hint: Complete the square and use the fact that $\int_{-\infty}^\infty e^{-x^2/2}\,dx = \sqrt{2\pi}$.]

159. A population, P, is said to be growing logistically if the time, T, taken for it to increase from P_1 to P_2 is given by

$$T = \int_{P_1}^{P_2} \frac{k\,dP}{P(L-P)},$$

where k and L are positive constants and $P_1 < P_2 < L$.
  (a) Calculate the time taken for the population to grow from $P_1 = L/4$ to $P_2 = L/2$.
  (b) What happens to T as $P_2 \to L$?

160. In 1987 the average per capita income in the US was \$26,000 and increasing continuously at a rate of $r(t) = 480(1.024)^t$ dollars per year, where t is the number of years since 1987.

    (a) Estimate the average per capita income in 1995.

    (b) Find a formula for the average per capita income as a function of time after 1987.

161. A patient is given an injection of Imitrex, a migraine medicine, at a rate of $r(t) = 2te^{-2t}$ ml/sec, where t is the number of seconds since the injection started.

    (a) By letting $t \to \infty$, estimate the total quantity of Imitrex injected.

    (b) What fraction of this dose has the patient received at the end of 5 seconds?

162. In 1990 humans generated $1.4 \cdot 10^{20}$ joules of energy from petroleum. At the time, it was estimated that all of the earth's petroleum would generate approximately 10^{22} joules. Assuming the use of energy generated by petroleum increases by 2% each year, how long will it be before all of our petroleum resources are used up?

163. An organism has a development time of T days at a temperature $H = f(t)^{\circ}$C. The total the number of degree-days S required for development to maturity is a constant defined by

$$S = \int_0^T (f(t) - H_{\min}) dt.$$

    (a) Evaluate this integral for $T = 18$, $f(t) = 30^{\circ}$C, and $H_{\min} = 10^{\circ}$C. What are the units of S?

    (b) Illustrate this definite integral on a graph. Label the features corresponding to T, $f(t)$, $H_{\min}$, and S.

    (c) Now suppose $H = g(t) = 20 + 10\cos(2\pi t/6)^{\circ}$C. Assuming that S remains constant, write a definite integral which determines the new development time, T_2. Sketch a graph illustrating this new integral. Judging from the graph, how does T_2 compare to T? Find T_2.

CAS Challenge Problems

164. **(a)** Use a computer algebra system to find $\displaystyle\int \frac{(\ln x)}{x} \, dx$, $\displaystyle\int \frac{(\ln x)^2}{x} \, dx$, and $\displaystyle\int \frac{(\ln x)^3}{x} \, dx$.

    (b) Guess a formula for $\displaystyle\int \frac{\ln(x)^n}{x} \, dx$ that works for any positive integer n.

    (c) Use a substitution to check your formula.

165. **(a)** Using a computer algebra system, find $\int (\ln x)^n \, dx$ for $n = 1, 2, 3, 4$.

    (b) There is a formula relating $\int (\ln x)^n \, dx$ to $\int (\ln x)^{n-1} \, dx$ for any positive integer n. Guess this formula using your answer to part (a). Check your guess using integration by parts.

In Problems 166–168:

    (a) Use a computer algebra system to find the indefinite integral of the given function.

    (b) Use the computer algebra system again to differentiate the result of part (a). Do not simplify.

    (c) Use algebra to show that the result of part (b) is the same as the original function. Show all the steps in your calculation.

166. $\sin^3 x$

167. $\sin x \cos x \cos(2x)$

168. $\displaystyle\frac{x^4}{(1 + x^2)^2}$

CHECK YOUR UNDERSTANDING

The left and right Riemann sums of a function f on the interval $[2, 6]$ are denoted by LEFT(n) and RIGHT(n), respectively, when the interval is divided into n equal parts. In Problems 1–9, decide whether the statements are true for all continuous functions, f. Give an explanation for your answer.

    1. If $n = 10$, then the subdivision size is $\Delta x = 1/10$.

    2. If we double the value of n, we make Δx half as large.

    3. LEFT(10) $\leq$ RIGHT(10)

    4. As n approaches infinity, LEFT(n) approaches 0.

    5. LEFT(n) $-$ RIGHT(n) $= (f(2) - f(6))\Delta x$.

    6. Doubling n decreases the difference LEFT(n) $-$ RIGHT(n) by exactly the factor $1/2$.

    7. If LEFT(n) $=$ RIGHT(n) for all n, then f is a constant function.

    8. The trapezoid estimate TRAP(n) $=$ (LEFT(n) $+$ RIGHT(n))/2 is always closer to $\int_2^6 f(x)dx$ than LEFT(n) or RIGHT(n).

    9. $\int_2^6 f(x)dx$ lies between LEFT(n) and RIGHT(n).

In Problems 10–23, decide whether the statements are true or false. Give an explanation for your answer.

    10. $\int f'(x) \cos(f(x)) \, dx = \sin(f(x)) + C$

    11. $\int (1/f(x)) \, dx = \ln|f(x)| + C$

    12. $\int t \sin(5 - t^2) \, dt$ can be evaluated using substitution.

13. $\int \sin^7 \theta \cos^6 \theta \, d\theta$ can be written as a polynomial with $\cos \theta$ as the variable.

14. $\int 1/(x^2 + 4x + 5) \, dx$ involves a natural logarithm.

15. $\int 1/(x^2 + 4x - 5) \, dx$ involves an arctangent.

16. $\int x^{-1}((\ln x)^2 + (\ln x)^3) \, dx$ is a polynomial with $\ln x$ as the variable.

17. $\int t \sin(5 - t) \, dt$ can be evaluated by parts.

18. If f is continuous for all x and $\int_0^\infty f(x)dx$ converges, then so does $\int_a^\infty f(x)dx$ for all positive a.

19. If $f(x)$ is a positive periodic function, then $\int_0^\infty f(x)dx$ diverges.

20. If $f(x)$ is continuous and positive for $x > 0$ and if $\lim_{x \to \infty} f(x) = 0$, then $\int_0^\infty f(x)dx$ converges.

21. If $f(x)$ is continuous and positive for $x > 0$ and if $\lim_{x \to \infty} f(x) = \infty$, then $\int_0^\infty (1/f(x))dx$ converges.

22. If $\int_0^\infty f(x)dx$ and $\int_0^\infty g(x)dx$ both converge, then $\int_0^\infty (f(x) + g(x))dx$ converges.

23. If $\int_0^\infty f(x)dx$ and $\int_0^\infty g(x)dx$ both diverge, then $\int_0^\infty (f(x) + g(x))dx$ diverges.

Suppose that f is continuous for all real numbers and that $\int_0^\infty f(x) \, dx$ converges. Let a be any positive number. Decide which of the statements in Problems 24–27 are true and which are false. Give an explanation for your answer.

24. $\int_0^\infty a f(x) \, dx$ converges.

25. $\int_0^\infty f(ax) \, dx$ converges.

26. $\int_0^\infty f(a + x) \, dx$ converges.

27. $\int_0^\infty (a + f(x)) \, dx$ converges.

PROJECTS FOR CHAPTER SEVEN

1. Taylor Polynomial Inequalities

(a) Use the fact that $e^x \geq 1 + x$ for all values of x and the formula

$$e^x = 1 + \int_0^x e^t \, dt$$

to show that

$$e^x \geq 1 + x + \frac{x^2}{2}$$

for all positive values of x. Generalize this idea to get inequalities involving higher-degree polynomials.

(b) Use the fact that $\cos x \leq 1$ for all x and repeated integration to show that

$$\cos x \leq 1 - \frac{x^2}{2!} + \frac{x^4}{4!}.$$

2. Slope Fields

Suppose we want to sketch the antiderivative, F, of the function f. To get an accurate graph of F, we must be careful about making F have the right slope at every point. The slope of F at any point (x, y) on its graph should be $f(x)$, since $F'(x) = f(x)$. We arrange this as follows: at the point (x, y) in the plane, draw a small line segment with slope $f(x)$. Do this at many points. We call such a diagram a *slope field*. If $f(x) = x$, we get the slope field in Figure 7.29.

Figure 7.29: Slope field of $f(x) = x$

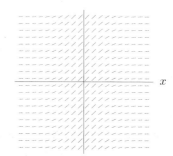

Figure 7.30: Slope field of $f(x) = e^{-x^2}$

Notice how the lines in Figure 7.29 seem to be arranged in a parabolic pattern. This is because the general antiderivative of x is $x^2/2 + C$, so the lines are all the tangent lines to the family of parabolas $y = x^2/2 + C$. This suggests a way of finding antiderivatives graphically even if we can't write down a formula for them: plot the slopes, and see if they suggest the graph of an antiderivative. For example, if you do this with $f(x) = e^{-x^2}$, which is one of the functions that does not have an elementary antiderivative, you get Figure 7.30.

You can see the ghost of the graph of a function lurking behind the slopes in Figure 7.30; in fact there is a whole stack of them. If you move across the plane in the direction suggested by the slope field at every point, you will trace out a curve. The slope field is tangent to the curve everywhere, so this is the graph of an antiderivative of e^{-x^2}.

(a) (i) Sketch a graph of $f(t) = \dfrac{\sin t}{t}$.

(ii) What does your graph tell you about the behavior of

$$\text{Si}(x) = \int_0^x \frac{\sin(t)}{t}\,dt$$

for $x > 0$? Is $\text{Si}(x)$ always increasing or always decreasing? Does $\text{Si}(x)$ cross the x-axis for $x > 0$?

(iii) By drawing the slope field for $f(t) = \dfrac{\sin t}{t}$, decide whether $\lim\limits_{x\to\infty} \text{Si}(x)$ exists.

(b) (i) Use your calculator or computer to sketch a graph of $y = x^{\sin x}$ for $0 < x \le 20$.

(ii) Using your answer to part (i), sketch by hand a graph of the function F, where

$$F(x) = \int_0^x t^{\sin t}\,dt.$$

(iii) Use a slope field program to check your answer to part (ii).

(c) Let $F(x)$ be the antiderivative of $\sin(x^2)$ satisfying $F(0) = 0$.

(i) Describe any general features of the graph of F that you can deduce by looking at the graph of $\sin(x^2)$ in Figure 7.31.

(ii) By drawing a slope field (using a calculator or computer), sketch a graph of F. Does F ever cross the x-axis in the region $x > 0$? Does $\lim\limits_{x\to\infty} F(x)$ exist?

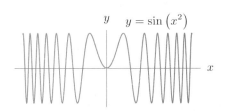

Figure 7.31

Chapter Eight

USING THE DEFINITE INTEGRAL

In Chapter 5, we saw how a definite integral can be used to compute an area, an average value, or a total change.

There are other quantities which can be expressed as definite integrals. To do this, we divide the quantity into small pieces and approximate it by a Riemann sum. In this chapter we use definite integrals to solve problems in geometry, physics, economics, and probability.

8.1 AREAS AND VOLUMES

In Chapter 5, we calculated areas using definite integrals. We obtained the integral by slicing up the region, constructing a Riemann sum, and then taking a limit.

In this section, we calculate volumes using definite integrals. To obtain the integral, we again slice up the region and construct a Riemann sum. The first two examples show how slicing can be used to find both areas and volumes.

Example 1   Use horizontal slices to set up a definite integral to calculate the area of the isosceles triangle in Figure 8.1.

Figure 8.1: Isosceles triangle

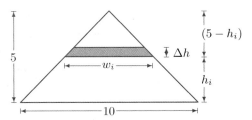

Figure 8.2: Horizontal slicing of isosceles triangle

Solution   Notice that we can find the area of a triangle without using an integral; we will use this to check the result from integration:

$$\text{Area} = \frac{1}{2} \text{ Base } \cdot \text{ Height } = 25 \text{ cm}^2.$$

To calculate the area using horizontal slices, see Figure 8.2. A typical strip is approximately a rectangle of length w_i and width Δh, so

$$\text{Area of strip } \approx w_i \Delta h \text{ cm}^2.$$

To get w_i in terms of h_i, the height above the base, use the similar triangles in Figure 8.2:

$$\frac{w_i}{10} = \frac{5 - h_i}{5}$$
$$w_i = 2(5 - h_i) = 10 - 2h_i.$$

Summing the areas of the strips gives the Riemann sum approximation

$$\text{Area of triangle } \approx \sum_{i=1}^{n} w_i \Delta h = \sum_{i=1}^{n} (10 - 2h_i) \Delta h \text{ cm}^2.$$

Taking the limit as $n \to \infty$ and $\Delta h \to 0$ gives the integral:

$$\text{Area of triangle } = \lim_{n \to \infty} \sum_{i=1}^{n} (10 - 2h_i) \Delta h = \int_0^5 (10 - 2h) \, dh \text{ cm}^2.$$

Evaluating the integral gives

$$\text{Area of triangle } = \int_0^5 (10 - 2h) \, dh = \left. (10h - h^2) \right|_0^5 = 25 \text{ cm}^2.$$

Notice that the limits in the definite integral are the limits for the variable h. Once we decided to slice the triangle horizontally, we knew that a typical slice has thickness Δh, so h is the variable in our definite integral, and the limits must be values of h.

Finding Volumes by Slicing

To calculate the volume of a solid using Riemann sums, we chop the solid into slices whose volumes we can estimate.

Figure 8.3: Cone cut into vertical slices

Figure 8.4: Cone cut into horizontal slices

Let's see how we might slice a cone standing with the vertex uppermost. We could divide the cone vertically into arch-shaped slices; see Figure 8.3. We could also divide the cone horizontally, giving coin-shaped slices; see Figure 8.4.

To calculate the volume of the cone, we choose the circular slices because it is easier to estimate the volumes of the coin-shaped slices.

Example 2   Use horizontal slicing to find the volume of the cone in Figure 8.5.

Figure 8.5: Cone

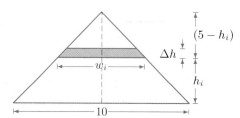

Figure 8.6: Vertical cross-section of cone in Figure 8.5

Solution   Each slice is a circular disk of thickness Δh. See Figure 8.5. The disk at height h_i above the base has radius $r_i = \frac{1}{2} w_i$. From Figure 8.6 and the previous example, we have

$$w_i = 10 - 2h_i \quad \text{so} \quad r_i = 5 - h_i.$$

Each slice is approximately a cylinder of radius r_i and thickness Δh, so

$$\text{Volume of slice} \approx \pi r_i^2 \Delta h = \pi (5 - h_i)^2 \Delta h \text{ cm}^3.$$

Summing over all slices, we have

$$\text{Volume of cone} \approx \sum_{i=1}^{n} \pi (5 - h_i)^2 \Delta h \text{ cm}^3.$$

Taking the limit as $n \to \infty$, so $\Delta h \to 0$, gives

$$\text{Volume of cone} = \lim_{n \to \infty} \sum_{i=1}^{n} \pi (5 - h_i)^2 \Delta h = \int_0^5 \pi (5 - h)^2 \, dh \text{ cm}^3.$$

The integral can be evaluated using the substitution $u = 5 - h$ or by multiplying out $(5 - h)^2$. Using the substitution, we have

$$\text{Volume of cone} = \int_0^5 \pi (5 - h)^2 dh = -\frac{\pi}{3} (5 - h)^3 \Big|_0^5 = \frac{125}{3} \pi \text{ cm}^3.$$

Note that the sum represented by the $\sum$ sign is over all the strips. In the future, we will not write down the points, $h_0, h_1, \ldots h_n$, that are needed to express the Riemann sum precisely. This is unnecessary if all we want is the final expression for the definite integral. We now calculate the area of a semicircle and the volume of a hemisphere by slicing.

Example 3   Use horizontal slices to set up a definite integral representing the area of the semicircle of radius 7 cm in Figure 8.7.

Figure 8.7: Semicircle

Figure 8.8: Horizontal slices of semicircle

Solution   As in Example 1, to calculate the area using horizontal slices, we divide the region into strips; see Figure 8.8. A typical strip at height h above the base has width w and thickness Δh, so

$$\text{Area of strip } \approx w\Delta h \text{ cm}^2.$$

To get w in terms of h, we use the Pythagorean Theorem in Figure 8.8, giving

$$h^2 + \left(\frac{w}{2}\right)^2 = 7^2,$$

so

$$w = \sqrt{4(7^2 - h^2)} = 2\sqrt{49 - h^2}.$$

Summing the areas of the strips gives the Riemann sum approximation

$$\text{Area of semicircle } \approx \sum w\Delta h = \sum 2\sqrt{49 - h^2}\Delta h \text{ cm}^2.$$

Taking the limit as $\Delta h \to 0$ gives the integral:

$$\text{Area of semicircle } = \lim_{\Delta h \to 0} \sum 2\sqrt{49 - h^2}\Delta h = 2\int_0^7 \sqrt{49 - h^2}\, dh \text{ cm}^2.$$

Using the table of integrals VI-28 and VI-30, or a calculator or computer, gives

$$\text{Area of semicircle } = 2\cdot\frac{1}{2}\left(h\sqrt{49 - h^2} + 49\arcsin\left(\frac{h}{7}\right)\right)\Bigg|_0^7 = 49\arcsin 1 = \frac{49}{2}\pi = 76.97 \text{ cm}^2.$$

As a check, notice that the area of the whole circle of radius 7 is $\pi \cdot 7^2 = 49\pi \text{ cm}^2$.

Example 4   Set up and evaluate an integral giving the volume of the hemisphere of radius 7 cm in Figure 8.9.

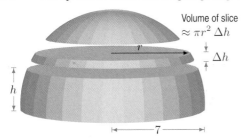

Figure 8.9: Slicing to find the volume
of a hemisphere

Figure 8.10: Vertical cut through center of hemisphere
showing relation between r_i and h_i

Solution   We will not use the formula $\frac{4}{3}\pi r^3$ for the volume of a sphere. However our approach can be used to derive that formula.

Divide the hemisphere into horizontal slices of thickness Δh cm. (See Figure 8.9.) Each slice is a circular slab. Let r be the radius of the slice at height h, so

$$\text{Volume of slice} \approx \pi r^2 \Delta h \text{ cm}^3.$$

We want to express r in terms of h. From Figure 8.10, we have $r^2 = 7^2 - h^2$, so

$$\text{Volume of slice} \approx \pi r^2 \,\Delta h = \pi(7^2 - h^2)\,\Delta h \text{ cm}^3.$$

Summing the volumes of all slices gives:

$$\text{Volume} \approx \sum \pi r^2 \,\Delta h = \sum \pi(7^2 - h^2)\,\Delta h \text{ cm}^3.$$

As the thickness of each slab tends to zero, the sum becomes a definite integral. Since the radius of the hemisphere is 7, we know that h varies from 0 to 7, so these are the limits of integration:

$$\text{Volume} = \int_0^7 \pi(7^2 - h^2)\,dh = \pi\left(7^2 h - \frac{1}{3}h^3\right)\bigg|_0^7 = \frac{2}{3}\pi 7^3 = 718.4 \text{ cm}^3.$$

Notice that the volume of the hemisphere is half of $\frac{4}{3}\pi 7^3$ cm^3, as we expected.

We now use slicing to find the volume of a pyramid. You may already know the formula $V = \frac{1}{3}b^2 \cdot h$ for the volume V of a pyramid with height h and a square base of length b. We will not use the formula, but our approach can be used to derive it.

Example 5   Compute the volume, in cubic feet, of the Great Pyramid of Egypt, whose base is a square 755 feet by 755 feet and whose height is 410 feet.

Solution   We slice the pyramid horizontally. Each slice is a square slab with thickness Δh. The bottom layer is a square slab 755 feet by 755 feet and volume about $(755)^2 \Delta h$ ft^3. As we move up the pyramid, the layers have shorter side lengths. We divide the height into n subintervals of length Δh. See Figure 8.11. Let s be the side length of the slice at height h, then

$$\text{Volume of slice} \approx s^2\,\Delta h \text{ ft}^3.$$

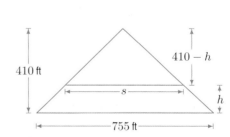

Figure 8.11: The Great Pyramid   Figure 8.12: Cross-section relating s and h

We express s as a function of h using the vertical cross-section in Figure 8.12. By similar triangles, we get

$$\frac{s}{755} = \frac{(410 - h)}{410}.$$

Thus,

$$s = \left(\frac{755}{410}\right)(410 - h),$$

and the total volume, V, is approximated by adding the volumes of the n layers:

$$V \approx \sum s^2 \, \Delta h = \sum \left[\left(\frac{755}{410} \right) (410 - h) \right]^2 \Delta h \text{ ft}^3.$$

As the thickness of each slice tends to zero, the sum becomes a definite integral. Finally, since h varies from 0 to 410, the height of the pyramid, we have

$$V = \int_0^{410} \left[\left(\frac{755}{410} \right) (410 - h) \right]^2 dh = \left(\frac{755}{410} \right)^2 \int_0^{410} (410 - h)^2 \, dh$$

$$= \left(\frac{755}{410} \right)^2 \left[-\frac{(410 - h)^3}{3} \right] \Bigg|_0^{410} = \left(\frac{755}{410} \right)^2 \frac{(410)^3}{3} = \frac{1}{3}(755)^2(410) \approx 78 \text{ million ft}^3.$$

Note that $V = \frac{1}{3}(755)^2(410) = \frac{1}{3}b^2 \cdot h$, as expected.

Exercises and Problems for Section 8.1

Exercises

In Exercises 1–8, write a Riemann sum and then a definite integral representing the area of the region, using the strip shown. Evaluate the integral exactly.

1.

2.

3.

4.

5.

6.

7.

8.

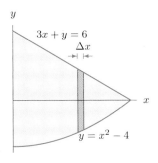

In Exercises 9–14, write a Riemann sum and then a definite integral representing the volume of the region, using the slice shown. Evaluate the integral exactly. (Regions are parts of cones, cylinders, spheres, and pyramids.)

9.

10.

11.

12.

13.

14.

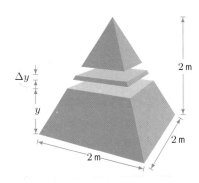

Problems

The integrals in Problems 15–18 represent the area of either a triangle or part of a circle, and the variable of integration measures a distance. In each case, say which shape is represented, and give the radius of the circle or the base and height of the triangle. Make a sketch to support your answer showing the variable and all other relevant quantities.

15. $\int_0^1 3x\,dx$

16. $\int_{-9}^9 \sqrt{81-x^2}\,dx$

17. $\int_0^{\sqrt{15}} \sqrt{15-h^2}\,dh$

18. $\int_0^7 5\left(1-\frac{h}{7}\right)dh$

The integrals in Problems 19–22 represent the volume of either a hemisphere or a cone, and the variable of integration measures a length. In each case, say which shape is represented, and give the radius of the hemisphere or the radius and height of the cone. Make a sketch to support your answer showing the variable and all other relevant quantities.

19. $\int_0^{12} \pi(144-h^2)\,dh$

20. $\int_0^{12} \pi(x/3)^2\,dx$

21. $\int_0^6 \pi(3-y/2)^2\,dy$

22. $\int_0^2 \pi(2^2-(2-y)^2)\,dy$

23. Find, by slicing, the volume of a cone whose height is 3 cm and whose base radius is 1 cm. Slice the cone as shown in Figure 8.4 on page 369.

24. Find the volume of a sphere of radius r by slicing.

25. Find, by slicing, a formula for the volume of a cone of height h and base radius r.

26. Figure 8.13 shows a solid with both rectangular and triangular cross sections.

    (a) Slice the solid parallel to the triangular faces. Sketch one slice and calculate its volume in terms of x, the distance of the slice from one end. Then write and evaluate an integral giving the volume of the solid.

    (b) Repeat part (a) for horizontal slices. Instead of x, use h, the distance of a slice from the top.

Figure 8.13

27. A rectangular lake is 150 km long and 3 km wide. The vertical cross-section through the lake in Figure 8.14 shows that the lake is 0.2 km deep at the center. (These are the approximate dimensions of Lake Mead, the largest reservoir in the US, which provides water to California, Nevada, and Arizona.) Set up and evaluate a definite integral giving the total volume of the lake.

Figure 8.14: Not to scale

28. A dam has a rectangular base 1400 meters long and 160 meters wide. Its cross-section is shown in Figure 8.15. (The Grand Coulee Dam in Washington state is roughly this size.) By slicing horizontally, set up and evaluate a definite integral giving the volume of material used to build this dam.

Figure 8.15: Not to scale

8.2   APPLICATIONS TO GEOMETRY

In Section 8.1, we calculated volumes using slicing and definite integrals. In this section, we use the same method to calculate the volumes of more complicated regions as well as the length of a curve. The method is summarized in the following steps:

> ### To Compute a Volume or Length Using an Integral
>
> - Divide the solid (or curve) into small pieces whose volume (or length) we can easily approximate;
> - Add the contributions of all the pieces, obtaining a Riemann sum that approximates the total volume (or length);
> - Take the limit as the number of terms in the sum tends to infinity, giving a definite integral for the total volume (or total length).

In the previous section, all the slices we created were disks or squares. We now look at different ways of generating volumes whose cross-sections include circles, rings, and squares.

Volumes of Revolution

One way to create a solid having circular cross-sections is to revolve a region in the plane around a line, giving a *solid of revolution*, as in the following examples.

Example 1   The region bounded by the curve $y = e^{-x}$ and the x-axis between $x = 0$ and $x = 1$ is revolved around the x-axis. Find the volume of this solid of revolution.

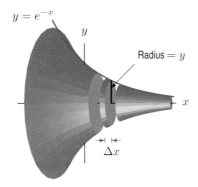

Figure 8.16: A thin strip rotated around the x-axis to form a circular slice

Solution   We slice the region perpendicular to the x-axis, giving circular disks of thickness Δx. See Figure 8.16. The radius of the disk is $y = e^{-x}$, so

$$\text{Volume of the slice} \approx \pi y^2 \, \Delta x = \pi (e^{-x})^2 \, \Delta x,$$

$$\text{Total volume} \approx \sum \pi y^2 \, \Delta x = \sum \pi \left(e^{-x}\right)^2 \, \Delta x.$$

As the thickness of each slice tends to zero, we get

$$\text{Total volume} = \int_0^1 \pi (e^{-x})^2 \, dx = \pi \int_0^1 e^{-2x} \, dx = \pi \left(-\frac{1}{2}\right) e^{-2x} \bigg|_0^1$$

$$= \pi \left(-\frac{1}{2}\right) (e^{-2} - e^0) = \frac{\pi}{2}(1 - e^{-2}) \approx 1.36.$$

Example 2   A table leg in Figure 8.17 has a circular cross section with radius r cm at a height of y cm above the ground given by $r = 3 + \cos(\pi y/25)$. Find the volume of the table leg.

100 cm

Figure 8.17

Solution   The table leg is formed by rotating the curve $r = 3 + \cos(\pi y / 25)$ around the y-axis. Slicing the table leg horizontally gives circular disks of thickness Δy and radius $r = 3 + \cos(\pi y / 25)$.

To set up a definite integral for the volume, we find the volume of a typical slice:

$$\text{Volume of slice} \approx \pi r^2 \Delta y = \pi \left(3 + \cos \left(\frac{\pi}{25} y \right) \right)^2 \Delta y.$$

Summing over all slices gives the Riemann sum approximation:

$$\text{Total volume} = \sum \pi \left(3 + \cos \left(\frac{\pi}{25} y \right) \right)^2 \Delta y.$$

Taking the limit as $\Delta y \to 0$ gives the definite integral

$$\text{Total volume} = \lim_{\Delta y \to 0} \sum \pi \left(3 + \cos \left(\frac{\pi}{25} y \right) \right)^2 \Delta y = \int_0^{100} \pi \left(3 + \cos \left(\frac{\pi}{25} y \right) \right)^2 dy.$$

Evaluating the integral numerically gives

$$\text{Total volume} = \int_0^{100} \pi \left(3 + \cos \left(\frac{\pi}{25} y \right) \right)^2 dy = 2984.5 \text{ cm}^3.$$

Example 3   The region bounded by the curves $y = x$ and $y = x^2$ is rotated about the line $y = 3$. Compute the volume of the resulting solid.

Figure 8.18: Cutaway view of volume showing inner and outer radii

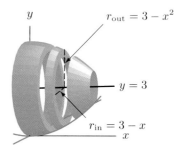

Figure 8.19: One slice (a disk-with-a-hole)

Solution   The solid is shaped like a bowl with the base removed. See Figure 8.18. To compute the volume, we divide the area in the xy-plane into thin vertical strips of thickness Δx, as in Figure 8.20.

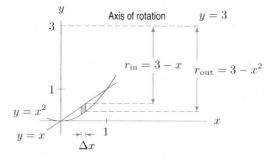

Figure 8.20: The region for Example 3

As each strip is rotated around the line $y = 3$, it sweeps out a slice shaped like a circular disk with a hole in it. See Figure 8.19. This disk-with-a-hole has an inner radius of $r_{\text{in}} = 3 - x$ and an outer radius of $r_{\text{out}} = 3 - x^2$. Think of the slice as a circular disk of radius r_{out} from which has been removed a smaller disk of radius r_{in}. Then,

$$\text{Volume of slice} \approx \pi r_{\text{out}}^2 \Delta x - \pi r_{\text{in}}^2 \Delta x = \pi (3 - x^2)^2 \Delta x - \pi (3 - x)^2 \Delta x.$$

Adding the volumes of all the slices, we have

$$\text{Total volume} = V \approx \sum \left(\pi r_{\text{out}}^2 - \pi r_{\text{in}}^2 \right) \Delta x = \sum \left(\pi (3 - x^2)^2 - \pi (3 - x)^2 \right) \Delta x.$$

We let Δx, the thickness of each slice, tend to zero to obtain a definite integral. Since the curves $y = x$ and $y = x^2$ intersect at $x = 0$ and $x = 1$, these are the limits of integration:

$$V = \int_0^1 \left(\pi (3 - x^2)^2 - \pi (3 - x)^2 \right) dx = \pi \int_0^1 \left((9 - 6x^2 + x^4) - (9 - 6x + x^2) \right) dx$$

$$= \pi \int_0^1 (6x - 7x^2 + x^4) \, dx = \pi \left(3x^2 - \frac{7x^3}{3} + \frac{x^5}{5} \right) \bigg|_0^1 \approx 2.72.$$

Volumes of Regions of Known Cross-Section

We now calculate the volume of a solid constructed by a different method. Starting with a region in the xy-plane as a base, the solid is built by standing squares, semicircles, or triangles vertically on edge in this region.

Example 4   Find the volume of the solid whose base is the region in the xy-plane bounded by the curves $y = x^2$ and $y = 8 - x^2$ and whose cross-sections perpendicular to the x-axis are squares with one side in the xy-plane. (See Figure 8.21.)

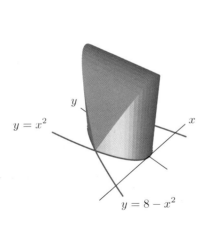

Figure 8.21: The solid for Example 4

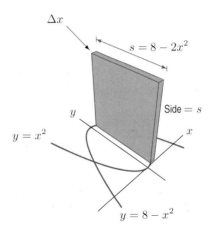

Figure 8.22: A slice of the solid for Example 4

Solution   We view the solid as a loaf of bread sitting on the xy-plane and made up of square slices. A typical slice of thickness Δx is shown in Figure 8.22. The side length, s, of the square is the distance (in the y direction) between the two curves, so $s = (8 - x^2) - x^2 = 8 - 2x^2$ giving

$$\text{Volume of slice} \approx s^2 \, \Delta x = (8 - 2x^2)^2 \, \Delta x.$$

Thus

$$\text{Total volume} = V \approx \sum s^2 \, \Delta x = \sum (8 - 2x^2)^2 \, \Delta x.$$

As the thickness Δx of each slice tends to zero, the sum becomes a definite integral. Since the curves $y = x^2$ and $y = 8 - x^2$ intersect at $x = -2$ and $x = 2$, these are the limits of integration. We have

$$V = \int_{-2}^2 (8 - 2x^2)^2 \, dx = \int_{-2}^2 (64 - 32x^2 + 4x^4) \, dx$$

$$= 64x - \frac{32}{3}x^3 + \frac{4}{5}x^5 \bigg|_{-2}^2 = \frac{2048}{15} \approx 136.5.$$

Arc Length

A definite integral can be used to compute the *arc length*, or length, of a curve. To compute the length of the curve $y = f(x)$ from $x = a$ to $x = b$, where $a < b$, we divide the curve into small pieces, each one approximately straight.

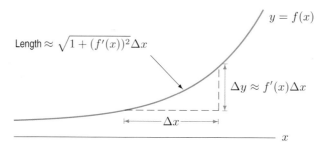

$$\text{Length} \approx \sqrt{1 + (f'(x))^2}\,\Delta x$$

$$\Delta y \approx f'(x)\,\Delta x$$

$$\Delta x$$

$$y = f(x)$$

$$x$$

Figure 8.23: Length of a small piece of curve approximated using Pythagoras' theorem

Figure 8.23 shows that a small change Δx corresponds to a small change $\Delta y \approx f'(x)\,\Delta x$. The length of the piece of the curve is approximated by

$$\text{Length} \approx \sqrt{(\Delta x)^2 + (\Delta y)^2} \approx \sqrt{(\Delta x)^2 + (f'(x)\,\Delta x)^2} = \sqrt{1 + (f'(x))^2}\,\Delta x.$$

Thus, the arc length of the entire curve is approximated by a Riemann sum:

$$\text{Arc length} \approx \sum \sqrt{1 + (f'(x))^2}\,\Delta x.$$

Since x varies between a and b, as we let Δx tend to zero, the sum becomes the definite integral:

For $a < b$, the arc length of the curve $y = f(x)$ from $x = a$ to $x = b$ is given by

$$\text{Arc length} = \int_a^b \sqrt{1 + (f'(x))^2}\,dx.$$

Example 5   Set up and evaluate an integral to compute the length of the curve $y = x^3$ from $x = 0$ to $x = 5$.

Solution   If $f(x) = x^3$, then $f'(x) = 3x^2$, so

$$\text{Arc length} = \int_0^5 \sqrt{1 + (3x^2)^2}\,dx.$$

Although the formula for the arc length of a curve is easy to apply, the integrands it generates often do not have elementary antiderivatives. Evaluating the integral numerically, we find the arc length to be 125.68. To check that the answer is reasonable, notice that the curve starts at $(0, 0)$ and goes to $(5, 125)$, so its length must be at least the length of a straight line between these points, or $\sqrt{5^2 + 125^2} = 125.10$. (See Figure 8.24.)

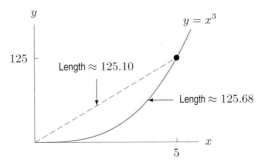

Figure 8.24: Arc length of $y = x^3$ (Note: The picture is distorted because the scales on the two axes are quite different)

Arc Length of a Parametric Curve

A particle moving along a curve in the plane given by the parametric equations $x = f(t)$, $y = g(t)$, where t is time, has speed given by:

$$v(t) = \sqrt{\left(\frac{dx}{dt}\right)^2 + \left(\frac{dy}{dt}\right)^2}.$$

We can find the distance traveled by a particle along a curve between $t = a$ and $t = b$ by integrating its speed. Thus,

$$\text{Distance traveled} = \int_a^b v(t)\,dt.$$

If the particle never stops or reverses its direction as it moves along the curve, the distance it travels is the same as the length of the curve. This suggests the following formula:

If a curve is given parametrically for $a \leq t \leq b$ by differentiable functions and if the velocity $v(t)$ is not 0 for $a < t < b$, then

$$\text{Arc length of curve} = \int_a^b v(t)\,dt = \int_a^b \sqrt{\left(\frac{dx}{dt}\right)^2 + \left(\frac{dy}{dt}\right)^2}\,dt.$$

Example 6   Find the circumference of the ellipse given by the parametric equations

$$x = 2\cos t, \quad y = \sin t, \quad 0 \leq t \leq 2\pi.$$

Solution   The circumference of this curve is given by an integral which must be calculated numerically:

$$\text{Circumference} = \int_0^{2\pi} \sqrt{\left(\frac{dx}{dt}\right)^2 + \left(\frac{dy}{dt}\right)^2}\,dt = \int_0^{2\pi} \sqrt{(-2\sin t)^2 + (\cos t)^2}\,dt$$

$$= \int_0^{2\pi} \sqrt{4\sin^2 t + \cos^2 t}\,dt = 9.69.$$

Since the ellipse is inscribed in a circle of radius 2 and circumscribes a circle of radius 1, we expect the length of the ellipse to be between $2\pi(2) \approx 12.57$ and $2\pi(1) \approx 6.28$, so the value of 9.69 is reasonable.

Exercises and Problems for Section 8.2

Exercises

In Exercises 1–10, the region is rotated around the x-axis. Find the volume.

1. Bounded by $y = x^2$, $y = 0$, $x = 0$, $x = 1$.

2. Bounded by $y = (x+1)^2$, $y = 0$, $x = 1$, $x = 2$.

3. Bounded by $y = 4 - x^2$, $y = 0$, $x = -2$, $x = 0$.

4. Bounded by $y = \sqrt{x+1}$, $y = 0$, $x = -1$, $x = 1$.

5. Bounded by $y = e^x$, $y = 0$, $x = -1$, $x = 1$.

6. Bounded by $y = \cos x$, $y = 0$, $x = 0$, $x = \pi/2$.

7. Bounded by $y = 1/(x+1)$, $y = 0$, $x = 0$, $x = 1$.

8. Bounded by $y = \sqrt{\cosh 2x}$, $y = 0$, $x = 0$, $x = 1$.

9. Bounded by $y = x^2$, $y = x$, $x = 0$, $x = 1$.

10. Bounded by $y = e^{3x}$, $y = e^x$, $x = 0$, $x = 1$.

For Exercises 11–13, find the arc length of the graph of the function from $x = 0$ to $x = 2$.

11. $f(x) = \sqrt{x^3}$ **12.** $f(x) = \sqrt{4 - x^2}$

13. $f(x) = \cosh x$

Find the length of the parametric curves in Exercises 14–16.

14. $x = 3 + 5t$, $y = 1 + 4t$ for $1 \le t \le 2$. Explain why your answer is reasonable.

15. $x = \cos(e^t)$, $y = \sin(e^t)$ for $0 \le t \le 1$. Explain why your answer is reasonable.

16. $x = \cos(3t)$, $y = \sin(5t)$ for $0 \le t \le 2\pi$.

Problems

17. Consider the hyperbola $x^2 - y^2 = 1$ in Figure 8.25.

    (a) The shaded region $2 \le x \le 3$ is rotated around the x-axis. What is the volume generated?

    (b) What is the arc length with $y \ge 0$ from $x = 2$ to $x = 3$?

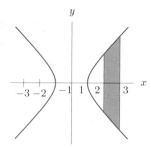

Figure 8.25

18. Rotating the ellipse $x^2/a^2 + y^2/b^2 = 1$ about the x-axis generates an ellipsoid. Compute its volume.

For Problems 19–21, sketch the solid obtained by rotating each region around the indicated axis. Using the sketch, show how to approximate the volume of the solid by a Riemann sum, and hence find the volume.

19. Bounded by $y = x^3$, $x = 1$, $y = -1$. Axis: $y = -1$.

20. Bounded by $y = \sqrt{x}$, $x = 1$, $y = 0$. Axis: $x = 1$.

21. Bounded by the first arch of $y = \sin x$, $y = 0$. Axis: x axis.

Problems 22–27 concern the region bounded by $y = x^2$, $y = 1$, and the y-axis. Find the volume of the following solids.

22. The solid obtained by rotating the region around the y-axis.

23. The solid obtained by rotating the region about the x-axis.

24. The solid obtained by rotating the region about the line $y = -2$.

25. The solid whose base is the region and whose cross-sections perpendicular to the x-axis are squares.

26. The solid whose base is the region and whose cross-sections perpendicular to the x-axis are semicircles.

27. The solid whose base is the region and whose cross-sections perpendicular to the y-axis are equilateral triangles.

For Problems 28–32 consider the region bounded by $y = e^x$, the x-axis, and the lines $x = 0$ and $x = 1$. Find the volume of the following solids.

28. The solid obtained by rotating the region about the x-axis.

29. The solid obtained by rotating the region about the horizontal line $y = -3$.

30. The solid obtained by rotating the region about the horizontal line $y = 7$.

31. The solid whose base is the given region and whose cross-sections perpendicular to the x-axis are squares.

32. The solid whose base is the given region and whose cross-sections perpendicular to the x-axis are semicircles.

33. (a) A pie dish is 9 inches across the top, 7 inches across the bottom, and 3 inches deep. See Figure 8.26. Compute the volume of this dish.

(b) Make a rough estimate of the volume in cubic inches of a single cut-up apple, and estimate the number of apples that is needed to make an apple pie that fills this dish.

Figure 8.26

34. A 100 cm long gutter is made of three strips of metal, each 5 cm wide; Figure 8.27 shows a cross-section.

(a) Find the volume of water in the gutter when the depth is h cm.

(b) What is the maximum value of h?

(c) What is the maximum volume of water that the gutter can hold?

(d) If the gutter is filled with half the maximum volume of water, is the depth larger or smaller than half of the answer to part (b)? Explain how you can answer without any calculation.

(e) Find the depth of the water when the gutter contains half the maximum possible volume.

Figure 8.27

35. The design of boats is based on Archimedes' Principle, which states that the buoyant force on an object in water is equal to the weight of the water displaced. Suppose you want to build a sailboat whose hull is parabolic with cross section $y = ax^2$, where a is a constant. Your boat will have length L and its maximum draft (the maximum vertical depth of any point of the boat beneath the water line) will be H. See Figure 8.28. Every cubic meter of water weighs 10,000 newtons. What is the maximum possible weight for your boat and cargo?

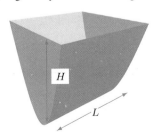

Figure 8.28

36. The circumference of a tree at different heights above the ground is given in the table below. Assume that all horizontal cross-sections of the tree are circles. Estimate the volume of the tree.

Height (inches)	0	20	40	60	80	100	120
Circumference (inches)	31	28	21	17	12	8	2

37. A bowl has the shape of the graph of $y = x^4$ between the points $(1, 1)$ and $(-1, 1)$ rotated about the y-axis. When the bowl contains water to a depth of h units, it flows out through a hole in the bottom at a rate (volume/time) proportional to $\sqrt{h}$, with constant of proportionality 6.

(a) Show that the water level falls at a constant rate.

(b) Find how long it takes to empty the bowl if it is originally full to the brim.

38. The hull of a boat has widths given by the following table. Reading across a row of the table gives widths at points 0, 10, ..., 60 feet from the front to the back at a certain level below waterline. Reading down a column of the table gives widths at levels 0, 2, 4, 6, 8 feet below waterline at a certain distance from the front. Use the trapezoidal rule to estimate the volume of the hull below waterline.

		Front of boat $\longrightarrow$ Back of boat						
		0	10	20	30	40	50	60
	0	2	8	13	16	17	16	10
Depth	2	1	4	8	10	11	10	8
below	4	0	3	4	6	7	6	4
waterline	6	0	1	2	3	4	3	2
(in feet)	8	0	0	1	1	1	1	1

39. (a) Write an integral which represents the circumference of a circle of radius r.

(b) Evaluate the integral, and show that you get the answer you expect.

40. Compute the perimeter of the region used for the base of the solids in Problems 28–32.

41. The catenary $\cosh x = \frac{1}{2}(e^x + e^{-x})$ represents the shape of a hanging cable. Find the exact length of this catenary between $x = -1$ and $x = 1$.

42. Set up an integral for the circumference of an ellipse with semi-major axis $a = 2$ and semi-minor axis $b = 1$. Evaluating this integral numerically (i.e., using a calculator or computer) poses a problem. Describe this problem. Formulate a way to solve it.

43. There are very few elementary functions $y = f(x)$ for which arc length can be computed in elementary terms using the formula

$$\int_a^b \sqrt{1 + \left(\frac{dy}{dx}\right)^2}\, dx.$$

You have seen some such functions f in Problems 11, 12, and 41, namely, $f(x) = \sqrt{4 - x^2}$, $f(x) = \sqrt{x^3}$, and $f(x) = \frac{1}{2}(e^x + e^{-x})$. Try to find some other function that "works," that is, a function whose arc length you can find using this formula and antidifferentiation.

44. After doing Problem 43, you may wonder what sort of functions can represent arc length. If $g(0) = 0$ and g is differentiable and increasing, then can $g(x)$, $x \geq 0$, represent arc length? That is, can we find a function $f(t)$ such that

$$\int_0^x \sqrt{1 + (f'(t))^2}\, dt = g(x)?$$

(a) Show that $f(x) = \int_0^x \sqrt{(g'(t))^2 - 1}\, dt$ works as long as $g'(x) \geq 1$. In other words, show that the arc length of the graph of f from 0 to x is $g(x)$.

(b) Show that if $g'(x) < 1$ for some x, then $g(x)$ cannot represent the arc length of the graph of any function.

(c) Find a function f whose arc length from 0 to x is $2x$.

8.3 AREA AND ARC LENGTH IN POLAR COORDINATES

Many curves and regions in the plane are easier to describe in polar coordinates than in Cartesian coordinates. Thus their areas and arc lengths are best found using integrals in polar coordinates.

A point, P, in the plane is often identified by its *Cartesian coordinates* (x, y), where x is the horizontal distance to the point from the origin and y is the vertical distance.[1] Alternatively, we can identify the point, P, by specifying its distance, r, from the origin and the angle, θ, shown in Figure 8.29. The angle θ is measured counterclockwise from the positive x-axis to the line joining P to the origin. The labels r and θ are called the *polar coordinates* of point P.

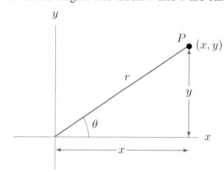

Figure 8.29: Cartesian and polar coordinates for the point P

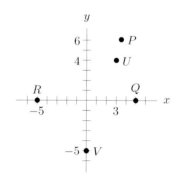

Figure 8.30

Relation Between Cartesian and Polar Coordinates

From the right triangle in Figure 8.29, we see that

- $x = r\cos\theta$ and $y = r\sin\theta$
- $r = \sqrt{x^2 + y^2}$ and $\tan\theta = \dfrac{y}{x}$, $x \neq 0$

The angle θ is determined by the equations $\cos\theta = x/\sqrt{x^2 + y^2}$ and $\sin\theta = y/\sqrt{x^2 + y^2}$.

Warning: In general $\theta \neq \tan^{-1}(y/x)$. It is not possible to determine which quadrant θ is in from the value of $\tan\theta$ alone.

Example 1   (a) Give Cartesian coordinates for the points with polar coordinates (r, θ) given by $P = (7, \pi/3)$, $Q = (5, 0)$, $R = (5, \pi)$.

(b) Give polar coordinates for the points with Cartesian coordinates (x, y) given by $U = (3, 4)$ and $V = (0, -5)$.

Solution   (a) See Figure 8.30. Point P is a distance of 7 from the origin. The angle θ is $\pi/3$ radians (60°). The Cartesian coordinates of P are

$$x = r\cos\theta = 7\cos\frac{\pi}{3} = \frac{7}{2} \quad \text{and} \quad y = r\sin\theta = 7\sin\frac{\pi}{3} = \frac{7\sqrt{3}}{2}.$$

[1]Cartesian coordinates can also be called rectangular coordinates.

Point Q is located a distance of 5 units along the positive x-axis with Cartesian coordinates

$$x = r\cos\theta = 5\cos 0 = 5 \quad\text{and}\quad y = r\sin\theta = 5\sin 0 = 0.$$

For point R, which is on the negative x-axis,

$$x = r\cos\theta = 5\cos\pi = -5 \quad\text{and}\quad y = r\sin\theta = 5\sin\pi = 0.$$

(b) For $U = (3,4)$, we have $r = \sqrt{3^2 + 4^2} = 5$ and $\tan\theta = 4/3$. A possible value for θ is $\theta = \arctan 4/3 = 0.927$ radians, or about $53°$. The polar coordinates of U are $(5, 0.927)$. The point V falls on the negative y-axis, so we can choose $r = 5, \theta = 3\pi/2$ for its polar coordinates. In this case, we cannot use $\tan\theta = y/x$ to find θ, because $\tan\theta = y/x = -5/0$ is undefined.

Because the angle θ can be allowed to wrap around the origin more than once, there are many possibilities for the polar coordinates of a point. For the point V in Example 1, we can also choose $\theta = -\pi/2$ or $\theta = 7\pi/2$, so that $(5, -\pi/2)$, $(5, 7\pi/2)$, and $(5, 3\pi/2)$ are all polar coordinates for V. However, we often choose θ between 0 and 2π.

Example 2   Give three possible sets of polar coordinates for the point P in Figure 8.31.

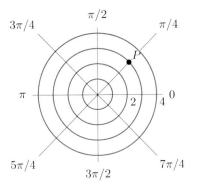

Figure 8.31

Solution   Because $r = 3$ and $\theta = \pi/4$, one set of polar coordinates for P is $(3, \pi/4)$. We can also use $\theta = \pi/4 + 2\pi = 9\pi/4$ and $\theta = \pi/4 - 2\pi = -7\pi/4$, to get $(3, 9\pi/4)$ and $(3, -7\pi/4)$.

Graphing Equations in Polar Coordinates

The equations for certain graphs are much simpler when expressed in polar coordinates than in Cartesian coordinates. On the other hand, some graphs that have simple equations in Cartesian coordinates have complicated equations in polar coordinates.

Example 3   (a) Describe in words the graphs of the equation $y = 1$ (in Cartesian coordinates) and the equation $r = 1$ (in polar coordinates).
(b) Write the equation $r = 1$ using Cartesian coordinates. Write the equation $y = 1$ using polar coordinates.

Solution   (a) The equation $y = 1$ describes a horizontal line. Since the equation $y = 1$ places no restrictions on the value of x, it describes every point having a y-value of 1, no matter what the value of its x-coordinate. Similarly, the equation $r = 1$ places no restrictions on the value of θ. Thus, it describes every point having an r-value of 1, that is, having a distance of 1 from the origin. This set of points is the unit circle. See Figure 8.32.

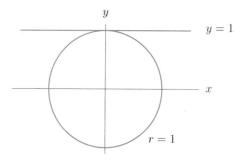

Figure 8.32: The graph of the equation $r = 1$ is the unit circle because $r = 1$ for every point regardless of the value of θ. The graph of $y = 1$ is a horizontal line since $y = 1$ for any x

Figure 8.33: A graph of the Archimedean spiral $r = \theta$

(b) Since $r = \sqrt{x^2 + y^2}$, we rewrite the equation $r = 1$ using Cartesian coordinates as $\sqrt{x^2 + y^2} = 1$, or, squaring both sides, as $x^2 + y^2 = 1$. We see that the equation for the unit circle is simpler in polar coordinates than it is in Cartesian coordinates.

On the other hand, since $y = r\sin\theta$, we can rewrite the equation $y = 1$ in polar coordinates as $r\sin\theta = 1$, or, dividing both sides by $\sin\theta$, as $r = 1/\sin\theta$. We see that the equation for this horizontal line is simpler in Cartesian coordinates than in polar coordinates.

Example 4   Graph the equation $r = \theta$. The graph is called an *Archimedean spiral* after the Greek mathematician Archimedes who described its properties (although not by using polar coordinates).

Solution   To construct this graph, use the values in Table 8.1. To help us visualize the shape of the spiral, we convert the angles in Table 8.1 to degrees and the r-values to decimals. See Table 8.2.

Table 8.1   *Points on the Archimedean spiral $r = \theta$, with θ in radians*

θ	0	$\frac{\pi}{6}$	$\frac{\pi}{3}$	$\frac{\pi}{2}$	$\frac{2\pi}{3}$	$\frac{5\pi}{6}$	π	$\frac{7\pi}{6}$	$\frac{4\pi}{3}$	$\frac{3\pi}{2}$
r	0	$\frac{\pi}{6}$	$\frac{\pi}{3}$	$\frac{\pi}{2}$	$\frac{2\pi}{3}$	$\frac{5\pi}{6}$	π	$\frac{7\pi}{6}$	$\frac{4\pi}{3}$	$\frac{3\pi}{2}$

Table 8.2   *Points on the Archimedean spiral $r = \theta$, with θ in degrees*

θ	0	30°	60°	90°	120°	150°	180°	210°	240°	270°
r	0.00	0.52	1.05	1.57	2.09	2.62	3.14	3.67	4.19	4.71

Notice that as the angle θ increases, points on the curve move farther from the origin. At 0°, the point is at the origin. At 30°, it is 0.52 units away from the origin, at 60° it is 1.05 units away, and at 90° it is 1.57 units away. As the angle winds around, the point traces out a curve that moves away from the origin, giving a spiral. (See Figure 8.33.)

In our definition, r is positive. However, graphs of curves in polar coordinates are traditionally drawn using negative values of r as well, because this makes the graphs symmetric. If an equation $r = f(\theta)$ gives a negative r-value, it is plotted in the opposite direction to θ. See Examples 5 and 6 and Figures 8.34 and 8.36.

Example 5   For $a > 0$ and n a positive integer, curves of the form $r = a\sin n\theta$ or $r = a\cos n\theta$ are called *roses*. Graph the roses

(a) $r = 3\sin 2\theta$

(b) $r = 4\cos 3\theta$

Solution   (a) Using a calculator or making a table of values, we see that the graph is a rose with four petals, each of length 3. See Figure 8.34. Negative values of r for $\pi/2 < \theta < \pi$ and $3\pi/2 < \theta < 2\pi$ give the petals in Quadrants II and IV. For example, $\theta = 3\pi/4$ gives $r = -3$, which is plotted 3 units from the origin in the direction opposite to $\theta = 3\pi/4$, namely in Quadrant IV.

(b) The graph is a rose with three petals, each of length 4. See Figure 8.35.

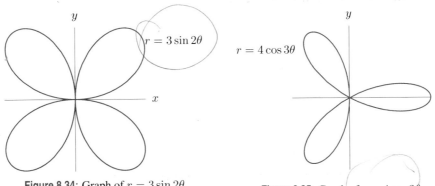

Figure 8.34: Graph of $r = 3 \sin 2\theta$
(Petals in Quadrants II and IV have $r < 0$)

Figure 8.35: Graph of $r = 4 \cos 3\theta$

Example 6   Curves of the form $r = a + b \sin \theta$ or $r = a + b \cos \theta$ are called limaçons. Graph $r = 1 + 2 \cos \theta$ and $r = 3 + 2 \cos \theta$.

Solution   See Figures 8.36 and 8.37. The equation $r = 1 + 2 \cos \theta$ leads to negative r values for some θ values between $\pi/2$ and $3\pi/2$; these values give the inner loop in Figure 8.36. For example, $\theta = \pi$ gives $r = -1$, which is plotted 1 unit from the origin in the direction opposite to $\theta = \pi$, namely on the positive x-axis. The equation $r = 3 + 2 \cos \theta$ does not lead to negative r-values.

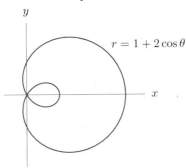

Figure 8.36: Graph of $r = 1 + 2 \cos \theta$
(Inner loop has $r < 0$)

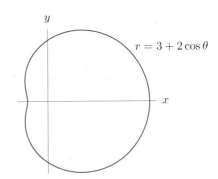

Figure 8.37: Graph of $r = 3 + 2 \cos \theta$

Polar coordinates can be used with inequalities to describe regions that are obtained from circles. Such regions are often much harder to represent in Cartesian coordinates.

Example 7   Using inequalities, describe a compact disk with outer diameter 120 mm and inner diameter 15 mm.

Solution   The compact disk lies between two circles of radius 7.5 mm and 60 mm. See Figure 8.38. Thus, if the origin is at the center, the disk is represented by

$$7.5 \leq r \leq 60 \quad \text{and} \quad 0 \leq \theta \leq 2\pi.$$

Figure 8.38: Compact disk

Figure 8.39: Pizza slice

Example 8   An 18 inch pizza is cut into 12 slices. Use inequalities to describe one of the slices.

Solution   The pizza has radius 9 inches; the angle at the center is $2\pi/12 = \pi/6$. See Figure 8.39. Thus, if the origin is at center of the original pizza, the slice is represented by

$$0 \le r \le 9 \qquad \text{and} \qquad 0 \le \theta \le \frac{\pi}{6}.$$

Area in Polar Coordinates

We can use a definite integral to find the area of a region described in polar coordinates. As previously, we slice the region into small pieces, construct a Riemann sum, and take a limit to obtain the definite integral. In this case, the slices are approximately circular sectors.

To calculate the area of the sector in Figure 8.40, we think of the area of the sector as a fraction $\theta/2\pi$ of the area of the entire circle (for θ in radians). Then

$$\text{Area of sector} = \frac{\theta}{2\pi} \cdot \pi r^2 = \frac{1}{2}r^2\theta.$$

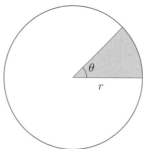

Figure 8.40: Area of shaded sector $= \frac{1}{2}r^2\theta$ (for θ in radians)

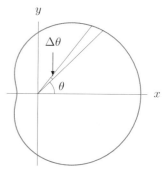

Figure 8.41: Finding the area of the limaçon $r = 3 + 2\cos\theta$

Example 9   Use circular sectors to set up a definite integral to calculate the area of the region bounded by the limaçon $r = 3 + 2\cos\theta$, for $0 \le \theta \le 2\pi$. See Figure 8.41.

Solution   The slices are not exactly circular sectors because the radius r depends on θ. However,

$$\text{Area of sector} \approx \frac{1}{2}r^2\Delta\theta = \frac{1}{2}(3 + 2\cos\theta)^2\,\Delta\theta.$$

Thus, for the whole area,

$$\text{Area of region} \approx \sum_{i=1}^{n} \frac{1}{2}(3 + 2\cos\theta)^2\,\Delta\theta.$$

Taking the limit as $n \to \infty$ and $\Delta\theta \to 0$ gives the integral

$$\text{Area} = \int_0^{2\pi} \frac{1}{2}(3 + 2\cos\theta)^2\,d\theta.$$

To compute this integral, we expand the integrand and use integration by parts or formula IV-18 from the table of integrals:

$$\text{Area} = \frac{1}{2}\int_0^{2\pi}\left(9 + 12\cos\theta + 4\cos^2\theta\right)d\theta$$

$$= \frac{1}{2}\left(9\theta + 12\sin\theta + \frac{4}{2}(\cos\theta\sin\theta + \theta)\right)\Bigg|_0^{2\pi}$$

$$= \frac{1}{2}(18\pi + 0 + 4\pi) = 11\pi.$$

The reasoning in Example 9 suggests a general formula for the area in Figure 8.42.

For a curve $r = f(\theta)$, with $\alpha \leq \theta \leq \beta$, which does not cross itself,

$$\text{Area of region enclosed} = \frac{1}{2} \int_\alpha^\beta f(\theta)^2 \, d\theta.$$

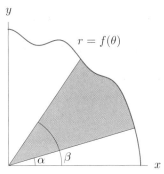

Figure 8.42: Area in polar coordinates

Example 10   Find the area of one petal of the four-petal rose $r = 3\sin 2\theta$ in Figure 8.43.

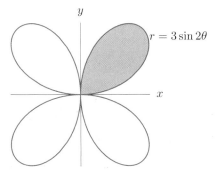

Figure 8.43: One petal of the rose
$r = 3\sin 2\theta$ with $0 \leq \theta \leq \pi/2$

Solution   The petal in the first quadrant is described by $r = 3\sin 2\theta$ for $0 \leq \theta \leq \pi/2$, so

$$\text{Area of shaded region} = \frac{1}{2}\int_0^{\pi/2} (3\sin 2\theta)^2 \, d\theta = \frac{9}{2}\int_0^{\pi/2} \sin^2 2\theta \, d\theta.$$

Using the substitution $w = 2\theta$, we rewrite the integral and use integration by parts or formula IV-17 from the table of integrals:

$$\text{Area} = \frac{9}{2}\int_0^{\pi/2} \sin^2 2\theta \, d\theta = \frac{9}{4}\int_0^{\pi} \sin^2 w \, dw$$

$$= \frac{9}{4}\left(-\frac{1}{2}\cos w \sin w + \frac{1}{2}w\right)\Big|_0^{\pi} = \frac{9}{4} \cdot \frac{\pi}{2} = \frac{9\pi}{8}.$$

Slope in Polar Coordinates

For a curve $r = f(\theta)$, we can express x and y in terms of θ as a parameter, giving

$$x = r \cos \theta = f(\theta) \cos \theta \quad \text{and} \quad y = r \sin \theta = f(\theta) \sin \theta.$$

To find the slope of the curve, we use the formula for the slope of a parametric curve

$$\boxed{\frac{dy}{dx} = \frac{dy/d\theta}{dx/d\theta}.}$$

Example 11   Find the slope of the curve $r = 3 \sin 2\theta$ at $\theta = \pi/3$.

Solution   Expressing x and y in terms of θ, we have

$$x = 3 \sin(2\theta) \cos \theta \quad \text{and} \quad y = 3 \sin(2\theta) \sin \theta.$$

The slope is given by

$$\frac{dy}{dx} = \frac{6 \cos(2\theta) \sin \theta + 3 \sin(2\theta) \cos \theta}{6 \cos(2\theta) \cos \theta - 3 \sin(2\theta) \sin \theta}.$$

At $\theta = \pi/3$, we have

$$\left.\frac{dy}{dx}\right|_{\theta=\pi/3} = \frac{6(-1/2)(\sqrt{3}/2) + 3(\sqrt{3}/2)(1/2)}{6(-1/2)(1/2) - 3(\sqrt{3}/2)(\sqrt{3}/2)} = \frac{\sqrt{3}}{5}.$$

Arc Length in Polar Coordinates

We can calculate the arc length of the curve $r = f(\theta)$ by expressing x and y in terms of θ as a parameter

$$x = f(\theta) \cos \theta \qquad y = f(\theta) \sin \theta$$

and using the formula for the arc length of a parametric curve:

$$\boxed{\text{Arc length} = \int_{\alpha}^{\beta} \sqrt{\left(\frac{dx}{d\theta}\right)^2 + \left(\frac{dy}{d\theta}\right)^2}\, d\theta.}$$

The calculations may be simplified by using the alternate form of the arc length integral in Problem 38.

Example 12   Find the arc length of one petal of the rose $r = 3 \sin 2\theta$ for $0 \leq \theta \leq \pi/2$. See Figure 8.43.

Solution   The curve is given parametrically by

$$x = 3 \sin(2\theta) \cos \theta \quad \text{and} \quad y = 3 \sin(2\theta) \sin \theta.$$

Thus, calculating $dx/d\theta$ and $dy/d\theta$ and evaluating the integral on a calculator, we have

$$\text{Arc length} = \int_{0}^{\pi/2} \sqrt{(6 \cos(2\theta) \cos \theta - 3 \sin(2\theta) \sin \theta)^2 + (6 \cos(2\theta) \sin \theta + 3 \sin(2\theta) \cos \theta)^2}\, d\theta$$

$$= 7.266.$$

Exercises and Problems for Section 8.3

Exercises

Convert the polar coordinates in Exercises 1–4 to Cartesian coordinates. Give exact answers.

1. $(1, 2\pi/3)$

2. $(\sqrt{3}, -3\pi/4)$

3. $(2\sqrt{3}, -\pi/6)$

4. $(2, 5\pi/6)$

Convert the Cartesian coordinates in Exercises 5–8 to polar coordinates.

5. $(1, 1)$

6. $(-1, 0)$

7. $(\sqrt{6}, -\sqrt{2})$

8. $(-\sqrt{3}, 1)$

9. (a) Make a table of values for the equation $r = 1 - \sin\theta$. Include $\theta = 0,\ \pi/3,\ \pi/2,\ 2\pi/3,\ \pi,\ \cdots$.
   (b) Use the table to graph the equation $r = 1 - \sin\theta$ in the xy-plane. This curve is called a *cardioid*.
   (c) At what point(s) does the cardioid $r = 1 - \sin\theta$ intersect a circle of radius 1/2 centered at the origin?
   (d) Graph the curve $r = 1 - \sin 2\theta$ in the xy-plane. Compare this graph to the cardioid $r = 1 - \sin\theta$.

10. Graph the equation $r = 1 - \sin(n\theta)$, for $n = 1, 2, 3, 4$. What is the relationship between the value of n and the shape of the graph?

11. Graph the equation $r = 1 - \sin\theta$, with $0 \le \theta \le n\pi$, for $n = 2, 3, 4$. What is the relationship between the value of n and the shape of the graph?

12. Graph the equation $r = 1 - n\sin\theta$, for $n = 2, 3, 4$. What is the relationship between the value of n and the shape of the graph?

13. Graph the equation $r = 1 - \cos\theta$. Describe its relationship to $r = 1 - \sin\theta$.

14. Give inequalities that describe the flat surface of a washer that is one inch in diameter and has an inner hole with a diameter of 3/8 inch.

15. Graph the equation $r = 1 - \sin(2\theta)$ for $0 \le \theta \le 2\pi$. There are two loops. For each loop, give a restriction on θ that shows all of that loop and none of the other loop.

16. A slice of pizza is one eighth of a circle of radius 1 foot. The slice is in the first quadrant, with one edge along the x-axis, and the center of the pizza at the origin. Give inequalities describing this region using:

   (a) Polar coordinates   **(b)** Rectangular coordinates

In Exercises 17–19, give inequalities for r and θ which describe the following regions in polar coordinates.

17.

18.

19.

Note: Region extends indefinitely in the y-direction.

Problems

20. Sketch the polar region described by the following integral expression for area:

$$\frac{1}{2} \int_0^{\pi/3} \sin^2(3\theta)\, d\theta.$$

21. Find the area inside the spiral $r = \theta$ for $0 \le \theta \le 2\pi$.

22. Find the area between the two spirals $r = \theta$ and $r = 2\theta$ for $0 \le \theta \le 2\pi$.

23. Find the area inside the cardioid $r = 1 + \cos\theta$ for $0 \le \theta \le 2\pi$.

24. (a) In polar coordinates, write equations for the line $x = 1$ and the circle of radius 2 centered at the origin.
   (b) Write an integral in polar coordinates representing the area of the region to the right of $x = 1$ and inside the circle.
   (c) Evaluate the integral.

25. Find the area inside the circle $r = 1$ and outside the cardioid $r = 1 + \sin\theta$.

26. Find the area inside the cardioid $r = 1 - \sin\theta$ and outside the circle $r = 1/2$.

27. Find the area lying outside $r = 2\cos\theta$ and inside $r = 1 + \cos\theta$.

28. **(a)** Graph $r = 2\cos\theta$ and $r = 2\sin\theta$ on the same axes.

(b) Using polar coordinates, find the area of the region shared by both curves.

29. For what value of a is the area enclosed by $r = \theta, \theta = 0$, and $\theta = a$ equal to 1?

30. **(a)** Sketch the bounded region inside the lemniscate $r^2 = 4\cos 2\theta$ and outside the circle $r = \sqrt{2}$.

(b) Compute the area of the region described in part (a).

31. Using Example 11 on page 388, find the equation of the tangent line to the curve $r = 3\sin 2\theta$ at $\theta = \pi/3$.

32. Using Example 11 on page 388 and Figure 8.34, find the points where the curve $r = 3\sin 2\theta$ has horizontal and vertical tangents.

33. For what values of θ on the polar curve $r = \theta$, with $0 \le \theta \le 2\pi$, are the tangent lines horizontal? Vertical?

34. **(a)** In Cartesian coordinates, write an equation for the tangent line to $r = 1/\theta$ at $\theta = \pi/2$.

(b) The graph of $r = 1/\theta$ has a horizontal asymptote as θ approaches 0. Find the equation of this asymptote.

35. Find the maximum value of the y-coordinate of points on the limaçon $r = 1 + 2\cos\theta$.

Find the arc length of the curves in Problems 36–37.

36. $r = \theta, 0 \le \theta \le 2\pi$

37. $r = 1/\theta, \pi \le \theta \le 2\pi$

38. For the curve $r = f(\theta)$ from $\theta = \alpha$ to $\theta = \beta$, show that

$$\text{Arc length} = \int_\alpha^\beta \sqrt{(f'(\theta))^2 + (f(\theta))^2}\, d\theta.$$

8.4 DENSITY AND CENTER OF MASS

Density and How to Slice a Region

The examples in this section involve the idea of *density*. For example,

- A population density is measured in, say, people per mile (along the edge of a road), or people per unit area (in a city), or bacteria per cubic centimeter (in a test tube).

- The density of a substance (e.g. air, wood, or metal), is the mass of a unit volume of the substance and is measured in, say, grams per cubic centimeter.

Suppose we want to calculate the total mass or total population, but the density is not constant over a region.

> **To find total quantity from density:** Divide the region into small pieces in such a way that the density is approximately constant on each piece, and add the contributions of the pieces.

Example 1   The Massachusetts Turnpike ("the Pike") starts in the middle of Boston and heads west. The number of people living next to it varies as it gets farther from the city. Suppose that, x miles out of town, the population density adjacent to the Pike is $P = f(x)$ people/mile. Express the total population living next to the Pike within 5 miles of Boston as a definite integral.

Solution   Divide the Pike up into segments of length Δx. The population density at the center of Boston is $f(0)$; let's use that density for the first segment. This gives an estimate of

$$\text{People living in first segment} \approx f(0)\text{ people/ mile} \cdot \Delta x\text{ mile} = f(0)\Delta x\text{ people.}$$

Population $\approx f(x)\Delta x$

Figure 8.44: Population along the Massachusetts Turnpike

Similarly, the population in a typical segment x miles from the center of Boston is the population density times the length of the interval, or roughly $f(x)\,\Delta x$. (See Figure 8.44.) The sum of all these estimates gives the estimate

$$\text{Total population} \approx \sum f(x)\,\Delta x.$$

Letting $\Delta x \to 0$ gives

$$\text{Total population} = \lim_{\Delta x \to 0} \sum f(x)\,\Delta x = \int_0^5 f(x)\,dx.$$

The 5 and 0 in the limits of the integral are the upper and lower limits of the interval over which we are integrating.

Example 2   The air density h meters above the earth's surface is $f(h)$ kg/m^3. Find the mass of a cylindrical column of air 2 meters in diameter and 25 kilometers high, with base on the surface of the earth.

Solution   The column of air is a circular cylinder 2 meters in diameter and 25 kilometers, or 25,000 meters, high. First we must decide how we are going to slice this column. Since the air density varies with altitude but remains constant horizontally, we take horizontal slices of air. That way, the density will be more or less constant over the whole slice, being close to its value at the bottom of the slice. (See Figure 8.45.)

Figure 8.45: Slicing a column of air horizontally

A slice is a cylinder of height Δh and diameter 2 m, so its radius is 1 m. We find the approximate mass of the slice by multiplying its volume by its density. If the thickness of the slice is Δh, then its volume is $\pi r^2 \cdot \Delta h = \pi 1^2 \cdot \Delta h = \pi\,\Delta h$ m^3. The density of the slice is given by $f(h)$. Thus,

$$\text{Mass of slice} \approx \text{Volume} \cdot \text{Density} = (\pi \Delta h \text{ m}^3)(f(h) \text{ kg/m}^3) = \pi\,\Delta h \cdot f(h) \text{ kg}.$$

Adding these slices up yields a Riemann sum:

$$\text{Total mass} \approx \sum \pi f(h)\,\Delta h \text{ kg}.$$

As $\Delta h \to 0$, this sum approximates the definite integral:

$$\text{Total mass} = \int_0^{25,000} \pi f(h)\,dh \text{ kg}.$$

In order to get a numerical value for the mass of air, we need an explicit formula for the density as a function of height, as in the next example.

Example 3   Find the mass of the column of air in Example 2 if the density of air at height h is given by

$$P = f(h) = 1.28e^{-0.000124h} \text{ kg/m}^3.$$

Solution   Using the result of the previous example, we have

$$\text{Mass} = \int_0^{25,000} \pi 1.28 e^{-0.000124h} \, dh = \frac{-1.28\pi}{0.000124} \left(e^{-0.000124h} \Big|_0^{25,000} \right)$$

$$= \frac{1.28\pi}{0.000124} \left(e^0 - e^{-0.000124(25,000)} \right) \approx 31,000 \text{ kg}.$$

It requires some thought to figure out how to slice a region. The key point is that you want the density to be nearly constant within each piece.

Example 4   The population density in Ringsburg is a function of the distance from the city center. At r miles from the center, the density is $P = f(r)$ people per square mile. Ringsburg is circular with radius 5 miles. Write a definite integral that expresses the total population of Ringsburg.

Solution   We want to slice Ringsburg up and estimate the population of each slice. If we were to take straight-line slices, the population density would vary on each slice, since it depends on the distance from the city center. We want the population density to be pretty close to constant on each slice. We therefore take slices that are thin rings around the center. (See Figure 8.46.) Since the ring is very thin, we can approximate its area by straightening it into a thin rectangle. (See Figure 8.47.) The width of the rectangle is Δr miles, and its length is approximately equal to the circumference of the ring, $2\pi r$ miles, so its area is about $2\pi r \, \Delta r$ mi^2. Since

$$\text{Population on ring} \approx \text{Density} \cdot \text{Area},$$

we get

$$\text{Population on ring} \approx (f(r) \text{ people/mi}^2)(2\pi r \Delta r \text{ mi}^2) = f(r) \cdot 2\pi r \, \Delta r \text{ people}.$$

Adding the contributions from each ring, we get

$$\text{Total population} \approx \sum 2\pi r f(r) \, \Delta r \text{ people}.$$

So

$$\text{Total population} = \int_0^5 2\pi r f(r) \, dr \text{ people}.$$

Figure 8.46: Ringsburg

Width $= \Delta r$

$2\pi r$

Figure 8.47: Ring from Ringsburg (straightened out)

Note: You may wonder what happens if we calculate the area of the ring by subtracting the area of the inner circle (πr^2) from the area of the outer circle [$\pi(r + \Delta r)^2$], giving

$$\text{Area} = \pi(r + \Delta r)^2 - \pi r^2.$$

Multiplying out and subtracting, we get

$$\begin{aligned}\text{Area} &= \pi[r^2 + 2r\,\Delta r + (\Delta r)^2] - \pi r^2 \\ &= 2\pi r\,\Delta r + \pi(\Delta r)^2.\end{aligned}$$

This expression differs from the one we used before by the $\pi(\Delta r)^2$ term. However, as Δr becomes very small, $\pi(\Delta r)^2$ becomes much, much smaller. We say its smallness is of *second order*, since the power of the small factor, Δr, is 2. In the limit as $\Delta r \to 0$, we can ignore $\pi(\Delta r)^2$.

Center of Mass

The center of mass of a mechanical system is important for studying its behavior when in motion. For example, some sport utility vehicles and light trucks tend to tip over in accidents, because of their high centers of mass.

In this section, we first define the center of mass for a system of point masses on a line. Then we use the definite integral to extend this definition.

Point Masses

Two children on a seesaw, one twice the weight of the other, will balance if the lighter child is twice as far from the pivot as the heavier child. Thus, the balance point is 2/3 of the way from the lighter child and 1/3 of the way from the heavier child. This balance point is the *center of mass* of the mechanical system consisting of the masses of the two children (we ignore the mass of the seesaw itself). See Figure 8.48.

To find the balance point, we use the *displacement* (signed distance) of each child from the pivot to calculate the *moment*, where

$$\text{Moment of mass about pivot} = \text{Mass} \times \text{Displacement from pivot}.$$

A moment represents the tendency of a child to turn the system about the pivot point; the seesaw balances if the total moment is zero. Thus, the center of mass is the point about which the total moment is zero.

Figure 8.48: Children on seesaw

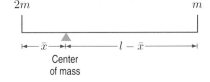

Figure 8.49: Center of mass of point masses

Example 5   Calculate the position of the center of mass of the children in Figure 8.48 using moments.

Solution   Suppose the center of mass in Figure 8.49 is at a distance of $\bar{x}$ from the left end. The moment of the left mass about the center of mass is $-2m\bar{x}$ (it is negative since it is to the left of the center of mass); the moment of the right mass about the center of mass is $m(l - \bar{x})$. The system balances if

$$-2m\bar{x} + m(l - \bar{x}) = 0 \quad \text{or} \quad ml - 3m\bar{x} = 0 \quad \text{so} \quad \bar{x} = \frac{1}{3}l.$$

Thus, the center of mass is $l/3$ from the left end.

$$m_1 \qquad\qquad m_2 \qquad m_3$$
$$x_1 \qquad\quad 0 \qquad x_2 \qquad x_3$$

Figure 8.50: Discrete masses, m_1, m_2, m_3

We use the same method to calculate the center of mass, $\bar{x}$, of the system in Figure 8.50. The sum of the moments of the three masses about $\bar{x}$ is 0, so

$$m_1(x_1 - \bar{x}) + m_2(x_2 - \bar{x}) + m_3(x_3 - \bar{x}) = 0.$$

Solving for $\bar{x}$, we get

$$m_1\bar{x} + m_2\bar{x} + m_3\bar{x} = m_1x_1 + m_2x_2 + m_3x_3$$

$$\bar{x} = \frac{m_1x_1 + m_2x_2 + m_3x_3}{m_1 + m_2 + m_3} = \frac{\sum_{i=1}^{3} m_ix_i}{\sum_{i=1}^{3} m_i}.$$

Generalizing leads to the following formula:

The **center of mass** of a system of n point masses $m_1, m_2, \ldots, m_n$ located at positions $x_1, x_2, \ldots, x_n$ along the x-axis is given by

$$\bar{x} = \frac{\sum x_i m_i}{\sum m_i}.$$

The numerator is the sum of the moments of the masses about the origin; the denominator is the total mass of the system.

Example 6   Show that the definition of $\bar{x}$ gives the same answer as we found in Example 5.

Solution   Suppose the origin is at the left end of the seesaw in Figure 8.48. The total mass of the system is $2m + m = 3m$. We compute

$$\bar{x} = \frac{\sum x_i m_i}{\sum m_i} = \frac{1}{3m}(2m \cdot 0 + m \cdot l) = \frac{ml}{3m} = \frac{l}{3}.$$

Continuous Mass Density

Instead of discrete masses arranged along the x-axis, suppose we have an object lying on the x-axis between $x = a$ and $x = b$. At point x, suppose the object has mass density (mass per unit length) of $\delta(x)$. To calculate the center of mass of such an object, divide it into n pieces, each of length Δx. On each piece, the density is nearly constant, so the mass of the piece is given by density times length. See Figure 8.51. Thus, if x_i is a point in the ith piece,

Mass of the ith piece, $m_i \approx \delta(x_i)\Delta x.$

Then the formula for the center of mass, $\bar{x} = \sum x_i m_i / \sum m_i$, applied to the n pieces of the object gives

$$\bar{x} = \frac{\sum x_i \delta(x_i)\Delta x}{\sum \delta(x_i)\Delta x}.$$

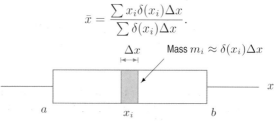

Figure 8.51: Calculating the center of mass of an object of variable density, $\delta(x)$

In the limit as $n \to \infty$ we have the following formula:

> The **center of mass** $\bar{x}$ of an object lying along the x-axis between $x = a$ and $x = b$ is
>
> $$\bar{x} = \frac{\int_a^b x\delta(x)\,dx}{\int_a^b \delta(x)\,dx},$$
>
> where $\delta(x)$ is the density (mass per unit length) of the object.

As in the discrete case, the denominator is the total mass of the object.

Example 7   Find the center of mass of a 2-meter rod lying on the x-axis with its left end at the origin if:
(a)   The density is constant and the total mass is 5 kg.   (b)   The density is $\delta(x) = 15x^2$ kg/m.

Solution   (a) Since the density is constant along the rod, we expect the balance point to be in the middle, that is, $\bar{x} = 1$. To check this, we compute $\bar{x}$. The density is the total mass divided by the length, so $\delta(x) = 5/2$ kg/m. Then

$$\bar{x} = \frac{\text{Moment}}{\text{Mass}} = \frac{\int_0^2 x \cdot \frac{5}{2}\,dx}{5} = \frac{1}{5} \cdot \frac{5}{2} \cdot \frac{x^2}{2}\Big|_0^2 = 1 \text{ meter.}$$

(b) Since more of the mass of the rod is closer to its right end (the density is greatest there), we expect the center of mass to be in the right half of the rod, that is, between $x = 1$ and $x = 2$. We have

$$\text{Total mass} = \int_0^2 15x^2\,dx = 5x^3\Big|_0^2 = 40 \text{ kg.}$$

Thus,

$$\bar{x} = \frac{\text{Moment}}{\text{Mass}} = \frac{\int_0^2 x \cdot 15x^2\,dx}{40} = \frac{15}{40} \cdot \frac{x^4}{4}\Big|_0^2 = 1.5 \text{ meter.}$$

Two- and Three-Dimensional Regions

For a system of masses that lies in the plane, the center of mass is a point with coordinates $(\bar{x}, \bar{y})$. In three dimensions, the center of mass is a point with coordinates $(\bar{x}, \bar{y}, \bar{z})$. To compute the center of mass in three dimensions, we use the following formulas in which $A_x(x)$ is the area of a slice perpendicular to the x-axis at x, and $A_y(y)$ and $A_z(z)$ are defined similarly. In two dimensions, we use the same formulas for $\bar{x}$ and $\bar{y}$, but we interpret $A_x(x)$ and $A_y(y)$ as the lengths of strips perpendicular to the x- and y-axes, respectively.

> For a region of constant density δ, the center of mass is given by
>
> $$\bar{x} = \frac{\int x\delta A_x(x)\,dx}{\text{Mass}} \quad \bar{y} = \frac{\int y\delta A_y(y)\,dy}{\text{Mass}} \quad \bar{z} = \frac{\int z\delta A_z(z)\,dz}{\text{Mass}}.$$

The expression $\delta A_x(x)\Delta x$ is the moment of a slice perpendicular to the x-axis. Thus, these formulas are extensions of that on page 395. In the two and three dimensional case, we are assuming that the density δ is constant. If the density is not constant, finding the center of mass may require a double or triple integral from multivariable calculus.

Example 8   Find the coordinates of the center of mass of the isosceles triangle in Figure 8.52. The triangle has constant density and mass m.

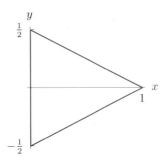

Figure 8.52: Find center of mass of this triangle

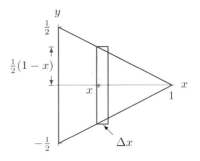

Figure 8.53: Sliced triangle

Solution   Because the mass of the triangle is symmetrically distributed with respect to the x-axis, $\bar{y} = 0$. We expect $\bar{x}$ to be closer to $x = 0$ than to $x = 1$, since the triangle is wider near the origin.

The area of the triangle is $\frac{1}{2} \cdot 1 \cdot 1 = \frac{1}{2}$. Thus Density = Mass/Area = $2m$. If we slice the triangle into strips of width Δx, then the strip at position x has length $A_x(x) = 2 \cdot \frac{1}{2}(1-x) = (1-x)$. (See Figure 8.53.) So

$$\text{Area of strip} = A_x(x)\Delta x \approx (1-x)\Delta x.$$

Since the density is $2m$, the center of mass is given by

$$\bar{x} = \frac{\int x\delta A_x(x)\, dx}{\text{Mass}} = \frac{\int_0^1 2mx(1-x)\, dx}{m} = 2\left(\frac{x^2}{2} - \frac{x^3}{3}\right)\bigg|_0^1 = \frac{1}{3}.$$

So the center of mass of this triangle is at the point $(\bar{x}, \bar{y}) = (1/3, 0)$.

Example 9   Find the center of mass of a hemisphere of radius 7 cm and constant density δ.

Solution   Stand the hemisphere with its base horizontal in the xy-plane, with the center at the origin. Symmetry tells us that its center of mass lies directly above the center of the base, so $\bar{x} = \bar{y} = 0$. Since the hemisphere is wider near its base, we expect the center of mass to be nearer to the base than the top.

To calculate the center of mass, slice the hemisphere into horizontal disks as in Figure 8.9 on page 375. A disk of thickness Δz at height z above the base has

$$\text{Volume of disk} = A_z(z)\Delta z \approx \pi(7^2 - z^2)\Delta z \text{ cm}^3.$$

So, since the density is δ,

$$\bar{z} = \frac{\int z\delta A_z(z)\, dz}{\text{Mass}} = \frac{\int_0^7 z\delta\pi(7^2 - z^2)\, dz}{\text{Mass}}.$$

Since the total mass of the hemisphere is $\left(\frac{2}{3}\pi 7^3\right)\delta$, we get

$$\bar{z} = \frac{\delta\pi \int_0^7 (7^2 z - z^3)\, dz}{\text{Mass}} = \frac{\delta\pi \left(7^2 z^2/2 - z^4/4\right)\big|_0^7}{\text{Mass}} = \frac{\frac{7^4}{4}\delta\pi}{\frac{2}{3}\pi 7^3\delta} = \frac{21}{8} = 2.625 \text{ cm}.$$

The center of mass of the hemisphere is 2.625 cm above the center of its base. As expected, it is closer to the base of the hemisphere than its top.

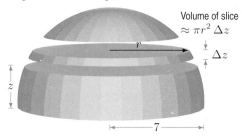

Figure 8.54: Slicing to find the center of mass of a hemisphere

Exercises and Problems for Section 8.4

Exercises

1. Find the mass of a rod of length 10 cm with density $\delta(x) = e^{-x}$ gm/cm at a distance of x cm from the left end.

2. A plate occupying the region $0 \le x \le 2, 0 \le y \le 3$ has density $\delta = 5$ gm/cm^2. Set up two integrals giving the mass of the plate, one corresponding to strips in the x-direction and one corresponding to strips in the y-direction.

3. A rod has length 2 meters. At a distance x meters from its left end, the density of the rod is given by

$$\delta(x) = 2 + 6x \text{ gm/m}.$$

    (a) Write a Riemann sum approximating the total mass of the rod.

    (b) Find the exact mass by converting the sum into an integral.

4. If a rod lies along the x-axis between a and b, the moment of the rod is $\int_a^b x\delta(x)\, dx$, where $\delta(x)$ is its density in grams/meter at a position x meters. Find the moment and center of mass of the rod in Problem 3.

5. The density of cars (in cars per mile) down a 20-mile stretch of the Pennsylvania Turnpike is approximated by

$$\delta(x) = 300 \left(2 + \sin\left(4\sqrt{x + 0.15}\right)\right),$$

at a distance x miles from the Breezewood toll plaza.

    (a) Sketch a graph of this function for $0 \le x \le 20$.

    (b) Write a Riemann sum that approximates the total number of cars on this 20-mile stretch.

    (c) Find the total number of cars on the 20-mile stretch.

6. (a) Find a Riemann sum which approximates the total mass of a 3×5 rectangular sheet, whose density per unit area at a distance x from one of the sides of length 5 is $1/(1 + x^4)$.

    (b) Calculate the mass.

7. A point mass of 2 grams located 3 centimeters to the left of the origin and a point mass of 5 grams located 4 centimeters to the right of the origin are connected by a thin, light rod. Find the center of mass of the system.

8. Find the center of mass of a system containing three point masses of 5 gm, 3 gm, and 1 gm located respectively at $x = -10$, $x = 1$, and $x = 2$.

Problems

9. Find the total mass of the triangular region in Figure 8.55, which has density $\delta(x) = 1 + x$ grams/cm^2.

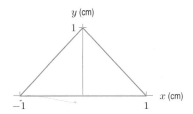

Figure 8.55

10. The density of oil in a circular oil slick on the surface of the ocean at a distance r meters from the center of the slick is given by $\delta(r) = 50/(1 + r)$ kg/m^2.

    (a) If the slick extends from $r = 0$ to $r = 10{,}000$ m, find a Riemann sum approximating the total mass of oil in the slick.

    (b) Find the exact value of the mass of oil in the slick by turning your sum into an integral and evaluating it.

    (c) Within what distance r is half the oil of the slick contained?

11. The soot produced by a garbage incinerator spreads out in a circular pattern. The depth, $H(r)$, in millimeters, of the soot deposited each month at a distance r kilometers from the incinerator is given by $H(r) = 0.115e^{-2r}$.

(a) Write a definite integral giving the total volume of soot deposited within 5 kilometers of the incinerator each month.

(b) Evaluate the integral you found in part (a), giving your answer in cubic meters.

12. A cardboard figure has the shape shown in Figure 8.56. The region is bounded on the left by the line $x = a$, on the right by the line $x = b$, above by $f(x)$, and below by $g(x)$. If the density $\delta(x)$ gm/cm^2 varies only with x, find an expression for the total mass of the figure, in terms of $f(x)$, $g(x)$, and $\delta(x)$.

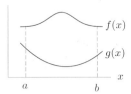

Figure 8.56

13. The storage shed in Figure 8.57 is the shape of a half-cylinder of radius r and length l.

(a) What is the volume of the shed?

(b) The shed is filled with sawdust whose density (mass/unit volume) at any point is proportional to the distance of that point from the floor. The constant of proportionality is k. Calculate the total mass of sawdust in the shed.

Figure 8.57

14. The following table gives the density D (in gm/cm^3) of the earth at a depth x km below the earth's surface. The radius of the earth is about 6370 km. Find an upper and a lower bound for the earth's mass such that the upper bound is less than twice the lower bound. Explain your reasoning; in particular, what assumptions have you made about the density?

x	0	1000	2000	2900	3000	4000	5000	6000	6370
D	3.3	4.5	5.1	5.6	10.1	11.4	12.6	13.0	13.0

15. Water leaks out of a tank through a square hole with 1-inch sides. At time t (in seconds) the velocity of water flowing through the hole is $v = g(t)$ ft/sec. Write a definite integral that represents the total amount of water lost in the first minute.

16. An exponential model for the density of the earth's atmosphere says that if the temperature of the atmosphere were constant, then the density of the atmosphere as a function of height, h (in meters), above the surface of the earth would be given by

$$\delta(h) = 1.28e^{-0.000124h} \text{ kg/m}^3.$$

(a) Write (but do not evaluate) a sum that approximates the mass of the portion of the atmosphere from $h = 0$ to $h = 100$ m (i.e., the first 100 meters above sea level). Assume the radius of the earth is 6400 km.

(b) Find the exact answer by turning your sum in part (a) into an integral. Evaluate the integral.

17. Three point masses of 4 gm each are placed at $x = -6, 1$ and 3. Where should a point fourth mass of 4 gm be placed to make the center of mass at the origin?

18. A rod of length 3 meters with density $\delta(x) = 1 + x^2$ grams/meter is positioned along the positive x-axis, with its left end at the origin. Find the total mass and the center of mass of the rod.

19. A rod with density $\delta(x) = 2 + \sin x$ lies on the x-axis between $x = 0$ and $x = \pi$. Find the center of mass of the rod.

20. A rod of length 1 meter has density $\delta(x) = 1 + kx^2$ grams/meter, where k is a positive constant. The rod is lying on the positive x-axis with one end at the origin.

(a) Find the center of mass as a function of k.

(b) Show that the center of mass of the rod satisfies $0.5 < \bar{x} < 0.75$.

21. A rod of length 2 meters and density $\delta(x) = 3 - e^{-x}$ kilograms per meter is placed on the x-axis with its ends at $x = \pm 1$.

(a) Will the center of mass of the rod be on the left or right of the origin? Explain.

(b) Find the coordinate of the center of mass.

22. One half of a uniform circular disk of radius 1 meter lies in the xy-plane with its diameter along the y-axis, its center at the origin, and $x > 0$. The mass of the half-disk is 3 kg. Find $(\bar{x}, \bar{y})$.

23. A metal plate, with constant density 2 gm/cm^2, has a shape bounded by the curve $y = x^2$ and the x-axis, with $0 \le x \le 1$ and x, y in cm.

(a) Find the total mass of the plate.

(b) Sketch the plate, and decide, on the basis of the shape, whether $\bar{x}$ is less than or greater than $1/2$.

(c) Find $\bar{x}$.

24. A metal plate, with constant density 5 gm/cm^2, has a shape bounded by the curve $y = \sqrt{x}$ and the x-axis, with $0 \le x \le 1$ and x, y in cm.

(a) Find the total mass of the plate.

(b) Find $\bar{x}$ and $\bar{y}$.

25. An isosceles triangle with uniform density, altitude a, and base b is placed in the xy-plane as in Figure 8.58. Show that the center of mass is at $\bar{x} = a/3$, $\bar{y} = 0$. Hence show that the center of mass is independent of the triangle's base.

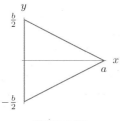

Figure 8.58

26. Find the center of mass of a cone of height 5 cm and base diameter 10 cm with constant density δ gm/cm^3.

27. A solid is formed by rotating the region bounded by the curve $y = e^{-x}$ and the x-axis between $x = 0$ and $x = 1$, around the x-axis. It was shown in Example 1 on page 375 that the volume of this solid is $\pi(1 - e^{-2})/2$. Assuming the solid has constant density δ, find $\bar{x}$ and $\bar{y}$.

28. **(a)** Find the mass of a pyramid of constant density δ gm/cm^3 with a square base of side 40 cm and height 10 cm. [That is, the vertex is 10 cm above the center of the base.]

(b) Find the center of mass of the pyramid.

8.5 APPLICATIONS TO PHYSICS

Although geometric problems were a driving force for the development of the calculus in the seventeenth century, it was Newton's spectacularly successful applications of the calculus to physics that most clearly demonstrated the power of this new mathematics.

Work

In physics the word "work" has a technical meaning which is different from its everyday meaning. Physicists say that if a constant force, F, is applied to some object to move it a distance, d, then the force has done work on the object. The force must be parallel to the motion (in the same or the opposite direction). We make the following definition:

$$\text{Work done} = \text{Force} \cdot \text{Distance} \qquad \text{or} \qquad W = F \cdot d.$$

Notice that if we walk across a room holding a book, we do no work on the book, since the force we exert on the book is vertical, but the motion of the book is horizontal. On the other hand, if we lift the book from the floor to a table, we accomplish work.

There are several sets of units in common use. To measure work, we will generally use the two sets of units, International (SI) and British, in the following table.

	Force	Distance	Work	Conversions
International (SI) units	newton (nt)	meter (m)	joule (j)	1 lb = 4.45 nt
British units	pound (lb)	foot (ft)	foot-pound (ft-lb)	1 ft = 0.305 m
				1 ft-lb = 1.36 joules

One joule of work is done when a force of 1 newton moves an object through 1 meter, so 1 joule = 1 newton-meter.

Example 1 Calculate the work done on an object when

(a) A force of 2 newtons moves it 12 meters.   (b) A 3-lb force moves it 4 feet.

Solution (a) Work done = 2 nt · 12 m = 24 joules.   (b) Work done = 3 lb · 4 ft = 12 ft-lb.

In the previous example, the force was constant and we calculated the work by multiplication. In the next example, the force varies, so we need an integral. We divide up the distance moved and sum to get a definite integral representing the work.

Example 2   Hooke's Law says that the force, F, required to compress the spring in Figure 8.59 by a distance x, in meters, is given by $F = kx$, for some constant k. Find the work done in compressing the spring by 0.1 m if $k = 8$ nt/m.

Figure 8.59: Compression of spring: Force is kx

Figure 8.60: Work done in compressing spring a small distance Δx is $kx\Delta x$

Solution   Since k is in newtons/meter and x is in meters, we have $F = 8x$ newtons. Since the force varies with x, we divide the distance moved into small increments, Δx, as in Figure 8.60. Then

$$\text{Work done in moving through an increment} \approx F\Delta x = 8x\Delta x \text{ joules.}$$

So, summing over all increments gives the Riemann sum approximation

$$\text{Total work done} \approx \sum 8x\Delta x.$$

Taking the limit as $\Delta x \to 0$ gives

$$\text{Total work done} = \int_0^{0.1} 8x \, dx = 4x^2 \Big|_0^{0.1} = 0.04 \text{ joules.}$$

In general, if force is a function $F(x)$ of position x, then in moving from $x = a$ to $x = b$,

$$\boxed{\text{Work done} = \int_a^b F(x) \, dx.}$$

The Force Due to Gravity: Mass versus Weight

When an object is lifted, work is done against the force exerted by gravity on the object. By Newton's Second Law, the downward gravitational force acting on a mass m is mg, where g is the acceleration due to gravity. To lift the object, we need to exert a force equal to the gravitational force but in the opposite direction.

In International units, $g = 9.8$ m/sec^2, and we usually measure mass, m, in *kilograms*. In British units, mass is seldom used. Instead, we usually talk about the *weight* of an object, which is the force exerted by gravity on the object. Roughly speaking, the mass represents the quantity of matter in an object, whereas the weight represents the force of gravity on it. The mass of an object is the same everywhere, whereas the weight can vary if, for example, the object is moved to outer space where gravitational forces are smaller.

When we are given the weight of an object, we do not multiply by g to find the gravitational force as it has already been done. In British units, a *pound* is a unit of weight. In International units, a kilogram is a unit of mass, and the unit of weight is a newton, where 1 newton $= 1$ kg $\cdot$ m/sec^2.

Example 3   How much work is done in lifting

(a)   A 5-pound book 3 feet off the floor?   (b)   A 1.5-kilogram book 2 meters off the floor?

Solution   (a) The force due to gravity is 5 lb, so $W = F \cdot d = (5 \text{ lb})(3 \text{ ft}) = 15$ foot-pounds.
                  (b) The force due to gravity is $mg = (1.5 \text{ kg})(g \text{ m/sec}^2)$, so

$$W = F \cdot d = [(1.5 \text{ kg})(9.8 \text{ m/sec}^2)] \cdot (2 \text{ m}) = 29.4 \text{ joules}.$$

In the previous example, work is found by multiplication. In the next example, different parts of the object move different distances, so an integral is needed.

Example 4   A 28-meter uniform chain with a mass 2 kilograms per meter is dangling from the roof of a building. How much work is needed to pull the chain up onto the top of the building?

Solution   Since 1 meter of the chain has mass density 2 kg, the gravitational force per meter of chain is $(2 \text{ kg})(9.8 \text{ m/sec}^2) = 19.6$ newtons. Let's divide the chain into small sections of length Δy, each requiring a force of $19.6\,\Delta y$ newtons to move it against gravity. See Figure 8.61. If Δy is small, all of this piece is hauled up approximately the same distance, namely y, so

$$\text{Work done on the small piece } \approx (19.6\,\Delta y \text{ newtons})(y \text{ meters}) = 19.6y\,\Delta y \text{ joules}.$$

The work done on the entire chain is given by the total of the work done on each piece:

$$\text{Work done } \approx \sum 19.6y\,\Delta y \text{ joules}.$$

As $\Delta y \to 0$, we obtain a definite integral. Since y varies from 0 to 28 meters, the total work is

$$\text{Work done } = \int_0^{28} (19.6y)\,dy = 9.8y^2 \Big|_0^{28} = 7683.2 \text{ joules}.$$

Top of building

Figure 8.61: Chain for Example 4

Example 5   Calculate the work done in pumping oil from the cone-shaped tank in Figure 8.62 to the rim. The oil has density 800 kg/m^3 and its vertical depth is 10 m.

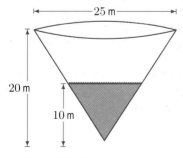

Figure 8.62: Cone-shaped tank
containing oil

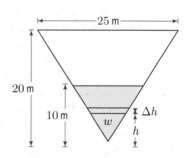

Figure 8.63: Slicing the oil horizontally to
compute work

Solution   We slice the oil horizontally because each part of such a slice moves the same vertical distance. Each slice is approximately a circular disk with radius $w/2$ m, so

$$\text{Volume of slice } \approx \pi \left(\frac{w}{2}\right)^2 \Delta h = \frac{\pi}{4} w^2 \Delta h \text{ m}^3.$$

$$\text{Force of gravity on slice } = \text{ Density } \cdot g \cdot \text{ Volume } = 800g \frac{\pi}{4} w^2 \Delta h = 200\pi g w^2 \Delta h \text{ nt.}$$

Since each part of the slice has to move a vertical distance of $(20 - h)$ m, we have

$$\text{Work done on slice } \approx \text{ Force } \cdot \text{ Distance } = 200\pi g w^2 \Delta h \text{ nt } \cdot (20 - h) \text{ m}$$
$$= 200\pi g w^2 (20 - h)\Delta h \text{ joules.}$$

To find w in terms of h, we use the similar triangles in Figure 8.63:

$$\frac{w}{h} = \frac{25}{20} \quad \text{so} \quad w = \frac{5}{4}h = 1.25h.$$

Thus,

$$\text{Work done on strip } \approx 200\pi g (1.25h)^2 (20 - h)\Delta h = 312.5\pi g h^2 (20 - h)\Delta h \text{ joules.}$$

Summing and taking the limit as $\Delta h \to 0$ gives an integral with upper limit $h = 10$, the depth of the oil.

$$\text{Total work } = \lim_{\Delta h \to 0} \sum 312.5\pi g h^2 (20 - h)\Delta h = \int_0^{10} 312.5\pi g h^2 (20 - h)\, dh \text{ joules.}$$

Evaluating the integral using $g = 9.8$ m/sec^2 gives

$$\text{Total work } = 312.5\pi g \left(20\frac{h^3}{3} - \frac{h^4}{4}\right)\Big|_0^{10} = 1{,}302{,}083\pi g \approx 4.0 \cdot 10^7 \text{ joules.}$$

In the following example, information is given in British units about the weight of the pyramid, so we do not need to multiply by g to find the gravitational force.

Example 6   It is reported that the Great Pyramid of Egypt was built in 20 years. If the stone making up the pyramid has density 200 pounds per cubic foot, find the total amount of work done in building the pyramid. The pyramid is 410 feet high and has a square base 755 feet by 755 feet. Estimate how many workers were needed to build the pyramid.

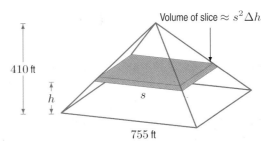

Figure 8.64: Pyramid for Example 6

Solution

We assume that the stones were originally located at the approximate height of the construction site. Imagine the pyramid built in layers as we did in Example 5 on page 371.

By similar triangles, the layer at height h has a side length $s = \frac{755}{410}(410 - h)$ ft. (See Figure 8.64.) The layer at height h has a volume of $s^2 \, \Delta h$ ft^3, so its weight is $200s^2 \, \Delta h$ lb. This layer is lifted through a height of h, so

$$\text{Work to lift layer} = (200s^2 \, \Delta h \text{ lb})(h \text{ ft}) = 200s^2 h \, \Delta h \, ft - lb.$$

Summing over all layers gives

$$\text{Total work} \approx \sum 200s^2 \, h\Delta h = \sum 200 \left(\frac{755}{410}\right)^2 (410 - h)^2 h \, \Delta h \text{ ft-lb.}$$

Since h varies from 0 to 410, as $\Delta h \to 0$, we obtain

$$\text{Total work} = \int_0^{410} 200 \left(\frac{755}{410}\right)^2 (410 - h)^2 h \, dh \approx 1.6 \cdot 10^{12} \text{ foot-pounds.}$$

We have calculated the total work done in building the pyramid; now we want to estimate the total number of workers needed. Let's assume every laborer worked 10 hours a day, 300 days a year, for 20 years. Assume that a typical worker lifted ten 50 pound blocks a distance of 4 feet every hour, thus performing 2000 foot-pounds of work per hour (this is a very rough estimate). Then each laborer performed $(10)(300)(20)(2000) = 1.2 \cdot 10^8$ foot-pounds of work over a twenty year period. Thus, the number of workers needed was about $(1.6 \cdot 10^{12})/(1.2 \cdot 10^8)$, or about 13,000.

Force and Pressure

We can use the definite integral to compute the force exerted by a liquid on a surface, for example, the force of water on a dam. The idea is to get the force from the *pressure*. The pressure in a liquid is the force per unit area exerted by the liquid. Two things you need to know about pressure are:

- At any point, pressure is exerted equally in all directions—up, down, sideways.
- Pressure increases with depth. (That is one of the reasons why deep sea divers have to take much greater precautions than scuba divers.)

At a depth of h meters, the pressure, p, exerted by the liquid, measured in newtons per square meter, is given by computing the total weight of a column of liquid h meters high with a base of 1 square meter. The volume of such a column of liquid is just h cubic meters. If the liquid has density δ (mass per unit volume), then its weight per unit volume is δg, where g is the acceleration due to gravity. The weight of the column of liquid is $\delta g h$, so

$$\boxed{\text{Pressure} = \text{Mass density} \cdot g \cdot \text{Depth} \quad \text{or} \quad p = \delta g h.}$$

Provided the pressure is constant over a given area, we also have the following relation:

$$\boxed{\text{Force} = \text{Pressure} \cdot \text{Area.}}$$

The units and data we will generally use are given in the following table:

	Density of water	Force	Area	Pressure	Conversions
SI units	1000 kg/m^3 (mass)	newton (nt)	meter^2	pascal (nt/m^2)	1 lb = 4.45 nt
British units	62.4 lb/ft^3 (weight)	pound (lb)	foot^2	lb/ft^2	1ft^2 = 0.093 m^2
					1 lb/ft^2 = 47.9 pa

In International units, the mass density of water is 1000 kg/m³, so the pressure at a depth of h meters is $\delta g h = 1000 \cdot 9.8h = 9800h$ nt/m². See Figure 8.65.

In British units, the density of the liquid is usually given as a weight per unit volume, rather than a mass per unit volume. In that case, we do not need to multiply by g because it has already been done. For example, water weighs 62.4 lb/ft³, so the pressure at depth h feet is $62.4h$ lb/ft². See Figure 8.66.

Figure 8.65: Pressure exerted by column of water (International units)

Figure 8.66: Pressure exerted by a column of water (British units)

If the pressure is constant over a surface, we calculate the force on the surface by multiplying the pressure by the area of the surface. If the pressure is not constant, we divide the surface into small pieces *in such a way that the pressure is nearly constant on each one* to obtain a definite integral for the force on the surface. Since the pressure varies with depth, we divide the surface into horizontal strips, each of which is at an approximately constant depth.

Example 7   In 1912, the ocean liner Titanic sank to the bottom of the Atlantic, 12,500 feet (nearly 2.5 miles) below the surface. Find the force on one side of a 100-foot square plate at the depth of the Titanic if the plate is: (a)   Lying horizontally   (b)   Standing vertically.

Solution   (a)   When the plate is horizontal, the pressure is the same at every point on the plate, so

$$\text{Pressure} \ = 62.4 \text{ lb/ft}^3 \cdot 12{,}500 \text{ ft} \ = 780{,}000 \text{ lb/ft}^2.$$

To imagine this pressure, convert to pounds per square inch, giving $780{,}000/144 \approx 5400$ lb/in². For the horizontal plate

$$\text{Force} \ = 780{,}000 \text{ lb/ft}^2 \cdot 100^2 \text{ ft}^2 = 7.8 \cdot 10^9 \text{ pounds.}$$

(b)   When the plate is vertical, only the bottom is at 12,500 feet; the top is at 12,400 feet. Dividing into horizontal strips of width Δh, as in Figure 8.67, we have

$$\text{Area of strip} \ = 100\Delta h \text{ ft}^2.$$

Since the pressure on a strip at a depth of h feet is $62.4h$ lb/ft²,

$$\text{Force on strip} \ \approx 62.4h \cdot 100\Delta h = 6240h\Delta h \text{ pounds.}$$

Summing over all strips and taking the limit as $\Delta h \to 0$ gives a definite integral. The strips vary in depth from 12,400 to 12,500 feet, so

$$\text{Total force} \ = \lim_{\Delta h \to 0} \sum 6240h\Delta h = \int_{12{,}400}^{12{,}500} 6240h \, dh \text{ pounds.}$$

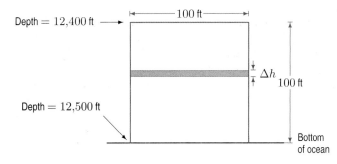

Figure 8.67: Square plate at bottom of ocean; h measured from the surface of water

Evaluating the integral gives

$$\text{Total force } = 6240 \frac{h^2}{2} \bigg|_{12,400}^{12,500} = 3120(12,500^2 - 12,400^2) = 7.77 \cdot 10^9 \text{ pounds.}$$

Notice that the answer to part (b) is smaller than the answer to part (a). This is because part of the plate is at a smaller depth in part (b) than in part (a).

Example 8   Figure 8.68 shows a dam approximately the size of Hoover Dam, which stores water for California, Nevada, and Arizona. Calculate: (a) The water pressure at the base of the dam.   (b) The total force of the water on the dam.

Figure 8.68: Trapezoid-shaped dam

Figure 8.69: Dividing dam into horizontal strips

Solution   (a) Since the density of water is $\delta = 1000$ kg/m³, at the base of the dam,

$$\text{Water pressure } = \delta g h = 1000 \cdot 9.8 \cdot 220 = 2.156 \cdot 10^6 \text{ nt/m}^2.$$

(b) To calculate the force on the dam, we divide the dam into horizontal strips because the pressure along each strip is approximately constant. See Figure 8.69. Since each strip is approximately rectangular,

$$\text{Area of strip } \approx w\Delta h \text{ m}^2.$$

The pressure at a depth of h meters is $\delta g h = 9800h$ nt/m². Thus,

$$\text{Force on strip } \approx \text{Pressure } \cdot \text{ Area } = 9800hw\Delta h \text{ nt.}$$

To find w in terms of h, we use the fact that w decreases linearly from $w = 400$ when $h = 0$ to $w = 200$ when $h = 220$. Thus w is a linear function of h, with slope $(200 - 400)/220 = -10/11$, so

$$w = 400 - \frac{10}{11}h.$$

Thus

$$\text{Force on strip} \approx 9800h \left(400 - \frac{10}{11}h \right) \Delta h \text{ nt.}$$

Summing over all strips and taking the limit as $\Delta h \to 0$ gives

$$\text{Total force on dam} = \lim_{\Delta h \to 0} \sum 9800h \left(400 - \frac{10}{11}h \right) \Delta h$$

$$= \int_0^{220} 9800h \left(400 - \frac{10}{11}h \right) dh \text{ newtons.}$$

Evaluating the integral gives

$$\text{Total force} = 9800 \left(200h^2 - \frac{10}{33}h^3 \right) \bigg|_0^{220} = 6.32 \cdot 10^{10} \text{ newtons.}$$

In fact, Hoover Dam is not flat, as the problem assumed, but arched, to better withstand the pressure.

Exercises and Problems for Section 8.5

Exercises

In Exercises 1–3, the force, F, required to compress a spring by a distance x meters is given by $F = 3x$ newtons.

1. Find the work done in compressing the spring from $x = 1$ to $x = 2$.

2. Find the work done to compress the spring to $x = 3$, starting at the equilibrium position, $x = 0$.

3. **(a)** Find the work done in compressing the spring from $x = 0$ to $x = 1$ and in compressing the spring from $x = 4$ to $x = 5$.
  (b) Which of the two answers is larger? Why?

4. The gravitational force on a 1 kg object at a distance r meters from the center of the earth is $F = 4 \cdot 10^{14}/r^2$ newtons. Find the work done in moving the object from the surface of the earth to a height of 10^6 meters above the surface. The radius of the earth is $6.4 \cdot 10^6$ meters.

5. How much work is required to lift a 1000-kg satellite to an altitude of $2 \cdot 10^6$ m above the surface of the earth? The gravitational force is $F = GMm/r^2$, where M is the mass of the earth, m is the mass of the satellite, and r is the distance between them. The radius of the earth is $6.4 \cdot 10^6$ m, its mass is $6 \cdot 10^{24}$ kg, and in these units the gravitational constant, G, is $6.67 \cdot 10^{-11}$.

Problems

6. A worker on a scaffolding 75 ft above the ground needs to lift a 500 lb bucket of cement from the ground to a point 30 ft above the ground by pulling on a rope weighing 0.5 lb/ft. How much work is required?

7. An anchor weighing 100 lb in water is attached to a chain weighing 3 lb/ft in water. Find the work done to haul the anchor and chain to the surface of the water from a depth of 25 ft.

8. A 1000-lb weight is being lifted to a height 10 feet off the ground. It is lifted using a rope which weighs 4 lb per foot and which is being pulled up by construction workers standing on a roof 30 feet off the ground. Find the work done to lift the weight.

9. A bucket of water of mass 20 kg is pulled at constant velocity up to a platform 40 meters above the ground. This takes 10 minutes, during which time 5 kg of water drips

out at a steady rate through a hole in the bottom. Find the work needed to raise the bucket to the platform.

10. A rectangular water tank has length 20 ft, width 10 ft, and depth 15 ft. If the tank is full, how much work does it take to pump all the water out?

11. A rectangular swimming pool 50 ft long, 20 ft wide, and 10 ft deep is filled with water to a depth of 9 ft. Use an integral to find the work required to pump all the water out over the top.

12. A water tank is in the form of a right circular cylinder with height 20 ft and radius 6 ft. If the tank is half full of water, find the work required to pump all of it over the top rim.

13. Suppose the tank in Problem 12 is full of water. Find the work required to pump all of it to a point 10 ft above the top of the tank.

14. Water in a cylinder of height 10 ft and radius 4 ft is to be pumped out. Find the work required if

    (a) The tank is full of water and the water is to pumped over the top of the tank.

    (b) The tank is full of water and the water must be pumped to a height 5 ft above the top of the tank.

    (c) The depth of water in the tank is 8 ft and the water must be pumped over the top of the tank.

15. A fuel oil tank is an upright cylinder, buried so that its circular top is 10 feet beneath ground level. The tank has a radius of 5 feet and is 15 feet high, although the current oil level is only 6 feet deep. Calculate the work required to pump all of the oil to the surface. Oil weighs 50 lb/ft^3.

16. A cone with height 12 ft and radius 4 ft, pointing downward, is filled with water to a depth of 9 ft. Find the work required to pump all the water out over the top.

17. A gas station stores its gasoline in a tank under the ground. The tank is a cylinder lying horizontally on its side. (In other words, the tank is not standing vertically on one of its flat ends.) If the radius of the cylinder is 4 feet, its length is 12 feet, and its top is 10 feet under the ground, find the total amount of work needed to pump the gasoline out of the tank. (Gasoline weighs 42 lb/ft^3.)

18. A cylindrical barrel, standing upright on its circular end, contains muddy water. The top of the barrel, which has diameter 1 meter, is open. The height of the barrel is 1.8 meter and it is filled to a depth of 1.5 meter. The density of the water at a depth of h meters below the surface is given by $\delta(h) = 1 + kh$ kg/m^3, where k is a positive constant. Find the total work done to pump the muddy water to the top rim of the barrel. (You can leave π, k, and g in your answer.)

19. **(a)** The trough in Figure 8.70 is full of water. Find the force of the water on a triangular end.

    (b) Find the work to pump all the water over the top.

Figure 8.70

20. **(a)** A reservoir has a dam at one end. The dam is a rectangular wall, 1000 feet long and 50 feet high. Approximate the total force of the water on the dam by a Riemann sum.

    (b) Write an integral which represents the force, and evaluate it.

21. Before any water was pumped out, what was the total force on the bottom and each side of the tank in Problem 10?

22. A rectangular dam is 100 ft long and 50 ft high. If the water is 40 ft deep, find the force of the water on the dam.

23. A lobster tank in a restaurant is 4 ft long by 3 ft wide by 2 ft deep. Find the water force on the bottom and on each of the four sides.

24. The Three Gorges Dam is currently being built in China. When it is finished in 2009, it will be the largest dam in the world: about 2000 m long and 180 m high, creating a lake the length of Lake Superior. Assume the dam is rectangular in shape.

    (a) Estimate the water pressure at the base of the dam.

    (b) Set up and evaluate a definite integral giving the total force of the water on the dam.

25. On August 12, 2000, the Russian submarine Kursk sank to the bottom of the sea, 350 feet below the surface. Find the following at the depth of the Kursk.

    (a) The water pressure in pounds per square foot and pounds per square inch.

    (b) The force on a 5-foot square metal sheet held

        (i) Horizontally.

        (ii) Vertically.

26. The ocean liner Titanic lies under 12,500 feet of water at the bottom of the Atlantic Ocean.

    (a) What is the water pressure at the Titanic? Give your answer in pounds per square foot and pounds per square inch.

    (b) Set up and calculate an integral giving the total force on a circular porthole (window) of diameter 6 feet standing vertically with its center at the depth of the Titanic.

27. Set up and calculate a definite integral giving the total force on the dam shown in Figure 8.71, which is about the size of the Aswan Dam in Egypt.

Figure 8.71

28. We define the electric potential at a distance r from an electric charge q by q/r. The electric potential of a charge distribution is obtained by adding up the potential from each point. Electric charge is sprayed (with constant density σ in units of charge/unit area) on to a circular disk of radius a. Consider the axis perpendicular to the disk and through its center. Find the electric potential at the point P on this axis at a distance R from the center. (See Figure 8.72.)

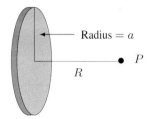

Radius $= a$

P

R

Figure 8.72

For Problems 29–30, find the kinetic energy of the rotating body. Use the fact that the kinetic energy of a particle of mass m moving at a speed v is $\frac{1}{2}mv^2$. Slice the object into pieces in such a way that the velocity is approximately constant on each piece.

29. Find the kinetic energy of a rod of mass 10 kg and length 6 m rotating about an axis perpendicular to the rod at its midpoint, with an angular velocity of 2 radians per second. (Imagine a helicopter blade of uniform thickness.)

30. Find the kinetic energy of a phonograph record of uniform density, mass 50 gm and radius 10 cm rotating at $33\frac{1}{3}$ revolutions per minute.

For Problems 31–33, find the gravitational force between two objects. Use the fact that the gravitational attraction between particles of mass m_1 and m_2 at a distance r apart is Gm_1m_2/r^2. Slice the objects into pieces, use this formula for the pieces, and sum using a definite integral.

31. What is the force of gravitational attraction between a thin uniform rod of mass M and length l and a particle of mass m lying in the same line as the rod at a distance a from one end?

32. Two long, thin, uniform rods of lengths l_1 and l_2 lie on a straight line with a gap between them of length a. Suppose their masses are M_1 and M_2, respectively, and the constant of the gravitation is G. What is the force of attraction between the rods? (Use the result of Problem 31.)

33. Find the gravitational force exerted by a thin uniform ring of mass M and radius a on a particle of mass m lying on a line perpendicular to the ring through its center. Assume m is at a distance y from the center of the ring.

34. A uniform, thin, circular disk of radius a and mass M lies on a horizontal plane. The point P lies a distance y directly above O, the center of the disk. Calculate the gravitational force on a mass m at the point P. (See Figure 8.73.) Use the fact that the gravitational force exerted on the mass m by a thin horizontal ring of radius r, mass μ, and center O is toward O and given by

$$F = \frac{G\mu my}{(r^2 + y^2)^{3/2}}, \quad \text{where } G \text{ is constant.}$$

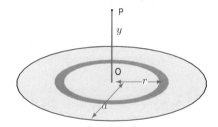

P

y

O

r

a

Figure 8.73

8.6 APPLICATIONS TO ECONOMICS

Present and Future Value

Many business deals involve payments in the future. If you buy a car or furniture, for example, you may buy it on credit and pay over a period of time. If you are going to accept payment in the future under such a deal, you obviously need to know how much you should be paid. Being paid $100 in the future is clearly worse than being paid $100 today for many reasons. If you are given the money today, you can do something else with it—for example, put it in the bank, invest it somewhere, or spend it. Thus, even without considering inflation, if you are to accept payment in the future, you would expect to be paid more to compensate for this loss of potential earnings. The question we will consider now is, how much more?

To simplify matters, we consider only what we would lose by not earning interest; we will not consider the effect of inflation. Let's look at some specific numbers. Suppose you deposit $100 in an account which earns 7% interest compounded annually, so that in a year's time you will have $107. Thus, $100 today will be worth $107 a year from now. We say that the $107 is the *future value* of the $100, and that the $100 is the *present value* of the $107. Observe that the present value is smaller than the future value. In general, we say the following:

- The **future value**, B, of a payment, P, is the amount to which the P would have grown if deposited in an interest bearing bank account.
- The **present value**, P, of a future payment, B, is the amount which would have to be deposited in a bank account today to produce exactly B in the account at the relevant time in the future.

With an interest rate of r, compounded annually, and a time period of t years, a deposit of P grows to a future balance of B, where

$$B = P(1+r)^t, \quad \text{or equivalently,} \quad P = \frac{B}{(1+r)^t}.$$

Note that for a 7% interest rate, $r = 0.07$. If instead of annual compounding, we have continuous compounding, we get the following result:

$$B = Pe^{rt}, \quad \text{or equivalently,} \quad P = \frac{B}{e^{rt}} = Be^{-rt}.$$

Example 1   You win the lottery and are offered the choice between $1 million in four yearly installments of $250,000 each, starting now, and a lump-sum payment of $920,000 now. Assuming a 6% interest rate, compounded continuously, and ignoring taxes, which should you choose?

Solution   We will do the problem in two ways. First, we assume that you pick the option with the largest present value. The first of the four $250,000 payments is made now, so

$$\text{Present value of first payment} = \$250,000.$$

The second payment is made one year from now, so

$$\text{Present value of second payment} = \$250,000e^{-0.06(1)}.$$

Calculating the present value of the third and fourth payments similarly, we find:

$$\text{Total present value} = \$250,000 + \$250,000e^{-0.06(1)} + \$250,000e^{-0.06(2)} + \$250,000e^{-0.06(3)}$$
$$\approx \$250,000 + \$235,441 + \$221,730 + \$208,818$$
$$= \$915,989.$$

Since the present value of the four payments is less than $920,000, you are better off taking the $920,000 right now.

Alternatively, we can compare the future values of the two pay schemes. The scheme with the highest future value is the best from a purely financial point of view. We calculate the future value of both schemes three years from now, on the date of the last $250,000 payment. At that time,

$$\text{Future value of the lump sum payment} = \$920,000e^{0.06(3)} \approx \$1,101,440.$$

Now we calculate the future value of the first $250,000 payment:

$$\text{Future value of the first payment} = \$250,000e^{0.06(3)}.$$

Calculating the future value of the other payments similarly, we find:

$$\text{Total future value} = \$250,000e^{0.06(3)} + \$250,000e^{0.06(2)} + \$250,000e^{0.06(1)} + \$250,000$$
$$\approx \$299,304 + \$281,874 + \$265,459 + \$250,000$$
$$= \$1,096,637.$$

The future value of the $920,000 payment is greater, so you are better off taking the $920,000 right now. Of course, since the present value of the $920,000 payment is greater than the present value of the four separate payments, you would expect the future value of the $920,000 payment to be greater than the future value of the four separate payments.

(Note: If you read the fine print, you will find that many lotteries do not make their payments right away, but often spread them out, sometimes far into the future. This is to reduce the present value of the payments made, so that the value of the prizes is much less than it might first appear!)

Income Stream

When we consider payments made to or by an individual, we usually think of *discrete* payments, that is, payments made at specific moments in time. However, we may think of payments made by a company as being *continuous*. The revenues earned by a huge corporation, for example, come in essentially all the time and can be represented by a continuous *income stream*, written

$$P(t) \text{ dollars/year.}$$

Notice that $P(t)$ is the *rate* at which deposits are made (its units are dollars per year, for example) and that this rate may vary with time, t.

Present and Future Values of an Income Stream

Just as we can find the present and future values of a single payment, so we can find the present and future values of a stream of payments. We will assume that interest is compounded continuously.

Suppose that we want to calculate the present value of the income stream described by a rate of $P(t)$ dollars per year, and that we are interested in the period from now until M years in the future. We divide the stream into many small deposits, each of which is made at approximately one instant. We divide the interval $0 \leq t \leq M$ into subintervals, each of length Δt:

Assuming Δt is small, the rate, $P(t)$, at which deposits are being made will not vary much within one subinterval. Thus, between t and $t + \Delta t$:

$$\text{Amount deposited} \approx \text{Rate of deposits} \times \text{Time}$$
$$\approx (P(t) \text{ dollars/year})(\Delta t \text{ years})$$
$$= P(t)\Delta t \text{ dollars.}$$

Measured from the present, the deposit of $P(t)\Delta t$ is made t years in the future. Thus,

$$\begin{array}{c}\text{Present value of money deposited} \\ \text{in interval } t \text{ to } t + \Delta t\end{array} \approx P(t)\Delta t e^{-rt}.$$

Summing over all subintervals gives

$$\text{Total present value} \approx \sum P(t)e^{-rt}\Delta t \text{ dollars.}$$

In the limit as $\Delta t \to 0$, we get the following integral:

$$\boxed{\text{Present value} = \int_0^M P(t)e^{-rt} dt \text{ dollars.}}$$

In computing future value, the deposit of $P(t)\Delta t$ has a period of $(M - t)$ years to earn interest, and therefore

$$\begin{array}{c}\text{Future value of money deposited} \\ \text{in interval } t \text{ to } t + \Delta t\end{array} \approx [P(t)\Delta t]\, e^{r(M-t)}.$$

Summing over all subintervals, we get:

$$\text{Total future value} \approx \sum P(t)\Delta t e^{r(M-t)} \text{ dollars.}$$

As the length of the subdivisions tends toward zero, the sum becomes an integral:

$$\text{Future value} = \int_0^M P(t)e^{r(M-t)}dt \text{ dollars.}$$

In addition, by writing $e^{r(M-t)} = e^{rM} \cdot e^{-rt}$ and factoring out e^{rM}, we see that

$$\text{Future value} = e^{rM} \cdot \text{Present value.}$$

Example 2   Find the present and future values of a constant income stream of \$1000 per year over a period of 20 years, assuming an interest rate of 10% compounded continuously.

Solution   Using $P(t) = 1000$ and $r = 0.1$, we have

$$\text{Present value} = \int_0^{20} 1000e^{-0.1t}dt - 1000\left(-\frac{e^{-0.1t}}{0.1}\right)\bigg|_0^{20} = 10{,}000(1-e^{-2}) \approx 8646.65 \text{ dollars.}$$

There are two ways to compute the future value. Using the present value of \$8646.65, we have

$$\text{Future value} = 8646.65e^{0.1(20)} = 63{,}890.58 \text{ dollars.}$$

Alternatively, we can use the integral formula:

$$\text{Future value} = \int_0^{20} 1000e^{0.1(20-t)}dt = \int_0^{20} 1000e^2 e^{-0.1t}dt$$

$$= 1000e^2\left(-\frac{e^{-0.1t}}{0.1}\right)\bigg|_0^{20} = 10{,}000e^2(1-e^{-2}) \approx 63890.58 \text{ dollars.}$$

Notice that the total amount deposited is \$1000 per year for 20 years, or \$20,000. The additional \$43,895.58 of the future value comes from interest earned.

Supply and Demand Curves

In a free market, the quantity of a certain item produced and sold can be described by the supply and demand curves of the item. The *supply curve* shows the quantity of the item the producers will supply at different price levels. It is usually assumed that as the price increases, the quantity supplied will increase. The consumers' behavior is reflected in the *demand curve*, which shows what quantity of goods are bought at various prices. An increase in price is usually assumed to cause a decrease in the quantity purchased. See Figure 8.74.

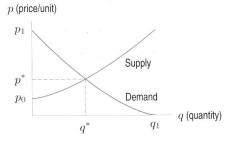

Figure 8.74: Supply and demand curves

It is assumed that the market settles to the *equilibrium price* and *quantity*, p^* and q^*, where the graphs cross. At equilibrium, a quantity q^* of an item is produced and sold for a price of p^* each.

Consumer and Producer Surplus

Notice that at equilibrium, a number of consumers have bought the item at a lower price than they would have been willing to pay. (For example, there are some consumers who would have been willing to pay prices up to p_1.) Similarly, there are some suppliers who would have been willing to produce the item at a lower price (down to p_0, in fact). We define the following terms:

> - The **consumer surplus** measures the consumers' gain from trade. It is the total amount gained by consumers by buying the item at the current price rather than at the price they would have been willing to pay.
> - The **producer surplus** measures the suppliers' gain from trade. It is the total amount gained by producers by selling at the current price, rather than at the price they would have been willing to accept.
> In the absence of price controls, the current price is assumed to be the equilibrium price.

Both consumers and producers are richer for having traded. The consumer and producer surplus measure how much richer they are.

Suppose that all consumers buy the good at the maximum price they are willing to pay. Divide the interval from 0 to q^* into subintervals of length Δq. Figure 8.75 shows that a quantity Δq of items are sold at a price of about p_1, another Δq are sold for a slightly lower price of about p_2, the next Δq for a price of about p_3, and so on. Thus,

$$\text{Consumers' total expenditure} \approx p_1\Delta q + p_2\Delta q + p_3\Delta q + \cdots = \sum p_i\Delta q.$$

If D is the demand function given by $p = D(q)$, and if all consumers who were willing to pay more than p^* paid as much as they were willing, then as $\Delta q \to 0$, we would have

$$\text{Consumer expenditure} = \int_0^{q^*} D(q)dq = \begin{array}{l}\text{Area under demand}\\\text{curve from 0 to } q^*.\end{array}$$

Now if all goods are sold at the equilibrium price, the consumers' actual expenditure is p^*q^*, the area of the rectangle between the axes and the lines $q = q^*$ and $p = p^*$. Thus, if p^* and q^* are equilibrium price and quantity, the consumer surplus is calculated as follows:

$$\text{Consumer surplus} = \left(\int_0^{q^*} D(q)dq\right) - p^*q^* = \begin{array}{l}\text{Area under demand}\\\text{curve above } p = p^*.\end{array}$$

Figure 8.75: Calculation of consumer surplus

Figure 8.76: Consumer and producer surplus

See Figure 8.76. Similarly, if the supply curve is given by the function $p = S(q)$ and p^* and q^* are equilibrium price and quantity, the producer surplus is calculated as follows:

$$\text{Producer surplus} = p^*q^* - \left(\int_0^{q^*} S(q)dq \right) = \frac{\text{Area between supply}}{\text{curve and line } p = p^*.}$$

Exercises and Problems for Section 8.6

Exercises

1. Find the future value of an income stream of $1000 per year, deposited into an account paying 8% interest, compounded continuously, over a 10-year period.

2. Find the present and future values of an income stream of $3000 per year over a 15-year period, assuming a 6% annual interest rate compounded continuously.

3. Find the present and future values of an income stream of $2000 a year, for a period of 5 years, if the continuous interest rate is 8%.

4. A person deposits money into a retirement account, which pays 7% interest compounded continuously, at a rate of $1000 per year for 20 years. Calculate:

   (a) The balance in the account at the end of the 20 years.
   (b) The amount of money actually deposited into the account.
   (c) The interest earned during the 20 years.

Problems

5. Draw a graph, with time in years on the horizontal axis, of what an income stream might look like for a company that sells sunscreen in the northeast United States.

6. On April 15, 1999, Maria Grasso won the largest lottery amount awarded up to that date. She was given her choice between $197 million, paid out continuously over 26 years, or a lump sum of $104 million, paid immediately.

   (a) Which option is better if the interest rate is 6%, compounded continuously? An interest rate of 5%?
   (b) The winner chose the lump sum option. What assumption was she making about interest rates?

7. (a) A bank account earns 10% interest compounded continuously. At what (constant, continuous) rate must a parent deposit money into such an account in order to save $100,000 in 10 years for a child's college expenses?
   (b) If the parent decides instead to deposit a lump sum now in order to attain the goal of $100,000 in 10 years, how much must be deposited now?

8. (a) If you deposit money continuously at a constant rate of $1000 per year into a bank account that earns 5% interest, how many years will it take for the balance to reach $10,000?
   (b) How many years would it take if the account had $2000 in it initially?

9. A business associate who owes you $3000 offers to pay you $2800 now, or else pay you three yearly installments of $1000 each, with the first installment paid now. If you

use only financial reasons to make your decision, which option should you choose? Justify your answer, assuming a 6% market interest rate, compounded continuously.

10. Big Tree McGee is negotiating his rookie contract with a professional basketball team. They have agreed to a three-year deal which will pay Big Tree a fixed amount at the end of each of the three years, plus a signing bonus at the beginning of his first year. They are still haggling about the amounts and Big Tree must decide between a big signing bonus and fixed payments per year, or a smaller bonus with payments increasing each year. The two options are summarized in the table. All values are payments in millions of dollars.

	Signing bonus	Year 1	Year 2	Year 3
Option #1	6.0	2.0	2.0	2.0
Option #2	1.0	2.0	4.0	6.0

   (a) Big Tree decides to invest all income in stock funds which he expects to grow at a rate of 10% per year, compounded continuously. He would like to choose the contract option which gives him the greater future value at the end of the three years when the last payment is made. Which option should he choose?
   (b) Calculate the present value of each contract offer.

11. Sales of Version 6.0 of a computer software package start out high and decrease exponentially. At time t, in years, the sales are $s(t) = 50e^{-t}$ thousands of dollars per

year. After two years, Version 7.0 of the software is released and replaces Version 6.0. Assume that all income from software sales is immediately invested in government bonds which pay interest at a 6% rate compounded continuously, calculate the total value of sales of Version 6.0 over the two year period.

12. The value of good wine increases with age. Thus, if you are a wine dealer, you have the problem of deciding whether to sell your wine now, at a price of $\$P$ a bottle, or to sell it later at a higher price. Suppose you know that the amount a wine-drinker is willing to pay for a bottle of this wine t years from now is $\$P(1 + 20\sqrt{t})$. Assuming continuous compounding and a prevailing interest rate of 5% per year, when is the best time to sell your wine?

13. An oil company discovered an oil reserve of 100 million barrels. For time $t > 0$, in years, the company's extraction plan is a linear declining function of time as follows:

$$q(t) = a - bt,$$

where $q(t)$ is the rate of extraction of oil in millions of barrels per year at time t and $b = 0.1$ and $a = 10$.

  (a) How long does it take to exhaust the entire reserve?
  (b) The oil price is a constant $20 per barrel, the extraction cost per barrel is a constant $10, and the market interest rate is 10% per year, compounded continuously. What is the present value of the company's profit?

14. In 1980 West Germany made a loan of 20 billion Deutsche Marks to the Soviet Union, to be used for the construction of a natural gas pipeline connecting Siberia to Western Russia, and continuing to West Germany (Urengoi–Uschgorod–Berlin). Assume that the deal was as follows: In 1985, upon completion of the pipeline, the Soviet Union would deliver natural gas to West Germany, at a constant rate, for all future times. Assuming a constant price of natural gas of 0.10 Deutsche Mark per cubic meter, and assuming West Germany expects 10% annual interest on its investment (compounded continuously), at what rate does the Soviet Union have to deliver the gas, in billions of cubic meters per year? Keep in mind that delivery of gas could not begin until the pipeline was completed. Thus, West Germany received no return on its investment until after five years had passed. (Note: A more complex deal of this type was actually made between the two countries.)

15. In May 1991 *Car and Driver* described a Jaguar that sold for $980,000. At that price only 50 have been sold. It

is estimated that 350 could have been sold if the price had been $560,000. Assuming that the demand curve is a straight line, and that $560,000 and 350 are the equilibrium price and quantity, find the consumer surplus at the equilibrium price.

16. Using Riemann sums, explain the economic significance of $\int_0^{q^*} S(q)\, dq$ to the producers.

17. Using Riemann sums, give an interpretation of producer surplus, $\int_0^{q^*} (p^* - S(q))\, dq$ analogous to the interpretation of consumer surplus.

18. In Figure 8.76, page 412, mark the regions representing the following quantities and explain their economic meaning:

  (a) $p^* q^*$
  (b) $\displaystyle\int_0^{q^*} D(q)\, dq$
  (c) $\displaystyle\int_0^{q^*} S(q)\, dq$
  (d) $\displaystyle\int_0^{q^*} D(q)\, dq - p^* q^*$
  (e) $p^* q^* - \displaystyle\int_0^{q^*} S(q)\, dq$
  (f) $\displaystyle\int_0^{q^*} (D(q) - S(q))\, dq$

19. The dairy industry is an example of cartel pricing: the government has set milk prices artificially high. On a supply and demand graph, label p^+, a price above the equilibrium price. Using the graph, describe the effect of forcing the price up to p^+ on:

  (a) The consumer surplus.
  (b) The producer surplus.
  (c) The total gains from trade (Consumer surplus + Producer surplus).

20. Rent controls on apartments are an example of price controls on a commodity. They keep the price artificially low (below the equilibrium price). Sketch a graph of supply and demand curves, and label on it a price p^- below the equilibrium price. What effect does forcing the price down to p^- have on:

  (a) The producer surplus?
  (b) The consumer surplus?
  (c) The total gains from trade (Consumer surplus + Producer surplus)?

8.7   DISTRIBUTION FUNCTIONS

Understanding the distribution of various quantities through the population is important to decision makers. For example, the income distribution gives useful information about the economic structure of a society. In this section we will look at the distribution of ages in the US. To allocate funding for education, health care, and social security, the government needs to know how many people are in each age group. We will see how to represent such information by a density function.

US Age Distribution

Table 8.3 *Distribution of ages in the US in 1995*

Age group	Fraction of total population
0–20	29% = 0.29
20–40	31% = 0.31
40–60	24% = 0.24
60–80	13% = 0.13
80–100	3% = 0.03

Figure 8.77: How ages were distributed in the US in 1995

The data in Table 8.3 shows how the ages of the US population were distributed in 1995. To represent this information graphically we use a type[2] of *histogram*, putting a vertical bar above each age group in such a way that the *area* of each bar represents the fraction of the population in that age group. The total area of all the rectangles is 100% = 1. We only consider people who are less than 100 years old.[3] For the 0–20 age group, the base of the rectangle is 20, and we want the area to be 0.29, so the height must be 0.29/20 = 0.0145. We treat ages as though they were continuously distributed. The category 0–20, for example, contains people who are just one day short of their twentieth birthday. (See Figure 8.77.)

Example 1 In 1995, estimate what fraction of the US population was:

(a) Between 20 and 60 years old.
(b) Less than 10 years old.
(c) Between 75 and 80 years old.
(d) Between 80 and 85 years old.

Solution

(a) We add the fractions, so 0.31 + 0.24 = 0.55; that is, 55% of the US population was in this age group.

(b) To find the fraction less than 10 years old, we could assume, for example, that the population was distributed evenly over the 0–20 group. (This means we are assuming that babies were born at a fairly constant rate over the last 20 years, which is probably reasonable.) If we make this assumption, then we can say that the population less than 10 years old was about half that in the 0–20 group, that is, 0.145 of the total population. Notice that we get the same result by computing the area of the rectangle from 0 to 10. (See Figure 8.78.)

(c) To find the population between 75 and 80 years old, since 0.13 of Americans in 1990 were in the 60-80 group, we might apply the same reasoning and say that $\frac{1}{4}(0.13) = 0.0325$ of the population was in this age group. This result is represented as an area in Figure 8.78. The assumption that the population was evenly distributed is not a good one here; certainly there were more people between the ages of 60 and 65 than between 75 and 80. Thus, the estimate of 0.0325 is certainly too high.

(d) Again using the (faulty) assumption that ages in each group were distributed uniformly, we would find that the fraction between 80 and 85 was $\frac{1}{4}(0.03) = 0.0075$. (See Figure 8.78.) This estimate is also poor—there were certainly more people in the 80–85 group than, say, the 95–100 group, and so the 0.0075 estimate is too low.

---

[2] There are other types of histogram which have frequency on the vertical axis.
[3] In fact, 0.02% of the population is over 100, but this is too small to be visible on the histogram.

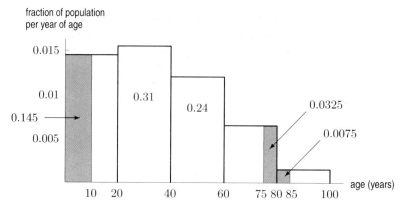

Figure 8.78: Ages in the US in 1995 — various subgroups (for Example 1)

Smoothing Out the Histogram

We could get better estimates if we had smaller age groups (each age group in Figure 8.77 is 20 years, which is quite large). The more detailed data in Table 8.4 leads to the new histogram in Figure 8.79. As we get more detailed information, the upper silhouette of the histogram becomes smoother, but the area of any of the bars still represents the percentage of the population in that age group. Imagine, in the limit, replacing the upper silhouette of the histogram by a smooth curve in such a way that area under the curve above one age group is the same as the area in the corresponding rectangle. The total area under the whole curve is again $100\% = 1$. (See Figure 8.79.)

The Age Density Function

If t is age in years, we define $p(t)$, the age *density function*, to be a function which "smooths out" the age histogram. This function has the property that

$$\begin{array}{c}\text{Fraction of population} \\ \text{between ages } a \text{ and } b\end{array} = \begin{array}{c}\text{Area under graph of } p \\ \text{between } a \text{ and } b\end{array} = \int_a^b p(t)\,dt.$$

If a and b are the smallest and largest possible ages (say, $a = 0$ and $b = 100$), so that the ages of all of the population are between a and b, then

$$\int_a^b p(t)\,dt = \int_0^{100} p(t)\,dt = 1.$$

Table 8.4 *Ages in the US in 1995 (more detailed)*

Age group	Fraction of total population
0–10	$15\% = 0.15$
10–20	$14\% = 0.14$
20–30	$14\% = 0.14$
30–40	$17\% = 0.17$
40–50	$14\% = 0.14$
50–60	$10\% = 0.10$
60–70	$8\% = 0.08$
70–80	$5\% = 0.05$
80–90	$2\% = 0.02$
90–100	$1\% = 0.01$

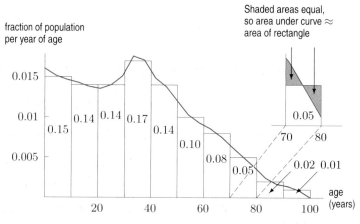

Figure 8.79: Smoothing out the age histogram

What does the age density function p tell us? Notice that we have not talked about the meaning of $p(t)$ itself, but *only* of the integral $\int_a^b p(t)\,dt$. Let's look at this in a bit more detail. Suppose, for example, that $p(10) = 0.015$ per year. This is *not* telling us that 0.015 of the population is precisely 10 years old (where 10 years old means exactly 10, not $10\frac{1}{2}$, not $10\frac{1}{4}$, not 10.1). However, $p(10) = 0.015$ does tell us that for some small interval Δt around 10, the fraction of the population with ages in this interval is approximately $p(10)\,\Delta t = 0.015\,\Delta t$.

The Density Function

Suppose we are interested in how a certain characteristic, x, is distributed through a population. For example, x might be height or age if the population is people, or might be wattage for a population of light bulbs. Then we define a general density function with the following properties:

The function, $p(x)$, is a **density function** if

$$\begin{array}{c} \text{Fraction of population for which} \\ x \text{ is between } a \text{ and } b \end{array} = \begin{array}{c} \text{Area under graph of } p \\ \text{between } a \text{ and } b \end{array} = \int_a^b p(x)\,dx.$$

$$\int_{-\infty}^{\infty} p(x)\,dx = 1 \quad \text{and} \quad p(x) \geq 0 \quad \text{for all } x.$$

The density function must be nonnegative because its integral always gives a fraction of the population. Also, the fraction of the population with x between $-\infty$ and ∞ is 1 because the entire population has the characteristic x between $-\infty$ and ∞. The function p that was used to smooth out the age histogram satisfies this definition of a density function. We do not assign a meaning to the value $p(x)$ directly, but rather interpret $p(x)\,\Delta x$ as the fraction of the population with the characteristic in a short interval of length Δx around x.

The density function is often approximated by formulas, as in the next example.

Example 2 Find formulas to approximate the density function, p, for the US age distribution. To reflect Figure 8.79, use a continuous function, constant at 0.015 up to age 40 and then dropping linearly.

Solution We have $p(t) = 0.015$ for $0 \leq t < 40$. For $t \geq 40$, we need a linear function sloping downward. Because p is continuous, we have $p(40) = 0.015$. Because p is a density function we have $\int_0^{100} p(t)\,dt = 1$. Suppose b is as in Figure 8.80 then

$$\int_0^{100} p(t)\,dt = \int_0^{40} p(t)\,dt + \int_{40}^{100} p(t)\,dt = 40(0.015) + \frac{1}{2}(0.015)b = 1,$$

where $\int_{40}^{100} p(t)\,dt$ is given by the area of the triangle. This gives

$$\frac{0.015}{2}b = 0.4, \quad \text{and so} \quad b \approx 53.3.$$

Thus the slope of the line is $-0.015/53.3 \approx -0.00028$, so for $40 \leq t \leq 40 + 53.3 = 93.3$, we have

$$p(t) - 0.015 = -0.00028(t - 40),$$
$$p(t) = 0.0262 - 0.00028t.$$

According to this way of smoothing the data, there is no one over 93.3 years old, so $p(t) = 0$ for $t > 93.3$.

fraction of population
per year of age

0.015

$p(t) = 0.015$ here

$p(t)$

100

40

b

t (age in years)

Figure 8.80: Age density function

Cumulative Distribution Function for Ages

Another way of showing how ages are distributed in the US is by using the *cumulative distribution function* $P(t)$, defined by

$$P(t) = \begin{array}{c}\text{Fraction of population} \\ \text{of age less than } t\end{array} = \int_0^t p(x)\,dx.$$

Thus, P is the antiderivative of p with $P(0) = 0$, and $P(t)$ gives the area under the density curve between 0 and t.

Notice that the cumulative distribution function is nonnegative and increasing (or at least non-decreasing), since the number of people younger than age t increases as t increases. Another way of seeing this is to notice that $P' = p$, and p is positive (or nonnegative). Thus the cumulative age distribution is a function which starts with $P(0) = 0$ and increases as t increases. $P(t) = 0$ for $t < 0$ because, when $t < 0$, there is no one whose age is less than t. The limiting value of P, as $t \to \infty$, is 1 since as t becomes very large (100 say), everyone is younger than age t, so the fraction of people with age less than t tends toward 1. (See Figure 8.81.) For t less than 40, the graph of P is a straight line, because p is constant there. For $t > 40$, the graph of P levels off as p tends to 0.

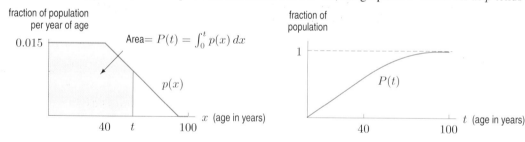

Figure 8.81: $P(t)$, the cumulative age distribution function, and its relation to $p(x)$, the age density function

Cumulative Distribution Function

A **cumulative distribution function**, $P(t)$, of a density function p, is defined by

$$P(t) = \int_{-\infty}^t p(x)\,dx = \begin{array}{c}\text{Fraction of population having} \\ \text{values of } x \text{ below } t.\end{array}$$

Thus, P is an antiderivative of p, that is, $P' = p$.
Any cumulative distribution has the following properties:
- P is increasing (or nondecreasing).
- $\displaystyle\lim_{t \to \infty} P(t) = 1$ and $\displaystyle\lim_{t \to -\infty} P(t) = 0.$
- $\begin{array}{c}\text{Fraction of population having} \\ \text{values of } x \text{ between } a \text{ and } b\end{array} = \int_a^b p(x)\,dx = P(b) - P(a).$

Exercises and Problems for Section 8.7

Exercises

In Exercises 1–3, graph a density function and a cumulative distribution function which could represent the distribution of income through a population with the given characteristics.

1. A large middle class.
2. Small middle and upper classes and many poor people.
3. Small middle class, many poor and many rich people.

Decide if the function graphed in Exercises 4–9 is a probability density function (pdf) or a cumulative distribution function (cdf). Give reasons. Find the value of c. Sketch and label the other function. (That is, sketch and label the cdf if the problem shows a pdf, and the pdf if the problem shows a cdf.)

4.

5.

6.

7.

8.

9.

Problems

10. A large number of people take a standardized test, receiving scores described by the density function p graphed in Figure 8.82. Does the density function imply that most people receive a score near 50? Explain why or why not.

Figure 8.82

11. An experiment is done to determine the effect of two new fertilizers A and B on the growth of a species of peas. The cumulative distribution functions of the heights of the mature peas without treatment and treated with each of A and B are graphed in Figure 8.83.

    (a) About what height are most of the unfertilized plants?
    (b) Explain in words the effect of the fertilizers A and B on the mature height of the plants.

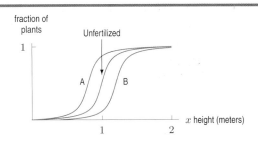

Figure 8.83

12. Suppose $F(x)$ is the cumulative distribution function for heights (in meters) of trees in a forest.

    (a) Explain in terms of trees the meaning of the statement $F(7) = 0.6$.
    (b) Which is greater, $F(6)$ or $F(7)$? Justify your answer in terms of trees.

13. Suppose that $p(x)$ is the density function for heights of American men, in inches. What is the meaning of the statement $p(68) = 0.2$?

14. Suppose $P(t)$ is the fraction of the US population of age less than t. Using Table 8.4 on page 416, make a table of values for $P(t)$.

15. Figure 8.84 shows a density function and the corresponding cumulative distribution function.[4]

---

[4] Adapted from *Calculus*, by David A. Smith and Lawerence C. Moore (Lexington: D.C. Heath, 1994).

(a) Which curve represents the density function and which represents the cumulative distribution function? Give a reason for your choice.

(b) Put reasonable values on the tick marks on each of the axes.

Figure 8.84

16. The density function and cumulative distribution function of heights of grass plants in a meadow are in Figures 8.85 and 8.86, respectively.

(a) There are two species of grass in the meadow, a short grass and a tall grass. Explain how the graph of the density function reflects this fact.

(b) Explain how the graph of the cumulative distribution functions reflects the fact that there are two species of grass in the meadow.

(c) About what percentage of the grasses in the meadow belong to the short grass species?

Figure 8.85

Figure 8.86

17. After measuring the duration of many telephone calls, the telephone company found their data was well-approximated by the density function $p(x) = 0.4e^{-0.4x}$, where x is the duration of a call, in minutes.

(a) What percentage of calls last between 1 and 2 minutes?

(b) What percentage of calls last 1 minute or less?

(c) What percentage of calls last 3 minutes or more?

(d) Find the cumulative distribution function.

18. Students at the University of California were surveyed and asked their grade point average . (The GPA ranges from 0 to 4, where 2 is just passing.) The distribution of GPAs is shown in Figure 8.87.[5]

(a) Roughly what fraction of students are passing?

(b) Roughly what fraction of the students have honor grades (GPAs above 3)?

(c) Why do you think there is a peak around 2?

(d) Sketch the cumulative distribution function.

Figure 8.87

19. Figure 8.88[5] shows the distribution of elevation, in miles, across the earth's surface. Positive elevation denotes land above sea level; negative elevation shows land below sea level (i.e., the ocean floor).

(a) Describe in words the elevation of most of the earth's surface.

(b) Approximately what fraction of the earth's surface is below sea level?

Figure 8.88

20. Consider a population of individuals with a disease. Suppose that t is the number of years since the onset of the disease. The death density function, $f(t) = cte^{-kt}$, approximates the fraction of the sick individuals who die in the time interval $[t, t + \Delta t]$ as follows:

$$\text{Fraction who die} \approx f(t)\Delta t = cte^{-kt}\Delta t$$

where c and k are positive constants whose values depend on the particular disease.

(a) Find the value of c in terms of k.

(b) If 40% of the population dies within 5 years, find c and k.

(c) Find the cumulative death distribution function, $C(t)$. Give your answer in terms of k.

---

[5] Adapted from *Statistics*, by Freedman, Pisani, Purves, and Adikhari (New York: Norton, 1991).

8.8   PROBABILITY, MEAN, AND MEDIAN

Probability

Suppose we pick a member of the US population at random and ask what is the probability that the person is between, say, the ages of 70 and 80. We saw in Table 8.4 on page 416 that $5\% = 0.05$ of the population is in this age group. We say that the probability, or chance, that the person is between 70 and 80 is 0.05. Using any age density function $p(t)$, we can define probabilities as follows:

$$\begin{array}{c} \text{Probability that a person is} \\ \text{between ages } a \text{ and } b \end{array} = \begin{array}{c} \text{Fraction of population} \\ \text{between ages } a \text{ and } b \end{array} = \int_a^b p(t)\,dt.$$

Since the cumulative distribution function gives the fraction of the population younger than age t, the cumulative distribution can also be used to calculate the probability that a randomly selected person is in a given age group.

$$\begin{array}{c} \text{Probability that a person is} \\ \text{younger than age } t \end{array} = \begin{array}{c} \text{Fraction of population} \\ \text{younger than age } t \end{array} = P(t) = \int_0^t p(x)\,dx.$$

In the next example, both a density function and a cumulative distribution function are used to describe the same situation.

Example 1   Suppose you want to analyze the fishing industry in a small town. Each day, the boats bring back at least 2 tons of fish, and never more than 8 tons.

(a) Using the density function describing the daily catch in Figure 8.89, find and graph the corresponding cumulative distribution function and explain its meaning.

(b) What is the probability that the catch will be between 5 and 7 tons?

Figure 8.89: Density function of daily catch

Solution   (a) The cumulative distribution function $P(t)$ is equal to the fraction of days on which the catch is less than t tons of fish. Since the catch is never less than 2 tons, we have $P(t) = 0$ for $t \leq 2$. Since the catch is always less than 8 tons, we have $P(t) = 1$ for $t \geq 8$. For t in the range $2 < t < 8$, we must evaluate the integral

$$P(t) = \int_{-\infty}^t p(x)dx = \int_2^t p(x)dx.$$

This integral equals the area under the graph of $p(x)$ between $x = 2$ and $x = t$. It can be calculated by noting that $p(x)$ is given by the formula

$$p(x) = \begin{cases} 0.04x & \text{for } 2 \leq x \leq 6 \\ -0.06x + 0.6 & \text{for } 6 < x \leq 8 \end{cases}$$

and $p(x) = 0$ for $x < 2$ or $x > 8$. Thus, for $2 \leq t \leq 6$,

$$P(t) = \int_2^t 0.04x\, dx = 0.04 \frac{x^2}{2}\bigg|_2^t = 0.02t^2 - 0.08.$$

And for $6 \leq t \leq 8$,

$$P(t) = \int_2^t p(x)\, dx = \int_2^6 p(x)\, dx + \int_6^t p(x)\, dx$$

$$= 0.64 + \int_6^t (-0.06x + 0.6)\, dx = 0.64 + \left(-0.06\frac{x^2}{2} + 0.6x\right)\bigg|_6^t$$

$$= -0.03t^2 + 0.6t - 1.88.$$

Thus

$$P(t) = \begin{cases} 0.02t^2 - 0.08 & \text{for } 2 \leq t \leq 6 \\ -0.03t^2 + 0.6t - 1.88 & \text{for } 6 < t \leq 8. \end{cases}$$

In addition $P(t) = 0$ for $t < 2$ and $P(t) = 1$ for $8 < t$. (See Figure 8.90.)

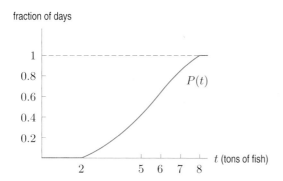

Figure 8.90: Cumulative distribution of daily catch

Figure 8.91: Shaded area represents the probability the catch is between 5 and 7 tons

(b) The probability that the catch is between 5 and 7 tons can be found using either the density function, p, or the cumulative distribution function, P. If we use the density function, this probability can be represented by the shaded area in Figure 8.91, which is about 0.43.

$$\begin{array}{c} \text{Probability catch is} \\ \text{between 5 and 7 tons} \end{array} = \int_5^7 p(x)\, dx = 0.43.$$

The probability can be found from the cumulative distribution as follows:

$$\begin{array}{c} \text{Probability catch is} \\ \text{between 5 and 7 tons} \end{array} = P(7) - P(5) = 0.85 - 0.42 = 0.43.$$

The Median and Mean

It is often useful to be able to give an "average" value for a distribution. Two measures that are in common use are the *median* and the *mean*.

The Median

> A **median** of a quantity x distributed through a population is a value T such that half the population has values of x less than (or equal to) T, and half the population has values of x greater than (or equal to) T. Thus, a median T satisfies
>
> $$\int_{-\infty}^{T} p(x)\, dx = 0.5,$$
>
> where p is the density function. In other words, half the area under the graph of p lies to the left of T.

Example 2   Find the median age in the US in 1995, using the age density function given by

$$p(t) = \begin{cases} 0.015 & \text{for } 0 \le t \le 40 \\ 0.0262 - 0.00028t & \text{for } 40 < t \le 93.3. \end{cases}$$

Solution   We want to find the value of T such that

$$\int_{-\infty}^{T} p(t)\, dt = \int_{0}^{T} p(t)\, dt = 0.5.$$

Since $p(t) = 0.015$ up to age 40, we have

$$\text{Median} = T = \frac{0.5}{0.015} \approx 33 \text{ years.}$$

(See Figure 8.92.)

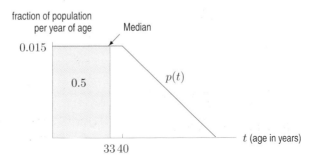

Figure 8.92: Median of age distribution

The Mean

Another commonly used average value is the *mean*. To find the mean of N numbers, you add the numbers and divide the sum by N. For example, the mean of the numbers 1, 2, 7, and 10 is $(1 + 2 + 7 + 10)/4 = 5$. The mean age of the entire US population is therefore defined as

$$\text{Mean age } = \frac{\sum \text{Ages of all people in the US}}{\text{Total number of people in the US}}.$$

Calculating the sum of all the ages directly would be an enormous task; we will approximate the sum by an integral. The idea is to "slice up" the age axis and consider the people whose age is between t and $t + \Delta t$. How many are there?

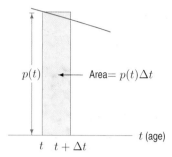

Figure 8.93: Shaded area is percentage of population with age between t and $t + \Delta t$

The fraction of the population between t and $t + \Delta t$ is the area under the graph of p between these points, which is well approximated by the area of the rectangle, $p(t)\Delta t$. (See Figure 8.93.) If the total number of people in the population is N, then

$$\begin{array}{c}\text{Number of people with age}\\ \text{between } t \text{ and } t + \Delta t\end{array} \approx p(t)\Delta t N.$$

The age of all of these people is approximately t:

$$\begin{array}{c}\text{Sum of ages of people}\\ \text{between age } t \text{ and } t + \Delta t\end{array} \approx tp(t)\Delta t N.$$

Therefore, adding and factoring out an N gives us

$$\text{Sum of ages of all people} \approx \left(\sum tp(t)\Delta t\right) N.$$

In the limit, as we allow Δt to shrink to 0, the sum becomes an integral, so

$$\text{Sum of ages of all people} = \left(\int_0^{100} tp(t)dt\right) N.$$

Therefore, with N equal to the total number of people in the US, and assuming no person is over 100 years old,

$$\text{Mean age} = \frac{\text{Sum of ages of all people in US}}{N} = \int_0^{100} tp(t)dt.$$

We can give the same argument for any[6] density function $p(x)$.

If a quantity has density function $p(x)$,

$$\textbf{Mean value} \text{ of the quantity} = \int_{-\infty}^{\infty} xp(x)\, dx.$$

It can be shown that the mean is the point on the horizontal axis where the region under the graph of the density function, if it were made out of cardboard, would balance.

Example 3   Find the mean age of the US population, using the density function of Example 2.

Solution   The formula for p is

$$p(t) = \begin{cases} 0 & \text{for } t < 0 \\ 0.015 & \text{for } 0 \leq t \leq 40 \\ 0.0262 - 0.00028t & \text{for } 40 < t \leq 93.3 \\ 0 & \text{for } t > 93.3. \end{cases}$$

---

[6]Provided all the relevant improper integrals converge.

Using these formulas, we compute

$$\text{Mean age} = \int_0^{100} tp(t)\,dt = \int_0^{40} t(0.015)\,dt + \int_{40}^{93.3} t(0.0262 - 0.00028t)\,dt$$

$$= 0.015\frac{t^2}{2}\Big|_0^{40} + 0.0262\frac{t^2}{2}\Big|_{40}^{93.3} - 0.00028\frac{t^3}{3}\Big|_{40}^{93.3} \approx 35 \text{ years.}$$

The mean is shown is Figure 8.94.

fraction of population
per year of age

$p(t)$

35

t

Mean = Balance point

Figure 8.94: Mean of age distribution

Normal Distributions

How much rain do you expect to fall in your home town this year? If you live in Anchorage, Alaska, the answer is something close to 15 inches (including the snow). Of course, you don't expect exactly 15 inches. Some years there are more than 15 inches, and some years there is less. Most years, however, the amount of rainfall is close to 15 inches; only rarely is it well above or well below 15 inches. What does the density function for the rainfall look like? To answer this question, we look at rainfall data over many years. Records show that the distribution of rainfall is well-approximated by a *normal distribution*. Its graph is a bell-shaped curve which peaks at 15 inches and slopes downward approximately symmetrically on either side.

Normal distributions are frequently used to model real phenomena, from grades on an exam to the number of airline passengers on a particular flight. A normal distribution is characterized by its *mean*, μ, and its *standard deviation*, σ. The mean tells us the location of the central peak. The standard deviation tells us how closely the data is clustered around the mean. A small value of σ tells us that the data is close to the mean; a large σ tells us the data is spread out. In the following formula for a normal distribution, the factor of $1/(\sigma\sqrt{2\pi})$ makes the area under the graph equal to 1.

> A **normal distribution** has a density function of the form
>
> $$p(x) = \frac{1}{\sigma\sqrt{2\pi}}e^{-(x-\mu)^2/(2\sigma^2)},$$
>
> where μ is the mean of the distribution and σ is the standard deviation, with $\sigma > 0$.

To model the rainfall in Anchorage, we use a normal distribution with $\mu = 15$ and $\sigma = 1$. (See Figure 8.95.)

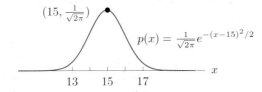

$\left(15, \frac{1}{\sqrt{2\pi}}\right)$

$p(x) = \frac{1}{\sqrt{2\pi}}e^{-(x-15)^2/2}$

13   15   17 x

Figure 8.95: Normal distribution with $\mu = 15$ and $\sigma = 1$

Example 4   For Anchorage's rainfall, use the normal distribution with the density function with $\mu = 15$ and $\sigma = 1$ to compute the fraction of the years with rainfall between
(a)   14 and 16 inches,   (b)   13 and 17 inches,   (c)   12 and 18 inches.

Solution   (a) The fraction of the years with annual rainfall between 14 and 16 inches is $\int_{14}^{16} \frac{1}{\sqrt{2\pi}} e^{-(x-15)^2/2} \, dx$.
Since there is no elementary antiderivative for $e^{-(x-15)^2/2}$, we find the integral numerically. Its value is about 0.68.

$$\begin{array}{l} \text{Fraction of years with rainfall} \\ \text{between 14 and 16 inches} \end{array} = \int_{14}^{16} \frac{1}{\sqrt{2\pi}} e^{-(x-15)^2/2} \, dx \approx 0.68.$$

(b) Finding the integral numerically again:

$$\begin{array}{l} \text{Fraction of years with rainfall} \\ \text{between 13 and 17 inches} \end{array} = \int_{13}^{17} \frac{1}{\sqrt{2\pi}} e^{-(x-15)^2/2} \, dx \approx 0.95.$$

(c)

$$\begin{array}{l} \text{Fraction of years with rainfall} \\ \text{between 12 and 18 inches} \end{array} = \int_{12}^{18} \frac{1}{\sqrt{2\pi}} e^{-(x-15)^2/2} \, dx \approx 0.997.$$

Since 0.95 is so close to 1, we expect that most of the time the rainfall will be between 13 and 17 inches a year.

Among the normal distributions, the one having $\mu = 0$, $\sigma = 1$ is called the *standard normal distribution*. Values of the corresponding cumulative distribution function are published in tables.

Exercises and Problems for Section 8.8

Exercises

1. Show that the area under the fishing density function in Figure 8.89 on page 421 is 1. Why is this to be expected?

2. Find the mean daily catch for the fishing data in Figure 8.89, page 421.

3. **(a)** Using a calculator or computer, sketch graphs of the density function of the normal distribution

$$p(x) = \frac{1}{\sigma\sqrt{2\pi}} e^{-(x-\mu)^2/(2\sigma^2)}.$$

   (i) For fixed μ (say, $\mu = 5$) and varying σ (say, $\sigma = 1, 2, 3$).
   (ii) For varying μ (say, $\mu = 4, 5, 6$) and fixed σ (say, $\sigma = 1$).

   (b) Explain how the graphs confirm that μ is the mean of the distribution and that σ is a measure of how closely the data is clustered around the mean.

Problems

4. The probability of a transistor failing between $t = a$ months and $t = b$ months is given by $c \int_a^b e^{-ct} dt$, for some constant c.

   (a) If the probability of failure within the first six months is 10%, what is c?
   (b) Given the value of c in part (a), what is the probability the transistor fails within the second six months?

5. Suppose that x measures the time (in hours) it takes for a student to complete an exam. All students are done

within two hours and the density function for x is

$$p(x) = \begin{cases} x^3/4 & \text{if } 0 < x < 2 \\ 0 & \text{otherwise.} \end{cases}$$

   (a) What proportion of students take between 1.5 and 2.0 hours to finish the exam?
   (b) What is the mean time for students to complete the exam?
   (c) Compute the median of this distribution.

6. In 1950 an experiment was done observing the time gaps between successive cars on the Arroyo Seco Freeway.[7] The data show that the density function of these time gaps was given approximately by

$$p(x) = ae^{-0.122x}$$

where x is the time in seconds and a is a constant.

(a) Find a.

(b) Find P, the cumulative distribution function.

(c) Find the median and mean time gap.

(d) Sketch rough graphs of p and P.

7. Consider a group of people who have received treatment for a disease such as cancer. Let t be the *survival time*, the number of years a person lives after receiving treatment. The density function giving the distribution of t is $p(t) = Ce^{-Ct}$ for some positive constant C.

(a) What is the practical meaning for the cumulative distribution function $P(t) = \int_0^t p(x)\,dx$?

(b) The survival function, $S(t)$, is the probability that a randomly selected person survives for at least t years. Find $S(t)$.

(c) Suppose a patient has a 70% probability of surviving at least two years. Find C.

8. While taking a walk along the road where you live, you accidentally drop your glove, but you don't know where. The probability density $p(x)$ for having dropped the glove x kilometers from home (along the road) is

$$p(x) = 2e^{-2x} \quad \text{for } x \geq 0.$$

(a) What is the probability that you dropped it within 1 kilometer of home?

(b) At what distance y from home is the probability that you dropped it within y km of home equal to 0.95?

9. The distribution of IQ scores is often modeled by the normal distribution with mean 100 and standard deviation 15.

(a) Write a formula for the density distribution of IQ scores.

(b) Estimate the fraction of the population with IQ between 115 and 120.

10. The speeds of cars on a road are approximately normally distributed with a mean $\mu = 58$ km/hr and standard deviation $\sigma = 4$ km/hr.

(a) What is the probability that a randomly selected car is going between 60 and 65 km/hr?

(b) What fraction of all cars are going slower than 52 km/hr?

11. Consider the normal distribution, $p(x)$.

(a) Show that $p(x)$ is a maximum when $x = \mu$. What is that maximum value?

(b) Show that $p(x)$ has points of inflection where $x = \mu + \sigma$ and $x = \mu - \sigma$.

(c) Describe in your own words what μ and σ tell you about the distribution.

12. For a normal population of mean 0, show that the fraction of the population within one standard deviation of the mean does not depend on the standard deviation.

[Hint: Use the substitution $w = x/\sigma$.]

13. Which of the following functions makes the most sense as a model for the probability density representing the time (in minutes, starting from $t = 0$) that the next customer walks into a store?

(a) $p(t) = \begin{cases} \cos t & 0 \leq t \leq 2\pi \\ e^{t-2\pi} & t \geq 2\pi \end{cases}$

(b) $p(t) = 3e^{-3t}$ for $t \geq 0$

(c) $p(t) = e^{-3t}$ for $t \geq 0$

(d) $p(t) = 1/4$ for $0 \leq t \leq 4$

14. Let $P(x)$ be the cumulative distribution function for the income distribution in the US in 1973 (income is measured in thousands of dollars). Some values of $P(x)$ are in the following table:

Income x (thousands)	1	4.4	7.8	12.6	20	50
$P(x)$ (%)	1	10	25	50	75	99

(a) What fraction of the population made between $20,000 and $50,000?

(b) What was the median income?

(c) Sketch a density function for this distribution. Where, approximately, does your density function have a maximum? What is the significance of this point, in terms of income distribution? How can you recognize this point on the graph of the density function and on the graph of the cumulative distribution?

15. If we think of an electron as a particle, the function

$$P(r) = 1 - (2r^2 + 2r + 1)e^{-2r}$$

is the cumulative distribution function of the distance, r, of the electron in a hydrogen atom from the center of the atom. The distance is measured in Bohr radii. (1 Bohr radius = 5.29×10^{-11} m. Niels Bohr (1885–1962) was a Danish physicist.)

For example, $P(1) = 1 - 5e^{-2} \approx 0.32$ means that the electron is within 1 Bohr radius from the center of the atom 32% of the time.

(a) Find a formula for the density function of this distribution. Sketch the density function and the cumulative distribution function.

(b) Find the median distance and the mean distance. Near what value of r is an electron most likely to be found?

(c) The Bohr radius is sometimes called the "radius of the hydrogen atom." Why?

---

[7]Reported by Daniel Furlough and Frank Barnes.

CHAPTER SUMMARY (see also Ready Reference at the end of the book)

- **Geometry**
  Area, volume, arc length.
- **Density**
  Finding total quantity from density, center of mass.
- **Physics**
  Work, force and pressure.

- **Economics**
  Present and future value of income stream, consumer and producer surplus.
- **Probability**
  Density function, cumulative distribution function, mean, median, normal distribution.

REVIEW EXERCISES AND PROBLEMS FOR CHAPTER EIGHT

Exercises

1. Imagine a hard-boiled egg lying on its side cut into thin slices. First think about vertical slices and then horizontal ones. What would these slices look like? Sketch them.

For each region in Exercises 2–4, write a definite integral which represents its area. Evaluate the integral to derive a formula for the area.

2. A rectangle with base b and height h:

3. A circle of radius r:

4. A right triangle of base b and height h:

In Exercises 5–9, the region is rotated about the x-axis. Find the volume.

5. Bounded by $y = x^2 + 1$, the x-axis, $x = 0$, $x = 4$.
6. Bounded by $y = \sqrt{x}$, x-axis, $x = 1$, $x = 2$.
7. Bounded by $y = e^{-2x}$, the x-axis, $x = 0$, $x = 1$.
8. Bounded by $y = 4 - x^2$ and the x-axis.
9. Bounded by $y = 2x$, $y = x$, $x = 0$, $x = 3$.
10. Find the volume of the region in Figure 8.96, given that the radius, r of the circular slice at h is $r = \sqrt{h}$.

Figure 8.96

11. Set up and evaluate an integral to find the volume of a cone of height 12 m and radius 3 m at the open end.

12. (a) Set up and evaluate an integral giving the volume of a pyramid of height 10 m and square base 8 m by 8 m.
    (b) The pyramid in part (a) is cut off at a height of 6 m. See Figure 8.97. Find the volume.

Figure 8.97

13. The exterior of a holding tank is a cylinder with radius 3 m and height 6 m; the interior is cone-shaped; Figure 8.98 shows its cross-section. Using an integral, find the volume of material needed to make the tank.

Figure 8.98

For the curves described in Exercises 14–15, write the integral that gives the exact length of the curve; do not evaluate it.

14. One arch of the sine curve, from $x = 0$ to $x = \pi$.
15. The ellipse with equation $(x^2/a^2) + (y^2/b^2) = 1$.

In Exercises 16–17, find the arclength of the function from $x = 0$ to $x = 3$. Use a graph to explain why your answer is reasonable.

16. $f(x) = \sin x$

17. $f(x) = 5x^2$

Problems

18. (a) Find the area of the region between $y = x^2$ and $y = 2x$.
   (b) Find the volume of the solid of revolution if this region is rotated about the x-axis.
   (c) Find the length of the perimeter of this region.

19. (a) Sketch the solid obtained by rotating the region bounded by $y = \sqrt{x}$, $x = 1$, and $y = 0$ around the line $y = 0$.
   (b) Approximate its volume by Riemann sums, showing the volume represented by each term in your sum on the sketch.
   (c) Now find the volume of this solid using an integral.

20. Using the region of Problem 19, find the volume when it is rotated around
   (a) The line $y = 1$.   **(b)** The y-axis.

21. (a) Find (in terms of a) the area of the region bounded by $y = ax^2$, the x-axis, and $x = 2$. Assume $a > 0$.
   (b) If this region is rotated about the x-axis, find the volume of the solid of revolution in terms of a.

22. (a) Find (in terms of b) the area of the region between $y = e^{-bx}$ and the x-axis, between $x = 0$ and $x = 1$. Assume $b > 0$.
   (b) If this region is rotated about the x-axis, find the volume of the solid of revolution in terms of b.

For Problems 23–25, set up and compute an integral giving the volume of the solid of revolution.

23. Bounded by $y = \sin x$, $y = 0.5x$, $x = 0$, $x = 3$;
   (a) Rotated about the x-axis.
   (b) Rotated about $y = 5$.

24. Bounded by $y = 2x$, the x-axis, $x = 0$, $x = 4$. Axis: $y = -5$.

25. Bounded by $y = x^2$, the x-axis, $x = 0$, $x = 3$;
   (a) Rotated about $y = -2$.
   (b) Rotated about $y = 10$.

Problems 26–31 concern the region bounded by the quarter circle $x^2 + y^2 = 1$, with $x \geq 0$, $y \geq 0$. Find the volume of the following solids.

26. The solid obtained by rotating the region about the x-axis.

27. The solid obtained by rotating the region about the line $x = -2$.

28. The solid obtained by rotating the region about the line $x = 1$.

29. The solid whose base is the region and whose cross-sections perpendicular to the x-axis are squares.

30. The solid whose base is the region and whose cross-sections perpendicular to the y-axis are semicircles.

31. The solid whose base is the region and whose cross-section perpendicular to the y-axis is an isosceles right triangle with one leg in the region.

32. The reflector behind a car headlight is made in the shape of the parabola, $x = \frac{4}{9}y^2$, with a circular cross-section, as shown in Figure 8.99.
   (a) Find a Riemann sum approximating the volume contained by this headlight.
   (b) Find the volume exactly.

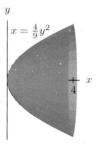

Figure 8.99

33. In this problem, you will derive the formula for the volume of a right circular cone with height l and base radius b by rotating the line $y = ax$ from $x = 0$ to $x = l$ around the x-axis. See Figure 8.100.
   (a) What value should you choose for a such that the cone will have height l and base radius b?
   (b) Given this value of a, find the volume of the cone.

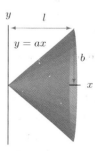

Figure 8.100

34. Figure 8.101 shows a cross section through an apple. (Scale: One division = 1/2 inch.)
   (a) Give a rough estimate for the volume of this apple (in cubic inches).
   (b) The density of these apples is about 0.03 lb/in^3 (a little less than the density of water—as you might expect, since apples float). Estimate how much this apple would cost. (They go for 80 cents a pound.)

Figure 8.101

35. The circle $x^2 + y^2 = 1$ is rotated about the line $y = 3$ forming a torus (a doughnut-shaped figure). Find the volume of this torus.

Problems 36–40 concern, C, the circle $r = 2a\cos\theta$, for $-\pi/2 \leq \theta \leq \pi/2$, of radius $a > 0$ centered at the point $(x, y) = (a, 0)$ on the x-axis.

36. By converting to Cartesian coordinates, show that $r = 2a\cos\theta$ gives the circle described.

37. Find the area of the circle C by integrating in polar coordinates.

38. Find the area of the region enclosed by C and outside the circle of radius a centered at the origin. What percent is this of the area of C?

39. **(a)** Find the slope of C at the angle θ.
   (b) At what value of θ does the maximum y-value occur?

40. Calculate the arc length of C using polar coordinates.

41. Find the center of mass of a system containing four identical point masses of 3 gm, located at $x = -5, -3, 2, 7$.

42. A metal plate, with constant density 2 gm/cm^2, has a shape bounded by the two curves $y = x^2$ and $y = \sqrt{x}$, with $0 \leq x \leq 1$, and x, y in cm.

   (a) Find the total mass of the plate.
   (b) Because of the symmetry of the plate about the line $y = x$, we have $\bar{x} = \bar{y}$. Sketch the plate, and decide, on the basis of the shape, whether $\bar{x}$ is less than or greater than $1/2$.
   (c) Find $\bar{x}$ and $\bar{y}$.

43. A 200-lb weight is hoisted by a rope from the ground to a point 20 ft above the ground. The rope weighs 2 lb/ft. Find the work done in lifting the weight.

44. Water is raised from a well 40 ft deep by a bucket attached to a rope. When the bucket is full, it weighs 30 lb. However, a leak in the bucket causes it to lose water at a rate of 1/4 lb for each foot that the bucket is raised. Neglecting the weight of the rope, find the work done in raising the bucket to the top.

45. A water tank is in the shape of a right circular cone with height 18 ft and radius 12 ft at the top. If it is filled with water to a depth of 15 ft, find the work done in pumping all of the water over the top of the tank. (The density of water is $\delta = 62.4$ lb/ft^3.)

46. An underground tank filled with gasoline of density 42 lb/ft^3 is a hemisphere of radius 5 ft, as in Figure 8.102. Use an integral to find the work to pump the gasoline over the top of the tank.

Figure 8.102

47. The dam in Hannawa Falls, NY, on the Raquette River is approximately 60 feet across and 25 feet high. Find the water force on the dam.

48. A crane lifts a 1000 lb object to a height of 20 ft using chain that weighs 2 lb/ft. If the crane arm is at a height of 50 ft, find the work required.

49. **(a)** Find the present and future values of a constant income stream of $100 per year over a period of 20 years, assuming a 10% annual interest rate compounded continuously.
   (b) How many years will it take for the balance to reach $5000?

50. You are manufacturing a particular item. After t years, the rate at which you earn a profit on the item is $(2 - 0.1t)$ thousand dollars per year. (A negative profit represents a loss.) Interest is 10%, compounded continuously,

   (a) Write a Riemann sum approximating the present value of the total profit earned up to a time M years in the future.
   (b) Write an integral representing the present value in part (a). (You need not evaluate this integral.)
   (c) For what M is the present value of the stream of profits on this item maximized? What is the present value of the total profit earned up to that time?

51. A nuclear power plant produces strontium-90 at a rate of 3 kg/yr. How much of the strontium produced since 1971 (when the plant opened) was still around in 1992? (The half-life of strontium-90 is 28 years.)

52. Mt. Shasta is a cone-like volcano whose radius at an elevation of h feet above sea level is approximately $(3.5 \cdot 10^5)/\sqrt{h + 600}$ feet. Its bottom is 400 feet above sea level, and its top is 14,400 feet above sea level. See Figure 8.103. (Note: Mt. Shasta is in northern California, and for some time was thought to be the highest point in the US outside Alaska.)

   (a) Give a Riemann sum approximating the volume of Mt. Shasta.

(b) Find the volume in cubic feet.

Figure 8.103: Mt. Shasta

53. Figure 8.104 shows an ancient Greek water clock called a clepsydra, which is designed so that the depth of the water decreases at a constant rate as the water runs out a hole in the bottom. This design allows the hours to be marked by a uniform scale. The tank of the clepsydra is a volume of revolution about a vertical axis. According to Torricelli's law, the exit speed of the water flowing through the hole is proportional to the square root of the depth of the water. Use this to find the formula $y = f(x)$ for this profile, assuming that $f(1) = 1$.

Figure 8.104

54. Figure 8.105 shows the distribution of kinetic energy of molecules in a gas at temperatures 300 kelvins and 500 kelvins. At higher temperatures, more of the molecules in a gas have higher kinetic energies. Which graph corresponds to which temperature?

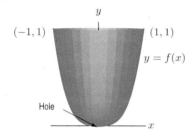

Figure 8.105

55. Figure 8.106 shows the distribution of the velocity of molecules in two gases. In which gas is the average velocity larger?

Figure 8.106

56. A radiation detector is a circular disk which registers photons which hit it. The probability that a photon hitting the disk at a distance r from the center is actually detected is given by $S(r)$. A radiation detector of radius R is bombarded by constant radiation of N photons per second per unit area. Write an integral representing the number of photons per second registered by the detector.

57. Find the volume of the snowman in Figure 8.107. If x and y are in meters, and the origin is on the ground, the x-axis is horizontal and the y-axis is vertical, then the body is approximated by rotating the curve $y = 1 - 4x^2$ about the y-axis. The neck is a cylinder of radius 0.1 meter and length 0.15 meter; the head is spherical with radius 0.2 meter.

Figure 8.107

58. A blood vessel is cylindrical with radius R and length l. The blood near the boundary moves slowly; blood at the center moves the fastest. The velocity, v, of the blood at a distance r from the center of the artery is given by

$$v = \frac{P}{4\eta l}(R^2 - r^2)$$

where P is the pressure difference between the ends of the blood vessel and η is the viscosity of blood.

(a) Find the rate at which the blood is flowing down the blood vessel. (Give your answer as a volume per unit time.)

(b) Show that your result agrees with Poiseuille's Law which says that the rate at which blood is flowing down the blood vessel is proportional to the radius of the blood vessel to the fourth power.

59. A car moving at a speed of v mph achieves $25 + 0.1v$ mpg (miles per gallon) for v between 20 and 60 mph. Your speed as a function of time, t, in hours, is given by

$$v = 50\frac{t}{t+1} \text{ mph}.$$

How many gallons of gas do you consume between $t = 2$ and $t = 3$?

60. A bowl is made by rotating the curve $y = ax^2$ around the y-axis (a is a constant).

(a) The bowl is filled with water to depth h. What is the volume of water in the bowl? (Your answer will contain a and h.)

(b) What is the area of the surface of the water if the bowl is filled to depth h? (Your answer will contain a and h.)

(c) Water is evaporating from the surface of the bowl at a rate proportional to the surface area, with proportionality constant k. Find a differential equation satisfied by h as a function of time, t. (That is, find an equation for dh/dt.)

(d) If the water starts at depth h_0, find the time taken for all the water to evaporate.

61. A cylindrical centrifuge of radius 1 m and height 2 m is filled with water to a depth of 1 meter (see Figure 8.108(I)). As the centrifuge accelerates, the water level rises along the wall and drops in the center; the cross-section will be a parabola. (See Figure 8.108(II).)

(a) Find the equation of the parabola in Figure 8.108(II) in terms of h, the depth of the water at its lowest point.

(b) As the centrifuge rotates faster and faster, either water will be spilled out the top, as in Figure 8.108(III), or the bottom of the centrifuge will be exposed, as in Figure 8.108(IV). Which happens first?

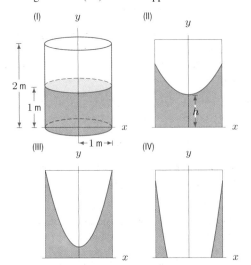

Figure 8.108

In Problems 62–63, you are given two objects which have the same mass M, the same radius R, and the same angular velocity about the indicated axes (say, one revolution per minute). For each problem, determine which of the two objects has the greater kinetic energy. (The kinetic energy of a particle of mass m with speed v is $\frac{1}{2}mv^2$.) Don't compute the kinetic energy of the objects to do this; just use reasoning.

62.

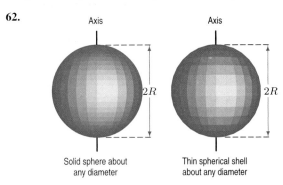

Solid sphere about   Thin spherical shell
any diameter   about any diameter

63.

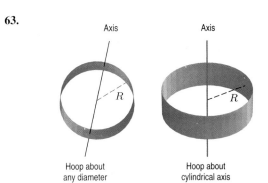

Hoop about   Hoop about
any diameter   cylindrical axis

CAS Challenge Problems

64. For a positive constant a, consider the curve

$$y = \sqrt{\frac{x^3}{a - x}}, \qquad 0 \le x < a.$$

(a) Using a computer algebra system, show that for $0 \le t < \pi/2$, the point with coordinates (x, y) lies on the curve if:

$$x = a \sin^2 t, \quad y = \frac{a \sin^3 t}{\cos t}.$$

(b) A solid is obtained by rotating the curve about its asymptote at $x = a$. Use horizontal slicing to write an integral in terms of x and y that represents the volume of this solid.

(c) Use part (a) to substitute in the integral for both x and y in terms of t. Use a computer algebra system or trigonometric identities to calculate the volume of the solid.

For Problems 65–66, define $A(t)$ to be the arc length of the graph of $y = f(x)$ from $x = 0$ to $x = t$, for $t \ge 0$.

(a) Use the integral expression for arc length and a computer algebra system to obtain a formula for $A(t)$.

(b) Graph $A(t)$ for $0 \le t \le 10$. What simple function does $A(t)$ look like? What does this tell you about the approximate value of $A(t)$ for large t?

(c) In order to estimate arc length visually, you need the same scales on both axes, so that the lengths are not distorted in one direction. Draw a graph of $f(x)$ with viewing window $0 \le x \le 100, 0 \le y \le 100$. Explain what you noticed in part (b) in terms of this graph.

65. $f(x) = x^2$

66. $f(x) = \sqrt{x}$

67. A bead is formed by drilling a cylindrical hole of circular cross section and radius a through a sphere of radius $r > a$, the axis of the hole passing through the center of the sphere.

(a) Write a definite integral expressing the volume of the bead.

(b) Find a formula for the bead by evaluating the definite integral in part (a).

CHECK YOUR UNDERSTANDING

In Problems 1–21, are the statements true or false? Give an explanation for your answer.

1. The integral $\int_{-3}^{3} \pi(9 - x^2) \, dx$ represents the volume of a sphere of radius 3.

2. The integral $\int_{0}^{h} \pi(r - y) \, dy$ gives the volume of a cone of radius r and height h.

3. The integral $\int_{0}^{r} \pi \sqrt{r^2 - y^2} \, dy$ gives the volume of a hemisphere of radius r.

4. A cylinder of radius r and length l is lying on its side. Horizontal slicing tells us that the volume is given by $\int_{-r}^{r} 2l\sqrt{r^2 - y^2} \, dy$.

5. If a region in the xy-plane lies below the x-axis, then revolving that region around the x-axis gives a solid of negative volume.

6. To find the total population in a circular city, we always slice it into concentric rings, no matter what the population density function.

7. A city occupies a region in the xy-plane, with population density $\delta(y) = 1 + y$. To set up an integral representing the total population in the city, we should slice the region parallel to the y-axis.

8. The population density in a circular city of radius 2 depends on the distance r from the center by $f(r) = 10 - 3r$, so that the density is greatest at the center. Then the population of the inner city, $0 \le r \le 1$, is greater than the population of the suburbs, $1 \le r \le 2$.

9. The location of the center of mass of a system of three masses on the x-axis does not change if all the three masses are doubled.

10. The center of mass of a region in the plane cannot be outside the region.

11. Particles are shot at a circular target. The density of particles hitting the target decreases with the distance from the center. To set up a definite integral to calculate the total number of particles hitting the target, we should slice the region into concentric rings.

12. A metal rod of density $f(x)$ lying along the x-axis from $x = 0$ to $x = 4$ has its center of mass at $x = 2$. Then the two halves of the rod on either side of $x = 2$ have equal mass.

13. It takes more work to lift a 20 lb weight 10 ft slowly than to lift it the same distance quickly.

14. Work can be negative or positive.

15. The force on a rectangular dam is doubled if its length stays the same and its depth is doubled.

16. To find the force of water on a vertical wall, we always slice the wall horizontally, no matter what the shape of the wall.

17. The force of a liquid on a wall can be negative or positive.

18. If the average value of the force $F(x)$ is 7 on the interval $1 \leq x \leq 4$, then the work done by the force in moving from $x = 1$ to $x = 4$ is 21.

19. The present value of an income stream is always less than its future value.

20. If $p(x) = xe^{-x^2}$ for all x, then $p(x)$ is a probability density function.

21. If $p(x) = xe^{-x^2}$ for all $x > 0$ and $p(x) = 0$ for $x \leq 0$, then $p(x)$ is a probability density function.

Are the statements in Problems 22–25 true or false? If a statement is true, explain how you know. If a statement is false, give a counterexample

22. Of two solids of revolution, the one with the greater volume is obtained by revolving the region in the plane with the greater area.

23. If f is differentiable on the interval $[0, 10]$, then the arc length of the graph of f on the interval $[0, 1]$ is less than the arc length of the graph of f on the interval $[1, 10]$.

24. If f is concave up for all x and $f'(0) = 3/4$, then the arc length of the graph of f on the interval $[0, 4]$ is at least 5.

25. If f is concave down for all x and $f'(0) = 3/4$, then the arc length of the graph of f on the interval $[0, 4]$ is at most 5.

In Problems 26–30, a quantity x is distributed through a population with probability density function $p(x)$ and cumulative distribution function $P(x)$. Decide if the statements in Problems 26–30 are true or false. Give an explanation for your answer.

26. If $p(10) = 1/2$, then half the population has $x < 10$.

27. If $P(10) = 1/2$, then half the population has $x < 10$.

28. If $p(10) = 1/2$, then the fraction of the population lying between $x = 9.98$ and $x = 10.04$ is about 0.03.

29. If $p(10) = p(20)$, then none of the population has x values lying between 10 and 20.

30. If $P(10) = P(20)$, then none of the population has x values lying between 10 and 20.

PROJECTS FOR CHAPTER EIGHT

1. Volume Enclosed by Two Cylinders

Two cylinders are inscribed in a cube of side length 2, as shown in Figure 8.109. What is the volume of the solid that the two cylinders enclose? [Hint: Use horizontal slices.] Note: The solution was known to Archimedes. The Chinese mathematician Liu Hui (third century A.D.) tried to find this volume, but he failed; he wrote a poem about his efforts calling the enclosed volume a "box-lid":

Look inside the cube
And outside the box-lid;
Though the dimension increases,
It doesn't quite fit.
The marriage preparations are complete;
But square and circle wrangle,
Thick and thin are treacherous plots,
They are incompatible.
I wish to give my humble reflections,
But fear that I will miss the correct principle;
I dare to let the doubtful points stand,
Waiting
For one who can expound them.

Figure 8.109

2. Length of a Hanging Cable

The distance between the towers of the main span of the Golden Gate Bridge is about 1280 m; the sag of the cable halfway between the towers on a cold winter day is about 143 m. See Figure 8.110.

Figure 8.110

(a) How long is the cable, assuming it has an approximately parabolic shape? (Represent the cable as a parabola of the form $y = kx^2$ and determine k to at least 1 decimal place.)

(b) On a hot summer day the cable is about 0.05% longer, due to thermal expansion. By how much does the sag increase? Assume no movement of the towers.

3. Surface Area of an Unpaintable Can of Paint

This project introduces the formula for the surface area of a volume of revolution and shows that it is possible to have a solid with finite volume but infinite surface area.

We know that the arclength of the curve $y = f(x)$ from a to b can be found using the integral

$$\text{Arc length} = \int_a^b \sqrt{1 + (f'(x))^2}\, dx.$$

Figure 8.111 shows the corresponding arc length of a small piece of the curve. Similarly, it can be shown that the surface area from a to b of the solid obtained by revolving $y = f(x)$ around the x-axis is given by

$$\text{Surface area } = 2\pi \int_a^b f(x)\sqrt{1 + (f'(x))^2}\, dx.$$

We can see this why this might be true by looking at Figure 8.112. We approximate the small piece of the surface by a slanted cylinder of radius y and "height" equal to the arclength of the curve, so that

$$\text{Surface area of edge of slice } \approx 2\pi y \sqrt{1 + (f'(x))^2}\, \Delta x.$$

Integrating this expression from $x = a$ to $x = b$ gives the surface area of the solid.

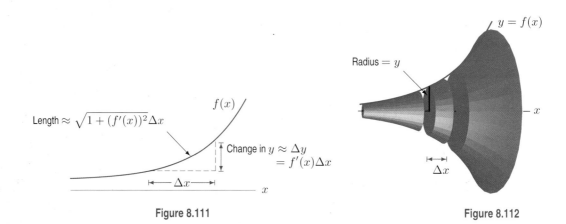

Figure 8.111   Figure 8.112

(a) Calculate the surface area of a sphere of radius r.

(b) Calculate the surface area of a cone of radius r and height h.

(c) Rotate the curve $y = 1/x$ for $x \geq 1$ around the x-axis. Find the volume of this solid.

(d) Show that the surface area of the solid in part (c) is infinite. [Hint: You might not be able to find an antiderivative of the integrand in the surface area formula; can you get a lower bound on the integral?]

(e) (Optional. Requires Chapter 11.) Find a curve such that when the portion of the curve from $x = a$ to $x = b$ is rotated around the x-axis (for any a and b), the volume of the solid of revolution is equal to its surface area. You may assume $dy/dx \geq 0$.

4. **Maxwell's Distribution of Molecular Velocities**

Let v be the speed, in meters/second, of an oxygen molecule, and let $p(v)$ be the density function of the speed distribution of oxygen molecules at room temperature. Maxwell showed that

$$p(v) = av^2 e^{-mv^2/(2kT)},$$

where $k = 1.4 \times 10^{-23}$ is the Boltzmann constant, T is the temperature in Kelvin (at room temperature, $T = 293$), and $m = 5 \times 10^{-26}$ is the mass of the oxygen molecule in kilograms.

(a) Find the value of a.

(b) Estimate the median and the mean speed. Find the maximum of $p(v)$.

(c) How do your answers in part (b) for the mean and the maximum of $p(v)$ change as T changes?

Chapter Nine

SEQUENCES AND SERIES

In this chapter, we look at infinite lists of numbers, called sequences, and infinite sums, called series. In Section 9.1, we study sequences. In Section 9.2, we begin with a particular type of series, called a geometric series. In Section 9.3, we consider general series of constants, what it means for such a series to converge, and how series are related to improper integrals. Tests that allow us to determine convergence are in Sections 9.3 and 9.4.

Section 9.5 introduces power series, in which the terms are constants multiplied by powers of x. These series converge for some x-values and not for others; the radius of convergence is introduced to identify the interval on which the series converges.

9.1 SEQUENCES

A sequence[1] is an infinite list of numbers $s_1, s_2, s_3, \ldots, s_n, \ldots$. We call s_1 the first term, s_2 the second term; s_n is the general term. For example, we can denote the sequence of squares, $1, 4, 9, \ldots, n^2, \ldots$ by the general term, $s_n = n^2$. Thus, a sequence is a function whose domain is the positive integers, but it is traditional to denote the terms of a sequence using subscripts, s_n, rather than function notation, $s(n)$. In addition, we may talk about sequences whose general term has no simple formula, such as the sequence $3, 3.1, 3.14, 3.141, 3.1415, \ldots$, in which s_n gives the first n digits of π.

The Numerical, Algebraic, and Graphical Viewpoint

Just as we can view a function algebraically, numerically, graphically, or verbally, we can view sequences in different ways. We may give an algebraic formula for the general term. We may give the numerical values of the first few terms of the sequence, suggesting a pattern for the later terms.

Example 1   Give the first six terms of the following sequences:

(a) $s_n = \dfrac{n(n+1)}{2}$

(b) $s_n = \dfrac{n + (-1)^n}{n}$

Solution   (a) Substituting $n = 1, 2, 3, 4, 5, 6$ into the formula for the general term, we get

$$\frac{1 \cdot 2}{2}, \frac{2 \cdot 3}{2}, \frac{3 \cdot 4}{2}, \frac{4 \cdot 5}{2}, \frac{5 \cdot 6}{2}, \frac{6 \cdot 7}{2} = 1, 3, 6, 10, 15, 21.$$

(b) Substituting $n = 1, 2, 3, 4, 5, 6$ into the formula for the general term, we get

$$\frac{1-1}{1}, \frac{2+1}{2}, \frac{3-1}{3}, \frac{4+1}{4}, \frac{5-1}{5}, \frac{6+1}{6} = 0, \frac{3}{2}, \frac{2}{3}, \frac{5}{4}, \frac{4}{5}, \frac{7}{6}.$$

Example 2   Give a general term for the following sequences:

(a) $1, 2, 4, 8, 16, 32, \ldots$

(b) $\dfrac{7}{2}, \dfrac{7}{5}, \dfrac{7}{8}, \dfrac{7}{11}, \dfrac{1}{2}, \dfrac{7}{17}, \ldots$

Solution   Although the first six terms do not determine the sequence, we can sometimes use them to guess a possible formula for the general term.

(a) We have powers of 2, so we guess $s_n = 2^n$. When we check by substituting in $n = 1, 2, 3, 4, 5, 6$, we get $2, 4, 8, 16, 32, 64$, instead of $1, 2, 4, 8, 16, 32$. We fix our guess by subtracting 1 from the exponent, so the general term is

$$s_n = 2^{n-1}.$$

Substituting the first six values of n shows that the formula checks.

(b) In this sequence, the fifth term looks different from the others, whose numerators are all 7. We can fix this by rewriting $1/2 = 7/14$. The sequence of denominators is then $2, 5, 8, 11, 14, 17$. This looks like a linear function with slope 3, so we expect the denominator has formula $3n + k$ for some k. When $n = 1$, the denominator is 2, so

$$2 = 3 \cdot 1 + k \quad \text{giving} \quad k = -1$$

and the denominator of s_n is $3n - 1$. Our general term is then

$$s_n = \frac{7}{3n - 1}.$$

To check this, evaluate s_n for $n = 1, \ldots, 6$.

---

[1]In everyday English, the words "sequence" and "series" are used interchangeably. In mathematics, they have different meanings and cannot be interchanged.

There are two ways to visualize a sequence. One is to plot points with n on the horizontal axis and s_n on the vertical axis. The other is to label points on a number line $s_1, s_2, s_3, \ldots$. See Figure 9.1 for the sequence $s_n = 1 + (-1)^n/n$.

Figure 9.1: The sequence $s_n = 1 + (-1)^n/n$

Defining Sequences Recursively

Sequences can also be defined *recursively*, by giving an equation relating the nth term to the previous terms and as many of the first terms as are needed to get started.

Example 3   Give the first six terms of the recursively defined sequences.

(a) $s_n = s_{n-1} + 3$ for $n > 1$ and $s_1 = 4$
(b) $s_n = -3s_{n-1}$ for $n > 1$ and $s_1 = 2$
(c) $s_n = \frac{1}{2}(s_{n-1} + s_{n-2})$ for $n > 2$ and $s_1 = 0$, $s_2 = 1$
(d) $s_n = ns_{n-1}$ for $n > 1$ and $s_1 = 1$

Solution

(a) When $n = 2$, we obtain $s_2 = s_1 + 3 = 4 + 3 = 7$. When $n = 3$, we obtain $s_3 = s_2 + 3 = 7 + 3 = 10$. In words, we obtain each term by adding 3 to the previous term. The first six terms are
$$4, 7, 10, 13, 16, 19.$$

(b) Each term is -3 times the previous term, starting with $s_1 = 2$. We have $s_2 = -3s_1 = -3 \cdot 2 = -6$ and $s_3 = -3s_2 = -3(-6) = 18$. Continuing, we get
$$2, -6, 18, -54, 162, -486.$$

(c) Each term is the average of the previous two terms, starting with $s_1 = 0$ and $s_2 = 1$. We get $s_3 = (s_2 + s_1)/2 = (1 + 0)/2 = 1/2$. Then $s_4 = (s_3 + s_2)/2 = ((1/2) + 1)/2 = 3/4$. Continuing, we get
$$0, 1, \frac{1}{2}, \frac{3}{4}, \frac{5}{8}, \frac{11}{16}.$$

(d) Here $s_2 = 2s_1 = 2 \cdot 1 = 2$ so $s_3 = 3s_2 = 3 \cdot 2 = 6$ and $s_4 = 4s_3 = 4 \cdot 6 = 24$. Continuing gives
$$1, 2, 6, 24, 120, 720.$$

The general term of part (d) of the previous example is given by $s_n = n(n-1)(n-2)\ldots 3 \cdot 2 \cdot 1$, which is denoted $s_n = n!$ and is called n factorial .

We can also look at the first few terms of a sequence and try to guess a recursive definition by looking for a pattern.

Example 4   Give a recursive definition of the following sequences.

(a) $1, 3, 7, 15, 31, 63, \ldots$ (b) $1, 4, 9, 16, 25, 36, \ldots$

Solution

(a) Each term is twice the previous term plus one; for example $7 = 2 \cdot 3 + 1$ and $63 = 2 \cdot 31 + 1$. Thus, a recursive definition is
$$s_n = 2s_{n-1} + 1 \text{ for } n > 1 \text{ and } s_1 = 1.$$

There are other ways to define the sequence recursively. We might notice, for example, that the differences of consecutive terms are powers of 2. Thus, we could also use

$$s_n = s_{n-1} + 2^{n-1} \quad \text{for } n > 1 \text{ and } s_1 = 1.$$

(b) We recognize the terms as the squares of the positive integers, but we are looking for a recursive definition which relates consecutive terms. We see that

$$s_2 = s_1 + 3$$
$$s_3 = s_2 + 5$$
$$s_4 = s_3 + 7$$
$$s_5 = s_4 + 9,$$

so the differences between consecutive terms are consecutive odd integers. The difference between s_n and s_{n-1} is $2n - 1$, so a recursive definition is

$$s_n = s_{n-1} + 2n - 1, \text{ for } n > 1 \text{ and } s_1 = 1.$$

Recursively defined sequences, sometimes called recurrence relations, are powerful tools used frequently in computer science, as well as differential equations. Finding a formula for the general term—which can be surprisingly difficult—uses the principle of mathematical induction.

Convergence of Sequences

The limit of a sequence s_n as $n \to \infty$ is defined the same way as the limit of a function $f(x)$ as $x \to \infty$; see also Problem 44.

> The sequence $s_1, s_2, s_3, \ldots, s_n, \ldots$ has a **limit** L, written $\lim_{n \to \infty} s_n = L$, if s_n is as close to L as we please whenever n is sufficiently large. If a limit, L, exists, we say the sequence **converges** to its limit L. If no limit exists, we say the sequence **diverges**.

To calculate the limit of a sequence, we use what we know about the limits of functions, including the properties in Theorem 1.2 and the following facts:

- The sequence $s_n = x^n$ converges to 0 if $|x| < 1$ and diverges if $|x| > 1$
- The sequence $s_n = 1/n^p$ converges to 0 if $p > 0$

Example 5 Do the following sequences converge or diverge? If a sequence converges, find its limit.

(a) $s_n = (0.8)^n$ (b) $s_n = \dfrac{1 - e^{-n}}{1 + e^{-n}}$ (c) $s_n = 1 + (-1)^n$

Solution (a) Since $0.8 < 1$, the sequence converges by the first fact and the limit is 0.
(b) Since $e^{-1} < 1$, we have $\lim_{n \to \infty} e^{-n} = \lim_{n \to \infty} (e^{-1})^n = 0$ by the first fact, so $\lim_{n \to \infty} s_n = 1$.
(c) Since $(-1)^n$ alternates in sign, the sequence alternates between 0 and 2. Thus the sequence s_n diverges, since it does not get close to any fixed value.

Convergence and Bounded Sequences

A sequence s_n is *bounded* if there are numbers K and M such that $K \leq s_n \leq M$ for all terms. If $\lim_{n \to \infty} s_n = L$, then from some point on, the terms are bounded between $L - 1$ and $L + 1$. Thus we have the following fact:

> A convergent sequence is bounded.

On the other hand, a bounded sequence need not be convergent. In Example 5, we saw that $1 + (-1)^n$ diverges, but it is bounded between 0 and 2. To ensure that a bounded sequence converges we need to rule out this sort of oscillation. The following theorem gives a condition that ensures convergence for a bounded sequence. A sequence s_n is called *monotone* if it is either increasing, that is $s_n < s_{n+1}$ for all n, or decreasing, that is $s_n > s_{n+1}$ for all n.

Theorem 9.1: Convergence of a Monotone, Bounded Sequence

If a sequence s_n is bounded and monotone, it converges.

To understand this theorem graphically, see Figure 9.2. The sequence s_n is increasing and bounded above by M, so the values of s_n must "pile up" at some number less than or equal to M. This number is the limit.[2]

Figure 9.2: Values of s_n for $n = 1, 2, \cdots, 10$

As an example, the sequence $s_n = (1 + 1/n)^n$ can be shown to be increasing and bounded (see Project 1 on page 475). Theorem 9.1 then guarantees that this sequence has a limit, which turns out to be e. (In fact, the sequence can be used to define e.)

Example 6   If $s_n = (1 + 1/n)^n$, find s_{100} and s_{1000}. How many decimal places agree with e?

Solution   We have $s_{100} = (1.01)^{100} = 2.7048$ and $s_{1000} = (1.001)^{1000} = 2.7169$. Since $e = 2.7183\ldots$, we see that s_{100} agrees with e to one decimal place and s_{1000} agrees with e to two decimal places.

Exercises and Problems for Section 9.1

Exercises

For Exercises 1–6, find the first five terms of the sequence from the formula for s_n, $n \geq 1$.

1. $2^n + 1$

2. $n + (-1)^n$

3. $\dfrac{2n}{2n + 1}$

4. $(-1)^n \left(\dfrac{1}{2}\right)^n$

5. $(-1)^{n+1} \left(\dfrac{1}{2}\right)^{n-1}$

6. $\left(1 - \dfrac{1}{n + 1}\right)^{n+1}$

In Exercises 7–12, find a formula for s_n, $n \geq 1$.

7. $4, 8, 16, 32, 64, \ldots$

8. $1, 3, 7, 15, 31, \ldots$

9. $2, 5, 10, 17, 26, \ldots$

10. $1, -3, 5, -7, 9, \ldots$

11. $1/3, 2/5, 3/7, 4/9, 5/11, \ldots$

12. $1/2, -1/4, 1/6, -1/8, 1/10, \ldots$

In Exercises 13–16, find the first six terms of the recursively defined sequence.

13. $s_n = s_{n-1} + n$ for $n > 1$ and $s_1 = 1$

14. $s_n = 2s_{n-1} + 3$ for $n > 1$ and $s_1 = 1$

15. $s_n = s_{n-1} + \left(\frac{1}{2}\right)^{n-1}$ for $n > 1$ and $s_1 = 0$

16. $s_n = s_{n-1} + 2s_{n-2}$ for $n > 2$ and $s_1 = 1$, $s_2 = 5$

---

[2] See the online supplement for a proof.

Problems

17. Match formulas (a)–(d) with graphs (I)–(IV).

(a) $s_n = 1 - 1/n$ (b) $s_n = 1 + (-1)^n/n$

(c) $s_n = 1/n$ (d) $s_n = 1 + 1/n$

Do the sequences in Problems 20–31 converge or diverge? If a sequence converges, find its limit.

20. $(0.2)^n$ **21.** 2^n

22. $(-0.3)^n$ **23.** $3 + e^{-2n}$

24. $\dfrac{2^n}{3^n}$ **25.** $\dfrac{n}{10} + \dfrac{10}{n}$

26. $\dfrac{(-1)^n}{n}$ **27.** $\dfrac{2n+1}{n}$

28. $\cos(\pi n)$ **29.** $\dfrac{\sin n}{n}$

30. $\dfrac{2n + (-1)^n 5}{4n - (-1)^n 3}$ **31.** $\dfrac{2^n}{n^3}$

18. Match formulas (a)–(e) with graphs (I)–(V).

(a) $s_n = 2 - 1/n$

(b) $s_n = (-1)^n 2 + 1/n$

(c) $s_n = 2 + (-1)^n/n$

(d) $s_n = 2 + 1/n$

(e) $s_n = (-1)^n 2 + (-1)^n/n$

In electrical engineering, a continuous function like $f(t) = \sin t$, where t is time in seconds, is referred to as an analog signal. To digitize the signal, we sample $f(t)$ every Δt seconds to form the sequence $s_n = f(n\Delta t)$. For example, sampling f every $1/10$ second produces the sequence $\sin(1/10), \sin(2/10), \sin(3/10), \ldots$. In Problems 32–34, give the first 6 terms of a sampling of the signal every Δt seconds.

32. $f(t) = \cos 5t, \Delta t = 0.1$

33. $f(t) = (x-1)^2, \Delta t = 0.5$

34. $f(t) = \dfrac{\sin t}{t}, \Delta t = 1$

To smooth a sequence, $s_1, s_2, s_3, \ldots$, we replace each term s_n by t_n, the average of s_n with its neighboring terms

$$t_n = \frac{(s_{n-1} + s_n + s_{n+1})}{3} \text{ for } n > 1.$$

We start with $t_1 = (s_1 + s_2)/2$, since s_1 has only one neighbor. For Problems 35–37, smooth the sequence once and then smooth the resulting sequence. What do you notice?

35. $18, -18, 18, -18, 18, -18, 18 \ldots$

36. $0, 0, 0, 18, 0, 0, 0, 0 \ldots$

37. $1, 2, 3, 4, 5, 6, 7, 8 \ldots$

19. Match formulas (a)–(e) with descriptions (I)–(V) of the behavior of the sequence as $n \to \infty$.

(a) $s_n = n(n+1) - 1$

(b) $s_n = 1/(n+1)$

(c) $s_n = 1 - n^2$

(d) $s_n = \cos(1/n)$

(e) $s_n = (\sin n)/n$

(I) Diverges to $-\infty$

(II) Diverges to $+\infty$

(III) Converges to 0 through positive numbers

(IV) Converges to 1

(V) Converges to 0 through positive and negative numbers

38. Let V_n be the number of new SUVs sold in the US in month n, where $n = 1$ is January 2004. In terms of SUVs, what do the following represent?

(a) V_{10}

(b) $V_n - V_{n-1}$

(c) $\sum_{i=1}^{12} V_i$ and $\sum_{i=1}^{n} V_i$

[3] www.bp.com, accessed May 28, 2004.

39. World oil consumption was 75.747 million barrels per day in 2002 and is increasing by about 0.3% per year.[3] Let c_n be daily world oil consumption n years after 2002.

(a) Find a formula for c_n.

(b) Find and interpret $c_n - c_{n-1}$.

(c) What does the sum $\sum_{n=1}^{18} c_n$ represent? (You do not need to compute this sum.)

40. (a) Let s_n be the number of ancestors a person has n generations ago. What is s_1? s_2? Find a formula for s_n.

(b) For which n is s_n greater than 6 billion, the current world population? What does this tell you about your ancestors?

41. For $0 \leq n \leq 10$, find a formula for p_n, the payment in year n on a loan of $100,000. Interest is 5% per year, compounded annually, and payments are made at the end of each year for ten years. Each payment is $10,000 plus the interest on the amount of money outstanding.

42. Baby formula can contain bacteria which double in number every half hour at room temperature and every 10 hours in the refrigerator.[4] Suppose there are B_0 bacteria initially.

(a) Write a formula for

(i) R_n, the number of bacteria n hours later if the baby formula is kept at room temperature.

(ii) F_n, the number of bacteria n hours later if the baby formula is kept in the refrigerator.

(iii) Y_n, the ratio of the number of bacteria at room temperature to the number of bacteria in the refrigerator.

(b) How many hours does it take before there are a million times as many bacteria in baby formula kept at room temperature as in baby formula kept in the refrigerator?

43. You are deciding whether to buy a new or a two-year-old car (of the same make) based on which will have cost you less when you resell it at the end of three years. Your cost consists of two parts: the loss in value of the car and the repairs. A new car costs $20,000 and loses 12% of its value each year. Repairs are $400 the first year and increase by 18% each subsequent year.

(a) For a new car, find the first three terms of the sequence d_n giving the depreciation (loss of value) in dollars in year n. Give a formula for d_n.

(b) Find the first three terms of the sequence r_n, the repair cost in dollars for a new car in year n. Give a formula for r_n.

(c) Find the total cost of owning a new car for three years.

(d) Find the total cost of owning the two-year-old car for three years. Which should you buy?

44. Write a definition for $\lim_{n \to \infty} s_n = L$ similar to the ϵ, δ definition for $\lim_{x \to a} f(x) = L$ in Section 1.8. Instead of δ, you will need N, a value of n.

45. The sequence s_n is increasing, the sequence t_n converges, and $s_n \leq t_n$ for all n. Show that s_n converges.

In Exercises 46–51, find a recursive definition for the sequence.

46. $1, 3, 5, 7, 9, \ldots$

47. $2, 4, 6, 8, 10, \ldots$

48. $3, 5, 9, 17, 33, \ldots$

49. $1, 5, 14, 30, 55, \ldots$

50. $1, 3, 6, 10, 15, \ldots$

51. $1, 2, \dfrac{3}{2}, \dfrac{5}{3}, \dfrac{8}{5}, \dfrac{13}{8}, \ldots$

In Problems 52–54, show that the sequence s_n satisfies the recurrence relation.

52. $s_n = 3n - 2$
$s_n = s_{n-1} + 3$ for $n > 1$ and $s_1 = 1$

53. $s_n = n(n + 1)/2$
$s_n = s_{n-1} + n$ for $n > 1$ and $s_1 = 1$

54. $s_n = 2n^2 - n$
$s_n = s_{n-1} + 4n - 3$ for $n > 1$ and $s_1 = 1$

55. (a) Cans are stacked in a triangle on a shelf. The bottom row contains k cans, the row above contains one can fewer, and so on. How many rows are there? Find a_n, the number of cans in the nth row from the top, $1 \leq n \leq k$.

(b) Let T_n be the total number of cans in the top n rows. Find a recurrence relation for T_n in terms of T_{n-1}.

(c) Show that $T_n = \frac{1}{2}n(n + 1)$ satisfies the recurrence relation.

56. The Fibonacci sequence first studied by the thirteenth century Italian mathematician Leonardo di Pisa, also known as Fibonacci, is defined recursively by

$$F_n = F_{n-1} + F_{n-2} \text{ for } n > 2 \text{ and } F_1 = 1, F_2 = 1.$$

The Fibonacci sequence occurs in many branches of mathematics and can be found in patterns of plant growth (phyllotaxis).

(a) Find the first 12 terms.

(b) Show that the sequence of successive ratios F_{n+1}/F_n appears to converge to a number r satisfying the equation $r^2 = r + 1$. (The number r was known as the golden ratio to the ancient Greeks.)

(c) Let r satisfy $r^2 = r + 1$. Show that the sequence $s_n = A r^n$, where A is constant, satisfies the Fibonacci equation $s_n = s_{n-1} + s_{n-2}$ for $n > 2$.

---

[4]Iverson, C. and Forsythe, F., reported in "Baby Food Could Trigger Meningitis," www.newscientist.com, June 3, 2004.

For a function f, define a sequence recursively by $x_n = f(x_{n-1})$ for $n > 1$ and $x_1 = a$. Depending on f and the starting value a, this sequence may converge to a limit L. If L exists, it has the property that $f(L) = L$. For the functions and starting values in Problems 57–58, use a calculator to see if the sequence converges. [To obtain the terms of the sequence, repeatedly push the function button.]

57. $f(x) = \cos x, a = 0$ **58.** $f(x) = e^{-x}, a = 0$

9.2 GEOMETRIC SERIES

This section introduces infinite series of constants, which are sums of the form

$$1 + \tfrac{1}{2} + \tfrac{1}{3} + \tfrac{1}{4} + \cdots$$
$$0.4 + 0.04 + 0.004 + 0.0004 + \cdots.$$

The individual numbers, $1, \tfrac{1}{2}, \tfrac{1}{3}, \cdots$, or $0.4, 0.04, \cdots$, etc., are called *terms* in the series. To talk about the *sum* of the series, we must first explain how to add infinitely many numbers.

Let us look at the repeated administration of a drug. In this example, the terms in the series represent each dose; the sum of the series represents the drug level in the body in the long run.

Repeated Drug Dosage

A person with an ear infection is told to take antibiotic tablets regularly for several days. Since the drug is being excreted by the body between doses, how can we calculate the quantity of the drug remaining in the body at any particular time?

To be specific, let's suppose the drug is ampicillin (a common antibiotic) taken in 250 mg doses four times a day (that is, every six hours). It is known that at the end of six hours, about 4% of the drug is still in the body. What quantity of the drug is in the body right after the tenth tablet? The fortieth?

Let Q_n represent the quantity, in milligrams, of ampicillin in the blood right after the nth tablet. Then

$$Q_1 = 250 \qquad\qquad = 250 \text{ mg}$$

$$Q_2 = \underbrace{250(0.04)}_{\text{Remnants of first tablet}} + \underbrace{250}_{\text{New tablet}} \qquad\qquad = 260 \text{ mg}$$

$$Q_3 = Q_2(0.04) + 250 = (250(0.04) + 250)(0.04) + 250$$
$$= \underbrace{250(0.04)^2 + 250(0.04)}_{\text{Remnants of first and second tablets}} + \underbrace{250}_{\text{New tablet}} \qquad = 260.4 \text{ mg}$$

$$Q_4 = Q_3(0.04) + 250 = \left(250(0.04)^2 + 250(0.04) + 250\right)(0.04) + 250$$
$$= \underbrace{250(0.04)^3 + 250(0.04)^2 + 250(0.04)}_{\text{Remnants of first, second, and third tablets}} + \underbrace{250}_{\text{New tablet}} \qquad = 260.416 \text{ mg}.$$

Looking at the pattern that is emerging, we guess that

$$Q_5 = 250(0.04)^4 + 250(0.04)^3 + 250(0.04)^2 + 250(0.04) + 250$$
$$Q_{10} = 250(0.04)^9 + 250(0.04)^8 + \cdots + 250(0.04) + 250.$$

Notice that there are 10 terms in this sum—one for every tablet—but that the highest power of 0.04 is the ninth, because no tablet has been in the body for more than 9 six-hour time periods. (Do you see why?) Now suppose we actually want to find the numerical value of Q_{10}. It seems that we have to add 10 terms—and if we want the value of Q_{40}, we would be faced with adding 40 terms:

$$Q_{40} = 250(0.04)^{39} + 250(0.04)^{38} + \cdots + 250(0.04) + 250.$$

Fortunately, there's a better way. Let's start with Q_{10}.

$$Q_{10} = 250(0.04)^9 + 250(0.04)^8 + 250(0.04)^7 + \cdots + 250(0.04)^2 + 250(0.04) + 250.$$

Notice the remarkable fact that if you subtract $(0.04)Q_{10}$ from Q_{10}, a great many terms (all but two, in fact) drop out. First multiplying by 0.04, we get

$$(0.04)Q_{10} = 250(0.04)^{10} + 250(0.04)^9 + 250(0.04)^8 + \cdots + 250(0.04)^3 + 250(0.04)^2 + 250(0.04).$$

Subtracting gives

$$Q_{10} - (0.04)Q_{10} = 250 - 250(0.04)^{10}.$$

Factoring Q_{10} on the left and solving for Q_{10} gives

$$Q_{10}(1 - 0.04) = 250\left(1 - (0.04)^{10}\right)$$

$$Q_{10} = \frac{250\left(1 - (0.04)^{10}\right)}{1 - 0.04}.$$

This is called the *closed-form* expression for Q_{10}. It is easy to evaluate on a calculator, giving $Q_{10} = 260.42$ (to two decimal places). Similarly, Q_{40} is given in closed-form by

$$Q_{40} = \frac{250\left(1 - (0.04)^{40}\right)}{1 - 0.04}.$$

Evaluating this on a calculator shows $Q_{40} = 260.42$, which is the same (to two decimal places) as Q_{10}. Thus after ten tablets, the value of Q_n appears to have stabilized at just over 260 mg.

Looking at the closed-forms for Q_{10} and Q_{40}, we can see that, in general, Q_n must be given by

$$Q_n = \frac{250\left(1 - (0.04)^n\right)}{1 - 0.04}.$$

What Happens as $n \to \infty$?

What does this closed-form for Q_n predict about the long-run level of ampicillin in the body? As $n \to \infty$, the quantity $(0.04)^n \to 0$. In the long run, assuming that 250 mg continue to be taken every six hours, the level right after a tablet is taken is given by

$$Q_n = \frac{250\left(1 - (0.04)^n\right)}{1 - 0.04} \to \frac{250(1 - 0)}{1 - 0.04} = 260.42.$$

The Geometric Series in General

In the previous example we encountered sums of the form $a + ax + ax^2 + \cdots + ax^8 + ax^9$ (with $a = 250$ and $x = 0.04$). Such a sum is called a finite *geometric series*. A geometric series is one in which each term is a constant multiple of the one before. The first term is a, and the constant multiplier, or *common ratio* of successive terms, is x. (In our example, $a = 250$ and $x = 0.04$.)

A finite geometric series has the form

$$a + ax + ax^2 + \cdots + ax^{n-2} + ax^{n-1}.$$

An **infinite geometric series** has the form

$$a + ax + ax^2 + \cdots + ax^{n-2} + ax^{n-1} + ax^n + \cdots.$$

The "$\cdots$" at the end of the second series tells us that the series is going on forever—in other words, that it is infinite.

Sum of a Finite Geometric Series

The same procedure that enabled us to find the closed-form for Q_{10} can be used to find the sum of any finite geometric series. Suppose we write S_n for the sum of the first n terms, which means up to the term containing x^{n-1}:

$$S_n = a + ax + ax^2 + \cdots + ax^{n-2} + ax^{n-1}.$$

Multiply S_n by x:

$$xS_n = ax + ax^2 + ax^3 + \cdots + ax^{n-1} + ax^n.$$

Now subtract xS_n from S_n, which cancels out all terms except for two, giving

$$S_n - xS_n = a - ax^n$$
$$(1-x)S_n = a(1-x^n).$$

Provided $x \neq 1$, we can solve to find a closed form for S_n as follows:

The **sum of a finite geometric series** is given by

$$S_n = a + ax + ax^2 + \cdots + ax^{n-1} = \frac{a(1-x^n)}{1-x}, \qquad \text{provided } x \neq 1.$$

Note that the value of n in the formula for S_n is the number of terms in the sum S_n.

Sum of an Infinite Geometric Series

In the ampicillin example, we found the sum Q_n and then let $n \to \infty$. We do the same here. The sum Q_n, which shows the effect of the first n doses, is an example of a *partial sum*. The first three partial sums of the series $a + ax + ax^2 + \cdots + ax^{n-1} + ax^n + \cdots$ are

$$S_1 = a$$
$$S_2 = a + ax$$
$$S_3 = a + ax + ax^2.$$

To find the sum, S, of this infinite series, we consider the partial sum, S_n, of the first n terms. The formula for the sum of a finite geometric series gives

$$S_n = a + ax + ax^2 + \cdots + ax^{n-1} = \frac{a(1-x^n)}{1-x}.$$

What happens to S_n as $n \to \infty$? It depends on the value of x. If $|x| < 1$, then $x^n \to 0$ as $n \to \infty$, so

$$\lim_{n \to \infty} S_n = \lim_{n \to \infty} \frac{a(1-x^n)}{1-x} = \frac{a(1-0)}{1-x} = \frac{a}{1-x}.$$

Thus, provided $|x| < 1$, as $n \to \infty$ the partial sums S_n approach a limit of $a/(1-x)$. When this happens, we define the sum of the infinite geometric series to be that limit and say the series *converges* to $a/(1-x)$.

For $|x| < 1$, the **sum of the infinite geometric series** is given by

$$S = a + ax + ax^2 + \cdots + ax^{n-1} + ax^n + \cdots = \frac{a}{1-x}.$$

If, on the other hand, $|x| > 1$, then x^n and the partial sums have no limit as $n \to \infty$ (if $a \neq 0$). In this case, we say the series *diverges*. If $x > 1$, the terms in the series become larger and larger in magnitude, and the partial sums diverge to $+\infty$ (if $a > 0$) or $-\infty$ (if $a < 0$). When $x < -1$, the terms become larger in magnitude, the partial sums oscillate as $n \to \infty$, and the series diverges.

What happens when $x = 1$? The series is

$$a + a + a + a + \cdots,$$

and if $a \neq 0$, the partial sums grow without bound, and the series does not converge. When $x = -1$, the series is

$$a - a + a - a + a - \cdots,$$

and, if $a \neq 0$, the partial sums oscillate between a and 0, and the series does not converge.

Example 1   For each of the following infinite geometric series, find several partial sums and the sum (if it exists).

(a) $1 + \dfrac{1}{2} + \dfrac{1}{4} + \dfrac{1}{8} + \cdots$ (b) $1 + 2 + 4 + 8 + \cdots$ (c) $6 - 2 + \dfrac{2}{3} - \dfrac{2}{9} + \dfrac{2}{27} - \cdots$

Solution   (a) This series may be written

$$1 + \frac{1}{2} + \left(\frac{1}{2}\right)^2 + \left(\frac{1}{2}\right)^3 + \cdots$$

which we can identify as a geometric series with $a = 1$ and $x = \frac{1}{2}$, so $S = \dfrac{1}{1 - (1/2)} = 2.$

Let's check this by finding the partial sums:

$$S_1 = 1$$
$$S_2 = 1 + \frac{1}{2} = \frac{3}{2} = 2 - \frac{1}{2}$$
$$S_3 = 1 + \frac{1}{2} + \frac{1}{4} = \frac{7}{4} = 2 - \frac{1}{4}$$
$$S_4 = 1 + \frac{1}{2} + \frac{1}{4} + \frac{1}{8} = \frac{15}{8} = 2 - \frac{1}{8}$$
$$S_5 = 1 + \frac{1}{2} + \frac{1}{4} + \frac{1}{8} + \frac{1}{16} = \frac{31}{16} = 2 - \frac{1}{16}.$$

The formula for S_n gives

$$S_n = \frac{1 - (\frac{1}{2})^n}{1 - \frac{1}{2}} = 2 - \left(\frac{1}{2}\right)^{n-1}.$$

Thus, the partial sums are creeping up to the value $S = 2$, so $S_n \to 2$ as $n \to \infty$.

(b) The partial sums of this geometric series (with $a = 1$ and $x = 2$) grow without bound, so the series has no sum:

$$S_1 = 1$$
$$S_2 = 1 + 2 = 3$$
$$S_3 = 1 + 2 + 4 = 7$$
$$S_4 = 1 + 2 + 4 + 8 = 15$$
$$S_5 = 1 + 2 + 4 + 8 + 16 = 31.$$

The formula for S_n gives

$$S_n = \frac{1 - 2^n}{1 - 2} = 2^n - 1.$$

(c) This is an infinite geometric series with $a = 6$ and $x = -\frac{1}{3}$. The partial sums,

$$S_1 = 6.00, \quad S_2 = 4.00, \quad S_3 \approx 4.67, \quad S_4 \approx 4.44, \quad S_5 \approx 4.52, \quad S_6 \approx 4.49,$$

appear to be converging to 4.5. This turns out to be correct because the sum is

$$S = \frac{6}{1 - (-1/3)} = 4.5.$$

Regular Deposits into a Savings Account

People who save money often do so by putting some fixed amount aside regularly. To be specific, suppose $1000 is deposited every year in a savings account earning 5% a year, compounded annually. What is the balance, B_n, in dollars, in the savings account right after the nth deposit?

As before, let's start by looking at the first few years:

$$B_1 = 1000$$

$$B_2 = B_1(1.05) + 1000 = \underbrace{1000(1.05)}_{\text{Original deposit}} + \underbrace{1000}_{\text{New deposit}}$$

$$B_3 = B_2(1.05) + 1000 = \underbrace{1000(1.05)^2 + 1000(1.05)}_{\text{First two deposits}} + \underbrace{1000}_{\text{New deposit}}$$

$$B_4 = B_3(1.05) + 1000 = \underbrace{1000(1.05)^3 + 1000(1.05)^2 + 1000(1.05)}_{\text{First three deposits}} + \underbrace{1000}_{\text{New deposit}}$$

Observing the pattern, we see

$$B_n = 1000(1.05)^{n-1} + 1000(1.05)^{n-2} + \cdots + 1000(1.05) + 1000.$$

So B_n is a finite geometric series with $a = 1000$ and $x = 1.05$. Thus we have

$$B_n = \frac{1000\left(1 - (1.05)^n\right)}{1 - 1.05}.$$

We can rewrite this so that both the numerator and denominator of the fraction are positive:

$$B_n = \frac{1000\left((1.05)^n - 1\right)}{1.05 - 1}.$$

What Happens as $n \to \infty$?

Common sense tells you that if you keep depositing \$1000 in an account and it keeps earning interest, your balance grows without bound. This is what the formula for B_n shows also: $(1.05)^n \to \infty$ as $n \to \infty$, so B_n has no limit. (Alternatively, observe that the infinite geometric series of which B_n is a partial sum has $x = 1.05$, which is greater than 1, so the series does not converge.)

Exercises and Problems for Section 9.2

Exercises

In Exercises 1–10, decide which of the following are geometric series. For those which are, give the first term and the ratio between successive terms. For those which are not, explain why not.

1. $2 + 1 + \dfrac{1}{2} + \dfrac{1}{4} + \dfrac{1}{8} + \cdots$

2. $1 - \dfrac{1}{2} + \dfrac{1}{4} - \dfrac{1}{8} + \dfrac{1}{16} + \cdots$

3. $1 + \dfrac{1}{2} + \dfrac{1}{3} + \dfrac{1}{4} + \dfrac{1}{5} + \cdots$

4. $5 - 10 + 20 - 40 + 80 - \cdots$

5. $1 + x + 2x^2 + 3x^3 + 4x^4 + \cdots$

6. $3 + 3z + 6z^2 + 9z^3 + 12z^4 + \cdots$

7. $1 + 2z + (2z)^2 + (2z)^3 + \cdots$

8. $y^2 + y^3 + y^4 + y^5 + \cdots$

9. $1 - x + x^2 - x^3 + x^4 - \cdots$

10. $1 - y^2 + y^4 - y^6 + \cdots$

11. Find the sum of the series in Exercise 7.

12. Find the sum of the series in Exercise 8.

13. Find the sum of the series in Exercise 9.

14. Find the sum of the series in Exercise 10.

For each finite geometric series in Exercises 15–17, say how many terms are in the series and find its sum.

15. $2 + 2(0.1) + 2(0.1)^2 + \cdots + 2(0.1)^{25}$

16. $2(0.1) + 2(0.1)^2 + \cdots + 2(0.1)^{10}$

17. $2(0.1)^5 + 2(0.1)^6 + \cdots + 2(0.1)^{13}$

Find the sum of the series in Exercises 18–21.

18. $-2 + 1 - \dfrac{1}{2} + \dfrac{1}{4} - \dfrac{1}{8} + \dfrac{1}{16} - \cdots$

19. $3 + \dfrac{3}{2} + \dfrac{3}{4} + \dfrac{3}{8} + \cdots + \dfrac{3}{2^{10}}$

20. $\displaystyle\sum_{n=4}^{\infty} \left(\dfrac{1}{3}\right)^n$

21. $\displaystyle\sum_{n=4}^{20} \left(\dfrac{1}{3}\right)^n$

Problems

22. Every term of a geometric series (finite or infinite) is multiplied by the same nonzero constant c. Is the new series also a geometric series? Explain.

23. A new series is obtained from a geometric series (finite or infinite) by taking the reciprocal of each term. Is the new series that results also a geometric series? Explain.

24. This problem shows another way of deriving the long-run ampicillin level. (See page 444.) In the long run the ampicillin levels off to Q mg right after each tablet is taken. Six hours later, right before the next dose, there will be less ampicillin in the body. However, if stability has been reached, the amount of ampicillin that has been excreted is exactly 250 mg because taking one more tablet raises the level back to Q mg. Use this to solve for Q.

25. Figure 9.3 shows the quantity of the drug atenolol in the blood as a function of time, with the first dose at time $t = 0$. Atenolol is taken in 50 mg doses once a day to lower blood pressure.

Figure 9.3

(a) If the half-life of atenolol in the blood is 6.3 hours, what percentage of the atenolol present at the start of a 24-hour period is still there at the end?

(b) Find expressions for the quantities $Q_0, Q_1, Q_2, Q_3,$ ..., and Q_n shown in Figure 9.3. Write the expression for Q_n in closed-form.

(c) Find expressions for the quantities $P_1, P_2, P_3,$ ..., and P_n shown in Figure 9.3. Write the expression for P_n in closed-form.

26. On page 444, you saw how to compute the quantity Q_n mg of ampicillin in the body right after the nth tablet of 250 mg, taken once every six hours.

(a) Do a similar calculation for P_n, the quantity of ampicillin (in mg) in the body right *before* the nth tablet is taken.

(b) Express P_n in closed form.

(c) What is $\lim_{n\to\infty} P_n$? Is this limit the same as $\lim_{n\to\infty} Q_n$? Explain in practical terms why your answer makes sense.

27. Draw a graph like that in Figure 9.3 for 250 mg of ampicillin taken every 6 hours, starting at time $t = 0$. Put on the graph the values of $Q_1, Q_2, Q_3,$ ... introduced in the text on page 444 and the values of $P_1, P_2, P_3,$ ... calculated in Problem 26.

28. A ball is dropped from a height of 10 feet and bounces. Each bounce is $\frac{3}{4}$ of the height of the bounce before. Thus, after the ball hits the floor for the first time, the ball rises to a height of $10\left(\frac{3}{4}\right) = 7.5$ feet, and after it hits the floor for the second time, it rises to a height of $7.5\left(\frac{3}{4}\right) = 10\left(\frac{3}{4}\right)^2 = 5.625$ feet. (Assume $g = 32$ ft/sec^2 and that there is no air resistance.)

(a) Find an expression for the height to which the ball rises after it hits the floor for the nth time.

(b) Find an expression for the total vertical distance the ball has traveled when it hits the floor for the first, second, third, and fourth times.

(c) Find an expression for the total vertical distance the ball has traveled when it hits the floor for the nth time. Express your answer in closed-form.

29. You might think that the ball in Problem 28 keeps bouncing forever since it takes infinitely many bounces. This is not true!

(a) Show that a ball dropped from a height of h feet reaches the ground in $\frac{1}{4}\sqrt{h}$ seconds.

(b) Show that the ball in Problem 28 stops bouncing after

$$\frac{1}{4}\sqrt{10} + \frac{1}{2}\sqrt{10}\sqrt{\frac{3}{4}}\left(\frac{1}{1 - \sqrt{3/4}}\right) \approx 11 \text{ seconds.}$$

30. One way of valuing a company is to calculate the present value of all its future earnings. Suppose a farm expects to sell $1000 worth of Christmas trees once a year forever, with the first sale in the immediate future. What is the present value of this Christmas tree business? Assume that the interest rate is 4% per year, compounded continuously.

31. This problem deals with the question of estimating the cumulative effect of a tax cut on a country's economy. Suppose the government proposes a tax cut totaling $100 million. We assume that all the people who have extra money to spend would spend 80% of it and save 20%. Thus, of the extra income generated by the tax cut, $100(0.8)$ million $=$ $80 million would be spent and so become extra income to someone else. Assume that these people also spend 80% of their additional income, or $80(0.8)$ million, and so on. Calculate the total additional spending created by such a tax cut.

32. Suppose the government proposes a tax cut of $100 million as in Problem 31, but that economists now predict that people will spend 90% of their extra income and save only 10%. How much additional spending would be generated by the tax cut under these assumptions?

9.3   CONVERGENCE OF SERIES

We now consider general series in which each term a_n is a number. The series can be written compactly using a $\sum$ sign as follows

$$\sum_{n=1}^{\infty} a_n = a_1 + a_2 + a_3 + \cdots + a_n + \cdots .$$

For any particular values of a and x, the geometric series is such a series, with general term $a_n = ax^{n-1}$.

Partial Sums and Convergence of Series

As in Section 9.2, we define the *partial sum*, S_n, of the first n terms of a series as

$$S_n = \sum_{i=1}^{n} a_i = a_1 + a_2 + \cdots + a_n .$$

To investigate the convergence of the series, we consider the sequence of partial sums

$$S_1, S_2, S_3, \ldots, S_n, \ldots .$$

If S_n has a limit as $n \to \infty$, then we define the sum of the series to be that limit.

If the sequence S_n of partial sums converges to S, so $\lim_{n \to \infty} S_n = S$, then we say the series $\sum_{n=1}^{\infty} a_n$ **converges** and that its sum is S. We write $\sum_{n=1}^{\infty} a_n = S$. If $\lim_{n \to \infty} S_n$ does not exist, we say that the series **diverges**.

Visualizing Series

We can visualize the terms of a series as in Figure 9.4. In this figure, we assume $a_n \geq 0$ for all n, so each rectangle has area a_n. Then the series converges if the total area of the rectangles is finite and the sum of the series is the total area of the rectangles. This is similar to an improper integral $\int_0^{\infty} f(x)\, dx$, in which the area under the graph of f can be finite, even on an infinite interval.

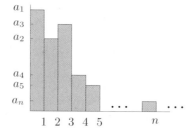

Figure 9.4: Height and area of the nth rectangle is a_n

Here are some properties that are useful in determining whether or not a series converges.

Theorem 9.2: Convergence Properties of Series

1. If $\displaystyle\sum_{n=1}^{\infty} a_n$ and $\displaystyle\sum_{n=1}^{\infty} b_n$ converge and if k is a constant, then

   - $\displaystyle\sum_{n=1}^{\infty} (a_n + b_n)$ converges to $\displaystyle\sum_{n=1}^{\infty} a_n + \sum_{n=1}^{\infty} b_n$.

   - $\displaystyle\sum_{n=1}^{\infty} ka_n$ converges to $k \displaystyle\sum_{n=1}^{\infty} a_n$.

2. Changing a finite number of terms in a series does not change whether or not it converges, although it may change the value of its sum if it does converge.

3. If $\displaystyle\lim_{n\to\infty} a_n \neq 0$ or $\displaystyle\lim_{n\to\infty} a_n$ does not exist, then $\displaystyle\sum_{n=1}^{\infty} a_n$ diverges.

4. If $\displaystyle\sum_{n=1}^{\infty} a_n$ diverges, then $\displaystyle\sum_{n=1}^{\infty} ka_n$ diverges if $k \neq 0$.

For proofs of these properties, see Problems 26–29. As for improper integrals, the convergence of a series is determined by its behavior for large n. (See the "behaves like" principle on page 357.) From Property 2 we see that, if N is a positive integer, then $\sum_{n=1}^{\infty} a_n$ and $\sum_{n=N}^{\infty} a_n$ either both converge or both diverge. Thus, if all we care about is the convergence of a series, we can omit the limits and write $\sum a_n$.

Example 1   Does the series $\displaystyle\sum (1 - e^{-n})$ converge?

Solution   Since the terms in the series, $a_n = 1 - e^{-n}$ tend to 1, not 0, as $n \to \infty$, the series diverges by Property 3 of Theorem 9.2.

Comparison of Series and Integrals

We investigate the convergence of some series by comparison with an improper integral. The *harmonic series* is the infinite series

$$\sum_{n=1}^{\infty} \frac{1}{n} = 1 + \frac{1}{2} + \frac{1}{3} + \frac{1}{4} + \cdots + \frac{1}{n} + \cdots.$$

Convergence of this sum would mean that the sequence of partial sums

$$S_1 = 1, \quad S_2 = 1 + \frac{1}{2}, \quad S_3 = 1 + \frac{1}{2} + \frac{1}{3}, \quad \cdots, \quad S_n = 1 + \frac{1}{2} + \frac{1}{3} + \cdots + \frac{1}{n}, \quad \cdots$$

tends to a limit as $n \to \infty$. Let's look at some values:

$$S_1 = 1, \quad S_{10} \approx 2.93, \quad S_{100} \approx 5.19, \quad S_{1000} \approx 7.49, \quad S_{10000} \approx 9.79.$$

The growth of these partial sums is slow, but they do in fact grow without bound, so the harmonic series diverges. This is justified in the following example and in Problem 33.

Example 2 Show that the harmonic series $1 + 1/2 + 1/3 + 1/4 + \dots$ diverges.

Solution The idea is to approximate $\int_1^\infty (1/x)\, dx$ by a left-hand sum, where the terms $1, 1/2, 1/3, \dots$ are heights of rectangles of base 1. In Figure 9.5, the sum of the areas of the 3 rectangles is larger than the area under the curve between $x = 1$ and $x = 4$, and the same kind of relationship holds for the first n rectangles. Thus, we have

$$S_n = 1 + \frac{1}{2} + \frac{1}{3} + \dots + \frac{1}{n} > \int_1^{n+1} \frac{1}{x}\, dx = \ln(n + 1).$$

Since $\ln(n + 1)$ gets arbitrarily large as $n \to \infty$, so do the partial sums, S_n. Thus, the partial sums have no limit, so the series diverges.

Figure 9.5: Comparing the harmonic series to $\int_1^\infty (1/x)\, dx$

Notice that the harmonic series diverges, even though $\displaystyle\lim_{n \to \infty} a_n = \lim_{n \to \infty} (1/n) = 0$.

Example 3 By comparison with the improper integral $\int_1^\infty (1/x^2)\, dx$, show that the following series converges:

$$\sum_{n=1}^\infty \frac{1}{n^2} = 1 + \frac{1}{4} + \frac{1}{9} + \cdots.$$

Figure 9.6: Comparing $\sum_{n=1}^\infty 1/n^2$ to $\int_1^\infty (1/x^2)\, dx$

Solution Since we want to show that $\displaystyle\sum_{n=1}^\infty 1/n^2$ converges, we want to show that the partial sums of this series tend to a limit. We do this by showing that the sequence of partial sums increases and is bounded above, so Theorem 9.1 applies.

Each successive partial sum is obtained from the previous one by adding one more term in the series. Since all the terms are positive, the sequence of partial sums is increasing.

To show that the partial sums of $\sum_{n=1}^{\infty} 1/n^2$ are bounded, we consider the right-hand sum represented by the area of the rectangles in Figure 9.6. We start at $x = 1$, since the area under the curve is infinite for $0 \leq x \leq 1$. The shaded rectangles in Figure 9.6 suggest that:

$$\frac{1}{4} + \frac{1}{9} + \frac{1}{16} + \cdots + \frac{1}{n^2} \leq \int_1^{\infty} \frac{1}{x^2}\, dx.$$

The area under the graph is finite, since

$$\int_1^{\infty} \frac{1}{x^2}\, dx = \lim_{b \to \infty} \int_1^b \frac{1}{x^2}\, dx = \lim_{b \to \infty}\left(-\frac{1}{b} + 1\right) = 1.$$

To get S_n, we add 1 to both sides, giving

$$S_n = 1 + \frac{1}{4} + \frac{1}{9} + \frac{1}{16} + \cdots + \frac{1}{n^2} \leq 1 + \int_1^{\infty} \frac{1}{x^2}\, dx = 2.$$

Thus, the sequence of partial sums is bounded above by 2. Hence, by Theorem 9.1 the sequence of partial sums converges, so the series converges.

Notice that we have shown that the series in the Example 3 converges, but we have not found its sum. The integral gives us a bound on the partial sums, but it does not give us the limit of the partial sums. Euler proved the remarkable fact that the sum is $\pi^2/6$.

The method of Examples 2 and 3 can be used to prove the following theorem. See Problem 32.

Theorem 9.3: The Integral Test

Suppose $a_n = f(n)$, where $f(x)$ is decreasing and positive for $x \geq c$.

- If $\int_c^{\infty} f(x)\, dx$ converges, then $\sum a_n$ converges.

- If $\int_c^{\infty} f(x)\, dx$ diverges, then $\sum a_n$ diverges.

The integral test allows us to analyze a family of series, the p-series, and see how convergence depends on the parameter p.

Example 4   For what values of p does the series $\sum_{n=1}^{\infty} 1/n^p$ converge?

Solution   If $p \leq 0$, the terms in the series $a_n = 1/n^p$ do not tend to 0 as $n \to \infty$. Thus the series diverges for $p \leq 0$.

If $p > 0$, we compare $\sum_{n=1}^{\infty} 1/n^p$ to the integral $\int_1^{\infty} 1/x^p\, dx$. In Example 3 of Section 7.7 we saw that the integral converges if $p > 1$ and diverges if $p \leq 1$. By the integral test, we conclude that $\sum 1/n^p$ converges if $p > 1$ and diverges if $p \leq 1$.

We can summarize Example 4 as follows:

The p-**series** $\sum_{n=1}^{\infty} 1/n^p$ converges if $p > 1$ and diverges if $p \leq 1$.

Exercises and Problems for Section 9.3

Exercises

Use the integral test to decide whether the series in Exercises 1–4 converge or diverge.

1. $\displaystyle\sum_{n=1}^{\infty} \frac{1}{n^3}$

2. $\displaystyle\sum_{n=1}^{\infty} \frac{n}{n^2 + 1}$

3. $\displaystyle\sum_{n=1}^{\infty} \frac{1}{e^n}$

4. $\displaystyle\sum_{n=2}^{\infty} \frac{1}{n(\ln n)^2}$

5. Use comparison with $\int_1^{\infty} x^{-3}\, dx$ to show that $\sum_{n=2}^{\infty} 1/n^3$ converges to a number less than or equal to $1/2$.

6. Use comparison with $\int_0^{\infty} 1/(x^2 + 1)\, dx$ to show that $\sum_{n=1}^{\infty} 1/(n^2 + 1)$ converges to a number less than or equal to $\pi/2$.

Explain why the integral test cannot be used to decide if the series in Exercises 7–9 converge or diverge.

7. $\displaystyle\sum_{n=1}^{\infty} n^2$

8. $\displaystyle\sum_{n=1}^{\infty} \frac{(-1)^n}{n}$

9. $\displaystyle\sum_{n=1}^{\infty} e^{-n} \sin n$

Problems

Do the series in Problems 10–19 converge or diverge?

10. $\displaystyle\sum_{n=0}^{\infty} \frac{3}{n + 2}$

11. $\displaystyle\sum_{n=1}^{\infty} \frac{3}{(2n - 1)^2}$

12. $\displaystyle\sum_{n=0}^{\infty} \frac{2}{\sqrt{2 + n}}$

13. $\displaystyle\sum_{n=1}^{\infty} \frac{n}{n + 1}$

14. $\displaystyle\sum_{n=1}^{\infty} \frac{4}{(2n + 1)^3}$

15. $\displaystyle\sum_{n=0}^{\infty} \frac{3}{n^2 + 4}$

16. $\displaystyle\sum_{n=1}^{\infty} \left(\left(\frac{3}{4}\right)^n + \frac{1}{n} \right)$

17. $\displaystyle\sum_{n=1}^{\infty} \frac{n + 2^n}{n 2^n}$

18. $\displaystyle\sum_{n=1}^{\infty} \frac{\ln n}{n}$

19. $\displaystyle\sum_{n=3}^{\infty} \frac{n + 1}{n^2 + 2n + 2}$

20. Show that $\displaystyle\sum_{n=1}^{\infty} \frac{1}{\ln(2^n)}$ diverges.

21. Show that $\displaystyle\sum_{n=1}^{\infty} \frac{1}{(\ln(2^n))^2}$ converges.

22. (a) Find the partial sum, S_n, of $\displaystyle\sum_{n=1}^{\infty} \ln\left(\frac{n + 1}{n}\right)$.

(b) Does the series in part (a) converge or diverge?

23. (a) Show $r^{\ln n} = n^{\ln r}$ for positive numbers n and r.

(b) For what values $r > 0$ does $\sum_{n=1}^{\infty} r^{\ln n}$ converge?

24. Consider the series $\displaystyle\sum_{k=1}^{\infty} \frac{1}{k(k + 1)} = \frac{1}{1 \cdot 2} + \frac{1}{2 \cdot 3} + \cdots$.

(a) Show that $\dfrac{1}{k} - \dfrac{1}{k + 1} = \dfrac{1}{k(k + 1)}$.

(b) Use part (a) to find the partial sums S_3, S_{10}, and S_n.

(c) Use part (b) to show that the sequence of partial sums S_n, and therefore the series, converges to 1.

25. Consider the series

$$\sum_{k=2}^{\infty} \ln\left(\frac{(k - 1)(k + 1)}{k^2} \right) = \ln\left(\frac{1 \cdot 3}{2 \cdot 2}\right) + \ln\left(\frac{2 \cdot 4}{3 \cdot 3}\right) + \cdots.$$

(a) Show that the partial sum $S_4 = \ln(5/8)$.

(b) Show that the partial sum $S_n = \ln\left(\dfrac{n + 1}{2n}\right)$.

(c) Use part (b) to show that the partial sums S_n, and therefore the series, converge to $\ln(1/2)$.

26. Show that if $\sum a_n$ and $\sum b_n$ converge and if k is a constant, then $\sum(a_n + b_n)$, $\sum(a_n - b_n)$, and $\sum ka_n$ converge.

27. Let N be a positive integer. Show that if $a_n = b_n$ for $n \geq N$, then $\sum a_n$ and $\sum b_n$ either both converge, or both diverge.

28. Show that if $\sum a_n$ converges, then $\lim_{n \to \infty} a_n = 0$. [Hint: Consider $\lim_{n \to \infty} (S_n - S_{n-1})$, where S_n is the nth partial sum.]

29. Show that if $\sum a_n$ diverges and $k \neq 0$, then $\sum ka_n$ diverges.

30. The series $\sum a_n$ converges. Explain, by looking at partial sums, why the series $\sum(a_{n+1} - a_n)$ also converges.

31. The series $\sum a_n$ diverges. Give examples that show the series $\sum(a_{n+1} - a_n)$ could converge or diverge.

32. In this problem, you will justify the integral test. Suppose $c \geq 0$ and $f(x)$ is a decreasing positive function, defined for all $x \geq c$, with $f(n) = a_n$ for all n.

(a) Suppose $\int_c^{\infty} f(x)\, dx$ converges. By considering rectangles under the graph of f, show that $\sum a_n$ converges. [Hint: See Example 3 on page 452.]

(b) Suppose that $\int_c^{\infty} f(x)\, dx$ diverges. By considering rectangles above the graph of f, show that $\sum a_n$ diverges. [Hint: See Example 2 on page 452.]

33. Consider the following grouping of terms in the harmonic series:

$$1 + \left(\frac{1}{2}\right) + \left(\frac{1}{3} + \frac{1}{4}\right) + \left(\frac{1}{5} + \frac{1}{6} + \frac{1}{7} + \frac{1}{8}\right) +$$
$$\left(\frac{1}{9} + \frac{1}{10} + \cdots + \frac{1}{16}\right) + \cdots$$

(a) Show that the sum of each group of fractions is more than $1/2$.

(b) Explain why this shows that the harmonic series does not converge.

34. Show that $\displaystyle\sum_{n=2}^{\infty} \frac{1}{n \ln n}$ diverges.

(a) Using the integral test.

(b) By considering the grouping of terms

$$\left(\frac{1}{2\ln 2}\right) + \left(\frac{1}{3\ln 3} + \frac{1}{4\ln 4}\right)$$
$$+ \left(\frac{1}{5\ln 5} + \frac{1}{6\ln 6} + \frac{1}{7\ln 7} + \frac{1}{8\ln 8}\right) + \cdots.$$

35. Consider the sequence given by

$$a_n = \left(1 + \frac{1}{2} + \frac{1}{3} + \cdots \frac{1}{n}\right) - \ln n.$$

(a) Show that a_n is positive for all n. [Use a left-sum approximation to $\int_1^n (1/x)\,dx$ with $\Delta x = 1$.]

(b) Show that $a_{n+1} < a_n$ for all n. [Use one rectangle to give a right approximation to $\int_n^{n+1} (1/x)\,dx$.]

(c) Explain why $\lim_{n\to\infty} a_n$ exists.

(d) The number $\gamma = \lim_{n\to\infty} a_n$ is called *Euler's constant*. Estimate γ to two decimal places by computing a_{200}.

36. On page 453, we gave Euler's result

$$\sum_{n=1}^{\infty} \frac{1}{n^2} = \frac{\pi^2}{6}.$$

(a) Find the sum of the first 20 terms of this series. Give your answer to three decimal places.

(b) Use your answer to estimate π. Give your answer to two decimal places.

(c) Repeat parts (a) and (b) with 100 terms.

(d) Use a right sum approximation to bound the error in approximating $\pi^2/6$ by $\sum_{n=1}^{20} (1/n^2)$ and by $\sum_{n=1}^{100} (1/n^2)$.

37. This problem approximates e using

$$e = \sum_{n=0}^{\infty} \frac{1}{n!}.$$

(a) Find a lower bound for e by evaluating the first five terms of the series.

(b) Show that $1/n! \le 1/2^{n-1}$ for $n \ge 1$.

(c) Find an upper bound for e using part (b).

38. In this problem we investigate how fast the partial sums $S_N = 1^5 + 2^5 + 3^5 + \cdots + N^5$ of the divergent series $\sum_{n=1}^{\infty} n^5$ grow as N gets larger and larger. Show that

(a) $S_N > N^6/6$ by considering the right-hand Riemann sum for $\int_0^N x^5\,dx$ with $\Delta x = 1$.

(b) $S_N < ((N+1)^6 - 1)/6$ by considering the left-hand Riemann sum for $\int_1^{N+1} x^5\,dx$ with $\Delta x = 1$.

(c) $\lim_{N\to\infty} S_N/(N^6/6) = 1$. We say that S_N is asymptotic to $N^6/6$ as N goes to infinity.

9.4 TESTS FOR CONVERGENCE

Comparison of Series

In Section 7.8, we compared two integrals to decide whether an improper integral converged. In Theorem 9.3 we compared an integral and a series. Now we compare two series.

Theorem 9.4: Comparison Test

Suppose $0 \le a_n \le b_n$ for all n.
- If $\sum b_n$ converges, then $\sum a_n$ converges.
- If $\sum a_n$ diverges, then $\sum b_n$ diverges.

Since $a_n \le b_n$, the plot of the a_n lies under the plot of the b_n. (See Figure 9.7.) The comparison test says that if the total area for $\sum b_n$ is finite, then the total area for $\sum a_n$ is finite also. If the total area for $\sum a_n$ is not finite, then neither is the total area for $\sum b_n$.

Figure 9.7: Each a_n is represented by the area of a dark rectangle, and each b_n by a dark plus a light rectangle

---

Example 1   Use the comparison test to determine whether the series $\displaystyle\sum_{n=1}^{\infty} \frac{1}{n^3 + 1}$ converges.

Solution   For $n \geq 1$, we know that $n^3 \leq n^3 + 1$, so

$$0 \leq \frac{1}{n^3 + 1} \leq \frac{1}{n^3}.$$

Thus, every term in the series $\sum_{n=1}^{\infty} 1/(n^3 + 1)$ is less than or equal to the corresponding term in $\sum_{n=1}^{\infty} 1/n^3$. Since we saw that $\sum_{n=1}^{\infty} 1/n^3$ converges as a p-series with $p > 1$, we know that $\sum_{n=1}^{\infty} 1/(n^3 + 1)$ converges.

Example 2   Decide whether the following series converge: (a) $\displaystyle\sum_{n=1}^{\infty} \frac{n-1}{n^3+3}$ (b) $\displaystyle\sum_{n=1}^{\infty} \frac{6n^2+1}{2n^3-1}$.

Solution   (a) Since the convergence is determined by the behavior of the terms for large n, we observe that

$$\frac{n-1}{n^3+3} \quad \text{behaves like} \quad \frac{n}{n^3} = \frac{1}{n^2} \quad \text{as} \quad n \to \infty.$$

Since $\sum 1/n^2$ converges, we guess that $\sum (n-1)/(n^3+3)$ converges. To confirm this, we use the comparison test. Since a fraction increases if its numerator is made larger or its denominator is made smaller, we have

$$0 \leq \frac{n-1}{n^3+3} \leq \frac{n}{n^3} = \frac{1}{n^2} \quad \text{for all} \quad n \geq 1.$$

Thus, the series $\sum (n-1)/(n^3+3)$ converges by comparison with $\sum 1/n^2$.

(b) First, we observe that

$$\frac{6n^2+1}{2n^3-1} \quad \text{behaves like} \quad \frac{6n^2}{2n^3} = \frac{3}{n} \quad \text{as} \quad n \to \infty.$$

Since $\sum 1/n$ diverges, so does $\sum 3/n$, and we guess that $\sum (6n^2+1)/(2n^3-1)$ diverges. To confirm this, we use the comparison test. Since a fraction decreases if its numerator is made smaller or its denominator is made larger, we have

$$0 \leq \frac{6n^2}{2n^3} \leq \frac{6n^2+1}{2n^3-1},$$

so

$$0 \leq \frac{3}{n} \leq \frac{6n^2+1}{2n^3-1}.$$

Thus, the series $\sum (6n^2+1)/(2n^3-1)$ diverges by comparison with $\sum 3/n$.

Limit Comparison Test

The convergence or divergence of a series $\sum a_n$ is determined by the values of a_n as $n \to \infty$. This often enables us to predict convergence or divergence by looking at the long-run behavior of a_n.

Example 3   Predict convergence or divergence of

$$\sum \frac{n^2 - 5}{n^3 + n + 2}.$$

Solution   As $n \to \infty$, the highest power terms in the numerator and denominator, n^2 and n^3, dominate. Thus the term

$$a_n = \frac{n^2 - 5}{n^3 + n + 2}$$

behaves, as $n \to \infty$, like

$$\frac{n^2}{n^3} = \frac{1}{n}.$$

Since the harmonic series $\sum 1/n$ diverges, we guess that $\sum a_n$ also diverges.

The comparison test is needed to confirm a prediction of convergence or divergence. Because the comparison test requires showing that $a_n \leq b_n$, we may prefer to use the following test, which avoids these inequalities.

Theorem 9.5: Limit Comparison Test

Suppose $a_n > 0$ and $b_n > 0$ for all n. If

$$\lim_{n \to \infty} \frac{a_n}{b_n} = c \qquad \text{where } c > 0,$$

then the two series $\sum a_n$ and $\sum b_n$ either both converge or both diverge.

The limit $\lim_{n \to \infty} a_n/b_n = c$ captures the idea that a_n "behaves like" cb_n as $n \to \infty$.

Example 4   Use the limit comparison test to determine if the following series converge or diverge.

(a) $\quad \sum \dfrac{n^2 + 6}{n^4 - 2n + 3}$ (b) $\quad \sum \sin\left(\dfrac{1}{n}\right)$

Solution   (a) We take $a_n = \dfrac{n^2 + 6}{n^4 - 2n + 3}$. Because a_n behaves like $\dfrac{n^2}{n^4} = \dfrac{1}{n^2}$ as $n \to \infty$ we take $b_n = 1/n^2$. We have

$$\lim_{n \to \infty} \frac{a_n}{b_n} = \lim_{n \to \infty} \frac{n^4 + 6n^2}{n^4 - 2n + 3} = 1.$$

The limit comparison test applies with $c = 1$. Since $\sum 1/n^2$ converges, the limit comparison test shows that $\sum \dfrac{n^2 + 6}{n^4 - 2n + 3}$ also converges.

(b) Since $\sin(x) \approx x$ for x near 0, we know that $\sin\left(\frac{1}{n}\right)$ behaves like $1/n$ as $n \to \infty$. We apply the limit comparison test with $a_n = \sin\left(\frac{1}{n}\right)$ and $b_n = 1/n$. We have

$$\lim_{n \to \infty} \frac{a_n}{b_n} = \lim_{n \to \infty} \frac{\sin(1/n)}{1/n} = 1.$$

Thus $c = 1$ and since $\sum 1/n$ diverges, the series $\sum \sin(1/n)$ also diverges.

Series of Both Positive and Negative Terms

If $\sum a_n$ has both positive and negative terms, then its plot has rectangles lying both above and below the horizontal axis. See Figure 9.8. The total area of the rectangles is no longer equal to $\sum a_n$. However, it is still true that if the total area of the rectangles above and below the axis is finite, then the series converges. The area of the nth rectangle is $|a_n|$, so we have:

Theorem 9.6: Convergence of Absolute Values Implies Convergence

If $\sum |a_n|$ converges, then so does $\sum a_n$.

Problem 67 shows how to prove this result.

Figure 9.8: Representing a series with positive and negative terms

---

Example 5   Explain how we know that the following series converges

$$\sum_{n=1}^{\infty} \frac{(-1)^{n-1}}{n^2} = 1 - \frac{1}{4} + \frac{1}{9} - \cdots.$$

Solution   Writing $a_n = (-1)^{n-1}/n^2$, we have

$$|a_n| = \left| \frac{(-1)^{n-1}}{n^2} \right| = \frac{1}{n^2}.$$

The p-series $\sum 1/n^2$ converges, since $p > 1$, so $\sum (-1)^{n-1}/n^2$ converges.

Comparison with a Geometric Series: The Ratio Test

A geometric series $\sum a_n$ has the property that the ratio a_{n+1}/a_n is constant for all n. For many other series, this ratio, although not constant, tends to a constant as n increases. In some ways, such series behave like geometric series. In particular, a geometric series converges if the ratio $|a_{n+1}/a_n| < 1$. A non-geometric series also converges if the ratio $|a_{n+1}/a_n|$ tends to a limit which is less than 1. This idea leads to the following test.

Theorem 9.7: The Ratio Test

For a series $\sum a_n$, suppose the sequence of ratios $|a_{n+1}|/|a_n|$ has a limit:

$$\lim_{n \to \infty} \frac{|a_{n+1}|}{|a_n|} = L.$$

- If $L < 1$, then $\sum a_n$ converges.
- If $L > 1$, or if L is infinite,[5] then $\sum a_n$ diverges.
- If $L = 1$, the test does not tell us anything about the convergence of $\sum a_n$.

---

[5]That is, the sequence $|a_{n+1}|/|a_n|$ grows without bound.

Proof   Here are the main steps in the proof. Suppose $\lim\limits_{n \to \infty} \dfrac{|a_{n+1}|}{|a_n|} = L < 1$. Let x be a number between L and 1. Then for all sufficiently large n, say for all $n \geq k$, we have

$$\frac{|a_{n+1}|}{|a_n|} < x.$$

Then,

$$|a_{k+1}| < |a_k|x,$$
$$|a_{k+2}| < |a_{k+1}|x < |a_k|x^2,$$
$$|a_{k+3}| < |a_{k+2}|x < |a_k|x^3,$$

and so on. Thus, writing $a = |a_k|$, we have for $i = 1, 2, 3, \ldots$,

$$|a_{k+i}| < ax^i.$$

Now we can use the comparison test: $\sum |a_{k+i}|$ converges by comparison with the geometric series $\sum ax^i$. Since $\sum |a_{k+i}|$ converges, Theorem 9.6 tells us that $\sum a_{k+i}$ converges. So, by property 2 of Theorem 9.2, we see that $\sum a_n$ converges too.

If $L > 1$, then for sufficiently large n, say $n \geq m$,

$$|a_{n+1}| > |a_n|,$$

so the sequence $|a_m|, |a_{m+1}|, |a_{m+2}|, \cdots$, is increasing. Thus, $\lim\limits_{n \to \infty} a_n \neq 0$, so $\sum a_n$ diverges (by Theorem 9.2, property 3). The argument in the case that $|a_{n+1}|/|a_n|$ is unbounded is similar.

Example 6   Show that the following series converges:[6]

$$\sum_{n=1}^{\infty} \frac{1}{n!} = 1 + \frac{1}{2!} + \frac{1}{3!} + \cdots.$$

Solution   Since $a_n = 1/n!$ and $a_{n+1} = 1/(n+1)!$, we have

$$\frac{|a_{n+1}|}{|a_n|} = \frac{1/(n+1)!}{1/n!} = \frac{n!}{(n+1)!} = \frac{n(n-1)(n-2)\cdots 2 \cdot 1}{(n+1)n(n-1)\cdots 2 \cdot 1}.$$

We cancel $n(n-1)(n-2)\cdots \cdots 2 \cdot 1$, giving

$$\lim_{n \to \infty} \frac{|a_{n+1}|}{|a_n|} = \lim_{n \to \infty} \frac{n!}{(n+1)!} = \lim_{n \to \infty} \frac{1}{n+1} = 0.$$

Because the limit is 0, which is less than 1, the ratio test tells us that $\sum\limits_{n=1}^{\infty} 1/n!$ converges. In Chapter 10, we show the limit is e.

Example 7   What does the ratio test tell us about the convergence of the following two series?

$$\sum_{n=1}^{\infty} \frac{1}{n} \quad \text{and} \quad \sum_{n=1}^{\infty} \frac{(-1)^{n-1}}{n}.$$

Solution   Because $|(-1)^n| = 1$, in both cases we have $\lim_{n \to \infty} |a_{n+1}/a_n| = \lim_{n \to \infty} n/(n+1) = 1$. Thus, the ratio test does not tell us anything about the convergence of either series. In fact, the first series is the harmonic series, which diverges. Example 8 will show that the second series converges.

---

[6]We define 2! to be $2 \cdot 1 = 2$. Similarly, $3! = 3 \cdot 2 \cdot 1 = 6$ and $n! = n \cdot (n-1) \cdots 2 \cdot 1$. We also define $0! = 1$.

Alternating Series

A series is called an *alternating series* if the terms alternate in sign. For example,

$$\sum_{n=1}^{\infty} \frac{(-1)^{n-1}}{n} = 1 - \frac{1}{2} + \frac{1}{3} - \frac{1}{4} + \cdots + \frac{(-1)^{n-1}}{n} + \cdots .$$

The convergence of an alternating series can often be determined using the following test:

Theorem 9.8: Alternating Series Test

A series of the form

$$\sum_{n=1}^{\infty} (-1)^{n-1} a_n = a_1 - a_2 + a_3 - a_4 + \cdots + (-1)^{n-1} a_n + \cdots$$

converges if

$$0 < a_{n+1} < a_n \quad \text{for all } n \qquad \text{and} \qquad \lim_{n \to \infty} a_n = 0.$$

Although we do not prove this result, we can see why it is reasonable. The first partial sum, $S_1 = a_1$, is positive. The second, $S_2 = a_1 - a_2$, is still positive, since $a_2 < a_1$, but S_2 is smaller than S_1. (See Figure 9.9.) The next sum, $S_3 = a_1 - a_2 + a_3$, is greater than S_2 but smaller than S_1. The partial sums oscillate back and forth, and since the distance between them tends to 0, they eventually converge.

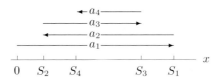

Figure 9.9: Partial sums, S_1, S_2, S_3, S_4 of an alternating series

---

Example 8

Show that the following alternating harmonic series converges:

$$\sum_{n=1}^{\infty} \frac{(-1)^{n-1}}{n} .$$

Solution

We have $a_n = 1/n$ and $a_{n+1} = 1/(n+1)$. Thus,

$$a_{n+1} = \frac{1}{n+1} < \frac{1}{n} = a_n \quad \text{for all } n, \quad \text{and} \quad \lim_{n \to \infty} 1/n = 0.$$

Thus, the hypothesis of Theorem 9.8 is satisfied, so the alternating harmonic series converges.

Suppose S is the sum of an alternating series, so $S = \lim_{n \to \infty} S_n$. Then S is trapped between any two consecutive partial sums, say S_3 and S_4 or S_4 and S_5 so

$$S_2 < S_4 < \cdots < S < \cdots < S_3 < S_1.$$

Thus, the error in using S_n to approximate the true sum S is less than the distance from S_n to S_{n+1}, which is a_{n+1}. Stated symbolically, we have the following result:

Theorem 9.9: Error Bounds for Alternating Series

Let $S_n = \sum_{i=1}^{n} (-1)^{i-1} a_i$ be the nth partial sum of an alternating series and let $S = \lim_{n \to \infty} S_n$. Suppose that $0 < a_{n+1} < a_n$ for all n and $\lim_{n \to \infty} a_n = 0$. Then

$$|S - S_n| < a_{n+1}.$$

Thus, provided S_n converges to S by the alternating series test, the error in using S_n to approximate S is less than the magnitude of the first term of the series which is omitted in the approximation.

Example 9 Estimate the error in approximating the sum of the alternating harmonic series $\sum_{n=1}^{\infty} (-1)^{n-1}/n$ by the sum of the first nine terms.

Solution The ninth partial sum is given by

$$S_9 = 1 - \frac{1}{2} + \frac{1}{3} - \cdots + \frac{1}{9} = 0.7456\ldots.$$

The first term omitted is $-1/10$, with magnitude 0.1. By Theorem 9.9, we know that the true value of the sum differs from $0.7456\ldots$ by less than 0.1.

Absolute and Conditional Convergence

We say that the series $\sum a_n$ is
- **absolutely convergent** if $\sum a_n$ and $\sum |a_n|$ both converge.
- **conditionally convergent** if $\sum a_n$ converges but $\sum |a_n|$ diverges.

Conditionally convergent series rely on cancelation between positive and negative terms for their convergence.

Example: The series $\sum_{n=1}^{\infty} \frac{(-1)^{n-1}}{n^2}$ is absolutely convergent because the series converges and the p-series $\sum 1/n^2$ also converges.

Example: The series $\sum_{n=1}^{\infty} \frac{(-1)^{n-1}}{n}$ is conditionally convergent because the series converges but the harmonic series $\sum 1/n$ diverges.

Exercises and Problems for Section 9.4

Exercises

Use the comparison test to confirm the statements in Exercises 1–3.

1. $\sum_{n=4}^{\infty} \frac{1}{n}$ diverges, so $\sum_{n=4}^{\infty} \frac{1}{n-3}$ diverges.

2. $\sum_{n=1}^{\infty} \frac{1}{n^2}$ converges, so $\sum_{n=1}^{\infty} \frac{1}{n^2+2}$ converges.

3. $\sum_{n=1}^{\infty} \frac{1}{n^2}$ converges, so $\sum_{n=1}^{\infty} \frac{e^{-n}}{n^2}$ converges.

Use the comparison test to determine whether the series in Exercises 4–9 converge.

4. $\displaystyle\sum_{n=1}^{\infty} \frac{1}{3^n + 1}$

5. $\displaystyle\sum_{n=1}^{\infty} \frac{1}{n^4 + e^n}$

6. $\displaystyle\sum_{n=2}^{\infty} \frac{1}{\ln n}$

7. $\displaystyle\sum_{n=1}^{\infty} \frac{n^2}{n^4 + 1}$

8. $\displaystyle\sum_{n=1}^{\infty} \frac{n \sin n}{n^3 + 1}$.

9. $\displaystyle\sum_{n=1}^{\infty} \frac{2^n + 1}{n2^n - 1}$

Use the ratio test to decide if the series in Exercises 10–15 converge or diverge.

10. $\displaystyle\sum_{n=1}^{\infty} \frac{1}{(2n)!}$

11. $\displaystyle\sum_{n=1}^{\infty} \frac{(n!)^2}{(2n)!}$

12. $\displaystyle\sum_{n=1}^{\infty} \frac{(2n)!}{n!(n+1)!}$

13. $\displaystyle\sum_{n=1}^{\infty} \frac{1}{r^n n!}, r > 0$

14. $\displaystyle\sum_{n=1}^{\infty} \frac{1}{ne^n}$

15. $\displaystyle\sum_{n=0}^{\infty} \frac{2^n}{n^3 + 1}$

Which of the the series in Exercises 16–19 are alternating?

16. $\displaystyle\sum_{n=1}^{\infty} (-1)^n \left(2 - \frac{1}{n}\right)$

17. $\displaystyle\sum_{n=1}^{\infty} \cos(n\pi)$

18. $\displaystyle\sum_{n=1}^{\infty} (-1)^n \cos(n\pi)$

19. $\displaystyle\sum_{n=1}^{\infty} (-1)^n \cos n$

Use the alternating series test to show that the series in Exercises 20–23 converge.

20. $\displaystyle\sum_{n=1}^{\infty} \frac{(-1)^{n-1}}{\sqrt{n}}$

21. $\displaystyle\sum_{n=1}^{\infty} \frac{(-1)^{n-1}}{2n+1}$

22. $\displaystyle\sum_{n=1}^{\infty} \frac{(-1)^{n-1}}{n^2 + 2n + 1}$

23. $\displaystyle\sum_{n=1}^{\infty} \frac{(-1)^{n-1}}{e^n}$

In Exercises 24–31, use the limit comparison test to determine whether the series converges or diverges.

24. $\displaystyle\sum_{n=1}^{\infty} \frac{5n+1}{3n^2}$, by comparing to $\displaystyle\sum_{n=1}^{\infty} \frac{1}{n}$

25. $\displaystyle\sum_{n=1}^{\infty} \left(\frac{1+n}{3n}\right)^n$, by comparing to $\displaystyle\sum_{n=1}^{\infty} \left(\frac{1}{3}\right)^n$

[Hint: $\lim_{n\to\infty}(1 + 1/n)^n = e$.]

26. $\displaystyle\sum \frac{1}{n^4 - 7}$

27. $\displaystyle\sum \frac{n+1}{n^2 + 2}$

28. $\displaystyle\sum \frac{n^3 - 2n^2 + n + 1}{n^4 - 2}$

29. $\displaystyle\sum \frac{2^n}{3^n - 1}$

30. $\displaystyle\sum \frac{1}{2\sqrt{n} + \sqrt{n+2}}$

31. $\displaystyle\sum \left(\frac{1}{2n-1} - \frac{1}{2n}\right)$

Problems

Explain why the comparison test cannot be used to decide if the series in Problems 32–33 converge or diverge.

32. $\displaystyle\sum_{n=1}^{\infty} \frac{(-1)^n}{\cdot n^2}$

33. $\displaystyle\sum_{n=1}^{\infty} \sin n$

Explain why the ratio test cannot be used to decide if the series in Problems 34–35 converge or diverge.

34. $\displaystyle\sum_{n=1}^{\infty} (-1)^n$

35. $\displaystyle\sum_{n=1}^{\infty} \sin n$

Explain why the alternating series test cannot be used to decide if the series in Problems 36–38 converge or diverge.

36. $\displaystyle\sum_{n=1}^{\infty} (-1)^{n-1} n$

37. $\displaystyle\sum_{n=1}^{\infty} (-1)^{n-1} \sin n$

38. $\displaystyle\sum_{n=1}^{\infty} (-1)^{n-1} \left(2 - \frac{1}{n}\right)$

Use a computer or calculator to investigate the behavior of the partial sums of the alternating series in Problems 39–41. Which ones converge? Confirm convergence using the alternating series test. If a series converges, estimate its sum.

39. $1 - 2 + 3 - 4 + 5 + \cdots + (-1)^n(n+1) + \cdots$

40. $1 - 0.1 + 0.01 - 0.001 + \cdots + (-1)^n 10^{-n} + \cdots$

41. $1 - \dfrac{1}{1!} + \dfrac{1}{2!} - \dfrac{1}{3!} + \cdots + (-1)^n \dfrac{1}{n!} + \cdots$

Determine which of the series in Problems 42–53 converge.

42. $\displaystyle\sum_{n=0}^{\infty} \frac{(0.1)^n}{n!}$

43. $\displaystyle\sum_{n=1}^{\infty} \frac{(n-1)!}{n^2}$

44. $\displaystyle\sum_{n=0}^{\infty} e^{-n}$

45. $\displaystyle\sum_{n=1}^{\infty} e^n$

46. $\displaystyle\sum_{n=1}^{\infty} \frac{(-1)^{n-1}}{\sqrt{3n-1}}$

47. $\displaystyle\sum_{n=1}^{\infty} \frac{(-1)^{n-1} 2^n}{n^2}$

48. $\displaystyle\sum_{n=1}^{\infty} \frac{\sin n}{n^2}$

49. $\displaystyle\sum_{n=1}^{\infty} \frac{\cos(n\pi)}{n}$

50. $\displaystyle\sum_{n=2}^{\infty} \frac{n+2}{n^2-1}$

51. $\displaystyle\sum_{n=2}^{\infty} \frac{3}{\ln n^2}$

52. $\displaystyle\sum_{n=1}^{\infty} \frac{1}{\sqrt{n^2(n+2)}}$

53. $\displaystyle\sum_{n=1}^{\infty} \frac{n(n+1)}{\sqrt{n^3+2n^2}}$

The series in Problems 54–56 converge by the alternating series test. How many terms give a partial sum, S_n, within 0.01 of the sum, S, of the series?

54. $\displaystyle\sum_{n=1}^{\infty} \frac{(-1)^{n-1}}{n}$

55. $\displaystyle\sum_{n=1}^{\infty} \left(-\frac{2}{3}\right)^{n-1}$

56. $\displaystyle\sum_{n=1}^{\infty} \frac{(-1)^{n-1}}{(2n)!}$

In Problems 57–60 determine whether the series are absolutely convergent, conditionally convergent, or divergent.

57. $\displaystyle\sum \frac{(-1)^n}{2^n}$

58. $\displaystyle\sum \frac{(-1)^n}{2n}$

59. $\displaystyle\sum (-1)^n \left(1 + \frac{1}{n^2}\right)$

60. $\displaystyle\sum \frac{(-1)^n}{n^4+7}$

61. Suppose $0 \leq b_n \leq 2^n \leq a_n$ and $0 \leq c_n \leq 2^{-n} \leq d_n$ for all n. Which of the series $\sum a_n$, $\sum b_n$, $\sum c_n$, and $\sum d_n$ definitely converge, which definitely diverge?

62. Given two convergent series $\sum a_n$ and $\sum b_n$, we know that the term-by-term sum $\sum(a_b + b_n)$ converges. What about the series formed by taking the products of the terms $\sum a_n \cdot b_n$? This problem shows that it may or may not converge.

(a) Show that if $\sum a_n = \sum 1/n^2$ and $\sum b_n = \sum 1/n^3$, then $\sum a_n \cdot b_n$ converges.

(b) Explain why $\sum(-1)^n/\sqrt{n}$ converges.

(c) Show that if $a_n = b_n = (-1)^n/\sqrt{n}$, then $\sum a_n \cdot b_n$ does not converge.

63. Suppose that $b_n > 0$ and $\sum b_n$ converges. Show that if $\displaystyle\lim_{n\to\infty} a_n/b_n = 0$ then $\sum a_n$ converges.

64. Suppose that $b_n > 0$ and $\sum b_n$ diverges. Show that if $\displaystyle\lim_{n\to\infty} a_n/b_n = \infty$ then $\sum a_n$ diverges.

65. A series $\sum a_n$ of positive terms (that is, $a_n > 0$) can be used to form another series $\sum b_n$ where each term b_n is the average of the first n terms of the original series, that is, $b_n = (a_1 + a_2 + \cdots + a_n)/n$. Show that $\sum b_n$ does not converge (even if $\sum a_n$ does). [Hint: Compare $\sum b_n$ to a multiple of the harmonic series.]

66. Show that if $\sum |a_n|$ converges, then $\sum(-1)^n a_n$ converges.

67. (a) For a series $\sum a_n$, show that $0 \leq a_n + |a_n| \leq 2|a_n|$.

(b) Use part (a) to show that if $\sum |a_n|$ converges, then $\sum a_n$ converges.

Problems 68–69 introduce the *root test* for convergence. Given a series $\sum a_n$ of positive terms (that is, $a_n > 0$) such that the root $\sqrt[n]{a_n}$ has a limit r as $n \to \infty$,

- if $r < 1$, then $\sum a_n$ converges
- if $r > 1$, then $\sum a_n$ diverges
- if $r = 1$, then $\sum a_n$ could converge or diverge.

(This test works since $\lim_{n\to\infty} \sqrt[n]{a_n} = r$ tells us that the series is comparable to a geometric series with ratio r.) Use this test to determine the behavior of the series.

68. $\displaystyle\sum_{n=1}^{\infty} \left(\frac{2}{n}\right)^n$

69. $\displaystyle\sum_{n=1}^{\infty} \left(\frac{5n+1}{3n^2}\right)^n$

9.5 POWER SERIES AND INTERVAL OF CONVERGENCE

In Section 9.2 we saw that the geometric series $\sum ax^n$ converges for $-1 < x < 1$ and diverges otherwise. This section studies the convergence of more general series constructed from powers. Chapter 10 shows how such power series are used to approximate functions such as e^x, $\sin x$, $\cos x$, and $\ln x$.

> A **power series** about $x = a$ is a sum of constants times powers of $(x-a)$:
>
> $$C_0 + C_1(x-a) + C_2(x-a)^2 + \cdots + C_n(x-a)^n + \cdots = \sum_{n=0}^{\infty} C_n(x-a)^n.$$

We think of a as a constant. For any fixed x, the power series $\sum C_n(x-a)^n$ is a series of numbers like those considered in Section 9.3. To investigate the convergence of a power series, we consider the partial sums, which in this case are the polynomials $S_n(x) = C_0 + C_1(x-a) + C_2(x-a)^2 + \cdots + C_n(x-a)^n$. As before, we consider the sequence[7]

$$S_0(x),\ S_1(x),\ S_2(x),\ \ldots, S_n(x), \ldots.$$

> For a fixed value of x, if this sequence of partial sums converges to a limit L, that is, if $\lim_{n \to \infty} S_n(x) = L$, then we say that the power series **converges** to L for this value of x.

A power series may converge for some values of x and not for others.

Example 1   Determine whether the power series $\displaystyle\sum_{n=0}^{\infty} \frac{x^n}{2^n}$ converges or diverges for

(a) $x = -1$ (b) $x = 3$

Solution   (a) Substituting $x = -1$, we have

$$\sum_{n=0}^{\infty} \frac{x^n}{2^n} = \sum_{n=0}^{\infty} \frac{(-1)^n}{2^n} = \sum_{n=0}^{\infty} \left(-\frac{1}{2}\right)^n.$$

This is a geometric series with ratio $-1/2$, so the series converges to $1/(1-(-\frac{1}{2})) = 2/3$.

(b) Substituting $x = 3$, we have

$$\sum_{n=0}^{\infty} \frac{x^n}{2^n} = \sum_{n=0}^{\infty} \frac{3^n}{2^n} = \sum_{n=0}^{\infty} \left(\frac{3}{2}\right)^n.$$

This is a geometric series with ratio greater than 1, so it diverges.

Example 2   Find an expression for the general term of the series and use it to write the series using $\sum$ notation:

$$\frac{(x-2)^4}{4} - \frac{(x-2)^6}{9} + \frac{(x-2)^8}{16} - \frac{(x-2)^{10}}{25} + \cdots.$$

Solution   The series is about $x = 2$ and the odd terms are zero. We use $(x-2)^{2n}$ and begin with $n = 2$. Since the series alternates and is positive for $n = 2$, we multiply by $(-1)^n$. For $n = 2$, we divide by 4, for $n = 3$ we divide by 9, and in general, we divide by n^2. One way to write this series is

$$\sum_{n=2}^{\infty} \frac{(-1)^n (x-2)^{2n}}{n^2}.$$

Numerical and Graphical View of Convergence

Consider the series

$$(x-1) - \frac{(x-1)^2}{2} + \frac{(x-1)^3}{3} - \frac{(x-1)^4}{4} + \cdots + (-1)^{n-1}\frac{(x-1)^n}{n} + \cdots.$$

To investigate the convergence of this series, we look at the sequence of partial sums graphed in Figure 9.10. The graph suggests that the partial sums converge for x in the interval $(0, 2)$. In Examples 4 and 6, we show that the series converges for $0 < x \leq 2$. This is called the *interval of convergence* of this series.

---

[7]Here we have chosen to call the first term in the sequence $S_0(x)$ rather than $S_1(x)$ to correspond to the power of $(x-a)$.

At $x = 1.4$, which is inside the interval, the series appears to converge quite rapidly:

$$S_5(1.4) = 0.33698\ldots \qquad S_7(1.4) = 0.33653\ldots$$
$$S_6(1.4) = 0.33630\ldots \qquad S_8(1.4) = 0.33645\ldots$$

Table 9.1 shows the results of using $x = 1.9$ and $x = 2.3$ in the power series. For $x = 1.9$, which is inside the interval of convergence but close to an endpoint, the series converges, though rather slowly. For $x = 2.3$, which is outside the interval of convergence, the series diverges: the larger the value of n, the more wildly the series oscillates. In fact, the contribution of the twenty-fifth term is about 28; the contribution of the hundredth term is about $-2,500,000,000$. Figure 9.10 shows the interval of convergence and the partial sums.

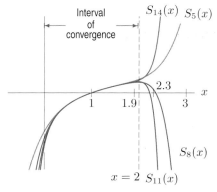

Figure 9.10: Partial sums for series in Example 4 converge for $0 < x < 2$

Table 9.1 *Partial sums for series in Example 4 with $x = 1.9$ inside interval of convergence and $x = 2.3$ outside*

n	$S_n(1.9)$	n	$S_n(2.3)$
2	0.495	2	0.455
5	0.69207	5	1.21589
8	0.61802	8	0.28817
11	0.65473	11	1.71710
14	0.63440	14	-0.70701

Notice that the interval of convergence, $0 < x \leq 2$, is centered on $x = 1$. Since the interval extends one unit on either side, we say the *radius of convergence* of this series is 1.

Intervals of Convergence

Each power series falls into one of the three following cases, characterized by its *radius of convergence*, R.

- The series converges only for $x = a$; the **radius of convergence** is defined to be $R = 0$.
- The series converges for all values of x; the **radius of convergence** is defined to be $R = \infty$.
- There is a positive number R, called the **radius of convergence**, such that the series converges for $|x - a| < R$ and diverges for $|x - a| > R$. See Figure 9.11.
- The **interval of convergence** is the interval between $a - R$ and $a + R$, including any endpoint where the series converges.

Figure 9.11: Radius of convergence, R, determines an interval, centered at $x = a$, in which the series converges

There are some series whose radius of convergence we already know. For example, the geometric series

$$1 + x + x^2 + \cdots + x^n + \cdots$$

converges for $|x| < 1$ and diverges for $|x| \geq 1$, so its radius of convergence is 1. Similarly, the series

$$1 + \frac{x}{3} + \left(\frac{x}{3}\right)^2 + \cdots + \left(\frac{x}{3}\right)^n + \cdots$$

converges for $|x/3| < 1$ and diverges for $|x/3| \geq 1$, so its radius of convergence is 3.

The next theorem gives a method of computing the radius of convergence for many series. To find the values of x for which the power series $\sum_{n=0}^{\infty} C_n(x-a)^n$ converges, we use the ratio test. Writing $a_n = C_n(x-a)^n$ and assuming $C_n \neq 0$ and $x \neq a$, we have

$$\lim_{n \to \infty} \frac{|a_{n+1}|}{|a_n|} = \lim_{n \to \infty} \frac{|C_{n+1}(x-a)^{n+1}|}{|C_n(x-a)^n|} = \lim_{n \to \infty} \frac{|C_{n+1}||x-a|}{|C_n|} = |x-a| \lim_{n \to \infty} \frac{|C_{n+1}|}{|C_n|}.$$

Case 1. Suppose $\lim_{n \to \infty} |a_{n+1}|/|a_n|$ is infinite. Then the ratio test shows that the power series converges only for $x = a$. The radius of convergence is $R = 0$.

Case 2. Suppose $\lim_{n \to \infty} |a_{n+1}|/|a_n| = 0$. Then the ratio test shows that the power series converges for all x. The radius of convergence is $R = \infty$.

Case 3. Suppose $\lim_{n \to \infty} |a_{n+1}|/|a_n| = K|x-a|$, where $\lim_{n \to \infty} |C_{n+1}|/|C_n| = K$. In Case 1, K does not exist; in Case 2, $K = 0$. Thus, we can assume K exists and $K \neq 0$, and we can define $R = 1/K$. Then we have

$$\lim_{n \to \infty} \frac{|a_{n+1}|}{|a_n|} = K|x-a| = \frac{|x-a|}{R},$$

so the ratio test tells us that the power series:

- Converges for $\dfrac{|x-a|}{R} < 1$; that is, for $|x-a| < R$

- Diverges for $\dfrac{|x-a|}{R} > 1$; that is, for $|x-a| > R$.

The results are summarized in the following theorem.

Theorem 9.10: Method for Computing Radius of Convergence

To calculate the radius of convergence, R, for the power series $\sum_{n=0}^{\infty} C_n(x-a)^n$, use the ratio test with $a_n = C_n(x-a)^n$.

- If $\lim_{n \to \infty} |a_{n+1}|/|a_n|$ is infinite, then $R = 0$.
- If $\lim_{n \to \infty} |a_{n+1}|/|a_n| = 0$, then $R = \infty$.
- If $\lim_{n \to \infty} |a_{n+1}|/|a_n| = K|x-a|$, where K is finite and nonzero, then $R = 1/K$.

Note that the ratio test does not tell us anything if $\lim_{n \to \infty} |a_{n+1}|/|a_n|$ fails to exist, which can occur, for example, if some of the C_ns are zero.

A proof that a power series has a radius of convergence and of Theorem 9.10 is given in the online theory supplement. To understand these facts informally, we can think of a power series as being like a geometric series whose coefficients vary from term to term. The radius of convergence depends on the behavior of the coefficients: if there are constants C and K such that for larger and larger n,

$$|C_n| \approx CK^n,$$

then it is plausible that $\sum C_n x^n$ and $\sum CK^n x^n = \sum C(Kx)^n$ converge or diverge together. The geometric series $\sum C(Kx)^n$ converges for $|Kx| < 1$, that is, for $|x| < 1/K$. We can find K using the ratio test, because

$$\frac{|a_{n+1}|}{|a_n|} = \frac{|C_{n+1}||(x-a)^{n+1}|}{|C_n||(x-a)^n|} \approx \frac{CK^{n+1}|(x-a)^{n+1}|}{CK^n|(x-a)^n|} = K|x-a|.$$

Example 3   Show that the following power series converges for all x:

$$1 + x + \frac{x^2}{2!} + \frac{x^3}{3!} + \cdots + \frac{x^n}{n!} + \cdots.$$

Solution   Because $C_n = 1/n!$, none of the C_ns are zero and we can use the ratio test:

$$\lim_{n\to\infty} \frac{|a_{n+1}|}{|a_n|} = |x| \lim_{n\to\infty} \frac{|C_{n+1}|}{|C_n|} = |x| \lim_{n\to\infty} \frac{1/(n+1)!}{1/n!} = |x| \lim_{n\to\infty} \frac{n!}{(n+1)!} = |x| \lim_{n\to\infty} \frac{1}{n+1} = 0.$$

This gives $R = \infty$, so the series converges for all x. We see in Chapter 10 that it converges to e^x.

Example 4   Determine the radius of convergence of the series

$$(x-1) - \frac{(x-1)^2}{2} + \frac{(x-1)^3}{3} - \frac{(x-1)^4}{4} + \cdots + (-1)^{n-1}\frac{(x-1)^n}{n} + \cdots.$$

What does this tell us about the interval of convergence of this series?

Solution   The general term of the series is $(x-1)^n/n$ if n is odd and $-(x-1)^n/n$ if n is even, so $C_n = (-1)^{n-1}/n$, and we can use the ratio test. We have

$$\lim_{n\to\infty} \frac{|a_{n+1}|}{|a_n|} = |x-1| \lim_{n\to\infty} \frac{|C_{n+1}|}{|C_n|} = |x-1| \lim_{n\to\infty} \frac{|(-1)^n/(n+1)|}{|(-1)^{n-1}/n|} = |x-1| \lim_{n\to\infty} \frac{n}{n+1} = |x-1|.$$

Thus, $K = 1$ in Theorem 9.10, so the radius of convergence is $R = 1$. The power series converges for $|x-1| < 1$ and diverges for $|x-1| > 1$, so the series converges for $0 < x < 2$. Notice that the radius of convergence does not tell us what happens at the endpoints, $x = 0$ and $x = 2$. We see in Chapter 10 that the series converges to $\ln x$ for $0 < x \le 2$.

The ratio test requires $\lim_{n\to\infty} |a_{n+1}|/|a_n|$ to exist for $a_n = C_n(x-a)^n$. What happens if some of the coefficients C_n are zero? Then we use the fact that an infinite series can be written in several ways and pick one in which the terms are nonzero. For example, we think of the power series

$$x - \frac{x^3}{3!} + \frac{x^5}{5!} - \frac{x^7}{7!} + \cdots + (-1)^{n-1}\frac{x^{2n-1}}{(2n-1)!} + \cdots$$

as the series with $a_1 = x$ and $a_2 = -x^3/3!, \ldots$, so $a_n = (-1)^{n-1}x^{2n-1}/(2n-1)!$. With this choice of a_n, all $a_n \ne 0$, so we can use the ratio test.[8]

[8]We do not take $a_1 = x, a_2 = 0, a_3 = -x^3/3!, a_4 = 0, \ldots$ because then $\lim_{n\to\infty} |a_{n+1}|/|a_n|$ does not exist.

Example 5   Find the radius and interval of convergence of the series

$$x - \frac{x^3}{3!} + \frac{x^5}{5!} - \frac{x^7}{7!} + \cdots + (-1)^{n-1} \frac{x^{2n-1}}{(2n-1)!} + \cdots$$

Solution   We take

$$a_n = (-1)^{n-1} \frac{x^{2n-1}}{(2n-1)!},$$

so that, replacing n by $n + 1$, we have

$$a_{n+1} = (-1)^{(n+1)-1} \frac{x^{2(n+1)-1}}{(2(n+1)-1)!} = (-1)^n \frac{x^{2n+1}}{(2n+1)!}.$$

Thus,

$$\frac{|a_{n+1}|}{|a_n|} = \frac{\left|(-1)^n \frac{x^{2n+1}}{(2n+1)!}\right|}{\left|(-1)^{n-1} \frac{x^{2n-1}}{(2n-1)!}\right|} = \left|\frac{(-1)^n x^{2n+1}(2n-1)!}{(-1)^{n-1} x^{2n-1}(2n+1)!}\right| = \left|\frac{(-1)x^2}{(2n+1)2n}\right| = \frac{x^2}{(2n+1)2n}.$$

Because

$$\lim_{n \to \infty} \frac{|a_{n+1}|}{|a_n|} = \lim_{n \to \infty} \frac{x^2}{(2n+1)2n} = 0,$$

we have $\lim_{n \to \infty} |a_{n+1}|/|a_n| = 0 < 1$ for all x. Thus, the ratio test guarantees that the power series converges for every x. The radius of convergence is infinite and the interval of convergence is all x. We see in Chapter 10 that the series converges to $\sin x$.

What Happens at the Endpoints of the Interval of Convergence?

The ratio test does not tell us whether a power series converges at the endpoints of its interval of convergence, $x = a \pm R$. There is no simple theorem that answers this question. Since substituting $x = a \pm R$ converts the power series to a series of numbers, the tests in Sections 9.3 and 9.4 are often useful. See Examples 6 and 7.

Example 6   Determine the interval of convergence of the series

$$(x-1) - \frac{(x-1)^2}{2} + \frac{(x-1)^3}{3} - \frac{(x-1)^4}{4} + \cdots + (-1)^{n-1} \frac{(x-1)^n}{n} + \cdots.$$

Solution   In Example 4 on page 467, we showed that this series has $R = 1$; it converges for $0 < x < 2$ and diverges for $x < 0$ or $x > 2$. We need to determine whether it converges at the endpoints of the interval of convergence, $x = 0$ and $x = 2$. At $x = 2$, we have the series

$$1 - \frac{1}{2} + \frac{1}{3} - \frac{1}{4} + \cdots + \frac{(-1)^{n-1}}{n} + \cdots.$$

This is an alternating series with $a_n = 1/(n+1)$, so by the alternating series test (Theorem 9.8), it converges. At $x = 0$, we have the series

$$-1 - \frac{1}{2} - \frac{1}{3} - \frac{1}{4} - \cdots - \frac{1}{n} - \cdots.$$

This is the negative of the harmonic series, so it diverges. Therefore, the interval of convergence is $0 < x \leq 2$. The right endpoint is included and the left endpoint is not.

Example 7   Find the radius and interval of convergence of the series

$$1 + 2^2 x^2 + 2^4 x^4 + 2^6 x^6 + \cdots + 2^{2n} x^{2n} + \cdots.$$

Solution   If we take $a_n = 2^n x^n$ for n even and $a_n = 0$ for n odd, $\lim_{n \to \infty} |a_{n+1}|/|a_n|$ does not exist. Therefore, for this series we take

$$a_n = 2^{2n} x^{2n},$$

so that, replacing n by $n + 1$, we have

$$a_{n+1} = 2^{2(n+1)} x^{2(n+1)} = 2^{2n+2} x^{2n+2}.$$

Thus,

$$\frac{|a_{n+1}|}{|a_n|} = \frac{|2^{2n+2} x^{2n+2}|}{|2^{2n} x^{2n}|} = |2^2 x^2| = 4x^2.$$

We have

$$\lim_{n \to \infty} \frac{|a_{n+1}|}{|a_n|} = 4x^2.$$

The ratio test guarantees that the power series converges if $4x^2 < 1$, that is, if $|x| < \frac{1}{2}$. The radius of convergence is $\frac{1}{2}$. The series converges for $-\frac{1}{2} < x < \frac{1}{2}$ and diverges for $x < \frac{1}{2}$ or $x > -\frac{1}{2}$. At $x = \pm \frac{1}{2}$, all the terms in the series are 1, so the series diverges (by Theorem 9.2 Property 3). Thus, the interval of convergence is $-\frac{1}{2} < x < \frac{1}{2}$.

Exercises and Problems for Section 9.5

Exercises

Which of the series in Exercises 1–4 are power series?

1. $x - x^3 + x^6 - x^{10} + x^{15} - \cdots$

2. $\dfrac{1}{x} + \dfrac{1}{x^2} + \dfrac{1}{x^3} + \dfrac{1}{x^4} + \cdots$

3. $1 + x + (x - 1)^2 + (x - 2)^3 + (x - 3)^4 + \cdots$

4. $x^7 + x + 2$

Find an expression for the general term of the series in Exercises 5–10. Give the starting value of the index (n or k for example).

5. $\dfrac{1}{2}x + \dfrac{1 \cdot 3}{2^2 \cdot 2!}x^2 + \dfrac{1 \cdot 3 \cdot 5}{2^3 \cdot 3!}x^3 + \cdots$

6. $px + \dfrac{p(p-1)}{2!}x^2 + \dfrac{p(p-1)(p-2)}{3!}x^3 + \cdots$

7. $1 - \dfrac{(x-1)^2}{2!} + \dfrac{(x-1)^4}{4!} - \dfrac{(x-1)^6}{6!} + \cdots$

8. $(x-1)^3 - \dfrac{(x-1)^5}{2!} + \dfrac{(x-1)^7}{4!} - \dfrac{(x-1)^9}{6!} + \cdots$

9. $\dfrac{x-a}{1} + \dfrac{(x-a)^2}{2 \cdot 2!} + \dfrac{(x-a)^3}{4 \cdot 3!} + \dfrac{(x-a)^4}{8 \cdot 4!} + \cdots$

10. $2(x+5)^3 + 3(x+5)^5 + \dfrac{4(x+5)^7}{2!} + \dfrac{5(x+5)^9}{3!} + \cdots$

Use the ratio test to find the radius of convergence of the power series in Exercises 11–21.

11. $\sum_{n=0}^{\infty} (5x)^n$

12. $\sum_{n=0}^{\infty} n^3 x^n$

13. $\sum_{n=0}^{\infty} \dfrac{(n+1)x^n}{2^n + n}$

14. $\sum_{n=1}^{\infty} \dfrac{2^n (x-1)^n}{n}$

15. $x + 4x^2 + 9x^3 + 16x^4 + 25x^5 + \cdots$

16. $x - \dfrac{x^2}{4} + \dfrac{x^3}{9} - \dfrac{x^4}{16} + \dfrac{x^5}{25} - \cdots$

17. $1 + 2x + \dfrac{4x^2}{2!} + \dfrac{8x^3}{3!} + \dfrac{16x^4}{4!} + \dfrac{32x^5}{5!} + \cdots$

18. $\dfrac{x}{3} + \dfrac{2x^2}{5} + \dfrac{3x^3}{7} + \dfrac{4x^4}{9} + \dfrac{5x^5}{11} + \cdots$

19. $1 + 2x + \dfrac{4!x^2}{(2!)^2} + \dfrac{6!x^3}{(3!)^2} + \dfrac{8!x^4}{(4!)^2} + \dfrac{10!x^5}{(5!)^2} + \cdots$

20. $3x + \dfrac{5}{2}x^2 + \dfrac{7}{3}x^3 + \dfrac{9}{4}x^4 + \dfrac{11}{5}x^5 + \cdots$

21. $x - \dfrac{x^3}{3} + \dfrac{x^5}{5} - \dfrac{x^7}{7} + \cdots$

22. (a) Determine the radius of convergence of the series

$$x - \frac{x^2}{2} + \frac{x^3}{3} - \frac{x^4}{4} + \cdots + (-1)^{n-1}\frac{x^n}{n} + \cdots.$$

What does this tell us about the interval of convergence of this series?

(b) Investigate convergence at the end points of the interval of convergence of this series.

23. Show that the series $\sum_{n=1}^{\infty} \dfrac{(2x)^n}{n}$ converges for $|x| < 1/2$.

Investigate whether the series converges for $x = 1/2$ and $x = -1/2$.

Problems

In Problems 24–27, find the interval of convergence.

24. $\displaystyle\sum_{n=0}^{\infty} \frac{x^n}{3^n}$

25. $\displaystyle\sum_{n=2}^{\infty} \frac{(x-3)^n}{n}$

26. $\displaystyle\sum_{n=1}^{\infty} \frac{n^2 x^{2n}}{2^{2n}}$

27. $\displaystyle\sum_{n=1}^{\infty} \frac{(-1)^n (x-5)^n}{2^n n^2}$

28. For constant p, find the radius of convergence of the binomial power series:[9]

$$1 + px + \frac{p(p-1)x^2}{2!} + \frac{p(p-1)(p-2)x^3}{3!} + \cdots.$$

29. Show that if $C_0 + C_1 x + C_2 x^2 + C_3 x^3 + \cdots$ converges for $|x| < R$ with R given by the ratio test, then so does $C_1 + 2C_2 x + 3C_3 x^2 + \cdots$.

30. The series $\sum C_n x^n$ converges at $x = -5$ and diverges at $x = 7$. What can you say about its radius of convergence?

31. The series $\sum C_n (x+7)^n$ converges at $x = 0$ and diverges at $x = -17$. What can you say about its radius of convergence?

32. Suppose that the power series $\sum C_n x^n$ converges when $x = -4$ and diverges when $x = 7$. Which of the following are true, false or not possible to determine?

   (a) The power series converges when $x = 10$.
   (b) The power series converges when $x = 3$.
   (c) The power series diverges when $x = 1$.
   (d) The power series diverges when $x = 6$.

33. If $\sum C_n (x-3)^n$ converges at $x = 7$ and diverges at $x = 10$, what can you say about the convergence at $x = 11$? At $x = 5$? At $x = 0$?

34. Bessel functions are important in such diverse areas as describing planetary motion and the shape of a vibrating drumhead. The Bessel function of order 0 is defined by

$$J(x) = \sum_{n=0}^{\infty} \frac{(-1)^n x^{2n}}{2^{2n}(n!)^2}.$$

(a) Find the domain of $J(x)$ by finding the interval of convergence for this power series.
(b) Find $J(0)$.
(c) Find the partial sum polynomials S_0, S_1, S_2, S_3, S_4.
(d) Estimate $J(1)$ to three decimal places.
(e) Use your answer to part (d) to estimate $J(-1)$.

35. For all x-values for which it converges, the function f is defined by the series

$$f(x) = \sum_{n=0}^{\infty} \frac{x^n}{n!}.$$

(a) What is $f(0)$?
(b) What is the domain of f?
(c) Assuming that f' can be calculated by differentiating the series term-by-term, find the series for $f'(x)$. What do you notice?
(d) Guess what well-known function f is.

36. From Example 5 on page 468, we know the following series converges for all x. We define $g(x)$ to be its sum:

$$g(x) = \sum_{n=1}^{\infty} (-1)^{n-1} \frac{x^{2n-1}}{(2n-1)!}.$$

(a) Is $g(x)$ odd, even, or neither? What is $g(0)$?
(b) Assuming that derivatives can be computed term-by-term, show that $g''(x) = -g(x)$.
(c) Guess what well-known function g might be. Check your guess using $g(0)$ and $g'(0)$.

37. The functions $p(x)$ and $q(x)$ are defined by the series

$$p(x) = \sum_{n=0}^{\infty} (-1)^n \frac{x^{2n}}{(2n)!}, \quad q(x) = \sum_{n=1}^{\infty} (-1)^{n-1} \frac{x^{2n-1}}{(2n-1)!}.$$

Assuming that these series converge for all x and that multiplication can be done term-by-term:

(a) Find the series for $(p(x))^2 + (q(x))^2$ up to the term in x^6.
(b) Guess what well-known functions p and q could be.

CHAPTER SUMMARY (see also Ready Reference at the end of the book)

- **Sequences**
  Recursive definition, monotone, bounded, convergence.
- **Geometric series**
  Finite sum, infinite sum.
- **Harmonic series**
- **Alternating series**

- **Tests for convergence of series**
  Integral test, p-series, comparison test, limit comparison test, ratio test, alternating series test.
  Absolute and conditional convergence.
- **Power series**
  Ratio test for radius of convergence, interval of convergence.

---

[9]For an explanation of the name, see Section 10.2.

REVIEW EXERCISES AND PROBLEMS FOR CHAPTER NINE

Exercises

Do the sequences in Exercises 1–4 converge or diverge? If a sequence converges, find its limit.

1. $\dfrac{3 + 4n}{5 + 7n}$

2. $(-1)^n \dfrac{(n+1)}{n}$

3. $\sin\left(\dfrac{\pi}{4} n\right)$

4. $\dfrac{1}{n} + \ln n$

5. Find the sum of the series $b^5 + b^6 + b^7 + b^8 + b^9 + b^{10}$.

6. Find the sum of the series $(0.5)^3 + (0.5)^4 + \cdots + (0.5)^k$.

7. Find the sum: $\displaystyle\sum_{n=0}^{\infty} \dfrac{3^n + 5}{4^n}$

Use the integral test to decide whether the series in Exercises 8–11 converge or diverge.

8. $\displaystyle\sum_{n=1}^{\infty} \dfrac{1}{(n+2)^2}$

9. $\displaystyle\sum_{n=1}^{\infty} \dfrac{3n^2 + 2n}{n^3 + n^2 + 1}$

10. $\displaystyle\sum_{n=0}^{\infty} ne^{-n^2}$

11. $\displaystyle\sum_{n=2}^{\infty} \dfrac{2}{n^2 - 1}$

Use the comparison test to confirm the statements in Exercises 12–13.

12. $\displaystyle\sum_{n=1}^{\infty} \left(\dfrac{1}{3}\right)^n$ converges, so $\displaystyle\sum_{n=1}^{\infty} \left(\dfrac{n^2}{3n^2 + 4}\right)^n$ converges.

13. $\displaystyle\sum_{n=1}^{\infty} \dfrac{1}{n}$ diverges, so $\displaystyle\sum_{n=1}^{\infty} \dfrac{1}{n \sin^2 n}$ diverges.

In Exercises 14–17 use the limit comparison test to determine whether the series converges or diverges.

14. $\displaystyle\sum \dfrac{\sqrt{n-1}}{n^2 + 3}$

15. $\displaystyle\sum \dfrac{n^3 - 2n^2 + n + 1}{n^5 - 2}$

16. $\displaystyle\sum \sin \dfrac{1}{n^2}$

17. $\displaystyle\sum \dfrac{1}{\sqrt{n^3 - 1}}$

Use the ratio test to decide if the series in Exercises 18–19 converge or diverge.

18. $\displaystyle\sum_{n=1}^{\infty} \dfrac{1}{2^n n!}$

19. $\displaystyle\sum_{n=1}^{\infty} \dfrac{n!(n+1)!}{(2n)!}$

Use the alternating series test to decide which of the series in Exercises 20–21 converge.

20. $\displaystyle\sum_{n=1}^{\infty} \dfrac{(-1)^n}{n^2 + 1}$

21. $\displaystyle\sum_{n=1}^{\infty} \dfrac{(-1)^{n-1}}{\sqrt{n^2 + 1}}$

Determine which of the series in Exercises 22–41 converge.

22. $\displaystyle\sum_{n=1}^{\infty} \dfrac{1}{n+1}$

23. $\displaystyle\sum_{n=1}^{\infty} \dfrac{1}{n^3}$

24. $\displaystyle\sum_{n=3}^{\infty} \dfrac{2}{\sqrt{n-2}}$

25. $\displaystyle\sum_{n=1}^{\infty} \dfrac{(-1)^{n-1}}{\sqrt{n}+1}$

26. $\displaystyle\sum_{n=1}^{\infty} \dfrac{n^2}{n^2 + 1}$

27. $\displaystyle\sum_{n=1}^{\infty} \dfrac{n^2}{n^3 + 1}$

28. $\displaystyle\sum_{n=1}^{\infty} \dfrac{3^n}{(2n)!}$

29. $\displaystyle\sum_{n=1}^{\infty} \dfrac{(2n)!}{(n!)^2}$

30. $\displaystyle\sum_{n=1}^{\infty} \dfrac{n^2 + 2^n}{n^2 2^n}$

31. $\displaystyle\sum_{n=1}^{\infty} 2^{-n} \dfrac{(n+1)}{(n+2)}$

32. $\displaystyle\sum_{n=1}^{\infty} (-1)^n \dfrac{2^n}{(2n+1)!}$

33. $\displaystyle\sum_{n=1}^{\infty} (-1)^n \dfrac{n+1}{\sqrt{n}}$

34. $\displaystyle\sum_{n=0}^{\infty} \dfrac{2 + 3^n}{5^n}$

35. $\displaystyle\sum_{n=1}^{\infty} \dfrac{1}{2 + \sin n}$

36. $\displaystyle\sum_{n=3}^{\infty} \dfrac{1}{(2n-5)^3}$

37. $\displaystyle\sum_{n=2}^{\infty} \dfrac{1}{n^3 - 3}$

38. $\displaystyle\sum_{n=1}^{\infty} \dfrac{\sin(n\pi/2)}{n^3}$

39. $\displaystyle\sum_{k=1}^{\infty} \ln\left(1 + \dfrac{1}{k}\right)$

40. $\displaystyle\sum_{n=1}^{\infty} \dfrac{n}{2^n}$

41. $\displaystyle\sum_{n=2}^{\infty} \dfrac{1}{(\ln n)^2}$

Find the radius of convergence in Exercises 42–45.

42. $\displaystyle\sum_{n=0}^{\infty} nx^n$

43. $\displaystyle\sum_{n=0}^{\infty} \dfrac{(2n)! x^n}{(n!)^2}$

44. $\displaystyle\sum_{n=0}^{\infty} (2^n + n^2) x^n$

45. $\displaystyle\sum_{n=0}^{\infty} \dfrac{x^n}{n! + 1}$

Problems

In Problems 46–47 determine whether the series are absolutely convergent, conditionally convergent, or divergent.

46. $\displaystyle\sum \frac{(-1)^n}{n^{1/2}}$

47. $\displaystyle\sum (-1)^n \frac{n}{n+1}$

48. For $r > 0$, how does the convergence of the following series depend on r?

$$\sum_{n=1}^{\infty} \frac{n^r + r^n}{n^r r^n}$$

In Problems 49–51, find the interval of convergence for the power series.

49. $\displaystyle\sum_{n=1}^{\infty} \frac{x^n}{3^n n^2}$

50. $\displaystyle\sum_{n=0}^{\infty} \frac{(-1)^n (x-2)^n}{5^n}$

51. $\displaystyle\sum_{n=1}^{\infty} \frac{(-1)^n x^n}{n}$

52. Suppose that $\displaystyle\sum_{n=0}^{\infty} C_n (x-2)^n$ converges when $x = 4$ and diverges when $x = 6$. Which of the following are true, false or not possible to determine? Give reasons for your answers.

(a) The power series converges when $x = 7$.
(b) The power series diverges when $x = 1$.
(c) The power series converges when $x = 0.5$.
(d) The power series diverges when $x = 5$.
(e) The power series converges when $x = -3$.

53. For all the t-values for which it converges, the function h is defined by the series

$$h(t) = \sum_{n=0}^{\infty} (-1)^n \frac{t^{2n}}{(2n)!}.$$

(a) What is the domain of h?
(b) Is h odd, even, or neither?
(c) Assuming that derivatives can be computed term-by-term, show that

$$h''(t) = -h(t).$$

54. A new car costs \$30,000; it loses 10% of its value each year. Maintenance is \$500 the first year and increases by 20% annually.

(a) Find a formula for l_n, the value lost by the car in year n.
(b) Find a formula for m_n, the maintenance expenses in year n.

(c) In what year do maintenance expenses first exceed the value lost by the car?

Problems 55–57 are about *bonds*, which are issued by a government to raise money. An individual who buys a \$1000 bond gives the government \$1000 and in return receives a fixed sum of money, called the *coupon*, every six months or every year for the life of the bond. At the time of the last coupon, the individual also gets the \$1000, or *principal* back.

55. What is the present value of a \$1000 bond which pays \$50 a year for 10 years, starting one year from now? Assume interest rate is 6% per year, compounded annually.

56. What is the present value of a \$1000 bond which pays \$50 a year for 10 years, starting one year from now? Assume the interest rate is 4% per year, compounded annually.

57. (a) What is the present value of a \$1000 bond which pays \$50 a year for 10 years, starting one year from now? Assume the interest rate is 5% per year, compounded annually.
(b) Since \$50 is 5% of \$1000, this bond is often called a 5% bond. What does your answer to part (a) tell you about the relationship between the principal and the present value of this bond when the interest rate is 5%?
(c) If the interest rate is more than 5% per year, compounded annually, which is larger: the principal or the present value of the bond? Why do you think the bond is then described as *trading at a discount*?
(d) If the interest rate is less than 5% per year, compounded annually, why is the bond described as *trading at a premium*?

58. Cephalexin is an antibiotic with a half-life in the body of 0.9 hours, taken in tablets of 250 mg every six hours.

(a) What percentage of the cephalexin in the body at the start of a six-hour period is still there at the end (assuming no tablets are taken during that time)?
(b) Write an expression for Q_1, Q_2, Q_3, Q_4, where Q_n mg, is the amount of cephalexin in the body right after the nth tablet is taken.
(c) Express Q_3, Q_4 in closed-form and evaluate them.
(d) Write an expression for Q_n and put it in closed-form.
(e) If the patient keeps taking the tablets, use your answer to part (d) to find the quantity of cephalexin in the body in the long run, right after taking a pill.

59. Around January 1, 1993, Barbra Streisand signed a contract with Sony Corporation for \$2 million a year for 10 years. Suppose the first payment was made on the day of signing and that all other payments are made on the first day of the year. Suppose also that all payments are made into a bank account earning 4% a year, compounded annually.

(a) How much money was in the account
  (i) On the night of December 31, 1999?
  (ii) On the day the last payment is made?

(b) What was the present value of the contract on the day it was signed?

60. Before World War I, the British government issued what are called *consols*, which pay the owner or his heirs a fixed amount of money every year forever. (Cartoonists of the time described aristocrats living off such payments as "pickled in consols.") What should a person expect to pay for consols which pay £10 a year forever? Assume the first payment is one year from the date of purchase and that interest remains 4% per year, compounded annually. (£ denotes pounds, the British unit of currency.)

61. This problem illustrates how banks create credit and can thereby lend out more money than has been deposited. Suppose that initially $100 is deposited in a bank. Experience has shown bankers that on average only 8% of the money deposited is withdrawn by the owner at any time. Consequently, bankers feel free to lend out 92% of their deposits. Thus $92 of the original $100 is loaned out to other customers (to start a business, for example). This $92 will become someone else's income and, sooner or later, will be redeposited in the bank. Then 92% of $92, or $92(0.92) = 84.64, is loaned out again and eventually redeposited. Of the $84.64, the bank again loans out 92%, and so on.

(a) Find the total amount of money deposited in the bank as a result of these transactions.

(b) The total amount of money deposited divided by the original deposit is called the *credit multiplier*. Calculate the credit multiplier for this example and explain what this number tells us.

62. In theory, drugs that decay exponentially always leave a residue in the body. However, in practice, once the drug has been in the body for 5 half-lives, it is regarded as being eliminated.[10] If a patient takes a tablet of the same drug every 5 half-lives forever, what is the upper limit to the amount of drug that can be in the body?

63. Estimate $\displaystyle\sum_{n=1}^{\infty} \frac{(-1)^{n-1}}{(2n-1)!}$ to within 0.01 of the actual sum of the series.

64. Is it possible to construct a convergent alternating series $\displaystyle\sum_{n=1}^{\infty}(-1)^{n-1}a_n$ for which $0 < a_{n+1} < a_n$ but $\displaystyle\lim_{n\to\infty} a_n \neq 0$?

65. Show that if $\sum a_n$ converges and $\sum b_n$ diverges, then $\sum(a_n + b_n)$ diverges. [Hint: Assume that $\sum(a_n + b_n)$ converges and consider $\sum(a_n + b_n) - \sum a_n$.]

Suppose that $\sum a_n$ converges with $a_n > 0$ for all n. Decide if the series in Problems 66–70 converge, diverge or if there is not enough information provided.

66. $\displaystyle\sum a_n/n$ **67.** $\displaystyle\sum 1/a_n$ **68.** $\displaystyle\sum n a_n$

69. $\displaystyle\sum (a_n + a_n/2)$ **70.** $\displaystyle\sum a_n^2$

71. Does $\displaystyle\sum_{n=1}^{\infty}\left(\frac{1}{n} + \frac{1}{n}\right)$ converge or diverge? Does $\displaystyle\sum_{n=1}^{\infty}\left(\frac{1}{n} - \frac{1}{n}\right)$ converge or diverge? Is the statement "If $\sum a_n$ and $\sum b_n$ diverge, then $\sum(a_n + b_n)$ may or may not diverge" true?

72. Estimate the sum of the first 100,000 terms of the harmonic series,

$$\sum_{k=1}^{100{,}000} \frac{1}{k},$$

to the closest integer. [Hint: Use left- and right-hand sums of the function $f(x) = 1/x$ on the interval from 1 to 100,000 with $\Delta x = 1$.]

73. Although the harmonic series does not converge, the partial sums grow very, very slowly. Take a right-hand sum approximating the integral of $f(x) = 1/x$ on the interval $[1, n]$, with $\Delta x = 1$, to show that

$$\frac{1}{2} + \frac{1}{3} + \frac{1}{4} + \cdots + \frac{1}{n} < \ln n.$$

If a computer could add a million terms of the harmonic series each second, estimate the sum after one year.

74. Is the following argument true or false? Give reasons for your answer.

Consider the infinite series $\displaystyle\sum_{n=2}^{\infty} \frac{1}{n(n-1)}$. Since

$$\frac{1}{n(n-1)} = \frac{1}{n-1} - \frac{1}{n}$$ we can write this series as

$$\sum_{n=2}^{\infty} \frac{1}{n-1} - \sum_{n=2}^{\infty} \frac{1}{n}.$$

For the first series $a_n = 1/(n-1)$. Since $n - 1 < n$ we have $1/(n-1) > 1/n$ and so this series diverges by comparison with the divergent harmonic series $\displaystyle\sum_{n=2}^{\infty} \frac{1}{n}$. The second series is the divergent harmonic series. Since both series diverge, their difference also diverges.

---

[10]http://dr.pierce1.net/PDF/half_life.pdf, accessed on May 10, 2003.

CHECK YOUR UNDERSTANDING

Decide if the statements in Problems 1–46 are true or false. Give an explanation for your answer.

1. You can tell if a sequence converges by looking at the first 1000 terms.

2. If the terms s_n of a convergent sequence are all positive then $\lim_{n \to \infty} s_n$ is positive.

3. If the sequence s_n of positive terms is unbounded, then the sequence has a term greater than a million.

4. If the sequence s_n of positive terms is unbounded, then the sequence has an infinite number of terms greater than a million.

5. If a sequence s_n is convergent, then the terms s_n tend to zero as n increases.

6. If a series converges, then the sequence of partial sums of the series also converges.

7. A monotone sequence can not have both positive and negative terms.

8. If a monotone sequence of positive terms does not converge, then it has a term greater than a million.

9. If the terms s_n of a sequence alternate in sign, then the sequence converges.

10. If all terms s_n of a sequence are less than a million, then the sequence is bounded.

11. A geometric series in x is a power series.

12. $\sum_{n=1}^{\infty} (x - n)^n$ is a power series.

13. If the power series $\sum C_n x^n$ converges for $x = 2$, then it converges for $x = 1$.

14. If the power series $\sum C_n x^n$ converges for $x = 1$, then the power series converges for $x = 2$.

15. If the power series $\sum C_n x^n$ does not converge for $x = 1$, then the power series does not converge for $x = 2$.

16. If $0 \le a_n \le b_n$ and $\sum a_n$ converges, then $\sum b_n$ converges.

17. If $0 \le a_n \le b_n$ and $\sum a_n$ diverges, then $\sum b_n$ diverges.

18. If $b_n \le a_n \le 0$ and $\sum b_n$ converges, then $\sum a_n$ converges.

19. If $\sum a_n$ converges, then $\sum |a_n|$ converges.

20. If $\sum |a_n + b_n|$ converges, then $\sum |a_n|$ and $\sum |b_n|$ converge.

21. If $\sum a_n$ converges, then $\lim_{n \to \infty} |a_{n+1}|/|a_n| \ne 1$.

22. $\sum_{n=0}^{\infty} (-1)^n \cos(\pi n)$ is an alternating series.

23. $\sum_{n=1}^{\infty} (1 + (-1)^n)$ is a series of nonnegative terms.

24. The series $\sum_{n=0}^{\infty} (-1)^n 2^n$ converges.

25. The series $\sum_{n=1}^{\infty} 2^{(-1)^n}$ converges.

26. If $\sum a_n$ converges, then $\sum (-1)^n a_n$ converges.

27. If an alternating series converges by the alternating series test, then the error in using the first n terms of the series to approximate the entire series is less in magnitude than the first term omitted.

28. If an alternating series converges, then the error in using the first n terms of the series to approximate the entire series is less in magnitude than the first term omitted.

29. If $\sum |a_n|$ converges, then $\sum (-1)^n |a_n|$ converges.

30. To find the sum of the alternating harmonic series $\sum (-1)^{n-1}/n$ to within 0.01 of the true value, we can sum the first 100 terms.

31. $\sum C_n (x - 1)^n$ and $\sum C_n x^n$ have the same radius of convergence.

32. If $\sum C_n x^n$ and $\sum B_n x^n$ have the same radius of convergence, then the coefficients, C_n and B_n, must be equal.

33. If a series $\sum a_n$ converges, then the terms, a_n, tend to zero as n increases.

34. If the terms, a_n, of a series tend to zero as n increases, then the series $\sum a_n$ converges.

35. If $\sum a_n$ does not converge and $\sum b_n$ does not converge, then $\sum a_n b_n$ does not converge.

36. If $\sum a_n b_n$ converges, then $\sum a_n$ and $\sum b_n$ converge.

37. If $\sum a_n$ is absolutely convergent, then it is convergent.

38. If $\sum a_n$ is conditionally convergent, then it is absolutely convergent.

39. If a power series converges at one endpoint of its interval of convergence, then it converges at the other.

40. If $a_n > 0.5 b_n > 0$ for all n and $\sum b_n$ diverges, then $\sum a_n$ diverges.

41. A power series always converges at at least one point.

42. If the power series $\sum C_n x^n$ converges at $x = 10$, then it converges at $x = -9$.

43. If the power series $\sum C_n x^n$ converges at $x = 10$, then it converges at $x = -10$.

44. $-5 < x \le 7$ is a possible interval of convergence of a power series.

45. $-3 < x < 2$ could be the interval of convergence of $\sum C_n x^n$.

46. If $-11 < x < 1$ is the interval of convergence of $\sum C_n (x - a)^n$, then $a = -5$.

PROJECTS FOR CHAPTER NINE

1. A Definition of e

We show that the sequence $s_n = \left(1 + \dfrac{1}{n}\right)^n$ converges; its limit can be used to define e.

(a) For a fixed integer $n > 0$, let $f(x) = (n+1)x^n - nx^{n+1}$. For $x > 1$, show f is decreasing and that $f(x) < 1$. Hence, for $x > 1$,

$$x^n(n+1-nx) < 1.$$

(b) Substitute the following x-value into the inequality from part (a)

$$x = \frac{1 + 1/n}{1 + 1/(n+1)},$$

and show that

$$x^n\left(\frac{n+1}{n+2}\right) < 1.$$

(c) Use the inequality from part (b) to show that $s_n < s_{n+1}$ for all $n > 0$. Conclude the sequence is increasing.

(d) Substitute $x = 1 + 1/2n$ into the inequality from part (a) to show that

$$\left(1 + \frac{1}{2n}\right)^n < 2.$$

(e) Use the inequality from part (d) to show $s_{2n} < 4$. Conclude the sequence is bounded.

(f) Use parts (c) and (e) to show that the sequence has a limit.

2. Probability of Winning in Sports

In certain sports, winning a game requires a lead of two points. That is, if the score is tied you have to score two points in a row to win.

(a) For some sports (e.g. tennis), a point is scored every play. Suppose your probability of scoring the next point is always p. Then, your opponent's probability of scoring the next point is always $1 - p$.

  (i) What is the probability that you win the next two points?

  (ii) What is the probability that you and your opponent split the next two points, that is, that neither of you wins both points?

  (iii) What is the probability that you split the next two points but you win the two after that?

  (iv) What is the probability that you either win the next two points or split the next two and then win the next two after that?

  (v) Give a formula for your probability w of winning a tied game.

  (vi) Compute your probability of winning a tied game when $p = 0.5$; when $p = 0.6$; when $p = 0.7$; when $p = 0.4$. Comment on your answers.

(b) In other sports (e.g. volleyball), you can score a point only if it is your turn, with turns alternating until a point is scored. Suppose your probability of scoring a point when it is your turn is p, and your opponent's probability of scoring a point when it is her turn is q.

  (i) Find a formula for the probability S that you are the first to score the next point, assuming it is currently your turn.

  (ii) Suppose that if you score a point, the next turn is yours. Using your answers to part (a) and your formula for S, compute the probability of winning a tied game (if you need two points in a row to win).

  • Assume $p = 0.5$ and $q = 0.5$ and it is your turn.
  • Assume $p = 0.6$ and $q = 0.5$ and it is your turn.

3. Quinine

Malaria is a parasitic infection transmitted by mosquito bites, mainly in tropical areas of the world. The disease has existed since ancient times, and currently there are hundreds of millions of cases each year, with millions of deaths. Around 1630, Jesuits in Peru introduced the bark of the cinchona tree to the West as the first treatment for malaria. The drug quinine is the active ingredient in the bark, and it is still used today. However, the concentration in the body of quinine, and of most drugs, must be kept within certain parameters. If the concentration is too low, the drug is ineffective. If the concentration is too high, toxic side effects can result.

Suppose you are a doctor who prescribes quinine for a 70 kg malaria patient. At 8 am each day, she is to receive 50 mg of the drug.[11] To be effective, the average concentration in the body must be at least 0.4 mg/kg. However, concentrations above 3.0 mg/kg can be fatal. The half-life of quinine in the body is 11.5 hours.

(a) What is the continuous rate of decay of quinine (in units of %/min)?

(b) How much quinine is in the patient's system just before and just after the first day's dose? After the second day's dose?

(c) How much is in her system after a few more doses? After the nth dose? Explain what happens in the long run.

(d) Graph the concentration of the drug versus time showing what the steady-state looks like. Find a formula for the repeating function.

(e) What is the average (over time) of the concentration of quinine in the patient's body?

(f) Is this treatment both effective and safe?

(g) Suppose instead that the patient were to receive two doses of 25 mg each day, one at 8 am and one at 8 pm. Is this a safe and effective treatment?

(h) Determine the average value of an exponentially decaying function between two points (x_0, y_0) and (x_1, y_1).

(i) Notice that the average concentration of quinine is the same for the 50 mg once per day and 25 mg twice per day treatments. Use part (h) to explain why. Would the average concentration be the same for a 100 mg dose once every two days?

(j) Suppose the original quinine regimen is stopped. How long after the last dose will the amount of quinine be less than 10^{-10} times the patient's body mass?

---

[11] This is a simplified model; actual treatments involve different drugs and more complicated dosage regimens. See, for example, *The Pharmacological Basis of Therapeutics*, 9th Ed., ed. Joel G. Hardman, Alfred Goodman Gilman, and Lee E. Limbird, (New York: McGraw Hill, 1996).

Chapter Ten

APPROXIMATING FUNCTIONS USING SERIES

In Section 9.2, for $|x| < 1$, we found the sum of a geometric series as a function of the common ratio x:

$$1 + x + x^2 + \cdots + x^n + \cdots = \frac{1}{1-x}.$$

Viewed from another perspective, this formula gives us a series whose sum is the function $f(x) = 1/(1-x)$ for $|x| < 1$. We now work in the opposite direction, starting with a function and looking for a series of simpler functions whose sum is that function. We first use polynomials, which lead us to Taylor series, and then trigonometric functions, which lead us to Fourier series.

Taylor approximations use polynomials, which may be considered the simplest functions. They are easy to use because they can be evaluated by simple arithmetic, unlike transcendental functions, such as e^x and $\ln x$.

Fourier approximations use sines and cosines, the simplest periodic functions, instead of polynomials. Taylor approximations are generally good approximations to the function locally (that is, near a specific point), whereas Fourier approximations are generally good approximations over an interval.

10.1   TAYLOR POLYNOMIALS

In this section, we see how to approximate a function by polynomials.

Linear Approximations

We already know how to approximate a function using a degree 1 polynomial, namely the tangent line approximation given in Section 3.9 :

$$f(x) \approx f(a) + f'(a)(x - a).$$

The tangent line and the curve have the same slope at $x = a$. As Figure 10.1 suggests, the tangent line approximation to the function is generally more accurate for values of x close to a.

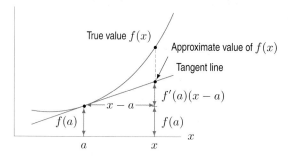

Figure 10.1: Tangent line approximation of $f(x)$ for x near a

We first focus on $a = 0$. The tangent line approximation at $x = 0$ is referred to as the *first Taylor approximation* at $x = 0$, or as follows:

Taylor Polynomial of Degree 1 Approximating $f(x)$ for x near 0

$$f(x) \approx P_1(x) = f(0) + f'(0)x$$

Example 1   Find the Taylor Polynomial of degree 1 for $g(x) = \cos x$, with x in radians, for x near 0.

Solution   The tangent line at $x = 0$ is just the horizontal line $y = 1$, as shown in Figure 10.2, so

$$g(x) = \cos x \approx 1, \quad \text{for } x \text{ near } 0.$$

If we take $x = 0.05$, then

$$g(0.05) = \cos(0.05) = 0.998\ldots,$$

which is quite close to the approximation $\cos x \approx 1$. Similarly, if $x = -0.1$, then

$$g(-0.1) = \cos(-0.1) = 0.995\ldots$$

is close to the approximation $\cos x \approx 1$. However, if $x = 0.4$, then

$$g(0.4) = \cos(0.4) = 0.921\ldots,$$

so the approximation $\cos x \approx 1$ is less accurate. The graph suggests that the farther a point x is away from 0, the worse the approximation is likely to be.

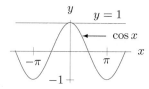

Figure 10.2: Graph of $\cos x$ and its tangent line at $x = 0$

The previous example shows that the Taylor polynomial of degree 1 might actually have degree less than 1.

Quadratic Approximations

To get a more accurate approximation, we use a quadratic function instead of a linear function.

Example 2   Find the quadratic approximation to $g(x) = \cos x$ for x near 0.

Solution   To ensure that the quadratic, $P_2(x)$, is a good approximation to $g(x) = \cos x$ at $x = 0$, we require that $\cos x$ and the quadratic have the same value, the same slope, and the same second derivative at $x = 0$. That is, we require $P_2(0) = g(0)$, $P_2'(0) = g'(0)$, and $P_2''(0) = g''(0)$. We take the quadratic polynomial

$$P_2(x) = C_0 + C_1 x + C_2 x^2,$$

and determine C_0, C_1, and C_2. Since

$$
\begin{aligned}
P_2(x) &= C_0 + C_1 x + C_2 x^2 &\quad\text{and}\quad& g(x) = \cos x \\
P_2'(x) &= C_1 + 2C_2 x & & g'(x) = -\sin x \\
P_2''(x) &= 2C_2 & & g''(x) = -\cos x,
\end{aligned}
$$

we have

$$
\begin{aligned}
C_0 &= P_2(0) = g(0) = \cos 0 = 1 &\quad\text{so}\quad& C_0 = 1 \\
C_1 &= P_2'(0) = g'(0) = -\sin 0 = 0 & & C_1 = 0 \\
2C_2 &= P_2''(0) = g''(0) = -\cos 0 = -1, & & C_2 = -\tfrac{1}{2}.
\end{aligned}
$$

Consequently, the quadratic approximation is

$$\cos x \approx P_2(x) = 1 + 0 \cdot x - \frac{1}{2}x^2 = 1 - \frac{x^2}{2}, \quad \text{for } x \text{ near } 0.$$

Figure 10.3 suggests that the quadratic approximation $\cos x \approx P_2(x)$ is better than the linear approximation $\cos x \approx P_1(x)$ for x near 0. Let's compare the accuracy of the two approximations. Recall that $P_1(x) = 1$ for all x. At $x = 0.4$, we have $\cos(0.4) = 0.921\ldots$ and $P_2(0.4) = 0.920$, so the quadratic approximation is a significant improvement over the linear approximation. The magnitude of the error is about 0.001 instead of 0.08.

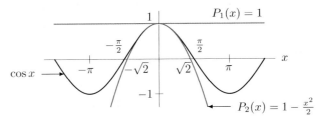

Figure 10.3: Graph of $\cos x$ and its linear, $P_1(x)$, and quadratic, $P_2(x)$, approximations for x near 0

Generalizing the computations in Example 2, we define the *second Taylor approximation* at $x = 0$.

Taylor Polynomial of Degree 2 Approximating $f(x)$ for x near 0

$$f(x) \approx P_2(x) = f(0) + f'(0)x + \frac{f''(0)}{2}x^2$$

Higher-Degree Polynomials

In a small interval around $x = 0$, the quadratic approximation to a function is usually a better approximation than the linear (tangent line) approximation. However, Figure 10.3 shows that the quadratic can still bend away from the original function for large x. We can attempt to fix this by using an approximating polynomial of higher degree. Suppose that we approximate a function $f(x)$ for x near 0 by a polynomial of degree n:

$$f(x) \approx P_n(x) = C_0 + C_1 x + C_2 x^2 + \cdots + C_{n-1} x^{n-1} + C_n x^n.$$

We need to find the values of the constants: $C_0, C_1, C_2, \ldots, C_n$. To do this, we require that the function $f(x)$ and each of its first n derivatives agree with those of the polynomial $P_n(x)$ at the point $x = 0$. In general, the more derivatives there are that agree at $x = 0$, the larger the interval on which the function and the polynomial remain close to each other.

To see how to find the constants, let's take $n = 3$ as an example:

$$f(x) \approx P_3(x) = C_0 + C_1 x + C_2 x^2 + C_3 x^3.$$

Substituting $x = 0$ gives

$$f(0) = P_3(0) = C_0.$$

Differentiating $P_3(x)$ yields

$$P_3'(x) = C_1 + 2C_2 x + 3C_3 x^2,$$

so substituting $x = 0$ shows that

$$f'(0) = P_3'(0) = C_1.$$

Differentiating and substituting again, we get

$$P_3''(x) = 2 \cdot 1 C_2 + 3 \cdot 2 \cdot 1 C_3 x,$$

which gives

$$f''(0) = P_3''(0) = 2 \cdot 1 C_2,$$

so that

$$C_2 = \frac{f''(0)}{2 \cdot 1}.$$

The third derivative, denoted by P_3''', is

$$P_3'''(x) = 3 \cdot 2 \cdot 1 C_3,$$

so

$$f'''(0) = P_3'''(0) = 3 \cdot 2 \cdot 1 C_3,$$

and then

$$C_3 = \frac{f'''(0)}{3 \cdot 2 \cdot 1}.$$

You can imagine a similar calculation starting with $P_4(x)$, using the fourth derivative $f^{(4)}$, which would give

$$C_4 = \frac{f^{(4)}(0)}{4 \cdot 3 \cdot 2 \cdot 1},$$

and so on. Using factorial notation,[1] we write these expressions as

$$C_3 = \frac{f'''(0)}{3!}, \quad C_4 = \frac{f^{(4)}(0)}{4!}.$$

Writing $f^{(n)}$ for the nth derivative of f, we have, for any positive integer n

$$C_n = \frac{f^{(n)}(0)}{n!}.$$

So we define the nth *Taylor approximation* at $x = 0$:

---

[1] Recall that $k! = k(k-1) \cdots 2 \cdot 1$. In addition, $1! = 1$, and $0! = 1$.

Taylor Polynomial of Degree n Approximating $f(x)$ for x near 0

$$f(x) \approx P_n(x)$$

$$= f(0) + f'(0)x + \frac{f''(0)}{2!}x^2 + \frac{f'''(0)}{3!}x^3 + \frac{f^{(4)}(0)}{4!}x^4 + \cdots + \frac{f^{(n)}(0)}{n!}x^n$$

We call $P_n(x)$ the Taylor polynomial of degree n centered at $x = 0$, or the Taylor polynomial about (or around) $x = 0$.

Example 3   Construct the Taylor polynomial of degree 7 approximating the function $f(x) = \sin x$ for x near 0. Compare the value of the Taylor approximation with the true value of f at $x = \pi/3$.

Solution   We have

$$
\begin{aligned}
f(x) &= \sin x &\text{giving} && f(0) &= 0 \\
f'(x) &= \cos x && & f'(0) &= 1 \\
f''(x) &= -\sin x && & f''(0) &= 0 \\
f'''(x) &= -\cos x && & f'''(0) &= -1 \\
f^{(4)}(x) &= \sin x && & f^{(4)}(0) &= 0 \\
f^{(5)}(x) &= \cos x && & f^{(5)}(0) &= 1 \\
f^{(6)}(x) &= -\sin x && & f^{(6)}(0) &= 0 \\
f^{(7)}(x) &= -\cos x, && & f^{(7)}(0) &= -1.
\end{aligned}
$$

Using these values, we see that the Taylor polynomial approximation of degree 7 is

$$\sin x \approx P_7(x) = 0 + 1 \cdot x + 0 \cdot \frac{x^2}{2!} - 1 \cdot \frac{x^3}{3!} + 0 \cdot \frac{x^4}{4!} + 1 \cdot \frac{x^5}{5!} + 0 \cdot \frac{x^6}{6!} - 1 \cdot \frac{x^7}{7!}$$

$$= x - \frac{x^3}{3!} + \frac{x^5}{5!} - \frac{x^7}{7!}, \quad \text{for } x \text{ near } 0.$$

Notice that since $f^{(8)}(0) = 0$, the seventh and eighth Taylor approximations to $\sin x$ are the same.

In Figure 10.4 we show the graphs of the sine function and the approximating polynomial of degree 7 for x near 0. They are indistinguishable where x is close to 0. However, as we look at values of x farther away from 0 in either direction, the two graphs move apart. To check the accuracy of this approximation numerically, we see how well it approximates $\sin(\pi/3) = \sqrt{3}/2 = 0.8660254\ldots$.

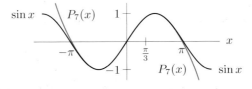

Figure 10.4: Graph of $\sin x$ and its seventh degree Taylor polynomial, $P_7(x)$, for x near 0

When we substitute $\pi/3 = 1.0471976\ldots$ into the polynomial approximation, we obtain $P_7(\pi/3) = 0.8660213\ldots$, which is extremely accurate—to about four parts in a million.

Example 4   Graph the Taylor polynomial of degree 8 approximating $g(x) = \cos x$ for x near 0.

Solution   We find the coefficients of the Taylor polynomial by the method of the preceding example, giving

$$\cos x \approx P_8(x) = 1 - \frac{x^2}{2!} + \frac{x^4}{4!} - \frac{x^6}{6!} + \frac{x^8}{8!}.$$

Figure 10.5 shows that $P_8(x)$ is close to the cosine function for a larger interval of x-values than the quadratic approximation $P_2(x) = 1 - x^2/2$ in Example 2 on page 479.

Figure 10.5: $P_8(x)$ approximates $\cos x$ better than $P_2(x)$ for x near 0

Example 5   Construct the Taylor polynomial of degree 10 about $x = 0$ for the function $f(x) = e^x$.

Solution   We have $f(0) = 1$. Since the derivative of e^x is equal to e^x, all the higher-order derivatives are equal to e^x. Consequently, for any $k = 1, 2, \ldots, 10$, $f^{(k)}(x) = e^x$ and $f^{(k)}(0) = e^0 = 1$. Therefore, the Taylor polynomial approximation of degree 10 is given by

$$e^x \approx P_{10}(x) = 1 + x + \frac{x^2}{2!} + \frac{x^3}{3!} + \frac{x^4}{4!} + \cdots + \frac{x^{10}}{10!}, \quad \text{for } x \text{ near } 0.$$

To check the accuracy of this approximation, we use it to approximate $e = e^1 = 2.718281828\ldots$. Substituting $x = 1$ gives $P_{10}(1) = 2.718281801$. Thus, P_{10} yields the first seven decimal places for e. For large values of x, however, the accuracy diminishes because e^x grows faster than any polynomial as $x \to \infty$. Figure 10.6 shows graphs of $f(x) = e^x$ and the Taylor polynomials of degree $n = 0, 1, 2, 3, 4$. Notice that each successive approximation remains close to the exponential curve for a larger interval of x-values.

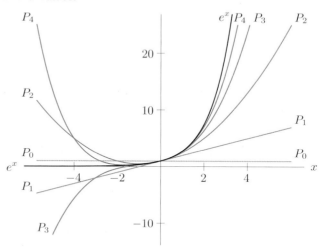

Figure 10.6: For x near 0, the value of e^x is more closely approximated by higher-degree Taylor polynomials

Example 6   Construct the Taylor polynomial of degree n approximating $f(x) = \dfrac{1}{1-x}$ for x near 0.

Solution   Differentiating gives $f(0) = 1$, $f'(0) = 1$, $f''(0) = 2$, $f'''(0) = 3!$, $f^{(4)}(0) = 4!$, and so on. This means

$$\frac{1}{1-x} \approx P_n(x) = 1 + x + x^2 + x^3 + x^4 + \cdots + x^n, \quad \text{for } x \text{ near } 0,$$

Let us compare the Taylor polynomial with the formula obtained from the sum of a finite geometric series on page 446:

$$\frac{1-x^{n+1}}{1-x} = 1 + x + x^2 + x^3 + x^4 + \cdots + x^n.$$

If x is close to 0 and x^{n+1} is small enough to neglect, the formula for the sum of a finite geometric series gives us the Taylor approximation of degree n:

$$\frac{1}{1-x} \approx 1 + x + x^2 + x^3 + x^4 + \cdots + x^n.$$

Taylor Polynomials Around $x = a$

Suppose we want to approximate $f(x) = \ln x$ by a Taylor polynomial. This function has no Taylor polynomial about $x = 0$ because the function is not defined for $x \le 0$. However, it turns out that we can construct a polynomial centered about some other point, $x = a$.

First, let's look at the equation of the tangent line at $x = a$:

$$y = f(a) + f'(a)(x-a).$$

This gives the first Taylor approximation

$$f(x) \approx f(a) + f'(a)(x-a) \quad \text{for } x \text{ near } a.$$

The $f'(a)(x-a)$ term is a correction term which approximates the change in f as x moves away from a.

Similarly, the Taylor polynomial $P_n(x)$ centered at $x = a$ is set up as $f(a)$ plus correction terms which are zero for $x = a$. This is achieved by writing the polynomial in powers of $(x-a)$ instead of powers of x:

$$f(x) \approx P_n(x) = C_0 + C_1(x-a) + C_2(x-a)^2 + \cdots + C_n(x-a)^n.$$

If we require n derivatives of the approximating polynomial $P_n(x)$ and the original function $f(x)$ to agree at $x = a$, we get the following result for the nth Taylor approximations at $x = a$:

Taylor Polynomial of Degree n Approximating $f(x)$ for x near a

$$f(x) \approx P_n(x)$$
$$= f(a) + f'(a)(x-a) + \frac{f''(a)}{2!}(x-a)^2 + \cdots + \frac{f^{(n)}(a)}{n!}(x-a)^n$$

We call $P_n(x)$ the Taylor polynomial of degree n centered at $x = a$, or the Taylor polynomial about $x = a$.

You can derive the formula for these coefficients in the same way that we did for $a = 0$. (See Problem 32, page 485.)

Example 7   Construct the Taylor polynomial of degree 4 approximating the function $f(x) = \ln x$ for x near 1.

Solution   We have

$$
\begin{array}{llll}
f(x) = & \ln x & \text{so} & f(1) = \ln(1) = 0 \\
f'(x) = & 1/x & & f'(1) = & 1 \\
f''(x) = & -1/x^2 & & f''(1) = & -1 \\
f'''(x) = & 2/x^3 & & f'''(1) = & 2 \\
f^{(4)}(x) = & -6/x^4, & & f^{(4)}(1) = & -6.
\end{array}
$$

The Taylor polynomial is therefore

$$\ln x \approx P_4(x) = 0 + (x-1) - \frac{(x-1)^2}{2!} + 2\frac{(x-1)^3}{3!} - 6\frac{(x-1)^4}{4!}$$

$$= (x-1) - \frac{(x-1)^2}{2} + \frac{(x-1)^3}{3} - \frac{(x-1)^4}{4}, \quad \text{for } x \text{ near } 1.$$

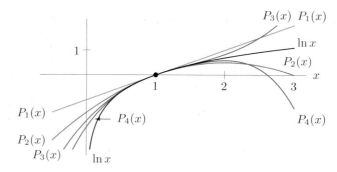

Figure 10.7: Taylor polynomials approximate $\ln x$ closely for x near 1, but not necessarily farther away

Graphs of $\ln x$ and several of its Taylor polynomials are shown in Figure 10.7. Notice that $P_4(x)$ stays reasonably close to $\ln x$ for x near 1, but bends away as x gets farther from 1. Also, note that the Taylor polynomials are defined for $x \leq 0$, but $\ln x$ is not.

The examples in this section suggest that the following results are true for common functions:
- Taylor polynomials centered at $x = a$ give good approximations to $f(x)$ for x near a. Farther away, they may or may not be good.
- The higher the degree of the Taylor polynomial, the larger the interval over which it fits the function closely.

Exercises and Problems for Section 10.1

Exercises

For Exercises 1–10, find the Taylor polynomials of degree n approximating the given functions for x near 0. (Assume p is a constant.)

1. $\dfrac{1}{1-x}$, $\quad n = 3, 5, 7$

2. $\dfrac{1}{1+x}$, $\quad n = 4, 6, 8$

3. $\sqrt{1+x}$, $\quad n = 2, 3, 4$

4. $\sqrt[3]{1-x}$, $\quad n = 2, 3, 4$

5. $\cos x$, $\quad n = 2, 4, 6$

6. $\ln(1+x)$, $\quad n = 5, 7, 9$

7. $\arctan x$, $\quad n = 3, 4$

8. $\tan x$, $\quad n = 3, 4$

9. $\dfrac{1}{\sqrt{1+x}}$, $\quad n = 2, 3, 4$

10. $(1+x)^p$, $\quad n = 2, 3, 4$

For Exercises 11–16, find the Taylor polynomial of degree n for x near the given point a.

11. e^x, $\quad a = 1$, $\quad n = 4$

12. $\sqrt{1+x}$, $\quad a = 1$, $\quad n = 3$

13. $\sin x$, $\quad a = \pi/2$, $\quad n = 4$

14. $\cos x$, $\quad a = \pi/4$, $\quad n = 3$

15. $\ln(x^2)$, $\quad a = 1$, $\quad n = 4$

16. $\sin 2x$, $\quad a = \pi/4$, $\quad n = 4$

Problems

For Problems 17–20, suppose $P_2(x) = a + bx + cx^2$ is the second degree Taylor polynomial for the function f about $x = 0$. What can you say about the signs of a, b, c if f has the graph given below?

17.

$f(x)$

x

18.

$f(x)$

x

19.

$f(x)$

x

20.

x

$f(x)$

21. The function $f(x)$ is approximated near $x = 0$ by the second degree Taylor polynomial

$$P_2(x) = 5 - 7x + 8x^2.$$

Give the value of

(a) $f(0)$ **(b)** $f'(0)$ **(c)** $f''(0)$

22. The function $f(x)$ is approximated near $x = 0$ by the sixth degree Taylor polynomial

$$P_6(x) = 3x - 4x^3 + 5x^6.$$

Give the value of

(a) $f(0)$ **(b)** $f'(0)$ **(c)** $f'''(0)$

(d) $f^{(5)}(0)$ **(e)** $f^{(6)}(0)$

23. Suppose g is a function which has continuous derivatives, and that $g(5) = 3, g'(5) = -2, g''(5) = 1,$ $g'''(5) = -3$.

   (a) What is the Taylor polynomial of degree 2 for g near 5? What is the Taylor polynomial of degree 3 for g near 5?

   (b) Use the two polynomials that you found in part (a) to approximate $g(4.9)$.

24. **(a)** Find the parabola best approximating the unit circle $x^2 + y^2 = 1$ near the point $(0, 1)$.

   (b) Use your answer to part (a) to estimate the coordinates of the point on the upper half of the unit circle with $x = 0.1$.

25. Find the second-degree Taylor polynomial for $f(x) = 4x^2 - 7x + 2$ about $x = 0$. What do you notice?

26. Find the third-degree Taylor polynomial for $f(x) = x^3 + 7x^2 - 5x + 1$ about $x = 0$. What do you notice?

27. **(a)** Based on your observations in Problems 25–26, make a conjecture about Taylor approximations in the case when f is itself a polynomial.

   (b) Show that your conjecture is true.

28. Show how you can use the Taylor approximation $\sin x \approx x - \dfrac{x^3}{3!}$, for x near 0, to explain why $\lim\limits_{x \to 0} \dfrac{\sin x}{x} = 1$.

29. Use the fourth-degree Taylor approximation $\cos x \approx 1 - \dfrac{x^2}{2!} + \dfrac{x^4}{4!}$ for x near 0 to explain why $\lim\limits_{x \to 0} \dfrac{1 - \cos x}{x^2} = \dfrac{1}{2}$.

30. Use a fourth degree Taylor approximation for e^h, for h near 0, to evaluate the following limits. Would your answer be different if you used a Taylor polynomial of higher degree?

   (a) $\lim\limits_{h \to 0} \dfrac{e^h - 1 - h}{h^2}$

   (b) $\lim\limits_{h \to 0} \dfrac{e^h - 1 - h - \frac{h^2}{2}}{h^3}$

31. If $f(2) = g(2) = h(2) = 0$, and $f'(2) = h'(2) = 0$, $g'(2) = 22$, and $f''(2) = 3$, $g''(2) = 5$, $h''(2) = 7$, calculate the following limits. Explain your reasoning.

   (a) $\lim\limits_{x \to 2} \dfrac{f(x)}{h(x)}$ **(b)** $\lim\limits_{x \to 2} \dfrac{f(x)}{g(x)}$

32. Derive the formulas given in the box on page 483 for the coefficients of the Taylor polynomial approximating a function f for x near a.

33. **(a)** Find the Taylor polynomial approximation of degree 4 about $x = 0$ for the function $f(x) = e^{x^2}$.

   (b) Compare this result to the Taylor polynomial approximation of degree 2 for the function $f(x) = e^x$ about $x = 0$. What do you notice?

   (c) Use your observation in part (b) to write out the Taylor polynomial approximation of degree 20 for the function in part (a).

   (d) What is the Taylor polynomial approximation of degree 5 for the function $f(x) = e^{-2x}$?

34. Consider the equations $\sin x = 0.2$ and $x - \dfrac{x^3}{3!} = 0.2$.

   (a) How many solutions does each equation have?

   (b) Which of the solutions of the two equations are approximately equal? Explain.

35. The integral $\int_0^1 (\sin t/t)\, dt$ is difficult to approximate using, for example, left Riemann sums or the trapezoid rule because the integrand $(\sin t)/t$ is not defined at $t = 0$. However, this integral converges; its value is $0.94608\ldots$. Estimate the integral using Taylor polynomials for $\sin t$ about $t = 0$ of

   (a) Degree 3   **(b)** Degree 5

36. One of the two sets of functions, f_1, f_2, f_3, or g_1, g_2, g_3, is graphed in Figure 10.8; the other set is graphed in Figure 10.9. Points A and B each have $x = 0$. Taylor polynomials of degree 2 approximating these functions near $x = 0$ are as follows:

$$f_1(x) \approx 2 + x + 2x^2 \qquad g_1(x) \approx 1 + x + 2x^2$$
$$f_2(x) \approx 2 + x - x^2 \qquad g_2(x) \approx 1 + x + x^2$$
$$f_3(x) \approx 2 + x + x^2 \qquad g_3(x) \approx 1 - x + x^2.$$

   (a) Which group of functions, the fs or the gs, is represented by each figure?

   (b) What are the coordinates of the points A and B?

   (c) Match each function with the graphs (I)–(III) in the appropriate figure.

Figure 10.8   **Figure 10.9**

10.2 TAYLOR SERIES

In the previous section we saw how to approximate a function near a point by Taylor polynomials. Now we define a Taylor series, which is a power series that can be thought of as a Taylor polynomial that goes on forever.

Taylor Series for $\cos x$, $\sin x$, e^x

We have the following Taylor polynomials centered at $x = 0$ for $\cos x$:

$$\cos x \approx P_0(x) = 1$$

$$\cos x \approx P_2(x) = 1 - \frac{x^2}{2!}$$

$$\cos x \approx P_4(x) = 1 - \frac{x^2}{2!} + \frac{x^4}{4!}$$

$$\cos x \approx P_6(x) = 1 - \frac{x^2}{2!} + \frac{x^4}{4!} - \frac{x^6}{6!}$$

$$\cos x \approx P_8(x) = 1 - \frac{x^2}{2!} + \frac{x^4}{4!} - \frac{x^6}{6!} + \frac{x^8}{8!}.$$

Here we have a sequence of polynomials, $P_0(x)$, $P_2(x)$, $P_4(x)$, $P_6(x)$, $P_8(x)$, ..., each of which is a better approximation to $\cos x$ than the last, for x near 0. When we go to a higher-degree polynomial (say from P_6 to P_8), we add more terms ($x^8/8!$, for example), but the terms of lower degree don't change. Thus, each polynomial includes the information from all the previous ones. We represent the whole sequence of Taylor polynomials by writing the *Taylor series* for $\cos x$:

$$1 - \frac{x^2}{2!} + \frac{x^4}{4!} - \frac{x^6}{6!} + \frac{x^8}{8!} - \cdots.$$

Notice that the partial sums of this series are the Taylor polynomials, $P_n(x)$.

We define the Taylor series for $\sin x$ and e^x similarly. It turns out that, for these functions, the Taylor series converges to the function for all x, so we can write the following:

$$\sin x = x - \frac{x^3}{3!} + \frac{x^5}{5!} - \frac{x^7}{7!} + \frac{x^9}{9!} - \cdots$$

$$\cos x = 1 - \frac{x^2}{2!} + \frac{x^4}{4!} - \frac{x^6}{6!} + \frac{x^8}{8!} - \cdots$$

$$e^x = 1 + x + \frac{x^2}{2!} + \frac{x^3}{3!} + \frac{x^4}{4!} + \cdots$$

These series are also called *Taylor expansions* of the functions $\sin x$, $\cos x$, and e^x about $x = 0$. The *general term* of a Taylor series is a formula which gives any term in the series. For example, $x^n/n!$ is the general term in the Taylor expansion for e^x, and $(-1)^k x^{2k}/(2k)!$ is the general term in the expansion for $\cos x$. We call n or k the *index*.

Taylor Series in General

Any function f, all of whose derivatives exist at 0, has a Taylor series. However, the Taylor series for f does not necessarily converge to $f(x)$ for all values of x. For the values of x for which the series does converge to $f(x)$, we have the following formula:

Taylor Series for $f(x)$ about $x = 0$

$$f(x) = f(0) + f'(0)x + \frac{f''(0)}{2!}x^2 + \frac{f'''(0)}{3!}x^3 + \cdots + \frac{f^{(n)}(0)}{n!}x^n + \cdots$$

In addition, just as we have Taylor polynomials centered at points other than 0, we can also have a Taylor series centered at $x = a$ (provided all the derivatives of f exist at $x = a$). For the values of x for which the series converges to $f(x)$, we have the following formula:

Taylor Series for $f(x)$ about $x = a$

$$f(x) = f(a) + f'(a)(x - a) + \frac{f''(a)}{2!}(x - a)^2 + \frac{f'''(a)}{3!}(x - a)^3 + \cdots + \frac{f^{(n)}(a)}{n!}(x - a)^n + \cdots$$

The Taylor series is a power series whose partial sums are the Taylor polynomials. As we saw in Section 9.5, power series generally converge on an interval centered at $x = a$. The Taylor series for such a function can be interpreted when x is replaced by a complex number. This extends the domain of the function. See Problem 45.

For a given function f and a given x, even if the Taylor series converges, it might not converge to $f(x)$. However, the Taylor series for most commonly encountered functions, including e^x, $\cos x$, and $\sin x$, do converge to the original function for all x. See Section 10.4.

Intervals of Convergence of Taylor Series

Let us look again at the Taylor polynomial for $\ln x$ about $x = 1$ that we derived in Example 7 on page 483. A similar calculation gives the Taylor Series

$$\ln x = (x - 1) - \frac{(x - 1)^2}{2} + \frac{(x - 1)^3}{3} - \frac{(x - 1)^4}{4} + \cdots + (-1)^{n-1}\frac{(x - 1)^n}{n} + \cdots.$$

Example 4 on page 467 and Example 6 on page 468 show that this power series has interval of convergence $0 < x \leq 2$. However, although we know that the series converges in this interval, we do not yet know that its sum is $\ln x$. The fact that in Figure 10.10 the polynomials fit the curve well for $0 < x < 2$ suggests that the Taylor series does converge to $\ln x$ for $0 < x \leq 2$. For such x-values, a higher-degree polynomial gives, in general, a better approximation.

However, when $x > 2$, the polynomials move away from the curve and the approximations get worse as the degree of the polynomial increases. Thus, the Taylor polynomials are effective only as approximations to $\ln x$ for values of x between 0 and 2; outside that interval, they should not be used. Inside the interval, but near the ends, 0 or 2, the polynomials converge very slowly. This means we might have to take a polynomial of very high degree to get an accurate value for $\ln x$.

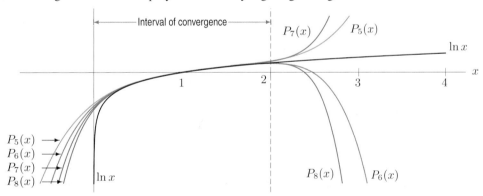

Figure 10.10: Taylor polynomials $P_5(x)$, $P_6(x)$, $P_7(x)$, $P_8(x)$, ... converge to $\ln x$ for $0 < x \leq 2$ and diverge outside that interval

To compute the interval of convergence exactly, we first compute the radius of convergence using the method on page 466. Convergence at the endpoints, $x = 0$ and $x = 2$, has to be determined separately. However, proving that the series converges to $\ln x$ on its interval of convergence, as Figure 10.10 suggests, requires the error term introduced in Section 10.4.

Example 1 Find the Taylor series for $\ln(1 + x)$ about $x = 0$, and calculate its interval of convergence.

Solution Taking derivatives of $\ln(1 + x)$ and substituting $x = 0$ leads to the Taylor series

$$\ln(1 + x) = x - \frac{x^2}{2} + \frac{x^3}{3} - \frac{x^4}{4} + \cdots.$$

Notice that this is the same series that we get by substituting $(1 + x)$ for x in the series for $\ln x$:

$$\ln x = (x - 1) - \frac{(x-1)^2}{2} + \frac{(x-1)^3}{3} - \frac{(x-1)^4}{4} + \cdots \qquad \text{for } 0 < x \le 2.$$

Since the series for $\ln x$ about $x = 1$ converges for $0 < x \le 2$, the interval of convergence for the Taylor series for $\ln(1 + x)$ about $x = 0$ is $-1 < x \le 1$. Thus we write

$$\ln(1 + x) = x - \frac{x^2}{2} + \frac{x^3}{3} - \frac{x^4}{4} + \cdots \qquad \text{for } -1 < x \le 1.$$

Notice that the series could not possibly converge to $\ln(1 + x)$ for $x \le -1$ since $\ln(1 + x)$ is not defined there.

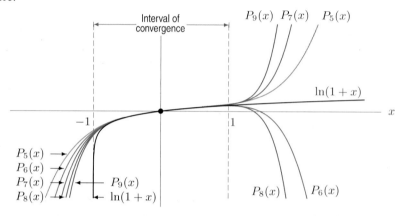

Figure 10.11: Interval of convergence for the Taylor series for $\ln(1 + x)$ is $-1 < x \le 1$

The Binomial Series Expansion

We now find the Taylor series about $x = 0$ for the function $f(x) = (1 + x)^p$, with p a constant, but not necessarily a positive integer. Taking derivatives:

$$
\begin{aligned}
f(x) &= (1 + x)^p & &\text{so} & f(0) &= 1 \\
f'(x) &= p(1 + x)^{p-1} & & & f'(0) &= p \\
f''(x) &= p(p - 1)(1 + x)^{p-2} & & & f''(0) &= p(p - 1) \\
f'''(x) &= p(p - 1)(p - 2)(1 + x)^{p-3}, & & & f'''(0) &= p(p - 1)(p - 2).
\end{aligned}
$$

Thus, the third-degree Taylor polynomial for x near 0 is

$$(1 + x)^p \approx P_3(x) = 1 + px + \frac{p(p - 1)}{2!}x^2 + \frac{p(p - 1)(p - 2)}{3!}x^3.$$

Graphing $P_3(x), P_4(x), \ldots$ for various specific values of p suggests that the Taylor polynomials converge to $f(x)$ for $-1 < x < 1$. (See Problems 26–28, page 489.) This can be confirmed using the radius of convergence test. The Taylor series for $f(x) = (1 + x)^p$ about $x = 0$ is as follows:

> **The Binomial Series**
>
> $$(1 + x)^p = 1 + px + \frac{p(p - 1)}{2!}x^2 + \frac{p(p - 1)(p - 2)}{3!}x^3 + \cdots \qquad \text{for } -1 < x < 1.$$

In fact the binomial series gives the same result as multiplying $(1 + x)^p$ out when p is a positive integer. (Newton discovered that the binomial series can be used for noninteger exponents.)

Example 2   Use the binomial series with $p = 3$ to expand $(1 + x)^3$.

Solution   The series is

$$(1+x)^3 = 1 + 3x + \frac{3 \cdot 2}{2!}x^2 + \frac{3 \cdot 2 \cdot 1}{3!}x^3 + \frac{3 \cdot 2 \cdot 1 \cdot 0}{4!}x^4 + \cdots.$$

The x^4 term and all terms beyond it turn out to be zero, because each coefficient contains a factor of 0. Simplifying gives

$$(1+x)^3 = 1 + 3x + 3x^2 + x^3,$$

which is the usual expansion obtained by multiplying out $(1+x)^3$.

Example 3   Find the Taylor series about $x = 0$ for $\dfrac{1}{1+x}$.

Solution   Since $\dfrac{1}{1+x} = (1+x)^{-1}$, use the binomial series with $p = -1$. Then

$$\frac{1}{1+x} = (1+x)^{-1} = 1 + (-1)x + \frac{(-1)(-2)}{2!}x^2 + \frac{(-1)(-2)(-3)}{3!}x^3 + \cdots$$

$$= 1 - x + x^2 - x^3 + \cdots \quad \text{for } -1 < x < 1.$$

This series is both a special case of the binomial series and an example of a geometric series. It converges for $-1 < x < 1$.

Exercises and Problems for Section 10.2

Exercises

For Exercises 1–8, find the first four nonzero terms of the Taylor series for the function about 0.

1. $(1+x)^{3/2}$

2. $\sqrt[4]{x+1}$

3. $\sin(-x)$

4. $\ln(1-x)$

5. $\dfrac{1}{1-x}$

6. $\dfrac{1}{\sqrt{1+x}}$

7. $\sqrt{1+x}$

8. $\sqrt[3]{1-y}$

For Exercises 9–15, find the first four terms of the Taylor series for the function about the point a.

9. $\sin x$, $a = \pi/4$

10. $\cos\theta$, $a = \pi/4$

11. $\sin\theta$, $a = -\pi/4$

12. $\tan x$, $a = \pi/4$

13. $1/x$, $a = 1$

14. $1/x$, $a = 2$

15. $1/x$, $a = -1$

Find an expression for the general term of the series in Exercises 16–23 and give the starting value of the index (n or k, for example).

16. $\dfrac{1}{1-x} = 1 + x + x^2 + x^3 + x^4 + \cdots$

17. $\dfrac{1}{1+x} = 1 - x + x^2 - x^3 + x^4 - \cdots$

18. $\ln(1-x) = -x - \dfrac{x^2}{2} - \dfrac{x^3}{3} - \dfrac{x^4}{4} - \cdots$

19. $\ln(1+x) = x - \dfrac{x^2}{2} + \dfrac{x^3}{3} - \dfrac{x^4}{4} + \dfrac{x^5}{5} - \cdots$

20. $\sin x = x - \dfrac{x^3}{3!} + \dfrac{x^5}{5!} - \dfrac{x^7}{7!} + \cdots$

21. $\arctan x = x - \dfrac{x^3}{3} + \dfrac{x^5}{5} - \dfrac{x^7}{7} + \cdots$

22. $e^{x^2} = 1 + x^2 + \dfrac{x^4}{2!} + \dfrac{x^6}{3!} + \dfrac{x^8}{4!} + \cdots$

23. $x^2\cos x^2 = x^2 - \dfrac{x^6}{2!} + \dfrac{x^{10}}{4!} - \dfrac{x^{14}}{6!} + \cdots$

Problems

24. (a) Find the terms up to degree 6 of the Taylor series for $f(x) = \sin(x^2)$ about $x = 0$ by taking derivatives.
(b) Compare your result in part (a) to the series for $\sin x$. How could you have obtained your answer to part (a) from the series for $\sin x$?

25. (a) Find the Taylor series for $f(x) = \ln(1 + 2x)$ about $x = 0$ by taking derivatives.

(b) Compare your result in part (a) to the series for $\ln(1 + x)$. How could you have obtained your answer to part (a) from the series for $\ln(1 + x)$?
(c) What do you expect the interval of convergence for the series for $\ln(1 + 2x)$ to be?

26. By graphing the function $f(x) = \sqrt{1+x}$ and several of its Taylor polynomials, estimate the interval of convergence of the series you found in Problem 7.

27. By graphing the function $f(x) = \dfrac{1}{\sqrt{1+x}}$ and several of its Taylor polynomials, estimate the interval of convergence of the series you found in Problem 6.

28. By graphing the function $f(x) = \dfrac{1}{1-x}$ and several of its Taylor polynomials, estimate the interval of convergence of the series you found in Problem 5. Compute the radius of convergence analytically.

29. Find the radius of convergence of the Taylor series around $x = 0$ for $\ln(1-x)$.

30. **(a)** Write the general term of the binomial series for $(1+x)^p$ about $x = 0$.
   (b) Find the radius of convergence of this series.

By recognizing each series in Problems 31–39 as a Taylor series evaluated at a particular value of x, find the sum of each of the following convergent series.

31. $1 + \dfrac{2}{1!} + \dfrac{4}{2!} + \dfrac{8}{3!} + \cdots + \dfrac{2^n}{n!} + \cdots$

32. $1 - \dfrac{1}{3!} + \dfrac{1}{5!} - \dfrac{1}{7!} + \cdots + \dfrac{(-1)^n}{(2n+1)!} + \cdots$

33. $1 + \dfrac{1}{4} + \left(\dfrac{1}{4}\right)^2 + \left(\dfrac{1}{4}\right)^3 + \cdots + \left(\dfrac{1}{4}\right)^n + \cdots$

34. $1 - \dfrac{100}{2!} + \dfrac{10000}{4!} + \cdots + \dfrac{(-1)^n \cdot 10^{2n}}{(2n)!} + \cdots$

35. $\dfrac{1}{2} - \dfrac{\left(\frac{1}{2}\right)^2}{2} + \dfrac{\left(\frac{1}{2}\right)^3}{3} - \dfrac{\left(\frac{1}{2}\right)^4}{4} + \cdots + \dfrac{(-1)^n \cdot \left(\frac{1}{2}\right)^{n+1}}{(n+1)} + \cdots$

36. $1 - 0.1 + 0.1^2 - 0.1^3 + \cdots$

37. $1 + 3 + \dfrac{9}{2!} + \dfrac{27}{3!} + \dfrac{81}{4!} + \cdots$

38. $1 - \dfrac{1}{2!} + \dfrac{1}{4!} - \dfrac{1}{6!} + \cdots$

39. $1 - 0.1 + \dfrac{0.01}{2!} - \dfrac{0.001}{3!} + \cdots$

In Problems 40–41 solve exactly for the variable.

40. $1 + x + x^2 + x^3 + \cdots = 5$

41. $x - \dfrac{1}{2}x^2 + \dfrac{1}{3}x^3 + \cdots = 0.2$

42. One of the two sets of functions, f_1, f_2, f_3, or g_1, g_2, g_3 is graphed in Figure 10.12; the other set is graphed in Figure 10.13. Taylor series for the functions about a point corresponding to either A or B are as follows:

$$f_1(x) = 3 + (x-1) - (x-1)^2 + \cdots$$

$$f_2(x) = 3 - (x-1) + (x-1)^2 + \cdots$$
$$f_3(x) = 3 - 2(x-1) + (x-1)^2 + \cdots$$
$$g_1(x) = 5 - (x-4) - (x-4)^2 + \cdots$$
$$g_2(x) = 5 - (x-4) + (x-4)^2 + \cdots$$
$$g_3(x) = 5 + (x-4) + (x-4)^2 + \cdots$$

(a) Which group of functions is represented in each figure?
(b) What are the coordinates of the points A and B?
(c) Match each function with the graphs (I)–(III) in the appropriate figure.

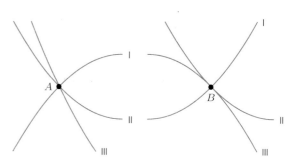

Figure 10.12   Figure 10.13

43. Suppose that you are told that the Taylor series of $f(x) = x^2 e^{x^2}$ about $x = 0$ is

$$x^2 + x^4 + \dfrac{x^6}{2!} + \dfrac{x^8}{3!} + \dfrac{x^{10}}{4!} + \cdots.$$

Find $\dfrac{d}{dx}\left(x^2 e^{x^2}\right)\bigg|_{x=0}$ and $\dfrac{d^6}{dx^6}\left(x^2 e^{x^2}\right)\bigg|_{x=0}$.

44. Suppose all the derivatives of some function f exist at 0, and that Taylor series for f about $x = 0$ is

$$x + \dfrac{x^2}{2} + \dfrac{x^3}{3} + \dfrac{x^4}{4} + \cdots + \dfrac{x^n}{n} + \cdots.$$

Find $f'(0)$, $f''(0)$, $f'''(0)$, and $f^{(10)}(0)$.

45. Let $i = \sqrt{-1}$. We define $e^{i\theta}$ by substituting $i\theta$ in the Taylor series for e^x. Use this definition[2] to explain Euler's formula

$$e^{i\theta} = \cos\theta + i\sin\theta.$$

10.3 FINDING AND USING TAYLOR SERIES

Finding a Taylor series for a function means finding the coefficients. Assuming the function has all its derivatives defined, finding the coefficients can always be done, in theory at least, by differentiation. That is how we derived the four most important Taylor series, those for the functions e^x, $\sin x$, $\cos x$, and $(1+x)^p$.

---

[2]Complex numbers are discussed in Appendix B.

For many functions, however, computing Taylor series coefficients by differentiation can be a very laborious business. We now introduce easier ways of finding Taylor series, if the series we want is closely related to a series that we already know.

New Series by Substitution

Suppose we want to find the Taylor series for e^{-x^2} about $x = 0$. We could find the coefficients by differentiation. Differentiating e^{-x^2} by the chain rule gives $-2xe^{-x^2}$, and differentiating again gives $-2e^{-x^2} + 4x^2e^{-x^2}$. Each time we differentiate we use the product rule, and the number of terms grows. Finding the tenth or twentieth derivative of e^{-x^2}, and thus the series for e^{-x^2} up to the x^{10} or x^{20} terms, by this method is tiresome (at least without a computer or calculator that can differentiate).

Fortunately, there's a quicker way. Recall that

$$e^y = 1 + y + \frac{y^2}{2!} + \frac{y^3}{3!} + \frac{y^4}{4!} + \cdots \quad \text{for all } y.$$

Substituting $y = -x^2$ tells us that

$$e^{-x^2} = 1 + (-x^2) + \frac{(-x^2)^2}{2!} + \frac{(-x^2)^3}{3!} + \frac{(-x^2)^4}{4!} + \cdots$$

$$= 1 - x^2 + \frac{x^4}{2!} - \frac{x^6}{3!} + \frac{x^8}{4!} + \cdots \quad \text{for all } x.$$

Using this method, it is easy to find the series up to the x^{10} or x^{20} terms. It can be shown that this is the Taylor series for e^{-x^2}.

Example 1   Find the Taylor series about $x = 0$ for $f(x) = \dfrac{1}{1 + x^2}$.

Solution   The binomial series tells us that

$$\frac{1}{1 + y} = (1 + y)^{-1} = 1 - y + y^2 - y^3 + y^4 + \cdots \quad \text{for } -1 < y < 1.$$

Substituting $y = x^2$ gives

$$\frac{1}{1 + x^2} = 1 - x^2 + x^4 - x^6 + x^8 + \cdots \quad \text{for } -1 < x < 1,$$

which is the Taylor series for $\dfrac{1}{1 + x^2}$.

These examples demonstrate that we can get new series from old ones by substitution. More advanced texts show that series obtained by substituting $g(x)$ into a Taylor series for $f(x)$ gives a Taylor series for $f(g(x))$.

In Example 1, we made the substitution $y = x^2$. We can also substitute an entire series into another one, as in the next example.

Example 2   Find the Taylor series about $\theta = 0$ for $g(\theta) = e^{\sin \theta}$.

Solution   For all y and θ, we know that

$$e^y = 1 + y + \frac{y^2}{2!} + \frac{y^3}{3!} + \frac{y^4}{4!} + \cdots$$

and

$$\sin \theta = \theta - \frac{\theta^3}{3!} + \frac{\theta^5}{5!} - \cdots.$$

Let's substitute the series for $\sin \theta$ for y:

$$e^{\sin \theta} = 1 + \left(\theta - \frac{\theta^3}{3!} + \frac{\theta^5}{5!} - \cdots \right) + \frac{1}{2!} \left(\theta - \frac{\theta^3}{3!} + \frac{\theta^5}{5!} - \cdots \right)^2 + \frac{1}{3!} \left(\theta - \frac{\theta^3}{3!} + \frac{\theta^5}{5!} - \cdots \right)^3 + \cdots.$$

To simplify, we multiply out and collect terms. The only constant term is the 1, and there's only one θ term. The only θ^2 term is the first term we get by multiplying out the square, namely $\theta^2/2!$. There are two contributors to the θ^3 term: the $-\theta^3/3!$ from within the first parentheses, and the first term we get from multiplying out the cube, which is $\theta^3/3!$. Thus the series starts

$$e^{\sin \theta} = 1 + \theta + \frac{\theta^2}{2!} + \left(-\frac{\theta^3}{3!} + \frac{\theta^3}{3!} \right) + \cdots$$

$$= 1 + \theta + \frac{\theta^2}{2!} + 0 \cdot \theta^3 + \cdots \quad \text{for all } \theta.$$

New Series by Differentiation and Integration

Just as we can get new series by substitution, we can also get new series by differentiation and integration. It can be shown that term-by-term differentiation of a Taylor series for $f(x)$ gives a Taylor series for $f'(x)$ and that the two series have the same radius of convergence. Integration works similarly.

Example 3   Find the Taylor Series about $x = 0$ for $\dfrac{1}{(1 - x)^2}$ from the series for $\dfrac{1}{1 - x}$.

Solution   We know that $\dfrac{d}{dx} \left(\dfrac{1}{1 - x} \right) = \dfrac{1}{(1 - x)^2}$, so we start with the geometric series

$$\frac{1}{1 - x} = 1 + x + x^2 + x^3 + x^4 + \cdots \quad \text{for } -1 < x < 1.$$

Differentiation term by term gives the binomial series

$$\frac{1}{(1 - x)^2} = \frac{d}{dx} \left(\frac{1}{1 - x} \right) = 1 + 2x + 3x^2 + 4x^3 + \cdots \quad \text{for } -1 < x < 1.$$

Example 4   Find the Taylor series about $x = 0$ for $\arctan x$ from the series for $\dfrac{1}{1 + x^2}$.

Solution   We know that $\dfrac{d(\arctan x)}{dx} = \dfrac{1}{1 + x^2}$, so we use the series from Example 1 on page 491:

$$\frac{d(\arctan x)}{dx} = \frac{1}{1 + x^2} = 1 - x^2 + x^4 - x^6 + x^8 - \cdots \quad \text{for } -1 < x < 1.$$

Antidifferentiating term by term gives

$$\arctan x = \int \frac{1}{1 + x^2} \, dx = C + x - \frac{x^3}{3} + \frac{x^5}{5} - \frac{x^7}{7} + \frac{x^9}{9} - \cdots \quad \text{for } -1 < x < 1,$$

where C is the constant of integration. Since $\arctan 0 = 0$, we have $C = 0$, so

$$\arctan x = x - \frac{x^3}{3} + \frac{x^5}{5} - \frac{x^7}{7} + \frac{x^9}{9} - \cdots \quad \text{for } -1 < x < 1.$$

The series for $\arctan x$ was discovered by James Gregory (1638–1675).

Applications of Taylor Series

Example 5   Use the series for $\arctan x$ to estimate the numerical value of π.

Solution   Since $\arctan 1 = \pi/4$, we use the series for $\arctan x$ from Example 4. We assume—as is the case—that the series does converge to $\pi/4$ at $x = 1$, the endpoint of its interval of convergence.

Substituting $x = 1$ into the series for $\arctan x$ gives

$$\pi = 4 \arctan 1 = 4 \left(1 - \frac{1}{3} + \frac{1}{5} - \frac{1}{7} + \frac{1}{9} - \cdots \right).$$

Table 10.1   *Approximating π using the series for $\arctan x$*

n	4	5	25	100	500	1000	10,000
S_n	2.895	3.340	3.182	3.132	3.140	3.141	3.141

Table 10.1 shows the value of the nth partial sum, S_n, obtained by summing the nonzero terms from 1 through n. The values of S_n do seem to converge to $\pi = 3.141\ldots$. However, this series converges very slowly, meaning that we have to take a large number of terms to get an accurate estimate for π. So this way of calculating π is not particularly practical. (A better one is given in Problem 1, page 519.) However, the expression for π given by this series is surprising and elegant.

A basic question we can ask about two functions is which one gives larger values. Taylor series can often be used to answer this question over a small interval. If the constant terms of the series for two functions are the same, compare the linear terms; if the linear terms are the same, compare the quadratic terms, and so on.

Example 6   By looking at their Taylor series, decide which of the following functions is largest, and which is smallest, for θ near 0.   (a) $1 + \sin \theta$ (b) e^θ (c) $\dfrac{1}{\sqrt{1 - 2\theta}}$

Solution   The Taylor expansion about $\theta = 0$ for $\sin \theta$ is

$$\sin \theta = \theta - \frac{\theta^3}{3!} + \frac{\theta^5}{5!} - \frac{\theta^7}{7!} + \cdots.$$

So

$$1 + \sin \theta = 1 + \theta - \frac{\theta^3}{3!} + \frac{\theta^5}{5!} - \frac{\theta^7}{7!} + \cdots.$$

The Taylor expansion about $\theta = 0$ for e^θ is

$$e^\theta = 1 + \theta + \frac{\theta^2}{2!} + \frac{\theta^3}{3!} + \frac{\theta^4}{4!} + \cdots.$$

The Taylor expansion about $\theta = 0$ for $1/\sqrt{1 + \theta}$ is

$$\frac{1}{\sqrt{1 + \theta}} = (1 + \theta)^{-1/2} = 1 - \frac{1}{2}\theta + \frac{(-\frac{1}{2})(-\frac{3}{2})}{2!}\theta^2 + \frac{(-\frac{1}{2})(-\frac{3}{2})(-\frac{5}{2})}{3!}\theta^3 + \cdots$$

$$= 1 - \frac{1}{2}\theta + \frac{3}{8}\theta^2 - \frac{5}{16}\theta^3 + \cdots.$$

So, substituting -2θ for θ:

$$\frac{1}{\sqrt{1 - 2\theta}} = 1 - \frac{1}{2}(-2\theta) + \frac{3}{8}(-2\theta)^2 - \frac{5}{16}(-2\theta)^3 + \cdots$$

$$= 1 + \theta + \frac{3}{2}\theta^2 + \frac{5}{2}\theta^3 + \cdots.$$

For θ near 0, we can neglect terms beyond the second degree. We are left with the approximations:

$$1 + \sin \theta \approx 1 + \theta$$

$$e^\theta \approx 1 + \theta + \frac{\theta^2}{2}$$

$$\frac{1}{\sqrt{1 - 2\theta}} \approx 1 + \theta + \frac{3}{2}\theta^2.$$

Since

$$1 + \theta < 1 + \theta + \frac{1}{2}\theta^2 < 1 + \theta + \frac{3}{2}\theta^2,$$

and since the approximations are valid for θ near 0, we conclude that, for θ near 0,

$$1 + \sin\theta < e^{\theta} < \frac{1}{\sqrt{1 - 2\theta}}.$$

Example 7   Two electrical charges of equal magnitude and opposite signs located near one another are called an electrical dipole. The charges Q and $-Q$ are a distance r apart. (See Figure 10.14.) The electric field, E, at the point P is given by

$$E = \frac{Q}{R^2} - \frac{Q}{(R + r)^2}.$$

Use series to investigate the behavior of the electric field far away from the dipole. Show that when R is large in comparison to r, the electric field is approximately proportional to $1/R^3$.

Figure 10.14: Approximating the electric field at P due to a dipole consisting of charges Q and $-Q$ a distance r apart

Solution   In order to use a series approximation, we need a variable whose value is small. Although we know that r is much smaller than R, we do not know that r itself is small. The quantity r/R is, however, very small. Hence we expand $1/(R + r)^2$ in powers of r/R so that we can safely use only the first few terms of the Taylor series. First we rewrite using algebra:

$$\frac{1}{(R + r)^2} = \frac{1}{R^2(1 + r/R)^2} = \frac{1}{R^2}\left(1 + \frac{r}{R}\right)^{-2}.$$

Now we use the binomial expansion for $(1 + x)^p$ with $x = r/R$ and $p = -2$:

$$\frac{1}{R^2}\left(1 + \frac{r}{R}\right)^{-2} = \frac{1}{R^2}\left(1 + (-2)\left(\frac{r}{R}\right) + \frac{(-2)(-3)}{2!}\left(\frac{r}{R}\right)^2 + \frac{(-2)(-3)(-4)}{3!}\left(\frac{r}{R}\right)^3 + \cdots\right)$$

$$= \frac{1}{R^2}\left(1 - 2\frac{r}{R} + 3\frac{r^2}{R^2} - 4\frac{r^3}{R^3} + \cdots\right).$$

So, substituting the series into the expression for E, we have

$$E = \frac{Q}{R^2} - \frac{Q}{(R + r)^2} = Q\left[\frac{1}{R^2} - \frac{1}{R^2}\left(1 - 2\frac{r}{R} + 3\frac{r^2}{R^2} - 4\frac{r^3}{R^3} + \cdots\right)\right]$$

$$= \frac{Q}{R^2}\left(2\frac{r}{R} - 3\frac{r^2}{R^2} + 4\frac{r^3}{R^3} - \cdots\right).$$

Since r/R is smaller than 1, the binomial expansion for $(1 + r/R)^{-2}$ converges. We are interested in the electric field far away from the dipole. The quantity r/R is small there, and $(r/R)^2$ and higher powers are smaller still. Thus, we approximate by disregarding all terms except the first, giving

$$E \approx \frac{Q}{R^2}\left(\frac{2r}{R}\right), \quad \text{so} \quad E \approx \frac{2Qr}{R^3}.$$

Since Q and r are constants, this means that E is approximately proportional to $1/R^3$.

In the previous example, we say that E is *expanded in terms of* r/R, meaning that the variable in the expansion is r/R.

Exercises and Problems for Section 10.3

Exercises

Find the first four nonzero terms of the Taylor series about 0 for the functions in Exercises 1–12.

1. e^{-x}

2. $\cos(\theta^2)$

3. $\sqrt{1-2x}$

4. $\ln(1-2y)$

5. $\arcsin x$

6. $\phi^3 \cos(\phi^2)$

7. $\dfrac{1}{\sqrt{1-z^2}}$

8. $\dfrac{z}{e^{z^2}}$

9. $t\sin(3t)$

10. $\sqrt{1+\sin\theta}$

11. $\sqrt{(1+t)}\sin t$

12. $e^t \cos t$

Find the Taylor series about 0 for the functions in Exercises 13–15, including the general term.

13. $(1+x)^3$

14. $t\sin(t^2)-t^3$

15. $\dfrac{1}{\sqrt{1-y^2}}$

For Exercises 16–21, expand the quantity about 0 in terms of the variable given. Give four nonzero terms.

16. $\dfrac{1}{2+x}$ in terms of $\dfrac{x}{2}$

17. $\dfrac{1}{a-r}$ in terms of $\dfrac{r}{a}$

18. $\sqrt{T+h}$ in terms of $\dfrac{h}{T}$

19. $\sqrt[3]{P+t}$ in terms of $\dfrac{t}{P}$

20. $\dfrac{1}{(a+r)^2}$ in terms of $\dfrac{r}{a}$

21. $\dfrac{a}{\sqrt{a^2+x^2}}$ in terms of $\dfrac{x}{a}$, where $a>0$

Problems

22. **(a)** Find the first three nonzero terms of the Taylor series for $e^x + e^{-x}$.
   (b) Explain why the graph of $e^x + e^{-x}$ looks like a parabola near $x = 0$. What is the equation of this parabola?

23. **(a)** Find the first three nonzero terms of the Taylor series for $e^x - e^{-x}$.
   (b) Explain why the graph of $e^x - e^{-x}$ near $x = 0$ looks like the graph of a cubic polynomial symmetric about the origin. What is the equation for this cubic?

24. Find the sum of $\displaystyle\sum_{n=1}^{\infty} \frac{k^{n-1}}{(n-1)!}e^{-k}$.

25. Find the sum of $\displaystyle\sum_{p=1}^{\infty} px^{p-1}$ for $|x| < 1$.

26. By looking at the Taylor series, decide which of the following functions is largest, and which is smallest, for small positive θ.
   (a) $1 + \sin\theta$ **(b)** $\cos\theta$ **(c)** $\dfrac{1}{1-\theta^2}$

27. For values of y near 0, put the following functions in increasing order, using their Taylor expansions.
   (a) $\ln(1+y^2)$ **(b)** $\sin(y^2)$ **(c)** $1 - \cos y$

28. Figure 10.15 shows the graphs of the four functions below for values of x near 0. Use Taylor series to match graphs and formulas.
   (a) $\dfrac{1}{1-x^2}$ **(b)** $(1+x)^{1/4}$
   (c) $\sqrt{1+\dfrac{x}{2}}$ **(d)** $\dfrac{1}{\sqrt{1-x}}$

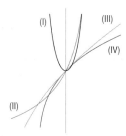

Figure 10.15

29. Consider the functions $y = e^{-x^2}$ and $y = 1/(1+x^2)$.
   (a) Write the Taylor expansions for the two functions about $x = 0$. What is similar about the two series? What is different?
   (b) Looking at the series, which function do you predict will be greater over the interval $(-1, 1)$? Graph both and see.
   (c) Are these functions even or odd? How might you see this by looking at the series expansions?
   (d) By looking at the coefficients, explain why it is reasonable that the series for $y = e^{-x^2}$ converges for all values of x, but the series for $y = 1/(1+x^2)$ converges only on $(-1, 1)$.

30. The hyperbolic sine and cosine are differentiable and satisfy the conditions $\cosh 0 = 1$ and $\sinh 0 = 0$, and

$$\frac{d}{dx}(\cosh x) = \sinh x \qquad \frac{d}{dx}(\sinh x) = \cosh x.$$

   (a) Using only this information, find the Taylor approximation of degree $n = 8$ about $x = 0$ for $f(x) = \cosh x$.

(b) Estimate the value of cosh 1.

(c) Use the result from part (a) to find a Taylor polynomial approximation of degree $n = 7$ about $x = 0$ for $g(x) = \sinh x$.

31. Use the series for e^x to find the Taylor series for $\sinh 2x$ and $\cosh 2x$.

32. Padé approximants are rational functions used to approximate more complicated functions. In this problem, you will derive the Padé approximant to the exponential function.

   (a) Let $f(x) = (1 + ax)/(1 + bx)$, where a and b are constants. Write down the first three terms of the Taylor series for $f(x)$ about $x = 0$.

   (b) By equating the first three terms of the Taylor series about $x = 0$ for $f(x)$ and for e^x, find a and b so that $f(x)$ approximates e^x as closely as possible near $x = 0$.

33. An electric dipole on the x-axis consists of a charge Q at $x = 1$ and a charge $-Q$ at $x = -1$. The electric field, E, at the point $x = R$ on the x-axis is given (for $R > 1$) by

$$E = \frac{kQ}{(R-1)^2} - \frac{kQ}{(R+1)^2}$$

where k is a positive constant whose value depends on the units. Expand E as a series in $1/R$, giving the first two nonzero terms.

34. Assume a is a positive constant. Suppose z is given by the expression

$$z = \sqrt{a^2 + x^2} - \sqrt{a^2 - x^2}.$$

Expand z as a series in x as far as the second nonzero term.

35. The electric potential, V, at a distance R along the axis perpendicular to the center of a charged disc with radius a and constant charge density σ, is given by

$$V = 2\pi\sigma(\sqrt{R^2 + a^2} - R).$$

Show that, for large R,

$$V \approx \frac{\pi a^2 \sigma}{R}.$$

36. One of Einstein's most amazing predictions was that light traveling from distant stars would bend around the sun on the way to earth. His calculations involved solving for ϕ in the equation

$$\sin\phi + b(1 + \cos^2\phi + \cos\phi) = 0$$

where b is a very small positive constant.

   (a) Explain why the equation could have a solution for ϕ which is near 0.

   (b) Expand the left-hand side of the equation in Taylor series about $\phi = 0$, disregarding terms of order ϕ^2 and higher. Solve for ϕ. (Your answer will involve b.)

37. The Michelson-Morley experiment, which contributed to the formulation of the Theory of Relativity, involved the difference between the two times t_1 and t_2 that light took to travel between two points. If v is the velocity of light; l_1, l_2, and c are constants; and $v < c$, then the times t_1 and t_2 are given by

$$t_1 = \frac{2l_2}{c(1 - v^2/c^2)} - \frac{2l_1}{c\sqrt{1 - v^2/c^2}}$$

$$t_2 = \frac{2l_2}{c\sqrt{1 - v^2/c^2}} - \frac{2l_1}{c(1 - v^2/c^2)}.$$

   (a) Find an expression for $\Delta t = t_1 - t_2$, and give its Taylor expansion in terms of v^2/c^2 up to the second nonzero term.

   (b) For small v, to what power of v is Δt proportional? What is the constant of proportionality?

38. A hydrogen atom consists of an electron, of mass m, orbiting a proton, of mass M, where m is much smaller than M. The reduced mass, μ, of the hydrogen atom is defined by

$$\mu = \frac{mM}{m + M}.$$

   (a) Show that $\mu \approx m$.

   (b) To get a more accurate approximation for μ, express μ as m times a series in m/M.

   (c) The approximation $\mu \approx m$ is obtained by disregarding all but the constant term in the series. The first-order correction is obtained by including the linear term but no higher terms. If $m \approx M/1836$, by what percentage does including the first-order correction change the estimate $\mu \approx m$?

39. A thin disk of radius a and mass M lies horizontally; a particle of mass m is at a height h directly above the center of the disk. The gravitational force, F, exerted by the disk on the mass m is given by

$$F = \frac{2GMmh}{a^2}\left(\frac{1}{h} - \frac{1}{(a^2 + h^2)^{1/2}}\right).$$

Assume $a < h$ and think of F as a function of a, with the other quantities constant.

   (a) Expand F as a series in a/h. Give the first two nonzero terms.

   (b) Show that the approximation for F obtained by using only the first nonzero term in the series is independent of the radius, a.

   (c) If $a = 0.02h$, by what percentage does the approximation in part (a) differ from the approximation in part (b)?

40. When a body is near the surface of the earth, we usually assume that the force due to gravity on it is a constant mg, where m is the mass of the body and g is the acceleration due to gravity at sea level. For a body at a distance h above the surface of the earth, a more accurate expression for the force F is

$$F = \frac{mgR^2}{(R+h)^2}$$

where R is the radius of the earth. We will consider the situation in which the body is close to the surface of the earth so that h is much smaller than R.

(a) Show that $F \approx mg$.
(b) Express F as mg multiplied by a series in h/R.
(c) The first-order correction to the approximation $F \approx mg$ is obtained by taking the linear term in the series but no higher terms. How far above the surface of the earth can you go before the first-order correc-

tion changes the estimate $F \approx mg$ by more than 10%? (Assume $R = 6400$ km.)

41. (a) Estimate the value of $\int_0^1 e^{-x^2} dx$ using Riemann sums for both left-hand and right-hand sums with $n = 5$ subdivisions.
(b) Approximate the function $f(x) = e^{-x^2}$ with a Taylor polynomial of degree 6.
(c) Estimate the integral in part (a) by integrating the Taylor polynomial from part (b).
(d) Indicate briefly how you could improve the results in each case.

42. Use Taylor series to explain the patterns in the digits in the following expansions:

(a) $\dfrac{1}{0.98} = 1.020408163264\ldots$

(b) $\left(\dfrac{1}{0.99}\right)^2 = 1.020304050607\ldots$

10.4   THE ERROR IN TAYLOR POLYNOMIAL APPROXIMATIONS

In order to use an approximation with confidence, we need to be able to estimate the size of the error, which is the difference between the exact answer (which we do not know) and the approximate value.

When we use $P_n(x)$, the nth degree Taylor polynomial, to approximate $f(x)$, the error is the difference

$$E_n(x) = f(x) - P_n(x).$$

We are interested in finding a *bound* on the magnitude of the error, $|E_n|$; that is, we want a number which we are sure is bigger than $|E_n|$. In practice, we want a bound which is reasonably close to the maximum value of $|E_n|$.

Finding an Error Bound

Recall that we constructed $P_n(x)$, the Taylor polynomial of f about 0, so that its first n derivatives equal the corresponding derivatives of $f(x)$. Therefore, $E_n(0) = 0$, $E_n'(0) = 0$, $E_n''(0) = 0$, $\cdots$, $E_n^{(n)}(0) = 0$. Since $P_n(x)$ is an nth degree polynomial, its $(n+1)^{\text{st}}$ derivative is 0, so $E_n^{(n+1)}(x) = f^{(n+1)}(x)$. In addition, suppose that $\left|f^{(n+1)}(x)\right|$ is bounded by a positive constant M, for all positive values of x near 0, say for $0 \le x \le d$, so that

$$-M \le f^{(n+1)}(x) \le M \quad \text{for } 0 \le x \le d.$$

This means that

$$-M \le E_n^{(n+1)}(x) \le M \quad \text{for } 0 \le x \le d.$$

Writing t for the variable, we integrate this inequality from 0 to x, giving

$$-\int_0^x M\,dt \le \int_0^x E_n^{(n+1)}(t)\,dt \le \int_0^x M\,dt \quad \text{for } 0 \le x \le d,$$

so

$$-Mx \le E_n^{(n)}(x) \le Mx \quad \text{for } 0 \le x \le d.$$

We integrate this inequality again from 0 to x, giving

$$-\int_0^x Mt\,dt \le \int_0^x E_n^{(n)}(t)\,dt \le \int_0^x Mt\,dt \quad \text{for } 0 \le x \le d,$$

so

$$-\frac{1}{2}Mx^2 \leq E_n^{(n-1)}(x) \leq \frac{1}{2}Mx^2 \quad \text{for } 0 \leq x \leq d.$$

By repeated integration, we obtain the following bound:

$$-\frac{1}{(n+1)!}Mx^{n+1} \leq E_n(x) \leq \frac{1}{(n+1)!}Mx^{n+1} \quad \text{for } 0 \leq x \leq d,$$

which means that

$$|E_n(x)| = |f(x) - P_n(x)| \leq \frac{1}{(n+1)!}Mx^{n+1} \quad \text{for } 0 \leq x \leq d.$$

When x is to the left of 0, so $-d \leq x \leq 0$, and when the Taylor series is centered at $a \neq 0$, similar calculations lead to the following result:

Theorem 10.1: The Lagrange Error Bound for $P_n(x)$

Suppose f and all its derivatives are continuous. If $P_n(x)$ is the nth Taylor polynomial for to $f(x)$ about a, then

$$|E_n(x)| = |f(x) - P_n(x)| \leq \frac{M}{(n+1)!}|x - a|^{n+1},$$

where $\max\left|f^{(n+1)}\right| \leq M$ on the interval between a and x.

Using the Lagrange Error Bound for Taylor Polynomials

Example 1   Give a bound on the error, E_4, when e^x is approximated by its fourth-degree Taylor polynomial about 0 for $-0.5 \leq x \leq 0.5$.

Solution   Let $f(x) = e^x$. We need to find a bound for the fifth derivative, $f^{(5)}(x) = e^x$. Since e^x is increasing,

$$|f^{(5)}(x)| \leq e^{0.5} = \sqrt{e} < 2 \quad \text{for} \quad -0.5 \leq x \leq 0.5.$$

So we can take $M = 2$. Then

$$|E_4| = |f(x) - P_4(x)| \leq \frac{2}{5!}|x|^5.$$

This means, for example, that on $-0.5 \leq x \leq 0.5$, the approximation

$$e^x \approx 1 + x + \frac{x^2}{2!} + \frac{x^3}{3!} + \frac{x^4}{4!}$$

has an error of at most $\frac{2}{120}(0.5)^5 < 0.0006$.

The Lagrange error bound for Taylor polynomials can be used to see how the accuracy of the approximation depends on the value of x or the value of n. Observe that the error for a Taylor polynomial of degree n depends on the $(n+1)^{\text{st}}$ power of $(x - a)$. That means, for example, with a Taylor polynomial of degree n centered at 0, if we decrease x by a factor of 2, the error bound decreases by a factor of 2^{n+1}.

Example 2   Compare the errors in the approximations

$$e^{0.1} \approx 1 + 0.1 + \frac{1}{2!}(0.1)^2 \quad \text{and} \quad e^{0.05} \approx 1 + (0.05) + \frac{1}{2!}(0.05)^2.$$

Solution   We are approximating e^x by its second-degree Taylor polynomial about 0. We evaluate the polynomial first at $x = 0.1$, and then at $x = 0.05$. Since we have decreased x by a factor of 2, the error bound decreases by a factor of about $2^3 = 8$. To see what actually happens to the errors, we compute them:

$$e^{0.1} - \left(1 + 0.1 + \frac{1}{2!}(0.1)^2\right) = 1.105171 - 1.105000 = 0.000171$$

$$e^{0.05} - \left(1 + 0.05 + \frac{1}{2!}(0.05)^2\right) = 1.051271 - 1.051250 = 0.000021$$

Since $(0.000171)/(0.000021) = 8.1$, the error has also decreased by a factor of about 8.

Convergence of Taylor Series for $\cos x$

We have already seen that the Taylor polynomials centered at $x = 0$ for $\cos x$ are good approximations for x near 0. (See Figure 10.16.) In fact, for any value of x, if we take a Taylor polynomial centered at $x = 0$ of high enough degree, its graph is nearly indistinguishable from the graph of the cosine near that point.

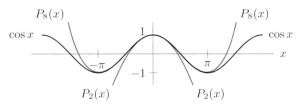

Figure 10.16: Graph of $\cos x$ and two Taylor polynomials for x near 0

Let's see what happens numerically. Let $x = \pi/2$. The successive Taylor polynomial approximations to $\cos(\pi/2) = 0$ about $x = 0$ are

$$\begin{aligned}
P_2(\pi/2) &= & 1 - (\pi/2)^2/2! & & = -0.23370\ldots \\
P_4(\pi/2) &= 1 - (\pi/2)^2/2! + (\pi/2)^4/4! &= & 0.01997\ldots \\
P_6(\pi/2) &= & \cdots & & = -0.00089\ldots \\
P_8(\pi/2) &= & \cdots & & = 0.00002\ldots.
\end{aligned}$$

It appears that the approximations converge to the true value, $\cos(\pi/2) = 0$, very rapidly. Now take a value of x somewhat farther away from 0, say $x = \pi$, then $\cos \pi = -1$ and

$$\begin{aligned}
P_2(\pi) &= 1 - (\pi)^2/2! = & -3.93480\ldots \\
P_4(\pi) &= \cdots & = 0.12391\ldots \\
P_6(\pi) &= \cdots & = -1.21135\ldots \\
P_8(\pi) &= \cdots & = -0.97602\ldots \\
P_{10}(\pi) &= \cdots & = -1.00183\ldots \\
P_{12}(\pi) &= \cdots & = -0.99990\ldots \\
P_{14}(\pi) &= \cdots & = -1.000004\ldots
\end{aligned}$$

We see that the rate of convergence is somewhat slower; it takes a 14th degree polynomial to approximate $\cos \pi$ as accurately as an 8th degree polynomial approximates $\cos(\pi/2)$. If x were taken still farther away from 0, then we would need still more terms to obtain as accurate an approximation of $\cos x$.

Using the ratio test, we can show the Taylor series for $\cos x$ converges for all values of x. In addition, we will prove that it converges to $\cos x$ using Theorem 10.1. Thus, we are justified in writing the equality:

$$\cos x = 1 - \frac{x^2}{2!} + \frac{x^4}{4!} - \frac{x^6}{6!} + \frac{x^8}{8!} - \cdots \qquad \text{for all } x.$$

Showing the Taylor Series for cos x Converges to cos x

The Lagrange error bound in Theorem 10.1 allows us to see if the Taylor series for a function converges to that function. In the series for $\cos x$, the odd powers are missing, so we assume n is even and write

$$E_n(x) = \cos x - P_n(x) = \cos x - \left(1 - \frac{x^2}{2!} + \cdots + (-1)^{n/2}\frac{x^n}{n!}\right),$$

giving

$$\cos x = 1 - \frac{x^2}{2!} + \cdots + (-1)^{n/2}\frac{x^n}{n!} + E_n(x).$$

Thus, for the Taylor series to converge to $\cos x$, we must have $E_n(x) \to 0$ as $n \to \infty$.

Showing $E_n(x) \to 0$ as $n \to \infty$

Proof Since $f(x) = \cos x$, the $(n+1)^{\text{st}}$ derivative $f^{(n+1)}(x)$ is $\pm \cos x$ or $\pm \sin x$, no matter what n is. So for all n, we have $|f^{(n+1)}(x)| \leq 1$ on the interval between 0 and x.

By the Lagrange error bound with $M = 1$, we have

$$|E_n(x)| = |\cos x - P_n(x)| \leq \frac{|x|^{n+1}}{(n+1)!} \quad \text{for every } n.$$

To show that the errors go to zero, we must show that for a fixed x,

$$\frac{|x|^{n+1}}{(n+1)!} \to 0 \quad \text{as} \quad n \to \infty.$$

To see why this is true, think about what happens when n is much larger than x. Suppose, for example, that $x = 17.3$. Let's look at the value of the sequence for n more than twice as big as 17.3, say $n = 36$, or $n = 37$, or $n = 38$:

$$\text{For } n = 36: \quad \frac{1}{37!}(17.3)^{37}$$

$$\text{For } n = 37: \quad \frac{1}{38!}(17.3)^{38} = \frac{17.3}{38} \cdot \frac{1}{37!}(17.3)^{37},$$

$$\text{For } n = 38: \quad \frac{1}{39!}(17.3)^{39} = \frac{17.3}{39} \cdot \frac{17.3}{38} \cdot \frac{1}{37!}(17.3)^{37}, \quad \cdots$$

Since $17.3/36$ is less than $\frac{1}{2}$, each time we increase n by 1, the term is multiplied by a number less than $\frac{1}{2}$. No matter what the value of $\frac{1}{37!}(17.3)^{37}$ is, if we keep on dividing it by two, the result gets closer to zero. Thus $\frac{1}{(n+1)!}(17.3)^{n+1}$ goes to 0 as n goes to infinity.

We can generalize this by replacing 17.3 by an arbitrary $|x|$. For $n > 2|x|$, the following sequence converges to 0 because each term is obtained from its predecessor by multiplying by a number less than $\frac{1}{2}$:

$$\frac{x^{n+1}}{(n+1)!}, \quad \frac{x^{n+2}}{(n+2)!}, \quad \frac{x^{n+3}}{(n+3)!}, \cdots.$$

Therefore, the Taylor series $1 - x^2/2! + x^4/4! - \cdots$ does converge to $\cos x$.

___

Problems 20 and 19 ask you to show that the Taylor series for $\sin x$ and e^x converge to the original function for all x. In each case, you again need the following limit:

$$\lim_{n \to \infty} \frac{x^n}{n!} = 0.$$

Exercises and Problems for Section 10.4

Exercises

Use Theorem 10.1 to find a reasonable bound for the error in approximating the quantity in Exercises 1–8 with a third-degree Taylor polynomial about $x = 0$.

1. $e^{0.1}$

2. $\sin(0.2)$

3. $\cos(-0.3)$

4. $\sqrt{0.9}$

5. $\ln(1.5)$

6. $1/\sqrt{3}$

7. $\tan 1$

8. $0.5^{1/3}$

Problems

9. You approximate $f(t) = e^t$ by a Taylor polynomial of degree 0 about $t = 0$ on the interval $[0, 0.5]$.

   (a) Reasoning informally, say whether the approximation is an overestimate or an underestimate.

   (b) Use Theorem 10.1 to bound the error. Check your answer graphically on a computer or calculator.

10. Repeat Problem 9 using the second-degree Taylor approximation, $P_2(t)$, to e^t.

11. Consider the error in using the approximation $\sin \theta \approx \theta$ on the interval $[-1, 1]$.

   (a) Reasoning informally, say where the approximation is an overestimate, and where is it an underestimate.

   (b) Use Theorem 10.1 to bound the error. Check your answer graphically on a computer or calculator.

12. Repeat Problem 11 for the approximation $\sin \theta \approx \theta - \theta^3/3!$.

13. **(a)** Use the graphs of $y = \cos x$ and its Taylor polynomials, $P_{10}(x)$ and $P_{20}(x)$, in Figure 10.17 to bound:

     (i) The error in approximating $\cos 6$ by $P_{10}(6)$ and by $P_{20}(6)$.

     (ii) The error in approximating $\cos x$ by $P_{20}(x)$ for $|x| \leq 9$.

   (b) If we want to approximate $\cos x$ by $P_{10}(x)$ to an accuracy of within 0.1, what is the largest interval of x-values on which we can work? Give your answer to the nearest integer.

Figure 10.17

14. Give a bound for the error for the nth degree Taylor polynomial about $x = 0$ approximating $\cos x$ on the interval $[0, 1]$. What is the bound for $\sin x$?

15. What degree Taylor polynomial about $x = 0$ do you need to calculate $\cos 1$ to four decimal places? To six decimal places? Justify your answer using the results of Problem 14.

16. **(a)** Using a calculator, make a table of the values to four decimal places of $\sin x$ for

$$x = -0.5, -0.4, \ldots, -0.1, 0, 0.1, \ldots, 0.4, 0.5.$$

   (b) Add to your table the values of the error $E_1 = \sin x - x$ for these x-values.

   (c) Using a calculator or computer, draw a graph of the quantity $E_1 = \sin x - x$ showing that

$$|E_1| < 0.03 \quad \text{for} \quad -0.5 \leq x \leq 0.5.$$

17. In this problem, you will investigate the error in the nth degree Taylor approximation to e^x about 0 for various values of n.

   (a) Let $E_1 = e^x - P_1(x) = e^x - (1 + x)$. Using a calculator or computer, graph E_1 for $-0.1 \leq x \leq 0.1$. What shape is the graph of E_1? Use the graph to confirm that

$$|E_1| \leq x^2 \quad \text{for} \quad -0.1 \leq x \leq 0.1.$$

   (b) Let $E_2 = e^x - P_2(x) = e^x - (1 + x + x^2/2)$. Choose a suitable range and graph E_2 for $-0.1 \leq x \leq 0.1$. What shape is the graph of E_2? Use the graph to confirm that

$$|E_2| \leq x^3 \quad \text{for} \quad -0.1 \leq x \leq 0.1.$$

   (c) Explain why the graphs of E_1 and E_2 have the shapes they do.

18. For $|x| \leq 0.1$, graph the error

$$E_0 = \cos x - P_0(x) = \cos x - 1.$$

Explain the shape of the graph, using the Taylor expansion of $\cos x$. Find a bound for $|E_0|$ for $|x| \leq 0.1$.

19. Show that the Taylor series about 0 for e^x converges to e^x for every x. Do this by showing that the error $E_n(x) \to 0$ as $n \to \infty$.

20. Show that the Taylor series about 0 for $\sin x$ converges to $\sin x$ for every x.

10.5 FOURIER SERIES

We have seen how to approximate a function by a Taylor polynomial of fixed degree. Such a polynomial is usually very close to the true value of the function near one point (the point at which the Taylor polynomial is centered), but not necessarily at all close anywhere else. In other words, Taylor polynomials are good approximations of a function *locally*, but not necessarily *globally*. In this section, we take another approach: we approximate the function by trigonometric functions, called *Fourier approximations*. The resulting approximation may not be as close to the original function at some points as the Taylor polynomial. However, the Fourier approximation is, in general, close over a larger interval. In other words, a Fourier approximation can be a better approximation globally. In addition, Fourier approximations are useful even for functions that are not continuous. Unlike Taylor approximations, Fourier approximations are periodic, so they are particularly useful for approximating periodic functions.

Many processes in nature are periodic or repeating, so it makes sense to approximate them by periodic functions. For example, sound waves are made up of periodic oscillations of air molecules. Heartbeats, the movement of the lungs, and the electrical current that powers our homes are all periodic phenomena. Two of the simplest periodic functions are the square wave in Figure 10.18 and the triangular wave in Figure 10.19. Electrical engineers use the square wave as the model for the flow of electricity when a switch is repeatedly flicked on and off.

Figure 10.18: Square wave

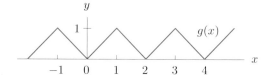

Figure 10.19: Triangular wave

Fourier Polynomials

We can express the square wave and the triangular wave by the formulas

$$f(x) = \begin{cases} \vdots & \vdots \\ 0 & -1 \le x < 0 \\ 1 & 0 \le x < 1 \\ 0 & 1 \le x < 2 \\ 1 & 2 \le x < 3 \\ 0 & 3 \le x < 4 \\ \vdots & \vdots \end{cases} \qquad g(x) = \begin{cases} \vdots & \vdots \\ -x & -1 \le x < 0 \\ x & 0 \le x < 1 \\ 2 - x & 1 \le x < 2 \\ x - 2 & 2 \le x < 3 \\ 4 - x & 3 \le x < 4 \\ \vdots & \vdots \end{cases}$$

However, these formulas are not particularly easy to work with. Worse, the functions are not differentiable at various points. Here we show how to approximate such functions by differentiable, periodic functions.

Since the sine and cosine are the simplest periodic functions, they are the building blocks we use. Because they repeat every 2π, we assume that the function f we want to approximate repeats every 2π. (Later, we deal with the case where f has some other period.) We start by considering the square wave in Figure 10.20. Because of the periodicity of all the functions concerned, we only have to consider what happens in the course of a single period; the same behavior repeats in any other period.

$$f(x) = \begin{cases} 0 & -\pi \le x < 0 \\ 1 & 0 \le x < \pi \end{cases}$$

Figure 10.20: Square wave on $[-\pi, \pi]$

We will attempt to approximate f with a sum of trigonometric functions of the form

$$f(x) \approx F_n(x)$$
$$= a_0 + a_1 \cos x + a_2 \cos(2x) + a_3 \cos(3x) + \cdots + a_n \cos(nx)$$
$$+ b_1 \sin x + b_2 \sin(2x) + b_3 \sin(3x) + \cdots + b_n \sin(nx)$$
$$= a_0 + \sum_{k=1}^{n} a_k \cos(kx) + \sum_{k=1}^{n} b_k \sin(kx).$$

$F_n(x)$ is known as a *Fourier polynomial of degree* n, named after the French mathematician Joseph Fourier (1768-1830), who was one of the first to investigate it.[3] The coefficients a_k and b_k are called *Fourier coefficients*. Since each of the component functions $\cos(kx)$ and $\sin(kx)$, $k = 1, 2, \ldots, n$, repeats every 2π, $F_n(x)$ must repeat every 2π and so is a potentially good match for $f(x)$, which also repeats every 2π. The problem is to determine values for the Fourier coefficients that achieve a close match between $f(x)$ and $F_n(x)$. We choose the following values:

The Fourier Coefficients for a Periodic Function f of Period 2π

$$a_0 = \frac{1}{2\pi} \int_{-\pi}^{\pi} f(x)\, dx,$$

$$a_k = \frac{1}{\pi} \int_{-\pi}^{\pi} f(x) \cos(kx)\, dx \quad \text{for } k > 0,$$

$$b_k = \frac{1}{\pi} \int_{-\pi}^{\pi} f(x) \sin(kx)\, dx \quad \text{for } k > 0.$$

Notice that a_0 is just the average value of f over the interval $[-\pi, \pi]$.

For an informal justification for the use of these values, see page 510. In addition, the integrals over $[-\pi, \pi]$ for a_k and b_k can be replaced by integrals over any interval of length 2π.

Example 1   Construct successive Fourier polynomials for the square wave function f, with period 2π, given by

$$f(x) = \begin{cases} 0 & -\pi \leq x < 0 \\ 1 & 0 \leq x < \pi. \end{cases}$$

Solution   Since a_0 is the average value of f on $[-\pi, \pi]$, we suspect from the graph of f that $a_0 = \frac{1}{2}$. We can verify this analytically:

$$a_0 = \frac{1}{2\pi} \int_{-\pi}^{\pi} f(x)\, dx = \frac{1}{2\pi} \int_{-\pi}^{0} 0\, dx + \frac{1}{2\pi} \int_{0}^{\pi} 1\, dx = 0 + \frac{1}{2\pi}(\pi) = \frac{1}{2}.$$

Furthermore,

$$a_1 = \frac{1}{\pi} \int_{-\pi}^{\pi} f(x) \cos x\, dx = \frac{1}{\pi} \int_{0}^{\pi} 1 \cos x\, dx = 0$$

and

$$b_1 = \frac{1}{\pi} \int_{-\pi}^{\pi} f(x) \sin x\, dx = \frac{1}{\pi} \int_{0}^{\pi} 1 \sin x\, dx = \frac{2}{\pi}.$$

Therefore, the Fourier polynomial of degree 1 is given by

$$f(x) \approx F_1(x) = \frac{1}{2} + \frac{2}{\pi} \sin x$$

---

[3]The Fourier polynomials are not polynomials in the usual sense of the word.

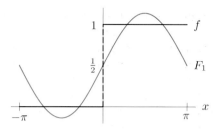

Figure 10.21: First Fourier approximation to the square wave

Figure 10.22: Third Fourier approximation to the square wave

and the graphs of the function and the first Fourier approximation are shown in Figure 10.21.

We next construct the Fourier polynomial of degree 2. The coefficients a_0, a_1, b_1 are the same as before. In addition,

$$a_2 = \frac{1}{\pi} \int_{-\pi}^{\pi} f(x) \cos(2x)\, dx = \frac{1}{\pi} \int_0^{\pi} 1 \cos(2x)\, dx = 0$$

and

$$b_2 = \frac{1}{\pi} \int_{-\pi}^{\pi} f(x) \sin(2x)\, dx = \frac{1}{\pi} \int_0^{\pi} 1 \sin(2x)\, dx = 0.$$

Since $a_2 = b_2 = 0$, the Fourier polynomial of degree 2 is identical to the Fourier polynomial of degree 1. Let's look at the Fourier polynomial of degree 3:

$$a_3 = \frac{1}{\pi} \int_{-\pi}^{\pi} f(x) \cos(3x)\, dx = \frac{1}{\pi} \int_0^{\pi} 1 \cos(3x)\, dx = 0$$

and

$$b_3 = \frac{1}{\pi} \int_{-\pi}^{\pi} f(x) \sin(3x)\, dx = \frac{1}{\pi} \int_0^{\pi} 1 \sin(3x)\, dx = \frac{2}{3\pi}.$$

So the approximation is given by

$$f(x) \approx F_3(x) = \frac{1}{2} + \frac{2}{\pi} \sin x + \frac{2}{3\pi} \sin(3x).$$

The graph of F_3 is shown in Figure 10.22. This approximation is better than $F_1(x) = \frac{1}{2} + \frac{2}{\pi} \sin x$, as comparing Figure 10.22 to Figure 10.21 shows.

Without going through the details, we calculate the coefficients for higher-degree Fourier approximations:

$$F_5(x) = \frac{1}{2} + \frac{2}{\pi} \sin x + \frac{2}{3\pi} \sin(3x) + \frac{2}{5\pi} \sin(5x)$$
$$F_7(x) = \frac{1}{2} + \frac{2}{\pi} \sin x + \frac{2}{3\pi} \sin(3x) + \frac{2}{5\pi} \sin(5x) + \frac{2}{7\pi} \sin(7x).$$

Figure 10.23 shows that higher-degree approximations match the step-like nature of the square wave function more and more closely.

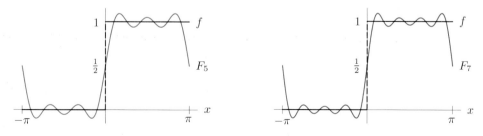

Figure 10.23: Fifth and seventh Fourier approximations to the square wave

We could have used a Taylor series to approximate the square wave, provided we did not center the series at a point of discontinuity. Since the square wave is a constant function on each interval, all its derivatives are zero, and so its Taylor series approximations are the constant functions: 0 or 1, depending on where the Taylor series is centered. They approximate the square wave perfectly on each piece, but they do not do a good job over the whole interval of length 2π. That is what Fourier polynomials succeed in doing: they approximate a curve fairly well everywhere, rather than just near a particular point. The Fourier approximations above look a lot like square waves, so they approximate well *globally*. However, they may not give good values near points of discontinuity. (For example, near $x = 0$, they all give values near $1/2$, which are incorrect.) Thus Fourier polynomials may not be good *local* approximations.

> Taylor polynomials give good *local* approximations to a function;
> Fourier polynomials give good *global* approximations to a function.

Fourier Series

As with Taylor polynomials, the higher the degree of the Fourier approximation, the more accurate it is. Therefore, we carry this procedure on indefinitely by letting $n \to \infty$, and we call the resulting infinite series a *Fourier series*.

> ### The Fourier Series for f on $[-\pi, \pi]$
>
> $$f(x) = a_0 + a_1 \cos x + a_2 \cos 2x + a_3 \cos 3x + \cdots$$
> $$+ b_1 \sin x + b_2 \sin 2x + b_3 \sin 3x + \cdots$$
>
> where a_k and b_k are the Fourier coefficients.

Thus, the Fourier series for the square wave is

$$f(x) = \frac{1}{2} + \frac{2}{\pi} \sin x + \frac{2}{3\pi} \sin 3x + \frac{2}{5\pi} \sin 5x + \frac{2}{7\pi} \sin 7x + \cdots.$$

Harmonics

Let us start with a function $f(x)$ that is periodic with period 2π, expanded in a Fourier series:

$$f(x) = a_0 + a_1 \cos x + a_2 \cos 2x + a_3 \cos 3x + \cdots$$
$$+ b_1 \sin x + b_2 \sin 2x + b_3 \sin 3x + \cdots$$

The function

$$a_k \cos kx + b_k \sin kx$$

is referred to as the kth *harmonic* of f, and it is customary to say that the Fourier series expresses f in terms of its harmonics. The first harmonic, $a_1 \cos x + b_1 \sin x$, is sometimes called the *fundamental harmonic* of f.

Example 2   Find a_0 and the first four harmonics of a *pulse train* function f of period 2π shown in Figure 10.24:

$$f(x) = \begin{cases} 1 & 0 \le x < \pi/2 \\ 0 & \pi/2 \le x < 2\pi \end{cases}$$

Figure 10.24: A train of pulses with period 2π

Solution   First, a_0 is the average value of the function, so

$$a_0 = \frac{1}{2\pi} \int_{-\pi}^{\pi} f(x)\,dx = \frac{1}{2\pi} \int_0^{\pi/2} 1\,dx = \frac{1}{4}.$$

Next, we compute a_k and b_k, $k = 1, 2, 3,$ and 4. The formulas

$$a_k = \frac{1}{\pi} \int_{-\pi}^{\pi} f(x) \cos(kx)\,dx = \frac{1}{\pi} \int_0^{\pi/2} \cos(kx)\,dx$$

$$b_k = \frac{1}{\pi} \int_{-\pi}^{\pi} f(x) \sin(kx)\,dx = \frac{1}{\pi} \int_0^{\pi/2} \sin(kx)\,dx$$

lead to the harmonics

$$a_1 \cos x + b_1 \sin x = \frac{1}{\pi} \cos x + \frac{1}{\pi} \sin x$$

$$a_2 \cos(2x) + b_2 \sin(2x) = \frac{1}{\pi} \sin(2x)$$

$$a_3 \cos(3x) + b_3 \sin(3x) = -\frac{1}{3\pi} \cos(3x) + \frac{1}{3\pi} \sin(3x)$$

$$a_4 \cos(4x) + b_4 \sin(4x) = 0.$$

Figure 10.25 shows the graph of the sum of a_0 and these harmonics, which is the fourth Fourier approximation of f.

Figure 10.25: Fourth Fourier approximation to pulse train f equals the sum of a_0 and the first four harmonics

Energy and the Energy Theorem

The quantity $A_k = \sqrt{a_k^2 + b_k^2}$ is called the amplitude of the kth harmonic. The square of the amplitude has a useful interpretation. Adopting terminology from the study of periodic waves, we define the *energy* E of a periodic function f of period 2π to be the number

$$E = \frac{1}{\pi} \int_{-\pi}^{\pi} [f(x)]^2\,dx.$$

Problem 16 on page 513 asks you to check that for all positive integers k,

$$\frac{1}{\pi} \int_{-\pi}^{\pi} (a_k \cos(kx) + b_k \sin(kx))^2\,dx = a_k^2 + b_k^2 = A_k^2.$$

This shows that the kth harmonic of f has energy A_k^2. The energy of the constant term a_0 of the Fourier series is $\frac{1}{\pi} \int_{-\pi}^{\pi} a_0^2\,dx = 2a_0^2$, so we make the definition

$$A_0 = \sqrt{2}a_0.$$

It turns out that for all reasonable periodic functions f, the energy of f equals the sum of the energy of its harmonics:

The Energy Theorem for a Periodic Function f of Period 2π

$$E = \frac{1}{\pi} \int_{-\pi}^{\pi} [f(x)]^2\,dx = A_0^2 + A_1^2 + A_2^2 + \cdots$$

where $A_0 = \sqrt{2}a_0$ and $A_k = \sqrt{a_k^2 + b_k^2}$ (for all integers $k \geq 1$).

The graph of A_k^2 against k is called the *energy spectrum* of f. It shows how the energy of f is distributed among its harmonics.

Example 3   (a) Graph the energy spectrum of the square wave of Example 1.

(b) What fraction of the energy of the square wave is contained in the constant term and first three harmonics of its Fourier series?

Solution   (a) We know from Example 1 that $a_0 = 1/2$, $a_k = 0$ for $k \geq 1$, $b_k = 0$ for k even, and $b_k = 2/(k\pi)$ for k odd. Thus

$$A_0^2 = 2a_0^2 = \frac{1}{2}$$

$$A_k^2 = 0 \quad \text{if } k \text{ is even}, \quad k \geq 1,$$

$$A_k^2 = \left(\frac{2}{k\pi}\right)^2 = \frac{4}{k^2\pi^2} \quad \text{if } k \text{ is odd}, \quad k \geq 1.$$

The energy spectrum is graphed in Figure 10.26. Notice that it is customary to represent the energy A_k^2 of the kth harmonic by a vertical line of length A_k^2. The graph shows that the constant term and first harmonic carry most of the energy of f.

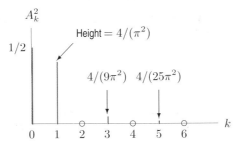

Figure 10.26: The energy spectrum of a square wave

(b) The energy of the square wave $f(x)$ is

$$E = \frac{1}{\pi} \int_{-\pi}^{\pi} [f(x)]^2 \, dx = \frac{1}{\pi} \int_{0}^{\pi} 1 \, dx = 1.$$

The energy in the constant term and the first three harmonics of the Fourier series is

$$A_0^2 + A_1^2 + A_2^2 + A_3^2 = \frac{1}{2} + \frac{4}{\pi^2} + 0 + \frac{4}{9\pi^2} = 0.950.$$

The fraction of energy carried by the constant term and the first three harmonics is

$$0.95/1 = 0.95, \text{ or } 95\%.$$

Musical Instruments

You may have wondered why different musical instruments sound different, even when playing the same note. A first step might be to graph the periodic deviations from the average air pressure that form the sound waves they produce. This has been done for clarinet and trumpet in Figure 10.27.[4] However, it is more revealing to graph the energy spectra of these functions, as in Figure 10.28. The most striking difference is the relative weakness of the second, fourth, and sixth harmonics for the clarinet, with the second harmonic completely absent. The trumpet sounds the second harmonic with as much energy as it does the fundamental.

[4] Adapted from C.A. Culver, *Musical Acoustics* (New York: McGraw-Hill, 1956) pp. 204, 220.

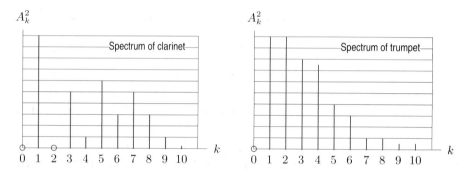

Figure 10.27: Sound waves of a clarinet and trumpet

Figure 10.28: Energy spectra of a clarinet and trumpet

What Do We Do if Our Function Does Not Have Period 2π?

We can easily adapt what we have already done by changing variables. Suppose we have a function $f(x)$ that is periodic with period b. If we define $x = bt/2\pi$, then x varies over the interval $[-b/2, b/2]$ when t varies over the interval $[-\pi, \pi]$. Thus, if we substitute $x = bt/2\pi$ into f and define a new function g by

$$g(t) = f\left(\frac{bt}{2\pi}\right) = f(x),$$

then g has period 2π. We can find the Fourier series for g as before, giving

$$g(t) = a_0 + a_1 \cos t + a_2 \cos(2t) + a_3 \cos(3t) + \cdots$$
$$+ b_1 \sin t + b_2 \sin(2t) + b_3 \sin(3t) + \cdots.$$

Substituting $t = 2\pi x/b$ allows us to convert back:

$$f(x) = g\left(\frac{2\pi x}{b}\right) = a_0 + a_1 \cos\left(\frac{2\pi x}{b}\right) + a_2 \cos\left(\frac{4\pi x}{b}\right) + a_3 \cos\left(\frac{6\pi x}{b}\right) + \cdots$$
$$+ b_1 \sin\left(\frac{2\pi x}{b}\right) + b_2 \sin\left(\frac{4\pi x}{b}\right) + b_3 \sin\left(\frac{6\pi x}{b}\right) + \cdots$$

Note that the terms in the Fourier series of a periodic function of period b are not $\cos(kx)$ and $\sin(kx)$, but instead are $\cos(2\pi kx/b)$ and $\sin(2\pi kx/b)$. The leads to the following result:

The Fourier Series for f on $[-b/2, b/2]$

$$f(x) = a_0 + \sum_{k=1}^{\infty} \left(a_k \cos\left(\frac{2\pi k x}{b}\right) + b_k \sin\left(\frac{2\pi k x}{b}\right) \right)$$

where $a_0 = \dfrac{1}{b} \displaystyle\int_{-b/2}^{b/2} f(x)\, dx$ and, for $k \geq 1$,

$$a_k = \frac{2}{b} \int_{-b/2}^{b/2} f(x) \cos\left(\frac{2\pi k x}{b}\right) dx \qquad b_k = \frac{2}{b} \int_{-b/2}^{b/2} f(x) \sin\left(\frac{2\pi k x}{b}\right) dx$$

Example 4   Find the fifth-degree Fourier polynomial of the square wave $f(x)$ graphed in Figure 10.29.

Figure 10.29: Square wave f and its fifth Fourier approximation F_5

Solution   Since f has period $b = 2$, we let $x = bt/2\pi = t/\pi$ and consider the function

$$g(t) = f\left(\frac{t}{\pi}\right) = f(x)$$

which has period 2π. Then

$$g(t) = \begin{cases} 0 & -\pi \leq t < 0 \\ 1 & 0 \leq t < \pi, \end{cases}$$

so $g(t)$ is the square wave of Example 1. We have, therefore,

$$g(t) \approx \frac{1}{2} + \frac{2}{\pi} \sin t + \frac{2}{3\pi} \sin(3t) + \frac{2}{5\pi} \sin(5t).$$

Hence, substituting $t = \pi x$,

$$f(x) = g(t) = g(\pi x) \approx \frac{1}{2} + \frac{2}{\pi} \sin(\pi x) + \frac{2}{3\pi} \sin(3\pi x) + \frac{2}{5\pi} \sin(5\pi x).$$

Seasonal Variation in the Incidence of Measles

Example 5   Fourier approximations have been used to analyze the seasonal variation in the incidence of diseases. One study[5] done in Baltimore, Maryland, for the years 1901–1931, studied $I(t)$, the average

---

[5]From C. I. Bliss and D. L. Blevins, *The Analysis of Seasonal Variation in Measles* (Am. J. Hyg. 70, 1959), reported by Edward Batschelet, *Introduction to Mathematics for the Life Sciences* (Springer-Verlag, Berlin, 1979).

number of cases of measles per 10,000 susceptible children in the tth month of the year. The data points in Figure 10.30 show $f(t) = \log I(t)$. The curve in Figure 10.30 shows the second Fourier approximation of $f(t)$. Figure 10.31 contains the graphs of the first and second harmonics of $f(t)$, plotted separately as deviations about a_0, the average logarithmic incidence rate. Describe what these two harmonics tell you about incidence of measles.

Figure 10.30: Logarithm of incidence of measles per month (dots) and second Fourier approximation (curve)

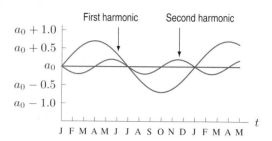

Figure 10.31: First and second harmonics of $f(t)$ plotted as deviations from average log incidence rate, a_0

Solution

Taking the log of $I(t)$ has the effect of reducing the amplitude of the oscillations. However, since the log of a function increases when the function increases, and decreases when it decreases, oscillations in $f(t)$ correspond to oscillations in $I(t)$.

Figure 10.31 shows that the first harmonic in the Fourier series has a period of one year (the same period as the original function); the second harmonic has a period of six months. The graph in Figure 10.31 shows that the first harmonic is approximately a sine with amplitude about 0.7; the second harmonic is approximately the negative of a sine with amplitude about 0.2. Thus, for t in months ($t = 0$ in January),

$$\log I(t) = f(t) \approx a_0 + 0.7 \sin\left(\frac{\pi}{6}t\right) - 0.2 \sin\left(\frac{\pi}{3}t\right),$$

where $\pi/6$ and $\pi/3$ are introduced to make the periods 12 and 6 months, respectively. We can estimate a_0 from the original graph of f: it is the average value of f, approximately 1.5. Thus

$$f(t) \approx 1.5 + 0.7 \sin\left(\frac{\pi}{6}t\right) - 0.2 \sin\left(\frac{\pi}{3}t\right).$$

Figure 10.30 shows that the second Fourier approximation of $f(t)$ is quite good. The harmonics of $f(t)$ beyond the second must be rather insignificant. This suggests that the variation in incidence in measles comes from two sources, one with a yearly cycle that is reflected in the first harmonic and one with a half-yearly cycle reflected in the second harmonic. At this point the mathematics can tell us no more; we must turn to the epidemiologists for further explanation.

Informal Justification of the Formulas for the Fourier Coefficients

Recall that the coefficients in a Taylor series (which is a good approximation locally) are found by differentiation. In contrast, the coefficients in a Fourier series (which is a good approximation globally) are found by integration.

We want to find the constants $a_0, a_1, a_2, \ldots$ and $b_1, b_2, \ldots$ in the expression

$$f(x) = a_0 + \sum_{k=1}^{\infty} a_k \cos(kx) + \sum_{k=1}^{\infty} b_k \sin(kx).$$

Consider the integral

$$\int_{-\pi}^{\pi} f(x)\,dx = \int_{-\pi}^{\pi} \left[a_0 + \sum_{k=1}^{\infty} a_k \cos(kx) + \sum_{k=1}^{\infty} b_k \sin(kx) \right] dx.$$

Splitting the integral into separate terms, and assuming we can interchange integration and summation, we get

$$\int_{-\pi}^{\pi} f(x)\,dx = \int_{-\pi}^{\pi} a_0\,dx + \int_{-\pi}^{\pi} \sum_{k=1}^{\infty} a_k \cos(kx)\,dx + \int_{-\pi}^{\pi} \sum_{k=1}^{\infty} b_k \sin(kx)\,dx$$

$$= \int_{-\pi}^{\pi} a_0\,dx + \sum_{k=1}^{\infty} \int_{-\pi}^{\pi} a_k \cos(kx)\,dx + \sum_{k=1}^{\infty} \int_{-\pi}^{\pi} b_k \sin(kx)\,dx.$$

But for $k \geq 1$, thinking of the integral as an area shows that

$$\int_{-\pi}^{\pi} \sin(kx)\,dx = 0 \text{ and } \int_{-\pi}^{\pi} \cos(kx)\,dx = 0,$$

so all terms drop out except the first, giving

$$\int_{-\pi}^{\pi} f(x)\,dx = \int_{-\pi}^{\pi} a_0\,dx = a_0 x \Big|_{-\pi}^{\pi} = 2\pi a_0$$

and so we get the following result:

$$\boxed{a_0 = \frac{1}{2\pi} \int_{-\pi}^{\pi} f(x)\,dx.}$$

Thus a_0 is the average value of f on the interval $[-\pi, \pi]$.

To determine the values of any of the other a_k or b_k (for positive k), we use a rather clever method which depends on the following facts. For all integers k and m,

$$\int_{-\pi}^{\pi} \sin(kx) \cos(mx)\,dx = 0,$$

and, provided $k \neq m$,

$$\int_{-\pi}^{\pi} \cos(kx) \cos(mx)\,dx = 0 \text{ and } \int_{-\pi}^{\pi} \sin(kx) \sin(mx)\,dx = 0.$$

(See Problems 23–27 on page 514.) In addition, provided $m \neq 0$, we have

$$\int_{-\pi}^{\pi} \cos^2(mx)\,dx = \pi \text{ and } \int_{-\pi}^{\pi} \sin^2(mx)\,dx = \pi.$$

To determine a_k, we multiply the Fourier series by $\cos(mx)$, where m is any positive integer:

$$f(x)\cos(mx) = a_0 \cos(mx) + \sum_{k=1}^{\infty} a_k \cos(kx)\cos(mx) + \sum_{k=1}^{\infty} b_k \sin(kx)\cos(mx).$$

We integrate this between $-\pi$ and π, term by term:

$$\int_{-\pi}^{\pi} f(x) \cos(mx)\, dx = \int_{-\pi}^{\pi} \left(a_0 \cos(mx) + \sum_{k=1}^{\infty} a_k \cos(kx) \cos(mx) + \sum_{k=1}^{\infty} b_k \sin(kx) \cos(mx) \right) dx$$

$$= a_0 \int_{-\pi}^{\pi} \cos(mx)\, dx + \sum_{k=1}^{\infty} \left(a_k \int_{-\pi}^{\pi} \cos(kx) \cos(mx)\, dx \right)$$

$$+ \sum_{k=1}^{\infty} \left(b_k \int_{-\pi}^{\pi} \sin(kx) \cos(mx)\, dx \right).$$

Provided $m \neq 0$, we have $\int_{-\pi}^{\pi} \cos(mx)\, dx = 0$. Since the integral $\int_{-\pi}^{\pi} \sin(kx) \cos(mx)\, dx = 0$, all the terms in the second sum are zero. Since $\int_{-\pi}^{\pi} \cos kx \cos mx\, dx = 0$ provided $k \neq m$, all the terms in the first sum are zero except where $k = m$. Thus the right-hand side reduces to one term:

$$\int_{-\pi}^{\pi} f(x) \cos(mx)\, dx = a_m \int_{-\pi}^{\pi} \cos(mx) \cos(mx)\, dx = \pi a_m.$$

This leads, for each value of $m = 1, 2, 3 \ldots$, to the following formula:

$$a_m = \frac{1}{\pi} \int_{-\pi}^{\pi} f(x) \cos(mx)\, dx.$$

To determine b_k, we multiply through by $\sin(mx)$ instead of $\cos(mx)$ and eventually obtain, for each value of $m = 1, 2, 3 \ldots$, the following result:

$$b_m = \frac{1}{\pi} \int_{-\pi}^{\pi} f(x) \sin(mx)\, dx.$$

Exercises and Problems for Section 10.5

Exercises

Which of the series in Exercises 1–4 are Fourier series?

1. $1 + \cos x + \cos^2 x + \cos^3 x + \cos^4 x + \cdots$

2. $\sin x + \sin(x+1) + \sin(x+2) + \cdots$

3. $\dfrac{\cos x}{2} + \sin x - \dfrac{\cos(2x)}{4} - \dfrac{\sin(2x)}{2} + \dfrac{\cos(3x)}{8} + \dfrac{\sin(3x)}{3} - \cdots$

4. $\dfrac{1}{2} - \dfrac{1}{3} \sin x + \dfrac{1}{4} \sin(2x) - \dfrac{1}{5} \sin(3x) + \cdots$

5. Construct the first three Fourier approximations to the square wave function

$$f(x) = \begin{cases} -1 & -\pi \leq x < 0 \\ 1 & 0 \leq x < \pi. \end{cases}$$

Use a calculator or computer to draw the graph of each approximation.

6. Repeat Problem 5 with the function

$$f(x) = \begin{cases} -x & -\pi \leq x < 0 \\ x & 0 \leq x < \pi. \end{cases}$$

7. What fraction of the energy of the function in Problem 6 is contained in the constant term and first three harmonics of its Fourier series?

For Exercises 8–10, find the nth Fourier polynomial for the given functions, assuming them to be periodic with period 2π. Graph the first three approximations with the original function.

8. $f(x) = x^2$, $-\pi < x \leq \pi$.

9. $h(x) = \begin{cases} 0 & -\pi < x \leq 0 \\ x & 0 < x \leq \pi. \end{cases}$

10. $g(x) = x$, $-\pi < x \leq \pi$.

Problems

11. (a) Find and graph the third Fourier approximation of the square wave $g(x)$ of period 2π:

$$g(x) = \begin{cases} 0 & -\pi \leq x < -\pi/2 \\ 1 & -\pi/2 \leq x < \pi/2 \\ 0 & \pi/2 \leq x < \pi. \end{cases}$$

(b) How does the result of part (a) differ from that of the square wave in Example 1?

12. Suppose we have a periodic function f with period 1 defined by $f(x) = x$ for $0 \leq x < 1$. Find the fourth degree Fourier polynomial for f and graph it on the interval $0 \leq x < 1$. [Hint: Remember that since the period is not 2π, you will have to start by doing a substitution. Notice that the terms in the sum are not $\sin(nx)$ and $\cos(nx)$, but instead turn out to be $\sin(2\pi nx)$ and $\cos(2\pi nx)$.]

13. Suppose f has period 2 and $f(x) = x$ for $0 \leq x < 2$. Find the fourth-degree Fourier polynomial and graph it on $0 \leq x < 2$. [Hint: See Problem 12.]

14. Suppose that a spacecraft near Neptune has measured a quantity A and sent it to earth in the form of a periodic signal $A \cos t$ of amplitude A. On its way to earth, the signal picks up periodic noise, containing only second and higher harmonics. Suppose that the signal $h(t)$ actually received on earth is graphed below in Figure 10.32. Determine the signal that the spacecraft originally sent and hence the value A of the measurement.

Figure 10.32

15. Figures 10.33 and 10.34 show the waveforms and energy spectra for notes produced by flute and bassoon.[6] Describe the principal differences between the two spectra.

Figure 10.33

Figure 10.34

16. Show that for positive integers k, the periodic function $f(x) = a_k \cos kx + b_k \sin kx$ of period 2π has energy $a_k^2 + b_k^2$.

17. Given the graph of f in Figure 10.35, find the first two Fourier approximations numerically.

Figure 10.35

18. Justify the formula $b_k = \frac{1}{\pi} \int_{-\pi}^{\pi} f(x) \sin(kx) \, dx$ for the Fourier coefficients, b_k, of a periodic function of period 2π. The argument is similar to that in the text for a_k. In addition to some of the formulas used there, you will need the formulas: $\int_{-\pi}^{\pi} \sin^2(kx) \, dx = \pi$ and $\int_{-\pi}^{\pi} \sin(kx) \sin(mx) \, dx = 0$ for $k \neq m$.

In Problems 19–22, the pulse train of width c is the periodic function f of period 2π given by

$$f(x) = \begin{cases} 0 & -\pi \leq x < -c/2 \\ 1 & -c/2 \leq x < c/2 \\ 0 & c/2 \leq x < \pi. \end{cases}$$

19. Suppose that f is the pulse train of width 1.

(a) What fraction of the energy of f is contained in the constant term of its Fourier series? In the constant term and the first harmonic together?

---

[6] Adapted from C.A. Culver, *Musical Acoustics* (New York: McGraw-Hill, 1956), pp. 200, 213.

(b) Find a formula for the energy of the kth harmonic of f. Use it to sketch the energy spectrum of f.

(c) How many terms of the Fourier series of f are needed to capture 90% of the energy of f?

(d) Graph f and its fifth Fourier approximation on the interval $[-3\pi, 3\pi]$.

20. Suppose that f is the pulse train of width 0.4.

(a) What fraction of the energy of f is contained in the constant term of its Fourier series? In the constant term and the first harmonic together?

(b) Find a formula for the energy of the kth harmonic of f. Use it to sketch the energy spectrum of f.

(c) What fraction of the energy of f is contained in the constant term and the first five harmonics of f? (The constant term and the first thirteen harmonics are needed to capture 90% of the energy of f.)

(d) Graph f and its fifth Fourier approximation on the interval $[-3\pi, 3\pi]$.

21. Suppose that f is the pulse train of width 2.

(a) What fraction of the energy of f is contained in the constant term of its Fourier series? In the constant term and the first harmonic together?

(b) How many terms of the Fourier series of f are needed to capture 90% of the energy of f?

(c) Graph f and its third Fourier approximation on the interval $[-3\pi, 3\pi]$.

22. After working Problems 19– 21, write a paragraph about the approximation of pulse trains by Fourier polynomials. Explain how the energy spectrum of a pulse train of width c changes as c gets closer and closer to 0 and how this affects the number of terms required for an accurate approximation.

For Problems 23–27, use the table of integrals inside the back cover to show that the following statements are true for positive integers k and m.

23. $\displaystyle\int_{-\pi}^{\pi} \cos(kx)\cos(mx)\,dx = 0, \quad \text{if } k \neq m.$

24. $\displaystyle\int_{-\pi}^{\pi} \cos^2(mx)\,dx = \pi.$

25. $\displaystyle\int_{-\pi}^{\pi} \sin^2(mx)\,dx = \pi.$

26. $\displaystyle\int_{-\pi}^{\pi} \sin(kx)\cos(mx)\,dx = 0.$

27. $\displaystyle\int_{-\pi}^{\pi} \sin(kx)\sin(mx)\,dx = 0, \quad \text{if } k \neq m.$

28. Suppose that $f(x)$ is a periodic function with period b. Show that

(a) $g(t) = f(bt/2\pi)$ is periodic with period 2π and $f(x) = g(2\pi x/b)$.

(b) The Fourier series for g is given by

$$g(t) = a_0 + \sum_{k=1}^{\infty} \left(a_k \cos(kt) + b_k \sin(kt) \right)$$

where the coefficients a_0, a_k, b_k are given in the box on page 509.

(c) The Fourier series for f is given by

$$f(x) = a_0 + \sum_{k=1}^{\infty} \left(a_k \cos\left(\frac{2\pi kx}{b}\right) + b_k \sin\left(\frac{2\pi kx}{b}\right) \right)$$

where the coefficients are the same as in part (c).

CHAPTER SUMMARY (see also Ready Reference at the end of the book)

- **Taylor series and polynomials**
  General expansion about $x = 0$ or $x = a$; specific series for $e^x, \sin x, \cos x, (1 + x)^p$; using known Taylor series to find others by substitution, integration, and differentiation; interval of convergence; error in Taylor polynomial

  expansion
- **Fourier series**
  Formula for coefficients on $[-\pi, \pi], [-b, b]$; Energy theorem

REVIEW EXERCISES AND PROBLEMS FOR CHAPTER TEN

Exercises

For Exercises 1–4, find the second-degree Taylor polynomial about the given point.

1. $e^x, \qquad x = 1$

2. $\ln x, \qquad x = 2$

3. $\sin x, \qquad x = -\pi/4$

4. $\tan\theta, \quad \theta = \pi/4$

5. Find the third-degree Taylor polynomial for $f(x) = x^3 + 7x^2 - 5x + 1$ at $x = 1$.

In Exercises 6–13, find the first four nonzero terms of the Taylor series about the origin of the given functions.

6. $t^2 e^t$

7. $\cos(3y)$

8. $\theta^2 \cos\theta^2$

9. $\sin t^2$

10. $\dfrac{1}{\sqrt{4 - x}}$

11. $\dfrac{1}{1 - 4z^2}$

12. $\dfrac{t}{1 + t}$

13. $\dfrac{z^2}{\sqrt{1 - z^2}}$

For Exercises 14–17, expand the quantity in a Taylor series around 0 in terms of the variable given. Give four nonzero terms.

14. $\sqrt{R - r}$ in terms of $\dfrac{r}{R}$ **15.** $\dfrac{a}{a + b}$ in terms of $\dfrac{b}{a}$

16. $(B^2 + y^2)^{3/2}$ in terms of $\dfrac{y}{B}$, where $B > 0$

17. $\dfrac{1}{(a + r)^{3/2}}$ in terms of $\dfrac{r}{a}$

Problems

18. Find an exact value for each of the following sums.

 (a) $7(1.02)^3 + 7(1.02)^2 + 7(1.02) + 7 + \dfrac{7}{(1.02)} + \dfrac{7}{(1.02)^2} + \cdots + \dfrac{7}{(1.02)^{100}}.$

 (b) $7 + 7(0.1)^2 + \dfrac{7(0.1)^4}{2!} + \dfrac{7(0.1)^6}{3!} + \cdots .$

Find the exact value of the sums in Problems 19–24.

19. $1 - \dfrac{1}{3} + \dfrac{1}{9} - \dfrac{1}{27} + \dfrac{1}{81} - \cdots$

20. $8 + 4 + 2 + 1 + \dfrac{1}{2} + \dfrac{1}{4} + \cdots + \dfrac{1}{2^{10}}$

21. $1 - 2 + \dfrac{4}{2!} - \dfrac{8}{3!} + \dfrac{16}{4!} + \cdots$

22. $2 - \dfrac{8}{3!} + \dfrac{32}{5!} - \dfrac{128}{7!} + \cdots$

23. $3 + 3 + \dfrac{3}{2!} + \dfrac{3}{3!} + \dfrac{3}{4!} + \dfrac{3}{5!} + \cdots$

24. $(0.1)^2 - \dfrac{(0.1)^4}{3!} + \dfrac{(0.1)^6}{5!} - \dfrac{(0.1)^8}{7!} + \cdots$

25. The function $f(x)$ is approximated near $x = 0$ by the third degree Taylor polynomial

$$P_3(x) = 2 - x - x^2/3 + 2x^3.$$

 Give the value of

 (a) $f(0)$ (b) $f'(0)$
 (c) $f''(0)$ (d) $f'''(0)$

26. A function f has $f(3) = 1$, $f'(3) = 5$ and $f''(3) = -10$. Find the best estimate you can for $f(3.1)$.

27. For a, b positive constants, the upper half of an ellipse has equation

$$y = f(x) = b\sqrt{1 - \dfrac{x^2}{a^2}},$$

 (a) Find the second degree Taylor polynomial of $f(x)$ about 0.
 (b) Explain how you could predict the coefficient of x in the Taylor polynomial from a graph of f.
 (c) For x near 0, the ellipse is approximated by a parabola. What is the equation of the parabola? What are its x-intercepts?
 (d) Taking $a = 3$ and $b = 2$, estimate the maximum difference between $f(x)$ and the second-degree Taylor polynomial for $-0.1 \le x \le 0.1$.

28. Suppose x is positive but very small. Arrange the following expressions in increasing order:

$$x , \quad \sin x, \quad \ln(1 + x), \quad 1 - \cos x,$$
$$e^x - 1, \quad \arctan x, \quad x\sqrt{1 - x}.$$

29. By plotting several of its Taylor polynomials and the function $f(x) = 1/(1 + x)$, estimate graphically the interval of convergence of the series expansions for this function about $x = 0$. Compute the radius of convergence analytically.

30. Use Taylor series to evaluate $\lim\limits_{x \to 0} \dfrac{\ln(1 + x + x^2) - x}{\sin^2 x}$.

31. (a) Find $\lim\limits_{\theta \to 0} \dfrac{\sin(2\theta)}{\theta}$. Explain your reasoning.

 (b) Use series to explain why $f(\theta) = \dfrac{\sin(2\theta)}{\theta}$ looks like a parabola near $\theta = 0$. What is the equation of the parabola?

32. (a) Find the Taylor series for $f(t) = te^t$ about $t = 0$.
 (b) Using your answer to part (a), find a Taylor series expansion about $x = 0$ for

$$\int_0^x te^t \, dt.$$

 (c) Using your answer to part (b), show that

$$\dfrac{1}{2} + \dfrac{1}{3} + \dfrac{1}{4(2!)} + \dfrac{1}{5(3!)} + \dfrac{1}{6(4!)} + \cdots = 1.$$

33. (a) Find the first two nonzero terms of the Taylor series for $f(x) = \sqrt{4 - x^2}$ about $x = 0$
 (b) Use your answer to part (a) and the Fundamental Theorem of Calculus to calculate an approximate value for $\displaystyle\int_0^1 \sqrt{4 - x^2} \, dx.$
 (c) Use the substitution $x = 2\sin t$ to calculate the exact value of the integral in part (b).
 (d) To how many decimal places do your answers to parts (b) and (c) agree?

34. (a) Find the Taylor series expansion of $\arcsin x$.
 (b) Use Taylor series to find the limit of $\dfrac{\arctan x}{\arcsin x}$ as $x \to 0$.

35. A particle moving along the x-axis has potential energy at the point x given by $V(x)$. The potential energy has a minimum at $x = 0$.

   (a) Write the Taylor polynomial of degree 2 for V about $x = 0$. What can you say about the signs of the co-efficients of each of the terms of the Taylor polynomial?

   (b) The force on the particle at the point x is given by $-V'(x)$. For small x, show that the force on the particle is approximately proportional to its distance from the origin. What is the sign of the proportionality constant? Describe the direction in which the force points.

36. The theory of relativity predicts that when an object moves at speeds close to the speed of light, the object appears heavier. The apparent, or relativistic, mass, m, of the object when it is moving at speed v is given by the formula

$$m = \frac{m_0}{\sqrt{1 - v^2/c^2}}$$

where c is the speed of light and m_0 is the mass of the object when it is at rest.

   (a) Use the formula for m to decide what values of v are possible.

   (b) Sketch a rough graph of m against v, labeling intercepts and asymptotes.

   (c) Write the first three nonzero terms of the Taylor series for m in terms of v.

   (d) For what values of v do you expect the series to converge?

37. The potential energy, V, of two gas molecules separated by a distance r is given by

$$V = -V_0 \left(2 \left(\frac{r_0}{r} \right)^6 - \left(\frac{r_0}{r} \right)^{12} \right),$$

where V_0 and r_0 are positive constants.

   (a) Show that if $r = r_0$, then V takes on its minimum value, $-V_0$.

   (b) Write V as a series in $(r - r_0)$ up through the quadratic term.

   (c) For r near r_0, show that the difference between V and its minimum value is approximately proportional to $(r - r_0)^2$. In other words, show that $V - (-V_0) = V + V_0$ is approximately proportional to $(r - r_0)^2$.

   (d) The force, F, between the molecules is given by $F = -dV/dr$. What is F when $r = r_0$? For r near r_0, show that F is approximately proportional to $(r - r_0)$.

38. The *gravitational field* at a point in space is the gravitational force that would be exerted on a unit mass placed there. We will assume that the gravitational field strength at a distance d away from a mass M is

$$\frac{GM}{d^2}$$

where G is constant. In this problem you will investigate the gravitational field strength, F, exerted by a system consisting of a large mass M and a small mass m, with a distance r between them. (See Figure 10.36.)

Figure 10.36

   (a) Write an expression for the gravitational field strength, F, at the point P.

   (b) Assuming r is small in comparison to R, expand F in a series in r/R.

   (c) By discarding terms in $(r/R)^2$ and higher powers, explain why you can view the field as resulting from a single particle of mass $M + m$, plus a correction term. What is the position of the particle of mass $M + m$? Explain the sign of the correction term.

39. Expand $f(x + h)$ and $g(x + h)$ in Taylor series and take a limit to confirm the product rule:

$$\frac{d}{dx}(f(x)g(x)) = f'(x)g(x) + f(x)g'(x).$$

40. Use Taylor expansions for $f(y + k)$ and $g(x + h)$ to confirm the chain rule:

$$\frac{d}{dx}(f(g(x))) = f'(g(x)) \cdot g'(x).$$

41. Suppose all the derivatives of g exist at $x = 0$ and that g has a critical point at $x = 0$.

   (a) Write the nth Taylor polynomial for g at $x = 0$.

   (b) What does the Second Derivative test for local maxima and minima say?

   (c) Use the Taylor polynomial to explain why the Second Derivative test works.

42. (Continuation of Problem 41) You may remember that the Second Derivative test tells us nothing when the second derivative is zero at the critical point. In this problem you will investigate that special case.

   Assume g has the same properties as in Problem 41, and that, in addition, $g''(0) = 0$. What does the Taylor polynomial tell you about whether g has a local maximum or minimum at $x = 0$?

43. Use the Fourier polynomials for the square wave

$$f(x) = \begin{cases} -1 & -\pi < x \leq 0 \\ 1 & 0 < x \leq \pi \end{cases}$$

to explain why the following sum must approach $\pi/4$ as $n \to \infty$:

$$1 - \frac{1}{3} + \frac{1}{5} - \frac{1}{7} + \cdots + (-1)^{2n+1} \frac{1}{2n+1}.$$

44. Find a Fourier polynomial of degree three for $f(x) = e^{2\pi x}$, for $0 \leq x < 1$.

45. Suppose that $f(x)$ is a differentiable periodic function of period 2π. Assume the Fourier series of f is differentiable term by term.

   (a) If the Fourier coefficients of f are a_k and b_k, show that the Fourier coefficients of its derivative f' are kb_k and $-ka_k$.

   (b) How are the amplitudes of the harmonics of f and f' related?

   (c) How are the energy spectra of f and f' related?

46. If the Fourier coefficients of f are a_k and b_k, and the Fourier coefficients of g are c_k and d_k, and if A and B are real, show that the Fourier coefficients of $Af + Bg$ are $Aa_k + Bc_k$ and $Ab_k + Bd_k$.

47. Suppose that f is a periodic function of period 2π and that g is a horizontal shift of f, say $g(x) = f(x + c)$. Show that f and g have the same energy.

CAS Challenge Problems

48. **(a)** Use a computer algebra system to find $P_{10}(x)$ and $Q_{10}(x)$, the Taylor polynomials of degree 10 about $x = 0$ for $\sin^2 x$ and $\cos^2 x$.

   (b) What similarities do you observe between the two polynomials? Explain your observation in terms of properties of sine and cosine.

49. **(a)** Use your computer algebra system to find $P_7(x)$ and $Q_7(x)$, the Taylor polynomials of degree 7 about $x = 0$ for $f(x) = \sin x$ and $g(x) = \sin x \cos x$.

   (b) Find the ratio between the coefficient of x^3 in the two polynomials. Do the same for the coefficients of x^5 and x^7.

   (c) Describe the pattern in the ratios that you computed in part (b). Explain it using the identity $\sin(2x) = 2\sin x \cos x$.

50. **(a)** Calculate the equation of the tangent line to the function $f(x) = x^2$ at $x = 2$. Do the same calculation for $g(x) = x^3 - 4x^2 + 8x - 7$ at $x = 1$ and for $h(x) = 2x^3 + 4x^2 - 3x + 7$ at $x = -1$.

   (b) Use a computer algebra system to divide $f(x)$ by $(x - 2)^2$, giving your result in the form

$$\frac{f(x)}{(x - 2)^2} = q(x) + \frac{r(x)}{(x - 2)^2},$$

   where $q(x)$ is the quotient and $r(x)$ is the remainder. In addition, divide $g(x)$ by $(x - 1)^2$ and $h(x)$ by $(x + 1)^2$.

   (c) For each of the functions, f, g, h, compare your answers to part (a) with the remainder, $r(x)$. What do you notice? Make a conjecture about the tangent line to a polynomial $p(x)$ at the point $x = a$ and the remainder, $r(x)$, obtained from dividing $p(x)$ by $(x - a)^2$.

   (d) Use the Taylor expansion of $p(x)$ about $x = a$ to prove your conjecture.[7]

51. Let $f(x) = \dfrac{x}{e^x - 1} + \dfrac{x}{2}$. Although the formula for f is not defined at $x = 0$, we can make f continuous by setting $f(0) = 1$. If we do this, f has a Taylor series about $x = 0$.

   (a) Use a computer algebra system to find $P_{10}(x)$, the Taylor polynomial of degree 10 about $x = 0$ for f.

   (b) What do you notice about the degrees of the terms in the polynomial? What property of f does this suggest?

   (c) Prove that f has the property suggested by part (b).

52. Let $S(x) = \int_0^x \sin(t^2)\, dt$.

   (a) Use a computer algebra system to find $P_{11}(x)$, the Taylor polynomial of degree 11 about $x = 0$, for $S(x)$.

   (b) What is the percentage error in the approximation of $S(1)$ by $P_{11}(1)$? What about the approximation of $S(2)$ by $P_{11}(2)$?

CHECK YOUR UNDERSTANDING

Decide if the statements in Problems 1–24 are true or false. Assume that the Taylor series for a function converges to that function. Give an explanation for your answer.

1. If $f(x)$ and $g(x)$ have the same Taylor polynomial of degree 2 near $x = 0$, then $f(x) = g(x)$.

2. Using $\sin\theta \approx \theta - \theta^3/3!$ with $\theta = 1°$, we have $\sin(1°) \approx 1 - 1^3/6 = 5/6$.

3. The Taylor polynomial of degree 2 for e^x near $x = 5$ is $1 + (x - 5) + (x - 5)^2/2$.

4. If the Taylor polynomial of degree 2 for $f(x)$ near $x = 0$ is $P_2(x) = 1 + x - x^2$, then $f(x)$ is concave up near $x = 0$.

5. The Taylor series for $\sin x$ about $x = \pi$ is

$$(x - \pi) - \frac{(x - \pi)^3}{3!} + \frac{(x - \pi)^5}{5!} - \cdots.$$

6. If f is an even function, then the Taylor series for f near $x = 0$ has only terms with even exponents.

---

[7]See "Tangents without Calculus" by Jorge Aarao, in *The College Mathematics Journal* Vol 31, No 5, Nov 2000 (Mathematical Association of America).

7. The Taylor series for $x^3 \cos x$ about $x = 0$ has only odd powers.

8. If f has the following Taylor series about $x = 0$, then $f^{(7)}(0) = -8$:

$$f(x) = 1 - 2x + \frac{3}{2!}x^2 - \frac{4}{3!}x^3 + \cdots.$$

(Assume the pattern of the coefficients continues.)

9. The Taylor series for $f(x)g(x)$ about $x = 0$ is

$$f(0)g(0) + f'(0)g'(0)x + \frac{f''(0)g''(0)}{2!}x^2 + \cdots.$$

10. To find the Taylor series for $\sin x + \cos x$ about any point, add the Taylor series for $\sin x$ and $\cos x$ about that point.

11. The quadratic approximation to $f(x)$ for x near 0 is better than the linear approximation for all values of x.

12. The Taylor series for f converges everywhere f is defined.

13. The graphs of e^x and its Taylor polynomial $P_{10}(x)$ get further and further apart as $x \to \infty$.

14. A Taylor polynomial for f near $x = a$ touches the graph of f only at $x = a$.

15. Let $P_n(x)$ be the nth Taylor polynomial for a function f near $x = a$. Although $P_n(x)$ is a good approximation to f near $x = a$, it is not possible to have $P_n(x) = f(x)$ for all x.

16. If $|f^{(n)}(x)| < 10$ for all $n > 0$ and all x, then the Taylor series for f about $x = 0$ converges to $f(x)$ for all x.

17. If f is an even function, then the Fourier series for f on $[-\pi, \pi]$ has only cosines.

18. The linear approximation to $f(x)$ near $x = -1$ shows that if $f(-1) = g(-1)$ and $f'(-1) < g'(-1)$, then $f(x) < g(x)$ for all x sufficiently close to -1 (but not equal to -1).

19. The quadratic approximation to $f(x)$ near $x = -1$ shows that if $f(-1) = g(-1)$, $f'(-1) = g'(-1)$, and $f''(-1) < g''(-1)$, then $f(x) < g(x)$ for all x sufficiently close to -1 (but not equal to -1).

20. If $L_1(x)$ is the linear approximation to $f_1(x)$ near $x = 0$ and $L_2(x)$ is the linear approximation to $f_2(x)$ near $x = 0$, then $L_1(x) + L_2(x)$ is the linear approximation to $f_1(x) + f_2(x)$ near $x = 0$

21. If $L_1(x)$ is the linear approximation to $f_1(x)$ near $x = 0$ and $L_2(x)$ is the linear approximation to $f_2(x)$ near $x = 0$, then $L_1(x)L_2(x)$ is the quadratic approximation to $f_1(x)f_2(x)$ near $x = 0$.

22. If $f^{(n)}(0) \geq n!$ for all n, then the Taylor series for f near $x = 0$ diverges at $x = 0$.

23. If $f^{(n)}(0) \geq n!$ for all n, then the Taylor series for f near $x = 0$ diverges at $x = 1$.

24. If $f^{(n)}(0) \geq n!$ for all n, then the Taylor series for f near $x = 0$ diverges at $x = 1/2$.

PROJECTS FOR CHAPTER TEN

1. Machin's Formula and the Value of π

(a) In the 17th century, Machin obtained the formula: $\pi/4 = 4\arctan(1/5) - \arctan(1/239)$. Use a calculator to check this formula.

(b) Use the Taylor polynomial approximation of degree 5 to the arctangent function to approximate the value of π. (Note: In 1873 William Shanks used this approach to calculate π to 707 decimal places. Unfortunately, in 1946 it was found that he made an error in the 528th place. Currently, several billion decimal places are known.)

(c) Why do the two series for arctangent converge so rapidly here while the series used in Example 5 on page 492 converges so slowly?

(d) Now we prove Machin's formula using the tangent addition formula

$$\tan(A + B) = \frac{\tan A + \tan B}{1 - \tan A \tan B}.$$

(i) Let $A = \arctan(120/119)$ and $B = -\arctan(1/239)$ and show that

$$\arctan\left(\frac{120}{119}\right) - \arctan\left(\frac{1}{239}\right) = \arctan 1.$$

(ii) Let $A = B = \arctan(1/5)$ and show that

$$2\arctan\left(\frac{1}{5}\right) = \arctan\left(\frac{5}{12}\right).$$

Use a similar method to show that

$$4 \arctan\left(\frac{1}{5}\right) = \arctan\left(\frac{120}{119}\right).$$

(iii) Derive Machin's formula.

2. Approximating the Derivative

In applications, the values of a function $f(x)$ are frequently known only at discrete values x_0, $x_0 \pm h$, $x_0 \pm 2h, \ldots$ Suppose we are interested in approximating the derivative $f'(x_0)$. The definition

$$f'(x_0) = \lim_{h \to 0} \frac{f(x_0 + h) - f(x_0)}{h}$$

suggests that for small h we can approximate $f'(x)$ as follows:

$$f'(x_0) \approx \frac{f(x_0 + h) - f(x_0)}{h}.$$

Such *finite-difference approximations* are used frequently in programming a computer to solve differential equations.[8]

Taylor series can be used to analyze the error in this approximation. Substituting

$$f(x_0 + h) = f(x_0) + f'(x_0)h + \frac{f''(x_0)}{2}h^2 + \cdots$$

into the approximation for $f'(x_0)$, we find

$$\frac{f(x_0 + h) - f(x_0)}{h} = f'(x_0) + \frac{f''(x_0)}{2}h + \cdots$$

This suggests (and it can be proved) that the error in the approximation is bounded as follows:

$$\left| \frac{f(x_0 + h) - f(x_0)}{h} - f'(x_0) \right| \leq \frac{Mh}{2},$$

where

$$|f''(x)| \leq M \qquad \text{for} \qquad |x - x_0| \leq |h|.$$

Notice that as $h \to 0$, the error also goes to zero, provided M is bounded.

As an example, we take $f(x) = e^x$ and $x_0 = 0$, so $f'(x_0) = 1$. The error for various values of h are given in Table 10.2. We see that decreasing h by a factor of 10 decreases the error by a factor of about 10, as predicted by the error bound $Mh/2$.

Table 10.2

h	$(f(x_0 + h) - f(x_0))/h$	Error
10^{-1}	1.05171	5.171×10^{-2}
10^{-2}	1.00502	5.02×10^{-3}
10^{-3}	1.00050	5.0×10^{-4}
10^{-4}	1.00005	5.0×10^{-5}

(a) Using Taylor series, suggest an error bound for each of the following finite-difference approximations.

(i) $f'(x_0) \approx \dfrac{f(x_0) - f(x_0 - h)}{h}$

[8]From Mark Kunka

(ii) $f'(x_0) \approx \dfrac{f(x_0 + h) - f(x_0 - h)}{2h}$

(iii) $f'(x_0) \approx \dfrac{-f(x_0 + 2h) + 8f(x_0 + h) - 8f(x_0 - h) + f(x_0 - 2h)}{12h}$

(b) Use each of the formulas in part (a) to approximate the first derivative of e^x at $x = 0$ for $h = 10^{-1}, 10^{-2}, 10^{-3}, 10^{-4}$. As h is decreased by a factor of 10, how does the error decrease? Does this agree with the error bounds found in part (a)? Which is the most accurate formula?

(c) Repeat part (b) using $f(x) = 1/x$ and $x_0 = 10^{-5}$. Why are these formulas not good approximations anymore? Continue to decrease h by factors of 10. How small does h have to be before formula (iii) is the best approximation? At these smaller values of h, what changed to make the formulas accurate again?

Chapter Eleven

DIFFERENTIAL EQUATIONS

The derivative was introduced as the rate of change of a function in Chapter 2. In Chapter 5, we saw how to use the definite integral to compute changes in the original function from the derivative. Chapter 6 introduced antiderivatives, differential equations, and going backward from the derivative to the original function.

Every time we antidifferentiate to find $\int f(x)\, dx$, we are actually solving the differential equation

$$\frac{dy}{dx} = f(x).$$

The new feature of this chapter is that the right-hand side of the differential equation may now be a function of y, or of both x and y, instead of just x.

We will see how to solve differential equations graphically first, then numerically, and, finally, in some of the special cases when it can be done, analytically. In the course of this chapter, we will treat first-order differential equations (involving first derivatives only) and some second-order differential equations (involving first and second derivatives).

11.1   WHAT IS A DIFFERENTIAL EQUATION?

How Fast Does a Person Learn?

Suppose we are interested in how fast an employee learns a new task. One theory claims that the more the employee already knows of the task, the slower he or she learns. In other words, if $y\%$ is the percentage of the task that has already been mastered, and dy/dt the rate at which the employee learns, then dy/dt decreases as y increases.

What can we say about y as a function of time, t? Figure 11.1 shows three graphs whose slope, dy/dt, decreases as y increases. Figure 11.1(a) represents an employee who starts learning at $t = 0$ and who eventually masters 100% of the task. Figure 11.1(b) represents an employee who starts later but eventually masters 100% of the task. Figure 11.1(c) represents an employee who starts learning at $t = 0$, but who does not master the whole task (since y levels off below 100%).

Figure 11.1: Possible graphs showing percentage of task learned, y, as a function of time, t

Setting up a Differential Equation to Model How a Person Learns

To describe more precisely how a person learns, we need more exact information about how dy/dt depends on y. Suppose, if time is measured in weeks, that

$$\text{Rate a person learns}\quad=\quad\text{Percentage of task not yet learned.}$$

Since y is the percentage learned by time t (in weeks), the percentage not yet learned by that time is $100 - y$. So we have

$$\frac{dy}{dt} = 100 - y.$$

Such an equation, which gives information about the rate of change of an unknown function, is called a *differential equation*.

Solving the Differential Equation Numerically

Suppose that the person starts learning at time zero, so $y = 0$ when $t = 0$. Then initially the person is learning at a rate

$$\frac{dy}{dt} = 100 - 0 = 100\% \text{ per week.}$$

In other words, if the person were to continue learning at this rate, the task would be mastered in a week. In fact, however, the rate at which the person learns decreases, so it takes more than a week to get close to mastering the task. Let's assume a five-day work week and that the 100% per week learning rate holds for the whole first day. (It doesn't, but we assume this for now.) One day is 1/5 of a week, so during the first day the person learns $100(1/5) = 20\%$ of the task. By the end of the first day the rate at which the person learns has therefore been reduced to

$$\frac{dy}{dt} = 100 - 20 = 80\% \text{ per week.}$$

Thus, during the second day the person learns $80(1/5) = 16\%$, so by the end of the second day the person knows $20 + 16 = 36\%$ of the task. Continuing in this fashion, we compute the approximate y-values[1] in Table 11.1.

---

[1]The values of y after 6, 7, ..., 9, 11, ..., 19 days were computed by the same method, but omitted from the table.

Table 11.1 *Approximate percentage of task learned as a function of time*

Time (working days)	0	1	2	3	4	5	10	20
Percentage learned	0	20	36	48.8	59.0	67.2	89.3	98.8

A Formula for the Solution to the Differential Equation

A function $y = f(t)$ which satisfies the differential equation is called a *solution*. Figure 11.1 contains graphs of possible solutions and Table 11.1 shows approximate numerical values of a solution to the equation

$$\frac{dy}{dt} = 100 - y.$$

Later in this chapter, we see how to obtain a formula for the solution:

$$y = 100 + Ce^{-t},$$

where C is a constant. To check that this formula is correct, we substitute into the differential equation, giving:

$$\text{Left side} = \frac{dy}{dt} = -Ce^{-t}$$

$$\text{Right side} = 100 - y = 100 - (100 + Ce^{-t}) = -Ce^{-t}.$$

Since we get the same result on both sides, $y = 100 + Ce^{-t}$ is a solution of this differential equation.

Finding the Arbitrary Constant: Initial Conditions

To find a value for the arbitrary constant C, we need an additional piece of information—usually the initial value of y. If, for example, we are told that $y = 0$ when $t = 0$, then substituting into

$$y = 100 + Ce^{-t}$$

shows us that

$$0 = 100 + Ce^0, \quad \text{so} \quad C = -100.$$

So the function $y = 100 - 100e^{-t}$ satisfies the differential equation *and* the condition that $y = 0$ when $t = 0$.

The Family of Solutions

Any solution to this differential equation is of the form $y = 100 + Ce^{-t}$ for some constant C. Like a family of antiderivatives, this family contains an arbitrary constant, C. We say that the *general solution* to the differential equation $dy/dt = 100 - y$ is the family of functions $y = 100 + Ce^{-t}$. The solution $y = 100 - 100e^{-t}$ that satisfies the differential equation together with the *initial condition* that $y = 0$ when $t = 0$ is called a *particular solution*. The differential equation and the initial condition together are called an *initial value problem*. Several members of the family of solutions are graphed in Figure 11.2.

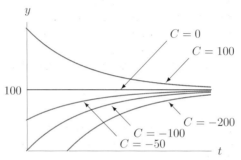

Figure 11.2: Solution curves for $dy/dt = 100 - y$:
Members of the family $y = 100 + Ce^{-t}$

First- and Second-Order Differential Equations

First, some more definitions. We often write y' to represent the derivative of y. The differential equation

$$y' = 100 - y$$

is called *first-order* because it involves the first derivative, but no higher derivatives. By contrast, if s is the height (in meters) of a body moving under the force of gravity and t is time (in seconds), then

$$\frac{d^2 s}{dt^2} = -9.8.$$

This is a *second-order* differential equation because it involves the second derivative of the unknown function, $s = f(t)$, but no higher derivatives.

Example 1   Show that $y = e^{2t}$ is not a solution to the second-order differential equation

$$\frac{d^2 y}{dt^2} + 4y = 0.$$

Solution   To decide whether the function $y = e^{2t}$ is a solution, substitute it into the differential equation:

$$\frac{d^2 y}{dt^2} + 4y = 2(2e^{2t}) + 4e^{2t} = 8e^{2t}.$$

Since $8e^{2t}$ is not identically zero, $y = e^{2t}$ is not a solution.

How Many Arbitrary Constants Should We Expect in the Family of Solutions?

Since a differential equation involves the derivative of an unknown function, solving it usually involves antidifferentiation, which introduces arbitrary constants. The solution to a first-order differential equation usually involves one antidifferentiation and one arbitrary constant (for example, the C in $y = 100 + Ce^{-t}$). Picking out one particular solution involves knowing one additional piece of information, such as an initial condition. Solving a second-order differential equation generally involves two antidifferentiations and so two arbitrary constants. Consequently, finding a particular solution usually involves two initial conditions.

For example, if s is the height (in meters) of a body above the surface of the earth at time t (in seconds), then

$$\frac{d^2 s}{dt^2} = -9.8.$$

Integrating gives

$$\frac{ds}{dt} = -9.8t + C_1,$$

and integrating again gives

$$s = -4.9t^2 + C_1 t + C_2.$$

Thus the general solution for s involves the two arbitrary constants C_1 and C_2. We can find C_1 and C_2 if we are told, for example, that the initial velocity is 30 meters per second upward and that the initial position is 5 meters above the ground. In this case, $C_1 = 30$ and $C_2 = 5$.

Exercises and Problems for Section 11.1

Exercises

1. Match the graphs in Figure 11.3 with the following descriptions.

    (a) The population of a new species introduced onto a tropical island

    (b) The temperature of a metal ingot placed in a furnace and then removed

    (c) The speed of a car traveling at uniform speed and then braking uniformly

    (d) The mass of carbon-14 in a historical specimen

    (e) The concentration of tree pollen in the air over the course of a year.

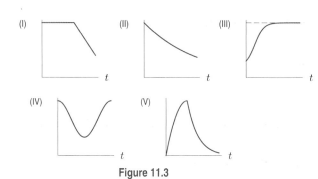

Figure 11.3

2. Is $y = x^3$ a solution to the differential equation

$$xy' - 3y = 0?$$

3. Show that $y = A + Ce^{kt}$ is a solution to the differential equation

$$\frac{dy}{dt} = k(y - A).$$

4. Show that, for any constant P_0, the function $P = P_0e^t$ satisfies the differential equation

$$\frac{dP}{dt} = P.$$

5. Fill in the missing values in the table given if you know that $dy/dt = 0.5y$. Assume the rate of growth given by dy/dt is approximately constant over each unit time interval and that the initial value of y is 8.

t	0	1	2	3	4
y	8				

6. Find the values of k for which $y = x^2 + k$ is a solution to the differential equation

$$2y - xy' = 10.$$

7. Suppose $Q = Ce^{kt}$ satisfies the differential equation

$$\frac{dQ}{dt} = -0.03Q.$$

What (if anything) does this tell you about the values of C and k?

8. For what values of k (if any) does $y = 5 + 3e^{kx}$ satisfy the differential equation:

$$\frac{dy}{dx} = 10 - 2y \ ?$$

9. Show that $y = \sin 2t$ satisfies

$$\frac{d^2y}{dt^2} + 4y = 0.$$

10. Find the value(s) of ω for which $y = \cos \omega t$ satisfies

$$\frac{d^2y}{dt^2} + 9y = 0.$$

11. Show that any function of the form

$$x = C_1 \cosh \omega t + C_2 \sinh \omega t$$

satisfies the differential equation

$$x'' - \omega^2 x = 0.$$

12. Use implicit differentiation to show that $x^2 + y^2 = r^2$ is a solution to the differential equation $dy/dx = -x/y$.

Problems

13. (a) For what values of C and n (if any) is $y = Cx^n$ a solution to the differential equation:

$$x\frac{dy}{dx} - 3y = 0 \ ?$$

    (b) If the solution satisfies $y = 40$ when $x = 2$, what more (if anything) can you say about C and n?

14. (a) Show that $P = 1/(1 + e^{-t})$ satisfies the logistic equation

$$\frac{dP}{dt} = P(1 - P).$$

    (b) What is the limiting value of P as $t \to \infty$?

15. Pick out which functions are solutions to which differential equations. (Note: Functions may be solutions to more than one equation or to none; an equation may have more than one solution.)

(a) $\dfrac{dy}{dx} = -2y$ (I) $y = 2\sin x$

(b) $\dfrac{dy}{dx} = 2y$ (II) $y = \sin 2x$

(c) $\dfrac{d^2y}{dx^2} = 4y$ (III) $y = e^{2x}$

(d) $\dfrac{d^2y}{dx^2} = -4y$ (IV) $y = e^{-2x}$

16. Match the following differential equations and possible solutions. (Note the given functions may satisfy more than one equation or none, and some equations may have more than one solution.)

(a) $y'' = y$ (I) $y = \cos x$

(b) $y' = -y$ (II) $y = \cos(-x)$

(c) $y' = 1/y$ (III) $y = x^2$

(d) $y'' = -y$ (IV) $y = e^x + e^{-x}$

(e) $x^2y'' - 2y = 0$ (V) $y = \sqrt{2x}$

11.2 SLOPE FIELDS

In this section, we see how to visualize a first-order differential equation. We start with the equation

$$\frac{dy}{dx} = y.$$

Any solution to this differential equation has the property that the slope at any point is equal to the y coordinate at that point. (That's what the equation $dy/dx = y$ is telling us!) If the solution goes through the point $(0, 0.5)$, its slope there is 0.5; if it goes through a point with $y = 1.5$, its slope there is 1.5. See Figure 11.4.

In Figure 11.4 a small line segment is drawn at each of the marked points showing the slope of the curve there. Imagine drawing many of these line segments, but leaving out the curves; this gives the *slope field* for the equation $dy/dx = y$ in Figure 11.5. From this picture, we can see that above the x-axis, the slopes are all positive (because y is positive there), and they increase as we move upward (as y increases). Below the x-axis, the slopes are all negative, and get more so as we move downward. On any horizontal line (where y is constant) the slopes are constant.

Figure 11.4: Solutions to $dy/dx = y$

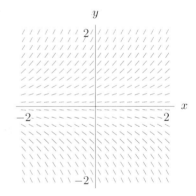

Figure 11.5: Slope field for $dy/dx = y$

In the slope field we can see the ghost of the solution curve lurking. Start anywhere on the plane and move so that the slope lines are tangent to our path; we trace out one of the solution curves. We think of the slope field as a set of signposts pointing in the direction we should go at each point. In this case, the slope field should trace out exponential curves of the form $y = Ce^x$, the solutions to the differential equation $dy/dx = y$.

Example 1   Figure 11.6 shows the slope field of the differential equation $dy/dx = 2x$.

(a) How does the slope field vary as we move around the xy-plane?

(b) Compare the solution curves sketched on the slope field with the formula for the solutions.

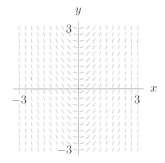

Figure 11.6: Slope field for $dy/dx = 2x$

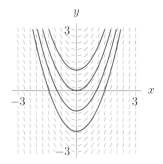

Figure 11.7: Some solutions to $dy/dx = 2x$

Solution

(a) In Figure 11.6 we notice that on a vertical line (where x is constant) the slopes are constant. This is because in this differential equation dy/dx depends on x only. (In the previous example, $dy/dx = y$, the slopes depended on y only.)

(b) The solution curves in Figure 11.7 look like parabolas. By antidifferentiation, we see that the solution to the differential equation $dy/dx = 2x$ is

$$y = \int 2x \, dx = x^2 + C,$$

so the solution curves really are parabolas.

Example 2   Using the slope field, guess the form of the solution curves of the differential equation

$$\frac{dy}{dx} = -\frac{x}{y}.$$

Solution   The slope field is shown in Figure 11.8. On the y-axis, where x is 0, the slope is 0. On the x-axis, where y is 0, the line segments are vertical and the slope is infinite. At the origin the slope is undefined, and there is no line segment.

The slope field suggests that the solution curves are circles centered at the origin. Later we see how to obtain the solution analytically, but even without this, we can check that the circle is a solution. We take the circle of radius r,

$$x^2 + y^2 = r^2,$$

and differentiate implicitly, thinking of y as a function of x. Using the chain rule, we get

$$2x + 2y \cdot \frac{dy}{dx} = 0.$$

Solving for dy/dx gives our differential equation,

$$\frac{dy}{dx} = -\frac{x}{y}.$$

This tells us that $x^2 + y^2 = r^2$ is a solution to the differential equation.

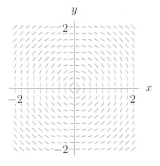

Figure 11.8: Slope field for $dy/dx = -x/y$

The previous example shows that the solution to a differential equation may be an implicit function.

Example 3   The slope fields of $dy/dt = 2 - y$ and $dy/dt = t/y$ are in Figures 11.9 and 11.10.
(a) On each slope field, sketch solution curves with initial conditions
   (i) $y = 1$ when $t = 0$ (ii) $y = 0$ when $t = 1$ (iii) $y = 3$ when $t = 0$
(b) For each solution curve, what can you say about the long-run behavior of y? For example, does $\lim\limits_{t\to\infty} y$ exist? If so, what is its value?

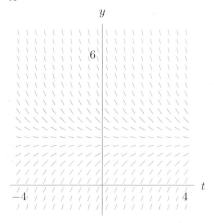

Figure 11.9: Slope field for $dy/dt = 2 - y$

Figure 11.10: Slope field for $dy/dt = t/y$

Solution   (a) See Figures 11.11 and 11.12.
(b) For $dy/dt = 2 - y$, all solution curves have $y = 2$ as a horizontal asymptote, so $\lim\limits_{t\to\infty} y = 2$.
   For $dy/dt = t/y$, as $t \to \infty$, it appears that either $y \to t$ or $y \to -t$.

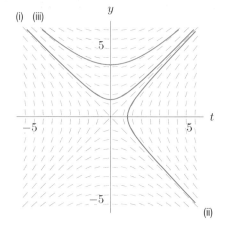

Figure 11.11: Solution curves for
$dy/dt = 2 - y$

Figure 11.12: Solution curves for
$dy/dt = t/y$

Existence and Uniqueness of Solutions

Since differential equations are used to model many real situations, the question of whether a solution is unique can have great practical importance. If we know how the velocity of a satellite is changing, can we know its velocity at any future time? If we know the initial population of a city, and we know how the population is changing, can we predict the population in the future? Common sense says yes: if we know the initial value of some quantity and we know exactly how it is changing, we should be able to figure out its future value.

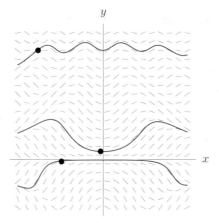

Figure 11.13: There's one and only one solution curve through each
point in the plane for this slope field (Dots represent initial conditions)

In the language of differential equations, an initial value problem (that is, a differential equation and an initial condition) almost always has a unique solution. One way to see this is by looking at the slope field. Imagine starting at the point representing the initial condition. Through that point there is usually a line segment pointing in the direction of the solution curve. By following these line segments, we trace out the solution curve. See Figure 11.13. In general, at each point there is one line segment and therefore only one direction for the solution curve to go. The solution curve *exists* and is *unique* provided we are given an initial point. Notice that even though we can draw the solution curves, we may have no simple formula for them.

It can be shown that if the slope field is continuous as we move from point to point in the plane, we can be sure that a solution curve exists everywhere. Ensuring that each point has only one solution curve through it requires a slightly stronger condition.

Exercises and Problems for Section 11.2

Exercises

1. Sketch three solution curves for each of the slope fields in Figures 11.14 and 11.15.

Figure 11.14   Figure 11.15

2. The slope field for the equation $y' = x(y - 1)$ is shown in Figure 11.16.

   (a) Sketch the solutions passing through the points
      (i) $(0, 1)$ (ii) $(0, -1)$ (iii) $(0, 0)$

   (b) From your sketch, write down the equation of the solution with $y(0) = 1$.

   (c) Check your solution to part (b) by substituting it into the differential equation.

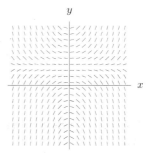

Figure 11.16

3. The slope field for the equation $y' = x + y$ is shown in Figure 11.17.

   (a) Sketch the solutions that pass through the points

   (i) $(0, 0)$ (ii) $(-3, 1)$ (iii) $(-1, 0)$

   (b) From your sketch, write the equation of the solution passing through $(-1, 0)$.

   (c) Check your solution to part (b) by substituting it into the differential equation.

Figure 11.17: $y' = x + y$

Problems

4. One of the slope fields in Figure 11.18 has the equation $y' = (x + y)/(x - y)$. Which one?

Figure 11.18

5. The slope field for the equation $dP/dt = 0.1P(10 - P)$, for $P \geq 0$, is in Figure 11.19.

   (a) Plot the solutions through the following points:

   (i) $(0, 0)$ (ii) $(1, 4)$ (iii) $(4, 1)$
   (iv) $(-5, 1)$ (v) $(-2, 12)$ (vi) $(-2, 10)$

   (b) For which positive values of P are the solutions increasing? Decreasing? What is the limiting value of P as t gets large?

Figure 11.19

6. The slope field for $y' = 0.5(1 + y)(2 - y)$ is shown in Figure 11.20.

   (a) Plot the following points on the slope field:

   (i) the origin   (ii) $(0, 1)$ (iii) $(1, 0)$
   (iv) $(0, -1)$ (v) $(0, -5/2)$ (vi) $(0, 5/2)$

   (b) Plot solution curves through the points in part (a).

   (c) For which regions are all solution curves increasing? For which regions are all solution curves decreasing? When can the solution curves have horizontal tangents? Explain why, using both the slope field and the differential equation.

Figure 11.20: Note: x and y scales are equal

7. Figure 11.21 shows the slope field for the equation $y' = (\sin x)(\sin y)$.

   (a) Sketch the solutions that pass through the points:
   (i) $(0, -2)$ (ii) $(0, \pi)$.

   (b) What is the equation of the solution that passes through $(0, n\pi)$, where n is any integer?

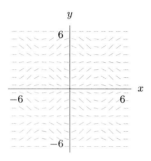

Figure 11.21: $y' = (\sin x)(\sin y)$

8. Match the slope fields in Figure 11.22 with their differential equations:

(a) $y' = -y$ (b) $y' = y$ (c) $y' = x$

(d) $y' = 1/y$ (e) $y' = y^2$

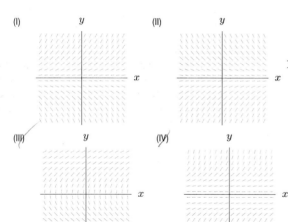

Figure 11.22

9. Match the slope fields in Figure 11.23 with their differential equations:

(a) $y' = 1 + y^2$ (b) $y' = x^2$ (c) $y' = \sin x$

(d) $y' = y$ (e) $y' = x - y$ (f) $y' = 4 - y$

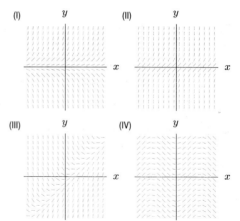

Figure 11.23: Each slope field is graphed for $-5 \le x \le 5, -5 \le y \le 5$

10. Match the slope fields in Figure 11.24 to the corresponding differential equations:

(a) $y' = xe^{-x}$ (b) $y' = \sin x$ (c) $y' = \cos x$

(d) $y' = x^2 e^{-x}$ (e) $y' = e^{-x^2}$ (f) $y' = e^{-x}$

Figure 11.24

11.3 EULER'S METHOD

In the preceding section we saw how to sketch a solution curve to a differential equation using its slope field. In this section we compute points on a solution curve numerically using *Euler's method*. (Leonhard Euler was an eighteenth-century Swiss mathematician.) In Section 11.4 we find formulas for some solution curves.

Here's the concept behind Euler's method. Think of the slope field as a set of signposts directing you across the plane. Pick a starting point (corresponding to the initial value), and calculate the slope at that point using the differential equation. This slope is a signpost telling you the direction to take. Head off a small distance in that direction. Stop and look at the new signpost. Recalculate the slope from the differential equation, using the coordinates of the new point. Change direction to correspond to the new slope, and move another small distance, and so on.

Example 1   Use Euler's method for $dy/dx = y$. Start at the point $P_0 = (0, 1)$ and take $\Delta x = 0.1$.

Solution   The slope at the point $P_0 = (0, 1)$ is $dy/dx = 1$. (See Figure 11.25.) As we move from P_0 to P_1, y increases by Δy, where

$$\Delta y = (\text{slope at } P_0)\Delta x = 1(0.1) = 0.1.$$

So we have

$$y\text{-value at } P_1 = (y \text{ value at } P_0) + \Delta y = 1 + 0.1 = 1.1.$$

Table 11.2   *Euler's method for $dy/dx = y$, starting at $(0, 1)$*

	x	y	$\Delta y = (\text{Slope})\Delta x$
P_0	0	1	$0.1 = (1)(0.1)$
P_1	0.1	1.1	$0.11 = (1.1)(0.1)$
P_2	0.2	1.21	$0.121 = (1.21)(0.1)$
P_3	0.3	1.331	$0.1331 = (1.331)(0.1)$
P_4	0.4	1.4641	$0.14641 = (1.4641)(0.1)$
P_5	0.5	1.61051	$0.161051 = (1.61051)(0.1)$

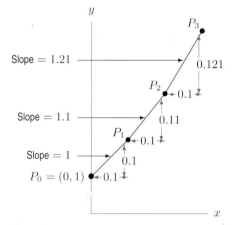

Figure 11.25: Euler's approximate solution to $dy/dx = y$

Thus the point P_1 is $(0.1, 1.1)$. Now, using the differential equation again, we see that

$$\text{Slope at } P_1 = 1.1,$$

so if we move to P_2, then y changes by

$$\Delta y = (\text{slope at } P_1)\Delta x = (1.1)(0.1) = 0.11.$$

This means

$$y\text{-value at } P_2 = (y \text{ value at } P_1) + \Delta y = 1.1 + 0.11 = 1.21.$$

Thus P_2 is $(0.2, 1.21)$. Continuing gives the results in Table 11.2.

Since the solution curves of $dy/dx = y$ are exponentials, they are concave up and bend upward away from the line segments of the slope field. Therefore, in this case, Euler's method produces y-values which are too small.

Notice that Euler's method calculates approximate y-values for points on a solution curve; it does not give a formula for y in terms of x.

Example 2   Show that Euler's method for $dy/dx = y$ starting at $(0, 1)$ and using two steps with $\Delta x = 0.05$ gives $y \approx 1.1025$ when $x = 0.1$.

Solution   At $(0, 1)$, the slope is 1 and $\Delta y = (1)(0.05) = 0.05$, so new $y = 1 + 0.05 = 1.05$. At $(0.05, 1.05)$, the slope is 1.05 and $\Delta y = (1.05)(0.05) = 0.0525$, so new $y = 1.05 + 0.0525 = 1.1025$ at $x = 0.1$.

In general, dy/dx may be a function of both x and y. Euler's method still works, as the next example shows.

Example 3   Approximate four points on the solution curve to $dy/dx = -x/y$ starting at $(0, 1)$; use $\Delta x = 0.1$. Are the approximate values overestimates or underestimates?

Solution   The results from Euler's method are in Table 11.3, along with the exact y-values (to two decimals) calculated from the equation of the circle $x^2 + y^2 = 1$, which is the solution curve through $(0, 1)$. Since the curve is concave down, the approximate y-values are above the exact ones. (See Figure 11.26.)

Table 11.3   *Euler's method for $dy/dx = -x/y$, starting at $(0, 1)$*

x	Approximate y-value	$\Delta y = (\text{Slope})\Delta x$	True y-value
0	1	$0 = (0)(0.1)$	1
0.1	1	$-0.01 = (-0.1/1)(0.1)$	0.99
0.2	0.99	$-0.02 = (-0.2/0.99)(0.1)$	0.98
0.3	0.97		0.95

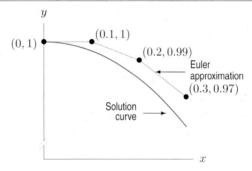

Figure 11.26: Euler's approximate solution to $dy/dx = -x/y$

The Accuracy of Euler's Method

To improve the accuracy of Euler's method, we choose Δx smaller. Let's go back to the differential equation $dy/dx = y$ and compare the exact and approximate values for different Δx's. The exact solution going through the point $(0, 1)$ is $y = e^x$, so the exact values are calculated using this function. (See Figure 11.27.) Where $x = 0.1$,

$$\text{Exact } y\text{-value } = e^{0.1} \approx 1.1051709.$$

In Example 1 we had $\Delta x = 0.1$, and where $x = 0.1$,

$$\text{Approximate } y\text{-value} = 1.1, \quad \text{so the error} \approx 0.005.$$

In Example 2 we decreased Δx to 0.05. After two steps, $x = 0.1$, and we had

$$\text{Approximate } y\text{-value} = 1.1025, \quad \text{so error} \approx 0.00267.$$

Thus, it appears that halving the step size has approximately halved the error.

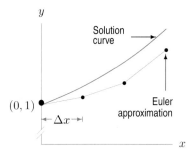

Figure 11.27: Euler's approximate solution to $dy/dx = y$

The *error* in using Euler's method is the difference between the approximate value and the exact value. If the number of steps used is n, the error is approximately proportional to $1/n$.

Just as there are more accurate numerical integration methods than left and right Riemann sums, there are more accurate methods than Euler's for approximating solution curves. However, Euler's method will be all we need.

Exercises and Problems for Section 11.3

Exercises

1. Consider the differential equation $y' = x + y$ whose slope field is in Figure 11.17 on page 530. Use Euler's method with $\Delta x = 0.1$ to estimate y when $x = 0.4$ for the solution curves satisfying

      (a) $y(0) = 1$ **(b)** $y(-1) = 0$.

2. (a) Use ten steps of Euler's method to determine an approximate solution for the differential equation $y' = x^3$, $y(0) = 0$, using a step size $\Delta x = 0.1$.
  (b) What is the exact solution? Compare it to the computed approximation.
  (c) Use a sketch of the slope field for this equation to explain the results of part (b).

3. Consider the solution of the differential equation $y' = y$ passing through $y(0) = 1$.

  (a) Sketch the slope field for this differential equation, and sketch the solution passing through the point $(0, 1)$.
  (b) Use Euler's method with step size $\Delta x = 0.1$ to estimate the solution at $x = 0.1, 0.2, \ldots, 1$.
  (c) Plot the estimated solution on the slope field; compare the solution and the slope field.
  (d) Check that $y = e^x$ is the solution of $y' = y$ with $y(0) = 1$.

4. (a) Use Euler's method to approximate the value of y at $x = 1$ on the solution curve to the differential equation

$$\frac{dy}{dx} = x^3 - y^3$$

that passes through $(0, 0)$. Use $\Delta x = 1/5$ (i.e., 5 steps).
  (b) Using Figure 11.28, sketch the solution that passes through $(0, 0)$. Show the approximation you made in part (a).
  (c) Using the the slope field, say whether your answer to part (a) is an overestimate or an underestimate.

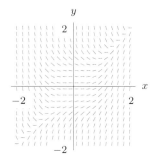

Figure 11.28: Slope field for $\frac{dy}{dx} = x^3 - y^3$

Problems

5. Consider the differential equation

$$\frac{dy}{dx} = 2x, \quad \text{with initial condition } y(0) = 1.$$

(a) Use Euler's method with two steps to estimate y when $x = 1$. Now use four steps.

(b) What is the formula for the exact value of y?

(c) Does the error in Euler's approximation behave as predicted in the box on page 534?

6. (a) Use ten steps of Euler's method to approximate y-values for $dy/dt = 1/t$, starting at $(1, 0)$ and using $\Delta t = 0.1$.

(b) Using integration, solve the differential equation to find the exact value of y at the end of these ten steps.

(c) Is your approximate value of y at the end of ten steps bigger or smaller than the exact value? Use a slope field to explain your answer.

7. Consider the differential equation $y' = (\sin x)(\sin y)$.

(a) Calculate approximate y-values using Euler's method with three steps and $\Delta x = 0.1$, starting at each of the following points:
    (i) $(0, 2)$ (ii) $(0, \pi)$.

(b) Use the slope field in Figure 11.21 on page 531 to explain your solution to part (a)(ii).

8. Consider the differential equation

$$\frac{dy}{dx} = \sin(xy), \quad \text{with initial condition } y(1) = 1.$$

Estimate $y(2)$, using Euler's method with step sizes $\Delta x = 0.2, 0.1, 0.05$.

Plot the computed approximations for $y(2)$ against Δx. What do you conclude? Use your observations to estimate the exact value of $y(2)$.

9. Use Euler's method to solve

$$\frac{dB}{dt} = 0.05B$$

with initial value $B = 1000$ when $t = 0$. Take:

(a) $\Delta t = 1$ and 1 step.   (b) $\Delta t = 0.5$ and 2 steps.

(c) $\Delta t = 0.25$ and 4 steps.

(d) Suppose B is the balance in a bank account earning interest. Explain why the result of your calculation in part (a) is equivalent to compounding the interest once a year instead of continuously.

(e) Interpret the result of your calculations in parts (b) and (c) in terms of compound interest.

10. Consider the differential equation $dy/dx = f(x)$ with initial value $y(0) = 0$. Explain why using Euler's method to approximate the solution curve gives the same results as using left Riemann sums to approximate $\int_0^x f(t)\,dt$.

11.4 SEPARATION OF VARIABLES

We have seen how to sketch solution curves of a differential equation using a slope field and how to calculate approximate numerical solutions. Now we see how solve certain differential equations analytically, finding an equation for the solution curve.

First, we look at a familiar example, the differential equation

$$\frac{dy}{dx} = -\frac{x}{y},$$

whose solution curves are the circles

$$x^2 + y^2 = C.$$

We can check that these circles are solutions by differentiation; the question now is how they were obtained. The method of *separation of variables* works by putting all the x's on one side of the equation and all the y's on the other, giving

$$y\,dy = -x\,dx.$$

We then integrate each side separately:

$$\int y\,dy = -\int x\,dx,$$

$$\frac{y^2}{2} = \frac{-x^2}{2} + k.$$

This gives the circles we were expecting:

$$x^2 + y^2 = C \qquad \text{where } C = 2k.$$

You might worry about whether it is legitimate to separate the dx and the dy. The reason it can be done is explained at the end of this section.

The Exponential Growth and Decay Equations

Let's use separation of variables on an equation that occurs frequently in practice:

$$\frac{dy}{dx} = ky.$$

Separating variables,

$$\frac{1}{y} dy = k\, dx,$$

and integrating,

$$\int \frac{1}{y}\, dy = \int k\, dx,$$

gives

$$\ln|y| = kx + C \quad \text{for some constant } C.$$

Solving for $|y|$ leads to

$$|y| = e^{kx+C} = e^{kx}e^{C} = Ae^{kx}$$

where $A = e^C$, so A is positive. Thus

$$y = (\pm A)e^{kx} = Be^{kx}$$

where $B = \pm A$, so B is any nonzero constant. Even though there's no C leading to $B = 0$, B can be 0 because $y = 0$ is a solution to the differential equation. We lost this solution when we divided through by y at the first step.

> The general solution to $\dfrac{dy}{dx} = ky$ is $y = Be^{kx}$ for any B.

The differential equation $dy/dx = ky$ always leads to exponential growth (if $k > 0$) or exponential decay (if $k < 0$). Graphs of solution curves for some fixed $k > 0$ are in Figure 11.29. For $k < 0$, the graphs are reflected in the y-axis.

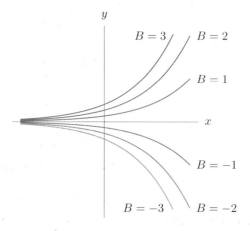

Figure 11.29: Graphs of $y = Be^{kx}$, which are solutions to $dy/dx = ky$, for some fixed $k > 0$

Example 1   For $k > 0$, find and graph solutions of

$$\frac{dH}{dt} = -k(H - 20).$$

Solution   The slope field in Figure 11.30 shows the qualitative behavior of the solutions. To find the equation of the solution curves, we separate variables and integrate:

$$\int \frac{1}{H - 20}\, dH = -\int k\, dt.$$

This gives

$$\ln|H - 20| = -kt + C.$$

Solving for H leads to:

$$|H - 20| = e^{-kt+C} = e^{-kt}e^{C} = Ae^{-kt}$$

or

$$H - 20 = (\pm A)e^{-kt} = Be^{-kt}$$
$$H = 20 + Be^{-kt}.$$

Again, $B = 0$ also gives a solution. Graphs for $k = 1$ and $B = -10, 0, 10$, with $t \geq 0$, are in Figure 11.30.

Figure 11.30: Slope field and some solution curves for $dH/dt = -k(H - 20)$, with $k = 1$

Example 2   Find and sketch the solution to

$$\frac{dP}{dt} = 2P - 2Pt$$

satisfying $P = 5$ when $t = 0$.

Solution   Factoring the right-hand side gives

$$\frac{dP}{dt} = P(2 - 2t).$$

Separating variables, we get

$$\int \frac{dP}{P} = \int (2 - 2t)\, dt,$$

so

$$\ln|P| = 2t - t^2 + C.$$

Solving for P leads to

$$|P| = e^{2t-t^2+C} = e^{C}e^{2t-t^2} = Ae^{2t-t^2}$$

with $A = e^{C}$, so $A > 0$. In addition, $A = 0$ gives a solution. Thus the general solution to the differential equation is

$$P = Be^{2t-t^2} \quad \text{for any } B.$$

To find the value of B, substitute $P = 5$ and $t = 0$ into the general solution, giving

$$5 = Be^{2(0)-0^2} = B$$

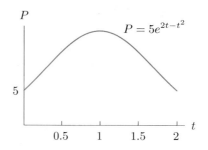

Figure 11.31: Bell-shaped solution curve

so

$$P = 5e^{2t-t^2}.$$

The graph of this function is in Figure 11.31. Since the solution can be rewritten as

$$P = 5e^{1-1+2t-t^2} = 5e^1 e^{-1+2t-t^2} = (5e)e^{-(t-1)^2},$$

the graph has the same shape as the graph of $y = e^{-t^2}$, the bell-shaped curve of statistics. Here the maximum, normally at $t = 0$, is shifted one unit to the right and occurs at $t = 1$.

Justification For Separation of Variables

Suppose a differential equation can be written in the form

$$\frac{dy}{dx} = g(x)f(y).$$

Provided $f(y) \neq 0$, we write $f(y) = 1/h(y)$ so the right-hand side can be thought of as a fraction,

$$\frac{dy}{dx} = \frac{g(x)}{h(y)}.$$

If we multiply through by $h(y)$ we get

$$h(y)\frac{dy}{dx} = g(x).$$

Thinking of y as a function of x, so $y = y(x)$, and $dy/dx = y'(x)$, we can rewrite the equation as

$$h(y(x)) \cdot y'(x) = g(x).$$

Now integrate both sides with respect to x:

$$\int h(y(x)) \cdot y'(x)\, dx = \int g(x)\, dx.$$

The form of the integral on the left suggests that we use the substitution $y = y(x)$. Since $dy = y'(x)\, dx$, we get

$$\int h(y)\, dy = \int g(x)\, dx.$$

If we can find antiderivatives of h and g, then this gives the equation of the solution curve.

Note that transforming the original differential equation,

$$\frac{dy}{dx} = \frac{g(x)}{h(y)},$$

into

$$\int h(y)\, dy = \int g(x)\, dx$$

looks as though we have treated dy/dx as a fraction, cross-multiplied and then integrated. Although that's not exactly what we've done, you may find this a helpful way of remembering the method. In fact, the dy/dx notation was introduced by Leibniz to allow shortcuts like this (more specifically, to make the chain rule look like cancellation).

Exercises and Problems for Section 11.4

Exercises

Find the solutions to the differential equations in Exercises 1–27, subject to the given initial conditions.

1. $\dfrac{dP}{dt} = -2P, \quad P(0) = 1$

2. $\dfrac{dP}{dt} = 0.02P, \quad P(0) = 20$

3. $\dfrac{dL}{dp} = \dfrac{L}{2}, \quad L(0) = 100$

4. $\dfrac{dQ}{dt} = \dfrac{Q}{5}, \quad Q = 50$ when $t = 0$

5. $\dfrac{dy}{dx} + \dfrac{y}{3} = 0, \quad y(0) = 10$

6. $P\dfrac{dP}{dt} = 1, \quad P(0) = 1$

7. $\dfrac{dm}{dt} = 3m, \quad m = 5$ when $t = 1$

8. $\dfrac{dI}{dx} = 0.2I, \quad I = 6$ where $x = -1$

9. $\dfrac{1}{z}\dfrac{dz}{dt} = 5, \quad z(1) = 5$

10. $\dfrac{dm}{ds} = m, \quad m(1) = 2$

11. $2\dfrac{du}{dt} = u^2, \quad u(0) = 1$

12. $\dfrac{dz}{dy} = zy, \quad z = 1$ when $y = 0$

13. $\dfrac{dy}{dt} = 0.5(y - 200), \quad y = 50$ when $t = 0$

14. $\dfrac{dP}{dt} = P + 4, \quad P = 100$ when $t = 0$

15. $\dfrac{dy}{dx} = 2y - 4, \quad$ through $(2, 5)$

16. $\dfrac{dQ}{dt} = 0.3Q - 120, \quad Q = 50$ when $t = 0$

17. $\dfrac{dm}{dt} = 0.1m + 200, \quad m(0) = 1000$

18. $\dfrac{dR}{dy} + R = 1, \quad R(1) = 0.1$

19. $\dfrac{dB}{dt} + 2B = 50, \quad B(1) = 100$

20. $\dfrac{dy}{dt} = \dfrac{y}{3 + t}, \quad y(0) = 1$

21. $\dfrac{dz}{dt} = te^z, \quad$ through the origin

22. $\dfrac{dy}{dx} = \dfrac{5y}{x}, \quad y = 3$ where $x = 1$

23. $\dfrac{dy}{dt} = y^2(1 + t), \quad y = 2$ when $t = 1$

24. $\dfrac{dz}{dt} = z + zt^2, \quad z = 5$ when $t = 0$

25. $\dfrac{dw}{d\theta} = \theta w^2 \sin \theta^2, \quad w(0) = 1$

26. $\dfrac{dw}{d\psi} = -w^2 \tan \psi, \quad w(0) = 2$

27. $x(x + 1)\dfrac{du}{dx} = u^2, \quad u(1) = 1$

28. Determine which of the following differential equations is separable. Do not solve the equations.

(a) $y' = y$ (b) $y' = x + y$
(c) $y' = xy$ (d) $y' = \sin(x + y)$
(e) $y' - xy = 0$ (f) $y' = y/x$
(g) $y' = \ln(xy)$ (h) $y' = (\sin x)(\cos y)$
(i) $y' = (\sin x)(\cos xy)$ (j) $y' = x/y$
(k) $y' = 2x$ (l) $y' = (x + y)/(x + 2y)$

Problems

Solve the differential equations in Problems 29–38. Assume a, b, and k are nonzero constants.

29. $\dfrac{dR}{dt} = kR$

30. $\dfrac{dQ}{dt} - \dfrac{Q}{k} = 0$

31. $\dfrac{dP}{dt} = P - a$

32. $\dfrac{dQ}{dt} = b - Q$

33. $\dfrac{dP}{dt} = k(P - a)$

34. $\dfrac{dR}{dt} = aR + b$

35. $\dfrac{dP}{dt} - aP = b$

36. $\dfrac{dy}{dt} = ky^2(1 + t^2)$

37. $\dfrac{dR}{dx} = a(R^2 + 1)$

38. $\dfrac{dL}{dx} = k(x + a)(L - b)$

Solve the differential equations in Problems 39–42. Assume $x, y, t > 0$.

39. $\dfrac{dy}{dt} = y(2 - y), \quad y(0) = 1$

40. $t\dfrac{dx}{dt} = (1 + 2\ln t)\tan x$

41. $\dfrac{dx}{dt} = \dfrac{x \ln x}{t}$

42. $\dfrac{dy}{dt} = -y \ln\left(\dfrac{y}{2}\right), \quad y(0) = 1$

43. (a) Find the general solution to the differential equation modeling how a person learns:

$$\dfrac{dy}{dt} = 100 - y.$$

(b) Plot the slope field of this differential equation and sketch solutions with $y(0) = 25$ and $y(0) = 110$.

(c) For each of the initial conditions in part (b), find the particular solution and add to your sketch.

(d) Which of these two particular solutions could represent how a person learns?

44. (a) Sketch the slope field for the differential equation $dy/dx = xy$.

(b) Sketch several solution curves.

(c) Solve the differential equation analytically.

45. (a) Sketch the slope field for $y' = x/y$.

(b) Sketch several solution curves.

(c) Solve the differential equation analytically.

46. (a) Sketch the slope field for $y' = -y/x$.

(b) Sketch several solution curves.

(c) Solve the differential equation analytically.

47. Compare the slope field for $y' = x/y$, Problem 45, with that for $y' = -y/x$, Problem 46. Show that the solution curves of Problem 45 intersect the solution curves of Problem 46 at right angles.

11.5   GROWTH AND DECAY

In this section we look at exponential growth and decay equations. Consider the population of a region. If there is no immigration or emigration, the rate at which the population is changing is often proportional to the population. In other words, the larger the population, the faster it is growing, because there are more people to have babies. If the population at time t is P, and its continuous growth rate is 2% per unit time, then we know

$$\text{Rate of growth of population} = 2\%(\text{Current population})$$

and we can write this as

$$\frac{dP}{dt} = 0.02P.$$

The 2% growth rate is called the *relative growth rate* to distinguish it from the *absolute growth rate*, dP/dt. Notice they measure different quantities. Since

$$\text{Relative growth rate} = 2\% = \frac{1}{P}\frac{dP}{dt},$$

the relative growth rate is a percent change per unit time, while

$$\text{Absolute growth rate} = \text{Rate of change of population} = \frac{dP}{dt}$$

is a change in population per unit time.

Since the equation $dP/dt = 0.02P$ is of the form $dP/dt = kP$ for $k = 0.02$, it has solution

$$P = P_0 e^{0.02t}.$$

Other processes are described by differential equations similar to that for population growth, but with negative values for k. In summary, we have the following result from the preceding section:

Every solution to the equation

$$\frac{dP}{dt} = kP$$

can be written in the form

$$P = P_0 e^{kt},$$

where P_0 is the value of P at $t = 0$, and $k > 0$ represents growth, whereas $k < 0$ represents decay.

Recall that the *doubling time* of an exponentially growing quantity is the time required for it to double. The *half-life* of an exponentially decaying quantity is the time for half of it to decay.

Continuously Compounded Interest

At a bank, continuous compounding means that interest is accrued at a rate that is a fixed percentage of the balance at that moment. Thus, the larger the balance, the faster interest is earned and the faster the balance grows.

Example 1   A bank account earns interest continuously at a rate of 5% of the current balance per year. Assume that the initial deposit is $1000, and that no other deposits or withdrawals are made.

(a) Write the differential equation satisfied by the balance in the account.
(b) Solve the differential equation and graph the solution.

Solution   (a) We are looking for B, the balance in the account in dollars, as a function of t, time in years. Interest is being added to the account continuously at a rate of 5% of the balance at that moment. Since no deposits or withdrawals are made, at any instant,

$$\text{Rate balance increasing} = \text{Rate interest earned} = 5\%(\text{Current balance}),$$

which we write as

$$\frac{dB}{dt} = 0.05B.$$

This is the differential equation that describes the process. It does not involve the initial condition $1000, because the initial deposit does not affect the process by which interest is earned.

(b) Solving the differential equation by separation of variables gives

$$B = B_0 e^{0.05t},$$

where B_0 is the value of B at $t = 0$, so $B_0 = 1000$. Thus

$$B = 1000 e^{0.05t}$$

and this function is graphed in Figure 11.32.

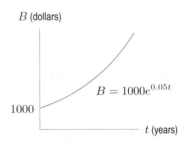

B (dollars)

$B = 1000 e^{0.05t}$

1000

t (years)

Figure 11.32: Bank balance against time

The Difference Between Continuous and Annual Percentage Growth Rates

If $P = P_0(1 + r)^t$ with t in years, we say that r is the *annual* growth rate, while if $P = P_0 e^{kt}$, we say that k is the *continuous* growth rate.

The constant k in the differential equation $dP/dt = kP$ is not the annual growth rate, but the continuous growth rate. In Example 1, with a continuous interest rate of 5%, we obtain a balance of $B = B_0 e^{0.05t}$, where time, t, is in years. At the end of one year the balance is $B_0 e^{0.05}$. In that one year, our balance has changed from B_0 to $B_0 e^{0.05}$; that is, by a factor of $e^{0.05} = 1.0513$.

Thus the annual growth rate is 5.13%. This is what the bank means when it says "5% compounded continuously for an effective annual yield of 5.13%." Since $P_0 e^{0.05t} = P_0(1.0513)^t$, we have two different ways to represent the same function.

Since most growth is measured over discrete time intervals, a continuous growth rate is an idealized concept. A demographer who says a population is growing at the rate of 2% per year usually means that after t years the population is $P = P_0(1.02)^t$. To find the continuous growth rate, k, we express the population as $P = P_0 e^{kt}$. At the end of one year $P = P_0 e^k$, so $e^k = 1.02$. Thus $k = \ln 1.02 \approx 0.0198$. The continuous growth rate, $k = 1.98\%$, is close to the annual growth rate of 2%, but it is not the same. Again, we have two different representations of the same function, since $P_0(1.02)^t = P_0 e^{0.0198t}$.

Pollution in the Great Lakes

In the 1960s pollution in the Great Lakes became an issue of public concern. We set up a model for how long it would take for the lakes to flush themselves clean, assuming no further pollutants are being dumped in the lake.

Suppose Q is the total quantity of pollutant in a lake of volume V at time t. Suppose that clean water is flowing into the lake at a constant rate r and that water flows out at the same rate. Assume that the pollutant is evenly spread throughout the lake, and that the clean water coming into the lake immediately mixes with the rest of the water.

We investigate how Q varies with time. Since pollutants are being taken out of the lake but not added, Q decreases, and the water leaving the lake becomes less polluted, so the rate at which the pollutants leave decreases. This tells us that Q is decreasing and concave up. In addition, the pollutants are never completely removed from the lake though the quantity remaining becomes arbitrarily small: in other words, Q is asymptotic to the t-axis. (See Figure 11.33.)

Setting Up a Differential Equation for the Pollution

To model exactly how Q changes with time, we write an equation for the rate at which Q changes. We know that

$$\binom{\text{Rate } Q}{\text{changes}} = -\left(\begin{array}{c} \text{Rate pollutants} \\ \text{leave in outflow} \end{array}\right)$$

where the negative sign represents the fact that Q is decreasing. At time t, the concentration of pollutants is Q/V and water containing this concentration is leaving at rate r. Thus

$$\binom{\text{Rate pollutants}}{\text{leave in outflow}} = \binom{\text{Rate of}}{\text{outflow}} \times \text{Concentration} = r \cdot \frac{Q}{V}.$$

So the differential equation is

$$\frac{dQ}{dt} = -\frac{r}{V}Q$$

and its solution is

$$Q = Q_0 e^{-rt/V}.$$

Table 11.4 contains values of r and V for four of the Great Lakes.[2] We use this data to calculate how long it would take for certain fractions of the pollution to be removed.

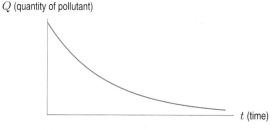

Figure 11.33: Pollutant in lake versus time

[2]Data from William E. Boyce and Richard C. DiPrima, *Elementary Differential Equations* (New York: Wiley, 1977).

Table 11.4   *Volume and outflow in Great Lakes*

	V (thousands of km^3)	r (km^3/year)
Superior	12.2	65.2
Michigan	4.9	158
Erie	0.46	175
Ontario	1.6	209

Example 2   According to this model, how long will it take for 90% of the pollution to be removed from Lake Erie? For 99% to be removed?

Solution   Substituting r and V for Lake Erie into the differential equation for Q gives

$$\frac{dQ}{dt} = -\frac{r}{V}Q = \frac{-175}{0.46 \times 10^3}Q = -0.38Q$$

where t is measured in years. Thus Q is given by

$$Q = Q_0 e^{-0.38t}.$$

When 90% of the pollution has been removed, 10% remains, so $Q = 0.1Q_0$. Substituting gives

$$0.1Q_0 = Q_0 e^{-0.38t}.$$

Canceling Q_0 and solving for t, we get

$$t = \frac{-\ln(0.1)}{0.38} \approx 6 \text{ years.}$$

When 99% of the pollution has been removed, $Q = 0.01Q_0$, so t satisfies

$$0.01Q_0 = Q_0 e^{-0.38t}.$$

Solving for t gives

$$t = \frac{-\ln(0.01)}{0.38} \approx 12 \text{ years.}$$

Newton's Law of Heating and Cooling

Newton proposed that the temperature of a hot object decreases at a rate proportional to the difference between its temperature and that of its surroundings. Similarly, a cold object heats up at a rate proportional to the temperature difference between the object and its surroundings.

For example, a hot cup of coffee standing on the kitchen table cools at a rate proportional to the temperature difference between the coffee and the surrounding air. As the coffee cools, the rate at which it cools decreases, because the temperature difference between the coffee and the air decreases. In the long run, the rate of cooling tends to zero, and the temperature of the coffee approaches room temperature. See Figure 11.34.

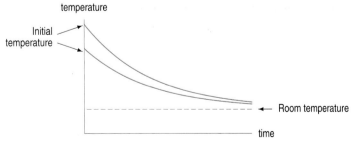

Figure 11.34: Temperature of two cups of coffee with different initial temperatures

Example 3   When a murder is committed, the body, originally at $37°C$, cools according to Newton's Law of Cooling. Suppose that after two hours the temperature is $35°C$, and that the temperature of the surrounding air is a constant $20°C$.

(a) Find the temperature, H, of the body as a function of t, the time in hours since the murder was committed.
(b) Sketch a graph of temperature against time.
(c) What happens to the temperature in the long run? Show this on the graph and algebraically.
(d) If the body is found at 4 pm at a temperature of $30°C$, when was the murder committed?

Solution   (a) We first find a differential equation for the temperature of the body as a function of time. Newton's Law of Cooling says that for some constant α,

$$\text{Rate of change of temperature} = \alpha(\text{Temperature difference}).$$

If H is the temperature of the body, then

$$\text{Temperature difference} = H - 20$$

so

$$\frac{dH}{dt} = \alpha(H - 20).$$

What about the sign of α? If the temperature difference is positive (i.e., $H > 20$), then H is falling, so the rate of change must be negative. Thus α should be negative, so we write:

$$\frac{dH}{dt} = -k(H - 20), \qquad \text{for some } k > 0.$$

Separating variables and solving, as in Example 1 on page 537, gives:

$$H - 20 = Be^{-kt}.$$

To find B, substitute the initial condition that $H = 37$ when $t = 0$:

$$37 - 20 = Be^{-k(0)} = B,$$

so $B = 17$. Thus,

$$H - 20 = 17e^{-kt}.$$

To find k, we use the fact that after 2 hours, the temperature is $35°C$, so

$$35 - 20 = 17e^{-k(2)}.$$

Dividing by 17 and taking natural logs, we get:

$$\ln\left(\frac{15}{17}\right) = \ln(e^{-2k})$$
$$-0.125 = -2k$$
$$k \approx 0.063.$$

Therefore, the temperature is given by

$$H - 20 = 17e^{-0.063t}$$

or

$$H = 20 + 17e^{-0.063t}.$$

(b) The graph of $H = 20 + 17e^{-0.063t}$ has a vertical intercept of $H = 37$, because the temperature of the body starts at $37°C$. The temperature decays exponentially with $H = 20$ as the horizontal asymptote. (See Figure 11.35.)

Figure 11.35: Temperature of dead body

(c) "In the long run" means as $t \to \infty$. The graph shows that as $t \to \infty$, $H \to 20$. Algebraically, since $e^{-0.063t} \to 0$ as $t \to \infty$, we have

$$H = 20 + \underbrace{17e^{-0.063t}}_{\text{goes to 0 as } t \to \infty} \longrightarrow 20 \quad \text{as } t \to \infty.$$

(d) We want to know when the temperature reaches $30°C$. Substitute $H = 30$ and solve for t:

$$30 = 20 + 17e^{-0.063t}$$
$$\frac{10}{17} = e^{-0.063t}.$$

Taking natural logs:

$$-0.531 = -0.063t,$$

which gives

$$t \approx 8.4 \, \text{hours}.$$

Thus the murder must have been committed about 8.4 hours before 4 pm. Since 8.4 hours $= 8$ hours 24 minutes, the murder was committed at about 7:30 am.

Equilibrium Solutions

Figure 11.36 shows the temperature of several objects in a $20°C$ room. One is initially hotter than $20°C$ and cools down toward $20°C$; another is initially cooler and warms up toward $20°C$. All these curves are solutions to the differential equation

$$\frac{dH}{dt} = -k(H - 20)$$

for some fixed $k > 0$, and all the solutions have the form

$$H = 20 + Ae^{-kt}$$

for some A. Notice that $H \to 20$ as $t \to \infty$ because $e^{-kt} \to 0$ as $t \to \infty$. In other words, in the long run, the temperature of the object always tends toward $20°C$, the temperature of the room. This means that what happens in the long run is independent of the initial condition.

In the special case when $A = 0$, we have the *equilibrium solution*

$$H = 20$$

for all t. This means that if the object starts at $20°C$, it remains at $20°C$ for all time. Notice that such a solution can be found directly from the differential equation by solving $dH/dt = 0$:

$$\frac{dH}{dt} = -k(H - 20) = 0$$

giving

$$H = 20.$$

Regardless of the initial temperature, H always gets closer and closer to 20 as $t \to \infty$. As a result, $H = 20$ is called a *stable* equilibrium[3] for H.

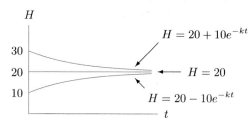

Figure 11.36: $H = 20$ is stable equilibrium ($k > 0$)

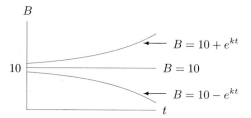

Figure 11.37: $B = 10$ is unstable equilibrium ($k > 0$)

A different situation is displayed in Figure 11.37, which shows solutions to the differential equation

$$\frac{dB}{dt} = k(B - 10)$$

for some fixed $k > 0$. Solving $dB/dt = 0$ gives the equilibrium $B = 10$, which is *unstable* because if B starts near 10, it moves away as $t \to \infty$.

In general, we have the following definitions.

- An **equilibrium solution** is constant for all values of the independent variable. The graph is a horizontal line.
- An equilibrium is **stable** if a small change in the initial conditions gives a solution which tends toward the equilibrium as the independent variable tends to positive infinity.
- An equilibrium is **unstable** if a small change in the initial conditions gives a solution curve which veers away from the equilibrium as the independent variable tends to positive infinity.

Solutions which do not veer away from an equilibrium solution are also called stable. If the differential equation is of the form $y' = f(y)$, equilibrium solutions can be found by setting y' to zero.

Exercises and Problems for Section 11.5

Exercises

1. Each curve in Figure 11.38 represents the balance in a bank account into which a single deposit was made at time zero. Assuming continuously compounded interest, find:

(a) The curve representing the largest initial deposit.
(b) The curve representing the largest interest rate.
(c) Two curves representing the same initial deposit.
(d) Two curves representing the same interest rate.

Figure 11.38

---

[3]In more advanced work, this behavior is described as asymptotic stability.

2. The graphs in Figure 11.39 represent the temperature, $H(°C)$, of four eggs as a function of time, t, in minutes. Match three of the graphs with the descriptions (a)–(c). Write a similar description for the fourth graph, including an interpretation of any intercepts and asymptotes.

(a) An egg is taken out of the refrigerator (just above $0°C$) and put into boiling water.

(b) Twenty minutes after the egg in part (a) is taken out of the fridge and put into boiling water, the same thing is done with another egg.

(c) An egg is taken out of the refrigerator at the same time as the egg in part (a) and left to sit on the kitchen table.

Figure 11.39

3. A yam is put in a $200°C$ oven and heats up according to the differential equation

$$\frac{dH}{dt} = -k(H - 200), \quad \text{for } k \text{ a positive constant.}$$

(a) If the yam is at $20°C$ when it is put in the oven, solve the differential equation.

(b) Find k using the fact that after 30 minutes the temperature of the yam is $120°C$.

4. (a) Find all equilibrium solutions for the differential equation

$$\frac{dy}{dx} = 0.5y(y - 4)(2 + y).$$

(b) Draw a slope field and use it to determine whether each equilibrium solution is stable or unstable.

Problems

7. The rate of growth of a tumor is proportional to the size of the tumor.

(a) Write a differential equation satisfied by S, the size of the tumor, in mm, as a function of time, t.

(b) Find the general solution to the differential equation.

(c) If the tumor is 5 mm across at time $t = 0$, what does that tell you about the solution?

(d) If, in addition, the tumor is 8 mm across at time $t = 3$, what does that tell you about the solution?

5. The slope field for $y' = 0.5(1 + y)(2 - y)$ is given in Figure 11.40.

(a) List equilibrium solutions and state whether each is stable or unstable.

(b) Draw solution curves on the slope field through each of the three marked points.

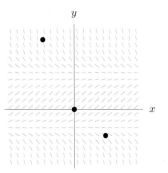

Figure 11.40

6. The slope field for a differential equation is given in Figure 11.41. Estimate all equilibrium solutions for this differential equation, and indicate whether each is stable or unstable.

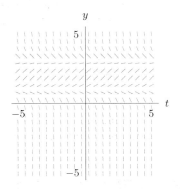

Figure 11.41

8. Money in a bank account grows continuously at an annual rate of r (when the interest rate is 5%, $r = 0.05$, and so on). Suppose $1000 is put into the account in 2000.

(a) Write a differential equation satisfied by M, the amount of money in the account at time t, measured in years since 2000.

(b) Solve the differential equation.

(c) Sketch the solution until the year 2030 for interest rates of 5% and 10%.

9. Hydrocodone bitartrate is used as a cough suppressant. After the drug is fully absorbed, the quantity of drug in the body decreases at a rate proportional to the amount left in the body. The half-life of hydrocodone bitartrate in the body is 3.8 hours, and the usual oral dose is 10 mg.

   (a) Write a differential equation for the quantity, Q, of hydrocodone bitartrate in the body at time t, in hours, since the drug was fully absorbed.

   (b) Solve the differential equation given in part (a).

   (c) Use the half-life to find the constant of proportionality, k.

   (d) How much of the 10 mg dose is still in the body after 12 hours?

10. (a) A cup of coffee is made with boiling water and stands in a room where the temperature is $20°$ C. If $T(t)$ is the temperature of the coffee at time t, explain what the differential equation

$$\frac{dT}{dt} = -k(T - 20)$$

   says in everyday terms. What is the sign of k?

   (b) Solve this differential equation. If the coffee cools to $90°$ C in 2 minutes, how long will it take to cool to $60°$ C degrees?

11. (a) An object is placed in a $68°$F room. Write a differential equation for H, the temperature of the object at time t.

   (b) Give the general solution for the differential equation.

   (c) The temperature of the object is $40°$F initially, and $48°$F one hour later. Find the temperature of the object after 3 hours.

12. Warfarin is a drug used as an anticoagulant. After administration of the drug is stopped, the quantity remaining in a patient's body decreases at a rate proportional to the quantity remaining. The half-life of warfarin in the body is 37 hours.

   (a) Sketch the quantity, Q, of warfarin in a patient's body as function of the time, t, since stopping administration of the drug. Mark the 37 hours on your graph.

   (b) Write a differential equation satisfied by Q.

   (c) How many days does it take for the drug level in the body to be reduced to 25% of the original level?

13. A roast is taken from the refrigerator (where the temperature is $40°$F) and put in a $350°$F oven. One hour later, the meat thermometer shows a temperature of $90°$F. If the roast is done when its temperature reaches $140°$F, what is the total time the roast should be in the oven? Newton's Law of Cooling says that the rate of change of temperature is proportional to the difference between the temperature of the object and the temperature of the surrounding air.

14. The rate at which a drug leaves the bloodstream and passes into the urine is proportional to the quantity of

the drug in the blood at that time. If an initial dose of Q_0 is injected directly into the blood, 20% is left in the blood after 3 hours.

   (a) Write and solve a differential equation for the quantity, Q, of the drug in the blood after t hours.

   (b) How much of this drug is in a patient's body after 6 hours if the patient is given 100 mg initially?

15. Morphine is often used as a pain-relieving drug. The half-life of morphine in the body is 2 hours. Suppose morphine is administered to a patient intravenously at a rate of 2.5 mg per hour, and the rate at which the morphine is eliminated is proportional to the amount present.

   (a) Show that, to three decimal places, the constant of proportionality for the rate at which morphine leaves the body (in mg/hour) is $k = -0.347$.

   (b) Write a differential equation for the quantity, Q, of morphine in the blood after t hours.

   (c) Use the differential equation to find the equilibrium solution. (This is the long-term amount of morphine in the body, once the system has stabilized.)

16. In Example 2 on page 543 we saw that it would take about 6 years for 90% of the pollution in Lake Erie to be removed, and about 12 years for 99% to be removed. Explain why one time is double the other.

17. Using the model in the text and the data in Table 11.4 on page 543, find how long it would take for 90% of the pollution to be removed from Lake Michigan and from Lake Ontario, assuming no new pollutants are added. Explain how you can tell which lake will take longer to be purified just by looking at the data in the table.

18. Use the model in the text and the data in Table 11.4 on page 543 to determine which of the Great Lakes would require the longest time and which would require the shortest time for 80% of the pollution to be removed, assuming no new pollutants are being added. Find the ratio of these two times.

19. Oil is pumped continuously from a well at a rate proportional to the amount of oil left in the well. Initially there were 1 million barrels of oil in the well; six years later 500,000 barrels remain.

   (a) At what rate was the amount of oil in the well decreasing when there were 600,000 barrels remaining?

   (b) When will there be 50,000 barrels remaining?

20. The amount of land in use for growing crops increases as the world's population increases. Suppose $A(t)$ represents the total number of hectares of land in use in year t. (A hectare is about $2\frac{1}{2}$ acres.)

   (a) Explain why it is plausible that $A(t)$ satisfies the equation $A'(t) = kA(t)$. What assumptions are you making about the world's population and its relation to the amount of land used?

(b) In 1950 about $1 \cdot 10^9$ hectares of land were in use; in 1980 the figure was $2 \cdot 10^9$. If the total amount of land available for growing crops is thought to be $3.2 \cdot 10^9$ hectares, when does this model predict it is exhausted? (Let $t = 0$ in 1950.)

21. A detective finds a murder victim at 9 am. The temperature of the body is measured at $90.3°F$. One hour later, the temperature of the body is $89.0°F$. The temperature of the room has been maintained at a constant $68°F$.

  (a) Assuming the temperature, T, of the body obeys Newton's Law of Cooling, write a differential equation for T.

  (b) Solve the differential equation to estimate the time the murder occurred.

22. At 1:00 pm one winter afternoon, there is a power failure at your house in Wisconsin, and your heat does not work without electricity. When the power goes out, it is $68°F$ in your house. At 10:00 pm, it is $57°F$ in the house, and you notice that it is $10°F$ outside.

  (a) Assuming that the temperature, T, in your home obeys Newton's Law of Cooling, write the differential equation satisfied by T.

  (b) Solve the differential equation to estimate the temperature in the house when you get up at 7:00 am the next morning. Should you worry about your water pipes freezing?

  (c) What assumption did you make in part (a) about the temperature outside? Given this (probably incorrect)

assumption, would you revise your estimate up or down? Why?

23. The radioactive isotope carbon-14 is present in small quantities in all life forms, and it is constantly replenished until the organism dies, after which it decays to stable carbon-12 at a rate proportional to the amount of carbon-14 present, with a half-life of 5730 years. Suppose $C(t)$ is the amount of carbon-14 present at time t.

  (a) Find the value of the constant k in the differential equation $C' = -kC$.

  (b) In 1988 three teams of scientists found that the Shroud of Turin, which was reputed to be the burial cloth of Jesus, contained 91% of the amount of carbon-14 contained in freshly made cloth of the same material.[4] How old is the Shroud of Turin, according to these data?

24. Before Galileo discovered that the speed of a falling body with no air resistance is proportional to the time since it was dropped, he mistakenly conjectured that the speed was proportional to the distance it had fallen.

  (a) Assume the mistaken conjecture to be true and write an equation relating the distance fallen, $D(t)$, at time t, and its derivative.

  (b) Using your answer to part (a) and the correct initial conditions, show that D would have to be equal to 0 for all t, and therefore the conjecture must be wrong.

11.6 APPLICATIONS AND MODELING

Much of this book involves functions that represent real processes, such as how the temperature of a yam or the population of Mexico is changing with time. You may wonder where such functions come from. In some cases, we fit functions to experimental data by trial and error. In other cases, we take a more theoretical approach, leading to a differential equation whose solution is the function we want. In this section we give examples of the more theoretical approach.

How a Layer of Ice Forms

When ice forms on a lake, the water on the surface freezes first. As heat from the water travels up through the ice and is lost to the air, more ice is formed. The question we will consider is: How thick is the layer of ice as a function of time? Since the thickness of the ice increases with time, the thickness function is increasing. In addition, as the ice gets thicker, it insulates better, therefore we expect the layer of ice to form more slowly as time goes on. Hence the thickness function is increasing at a decreasing rate, so its graph is concave down.

A Differential Equation for the Thickness of the Ice

To get more detailed information about the thickness function, we have to make some assumptions. Suppose y represents the thickness of the ice as a function of time, t. Since the thicker the ice, the longer it takes the heat to get through it, we will assume that the rate at which ice is formed is

---

[4]*The New York Times*, October 18, 1988

inversely proportional to the thickness. In other words, we assume that for some constant k,

$$\text{Rate thickness is increasing} = \frac{k}{\text{Thickness}},$$

so

$$\frac{dy}{dt} = \frac{k}{y} \quad \text{where} \quad k > 0.$$

This differential equation enables us to find a formula for y. Using separation of variables:

$$\int y \, dy = \int k \, dt$$

$$\frac{y^2}{2} = kt + C.$$

If we measure time so that $y = 0$ when $t = 0$, then $C = 0$. Since y must be non-negative, we have

$$y = \sqrt{2kt}\,.$$

Graphs of y against t are in Figure 11.42. Notice that the larger y is, the more slowly y increases. In addition, this model suggests that y increases indefinitely as time passes. (Of course, the value of y cannot increase beyond the depth of the lake.)

y (thickness)

Large k

$y = \sqrt{2kt}$

Small k

t (time)

Figure 11.42: Thickness of ice as a function of time

The Net Worth of a Company

In the preceding section, we saw an example in which money in a bank account was earning interest (Example 1, page 541). Consider a company whose revenues are proportional to its net worth (like interest on a bank account) but which must also make payroll payments. The question is: under what circumstances does the company make money and under what circumstances does it go bankrupt?

Common sense says that if the payroll exceeds the rate at which revenue is earned, the company will eventually be in trouble, whereas if revenue exceeds payroll, the company should do well. We assume that revenue is earned continuously and that payments are made continuously. (For a large company, this is a good approximation.) We also assume that the only factors affecting net worth are revenue and payroll.

Example 1 A company's revenue is earned at a continuous annual rate of 5% of its net worth. At the same time, the company's payroll obligations amount to $200 million a year, paid out continuously.
(a) Write a differential equation that governs the net worth of the company, W million dollars.
(b) Solve the differential equation, assuming an initial net worth of W_0 million dollars.
(c) Sketch the solution for $W_0 = 3000$, 4000, and 5000.

Solution First, let's see what we can learn without writing a differential equation. For example, we can ask if there is any initial net worth W_0 which will exactly keep the net worth constant. If there's such an equilibrium, the rate at which revenue is earned must exactly balance the payments made, so

Rate revenue is earned $=$ Rate payments are made.

If net worth is a constant W_0, revenue is earned at a constant rate of $0.05W_0$ per year, so we have

$$0.05W_0 = 200 \quad \text{giving} \quad W_0 = 4000.$$

Therefore, if the net worth starts at \$4000 million, the revenue and payments are equal, and the net worth remains constant. Therefore \$4000 million is an equilibrium solution.

Suppose, however, the initial net worth is above \$4000 million. Then, the revenue earned is more than the payroll expenses, and the net worth of the company increases, thereby increasing the revenue still further. Thus the net worth increases faster and faster. On the other hand, if the initial net worth is below \$4000 million, the revenue is not enough to meet the payments, and the net worth of the company declines. This decreases the revenue, making the net worth decrease still faster. The net worth will eventually go to zero, and the company goes bankrupt.

(a) Now we set up a differential equation for the net worth, using the fact that

$$\begin{array}{ccc} \text{Rate net worth} & & \text{Rate revenue} & & \text{Rate payroll payments} \\ \text{is increasing} & = & \text{is earned} & - & \text{are made} \end{array}.$$

In millions of dollars per year, revenue is earned at a rate of $0.05W$, and payments are made at a rate of 200 per year, so, for t in years,

$$\frac{dW}{dt} = 0.05W - 200.$$

The equilibrium solution, $W = 4000$, is obtained by setting $dW/dt = 0$.

(b) We solve this equation by separation of variables. It is helpful to factor out 0.05 before separating, so that the W moves over to the left-hand side without a coefficient:

$$\frac{dW}{dt} = 0.05(W - 4000).$$

Separating and integrating gives

$$\int \frac{dW}{W - 4000} = \int 0.05\, dt,$$

so

$$\ln|W - 4000| = 0.05t + C,$$

or

$$|W - 4000| = e^{0.05t+C} = e^C e^{0.05t}.$$

This means

$$W - 4000 = Ae^{0.05t} \qquad \text{where } A = \pm e^C.$$

To find A we use the initial condition that $W = W_0$ when $t = 0$.

$$W_0 - 4000 = Ae^0 = A.$$

Substituting this value for A into $W = 4000 + Ae^{0.05t}$ gives

$$W = 4000 + (W_0 - 4000)e^{0.05t}.$$

(c) If $W_0 = 4000$, then $W = 4000$, the equilibrium solution.

If $W_0 = 5000$, then $W = 4000 + 1000e^{0.05t}$.

If $W_0 = 3000$, then $W = 4000 - 1000e^{0.05t}$. Substituting $t \approx 27.7$ gives $W = 0$, so the company goes bankrupt in its twenty-eighth year.

In Figure 11.43, notice that if the net worth starts with W_0 near, but not equal to, \$4000 million, then W moves further away. Thus, $W = 4000$ is an unstable equilibrium.

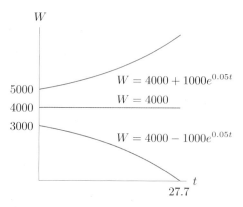

Figure 11.43: Solutions to $dW/dt = 0.05W - 200$

The Velocity of a Falling Body: Terminal Velocity

Think about the velocity of a sky-diver jumping out of a plane. When the sky-diver first jumps, his velocity is zero. The pull of gravity then makes his velocity increase. As the sky-diver speeds up, the air resistance also increases. Since the air resistance partly balances the pull of gravity, the force causing him to accelerate decreases. Thus, the velocity is an increasing function of time, but it is increasing at a decreasing rate. The air resistance increases until it balances gravity, when the sky-diver's velocity levels off. Thus, we expect the the graph of velocity against time to be concave down with a horizontal asymptote.

A Differential Equation: Air Resistance Proportional to Velocity

In order to compute the velocity function, we need to know exactly how air resistance depends on velocity. To decide whether air resistance is proportional to the velocity, say, or is some other function of velocity, requires either lab experiments or a theoretical idea of how the air resistance is created. We consider a very small object, such as a dust particle settling on a computer component during manufacturing,[5] and assume that air resistance is proportional to velocity. Thus, the net force on the object is $F = mg - kv$, where mg is the gravitational force, which acts downward, and kv is the air resistance, which acts upward, so $k > 0$. (See Figure 11.44.) Then, by Newton's Second Law of Motion,

$$\text{Force} = \text{Mass} \cdot \text{Acceleration},$$

we have

$$mg - kv = m\frac{dv}{dt}.$$

This differential equation can be solved by separation of variables. It is easier if we factor out $-k/m$ before separating, giving

$$\frac{dv}{dt} = -\frac{k}{m}\left(v - \frac{mg}{k}\right).$$

---

[5]Example suggested by Howard Stone.

Separating and integrating gives

$$\int \frac{dv}{v - mg/k} = -\frac{k}{m} \int dt$$

$$\ln \left| v - \frac{mg}{k} \right| = -\frac{k}{m}t + C.$$

Solving for v:

$$\left| v - \frac{mg}{k} \right| = e^{-kt/m+C} = e^C e^{-kt/m}$$

$$v - \frac{mg}{k} = Ae^{-kt/m},$$

where A is an arbitrary constant. We find A from the initial condition that the object starts from rest, so $v = 0$ when $t = 0$. Substituting

$$0 - \frac{mg}{k} = Ae^0$$

gives

$$A = -\frac{mg}{k}.$$

Thus

$$v = \frac{mg}{k} - \frac{mg}{k}e^{-kt/m} = \frac{mg}{k}(1 - e^{-kt/m}).$$

The graph of this function is in Figure 11.45. The horizontal asymptote represents the *terminal velocity*, mg/k.

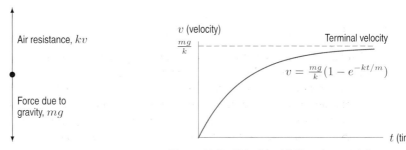

Figure 11.44: Forces acting on a falling object

Figure 11.45: Velocity of falling dust particle assuming that air resistance is kv

Notice that the terminal velocity can also be obtained from the differential equation by setting $dv/dt = 0$ and solving for v:

$$m\frac{dv}{dt} = mg - kv = 0 \quad \text{so} \quad v = \frac{mg}{k}.$$

Compartmental Analysis: A Reservoir

Many processes can be modeled as a container with various solutions flowing in and out—for example, drugs given intravenously or the discharge of pollutants into a lake. We consider a city's water reservoir, fed partly by clean water from a spring and partly by run-off from the surrounding land. In New England, and many other areas with much snow in the winter, the run-off contains salt which has been put on the roads to make them safe for driving. We consider the concentration of salt in the reservoir. If there is no salt in the reservoir initially, the concentration builds up until the rate at which the salt is entering into the reservoir balances the rate at which salt flows out. If, on the other hand, the reservoir starts with a great deal of salt in it, then initially the rate at which the salt is entering is less than the rate at which it is flowing out, and the quantity of salt in the lake decreases. In either case, the salt concentration levels off at an equilibrium value.

A Differential Equation for Salt Concentration

A water reservoir holds 100 million gallons of water and supplies a city with 1 million gallons a day. The reservoir is partly refilled by a spring which provides 0.9 million gallons a day, and the rest of the water, 0.1 million gallons a day, comes from run-off from the surrounding land. The spring is clean, but the run-off contains salt with a concentration of 0.0001 pound per gallon. There was no salt in the reservoir initially and the water is well mixed (that is, the out-flow contains the concentration of salt in the tank at that instant). We find the concentration of salt in the reservoir as a function of time.

It is important to distinguish between the total quantity, Q, of salt in pounds, and the concentration, C, of salt, in pounds/gallon, where

$$\text{Concentration} = C = \frac{\text{Quantity of salt}}{\text{Volume of water}} = \frac{Q}{100 \text{ million}} \left(\frac{\text{lb}}{\text{gal}}\right).$$

(The volume of the reservoir is 100 million pounds.) We will find Q first, and then C. We know that

$$\begin{array}{c}\text{Rate of change of}\\ \text{quantity of salt}\end{array} = \text{Rate salt entering} - \text{Rate salt leaving}.$$

Salt is entering through the run-off of 0.1 million gallons per day, with each gallon containing 0.0001 pound of salt. Therefore

$$\begin{aligned} \text{Rate salt entering} &= \text{Concentration} \cdot \text{Volume per day} \\ &= 0.0001 \left(\frac{\text{lb}}{\text{gal}}\right) \cdot 0.1 \left(\frac{\text{million gal}}{\text{day}}\right) \\ &= 0.00001 \left(\frac{\text{million lb}}{\text{day}}\right) = 10 \text{ lb/day}. \end{aligned}$$

Salt is leaving in the million gallons of water used by the city each day . Thus

$$\begin{aligned} \text{Rate salt leaving} &= \text{Concentration} \cdot \text{Volume per day} \\ &= \frac{Q}{100 \text{ million}} \left(\frac{\text{lb}}{\text{gal}}\right) \cdot 1 \left(\frac{\text{million gal}}{\text{day}}\right) = \frac{Q}{100} \text{ lb/day}. \end{aligned}$$

Therefore Q satisfies the differential equation

$$\frac{dQ}{dt} = 10 - \frac{Q}{100}.$$

We factor out $-1/100 = -0.1$ and separate variables, giving

$$\frac{dQ}{dt} = -0.01(Q - 1000),$$

$$\int \frac{dQ}{Q - 1000} = -\int 0.01 \, dt$$

$$\ln|Q - 1000| = -0.01t + k,$$

$$Q - 1000 = Ae^{-0.01t}.$$

There is no salt initially, so we substitute $Q = 0$ when $t = 0$:

$$0 - 1000 = Ae^0 \qquad \text{giving} \qquad A = -1000.$$

Thus

$$Q - 1000 = -1000e^{-0.01t}$$

so

$$Q = 1000(1 - e^{-0.01t}) \quad \text{pounds}.$$

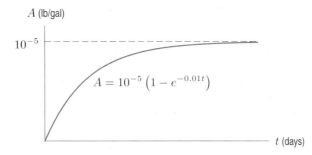

Figure 11.46: Concentration of salt in reservoir

Therefore

$$\text{Concentration} = C = \frac{Q}{100 \text{ million}} = \frac{1000}{10^8}(1 - e^{-0.01t}) = 10^{-5}(1 - e^{-0.01t}) \text{ lbs/gal.}$$

A sketch of concentration against time is in Figure 11.46.

Exercises and Problems for Section 11.6

Exercises

1. A bank account that earns 10% interest compounded continuously has an initial balance of zero. Money is deposited into the account at a constant rate of $1000 per year.

(a) Write a differential equation that describes the rate of change of the balance $B = f(t)$.

(b) Solve the differential equation to find the balance as a function of time.

2. A deposit is made to a bank account paying 8% interest compounded continuously. Payments totaling $2000 per year are made from this account.

(a) Write a differential equation for the balance, B, in the account after t years.

(b) Write the solution to the differential equation.

(c) How much is in the account after 5 years if the initial deposit is   (i) $20,000?   (ii) $30,000?

3. At time $t = 0$, a bottle of juice at $90°$F is stood in a mountain stream whose temperature is $50°$F. After 5 minutes, its temperature is $80°$F. Let $H(t)$ denote the temperature of the juice at time t, in minutes.

(a) Write a differential equation for $H(t)$ using Newton's Law of Cooling.

(b) Solve the differential equation.

(c) When will the temperature of the juice have dropped to $60°$F?

4. The velocity, v, of a dust particle of mass m and acceleration a satisfies the equation

$$ma = m\frac{dv}{dt} = mg - kv, \quad \text{where } g, k \text{ are constant.}$$

By differentiating this equation, find a differential equation satisfied by a. (Your answer may contain m, g, k, but not v.) Solve for a, given that $a(0) = g$.

Problems

5. In some chemical reactions, the rate at which the amount of a substance changes with time is proportional to the amount present. For example, this is the case as δ-glucono-lactone changes into gluconic acid.

(a) Write a differential equation satisfied by y, the quantity of δ-glucono-lactone present at time t.

(b) If 100 grams of δ-glucono-lactone is reduced to 54.9 grams in one hour, how many grams will remain after 10 hours?

6. When the electromotive force (emf) is removed from a circuit containing inductance and resistance but no capacitors, the rate of decrease of current is proportional to the current. If the initial current is 30 amps but decays to 11 amps after 0.01 seconds, find an expression for the current as a function of time.

7. Dead leaves accumulate on the ground in a forest at a rate of 3 grams per square centimeter per year. At the same time, these leaves decompose at a continuous rate of 75% per year. Write a differential equation for the total quantity of dead leaves (per square centimeter) at time t. Sketch a solution showing that the quantity of dead leaves tends toward an equilibrium level. What is that equilibrium level?

8. The rate (per foot) at which light is absorbed as it passes through water is proportional to the intensity, or brightness, at that point.

(a) Find the intensity as a function of the distance the light has traveled through the water.

(b) If 50% of the light is absorbed in 10 feet, how much is absorbed in 20 feet? 25 feet?

9. Water leaks out of a barrel at a rate proportional to the square root of the depth of the water at that time. If the water level starts at 36 inches and drops to 35 inches in 1 hour, how long will it take for all of the water to leak out of the barrel?

10. When a gas expands without gain or loss of heat, the rate of change of pressure with respect to volume is proportional to pressure divided by volume. Find a law connecting pressure and volume in this case.

11. According to a simple physiological model, an athletic adult needs 20 calories per day per pound of body weight to maintain his weight. If he consumes more or fewer calories than those required to maintain his weight, his weight changes at a rate proportional to the difference between the number of calories consumed and the number needed to maintain his current weight; the constant of proportionality is $1/3500$ pounds per calorie. Suppose that a particular person has a constant caloric intake of I calories per day. Let $W(t)$ be the person's weight in pounds at time t (measured in days).

    (a) What differential equation has solution $W(t)$?
    (b) Solve this differential equation.
    (c) Graph $W(t)$ if the person starts out weighing 160 pounds and consumes 3000 calories a day.

12. A mothball is in the shape of a sphere and starts with radius 1 cm. The material in the mothball evaporates at a rate proportional to the surface area. After one month, the radius is 0.5 cm. How many months (from the start) is it before the radius is 0.2 cm?

13. Water leaks from a vertical cylindrical tank through a small hole in its base at a rate proportional to the square root of the volume of water remaining. If the tank initially contains 200 liters and 20 liters leak out during the first day, when will the tank be half empty? How much water will there be after 4 days?

14. The rate at which barometric pressure decreases with altitude is proportional to the barometric pressure at that altitude. If the barometric pressure is measured in inches of mercury, and the altitude in feet, then the constant of proportionality is $3.7 \cdot 10^{-5}$. The barometric pressure at sea level is 29.92 inches of mercury.

    (a) Calculate the barometric pressure at the top of Mount Whitney, 14,500 feet (the highest mountain in the US outside Alaska), and at the top of Mount Everest, 29,000 feet (the highest mountain in the world).
    (b) People cannot easily survive at a pressure below 15 inches of mercury. What is the highest altitude to which people can safely go?

15. In 1692, Johann Bernoulli was teaching the Marquis de l'Hopital calculus in Paris. Solve the following problem, which is similar to the one that they did. What is the equation of the curve which has subtangent (distance BC in Figure 11.47) equal to twice its abscissa (distance OC)?

Figure 11.47

16. A bank account earns 10% annual interest, compounded continuously. Money is deposited in a continuous cash flow at a rate of $1200 per year into the account.

    (a) Write a differential equation that describes the rate at which the balance $B = f(t)$ is changing.
    (b) Solve the differential equation given an initial balance $B_0 = 0$.
    (c) Find the balance after 5 years.

17. (Continuation of Problem 16.) Now suppose the money is deposited once a month (instead of continuously) but still at a rate of $1200 per year.

    (a) Write down the sum that gives the balance after 5 years, assuming the first deposit is made one month from today, and today is $t = 0$.
    (b) The sum you wrote in part (a) is a Riemann sum approximation to the integral

    $$\int_0^5 1200 e^{0.1t} dt.$$

    Determine whether it is a left sum or right sum, and determine what Δt and N are. Then use your calculator to evaluate the sum.
    (c) Compare your answer in part (b) to your answer to Problem 16(c).

18. A patient is given the drug theophylline intravenously at a rate of 43.2 mg/hour to relieve acute asthma. You can imagine the drug as entering a compartment of volume 35,000 ml. (This is the volume of the part of the body through which the drug circulates.) The rate at which the quantity of drug leaves the body (mg/hour) is proportional to the quantity there, with proportionality constant 0.082 per hour. Assume the patient's body contains none of the drug initially. [Note: Concentration = Quantity/Volume.]

    (a) Describe in words how you would expect the concentration of theophylline in the patient to vary with time and sketch an approximate graph.
    (b) Write a differential equation satisfied by the concentration of the drug, $c(t)$.
    (c) Solve the differential equation and graph the solution. What happens to the concentration in the long run?

19. As you know, when a course ends, students start to forget the material they have learned. One model (called the Ebbinghaus model) assumes that the rate at which a student forgets material is proportional to the difference between the material currently remembered and some positive constant, a.

(a) Let $y = f(t)$ be the fraction of the original material remembered t weeks after the course has ended. Set up a differential equation for y. Your equation will contain two constants; the constant a is less than y for all t.

(b) Solve the differential equation.

(c) Describe the practical meaning (in terms of the amount remembered) of the constants in the solution $y = f(t)$.

20. An item is initially sold at a price of $\$p$ per unit. Over time, market forces push the price toward the equilibrium price, $\$p^*$, at which supply balances demand. The Evans Price Adjustment model says that the rate of change in the market price, $\$p$, is proportional to the difference between the market price and the equilibrium price.

(a) Write a differential equation for p as a function of t.

(b) Solve for p.

(c) Sketch solutions for various different initial prices, both above and below the equilibrium price.

(d) What happens to p as $t \to \infty$?

21. A drug is administered intravenously at a constant rate of r mg/hour and is excreted at a rate proportional to the quantity present, with constant of proportionality $\alpha > 0$.

(a) Solve a differential equation for the quantity, Q, in milligrams, of the drug in the body at time t hours. Assume there is no drug in the body initially. Your answer will contain r and α. Graph Q against t. What is Q_∞, the limiting long-run value of Q?

(b) What effect does doubling r have on Q_∞? What effect does doubling r have on the time to reach half the limiting value, $\frac{1}{2}Q_\infty$?

(c) What effect does doubling α have on Q_∞? On the time to reach $\frac{1}{2}Q_\infty$?

22. When people smoke, carbon monoxide is released into the air. In a room of volume 60 m^3, air containing 5% carbon monoxide is introduced at a rate of 0.002 m^3/min. (This means that 5% of the volume of the incoming air is carbon monoxide.) The carbon monoxide mixes immediately with the rest of the air and the mixture leaves the room at the same rate as it enters.

(a) Write a differential equation for $c(t)$, the concentration of carbon monoxide at time t, in minutes.

(b) Solve the differential equation, assuming there is no carbon monoxide in the room initially.

(c) What happens to the value of $c(t)$ in the long run?

23. (Continuation of Problem 22.) Medical texts[6] warn that exposure to air containing 0.1% carbon monoxide for some time can lead to a coma. How long does it take for the concentration of carbon monoxide in the room in Problem 22 to reach this level?

24. An aquarium pool has volume $2 \cdot 10^6$ liters. The pool initially contains pure fresh water. At $t = 0$ minutes, water containing 10 grams/liter of salt is poured into the pool at a rate of 60 liters/minute. The salt water instantly mixes with the fresh water, and the excess mixture is drained out of the pool at the same rate (60 liters/minute).

(a) Write a differential equation for $S(t)$, the mass of salt in the pool at time t.

(b) Solve the differential equation to find $S(t)$.

(c) What happens to $S(t)$ as $t \to \infty$?

25. An object of mass m is thrown vertically upward from the surface of the earth with initial velocity v_0. We will calculate the value of v_0, called the escape velocity, with which the object can escape the pull of the gravity and never return to earth. Since the object is moving far from the surface of the earth, we must take into account the variation of gravity with altitude. If the acceleration due to gravity at sea level is g, and R is the radius of the earth, the gravitational force, F, on the object of mass m at an altitude h above the surface of the earth is given by

$$F = \frac{mgR^2}{(R+h)^2},$$

(a) Suppose v is the velocity of the object (measured upward) at time t. Use Newton's Law of Motion to show that

$$\frac{dv}{dt} = -\frac{gR^2}{(R+h)^2}.$$

(b) Rewrite this equation with h instead of t as the independent variable using the chain rule $\frac{dv}{dt} = \frac{dv}{dh} \cdot \frac{dh}{dt}$. Hence show that

$$v\frac{dv}{dh} = -\frac{gR^2}{(R+h)^2}.$$

(c) Solve the differential equation in part (b).

(d) Find the escape velocity, the smallest value of v_0 such that v is never zero.

---

[6] See, for example, R.G. Petersdorf et al., *Harrison's Principles of Internal Medicine* (New York: McGraw Hill, 1983).

11.7 MODELS OF POPULATION GROWTH

Population projections have been important to political philosophers since at least the late eighteenth century. As concern for scarce resources has grown, so has interest in accurate population projections. In this section we look at two differential equations which are used to model both human and animal population growth. These differential equations have applications in economics and medicine, such as modeling the spread of an innovation or the growth of a tumor.

The US Population: 1790–1860

Every ten years the population of the United States is recorded by a census. The first such census was in 1790. Table 11.5 contains the census data from 1790 to 1940.

Table 11.5   *US Population in millions, 1790–1940*

Year	Population	Year	Population	Year	Population
1790	3.9	1850	23.1	1910	92.0
1800	5.3	1860	31.4	1920	105.7
1810	7.2	1870	38.6	1930	122.8
1820	9.6	1880	50.2	1940	131.7
1830	12.9	1890	62.9		
1840	17.1	1900	76.0		

The population is given only to the nearest 0.1 million (or 100,000), although the population is reported by the US Census Bureau down to the last digit (for example, 131,669,275 in 1940). We have rounded off because census figures are notoriously inaccurate. For example, in the census of 1990, New York City claimed the census had missed a million people in that city alone. Thus, giving more digits in the population does not necessarily give more accuracy.

We concentrate first on the relative growth rate, $(dP/dt)/P$, of the US population from 1790 to 1860. To estimate $(dP/dt)/P$ in 1830, we take the rate of change in the population and divide it by the population itself:

$$\frac{1}{P}\frac{dP}{dt} \approx \frac{1}{\text{Population in 1830}} \cdot \frac{\text{Population in 1840} - \text{Population in 1830}}{10 \text{ years}}$$
$$= \frac{1}{12.9} \cdot \frac{17.1 - 12.9}{10} = 0.0326 = 3.26\%.$$

Similar calculations for 1790, 1800, ..., 1850 give the percentages in Table 11.6:

Table 11.6   *Estimates of yearly relative growth rate of US population*

Year	1790	1800	1810	1820	1830	1840	1850
Relative growth rate	3.59%	3.58%	3.33%	3.44%	3.26%	3.57%	3.53%

The relative growth rate, while not precisely constant in Table 11.6, is nearly so. Since political and economic events such as war or recession affect the population, we don't expect the growth rate to be exactly constant. The simplest model for population growth is to assume that the relative growth rate *is* a constant, k, giving

$$\frac{1}{P}\frac{dP}{dt} = k,$$

This is equivalent to assuming that the population grows exponentially:

$$P = P_0 e^{kt}.$$

What should we take for the value of k? One possibility would be the average of the percentages we just calculated, namely 3.47%. However, there is a serious objection to using this percentage as an estimate for k. Remember that k is a continuous growth rate, but the populations are given at 10-year intervals. A 10-year population growth of 34.7% doesn't come from a continuous yearly rate of 3.47%. (This ignores the effects of compounding.) If the population increases by a factor of 34.7% in 10 years, then $P(10)/P_0 = 1.347$, where $P(10)$ is the population when $t = 10$. Assuming that $P = P_0 e^{kt}$, we need k to satisfy

$$\frac{P(10)}{P_0} = e^{k \cdot 10} = 1.347.$$

The continuous growth rate, k, is given by

$$k = \frac{\ln(1.347)}{10} = 0.0298.$$

Let's compare actual and predicted values of the population. We take $k = 0.0298$ and the initial population $P_0 = 3.9$ in 1790. With t in years since 1790, the model is

$$P = 3.9 e^{0.0298t}.$$

Substituting $t = 0, 10, 20, \dots, 70$ gives the predictions in Table 11.7.

Table 11.7   *Predicted versus actual US population 1790–1860 (exponential model)*

Year	Actual	Predicted	Year	Actual	Predicted
1790	3.9	3.9	1830	12.9	12.8
1800	5.3	5.3	1840	17.1	17.3
1810	7.2	7.1	1850	23.1	23.3
1820	9.6	9.5	1860	31.4	31.4

The agreement between the predicted and actual values is remarkable. Of course, since we used the data from the entire 70 year period to estimate k, we should expect good agreement throughout that period. What is surprising is that if we had used only the populations in 1790 and 1800 to estimate k, the predictions are still quite good. Let's find a new value of k using only the 1790 and 1800 data and compare predictions. The 10-year growth from 1790 to 1800 is 35.9% so $k = \ln(1.359)/10 = 0.0307$. We would then predict the population in 1860 to be

$$3.9 e^{0.0307(70)} = 33.4,$$

which is within about 6% of the actual population of 31.4. It is remarkable that a person in 1800 could accurately predict what the population of the US would be 60 years later, especially considering all the wars, recessions, epidemics, additions of new territory, and immigration that took place from 1800 to 1860.

Predictions From The Exponential Model

The exponential model led to the pessimistic predictions of Thomas Malthus, an early nineteenth-century clergyman and political philosopher, who believed that, if unchecked, a population would grow exponentially, whereas the food supply would grow linearly, and therefore the population would eventually outstrip the food supply.

Interestingly enough, an exponential model has fit the growth of world population and the population of many regions remarkably well for decades, even centuries. However, the model must break down at some point because it predicts that the population will continue to grow without bound as time goes on—and this cannot be true forever. Eventually the effects of crowding, emigration, disease, war, and lack of food will have to curb growth. In searching for an improvement, then, we should look for a model whose solution is approximately an exponential function for small values of the population, but which levels off later.

The US Population: 1790–1940

If we try to predict the population of the US in 2000 using our exponential model with $k = 0.0298$, we get

$$\text{US population in 2000} = 3.9e^{(0.0298)(210)} = 2{,}036 \text{ million},$$

which is far from the actual figure of around 280 million.

The problem is that growth rate in the US population between 1790 and 1860 did not stay constant in later decades. The 10-year percentage growths[7] from 1860 to 1930 are listed in Table 11.8:

Table 11.8 *Estimated 10-year growth rate of US population, 1860–1930*

Year	1860	1870	1880	1890	1900	1910	1920	1930
Relative growth rate	22.9%	30.1%	25.3%	20.8%	21.1%	14.9%	16.2%	7.2%

These figures are nothing like those during the period 1790 to 1860, where they hovered around 34%. The dramatic drop to 22.9% for the decade 1860–1870 is explainable by the Civil War (but don't ascribe the entire drop to deaths of the war—see Problem 4). The growth rate goes back up during the decade 1870–1880, but by 1890–1900 it has dropped below the rate during the Civil War. There is a slight increase in the rate during the immigrations of 1900 to 1910, another drop during World War I, a small bounce back, and finally the rate plummets during the recession of the 1930s. Notice that the effect of the recession is more dramatic than the effect of wars, which suggests that it is a decrease in birth rate rather than an increase in death rate that is a major factor in declines in the relative population growth rate.

The exponential model gives accurate predictions up to 1860. But it is inadequate for the following years. We look for a new model that takes into account the effects of overcrowding by having a relative growth rate which decreases as the population increases. Using the data in Table 11.8, it can be shown that, approximately,

$$\frac{1}{P}\frac{dP}{dt} = 0.0318 - 0.000170P.$$

Therefore, in our new model, P satisfies the *logistic* differential equation

$$\frac{dP}{dt} = 0.0318P - 0.000170P^2.$$

Its slope field is shown in Figure 11.48, together with the solution with $P(0) = 3.9$.

The most striking difference between this model and the exponential model is that the logistic model predicts that the US population will level off. The logistic model predicts that the population will continue to grow until

$$\frac{dP}{dt} = 0.0318P - 0.000170P^2 = 0,$$

that is, when $P = 0$ or $P \approx 187$ million.

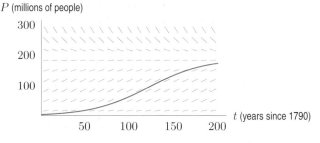

Figure 11.48: Solution to $dP/dt = 0.0318P - 0.000170P^2$ with $P(0) = 3.9$

[7]Calculated using estimates of the form $\frac{1}{P}\frac{dP}{dt}$ at 1860 $\approx \frac{\text{Population in 1870} - \text{Population in 1860}}{10(\text{Population in 1860})}$.

The shape of the solution curve in Figure 11.48 suggests that initially the population grows faster and faster and then slows down as the limiting value of 187 is approached; the fastest growth rate appears to be about half-way to the limiting value.

Later in this section we derive the formula for the solution to the logistic equation. For the US population, it turns out to be

$$P = \frac{187}{1 + 47e^{-0.0318t}}.$$

(The numbers 187 and the 47 are not exact values, but have been rounded.) The values predicted by this formula for P agree with the actual populations up to 1940. See Table 11.9. During the period from 1700 to 1940, the largest deviation is about 3% in 1840 and 1870 (the Civil War accounts for the second one). All other errors are less than 2%.

Table 11.9   *Predicted versus actual US population in millions, 1790–1980 (logistic model)*

Year	Actual	Predicted	Year	Actual	Predicted	Year	Actual	Predicted
1790	3.9	3.9	1870	38.6	39.9	1950	150.7	145.0
1800	5.3	5.3	1880	50.2	50.7	1960	179.3	154.4
1810	7.2	7.2	1890	62.9	63.3	1970	203.3	162.1
1820	9.6	9.8	1900	76.0	77.2	1980	226.5	168.2
1830	12.9	13.2	1910	92.0	91.9	1990	248.7	172.9
1840	17.1	17.7	1920	105.7	106.7	2000	281.4	176.6
1850	23.1	23.5	1930	122.8	120.8			
1860	31.4	30.8	1940	131.7	133.7			

Of course, the real test is how well our model, based on data from 1790 to 1940, predicts the population in the "future," 1950 to 2000. The fit between predicted and actual population in Table 11.9 is not good from 1950 on. Despite World War II, which undoubtedly depressed population growth between 1942 and 1945, in the last half of the 1940s the US population surged, wiping out in five years a deficit caused by 15 years of depression and war. The 1950s saw a population growth of 28 million, leaving our logistic model in the dust. This surge in population is referred to as the baby boom.

Once again we have reached a point where our model is no longer useful. This should not lead you to believe that a reasonable mathematical model cannot be found. No model is perfect and when one model fails, we seek a better one. Just as we abandoned the exponential model in favor of the logistic model for the US population, we could look further.

The Logistic Model

The logistic model we used to model the US population from 1790 to 1940 assumed that the relative growth rate of the population was a linearly decreasing function of P:

$$\frac{1}{P}\frac{dP}{dt} = k - aP.$$

(We took $k = 0.0318$ and $a = 0.000170$). For small P, we have approximately

$$\frac{1}{P}\frac{dP}{dt} \approx k.$$

The solution to this equation is an exponential function. This is why an exponential model fit the US population well during the years 1790-1860 when the population was relatively small. In the logistic model, as P increases, the relative growth rate decreases to zero; it reaches zero when P is given by

$$\frac{1}{P}\frac{dP}{dt} = k - aP = 0.$$

Solving gives $P = k/a$, the limiting value of the population, which we call L. The value L is the *carrying capacity* of the environment, and represents the largest population the environment can support. Writing $a = k/L$, the logistic equation becomes

$$\frac{1}{P}\frac{dP}{dt} = k - \frac{k}{L}P$$
$$\frac{dP}{dt} = kP\left(1 - \frac{P}{L}\right).$$

This is the general logistic differential equation, first proposed as a model for population growth by the Belgian mathematician P. F. Verhulst in the 1830s.

Qualitative Solution to the Logistic Equation

Figure 11.49 shows the slope field and characteristic *sigmoid*, or *S*-shaped, solution curve for the logistic model. Notice that for each fixed value of P, that is, along each horizontal line, the slopes are constant because dP/dt depends only on P and not on t. The slopes are small near $P = 0$ and near $P = L$; they are steepest around $P = L/2$. For $P > L$, the slopes are negative, so if the population is above the carrying capacity, the population decreases.

Figure 11.49: Slope field for $dP/dt = kP(1 - P/L)$

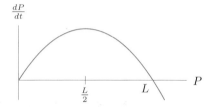

Figure 11.50: $dP/dt = kP(1 - P/L)$

We can locate precisely the inflection point where the slopes are greatest using the graph of dP/dt against P in Figure 11.50. The graph is a parabola because dP/dt is a quadratic function of P. The horizontal intercepts are at $P = 0$ and $P = L$, so the maximum, where the slope is greatest, is at $P = L/2$. The graph in Figure 11.50 also tells us that for $0 < P < L/2$, the slope dP/dt is positive and increasing, so the graph of P against t is concave up. (See Figure 11.51.) For $L/2 < P < L$, the slope dP/dt is positive and decreasing, so the graph of P against t is concave down. For $P > L$, the slope dP/dt is negative, so the graph of P against t is decreasing.

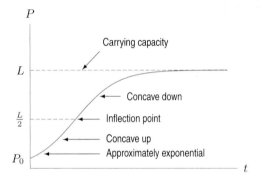

Figure 11.51: Logistic growth with inflection point

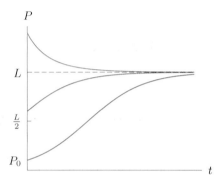

Figure 11.52: Solutions to the logistic equation

If $P = 0$ or $P = L$, there is an equilibrium solution (not a very interesting one if $P = 0$). Figure 11.52 shows that $P = 0$ is an unstable equilibrium because solutions which start near 0 move away. However, $P = L$ is a stable equilibrium.

The Analytic Solution to the Logistic Equation

We have already obtained a lot of information about logistic growth without finding a formula for the solution. However, the equation can be solved analytically by separating variables:

$$\frac{dP}{dt} = kP \left(1 - \frac{P}{L} \right) = kP \left(\frac{L - P}{L} \right)$$

giving

$$\int \frac{dP}{P(L - P)} = \int \frac{k}{L}\, dt.$$

We can integrate the left side using the integral tables (Formula 26), or by partial fractions:

$$\int \frac{1}{L} \left(\frac{1}{P} + \frac{1}{L - P} \right) dP = \int \frac{k}{L}\, dt.$$

Canceling the constant L, we integrate to get

$$\ln |P| - \ln |L - P| = kt + C.$$

Multiplying through by (-1) and using properties of logarithms, we have

$$\ln \left| \frac{L - P}{P} \right| = -kt - C.$$

Exponentiating both sides gives

$$\left| \frac{L - P}{P} \right| = e^{-kt-C} = e^{-C} e^{-kt},$$

so, writing $A = \pm e^{-C}$, we have

$$\frac{L - P}{P} = Ae^{-kt}.$$

We find A by substituting $P = P_0$ when $t = 0$, which gives

$$\frac{L - P_0}{P_0} = Ae^0 = A.$$

Since $(L - P)/P = (L/P) - 1$, we have

$$\frac{L}{P} = 1 + Ae^{-kt}$$

giving the formula for the logistic curve:

$$P = \frac{L}{1 + Ae^{-kt}} \qquad \text{where} \quad A = \frac{L - P_0}{P_0}.$$

Exercises and Problems for Section 11.7

Exercises

1. Assuming that Switzerland's population is growing exponentially at a continuous rate of 0.2% a year and that its 1988 population was 6.6 million, write an expression for the population as a function of time in years. (Let $t = 0$ in 1988.)

2. The Tojolobal Mayan Indian community in Southern Mexico has available a fixed amount of land. The proportion, P, of land in use for farming t years after 1935 is modeled with the logistic function in Figure 11.53:[8]

$$P = \frac{1}{1 + 2.968e^{-0.0275t}}.$$

(a) What proportion of the land was in use for farming in 1935?

(b) What is the long run prediction of this model?

(c) When was half the land in use for farming?

(d) When is the proportion of land used for farming increasing most rapidly?

P
(proportion of land in use)

Figure 11.53

3. (a) On the slope field for $dP/dt = 3P - 3P^2$ in Figure 11.54, sketch three solution curves showing different types of behavior for the population, P.

(b) Is there a stable value of the population? If so, what is it?

(c) Describe the meaning of the shape of the solution curves for the population: Where is P increasing? Decreasing? What happens in the long run? Are there any inflection points? Where? What do they mean for the population?

(d) Sketch a graph of dP/dt against P. Where is dP/dt positive? Negative? Zero? Maximum? How do your observations about dP/dt explain the shapes of your solution curves?

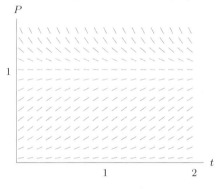

P

Figure 11.54

Problems

4. Estimate the US population in 1870 from the 1860 population of 31.4 million, assuming that the population increased at the same percentage rate during the 1860s as it did in the decades previous to 1860 (about 34.7% each decade). Compare your estimate with the actual 1870 population of 38.6 million. Find some estimate of the number of people who died in the Civil War. Does the number of deaths in the Civil War explain the shortfall in the actual population in 1870? What else might influence the shortfall?

5. Table 11.10 gives the percentage, P, of households with a VCR, as a function of year.

(a) Explain why a logistic model is a reasonable one to use for this data.

(b) Use the data to estimate the point of inflection of P.

What limiting value L does this point of inflection predict? Does this limiting value appear to be accurate given the percentages for 1990 and 1991?

(c) The best logistic equation for this data turns out to be the following. What limiting value does this model predict?

$$P = \frac{75}{1 + 316.75e^{-0.699t}}.$$

Table 11.10 *Percentage of households with a VCR*

Year	1978	1979	1980	1981	1982	1983	1984
$P(\%)$	0.3	0.5	1.1	1.8	3.1	5.5	10.6
Year	1985	1986	1987	1988	1989	1990	1991
$P(\%)$	20.8	36.0	48.7	58.0	64.6	71.9	71.9

[8]Adapted from J. S. Thomas and M. C. Robbins, "The Limits to Growth in a Tojolobal Maya Ejido", *Geoscience and Man 26* (Baton Rouge: Geoscience Publications, 1988), pp. 9–16.

6. The growth of an animal population is governed by the equation

$$\frac{1000}{P} \frac{dP}{dt} = 100 - P,$$

where $P(t)$ is the number of individuals in the colony at time t. The initial population is known to be 200 individuals. Sketch a graph of $P(t)$. Will there ever be more than 200 individuals in the colony? Will there ever be fewer than 100 individuals? Explain.

7. Another way to estimate the limiting population L for the logistic differential equation is to note that the derivative dP/dt is largest when $P = L/2$. Estimate dP/dt for the US census data for 1790–1940 using the two-sided difference quotient $(P(t + 10) - P(t - 10))/20$. Estimate L by doubling the population when dP/dt is largest. Compare this estimate with the text's estimate of $L = 187$.

8. A model for the population, $P(t)$, of carp in a landlocked lake is given by the differential equation

$$\frac{dP}{dt} = 0.25P(1 - 0.0004P).$$

(a) What is the long-term equilibrium population of carp in the lake?

(b) Ten years ago a census was taken and there were found to be 1000 carp in the lake. Estimate the current size of the population.

(c) It is planned to join the lake to a nearby river so that the fish will be able to leave the lake. It is estimated that there will be a net loss of 10% of the carp each year but that the patterns of birth and death are not expected to change. Revise the differential equation to take this into account. Use the revised differential equation to predict the future development of the carp population.

9. The total number of people infected with a virus often grows like a logistic curve. Suppose that 10 people originally have the virus, and that in the early stages of the virus (with time, t, measured in weeks), the number of people infected is increasing exponentially with $k = 1.78$. It is estimated that, in the long run, approximately 5000 people become infected.

(a) Use this information to find a logistic function to model this situation.

(b) Sketch a graph of your answer to part (a).

(c) Use your graph to estimate the length of time until the rate at which people are becoming infected starts to decrease. What is the vertical coordinate at this point?

10. The population of a species of elk on Reading Island in Canada has been monitored for some years. When the population was 600, the relative birth rate was 35% and the relative death rate was 15%. As the population grew to 800, the corresponding figures were 30% and 20%. The island is isolated so there is no hunting or migration.

(a) Write a differential equation to model the population as a function of time. Assume that relative growth rate is a linear function of population.

(b) Find the equilibrium size of the population. Today there are 900 elk on Reading Island. How do you expect the population to change in the future?

(c) Oil has been discovered on a neighboring island and the oil companies want to move 450 elk of the same species to Reading Island. What effect would this move have on the elk population on Reading Island in the future?

(d) Assuming the elk are moved to Reading Island, sketch the population on Reading Island as a function of time. Start before the elk are transferred and continue for some time into the future. Comment on the significance of your results.

11. It is of considerable interest to policy makers to model the spread of information through a population. For example, various agricultural ministries use models to help them understand the spread of technical innovations or new seed types through their countries. Two models, based on how the information is spread, are given below. Assume the population is of a constant size M.

(a) If the information is spread by mass media (TV, radio, newspapers), the rate at which information is spread is believed to be proportional to the number of people not having the information at that time. Write a differential equation for the number of people having the information by time t. Sketch a solution assuming that no one (except the mass media) has the information initially.

(b) If the information is spread by word of mouth, the rate of spread of information is believed to be proportional to the product of the number of people who know and the number who don't. Write a differential equation for the number of people having the information by time t. Sketch the solution for the cases in which

| (i) | no one | (ii) | 5% of the population |
| (iii) | 75% of the population |

knows initially. In each case, when is the information spreading fastest?

12. An alternative method of finding the analytic solution to the logistic equation

$$\frac{dP}{dt} = kP\left(1 - \frac{P}{L}\right)$$

uses the substitution $P = 1/u$.

(a) Show that

$$\frac{dP}{dt} = -\frac{1}{u^2}\frac{du}{dt}.$$

(b) Rewrite the logistic equation in terms of u and t, and solve for u in terms of t.

(c) Using your answer to part (b), find P as a function of t.

Any population, P, for which we can ignore immigration, satisfies

$$\frac{dP}{dt} = \text{Birth rate} - \text{Death rate}.$$

For organisms which need a partner for reproduction but rely on a chance encounter for meeting a mate, the birth rate is proportional to the square of the population. Thus, the population of such a type of organism satisfies a differential equation of the form

$$\frac{dP}{dt} = aP^2 - bP \quad \text{with } a, b > 0.$$

Problems 13–15 investigate the solutions to such an equation.

13. Consider the equation

$$\frac{dP}{dt} = 0.02P^2 - 0.08P.$$

  (a) Sketch the slope field for this differential equation for $0 \leq t \leq 50$, $0 \leq P \leq 8$.
  (b) Use your slope field to sketch the general shape of the solutions to the differential equation satisfying the following initial conditions:
   (i) $P(0) = 1$ (ii) $P(0) = 3$
   (iii) $P(0) = 4$ (iv) $P(0) = 5$
  (c) Are there any equilibrium values of the population? If so, are they stable?

14. Consider the equation

$$\frac{dP}{dt} = P^2 - 6P.$$

  (a) Sketch a graph of dP/dt against P for positive P.

  (b) Use the graph you drew in part (a) to sketch the approximate shape of the solution curve with $P(0) = 5$. To do this, consider the following question. For $0 < P < 6$, is dP/dt positive or negative? What does this tell you about the graph of P against t? As you move along the solution curve with $P(0) = 5$, how does the value of dP/dt change? What does this tell you about the concavity of the graph of P against t?
  (c) Use the graph you drew in part (a) to sketch the solution curve with $P(0) = 8$.
  (d) Describe the qualitative differences in the behavior of populations with initial value less than 6 and initial value more than 6. Why do you think $P = 6$ is called the *threshold population*?

15. Consider a population satisfying

$$\frac{dP}{dt} = aP^2 - bP \quad \text{with constants } a, b > 0.$$

  (a) Sketch a graph of dP/dt against P.
  (b) Use this graph to sketch the shape of solution curves with various initial values. Use your graph from part (a) to decide where dP/dt is positive or negative, and where it is increasing or decreasing. What does this tell you about the graph of P against t?
  (c) Why is $P = b/a$ called the threshold population? What happens if $P(0) = b/a$? What happens in the long-run if $P(0) > b/a$? What if $P(0) < b/a$?

11.8 SYSTEMS OF DIFFERENTIAL EQUATIONS

In Section 11.7 we modeled the growth of a single population over time. We now consider the growth of two populations which interact, such as a population of sick people infecting the healthy people around them. This involves not just one differential equation, but a system of two.

Diseases and Epidemics

Differential equations can be used to predict when an outbreak of a disease will become so severe that it is called an *epidemic*,[9] and to decide what level of vaccination is necessary to prevent an epidemic. Let's consider a specific example.

Flu in a British Boarding School

In January 1978, 763 students returned to a boys' boarding school after their winter vacation. A week later, one boy developed the flu, followed by two others the next day. By the end of the month, nearly half the boys were sick. Most of the school had been affected by the time the epidemic was over in mid-February.[10]

---

[9]Exactly when a disease should be called an epidemic is not always clear. The medical profession generally classifies a disease an epidemic when the frequency is higher than usually expected—leaving open the question of what is usually expected. See, for example, *Epidemiology in Medicine* by C. H. Hennekens and J. Buring (Boston: Little, Brown, 1987).

[10]Data from the Communicable Disease Surveillance Centre (UK); reported in "Influenza in a Boarding School," *British Medical Journal* March 4, 1978, and by J. D. Murray in *Mathematical Biology* (New York: Springer Verlag, 1990).

Being able to predict how many people will get sick, and when, is an important step toward controlling an epidemic. This is one of the responsibilities of Britain's Communicable Disease Surveillance Centre and the US's Center for Disease Control and Prevention.

The S-I-R model

We apply one of the most commonly used models for an epidemic, called the S-I-R model, to the boarding school flu example. The population of the school is divided into three groups:

S = the number of *susceptibles*, the people who are not yet sick but who could become sick

I = the number of *infecteds*, the people who are currently sick

R = the number of *recovered*, or *removed*, the people who have been sick and can no longer infect others or be reinfected.

The number of susceptibles decreases with time, as people become infected. We assume that the rate people become infected is proportional to the number of contacts between susceptible and infected people. We expect the number of contacts between the two groups to be proportional to both S and I. (If S doubles, we expect the number of contacts to double; similarly, if I doubles, we expect the number of contacts to double.) Thus we assume that the number of contacts is proportional to the product, SI. In other words, we assume that for some constant $a > 0$,

$$\frac{dS}{dt} = -\left(\begin{array}{c} \text{Rate susceptibles} \\ \text{get sick} \end{array} \right) = -aSI.$$

(The negative sign is used because S is decreasing.)

The number of infecteds is changing in two ways: newly sick people are added to the infected group, and others are removed. The newly sick people are exactly those people leaving the susceptible group and so accrue at a rate of aSI (with a positive sign this time). People leave the infected group either because they recover (or die), or because they are physically removed from the rest of the group and can no longer infect others. We assume that people are removed at a rate proportional to the number sick, or bI, where b is a positive constant. Thus,

$$\frac{dI}{dt} = \begin{array}{c} \text{Rate susceptibles} \\ \text{get sick} \end{array} - \begin{array}{c} \text{Rate infecteds} \\ \text{get removed} \end{array} = aSI - bI.$$

Assuming that those who have recovered from the disease are no longer susceptible, the recovered group increases at the rate of bI, so

$$\frac{dR}{dt} = bI.$$

We are assuming that having the flu confers immunity on a person, that is, that the person cannot get the flu again. (This is true for a given strain of flu, at least in the short run.)

In analyzing the flu, we can use the fact that the total population $S + I + R$ is not changing. (The total population, the total number of boys in the school, did not change during the epidemic.) Thus, once we know S and I, we can calculate R. So we restrict our attention to the two equations

$$\frac{dS}{dt} = -aSI$$
$$\frac{dI}{dt} = aSI - bI.$$

The Constants a and b

The constant a measures how infectious the disease is—that is, how quickly it is transmitted from the infecteds to the susceptibles. In the case of the flu, we know from medical accounts that the epidemic started with one sick boy, with two more becoming sick a day later. Thus, when $I = 1$ and $S = 762$, we have $dS/dt \approx -2$, enabling us to roughly[11] approximate a:

$$a = -\frac{dS/dt}{SI} = \frac{2}{(762)(1)} = 0.0026.$$

The constant b represents the rate at which infected people are removed from the infected population. In this case of the flu, boys were generally taken to the infirmary within one or two days of becoming sick. About half the infected population was removed each day, so we take $b \approx 0.5$. Thus, our equations are:

$$\frac{dS}{dt} = -0.0026SI$$

$$\frac{dI}{dt} = 0.0026SI - 0.5I.$$

The Phase Plane

We can get a good idea of the progress of the disease from graphs. You might expect that we would look for graphs of S and I against t, and eventually we will. However, we first look at a graph of I against S. If we plot a point (S, I) representing the number of susceptibles and the number of infecteds at any moment in time, then, as the numbers of susceptibles and infecteds change, the point moves. The SI-plane on which the point moves is called the *phase plane*. The path along which the point moves is called the *phase trajectory*, or *orbit*, of the point.

To find the phase trajectory, we need a differential equation relating S and I directly. Thinking of I as a function of S, and S as a function of t, we use the chain rule to get

$$\frac{dI}{dt} = \frac{dI}{dS} \cdot \frac{dS}{dt},$$

giving

$$\frac{dI}{dS} = \frac{dI/dt}{dS/dt}.$$

Substituting for dI/dt and dS/dt, we get

$$\frac{dI}{dS} = \frac{0.0026SI - 0.5I}{-0.0026SI}.$$

Assuming I is not zero, this equation simplifies to approximately

$$\frac{dI}{dS} = -1 + \frac{192}{S}.$$

The slope field of this differential equation is shown in Figure 11.55. The trajectory with initial condition $S_0 = 762$, $I_0 = 1$ is shown in Figure 11.56. Time is represented by the arrow showing the direction that a point moves on the trajectory. The disease starts at the point $S_0 = 762$, $I_0 = 1$. At first, more people become infected and fewer are susceptible. In other words, S decreases and I increases. Later, I decreases as S continues to decrease.

---

[11] The values of a and b are close to those obtained by J. D. Murray in *Mathematical Biology* (New York: Springer Verlag, 1990).

Figure 11.55: Slope field for $dI/dS = -1 + 192/S$ Figure 11.56: Trajectory for $S_0 = 762$, $I_0 = 1$

What does the SI-Phase Plane Tell Us?

To learn how the disease progresses, look at the shape of the curve in Figure 11.56. The value of I increases to about 300 (the maximum number infected and infectious at any one time); then I decreases to zero. This peak value of I occurs when $S \approx 200$. We can determine exactly when the peak value occurs by solving

$$\frac{dI}{dS} = -1 + \frac{192}{S} = 0,$$

which gives

$$S = 192.$$

Notice that the peak value for I always occurs at the same value of S, namely $S = 192$. The graph shows that if a trajectory starts with $S_0 > 192$, then I first increases and then decreases to zero. On the other hand, if $S_0 < 192$, there is no peak as I decreases right away.

For this example, the value $S_0 = 192$ is called a *threshold value*. If S_0 is around or below 192, there is no epidemic. If S_0 is significantly greater than 192, an epidemic occurs.[12]

The phase diagram makes clear that the maximum value of I is about 300. Another question answered by the phase plane diagram is the total number of students who are expected to get sick during the epidemic. (This is not the maximum value reached by I, which gives the maximum number infected at any one time.) The point at which the trajectory crosses the S-axis represents the time when the epidemic has passed (since $I = 0$). The S-intercept shows how many boys never get the flu and thus, how many do get it.

How Many People Should Be Vaccinated?

An epidemic can sometimes be avoided by vaccination. How many boys would have had to be vaccinated to prevent the flu epidemic? To answer this, think of vaccination as removing people from the S category (without increasing I), which amounts to moving the initial point on the trajectory to the left, parallel to the S-axis. To avoid an epidemic, the initial value of S_0 should be at or below the threshold value. Therefore, all but 192 boys would need to be vaccinated.

Graphs of S and I against t

To find out exactly when I reaches its maximum, we need numerical methods. A modification of Euler's method was used to generate the solution curves of S and I against t in Figure 11.57. Notice that the number of susceptibles drops throughout the disease as healthy people get sick. The number of infecteds peaks after about 6 days and then drops. The epidemic has run its course in 20 days.

---

[12]Here we are using J. D. Murray's definition of an epidemic as an outbreak in which the number of infecteds increases from the initial value, I_0. See *Mathematical Biology* (New York: Springer Verlag, 1990).

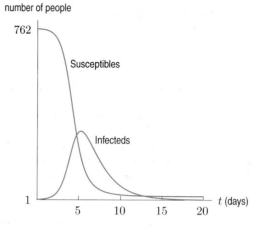

Figure 11.57: Progress of the flu over time

Analytical Solution for the SI-Phase Trajectory

The differential equation

$$\frac{dI}{dS} = -1 + \frac{192}{S}$$

can be integrated giving

$$I = -S + 192 \ln S + C.$$

Using $S_0 = 762$, $I_0 = 1$, gives $1 = -762 + 192 \ln 762 + C$, so we get $C = 763 - 192 \ln(762)$. Substituting this value for C, we get:

$$I = -S + 192 \ln S - 192 \ln(762) + 763$$
$$I = -S + 192 \ln \left(\frac{S}{762} \right) + 763.$$

This is the equation of the solution curve in Figure 11.56.

Two Interacting Populations: Predator-Prey

We now consider two populations which interact. They may compete for food, one may prey on the other, or they may enjoy a symbiotic relationship in which each helps the other. We model a predator-prey system using the *Lotka-Volterra equations*.

Robins and Worms

Let's look at an idealized case[13] in which robins are the predators and worms are the prey. There are r thousand robins and w million worms. If there were no robins, the worms would increase exponentially according to the equation

$$\frac{dw}{dt} = aw \quad \text{where } a \text{ is a constant and } a > 0.$$

If there were no worms, the robins would have no food and their population would decrease according to the equation[14]

$$\frac{dr}{dt} = -br \quad \text{where } b \text{ is a constant and } b > 0.$$

---

[13] Based on ideas from Thomas A. McMahon.

[14] You might criticize this assumption because it predicts that the number of robins will decay exponentially, rather than die out in finite time.

Now we account for the effect of the two populations on one another. Clearly, the presence of the robins is bad for the worms, so

$$\frac{dw}{dt} = aw - \text{Effect of robins on worms.}$$

On the other hand, the robins do better with the worms around, so

$$\frac{dr}{dt} = -br + \text{Effect of worms on robins.}$$

How exactly do the two populations interact? Let's assume the effect of one population on the other is proportional to the number of "encounters." (An encounter is when a robin eats a worm.) The number of encounters is likely to be proportional to the product of the populations because the more there are of either population, the more encounters there will be. So we assume

$$\frac{dw}{dt} = aw - cwr \quad \text{and} \quad \frac{dr}{dt} = -br + kwr,$$

where c and k are positive constants.

To analyze this system of equations, let's look at the specific example with $a = b = c = k = 1$:

$$\frac{dw}{dt} = w - wr \quad \text{and} \quad \frac{dr}{dt} = -r + wr.$$

To visualize the solutions to these equations, we look for trajectories in the phase plane. First we use the chain rule,

$$\frac{dr}{dw} = \frac{dr/dt}{dw/dt},$$

to obtain

$$\frac{dr}{dw} = \frac{-r + wr}{w - wr}.$$

The Slope Field and Equilibrium Points

We can get an idea of what solutions of this equation look like from the slope field in Figure 11.58. At the point $(1,1)$ there is no slope drawn because at this point the rate of change of the worm population with respect to time is zero:

$$\frac{dw}{dt} = 1 - (1)(1) = 0.$$

The rate of change of the robin population with respect to time is also zero:

$$\frac{dr}{dt} = -1 + (1)(1) = 0.$$

Thus dr/dw is undefined. In terms of worms and robins, this means that if at some moment $w = 1$ and $r = 1$ (that is, there are 1 million worms and 1 thousand robins), then w and r remain constant forever. The point $w = 1$, $r = 1$ is therefore an equilibrium solution. The slope field suggests that there are no other equilibrium points except the origin.

At an **equilibrium point**, both w and r are constant, so

$$\frac{dw}{dt} = 0 \quad \text{and} \quad \frac{dr}{dt} = 0.$$

Therefore, we look for equilibrium points by solving

$$\frac{dw}{dt} = w - wr = 0 \quad \text{and} \quad \frac{dr}{dt} = -r + rw = 0,$$

which has $w = 0$, $r = 0$ and $w = 1$, $r = 1$ as the only solutions.

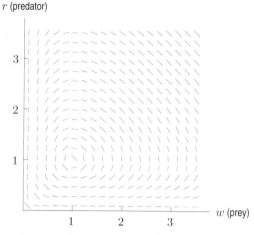

Figure 11.58: Slope field for $\dfrac{dr}{dw} = \dfrac{-r + wr}{w - wr}$

Trajectories in the wr-Phase Plane

Let's look at the trajectories in the phase plane. Remember that a point on a curve represents a pair of populations (w, r) existing at the same time t (though t is not shown on the graph). A short time later, the pair of populations is represented by a nearby point. As time passes, the point traces out a trajectory. The direction is marked on the curve by an arrow. (See Figure 11.59.)

How do we figure out which way to move on the trajectory? Approximating the solution numerically shows that the trajectory is traversed counterclockwise. Alternatively, look at the original pair of differential equations. At the point P_0 in Figure 11.60, where $w > 1$ and $r = 1$,

$$\frac{dr}{dt} = -r + wr = -1 + w > 0.$$

Therefore, r is increasing, so the point is moving counterclockwise around the closed curve.

Now let's think about why the solution curves are closed curves (that is, why they come back and meet themselves). Notice that the slope field is symmetric about the line $w = r$. We can confirm this by observing that interchanging w and r does not alter the differential equation for dr/dw. This means that if we start at point P on the line $w = r$ and travel once around the point $(1, 1)$, we arrive back at the same point P. The reason is that the second half of the path, from Q to P, is the reflection of the first half, from P to Q, in the line $w = r$. (See Figure 11.59.) If we did not end up at P again, the second half of our path would have a different shape from the first half.

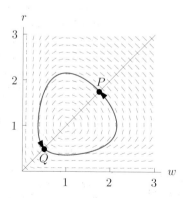

Figure 11.59: Solution curve is closed

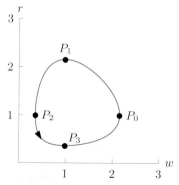

Figure 11.60: Trajectory in the phase plane

The Populations as Functions of Time

The shape of the trajectories tells us how the populations vary with time. We start at $t = 0$ at the point P_0 in Figure 11.60. Then we move to P_1 at time t_1, to P_2 at time t_2, to P_3 at time t_3, and so on. At time t_4 we are back at P_0, and the whole cycle repeats. Since the trajectory is a closed curve, both populations oscillate periodically with the same period. The worms (the prey) are at their maximum a quarter of a cycle before the robins. (See Figure 11.61.)

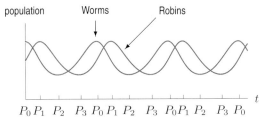

Figure 11.61: Populations of robins (in thousands) and worms (in millions) over time

Lynxes and Hares

A predator-prey system for which there are long-term data is the Canadian lynx and the hare. Since both animals were of interest to fur trappers, records of the Hudson Bay Company shed some light on their populations through much of the nineteenth century. These records show that both populations oscillated quite regularly, with a period of about ten years.

Exercises and Problems for Section 11.8

Exercises

1. Show that if S, I, and R satisfy the differential equations on page 567, the total population, $S + I + R$, is constant.

For Exercises 2–10, let w be the number of worms (in millions) and r the number of robins (in thousands) living on an island. Suppose w and r satisfy the following differential equations, which correspond to the slope field in Figure 11.62.

$$\frac{dw}{dt} = w - wr, \qquad \frac{dr}{dt} = -r + wr.$$

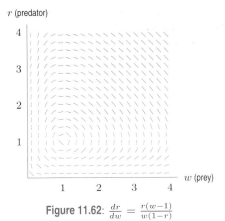

Figure 11.62: $\frac{dr}{dw} = \frac{r(w-1)}{w(1-r)}$

2. Explain why these differential equations are a reasonable model for interaction between the two populations. Why have the signs been chosen this way?

3. Solve these differential equations in the two special cases when there are no robins and when there are no worms living on the island.

4. Describe and explain the symmetry you observe in the slope field. What consequences does this symmetry have for the solution curves?

5. Assume $w = 2$ and $r = 2$ when $t = 0$. Do the numbers of robins and worms increase or decrease at first? What happens in the long run?

6. For the case discussed in Problem 5, estimate the maximum and the minimum values of the robin population. How many worms are there at the time when the robin population reaches its maximum?

7. On the same axes, graph w and r (the worm and the robin populations) against time. Use initial values of 1.5 for w and 1 for r. You may do this without units for t.

8. People on the island like robins so much that they decide to import 200 robins all the way from England, to increase the initial population to $r = 2.2$ when $t = 0$. Does this make sense? Why or why not?

9. Assume that $w = 3$ and $r = 1$ when $t = 0$. Do the numbers of robins and worms increase or decrease initially? What happens in the long run?

10. For the case discussed in Problem 9, estimate the maximum and minimum values of the robin population. Estimate the number of worms when the robin population reaches its minimum.

For Exercises 11–14, suppose x and y are the populations of two different species. Describe in words how each population changes with time.

11.

12.

13.

14.

Problems

The systems of differential equations in Problems 15–17 model the interaction of two populations x and y. In each case, answer the following two questions:

(a) What kinds of interaction (symbiosis[15], competition, predator-prey) do the equations describe?

(b) What happens in the long run? (For one of the systems, your answer will depend on the initial populations.) Use a calculator or computer to draw slope fields.

15. $\dfrac{1}{x}\dfrac{dx}{dt} = y - 1$

$\dfrac{1}{y}\dfrac{dy}{dt} = x - 1$

16. $\dfrac{1}{x}\dfrac{dx}{dt} = 1 - \dfrac{x}{2} - \dfrac{y}{2}$

$\dfrac{1}{y}\dfrac{dy}{dt} = 1 - x - y$

17. $\dfrac{1}{x}\dfrac{dx}{dt} = y - 1 - 0.05x$

$\dfrac{1}{y}\dfrac{dy}{dt} = 1 - x - 0.05y$

For Problems 18–20, consider a conflict between two armies of x and y soldiers, respectively. During World War I, F. W. Lanchester assumed that if both armies are fighting a conventional battle within sight of one another, the rate at which soldiers in one army are put out of action (killed or wounded) is proportional to the amount of fire the other army can concentrate on them, which is in turn proportional to the number of soldiers in the opposing army. Thus Lanchester assumed that if there are no reinforcements and t represents time since the start of the battle, then x and y obey the differential equations

$$\dfrac{dx}{dt} = -ay$$

$$\dfrac{dy}{dt} = -bx \quad a, b > 0.$$

18. Near the end of World War II a fierce battle took place between US and Japanese troops over the island of Iwo Jima, off the coast of Japan. Applying Lanchester's analysis to this battle, with x representing the number of US troops and y the number of Japanese troops, it has been estimated[16] that $a = 0.05$ and $b = 0.01$.

(a) Using these values for a and b, and ignoring reinforcements, write a differential equation involving dy/dx and sketch its slope field.

(b) Assuming that the initial strength of the US forces was 54,000 and that of the Japanese was 21,500, draw the trajectory which describes the battle. What outcome is predicted? (That is, which side do the differential equations predict will win?)

(c) Would knowing that the US in fact had 19,000 reinforcements, while the Japanese had none, alter the outcome predicted?

19. **(a)** For two armies of strengths x and y fighting a conventional battle governed by Lanchester's differential equations, write a differential equation involving dy/dx and the constants of attrition a and b.

(b) Solve the differential equation and hence show that the equation of the phase trajectory is

$$ay^2 - bx^2 = C$$

for some constant C. This equation is called *Lanchester's square law*. The value of C depends on the initial sizes of the two armies.

20. Consider the battle of Iwo Jima, described in Problem 18. Take $a = 0.05$, $b = 0.01$ and assume the initial strength of the US troops to be 54,000 and that of the Japanese troops to be 21,500. (Again, ignore reinforcements.)

(a) Using Lanchester's square law derived in Problem 19, find the equation of the trajectory describing the battle.

(b) Assuming that the Japanese fought without surrendering until they had all been killed, as was the case, how many US troops does this model predict would be left when the battle ended?

[15]Symbiosis takes place when the interaction of two species benefits both. An example is the pollination of plants by insects.

[16]See Martin Braun, *Differential Equations and their Applications*, 2nd ed. (New York: Springer and Verlag, 1975).

21. In this problem we adapt Lanchester's model for a conventional battle (described before Problem 18), to the case in which one or both of the armies is a guerrilla force. We assume that the rate at which a guerrilla force is put out of action is proportional to the product of the strengths of the two armies.

(a) Give a justification for the assumption that the rate at which a guerrilla force is put out of action is proportional to the product of the strengths of the two armies.

(b) Write the differential equations which describe a conflict between a guerrilla army of strength x and a conventional army of strength y, assuming all the constants of proportionality are 1.

(c) Find a differential equation involving dy/dx and solve it to find equations of phase trajectories.

(d) If $C > 0$, which side wins? If $C < 0$, which side wins? What if $C = 0$?

(e) Use your solution to part (d) to divide the phase plane into regions according to which side wins.

22. To model a conflict between two guerrilla armies, we assume that the rate that each one is put out of action is proportional to the product of the strengths of the two armies.

(a) Write the differential equations which describe a conflict between two guerrilla armies, of strengths x and y, respectively.

(b) Find a differential equation involving dy/dx and solve to find equations of phase trajectories.

(c) If $C > 0$, which side wins? If $C < 0$, which side wins? What if $C = 0$?

(d) Use your solution to part (b) to divide the phase plane into regions according to which side wins.

23. The concentrations of two chemicals A and B as functions of time are denoted by x and y respectively. Each alone decays at a rate proportional to its concentration. Put together, they also interact to form a third substance, at a rate proportional to the product of their concentrations. All this is expressed in the equations:

$$\frac{dx}{dt} = -2x - xy, \qquad \frac{dy}{dt} = -3y - xy.$$

(a) Find a differential equation describing the relationship between x and y, and solve it.

(b) Show that the only equilibrium state is $x = y = 0$. (Note that the concentrations are nonnegative.)

(c) Show that when x and y are positive and very small, y^2/x^3 is roughly constant. [Hint: When x is small, x is negligible compared to $\ln x$.]

If now the initial concentrations are $x(0) = 4$, $y(0) = 8$:

(d) Find the equation of the phase trajectory.

(e) What would be the concentrations of each substance if they become equal?

(f) If $x = e^{-10}$, find an approximate value for y.

11.9 ANALYZING THE PHASE PLANE

In the previous section we analyzed a system of differential equations using a slope field. In this section we analyze a system of differential equations using *nullclines*. We will consider two species having similar *niches*, or ways of living, and in competition for food and space. In such cases, one species often becomes extinct. This phenomenon is called the *Principle of Competitive Exclusion*. We see how differential equations predict this in a particular case.

Competitive Exclusion: Citrus Tree Parasites

The citrus farmers of Southern California are interested in controlling the insects that live on their trees. Some of these insects can be controlled by parasites which live on the trees too. Scientists are therefore interested in understanding under what circumstances these parasites flourish or die out. One such parasite was introduced accidentally from the Mediterranean; later, other parasites were introduced from China and India; in each case the previous parasite became extinct over part of its habitat. In 1963 a lab experiment was carried out to determine which one of a pair of species became extinct when they were in competition with each other. The data on one pair of species, called A. *fisheri* and A. *melinus*, with populations P_1 and P_2 respectively, is given in Table 11.11 and shows that A. *melinus* (P_2) became extinct after 8 generations.[17]

Table 11.11 *Population (in thousands) of two species of parasite as a function of time*

Generation number	1	2	3	4	5	6	7	8
Population P_1 (thousands)	0.193	1.093	1.834	5.819	13.705	16.965	18.381	16.234
Population P_2 (thousands)	0.083	0.229	0.282	0.378	0.737	0.507	0.13	0

[17]Data adapted from Paul DeBach and Ragnhild Sundby, "Competitive Displacement between Ecological Homologues," *Hilgardia 34*:17 (1963).

Data from the same experimenters indicates that, when alone, each population grows logistically. In fact, their data suggests that, when alone, the population of P_1 might grow according to the equation

$$\frac{dP_1}{dt} = 0.05P_1 \left(1 - \frac{P_1}{20} \right),$$

and when alone, the population of P_2 might grow according to the equation

$$\frac{dP_2}{dt} = 0.09P_2 \left(1 - \frac{P_2}{15} \right).$$

Now suppose both parasites are present. Each tends to reduce the growth rate of the other, so each differential equation is modified by subtracting a term on the right. The experimental data shows that together P_1 and P_2 can be well described by the equations

$$\frac{dP_1}{dt} = 0.05P_1 \left(1 - \frac{P_1}{20} \right) - 0.002P_1P_2$$

$$\frac{dP_2}{dt} = 0.09P_2 \left(1 - \frac{P_2}{15} \right) - 0.15P_1P_2.$$

The fact that P_2 dies out with time is reflected in these equations: the coefficient of P_1P_2 is so much larger in the equation for P_2 than in the equation for P_1. This indicates that the interaction has a much more devastating effect upon the growth of P_2 than on the growth of P_1.

The Phase Plane and Nullclines

We consider the phase plane with the P_1 axis horizontal and the P_2 axis vertical. To find the trajectories in the P_1P_2 phase plane, we could draw a slope field as in the previous section. Instead, we use a method which gives a good qualitative picture of the behavior of the trajectories even without a calculator or computer. We find the *nullclines* or curves along which $dP_1/dt = 0$ or $dP_2/dt = 0$. At points where $dP_2/dt = 0$, the population P_2 is momentarily constant, so only population P_1 is changing with time. Therefore, at this point the trajectory is horizontal. (See Figure 11.63.) Similarly, at points where $dP_1/dt = 0$, the population P_1 is momentarily constant and population P_2 is the only one changing, so the trajectory is vertical there. A point where both $dP_1/dt = 0$ and $dP_2/dt = 0$ is called an *equilibrium point* because P_1 and P_2 both remain constant if they reach these values.

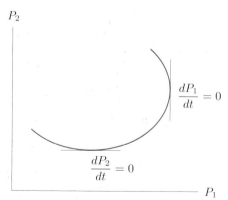

Figure 11.63: Points on a trajectory where
$dP_1/dt = 0$ or $dP_2/dt = 0$

On the $P_1 P_2$ phase plane:

- If $\dfrac{dP_1}{dt} = 0$, the trajectory is vertical.

- If $\dfrac{dP_2}{dt} = 0$, the trajectory is horizontal.

- If $\dfrac{dP_1}{dt} = \dfrac{dP_2}{dt} = 0$, there is an equilibrium point.

Using Nullclines to Analyze the Parasite Populations

In order to see where $dP_1/dt = 0$ or $dP_2/dt = 0$, we factor the right side of our differential equations:

$$\frac{dP_1}{dt} = 0.05 P_1 \left(1 - \frac{P_1}{20}\right) - 0.002 P_1 P_2 = 0.001 P_1 (50 - 2.5 P_1 - 2 P_2)$$

$$\frac{dP_2}{dt} = 0.09 P_2 \left(1 - \frac{P_2}{15}\right) - 0.15 P_1 P_2 = 0.001 P_2 (90 - 150 P_1 - 6 P_2).$$

Thus $dP_1/dt = 0$ where $P_1 = 0$ or where $50 - 2.5 P_1 - 2 P_2 = 0$. Graphing these equations in the phase plane gives two lines, which are nullclines. Since the trajectory is vertical where $dP_1/dt = 0$, in Figure 11.64 we draw small vertical line segments on these nullclines to represent the direction of the trajectories as they cross the nullcline. Similarly $dP_2/dt = 0$ where $P_2 = 0$ or where $90 - 150 P_1 - 6 P_2 = 0$. These equations are graphed in Figure 11.64 with small horizontal line segments on them.

The equilibrium points are where both $dP_1/dt = 0$ and $dP_2/dt = 0$, namely the points $P_1 = 0, P_2 = 0$ (meaning that both species die out); $P_1 = 0, P_2 = 15$ (where P_1 is extinct); and $P_1 = 20, P_2 = 0$ (where P_2 is extinct).

What Happens in the Regions Between the Nullclines?

Nullclines are useful because they divide the plane into regions in which the signs of dP_1/dt and dP_2/dt are constant. In each region, the direction of every trajectory remains roughly the same.

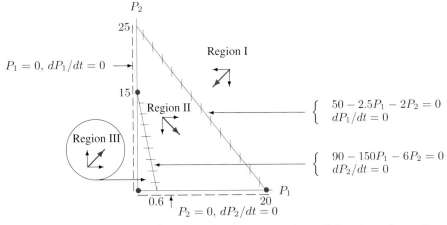

Figure 11.64: Analyzing three regions in the phase plane using nullclines (axes distorted) with equilibrium points represented by dots

Figure 11.65: Trajectories showing exclusion of population P_2 (not to scale)

In region I, for example, we might try the point $P_1 = 20, P_2 = 25$. Then

$$\frac{dP_1}{dt} = 0.001(20)(50 - 2.5(20) - 2(25)) < 0$$

$$\frac{dP_2}{dt} = 0.001(25)(90 - 150(20) - 6(25)) < 0$$

Now $dP_1/dt < 0$, so P_1 is decreasing, which can be represented by an arrow in the direction $\leftarrow$. Also, $dP_2/dt < 0$, so P_2 is decreasing, as represented by the arrow $\downarrow$. Combining these directions, we know that the trajectories in this region go approximately in the diagonal direction $\swarrow$ (See region I in Figure 11.64.)

In region II, try, for example, $P_1 = 1, P_2 = 1$. Then we have

$$\frac{dP_1}{dt} = 0.001(1)(50 - 2.5 - 2) > 0$$

$$\frac{dP_2}{dt} = 0.001(1)(90 - 150 - 6) < 0$$

So here, P_1 is increasing while P_2 is decreasing. (See region II in Figure 11.64.)

In region III, try $P_1 = 0.1, P_2 = 0.1$:

$$\frac{dP_1}{dt} = 0.001(0.1)(50 - 2.5(0.1) - 2(0.1)) > 0$$

$$\frac{dP_2}{dt} = 0.001(0.1)(90 - 150(0.1) - 6(0.1)) > 0$$

So here, both P_1 and P_2 are increasing. (See region III in Figure 11.64.)

Notice that the behavior of the populations in each region makes biological sense. In region I both populations are so large that overpopulation is a problem, so both populations decrease. In region III both populations are so small that they are effectively not in competition, so both grow. In region II competition between the species comes into play. The fact that P_1 increases while P_2 decreases in region II means that P_1 wins.

Solution Trajectories

Suppose the system starts with some of each population. This means that the initial point of the trajectory is not on one of the axes, and so it is in region I, II, or III. Then the point moves on a trajectory like one of those computed numerically and shown in Figure 11.65. Notice that *all* these trajectories tend toward the point $P_1 = 20, P_2 = 0$, corresponding to a population of 20,000 for P_1 and extinction for P_2. Consequently, this model predicts that no matter what the initial populations are, provided $P_1 \neq 0$, the population of P_2 is excluded by P_1, and P_1 tends to a constant value. This makes biological sense: in the absence of P_2, we would expect P_1 to settle down to the carrying capacity of the niche, which is 20,000.

Exercises and Problems for Section 11.9

Exercises

1. **(a)** Find the equilibrium points for the following system of equations

$$\frac{dx}{dt} = 20x - 10xy$$

$$\frac{dy}{dt} = 25y - 5xy.$$

   (b) Explain why $x = 2$, $y = 4$ is not an equilibrium point for this system.

2. The equations describing the flu epidemic in a boarding school are

$$\frac{dS}{dt} = -0.0026SI$$

$$\frac{dI}{dt} = 0.0026SI - 0.5I.$$

   (a) Find the nullclines and equilibrium points in the SI phase plane.
   (b) Find the direction of the trajectories in each region.
   (c) Sketch some typical trajectories and describe their behavior in words.

3. Use the idea of nullclines dividing the plane into sectors to analyze the equations describing the interactions of robins and worms.

$$\frac{dw}{dt} = w - wr$$

$$\frac{dr}{dt} = -r + rw.$$

Problems

For Problems 4–9, analyze the phase plane of the differential equations for x, $y \geq 0$. Show the nullclines and equilibrium points, and sketch the direction of the trajectories in each region.

4. $\dfrac{dx}{dt} = x(2 - x - y)$
   $\dfrac{dy}{dt} = y(1 - x - y)$

5. $\dfrac{dx}{dt} = x(2 - x - 3y)$
   $\dfrac{dy}{dt} = y(1 - 2x)$

6. $\dfrac{dx}{dt} = x(2 - x - 2y)$
   $\dfrac{dy}{dt} = y(1 - 2x - y)$

7. $\dfrac{dx}{dt} = x\left(1 - y - \dfrac{x}{3}\right)$
   $\dfrac{dy}{dt} = y\left(1 - \dfrac{y}{2} - x\right)$

8. $\dfrac{dx}{dt} = x\left(1 - x - \dfrac{y}{3}\right)$
   $\dfrac{dy}{dt} = y\left(1 - y - \dfrac{x}{2}\right)$

9. $\dfrac{dx}{dt} = x\left(1 - \dfrac{x}{2} - y\right)$
   $\dfrac{dy}{dt} = y\left(1 - \dfrac{y}{3} - x\right)$

10. Two companies share the market for a new technology. They have no competition except each other. Let $A(t)$ be the net worth of one company and $B(t)$ the net worth of the other at time t. Assume that net worth cannot be negative, because a company goes out of business if its net worth is zero. Suppose A and B satisfy the differential equations

$$A' = 2A - AB$$
$$B' = B - AB.$$

   (a) What do these equations predict about the net worth of each company if the other were not present? What effect do the companies have on each other?
   (b) Are there any equilibrium points? If so, what are they?

   (c) Sketch a slope field for these equations (using a computer or calculator), and hence describe the different possible long-run behaviors.

11. In the 1930s, the Soviet ecologist G. F. Gause performed a series of experiments on competition among two yeasts with populations P_1 and P_2, respectively. By performing population studies at low density in large volumes, he determined what he called the *coefficients of geometric increase* (and we would call continuous exponential growth rates). These coefficients described the growth of each yeast alone:

$$\frac{1}{P_1}\frac{dP_1}{dt} = 0.2$$

$$\frac{1}{P_2}\frac{dP_2}{dt} = 0.06$$

where P_1 and P_2 are measured in units that Gause established.

He also determined that, in his units, the carrying capacity of P_1 was 13 and the carrying capacity of P_2 was 6. He then observed that one P_2 occupies the niche space of 3 P_1 and that one P_1 occupied the niche space of 0.4 P_2. This led him to the following differential equations to describe the interaction of P_1 and P_2:

$$\frac{dP_1}{dt} = 0.2P_1\left(\frac{13 - (P_1 + 3P_2)}{13}\right),$$

$$\frac{dP_2}{dt} = 0.06P_2\left(\frac{6 - (P_2 + 0.4P_1)}{6}\right).$$

When both yeasts were growing together, Gause recorded the data in Table 11.12.

Table 11.12 *Gause's yeast populations*

Time (hours)	6	16	24	29	48	53
P_1	0.375	3.99	4.69	6.15	7.27	8.30
P_2	0.29	0.98	1.47	1.46	1.71	1.84

(a) Carry out a phase plane analysis of Gause's equations.

(b) Mark the data points on the phase plane and describe what would have happened had Gause continued the experiment.

12. When two countries are in an arms race, the rate at which each one spends money on arms is determined by its own current level of spending and by its opponent's level of spending. We expect that the more a country is already spending on armaments, the less willing it will be to increase its military expenditures. On the other hand, the more a country's opponent spends on armaments, the more rapidly the country will arm. If $x billion is the country's yearly expenditure on arms, and $y billion is its opponent's, then the *Richardson arms race model* proposes that x and y are determined by differential equations. Suppose that for some particular arms race the equations are

$$\frac{dx}{dt} = -0.2x + 0.15y + 20$$
$$\frac{dy}{dt} = 0.1x - 0.2y + 40.$$

(a) Explain the signs of the three terms on the right side of the equations for dx/dt.

(b) Find the equilibrium points for this system.

(c) Analyze the direction of the trajectories in each region.

(d) Are the equilibrium points stable or unstable?

(e) What does this model predict will happen if both countries disarm?

(f) What does this model predict will happen in the case of unilateral disarmament (one country disarms, and the other country does not)?

(g) What does the model predict will happen in the long run?

13. In the 1930's L. F. Richardson proposed that an arms race between two countries could be modeled by a system of differential equations. One arms race which can be reasonably well described by differential equations is the US-Soviet Union arms race between 1945 and 1960. If $x represents the annual Soviet expenditures on armaments (in billions of dollars) and $y represents the corresponding US expenditures, it has been suggested[18] that x and y obey the following differential equations

$$\frac{dx}{dt} = -0.45x + 10.5,$$
$$\frac{dy}{dt} = 8.2x - 0.8y - 142.$$

(a) Find the nullclines and equilibrium points for these differential equations. Which direction do the trajectories go in each region?

(b) Sketch some typical trajectories in the phase plane.

(c) What do these differential equations predict will be the long term outcome of the US-Soviet arms race?

(d) Discuss these predictions in the light of the actual expenditures in Table 11.13.

Table 11.13 *Arms budgets of the United States and the Soviet Union for the years 1945–1960 (billions of dollars)*

	USSR	USA		USSR	USA
1945	14	97	1953	25.7	71.4
1946	14	80	1954	23.9	61.6
1947	15	29	1955	25.5	58.3
1948	20	20	1956	23.2	59.4
1949	20	22	1957	23.0	61.4
1950	21	23	1958	22.3	61.4
1951	22.7	49.6	1959	22.3	61.7
1952	26.0	69.6	1960	22.1	59.6

11.10 SECOND-ORDER DIFFERENTIAL EQUATIONS: OSCILLATIONS

A Second-Order Differential Equation

When a body moves freely under gravity, we know that

$$\frac{d^2 s}{dt^2} = -g,$$

where s is the height of the body above ground at time t and g is the acceleration due to gravity. To solve this equation, we first integrate to get the velocity, $v = ds/dt$:

$$\frac{ds}{dt} = -gt + v_0,$$

---

[18]R. Taagepera, G. M. Schiffler, R. T. Perkins and D. L. Wagner, *Soviet-American and Israeli-Arab Arms Races and the Richardson Model* (General Systems, XX, 1975).

where v_0 is the initial velocity. Then we integrate again, giving

$$s = -\frac{1}{2}gt^2 + v_0 t + s_0,$$

where s_0 is the initial height.

The differential equation $d^2 s/dt^2 = -g$ is called *second order* because the equation contains a second derivative but no higher derivatives. The general solution to a second-order differential equation is a family of functions with two parameters, here v_0 and s_0. Finding values for the two constants corresponds to picking a particular function out of this family.

A Mass on a Spring

Not every second-order differential equation can be solved simply by integrating twice. Consider a mass m attached to the end of a spring hanging from the ceiling. We assume that the mass of the spring itself is negligible in comparison with the mass m. (See Figure 11.66.)

Figure 11.66: Spring and mass in equilibrium position and after upward displacement

When the system is left undisturbed, no net force acts on the mass. The force of gravity is balanced by the force the spring exerts on the mass, and the spring is in the *equilibrium position*. If we pull down on the mass, we feel a force pulling upward. If instead, we push upward on the mass, the opposite happens: a force pushes the mass down.[19]

What happens if we push the mass upward and then release it? We expect the mass to oscillate up and down around the equilibrium position.

Springs and Hooke's Law

In order to figure out how the mass moves, we need to know the exact relationship between its displacement, s, from the equilibrium position and the net force, F, exerted on the mass. (See Figure 11.66.) We would expect that the further the mass is from the equilibrium position, and the more the spring is stretched or compressed, the larger the force. In fact, provided that the displacement is not large enough to deform the spring permanently, experiments show that the net force, F, is approximately proportional to the displacement, s:

$$F = -ks,$$

where k is the *spring constant* ($k > 0$) and the negative sign means that the net force is in the opposite direction to the displacement. The value of k depends on the physical properties of the particular spring. This relationship is known as *Hooke's Law*. Suppose we push the mass upward some distance and then release it. After we let go the net force causes the mass to accelerate toward the equilibrium position. By Newton's Second Law of Motion we have

$$\text{Force} = \text{Mass} \times \text{Acceleration}.$$

---

[19]Pulling down on the mass stretches the spring, increasing the tension, so the combination of gravity and spring force is upward. Pushing up the spring decreases tension in the spring, so the combination of gravity and spring force is downward.

Since acceleration is d^2s/dt^2 and force is $F = -ks$ by Hooke's law, we have

$$-ks = m\frac{d^2s}{dt^2},$$

which is equivalent to the following differential equation.

Equation for Oscillations of a Mass on a Spring

$$\frac{d^2s}{dt^2} = -\frac{k}{m}s$$

Thus, the motion of the mass is described by a second-order differential equation. Since we expect the mass to oscillate, we guess that the solution to this equation involves trigonometric functions.

Solving the Differential Equation by Guess-and-Check

Guessing is the tried-and-true method for solving differential equations. It may surprise you to learn that there is no systematic method for solving most differential equations analytically, so guesswork is often extremely important.

Example 1   By guessing, find the general solution to the equation

$$\frac{d^2s}{dt^2} = -s.$$

Solution   We want to find functions whose second derivative is the negative of the original function. We are already familiar with two functions that have this property: $s(t) = \cos t$ and $s(t) = \sin t$. We check that they are solutions by substituting:

$$\frac{d^2}{dt^2}(\cos t) = \frac{d}{dt}(-\sin t) = -\cos t,$$

and

$$\frac{d^2}{dt^2}(\sin t) = \frac{d}{dt}(\cos t) = -\sin t.$$

The remarkable thing is that starting from these two particular solutions, we can build up *all* the solutions to our equation. Given two constants C_1 and C_2, the function

$$s(t) = C_1 \cos t + C_2 \sin t$$

satisfies the differential equation, since

$$\frac{d^2}{dt^2}(C_1 \cos t + C_2 \sin t) = \frac{d}{dt}(-C_1 \sin t + C_2 \cos t)$$
$$= -C_1 \cos t - C_2 \sin t$$
$$= -(C_1 \cos t + C_2 \sin t).$$

It can be shown (though we will not do so) that $s(t) = C_1 \cos t + C_2 \sin t$ is the most general form of the solution. As expected, it contains two constants, C_1 and C_2.

If the differential equation represents a physical problem, then C_1 and C_2 can often be computed, as shown in the next example.

Example 2   If the mass is displaced by a distance of s_0 and then released, find the solution to

$$\frac{d^2 s}{dt^2} = -s$$

Solution   The position of the mass is given by the equation

$$s(t) = C_1 \cos t + C_2 \sin t.$$

We also know that the initial position is s_0; thus,

$$s(0) = C_1 \cos 0 + C_2 \sin 0 = C_1 \cdot 1 + C_2 \cdot 0 = s_0,$$

so $C_1 = s_0$, the initial displacement. What is C_2? To find it, we use the fact that at $t = 0$, when the mass has just been released, its velocity is 0. Velocity is the derivative of the displacement, so

$$\left.\frac{ds}{dt}\right|_{t=0} = \left.(-C_1 \sin t + C_2 \cos t)\right|_{t=0} = -C_1 \cdot 0 + C_2 \cdot 1 = 0,$$

so $C_2 = 0$. Therefore the solution is $s = s_0 \cos t$.

Solution to the General Spring Equation

Having found the general solution to the equation $d^2 s/dt^2 = -s$, let us return to the equation

$$\frac{d^2 s}{dt^2} = -\frac{k}{m} s.$$

We write $\omega = \sqrt{k/m}$ (why will be clear in a moment), so the differential equation becomes

$$\frac{d^2 s}{dt^2} = -\omega^2 s.$$

This equation no longer has $\sin t$ and $\cos t$ as solutions, since

$$\frac{d^2}{dt^2}(\sin t) = -\sin t \neq -\omega^2 \sin t.$$

However, since we want a factor of ω^2 after differentiating twice, the chain rule leads us to guess that $\sin \omega t$ may be a solution. Checking this:

$$\frac{d^2}{dt^2}(\sin \omega t) = \frac{d}{dt}(\omega \cos \omega t) = -\omega^2 \sin \omega t,$$

we see that $\sin \omega t$ is a solution, and you can check that $\cos \omega t$ is a solution, too.

The general solution to the equation

$$\frac{d^2 s}{dt^2} + \omega^2 s = 0$$

is of the form

$$s(t) = C_1 \cos \omega t + C_2 \sin \omega t,$$

where C_1 and C_2 are arbitrary constants. (We assume $\omega > 0$.) The period of this oscillation is

$$T = \frac{2\pi}{\omega}.$$

Such oscillations are called **simple harmonic motion**.

The solution to our original equation, $\dfrac{d^2 s}{dt^2} + \dfrac{k}{m} s = 0$, is thus $s = C_1 \cos \sqrt{\frac{k}{m}} t + C_2 \sin \sqrt{\frac{k}{m}} t$.

Initial value and Boundary-Value Problems

A problem in which the initial position and the initial velocity are used to determine the particular solution is called an *initial value problem*. (See Example 2.) Alternatively, we may be given the position at two known times. Such a problem is known as a *boundary-value problem*.

Example 3   Find a solution to the differential equation satisfying each set of conditions below

$$\frac{d^2 s}{dt^2} + 4s = 0.$$

(a) The boundary conditions $s(0) = 0$, $s(\pi/4) = 20$.
(b) The initial conditions $s(0) = 1$, $s'(0) = -6$.

Solution   Since $\omega^2 = 4$, $\omega = 2$, the general solution to the differential equation is

$$s(t) = C_1 \cos 2t + C_2 \sin 2t.$$

(a) Substituting the boundary condition $s(0) = 0$ into the general solution gives

$$s(0) = C_1 \cos(2 \cdot 0) + C_2 \sin(2 \cdot 0) = C_1 \cdot 1 + C_2 \cdot 0 = C_1 = 0.$$

Thus $s(t)$ must have the form $s(t) = C_2 \sin 2t$. The second condition yields the value of C_2:

$$s\left(\frac{\pi}{4}\right) = C_2 \sin\left(2 \cdot \frac{\pi}{4}\right) = C_2 = 20.$$

Therefore, the solution satisfying the boundary conditions is

$$s(t) = 20 \sin 2t.$$

(b) For the initial value problem, we start from the same general solution: $s(t) = C_1 \cos 2t + C_2 \sin 2t$. Substituting 0 for t once again, we find

$$s(0) = C_1 \cos(2 \cdot 0) + C_2 \sin(2 \cdot 0) = C_1 = 1.$$

Differentiating $s(t) = \cos 2t + C_2 \sin 2t$ gives

$$s'(t) = -2 \sin 2t + 2C_2 \cos 2t,$$

and applying the second initial condition gives us C_2:

$$s'(0) = -2 \sin(2 \cdot 0) + 2C_2 \cos(2 \cdot 0) = 2C_2 = -6,$$

so $C_2 = -3$ and our solution is

$$s(t) = \cos 2t - 3 \sin 2t.$$

What do the Graphs of Our Solutions Look Like?

Since the general solution of the equation $d^2 s/dt^2 + \omega^2 s = 0$ is of the form

$$s(t) = C_1 \cos \omega t + C_2 \sin \omega t$$

it would be useful to know what the graph of such a sum of sines and cosines looks like. We start with the example $s(t) = \cos t + \sin t$, which is graphed in Figure 11.67.

Interestingly, the graph in Figure 11.67 looks like another sine function, and in fact it is one. If we measure the graph carefully, we find that its period is 2π, the same as $\sin t$ and $\cos t$, but the amplitude is approximately 1.414 (in fact, it's $\sqrt{2}$), and the graph is shifted along the t-axis (in fact by $\pi/4$). If we plot $C_1 \cos t + C_2 \sin t$ for any C_1 and C_2, the resulting graph is always a sine function with period 2π, though the amplitude and shift can vary. For example, the graph of $s = 6 \cos t - 8 \sin t$ is in Figure 11.68; it has period 2π.

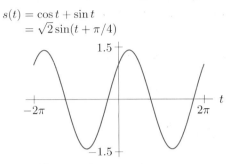

$s(t) = \cos t + \sin t$
$\quad = \sqrt{2} \sin(t + \pi/4)$

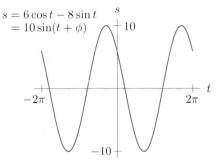

$s = 6 \cos t - 8 \sin t$
$\quad = 10 \sin(t + \phi)$

Figure 11.67: Graph of the sum:
$s(t) = \cos t + \sin t = \sqrt{2}\sin(t + \pi/4)$

Figure 11.68: Graph of a sum of sine and cosine. Amplitude $A = 10 = \sqrt{6^2 + 8^2}$

These graphs suggest that we can write the sum of a sine and a cosine of the same argument as one single sine function.[20] Problem 24 shows that this can always be done using the following relations.

If $C_1 \cos \omega t + C_2 \sin \omega t = A \sin(\omega t + \phi)$, the *amplitude*, A, is given by

$$A = \sqrt{C_1^2 + C_2^2}.$$

The angle ϕ is called the *phase shift* and satisfies

$$\tan \phi = \frac{C_1}{C_2}.$$

We choose ϕ in $(-\pi, \pi]$ such that if $C_1 > 0$, ϕ is positive, and if $C_1 < 0$, ϕ is negative.

The reason that it is often useful to write the solution to the differential equation as a single sine function $A\sin(\omega t + \phi)$, as opposed to the sum of a sine and a cosine, is that the amplitude A and the phase shift ϕ are easier to recognize in this form.

Warning: Phase Shift Is Not the Same as Horizontal Translation

Let's look at $s = \sin(3t + \pi/2)$. If we rewrite this as $s = \sin(3(t + \pi/6))$, we can see from Figure 11.69 that the graph of this function is the graph of $s = \sin 3t$ shifted to the left a distance of $\pi/6$. (Remember that replacing x by $x - 2$ shifts a graph by 2 to the right.) But $\pi/6$ is *not* the phase shift; the phase shift[21] is $\pi/2$. From the point of view of a scientist, the important question is often not the distance the curve has shifted, but the relation between the distance shifted and the period.

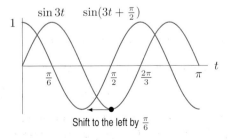

Figure 11.69: Phase shift is $\frac{\pi}{2}$; horizontal translation is $\frac{\pi}{6}$

---

[20] The sum of a sine and a cosine can also be written as a single cosine function.

[21] This definition of phase shift is the one usually used in the sciences.

Exercises and Problems for Section 11.10

Exercises

1. Check by differentiation that $y = 2\cos t + 3\sin t$ is a solution to $y'' + y = 0$.

2. Check by differentiation that $y(t) = 3\sin 2t + 2\cos 2t$ is a solution of $y'' + 4y = 0$.

3. Check by differentiation that $y = A\cos t + B\sin t$ is a solution to $y'' + y = 0$ for any constants A and B.

4. Check by differentiation that $y(t) = A\sin 2t + B\cos 2t$ is a solution of $y'' + 4y = 0$ for all values of A and B.

5. Check by differentiation that $y(t) = A\sin \omega t + B\cos \omega t$ is a solution of $y'' + \omega^2 y = 0$ for all values of A and B.

6. What values of α and A make $y = A\cos \alpha t$ a solution to $y'' + 5y = 0$ such that $y'(1) = 3$?

7. (a) Using a calculator or computer, graph $s = 4\cos t + 3\sin t$ for $-2\pi \le t \le 2\pi$.
  (b) If $4\cos t + 3\sin t = A\sin(t + \phi)$, use your graph to estimate the values of A and ϕ.
  (c) Calculate A and ϕ analytically.

8. Write the function $s(t) = \cos t - \sin t$ as a single sine function. Draw its graph.

9. Find the amplitude of $3\sin 2t + 7\cos 2t$.

10. Find the amplitude of the oscillation $y = 3\sin 2t + 4\cos x2t$.

11. Write $y(t) = 5\sin(2t) + 12\cos(2t)$ in the form $y(t) = A\sin(\omega t + \psi)$.

12. Write $7\sin \omega t + 24\cos \omega t$ in the form $A\sin(\omega t + \phi)$.

Problems

13. Each graph in Figure 11.70 represents a solution to one of the differential equations:
  (a) $x'' + x = 0$, (b) $x'' + 4x = 0$,
  (c) $x'' + 16x = 0$.

Assuming the t-scales on the four graphs are the same, which graph represents a solution to which equation? Find an equation for each graph.

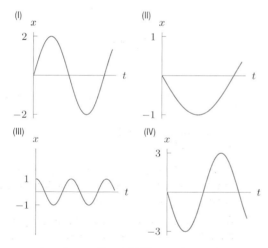

Figure 11.70

14. (a) Show that $y = c_1 \sinh wt + c_2 \cosh wt$ is a solution to $y'' - w^2 y = 0$.
  (b) Find a solution of this differential equation such that
    (i) $y(0) = 0, y(1) = 6$.
    (ii) $y'(0) = 0, y(1) = 6$.

The functions in Problems 15–17 describe the motion of a mass on a spring satisfying the differential equation $y'' = -9y$, where y is the displacement of the mass from the equilibrium position at time t, with upward as positive. In each case, describe in words how the motion starts when $t = 0$. For example, is the mass at the highest point, the lowest point, or in the middle? Is it moving up or down or is it at rest?

15. $y = 2\cos 3t$ **16.** $y = -0.5\sin 3t$

17. $y = -\cos 3t$

18. The following differential equations represent oscillating springs.

  (i) $s'' + 4s = 0$ $s(0) = 5,$ $s'(0) = 0$
  (ii) $4s'' + s = 0$ $s(0) = 10,$ $s'(0) = 0$
  (iii) $s'' + 6s = 0$ $s(0) = 4,$ $s'(0) = 0$
  (iv) $6s'' + s = 0$ $s(0) = 20,$ $s'(0) = 0$

  Which differential equation represents:

  (a) The spring oscillating most quickly (with the shortest period)?
  (b) The spring oscillating with the largest amplitude?
  (c) The spring oscillating most slowly (with the longest period)?
  (d) The spring with largest maximum velocity?

19. (a) Find the general solution of the differential equation

$$y'' + 9y = 0.$$

  (b) For each of the following initial conditions, find a particular solution.
    (i) $y(0) = 0, y'(0) = 1$
    (ii) $y(0) = 1, y'(0) = 0$
    (iii) $y(0) = 1, y(1) = 0$
    (iv) $y(0) = 0, y(1) = 1$
  (c) Sketch a graph of the solutions found in part (b).

20. Consider the motion described by the differential equations:

(a) $x'' + 16x = 0$, $x(0) = 5$, $x'(0) = 0$,
(b) $25x'' + x = 0$, $x(0) = -1$, $x'(0) = 2$.

In each case, find a formula for $x(t)$ and calculate the amplitude and period of the motion.

21. A brick of mass 3 kg hangs from the end of a spring. When the brick is at rest, the spring is stretched by 2 cm. The spring is then stretched an additional 5 cm and released. Assume there is no air resistance.

(a) Set up a differential equation with initial conditions describing the motion.
(b) Solve the differential equation.

22. A pendulum of length l makes an angle of x (radians) with the vertical (see Figure 11.71). When x is small, it can be shown that, approximately:

$$\frac{d^2x}{dt^2} = -\frac{g}{l}x,$$

where g is the acceleration due to gravity.

Figure 11.71

(a) Solve this equation assuming that $x(0) = 0$ and $x'(0) = v_0$.
(b) Solve this equation assuming that the pendulum is let go from the position where $x = x_0$. ("Let go" means that the velocity of the pendulum is zero when $x = x_0$. Measure t from the moment when the pendulum is let go.)

23. Look at the pendulum motion in Problem 22. What effect does it have on x as a function of time if:

(a) x_0 is increased?   (b) l is increased?

24. (a) Expand $A\sin(\omega t + \phi)$ using the trigonometric identity $\sin(x + y) = \sin x \cos y + \cos x \sin y$.
(b) Assume $A > 0$. If $A\sin(\omega t + \phi) = C_1 \cos \omega t + C_2 \sin \omega t$, show that we must have

$$A = \sqrt{C_1^2 + C_2^2} \quad \text{and} \quad \tan \phi = C_1/C_2.$$

Problems 25–27 concern the electric circuit in Figure 11.72. A charged capacitor connected to an inductor causes a current to flow through the inductor until the capacitor is fully discharged. The current in the inductor, in turn, charges up the capacitor until the capacitor is fully charged again. If $Q(t)$ is the charge on the capacitor at time t, and I is the current, then

$$I = \frac{dQ}{dt}.$$

If the circuit resistance is zero, then the charge Q and the current I in the circuit satisfy the differential equation

$$L\frac{dI}{dt} + \frac{Q}{C} = 0,$$

where C is the capacitance, and L is the inductance, so

$$L\frac{d^2Q}{dt^2} + \frac{Q}{C} = 0.$$

The unit of charge is the coulomb, the unit of capacitance the farad, the unit of inductance the henry, the unit of current is the ampere, and time is measured in seconds.

Capacitor

C

L

Inductor

Figure 11.72

25. If $L = 36$ henry and $C = 9$ farad, find a formula for $Q(t)$ if

(a) $Q(0) = 0$ $I(0) = 2$
(b) $Q(0) = 6$ $I(0) = 0$

26. Suppose $Q(0) = 0$, $Q'(0) = I(0) = 4$, and the maximum possible charge is $2\sqrt{2}$ coulombs. What is the capacitance if the inductance is 10 henry?

27. What happens to the charge and current as t goes to infinity? What does it mean that the charge and current are sometimes positive and sometimes negative?

11.11   LINEAR SECOND-ORDER DIFFERENTIAL EQUATIONS

A Spring with Friction: Damped Oscillations

The differential equation $d^2s/dt^2 = -(k/m)s$, which we used to describe the motion of a spring, disregards friction. But there is friction in every real system. For a mass on a spring, the frictional force from air resistance increases with the velocity of the mass. The frictional force is often approximately proportional to velocity, and so we introduce a *damping term* of the form $a(ds/dt)$, where a is a constant called the *damping coefficient* and ds/dt is the velocity of the mass.

Remember that without damping, the differential equation was obtained from

$$\text{Force} = \text{Mass} \cdot \text{Acceleration}.$$

With damping, the spring force $-ks$ is replaced by $-ks - a(ds/dt)$, where a is positive and the $a(ds/dt)$ term is subtracted because the frictional force is in the direction opposite to the motion. The new differential equation is therefore

$$-ks - a\frac{ds}{dt} = m\frac{d^2 s}{dt^2}$$

which is equivalent to the following differential equation:

Equation for Damped Oscillations of a Spring

$$\frac{d^2 s}{dt^2} + \frac{a}{m}\frac{ds}{dt} + \frac{k}{m}s = 0$$

We expect the solution to this equation to die away with time, as friction brings the motion to a stop.

The General Solution to a Linear Differential Equation

The equation for damped oscillations is an example of a *linear second-order differential equation with constant coefficients*. This section gives an analytic method of solving the equation,

$$\frac{d^2 y}{dt^2} + b\frac{dy}{dt} + cy = 0,$$

for constant b and c. As we have seen for the spring equation, if $f_1(t)$ and $f_2(t)$ satisfy the differential equation, then the *principle of superposition* says that, for any constants C_1 and C_2, the function

$$y(t) = C_1 f_1(t) + C_2 f_2(t)$$

is also a solution. It can be shown that the general solution is of this form, provided $f_1(t)$ is not a multiple of $f_2(t)$.

Finding Solutions: The Characteristic Equation

We now use complex numbers to solve the differential equation

$$\frac{d^2 y}{dt^2} + b\frac{dy}{dt} + cy = 0.$$

The method is a form of guess-and-check. We ask what kind of function might satisfy a differential equation in which the second derivative $d^2 y/dt^2$ is a sum of multiples of dy/dt and y. One possibility is an exponential function, so we try to find a solution of the form:

$$y = Ce^{rt},$$

where r may be a complex number.[22] To find r, we substitute into the differential equation:

$$\frac{d^2 y}{dt^2} + b\frac{dy}{dt} + cy = r^2 Ce^{rt} + b \cdot rCe^{rt} + c \cdot Ce^{rt} = Ce^{rt}(r^2 + br + c) = 0.$$

---

[22]See Appendix B on complex numbers.

We can divide by $y = Ce^{rt}$ provided $C \neq 0$, because the exponential function is never zero. If $C = 0$, then $y = 0$, which is not a very interesting solution (though it is a solution). So we assume $C \neq 0$. Then $y = Ce^{rt}$ is a solution to the differential equation if

$$r^2 + br + c = 0.$$

This quadratic is called the *characteristic equation* of the differential equation. Its solutions are

$$r = -\frac{1}{2}b \pm \frac{1}{2}\sqrt{b^2 - 4c}.$$

There are three different types of solutions to the differential equation, depending on whether the solutions to the characteristic equation are real and distinct, complex, or repeated. The sign of $b^2 - 4c$ determines the type of solutions.

The Case with $b^2 - 4c > 0$

There are two real solutions r_1 and r_2 to the characteristic equation, and the following two functions satisfy the differential equation:

$$C_1 e^{r_1 t} \quad \text{and} \quad C_2 e^{r_2 t}.$$

The sum of these two solutions is the general solution to the differential equation:

If $b^2 - 4c > 0$, the general solution to

$$\frac{d^2 y}{dt^2} + b\frac{dy}{dt} + cy = 0$$

is

$$y(t) = C_1 e^{r_1 t} + C_2 e^{r_2 t}$$

where r_1 and r_2 are the solutions to the characteristic equation.
If $r_1 < 0$ and $r_2 < 0$, the motion is called **overdamped**.

A physical system satisfying a differential equation of this type is said to be overdamped because it occurs when there is a lot of friction in the system. For example, a spring moving in a thick fluid such as oil or molasses is overdamped: it will not oscillate.

Example 1   A spring is placed in oil, where it satisfies the differential equation

$$\frac{d^2 s}{dt^2} + 3\frac{ds}{dt} + 2s = 0.$$

Solve this equation with the initial conditions $s = -0.5$ and $ds/dt = 3$ when $t = 0$.

Solution   The characteristic equation is

$$r^2 + 3r + 2 = 0,$$

with solutions $r = -1$ and $r = -2$, so the general solution to the differential equation is

$$s(t) = C_1 e^{-t} + C_2 e^{-2t}.$$

We use the initial conditions to find C_1 and C_2. At $t = 0$, we have

$$s = C_1 e^{-0} + C_2 e^{-2(0)} = C_1 + C_2 = -0.5.$$

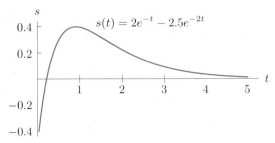

Figure 11.73: Solution to overdamped equation
$$\frac{d^2s}{dt^2} + 3\frac{ds}{dt} + 2s = 0$$

Furthermore, since $ds/dt = -C_1e^{-t} - 2C_2e^{-2t}$, we have

$$\frac{ds}{dt}\bigg|_{t=0} = -C_1e^{-0} - 2C_2e^{-2(0)} = -C_1 - 2C_2 = 3.$$

Solving these equations simultaneously, we find $C_1 = 2$ and $C_2 = -2.5$, so that the solution is

$$s(t) = 2e^{-t} - 2.5e^{-2t}.$$

The graph of this function is in Figure 11.73. The mass is so slowed by the oil that it passes through the equilibrium point only once (when $t \approx 1/4$) and for all practical purposes, it comes to rest after a short time. The motion has been "damped out" by the oil.

The Case with $b^2 - 4c = 0$

In this case, the characteristic equation has only one solution, $r = -b/2$. By substitution, we can check that both $y = e^{-bt/2}$ and $y = te^{-bt/2}$ are solutions.

If $b^2 - 4c = 0$,
$$\frac{d^2y}{dt^2} + b\frac{dy}{dt} + cy = 0$$

has general solution
$$y(t) = (C_1t + C_2)e^{-bt/2}.$$

If $b > 0$, the system is said to be **critically damped**.

The Case with $b^2 - 4c < 0$

In this case, the characteristic equation has complex roots. Using Euler's formula,[23] these complex roots lead to trigonometric functions which represent oscillations.

Example 2   An object of mass $m = 10$ kg is attached to a spring with spring constant $k = 20$ kg/sec^2, and the object experiences a frictional force proportional to the velocity, with constant of proportionality $a = 20$ kg/sec. At time $t = 0$, the object is released from rest 2 meters above the equilibrium position. Write the differential equation that describes the motion.

[23] See Appendix B on complex numbers.

Solution   The differential equation that describes the motion can be obtained from the following general expression for the damped motion of a spring:

$$m\underbrace{\frac{d^2 s}{dt^2}}_{\text{Mass·Acceleration}} = \underbrace{F}_{\text{Net force}} = -\underbrace{ks}_{\text{Spring force}} - \underbrace{a\frac{ds}{dt}}_{\text{Frictional force}}.$$

Substituting $m = 10$ kg, $k = 20$ kg/sec^2, and $a = 20$ kg/sec, we obtain the differential equation:

$$\frac{d^2 s}{dt^2} + 2\frac{ds}{dt} + 2s = 0.$$

At $t = 0$, the object is at rest 2 meters above equilibrium, so the initial conditions are $s(0) = 2$ and $s'(0) = 0$, where s is in meters and t in seconds.

Notice that this is the same differential equation as in Example 1 except that the coefficient of ds/dt has decreased from 3 to 2, which means that the frictional force has been reduced. This time, the roots of the characteristic equation have imaginary parts which lead to oscillations.

Example 3   Solve the differential equation

$$\frac{d^2 s}{dt^2} + 2\frac{ds}{dt} + 2s = 0,$$

subject to $s(0) = 2$, $s'(0) = 0$.

Solution   The characteristic equation is

$$r^2 + 2r + 2 = 0 \quad \text{giving} \quad r = -1 \pm i.$$

The solution to the differential equation is

$$s(t) = A_1 e^{(-1+i)t} + A_2 e^{(-1-i)t},$$

where A_1 and A_2 are arbitrary complex numbers. The initial condition $s(0) = 2$ gives

$$2 = A_1 e^{(-1+i)\cdot 0} + A_2 e^{(-1-i)\cdot 0} = A_1 + A_2.$$

Also,

$$s'(t) = A_1(-1 + i)e^{(-1+i)t} + A_2(-1 - i)e^{(-1-i)t},$$

so $s'(0) = 0$ gives

$$0 = A_1(-1 + i) + A_2(-1 - i).$$

Solving the simultaneous equations for A_1 and A_2 gives (after some algebra)

$$A_1 = 1 - i \quad \text{and} \quad A_2 = 1 + i.$$

The solution is therefore

$$s(t) = (1 - i)e^{(-1+i)t} + (1 + i)e^{(-1-i)t} = (1 - i)e^{-t}e^{it} + (1 + i)e^{-t}e^{-it}.$$

Using Euler's formula, $e^{it} = \cos t + i \sin t$ and $e^{-it} = \cos t - i \sin t$, we get

$$s(t) = (1 - i)e^{-t}(\cos t + i \sin t) + (1 + i)e^{-t}(\cos t - i \sin t).$$

Multiplying out and simplifying, all the complex terms drop out, giving

$$s(t) = e^{-t}\cos t + ie^{-t}\sin t - ie^{-t}\cos t + e^{-t}\sin t$$
$$+ e^{-t}\cos t - ie^{-t}\sin t + ie^{-t}\cos t + e^{-t}\sin t$$
$$= 2e^{-t}\cos t + 2e^{-t}\sin t.$$

The $\cos t$ and $\sin t$ terms tell us that the solution oscillates; the factor of e^{-t} tells us that the oscillations are damped. See Figure 11.74. However, the period of the oscillations does not change as the amplitude decreases. This is why a spring-driven clock can keep accurate time even as it is running down.

Figure 11.74: Solution to underdamped equation $\frac{d^2 s}{dt^2} + 2\frac{ds}{dt} + 2s = 0$

In Example 3, the coefficients A_1 and A_2 are complex, but the solution, $s(t)$, is real. (We expect this, since $s(t)$ represents a real displacement.) In general, provided the coefficients b and c in the original differential equation and the initial values are real, the solution is always real. The coefficients A_1 and A_2 are always complex conjugates (that is, of the form $\alpha \pm i\beta$).

If $b^2 - 4c < 0$, to solve

$$\frac{d^2 y}{dt^2} + b\frac{dy}{dt} + cy = 0,$$

- Find the solutions $r = \alpha \pm i\beta$ to the characteristic equation $r^2 + br + c = 0$.
- The general solution to the differential equation is, for some real C_1 and C_2,

$$y = C_1 e^{\alpha t}\cos\beta t + C_2 e^{\alpha t}\sin\beta t.$$

If $\alpha < 0$, such oscillations are called **underdamped**. If $\alpha = 0$, the oscillations are **undamped**.

Example 4   Find the general solution of the equations   (a) $y'' = 9y$ (b) $y'' = -9y$.

Solution   (a) The characteristic equation is $r^2 - 9 = 0$, so $r = \pm 3$. Thus the general solution is

$$y = C_1 e^{3t} + C_2 e^{-3t}.$$

(b) The characteristic equation is $r^2 + 9 = 0$, so $r = 0 \pm 3i$. The general solution is

$$y = C_1 e^{0t}\cos 3t + C_2 e^{0t}\sin 3t = C_1\cos 3t + C_2\sin 3t.$$

Notice that we have seen the solution to this equation in Section 11.10; it's the equation of undamped simple harmonic motion.

Example 5   Solve the initial value problem

$$y'' + 4y' + 13y = 0, \quad y(0) = 0, \ y'(0) = 30.$$

Solution   We solve the characteristic equation

$$r^2 + 4r + 13 = 0, \quad \text{getting} \quad r = -2 \pm 3i.$$

The general solution to the differential equation is

$$y = C_1 e^{-2t} \cos 3t + C_2 e^{-2t} \sin 3t.$$

Substituting $t = 0$ gives

$$y(0) = C_1 \cdot 1 \cdot 1 + C_2 \cdot 1 \cdot 0 = 0, \quad \text{so} \quad C_1 = 0.$$

Differentiating $y(t) = C_2 e^{-2t} \sin 3t$ gives

$$y'(t) = C_2(-2e^{-2t} \sin 3t + 3e^{-2t} \cos 3t).$$

Substituting $t = 0$ gives

$$y'(0) = C_2(-2 \cdot 1 \cdot 0 + 3 \cdot 1 \cdot 1) = 3C_2 = 30 \quad \text{so} \quad C_2 = 10.$$

The solution is therefore

$$y(t) = 10e^{-2t} \sin 3t.$$

Summary of Solutions to $y'' + by' + cy = 0$

If $b^2 - 4c > 0$, then $y = C_1 e^{r_1 t} + C_2 e^{r_2 t}$

If $b^2 - 4c = 0$, then $y = (C_1 t + C_2)e^{-bt/2}$

If $b^2 - 4c < 0$, then $y = C_1 e^{\alpha t} \cos \beta t + C_2 e^{\alpha t} \sin \beta t$

Exercises and Problems for Section 11.11

Exercises

For Exercises 1–12, find the general solution to the given differential equation.

1. $y'' + 4y' + 3y = 0$

2. $y'' + 4y' + 4y = 0$

3. $y'' + 4y' + 5y = 0$

4. $s'' - 7s = 0$

5. $s'' + 7s = 0$

6. $y'' - 3y' + 2y = 0$

7. $4z'' + 8z' + 3z = 0$

8. $\dfrac{d^2 x}{dt^2} + 4\dfrac{dx}{dt} + 8x = 0$

9. $\dfrac{d^2 p}{dt^2} + \dfrac{dp}{dt} + p = 0$

10. $z'' + 2z = 0$

11. $z'' + 2z' = 0$

12. $P'' + 2P' + P = 0$

For Exercises 13–20, solve the initial value problem.

13. $y'' + 5y' + 6y = 0, \qquad y(0) = 1, y'(0) = 0.$

14. $y'' + 5y' + 6y = 0, \qquad y(0) = 5, y'(0) = 1.$

15. $y'' - 3y' - 4y = 0, \qquad y(0) = 1, y'(0) = 0.$

16. $y'' - 3y' - 4y = 0, \qquad y(0) = 0, y'(0) = 0.5.$

17. $y'' + 6y' + 5y = 0, \qquad y(0) = 1, \quad y'(0) = 0$

18. $y'' + 6y' + 5y = 0, \qquad y(0) = 5, \quad y'(0) = 5$

19. $y'' + 6y' + 10y = 0, \qquad y(0) = 0, \quad y'(0) = 2$

20. $y'' + 6y' + 10y = 0, \qquad y(0) = 0, \quad y'(0) = 0$

For Exercises 21–24, solve the boundary value problem.

21. $y'' + 5y' + 6y = 0, \qquad y(0) = 1, y(1) = 0.$

22. $y'' + 5y' + 6y = 0, \qquad y(-2) = 0, y(2) = 3.$

23. $p'' + 2p' + 2p = 0, \qquad p(0) = 0, \quad p(\pi/2) = 20$

24. $p'' + 4p' + 5p = 0, \qquad p(0) = 1, \quad p(\pi/2) = 5$

Problems

25. Match the graphs of solutions in Figure 11.75 with the differential equations below.

(a) $x'' + 4x = 0$
(b) $x'' - 4x = 0$
(c) $x'' - 0.2x' + 1.01x = 0$
(d) $x'' + 0.2x' + 1.01x = 0$

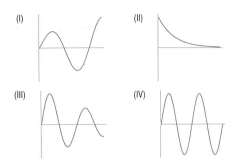

Figure 11.75

26. Match the differential equations to the solution graphs (I)–(IV). Use each graph only once.

(a) $y'' + 5y' + 6y = 0$ (b) $y'' + y' - 6y = 0$
(c) $y'' + 4y' + 9y = 0$ (d) $y'' = -9y$

27. If $y = e^{2t}$ is a solution to the differential equation

$$\frac{d^2y}{dt^2} - 5\frac{dy}{dt} + ky = 0,$$

find the value of the constant k and the general solution to this equation.

28. Assuming b, $c > 0$, explain how you know that the solutions of an underdamped differential equation must go to 0 as $t \to \infty$.

For each of the differential equations in Problems 29–30, find the values of b that make the general solution:

(a) overdamped,   (b) underdamped,
(c) critically damped.

29. $s'' + bs' + 5s = 0$ **30.** $s'' + bs' - 16s = 0$

Each of the differential equations (i)–(iv) represents the position of a 1 gram mass oscillating on the end of a damped spring. For Problems 31–35, pick the differential equation representing the system which answers the question.

(i) $s'' + s' + 4s = 0$ (ii) $s'' + 2s' + 5s = 0$
(iii) $s'' + 3s' + 3s = 0$ (iv) $s'' + 0.5s' + 2s = 0$

31. Which spring has the largest coefficient of damping?

32. Which spring exerts the smallest restoring force for a given displacement?

33. In which system does the mass experience the frictional force of smallest magnitude for a given velocity?

34. Which oscillation has the longest period?

35. Which spring is the stiffest? [Hint: You need to determine what it means for a spring to be stiff. Think of an industrial strength spring and a slinky.]

36. Find a solution to the following equation which satisfies $z(0) = 3$ and does not tend to infinity as $t \to \infty$:

$$\frac{d^2z}{dt^2} + \frac{dz}{dt} - 2z = 0.$$

37. Consider an overdamped differential equation with b, $c > 0$.

(a) Show that both roots of the characteristic equation are negative.

(b) Show that any solution to the differential equation goes to 0 as $t \to \infty$.

38. Could the graph in Figure 11.76 show the position of a mass oscillating at the end of an overdamped spring? Why or why not?

Figure 11.76

39. Consider the system of differential equations

$$\frac{dx}{dt} = -y \qquad \frac{dy}{dt} = -x.$$

(a) Convert this system to a second order differential equation in y by differentiating the second equation with respect to t and substituting for x from the first equation.

(b) Solve the equation you obtained for y as a function of t; hence find x as a function of t.

Recall the discussion of electric circuits on page 587. Just as a spring can have a damping force which affects its motion, so can a circuit. Problems 40–43 involve a damping force caused by the resistor in Figure 11.77. The charge Q on a capacitor in a circuit with inductance L, capacitance C, and resistance R, in ohms, satisfies the differential equation

$$L\frac{d^2Q}{dt^2} + R\frac{dQ}{dt} + \frac{1}{C}Q = 0.$$

Capacitor

Figure 11.77

40. If $L = 1$ henry, $R = 2$ ohms, and $C = 4$ farads, find a formula for the charge when

(a) $Q(0) = 0, Q'(0) = 2$.
(b) $Q(0) = 2, Q'(0) = 0$.

41. If $L = 1$ henry, $R = 1$ ohm, and $C = 4$ farads, find a formula for the charge when

(a) $Q(0) = 0, Q'(0) = 2$.
(b) $Q(0) = 2, Q'(0) = 0$.
(c) How did reducing the resistance affect the charge? Compare with your solution to Problem 40.

42. If $L = 8$ henry, $R = 2$ ohm, and $C = 4$ farads, find a formula for the charge when

(a) $Q(0) = 0, Q'(0) = 2$.
(b) $Q(0) = 2, Q'(0) = 0$.
(c) How did increasing the inductance affect the charge? Compare with your solution to Problem 40.

43. Given any positive values for R, L and C, what happens to the charge as t goes to infinity?

44. (a) If r_1 and r_2 satisfy $r^2 + br + c = 0$ show that

$$y = \frac{e^{r_1 t} - e^{r_2 t}}{r_1 - r_2}$$

is a solution to $y'' + by' + cy = 0$.
(b) If $r_1 = r_2 + h$, show that the solution in part (a) can be written

$$y = e^{r_2 t}\frac{(e^{ht} - 1)}{h}.$$

(c) Using the Taylor series, show that

$$\frac{(e^{ht} - 1)}{h} = t + \frac{ht^2}{2!} + \frac{h^2 t^3}{3!} + \cdots.$$

(d) Use the result of part (c) in the solution from part (b) to show that $\lim\limits_{h \to 0} e^{r_2 t}\frac{(e^{ht} - 1)}{h} = te^{r_2 t}$.
(e) If there is a double root $r_2 = r_1 = -b/2$. By direct substitution, show that $y = te^{-bt/2}$ satisfies $y'' + by' + cy = 0$.

CHAPTER SUMMARY (see also Ready Reference at the end of the book)

- **Differential equations terminology**
  Order, initial conditions, families of solutions, stable/unstable equilibrium solutions.
- **Solving first-order differential equations**
  Slope fields (graphical), Euler's method (numerical), separation of variables (analytical).
- **Modeling with differential equations**
  Growth and decay, Newton's law of heating and cooling, compartment models, logistic model.
- **Second-order differential equations**
  Initial and boundary conditions, oscillations (with and without damping), characteristic equation, solutions using complex numbers.
- **Systems of differential equations**
  S-I-R model, predator-prey model, phase plane.

REVIEW EXERCISES AND PROBLEMS FOR CHAPTER ELEVEN

Exercises

Find a general solution to the differential equations in Exercises 1–6.

1. $\dfrac{dP}{dt} = t$

2. $\dfrac{dy}{dx} = 0.2y - 8$

3. $\dfrac{dP}{dt} = 10 - 2P$

4. $\dfrac{dH}{dt} = 0.5H + 10$

5. $\dfrac{dR}{dt} = 2R - 6R^2$

6. $\dfrac{dP}{dt} = 0.4P(1 - 0.01P)$

For the differential equations in Exercises 7–22, find a solution which passes through the given point.

7. $\dfrac{dy}{dx} + xy^2 = 0$, $\quad y(1) = 1$

8. $\dfrac{dP}{dt} = 0.03P + 400$, $\quad P(0) = 0$

9. $1 + y^2 - \dfrac{dy}{dx} = 0$, $\quad y(0) = 0$

10. $2\sin x - y^2\dfrac{dy}{dx} = 0$, $\quad y(0) = 3$

11. $\dfrac{dk}{dt} = (1 + \ln t)k$, $\quad k(1) = 1$

12. $\dfrac{dy}{dx} = \dfrac{y(3 - x)}{x(0.5y - 4)}$, $\quad y(1) = 5$

13. $\dfrac{dy}{dx} = \dfrac{0.2y(18 + 0.1x)}{x(100 + 0.5y)}$, $\quad (10, 10)$

14. $\dfrac{dz}{dt} = z(z - 1)$, $\quad z(0) = 10$

15. $\dfrac{dy}{dt} = y(10 - y)$, $\quad y(0) = 1$

16. $\dfrac{dy}{dx} = \dfrac{y(100 - x)}{x(20 - y)}$, $\quad (1, 20)$

17. $\dfrac{df}{dx} = \sqrt{xf(x)}$, $\quad f(1) = 1$

18. $\dfrac{dy}{dx} = e^{x-y}$, $\quad y(0) = 1$

19. $\dfrac{dy}{dx} = e^{x+y}$, $\quad y = 0$ where $x = 1$

20. $e^{-\cos\theta}\dfrac{dz}{d\theta} = \sqrt{1 - z^2}\sin\theta$, $\quad z(0) = \frac{1}{2}$

21. $(1 + t^2)y\dfrac{dy}{dt} = 1 - y$, $\quad y(1) = 0$

22. $\dfrac{dy}{dt} = 2^y \sin^3 t$, $\quad y(0) = 0$

Find a general solution to the differential equations in Exercises 23–28.

23. $\dfrac{d^2z}{dt^2} + \pi^2 z = 0$

24. $9z'' - z = 0$

25. $9z'' + z = 0$

26. $y'' + 6y' + 8y = 0$

27. $y'' + 2y' + 3y = 0$

28. $x'' + 2x' + 10x = 0$

Problems

29. Match each of the slope field segments in (I)–(VI) with one or more of the differential equations in (a)–(f).

(a) $y' = e^{-x^2}$ **(b)** $y' = \cos y$ **(c)** $y' = \cos(4-y)$
(d) $y' = y(4-y)$ **(e)** $y' = y(3-y)$ **(f)** $y' = x(3-x)$

30. **(a)** Sketch the slope field for the differential equation $dy/dx = y/x$.
(b) Sketch several solution curves.
(c) Solve the differential equation analytically and compare the result with the sketches in part (b).

31. **(a)** What are the equilibrium solutions for the differential equation

$$\frac{dy}{dt} = 0.2(y - 3)(y + 2)?$$

(b) Use a slope field to determine whether each equilibrium solution is stable or unstable.

32. Consider the initial value problem

$$y' = 5 - y, \quad y(0) = 1.$$

(a) Use Euler's method with five steps to estimate $y(1)$.
(b) Sketch the slope field for this differential equation in the first quadrant, and use it to decide if your estimate is an over or underestimate.
(c) Find the exact solution to the differential equation and hence find $y(1)$ exactly.
(d) Without doing the calculation, roughly what would you expect the approximation for $y(1)$ to be with ten steps?

33. **(a)** If $B = f(t)$ is the balance at time t of a bank account that earns interest at a rate of $r\%$, compounded continuously, what is the differential equation describing the rate at which the balance changes? What is the constant of proportionality, in terms of r?

(b) What is the solution to this differential equation?

(c) Sketch the graph of $B = f(t)$ for an account that starts with \$1000 and earns interest at the following rates:

(i) 4%   (ii) 10%   (iii) 15%

34. The amount of radioactive carbon-14 in a sample is measured using a Geiger counter, which records each disintegration of an atom. Living tissue disintegrates at a rate of about 13.5 atoms per minute per gram of carbon. In 1977 a charcoal fragment found at Stonehenge, England, recorded 8.2 disintegrations per minute per gram of carbon. Assuming that the half-life of carbon-14 is 5730 years and that the charcoal was formed during the building of the site, estimate the date at which Stonehenge was built.

35. A spherical snowball melts at a rate proportional to its surface area.

(a) Write a differential equation for its volume, V.

(b) If the initial volume is V_0, solve the differential equation and graph the solution.

(c) When does the snowball disappear?

36. A chemical reaction involves one molecule of a substance A combining with one molecule of substance B to form one molecule of substance C, written $A + B \rightarrow C$. The Law of Mass Action states that the rate at which C is formed is proportional to the product of the quantities of A and B present. Assume a and b are the initial quantities of A and B, and x is the quantity of C present at time t.

(a) Write a differential equation for x.

(b) Solve the equation with $x(0) = 0$.

37. If the initial quantities, a and b, in Problem 36 are the same, write and solve a differential equation for x, with $x(0) = 0$.

For each of the differential equations in Problems 38–40, find the values of c that make the general solution:

(a) overdamped, **(b)** underdamped, **(c)** critically damped.

38. $s'' + 4s' + cs = 0$ **39.** $s'' + 2\sqrt{2}s' + cs = 0$

40. $s'' + 6s' + cs = 0$

41. Many organ pipes in old European churches are made of tin. In cold climates such pipes can be affected with *tin pest*, when the tin becomes brittle and crumbles into a gray powder. This transformation can appear to take place very suddenly because the presence of the gray powder encourages the reaction to proceed. At the start, when there is little gray powder, the reaction proceeds slowly. Similarly, toward the end, when there is little metallic tin left, the reaction is also slow. In between,

however, when there is plenty of both metallic tin and powder, the reaction can be alarmingly fast.

Suppose that the rate of the reaction is proportional to the product of the amount of tin left and the quantity of gray powder, p, present at time t. Assume also that when metallic tin is converted to gray powder, its mass does not change.

(a) Write a differential equation for p. Let the total quantity of metallic tin present originally be B.

(b) Sketch a graph of the solution $p = f(t)$ if there is a small quantity of powder initially. How much metallic tin has crumbled when it is crumbling fastest?

(c) Suppose there is no gray powder initially. (For example, suppose the tin is completely new.) What does this model predict will happen? How do you reconcile this with the fact that many organ pipes do get tin pest?

42. The spread of a non-fatal disease through a population of fixed size M can be modeled as follows. The rate that healthy people are infected, in people per day, is proportional to the product of the numbers of healthy and infected people. The constant of proportionality is $0.01/M$. The rate of recovery, in people per day, is 0.009 times the number of people infected. Construct a differential equation which models the spread of the disease. Assuming that initially only a small number of people are infected, plot a graph of the number of infected people against time. What fraction of the population is infected in the long run?

43. Juliet is in love with Romeo, who happens (in our version of this story) to be a fickle lover. The more Juliet loves him, the more he begins to dislike her. When she hates him, his feelings for her warm up. On the other hand, her love for him grows when he loves her and withers when he hates her. A model for their ill-fated romance is

$$\frac{dj}{dt} = Ar, \qquad \frac{dr}{dt} = -Bj,$$

where A and B are positive constants, $r(t)$ represents Romeo's love for Juliet at time t, and $j(t)$ represents Juliet's love for Romeo at time t. (Negative love is hate.)

(a) The constant on the right-hand side of Juliet's equation (the one including dj/dt) has a positive sign, whereas the constant in Romeo's equation is negative. Explain why these signs follow from the story.

(b) Derive a second-order differential equation for $r(t)$ and solve it. (Your equation should involve r and its derivatives, but not j and its derivatives.)

(c) Express $r(t)$ and $j(t)$ as functions of t, given $r(0) = 1$ and $j(0) = 0$. Your answer will contain A and B.

(d) As you may have discovered, the outcome of the relationship is a never-ending cycle of love and hate. Find what fraction of the time they both love one another.

44. The behavior of an electron is described by Schrödinger's equation. Under certain conditions, this equation becomes

$$\frac{-h^2}{8\pi^2 m}\frac{d^2\Psi}{dx^2} = E\Psi,$$

where Ψ, called a *wave function*, describes the electron's behavior as a function of x, the distance from a fixed point; the constants h, m, E are positive. In addition, Ψ satisfies the boundary conditions

$$\Psi(0) = \Psi(l) = 0$$

where l is another positive constant.

(a) Solutions to Schrödinger's equation are of the form $\Psi = C_1\cos(\omega x) + C_2\sin(\omega x)$ for some ω. Show

that if Ψ satisfies the boundary conditions, then Ψ must be of the form

$$\Psi = C_2\sin\left(\frac{n\pi x}{l}\right) \quad \text{for some positive integer } n.$$

(b) Using part (a), find E in terms of other constants.

(c) The constant E represents the energy level of the electron. Assuming that only positive integral values of n are possible, explain why not all values of E are possible. (This is described by saying that energy is quantized.) In addition, show that the possible energy levels are one of the following multiples of the lowest energy level: $1, 4, 9, 16\ldots$.

CAS Challenge Problems

45. Consider the differential equation

$$\frac{dP}{dt} = P(P-1)(2-P).$$

(a) Find all the equilibrium solutions.

(b) Show that the following two functions are solutions:

$$P_1(t) = 1 - \frac{e^t}{\sqrt{3 + e^{2t}}} \quad \text{and} \quad P_2(t) = 1 + \frac{e^t}{\sqrt{3 + e^{2t}}}.$$

[Hint: Use the computer algebra system to simplify the difference between the right and left-hand sides.]

(c) Find $P_1(0)$, $P_2(0)$, $\lim_{t\to\infty} P_1(t)$, and $\lim_{t\to\infty} P_2(t)$. Explain how you could have predicted the limits as $t \to \infty$ from the values at $t = 0$ without knowing the solutions explicitly.

46. In this problem we investigate *Picard's method* for approximating the solutions of differential equations. Consider the differential equation

$$y'(t) = y(t)^2 + t^2, \quad y(a) = b.$$

Integrating both sides with respect to t gives

$$y(s) - y(a) = \int_a^s y'(t)\,dt = \int_a^s (y(t)^2 + t^2)\,dt.$$

Since $y(a) = b$, we have

$$y(s) = b + \int_a^s (y(t)^2 + t^2)\,dt.$$

We have put the differential equation into the form of an *integral equation*. If we have an approximate solution $y_0(s)$, we can use the integral form to make a new approximation

$$y_1(s) = b + \int_a^s (y_0(t)^2 + t^2)\,dt.$$

Continuing this process, we get a sequence of approximations $y_0(s), y_1(s), \ldots, y_n(s), \ldots$ where each term in the sequence is defined in terms of the previous one by the equation

$$y_{n+1}(s) = b + \int_a^s (y_n(t)^2 + t^2)\,dt.$$

(a) Show that y_n satisfies the initial condition $y_n(a) = b$ for all n.

(b) Using the initial condition $y(1) = 0$, start with the approximation $y_0(s) = 0$ and use a computer algebra system to find the next three approximations y_1, y_2, and y_3.

(c) Use a computer algebra system to find the solution y satisfying $y(1) = 0$, and sketch y, y_1, y_2, y_3 on the same axes. On what domain do the approximations appear to be accurate? [The solution y cannot be expressed in terms of elementary functions. If your computer algebra system cannot solve the equation exactly, use a numerical method such as Euler's method.]

47. Figure 11.78 shows the slope field for the differential equation[24]

$$\frac{dy}{dx} = \sqrt{1 - y^2}.$$

Figure 11.78

(a) Sketch the solution with $y(0) = 0$ for $-3 \le x \le 3$.

(b) Use a computer algebra system to find the solution in part (a).

(c) Compare the computer algebra system's solution with your sketch. Do they agree? If not, which one is right?

[24]Adapted from David Lomen & David Lovelock, *Differential Equations* (New York: John Wiley & Sons, Inc., 1999).

CHECK YOUR UNDERSTANDING

In Problems 1–16, decide whether the statements are true or false. Give an explanation for your answer.

1. For all constants k, the equation $y' + ky = 0$ has exponential functions as solutions.

2. For all constants k, the equation $y'' + ky = 0$ has trigonometric functions as solutions.

3. Polynomials are never solutions to differential equations.

4. All solutions to the differential equation whose slope field is in Figure 11.79 have $\lim_{x \to \infty} y = \infty$.

Figure 11.79

5. All solutions to the differential equation whose slope field is in Figure 11.80 have $\lim_{x \to \infty} y = 0$.

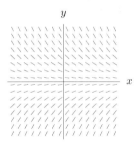

Figure 11.80

6. All solutions to the differential equation whose slope field is in Figure 11.81 have the same limiting value as $x \to \infty$.

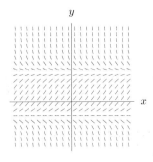

Figure 11.81

7. Euler's method gives the arc length of a solution curve.

8. The differential equation $dy/dx = x + y$ can be solved by separation of variables.

9. The differential equation $dy/dx - xy = x$ can be solved by separation of variables.

10. The only solution to the differential equation $dy/dx = 3y^{2/3}$ passing through the point $(0, 0)$ is $y = x^3$.

11. The solutions of the differential equation $dy/dx = x^2 + y^2 + 1$ are increasing at every point.

12. The solutions of the differential equation $dy/dx = x^2 + y^2 + 1$ are concave up at every point.

13. There is a solution curve for the logistic differential equation $dP/dt = P(2 - P)$ which goes through the points $(0, 1)$ and $(1, 3)$.

14. For any positive values of the constant k and any positive values of the initial value $P(0)$, the solution to the differential equation $dP/dt = kP(L - P)$ has limiting value L as $t \to \infty$.

15. The system of differential equations $dx/dt = -x + xy^2$ and $dy/dt = y - x^2 y$ requires initial conditions for both $x(0)$ and $y(0)$ to determine a unique solution.

16. Populations modeled by a system of differential equations never die out.

Suppose that $y = f(x)$ is a solution of the differential equation $dy/dx = g(x)$. Are the statements in Problems 17–24 true or false? If a statement is true, explain how you know. If a statement is false, give a counterexample.

17. If $g(x)$ is increasing for all x, then the graph of f is concave up for all x.

18. If $g(x)$ is increasing for $x > 0$, then so is $f(x)$.

19. If $g(0) = 1$ and $g(x)$ is increasing for $x \geq 0$, then $f(x)$ is also increasing for $x \geq 0$.

20. If $g(x)$ is periodic, then $f(x)$ is also periodic.

21. If $\lim_{x \to \infty} g(x) = 0$, then $\lim_{x \to \infty} f(x) = 0$.

22. If $\lim_{x \to \infty} g(x) = \infty$, then $\lim_{x \to \infty} f(x) = \infty$.

23. If $g(x)$ is even, then so is $f(x)$.

24. If $g(x)$ is even, then $f(x)$ is odd.

Suppose that $y = f(x)$ is a solution of the differential equation $dy/dx = 2x - y$. In Problems 25–33, decide whether the statements are true or false. Justify your answer.

25. If $f(a) = b$, the slope of the graph of f at (a, b) is $2a - b$.

26. $f'(x) = 2x - f(x)$.

27. There could be more than one value of x such that $f'(x) = 1$ and $f(x) = 5$.

28. If $y = f(x)$, then $d^2y/dx^2 = 2 - (2x - y)$.

29. If $f(1) = 5$, then $(1, 5)$ could be a critical point of f.

30. The graph of f is decreasing whenever it lies above the line $y = 2x$ and is increasing whenever it lies below the line $y = 2x$.

31. All the inflection points of f lie on the line $y = 2x - 2$.

32. If $g(x)$ is another solution to the differential equation $dy/dx = 2x - y$, then $g(x) = f(x) + C$.

33. If $g(x)$ is a different solution to the differential equation $dy/dx = 2x - y$, then $\lim_{x \to \infty}(g(x) - f(x)) = 0$. [Hint: Show that $w = g(x) - f(x)$ satisfies the differential equation $dw/dx = -w$.]

In Problems 34–37, give an example of a differential equation with the specified properties.

34. All of whose solutions are increasing and concave up.

35. All of whose solutions have their critical points on the parabola $y = x^2$.

36. All of whose solutions are the family of functions $f(x) = x^2 + C$.

37. All of whose solutions are the family of hyperbolas $x^2 - y^2 = C$.

PROJECTS FOR CHAPTER ELEVEN

The Spread of SARS

In the spring of 2003, SARS (Severe Acute Respiratory Syndrome) spread rapidly in several Asian countries and Canada. Predicting the course of the disease—how many people would be infected, how long it would last—was important to officials trying to minimize the impact of the disease.

1. SARS Predictions for Hong Kong

This project compares three predictions about the spread of SARS in Hong Kong. We measure time, t, in days since March 17, 2003, the date the World Health Organization (WHO) started to publish daily SARS reports.[25] Let P be the total number of cases reported in Hong Kong by day t. On March 17, Hong Kong reported 95 cases.

The constants in the three differential equations whose predictions are analyzed in this project were determined using WHO data available in March 2003. We compare predictions from the three models for June 12, 2003, the last day a new case was reported in Hong Kong.

(a) (i) *A Linear Model.* Suppose P satisfies

$$\frac{dP}{dt} = 30.2, \quad \text{with } P(0) = 95.$$

Solve the differential equation and use your solution to predict the number of cases of SARS in Hong Kong by June 12 ($t = 87$).

(ii) *An Exponential Model.* Suppose P satisfies

$$\frac{1}{P}\frac{dP}{dt} = 0.12, \quad \text{with } P(0) = 95.$$

Solve the differential equation and use your solution to predict the number of cases of SARS in Hong Kong by June 12 ($t = 87$).

(iii) *A Logistic Model.* Suppose P satisfies

$$\frac{1}{P}\frac{dP}{dt} = 0.19 - 0.0002P, \quad \text{with } P(0) = 95.$$

Solve the differential equation and use your solution to predict the number of cases of SARS in Hong Kong by June 12 ($t = 87$).

(b) (i) Comment on the June 12 predictions from the three models.

(ii) What do each of the three models in part (a) predict about the trend in the number of new cases each day?

---

[25] www.who.int/csr/country/en, accessed July 13, 2003

(iii) In May 2003, *The Lancet* published[26] a graph like Figure 11.82. What trend do you see in the data? What does this trend suggest about which model fits the data best?

(c) Assume that the data follow a logistic model, which is the only one in part (a) to predict that the total number of cases will level off:

   (i) Use your answer to part (a) (iii) to predict the maximum number of SARS cases.

   (ii) Assume that the data follow a logistic model, but not necessarily with the parameters given in part (a) (iii). Use Figure 11.82 and the fact that there had been a total of 998 cases of SARS by April 10 to predict the maximum number of SARS cases.

(d) To see how well the three models worked in practice, plot the data in Table 11.14 and each of the three solution curves from part (a).

Table 11.14 *Total number of SARS cases reported in Hong Kong by day t (where t = 0 is March 17, 2003)*

t	P	t	P	t	P	t	P
0	95	26	1108	54	1674	75	1739
5	222	33	1358	61	1710	81	1750
12	470	40	1527	68	1724	87	1755
19	800	47	1621				

Figure 11.82: Daily increase in Hong Kong SARS cases

2. A SIR Model for SARS

This project analyzes the spread of SARS through interaction between infected and susceptible people.

The variables are S, the number of susceptibles, I, the number of infecteds who can infect others, and R, the number removed (this group includes those in quarantine and those who die, as well as those who have recovered and acquired immunity). Time, t, is in days since March 17, 2003, the date the World Health Organization (WHO) started to publish daily SARS reports. On March 17, Hong Kong reported 95 cases. In this model

$$\frac{dS}{dt} = -aSI$$

$$\frac{dI}{dt} = aSI - bI,$$

and $S + I + R = 6.8$ million, the population of Hong Kong in 2003.[27] Estimates based on WHO data give $a = 1.25 \cdot 10^{-8}$.

(a) What are S_0 and I_0, the initial values of S and I?

(b) During March 2003, the value of b was about 0.06. Using a calculator or computer, sketch the slope field for this system of differential equations and the solution trajectory corresponding to the initial conditions. (Use $0 \leq S \leq 7 \cdot 10^6, 0 \leq I \leq 0.4 \cdot 10^6$.)

(c) What does your graph tell you about the total number of people infected over the course of the disease if $b = 0.06$? What is the threshold value? What does this value tell you?

(d) During April, as public health officials worked to get the disease under control, people who had been in contact with the disease were quarantined. Explain why quarantining has the effect of raising the value of b.

---

[26]SARS, Lay Epidemiology and Fear, by Razum, Beecher, Kapuan, Junghauss, *The Lancet*, May 2, 2003.

[27]www.census.gov, International Data Base (IDB), accessed June 8, 2004.

(e) Using the April value, $b = 0.24$, sketch the slope field. (Use the same value of a and the same window.)

(f) What is the threshold value for $b = 0.24$? What does this tell you? Comment on the quarantine policy.

(g) Comment on the effectiveness of each of the following policies intended to prevent an epidemic and protect a city from an outbreak of SARS in a nearby region.

    I Close off the city from contact with the infected region. Shut down roads, airports, trains, and other forms of direct contact.

    II Install a quarantine policy. Isolate anyone who has been in contact with a SARS patient or anyone who shows symptoms of SARS.

3. Pareto's Law

In analyzing a society, sociologists are often interested in how incomes are distributed through the society. Pareto's law asserts that each society has a constant k such that the average income of all people wealthier than you is k times your income ($k > 1$). If $p(x)$ is the number of people in the society with an income of x or above, we define $\Delta p = p(x + \Delta x) - p(x)$, for small Δx.

(a) Explain why the number of people with incomes between x and $x + \Delta x$ is represented by $-\Delta p$. Then show that the total amount of money earned by people with incomes between x and $x + \Delta x$ is approximated by $-x\Delta p$.

(b) Use Pareto's law to show that the total amount of money earned by people with incomes of x or above is $kxp(x)$. Then show that the total amount of money earned by people with incomes between x and $x + \Delta x$ is approximated by $-kp\Delta x - kx\Delta p$.

(c) Using your answers to parts (a) and (b), show that p satisfies the differential equation

$$(1 - k)xp' = kp.$$

(d) Solve this differential equation for $p(x)$.

(e) Sketch a graph of $p(x)$ for various values of k, some large and some near 1. How does the value of k alter the shape of the graph?

4. Vibrations in a Molecule

Suppose the force, F, between two atoms a distance r apart in a molecule is given by

$$F(r) = b\left(\frac{a^2}{r^3} - \frac{a}{r^2}\right)$$

where a and b are positive constants.

(a) Find the equilibrium distance, r, at which $F = 0$.

(b) Expand F in a series about the equilibrium position. Give the first two nonzero terms.

(c) Suppose one atom is fixed. Let x be the displacement of the other atom from the equilibrium position. Give the first two nonzero terms of the series for F in terms of x.

(d) Using Newton's Second Law, show that if x is small, the second atom oscillates about the equilibrium position. Find the period of the oscillation.

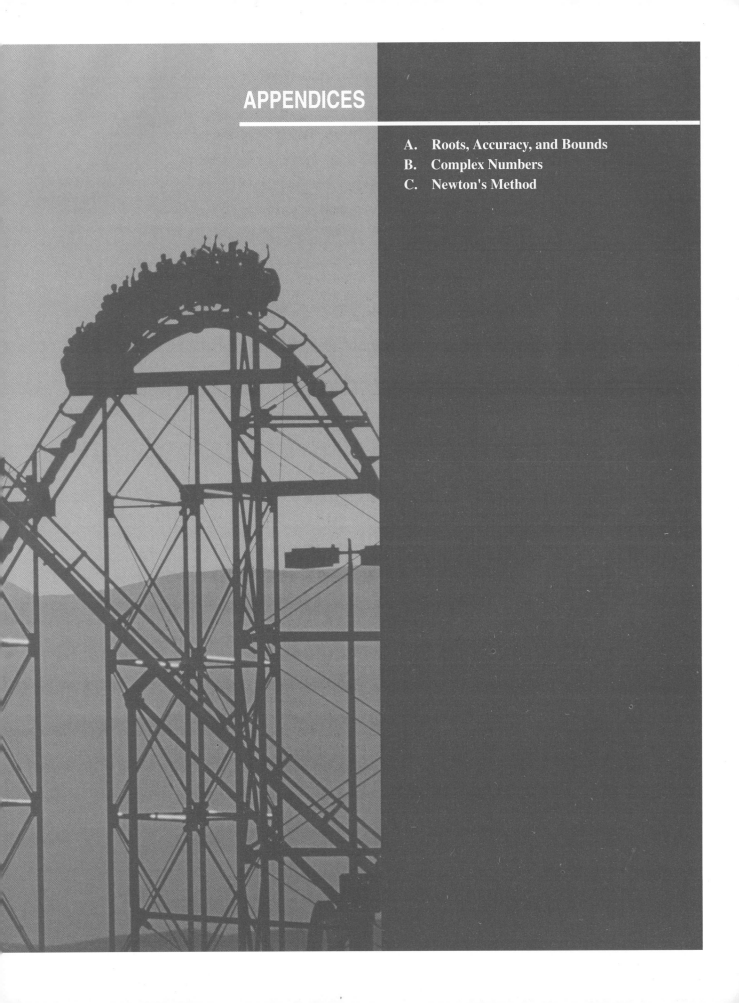

APPENDICES

A   ROOTS, ACCURACY, AND BOUNDS

It is often necessary to find the zeros of a polynomial or the points of intersection of two curves. So far, you have probably used algebraic methods, such as the quadratic formula, to solve such problems. Unfortunately, however, mathematicians' search for similar solutions to more complicated equations, has not been all that successful. The formulas for the solutions to third- and fourth-degree equations are so complicated that you'd never want to use them. Early in the nineteenth century, it was proved that there is no algebraic formula for the solutions to equations of degree 5 and higher. Most nonpolynomial equations cannot be solved using a formula either.

However, we can still find roots of equations, provided we use approximation methods, not formulas. In this section we will discuss three ways to find roots: algebraic, graphical, and numerical. Of these, only the algebraic method gives exact solutions.

First, let's get some terminology straight. Given the equation $x^2 = 4$, we call $x = -2$ and $x = 2$ the *roots*, or *solutions of the equation*. If we are given the function $f(x) = x^2 - 4$, then -2 and 2 are called the *zeros of the function*; that is, the zeros of the function f are the roots of the equation $f(x) = 0$.

The Algebraic Viewpoint: Roots by Factoring

If the product of two numbers is zero, then one or the other or both must be zero, that is, if $AB = 0$, then $A = 0$ or $B = 0$. This observation lies behind finding roots by factoring. You may have spent a lot of time factoring polynomials. Here you will also factor expressions involving trigonometric and exponential functions.

Example 1   Find the roots of $x^2 - 7x = 8$.

Solution   Rewrite the equation as $x^2 - 7x - 8 = 0$. Then factor the left side: $(x + 1)(x - 8) = 0$. By our observation about products, either $x + 1 = 0$ or $x - 8 = 0$, so the roots are $x = -1$ and $x = 8$.

Example 2   Find the roots of $\dfrac{1}{x} - \dfrac{x}{(x + 2)} = 0$.

Solution   Rewrite the left side with a common denominator:

$$\frac{x + 2 - x^2}{x(x + 2)} = 0.$$

Whenever a fraction is zero, the numerator must be zero. Therefore we must have

$$x + 2 - x^2 = (-1)(x^2 - x - 2) = (-1)(x - 2)(x + 1) = 0.$$

We conclude that $x - 2 = 0$ or $x + 1 = 0$, so 2 and -1 are the roots. They can be checked by substitution.

Example 3   Find the roots of $e^{-x} \sin x - e^{-x} \cos x = 0$.

Solution   Factor the left side: $e^{-x}(\sin x - \cos x) = 0$. The factor e^{-x} is never zero; it is impossible to raise e to a power and get zero. Therefore, the only possibility is that $\sin x - \cos x = 0$. This equation is equivalent to $\sin x = \cos x$. If we divide both sides by $\cos x$, we get

$$\frac{\sin x}{\cos x} = \frac{\cos x}{\cos x} \quad \text{so} \quad \tan x = 1.$$

The roots of this equation are

$$\dots, \frac{-7\pi}{4}, \frac{-3\pi}{4}, \frac{\pi}{4}, \frac{5\pi}{4}, \frac{9\pi}{4}, \frac{13\pi}{4}, \dots.$$

Warning: Using factoring to solve an equation only works when one side of the equation is 0. It is not true that if, say, $AB = 7$ then $A = 7$ or $B = 7$. For example, you *cannot* solve $x^2 - 4x = 2$ by factoring $x(x - 4) = 2$ and then assuming that either x or $x - 4$ equals 2.

The problem with factoring is that factors are not easy to find. For example, the left side of the quadratic equation $x^2 - 4x - 2 = 0$ does not factor, at least not into "nice" factors with integer coefficients. For the general quadratic equation

$$ax^2 + bx + c = 0,$$

there is the quadratic formula for the roots:

$$x = \frac{-b \pm \sqrt{b^2 - 4ac}}{2a}.$$

Thus the roots of $x^2 - 4x - 2 = 0$ are $(4 \pm \sqrt{24})/2$, or $2 + \sqrt{6}$ and $2 - \sqrt{6}$.

Notice that in each of these examples, we have found the roots exactly.

The Graphical Viewpoint: Roots by Zooming

To find the roots of an equation $f(x) = 0$, it helps to draw the graph of f. The roots of the equation, that is the zeros of f, are *the values of x where the graph of f crosses the x-axis*. Even a very rough sketch of the graph can be useful in determining how many zeros there are and their approximate values. If you have a computer or graphing calculator, then finding solutions by graphing is the easiest method, especially if you use the zoom feature. However, a graph can never tell you the exact value of a root, only an approximate one.

Example 4   Find the roots of $x^3 - 4x - 2 = 0$.

Solution   Attempting to factor the left side with integer coefficients will convince you it cannot be done, so we cannot easily find the roots by algebra. We know the graph of $f(x) = x^3 - 4x - 2$ will have the usual cubic shape; see Figure A.1.

There are clearly three roots: one between $x = -2$ and $x = -1$, another between $x = -1$ and $x = 0$, and a third between $x = 2$ and $x = 3$. Zooming in on the largest root with a graphing calculator or computer shows that it lies in the following interval:

$$2.213 < x < 2.215.$$

Thus, the root is $x = 2.21$, accurate to two decimal places. Zooming in on the other two roots shows them to be $x = -1.68$ and $x = -0.54$, accurate to two decimal places.

Useful trick: Suppose you want to solve the equation $\sin x - \cos x = 0$ graphically. Instead of graphing $f(x) = \sin x - \cos x$ and looking for zeros, you may find it easier to rewrite the equation as $\sin x = \cos x$ and graph $g(x) = \sin x$ and $h(x) = \cos x$. (After all, you already know what these two graphs look like. See Figure A.2.) The roots of the original equation are then precisely the x coordinates of the points of intersection of the graphs of $g(x)$ and $h(x)$.

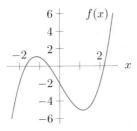

Figure A.1: The cubic
$f(x) = x^3 - 4x - 2$

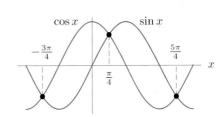

Figure A.2: Finding roots of
$\sin x - \cos x = 0$

Example 5   Find the roots of $2 \sin x - x = 0$.

Solution   Rewrite the equation as $2 \sin x = x$, and graph both sides. Since $g(x) = 2 \sin x$ is always between -2 and 2, there are no roots of $2 \sin x = x$ for $x > 2$ or for $x < -2$. We need only consider the graphs between -2 and 2 (or between $-\pi$ and π, which makes graphing the sine function easier). Figure A.3 shows the graphs. There are three points of intersection: one appears to be at $x = 0$, one between $x = \pi/2$ and $x = \pi$, and one between $x = -\pi/2$ and $x = -\pi$. You can tell that $x = 0$ is the exact value of one root because it satisfies the original equation exactly. Zooming in shows that there is a second root $x \approx 1.9$, and the third root is $x \approx -1.9$ by symmetry.

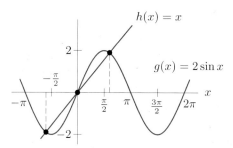

Figure A.3: Finding roots of $2 \sin x - x = 0$

The Numerical Viewpoint: Roots by Bisection

We now look at a numerical method of approximating the solutions to an equation. This method depends on the idea that if the value of a function $f(x)$ changes sign in an interval, and if we believe there is no break in the graph of the function there, then there is a root of the equation $f(x) = 0$ in that interval.

Let's go back to the problem of finding the root of $f(x) = x^3 - 4x - 2 = 0$ between 2 and 3. To locate the root, we close in on it by evaluating the function at the midpoint of the interval, $x = 2.5$. Since $f(2) = -2$, $f(2.5) = 3.625$, and $f(3) = 13$, the function changes sign between $x = 2$ and $x = 2.5$, so the root is between these points. Now we look at $x = 2.25$.

Since $f(2.25) = 0.39$, the function is negative at $x = 2$ and positive at $x = 2.25$, so there is a root between 2 and 2.25. Now we look at 2.125. We find $f(2.125) = -0.90$, so there is a root between 2.125 and 2.25, ... and so on. (You may want to round the decimals as you work.) See Figure A.4. The intervals containing the root are listed in Table A.1 and show that the root is $x = 2.21$ to two decimal places.

Figure A.4: Locating a root of $x^3 - 4x - 2 = 0$

Table A.1   *Intervals containing root of* $x^3 - 4x - 2 = 0$ *(Note $[2, 3]$ means $2 \le x \le 3$)*

$[2, 3]$
$[2, 2.5]$
$[2, 2.25]$
$[2.125, 2.25]$
$[2.1875, 2.25]$ — So $x = 2.2$ rounded to one decimal place
$[2.1875, 2.21875]$
$[2.203125, 2.21875]$
$[2.2109375, 2.21875]$
$[2.2109375, 2.2148438]$ — So $x = 2.21$ rounded to two decimal places

This method of estimating roots is called the **Bisection Method**:
- To solve an equation $f(x) = 0$ using the bisection method, we need two starting values for x, say, $x = a$ and $x = b$, such that $f(a)$ and $f(b)$ have opposite signs and f is continuous on $[a, b]$.
- Evaluate f at the midpoint of the interval $[a, b]$, and decide in which half-interval the root lies.
- Repeat, using the new half-interval instead of $[a, b]$.

There are some problems with the bisection method:
- The function may not change signs near the root. For example, $f(x) = x^2 - 2x + 1 = 0$ has a root at $x = 1$, but $f(x)$ is never negative because $f(x) = (x-1)^2$, and a square cannot be negative. (See Figure A.5.)
- The function f must be continuous between the starting values $x = a$ and $x = b$.
- If there is more than one root between the starting values $x = a$ and $x = b$, the method will find only one of the roots. For example, if we had tried to solve $x^3 - 4x - 2 = 0$ starting at $x = -12$ and $x = 10$, the bisection method would zero in on the root between $x = -2$ and $x = -1$, not the root between $x = 2$ and $x = 3$ that we found earlier. (Try it! Then see what happens if you use $x = -10$ instead of $x = -12$.)
- The bisection method is slow and not very efficient. Applying bisection three times in a row only traps the root in an interval $\left(\frac{1}{2}\right)^3 = \frac{1}{8}$ as large as the starting interval. Thus, if we initially know that a root is between, say, 2 and 3, then we would need to apply the bisection method at least four times to know the first digit after the decimal point.

There are much more powerful methods available for finding roots, such as Newton's method, which are more complicated but which avoid some of these difficulties.

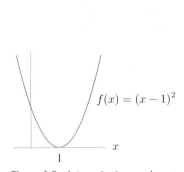

Figure A.5: f doesn't change sign at the root

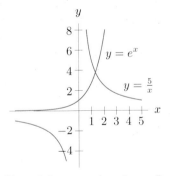

Figure A.6: Intersection of $y = e^x$ and $y = 5/x$

Table A.2   *Bisection method for* $f(x) = xe^x - 5 = 0$ *(Note that* $[1, 2]$ *means the interval* $1 \leq x \leq 2$*)*

Interval Containing Root
$[1, 2]$
$[1, 1.5]$
$[1.25, 1.5]$
$[1.25, 1.375]$
$[1.3125, 1.375]$
$[1.3125, 1.34375]$

Example 6   Find all the roots of $xe^x = 5$ to at least one decimal place.

Solution   If we rewrite the equation as $e^x = 5/x$ and graph both sides, as in Figure A.6, it is clear that there is exactly one root, and it is somewhere between 1 and 2. Table A.2 shows the intervals obtained by the bisection method. After five iterations, we have the root trapped between 1.3125 and 1.34375, so we can say the root is $x = 1.3$ to one decimal place.

Iteration

Both zooming in and bisection as discussed here are examples of *iterative* methods, in which a sequence of steps is repeated over and over again, using the results of one step as the input for the next. We can use such methods to locate a root to any degree of accuracy. In bisection, each iteration traps the root in an interval that is half the length of the previous one. Each time you zoom in on a calculator, you trap the root in a smaller interval; how much smaller depends on the settings on the calculator.

Accuracy and Error

In the previous discussion, we used the phrase "accurate to 2 decimal places." For an iterative process where we get closer and closer estimates of some quantity, we take a common-sense approach to accuracy: we watch the numbers carefully, and when a digit stays the same for several iterations, we assume it has stabilized and is correct, especially if the digits to the right of that digit also stay the same. For example, suppose 2.21429 and 2.21431 are two successive estimates for a zero of $f(x) = x^3 - 4x - 2$. Since these two estimates agree to the third digit after the decimal point, we probably have at least 3 decimal places correct.

There is a problem with this, however. Suppose we are estimating a root whose true value is 1, and the estimates are converging to the value from below—say, 0.985, 0.991, 0.997 and so on. In this case, not even the first decimal place is "correct," even though the difference between the estimates and the actual answer is very small—much less than 0.1. To avoid this difficulty, we say that an estimate a for some quantity r is *accurate to p decimal places* if the error, which is the absolute value of the difference between a and r, or $|r - a|$, is as follows:

Accuracy to p decimal places	means	Error less than
$p = 1$		0.05
2		0.005
3		0.0005
$\vdots$		$\vdots$
n		$0.\underbrace{000\ldots0}_{n}5$

This is the same as saying that r must lie in an interval of length twice the maximum error, centered on a. For example, if a is accurate to 1 decimal place, r must lie in the following interval:

$$a - 0.05 \qquad a \qquad a + 0.05$$

Since both the graphing calculator and the bisection method give us an interval in which the root is trapped, this definition of decimal accuracy is a natural one for these processes.

Example 7   Suppose the numbers $\sqrt{10}$, $22/7$, and 3.14 are given as approximations to $\pi = 3.1415\ldots$. To how many decimal places is each approximation accurate?

Solution   Using $\sqrt{10} = 3.1622\ldots$,

$$|\sqrt{10} - \pi| = |3.1622\ldots - 3.1415\ldots| = 0.0206\ldots < 0.05,$$

so $\sqrt{10}$ is accurate to one decimal place. Similarly, using $22/7 = 3.1428\ldots$,

$$\left|\frac{22}{7} - \pi\right| = |3.1428\ldots - 3.1415\ldots| = 0.0013\ldots < 0.005,$$

so 22/7 is accurate to two decimal places. Finally,

$$|3.14 - 3.1415\ldots| = 0.0015\ldots < 0.005,$$

so 3.14 is accurate to two decimal places.

Warning:

- Saying that an approximation is accurate to, say, 2 decimal places does *not* guarantee that its first two decimal places are "correct," that is, that the two digits of the approximation are the same as the corresponding two digits in the true value. For example, an approximate value of 5.997 is accurate to 2 decimal places if the true value is 6.001, but neither of the 9s in the approximation agrees with the 0s in the true value (nor does the digit 5 agree with the digit 6).

- When finding a root r of an equation, the number of decimal places of accuracy refers to the number of digits that have stabilized in the root. It does *not* refer to the number of digits of $f(r)$ that are zero. For example, Table A.1 on page 606 shows that $x = 2.2$ is a root of $f(x) = x^3 - 4x - 2 = 0$, accurate to one decimal place. Yet, $f(2.2) = -0.152$, so $f(2.2)$ does not have one zero after the decimal point. Similarly, $x = 2.21$ is the root accurate to two decimal places, but $f(2.21) = -0.046$ does not have two zeros after the decimal point.

Example 8   Is $x = 2.2143$ a zero of $f(x) = x^3 - 4x - 2$ accurate to four decimal places?

Solution   We want to know whether r, the exact value of the zero, lies in the interval

$$2.2143 - 0.00005 < r < 2.2143 + 0.00005$$

which is the same as

$$2.21425 < r < 2.21435.$$

Since $f(2.21425) < 0$ and $f(2.21435) > 0$, the zero does lie in this interval, and so $r = 2.2143$ is accurate to four decimal places.

How to Write a Decimal Answer

The graphing calculator and bisection method naturally give an interval for a root or a zero. However, other numerical techniques do not give a pair of numbers bounding the true value, but rather a single number near the true value. What should you do if you want a single number, rather than an interval, for an answer? In general, averaging the endpoint of the interval is the best solution.

When giving a single number as an answer and interpreting it, be careful about giving rounded answers. For example, suppose you know a root lies in the interval between 0.81 and 0.87. Averaging gives 0.84 as a single number estimating the root. But it would be wrong to round 0.84 to 0.8 and say that the answer is 0.8 accurate to one decimal place; the true value could be 0.86, which is not within 0.05 of 0.8. The right thing to say is that the answer is 0.84 accurate to one decimal place. Similarly, to give an answer accurate to, say, 2 decimal places, you may have to show 3 decimal places in your answer.

Bounds of a Function

Knowing how big or how small a function gets can sometimes be useful, especially when you can't easily find exact values of the function. You can say, for example, that $\sin x$ always stays between -1 and 1 and that $2\sin x + 10$ always stays between 8 and 12. But 2^x is not confined between any two numbers, because 2^x will exceed any number you can name if x is large enough. We say that $\sin x$ and $2\sin x + 10$ are *bounded* functions, and that 2^x is an *unbounded* function.

A function f is **bounded** on an interval if there are numbers L and U such that

$$L \le f(x) \le U$$

for all x in the interval. Otherwise, f is **unbounded** on the interval.

We say that L is a **lower bound** for f on the interval, and that U is an **upper bound** for f on the interval.

Example 9   Use Figures A.7 and A.8 to decide which of the following functions are bounded.

(a) x^3 on $-\infty < x < \infty$; on $0 \le x \le 100$.

(b) $2/x$ on $0 < x < \infty$; on $1 \le x < \infty$.

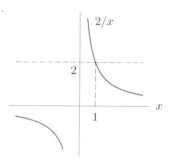

Figure A.7: Is x^3 bounded?   **Figure A.8**: Is $2/x$ bounded?

Solution   (a) The graph of x^3 in Figure A.7 shows that x^3 will exceed any number, no matter how large, if x is big enough, so x^3 does not have an upper bound on $-\infty < x < \infty$. Therefore, x^3 is unbounded on $-\infty < x < \infty$. But on the interval $0 \le x \le 100$, x^3 stays between 0 (a lower bound) and $100^3 = 1{,}000{,}000$ (an upper bound). Therefore, x^3 is bounded on the interval $0 \le x \le 100$. Notice that upper and lower bounds, when they exist, are not unique. For example, -100 is another lower bound and $2{,}000{,}000$ another upper bound for x^3 on $0 \le x \le 100$.

(b) $2/x$ is unbounded on $0 < x < \infty$, since it has no upper bound on that interval. But $0 \le 2/x \le 2$ for $1 \le x < \infty$, so $2/x$ is bounded, with lower bound 0 and upper bound 2, on $1 \le x < \infty$. (See Figure A.8.)

Best Possible Bounds

Consider a group of people whose height in feet, h, ranges from 5 feet to 6 feet. Then 5 feet is a lower bound for the people in the group and 6 feet is an upper bound:

$$5 \le h \le 6.$$

But the people in this group are also all between 4 feet and 7 feet, so it is also true that

$$4 \le h \le 7.$$

So, there are many lower bounds and many upper bounds. However, the 5 and the 6 are considered the best bounds because they are the closest together of all the possible pairs of bounds.

The **best possible bounds** for a function, f, over an interval are numbers A and B such that, for all x in the interval,

$$A \le f(x) \le B$$

and where A and B are as close together as possible. A is called the **greatest lower bound** and B is the **least upper bound**.

What Do Bounds Mean Graphically?

Upper and lower bounds can be represented on a graph by horizontal lines. See Figure A.9.

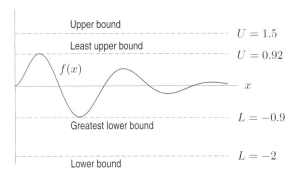

Figure A.9: Upper and lower bounds for the function f

Exercises for Appendix A

1. Use a calculator or computer graph of $f(x) = 13 - 20x - x^2 - 3x^4$ to determine:

   (a) The range of this function;
   (b) The number of zeros of this function.

For Exercises 2–12, determine the roots or points of intersection to an accuracy of one decimal place.

2. (a) The root of $x^3 - 3x + 1 = 0$ between 0 and 1
   (b) The root of $x^3 - 3x + 1 = 0$ between 1 and 2
   (c) The smallest root of $x^3 - 3x + 1 = 0$

3. The root of $x^4 - 5x^3 + 2x - 5 = 0$ between -2 and -1

4. The root of $x^5 + x^2 - 9x - 3 = 0$ between -2 and -1

5. The largest real root of $2x^3 - 4x^2 - 3x + 1 = 0$

6. All real roots of $x^4 - x - 2 = 0$

7. All real roots of $x^5 - 2x^2 + 4 = 0$

8. The smallest positive root of $x \sin x - \cos x = 0$

9. The left-most point of intersection between $y = 2x$ and $y = \cos x$

10. The left-most point of intersection between $y = 1/2^x$ and $y = \sin x$

11. The point of intersection between $y = e^{-x}$ and $y = \ln x$

12. All roots of $\cos t = t^2$

13. Estimate all real zeros of the following polynomials, accurate to 2 decimal places:

    (a) $f(x) = x^3 - 2x^2 - x + 3$
    (b) $f(x) = x^3 - x^2 - 2x + 2$

14. Find the largest zero of

    $$f(x) = 10xe^{-x} - 1$$

    to two decimal places, using the bisection method. Make sure to demonstrate that your approximation is as good as you claim.

15. (a) Find the smallest positive value of x where the graphs of $f(x) = \sin x$ and $g(x) = 2^{-x}$ intersect.
    (b) Repeat with $f(x) = \sin 2x$ and $g(x) = 2^{-x}$.

16. Use a graphing calculator to sketch $y = 2\cos x$ and $y = x^3 + x^2 + 1$ on the same set of axes. Find the positive zero of $f(x) = 2\cos x - x^3 - x^2 - 1$. A friend claims there is one more real zero. Is your friend correct? Explain.

17. Use the table below to investigate the zeros of the function

    $$f(\theta) = (\sin 3\theta)(\cos 4\theta) + 0.8$$

    in the interval $0 \le \theta \le 1.8$.

θ	0	0.2	0.4	0.6	0.8	1.0	1.2	1.4	1.6	1.8
$f(\theta)$	0.80	1.19	0.77	0.08	0.13	0.71	0.76	0.12	-0.19	0.33

   (a) Decide how many zeros the function has in the interval $0 \le \theta \le 1.8$.
   (b) Locate each zero, or a small interval containing each zero.
   (c) Are you sure you have found all the zeros in the interval $0 \le \theta \le 1.8$? Graph the function on a calculator or computer to decide.

18. (a) Use Table A.3 to locate approximate solution(s) to

$$(\sin 3x)(\cos 4x) = \frac{x^3}{\pi^3}$$

in the interval $1.07 \le x \le 1.15$. Give an interval of length 0.01 in which each solution lies.

Table A.3

x	x^3/π^3	$(\sin 3x)(\cos 4x)$
1.07	0.0395	0.0286
1.08	0.0406	0.0376
1.09	0.0418	0.0442
1.10	0.0429	0.0485
1.11	0.0441	0.0504
1.12	0.0453	0.0499
1.13	0.0465	0.0470
1.14	0.0478	0.0417
1.15	0.0491	0.0340

(b) Make an estimate for each solution accurate to two decimal places.

19. (a) With your calculator in radian mode, take the arctangent of 1 and multiply that number by 4. Now, take the arctangent of the result and multiply it by 4. Continue this process 10 times or so and record each result as in the accompanying table. At each step, you get 4 times the arctangent of the result of the previous step.

$$\begin{array}{c} \rule{3cm}{0.4pt} \\ 1 \\ 3.14159\ldots \\ 5.05050\ldots \\ 5.50129\ldots \\ \vdots \\ \rule{3cm}{0.4pt} \end{array}$$

(b) Your table allows you to find a solution of the equation

$$4 \arctan x = x.$$

Why? What is that solution?

(c) What does your table in part (a) have to do with Figure A.10?

[Hint: The coordinates of P_0 are $(1, 1)$. Find the coordinates of P_1, P_2, P_3,...]

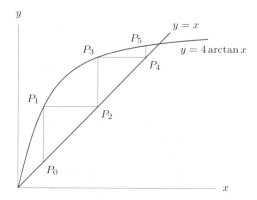

Figure A.10

(d) In part (a), what happens if you start with an initial guess of 10? Of -10? What types of behavior do you observe? (That is, for which initial guesses is the sequence increasing, and for which is it decreasing; does the sequence approach a limit?) Explain your answers graphically, as in part (c).

20. Using radians, apply the iteration method of Problem 19 to the equation

$$\cos x = x.$$

Represent your results graphically, as in Figure A.10.

For Exercises 21–23, draw a graph to decide if the function is bounded on the interval given. Give the best possible upper and lower bounds for any function which is bounded.

21. $f(x) = 4x - x^2$ on $[-1, 4]$

22. $h(\theta) = 5 + 3\sin\theta$ on $[-2\pi, 2\pi]$

23. $f(t) = \dfrac{\sin t}{t^2}$ on $[-10, 10]$

B   COMPLEX NUMBERS

The quadratic equation

$$x^2 - 2x + 2 = 0$$

is not satisfied by any real number x. If you try applying the quadratic formula, you get

$$x = \frac{2 \pm \sqrt{4 - 8}}{2} = 1 \pm \frac{\sqrt{-4}}{2}.$$

Apparently, you need to take a square root of -4. But -4 doesn't have a square root, at least, not one which is a real number. Let's give it a square root.

We define the imaginary number i to be a number such that

$$i^2 = -1.$$

Using this i, we see that $(2i)^2 = -4$, so

$$x = 1 \pm \frac{\sqrt{-4}}{2} = 1 \pm \frac{2i}{2} = 1 \pm i.$$

This solves our quadratic equation. The numbers $1 + i$ and $1 - i$ are examples of complex numbers.

A **complex number** is defined as any number that can be written in the form

$$z = a + bi,$$

where a and b are real numbers and $i^2 = -1$, so we say $i = \sqrt{-1}$.
The *real part* of z is the number a; the *imaginary part* is the number b.

Calling the number i imaginary makes it sound as if i doesn't exist in the same way that real numbers exist. In some cases, it is useful to make such a distinction between real and imaginary numbers. For example, if we measure mass or position, we want our answers to be real numbers. But the imaginary numbers are just as legitimate mathematically as the real numbers are.

As an analogy, consider the distinction between positive and negative numbers. Originally, people thought of numbers only as tools to count with; their concept of "five" or "ten" was not far removed from "five arrows" or "ten stones." They were unaware that negative numbers existed at all. When negative numbers were introduced, they were viewed only as a device for solving equations like $x + 2 = 1$. They were considered "false numbers," or, in Latin, "negative numbers." Thus, even though people started to use negative numbers, they did not view them as existing in the same way that positive numbers did. An early mathematician might have reasoned: "The number 5 exists because I can have 5 dollars in my hand. But how can I have -5 dollars in my hand?" Today we have an answer: "I have -5 dollars" means I owe somebody 5 dollars. We have realized that negative numbers are just as useful as positive ones, and it turns out that complex numbers are useful too. For example, they are used in studying wave motion in electric circuits.

Algebra of Complex Numbers

Numbers such as 0, 1, $\frac{1}{2}$, π, and $\sqrt{2}$ are called *purely real* because they contain no imaginary components. Numbers such as i, $2i$, and $\sqrt{2}i$ are called *purely imaginary* because they contain only the number i multiplied by a nonzero real coefficient.

Two complex numbers are called *conjugates* if their real parts are equal and if their imaginary parts are opposites. The complex conjugate of the complex number $z = a + bi$ is denoted $\bar{z}$, so we have

$$\bar{z} = a - bi.$$

(Note that z is real if and only if $z = \bar{z}$.) Complex conjugates have the following remarkable property: if $f(x)$ is any polynomial with real coefficients ($x^3 + 1$, say) and $f(z) = 0$, then $f(\bar{z}) = 0$. This means that if z is the solution to a polynomial equation with real coefficients, then so is $\bar{z}$.

- Two complex numbers are equal if and only if their real parts are equal and their imaginary parts are equal. Consequently, if $a + bi = c + di$, then $a = c$ and $b = d$.

- Adding two complex numbers is done by adding real and imaginary parts separately:

$$(a + bi) + (c + di) = (a + c) + (b + d)i.$$

- Subtracting is similar:

$$(a + bi) - (c + di) = (a - c) + (b - d)i.$$

- Multiplication works just as for polynomials, using $i^2 = -1$:

$$(a + bi)(c + di) = a(c + di) + bi(c + di)$$
$$= ac + adi + bci + bdi^2$$
$$= ac + adi + bci - bd = (ac - bd) + (ad + bc)i.$$

- Powers of i: We know that $i^2 = -1$; then $i^3 = i \cdot i^2 = -i$, and $i^4 = (i^2)^2 = (-1)^2 = 1$. Then $i^5 = i \cdot i^4 = i$, and so on. Thus we have

$$i^n = \begin{cases} i & \text{for } n = 1, 5, 9, 13, \ldots \\ -1 & \text{for } n = 2, 6, 10, 14, \ldots \\ -i & \text{for } n = 3, 7, 11, 15, \ldots \\ 1 & \text{for } n = 0, 4, 8, 12, 16, \ldots \end{cases}$$

- The product of a number and its conjugate is always real and nonnegative:

$$z \cdot \bar{z} = (a + bi)(a - bi) = a^2 - abi + abi - b^2 i^2 = a^2 + b^2.$$

- Dividing by a nonzero complex number is done by multiplying the denominator by its conjugate, thereby making the denominator real:

$$\frac{a + bi}{c + di} = \frac{a + bi}{c + di} \cdot \frac{c - di}{c - di} = \frac{ac - adi + bci - bdi^2}{c^2 + d^2} = \frac{ac + bd}{c^2 + d^2} + \frac{bc - ad}{c^2 + d^2} i.$$

Example 1   Compute $(2 + 7i)(4 - 6i) - i$.

Solution $(2 + 7i)(4 - 6i) - i = 8 + 28i - 12i - 42i^2 - i = 8 + 15i + 42 = 50 + 15i.$

Example 2   Compute $\dfrac{2 + 7i}{4 - 6i}$.

Solution

$$\frac{2 + 7i}{4 - 6i} = \frac{2 + 7i}{4 - 6i} \cdot \frac{4 + 6i}{4 + 6i} = \frac{8 + 12i + 28i + 42i^2}{4^2 + 6^2} = \frac{-34 + 40i}{52} = \frac{-17}{26} + \frac{10}{13} i.$$

You can check by multiplying out that $(-17/26 + 10i/13)(4 - 6i) = 2 + 7i$.

The Complex Plane and Polar Coordinates

It is often useful to picture a complex number $z = x + iy$ in the plane, with x along the horizontal axis and y along the vertical. The xy-plane is then called the *complex plane*. Figure B.11 shows the complex numbers $-2i$, $1 + i$, and $-2 + 3i$.

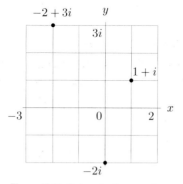

Figure B.11: Points in the complex plane

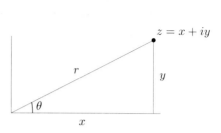

Figure B.12: The point $z = x + iy$ in the complex plane, showing polar coordinates

The triangle in Figure B.12 shows that a complex number can be written using polar coordinates as follows:

$$\boxed{z = x + iy = r\cos\theta + ir\sin\theta.}$$

Example 3   Express $z = -2i$ and $z = -2 + 3i$ using polar coordinates. (See Figure B.11.)

Solution   For $z = -2i$, the distance of z from the origin is 2, so $r = 2$. Also, one value for θ is $\theta = 3\pi/2$. Using polar coordinates, $-2i = 2\cos(3\pi/2) + i\,2(\sin 3\pi/2)$.

For $z = -2 + 3i$, we have $x = -2$, $y = 3$. So $r = \sqrt{(-2)^2 + 3^2} \approx 3.61$, and one solution of $\tan\theta = 3/(-2)$ with θ in quadrant II is $\theta \approx 2.16$. So $-2 + 3i \approx 3.61\cos(2.16) + i\,3.61\sin(2.16)$.

Example 4   Consider the point with polar coordinates $r = 5$ and $\theta = 3\pi/4$. What complex number does this point represent?

Solution   Since $x = r\cos\theta$ and $y = r\sin\theta$ we see that $x = 5\cos 3\pi/4 = -5/\sqrt{2}$, and $y = 5\sin 3\pi/4 = 5/\sqrt{2}$, so $z = -5/\sqrt{2} + i\,5/\sqrt{2}$.

Derivatives and Integrals of Complex-Valued Functions

Suppose $z(t) = x(t) + iy(t)$, where t is real, then we define $z'(t)$ and $\int z(t)\,dt$ by treating i like any other constant:

$$z'(t) = x'(t) + iy'(t)$$
$$\int z(t)\,dt = \int x(t)\,dt + i\int y(t)\,dt.$$

With these definitions, all the usual properties of differentiation and integration hold, such as

$$\int z'(t)\,dt = z(t) + C, \qquad \text{for } C \text{ is a complex constant.}$$

Euler's Formula

Consider the complex number z lying on the unit circle in Figure B.13. Writing z in polar coordinates, and using the fact that $r = 1$, we have

$$z = f(\theta) = \cos\theta + i\sin\theta.$$

It turns out that there is a particularly beautiful and compact way of rewriting $f(\theta)$ using complex exponentials. We take the derivative of f using the fact that $i^2 = -1$:

$$f'(\theta) = -\sin\theta + i\cos\theta = i\cos\theta + i^2\sin\theta.$$

Factoring out an i gives

$$f'(\theta) = i(\cos\theta + i\sin\theta) = i \cdot f(\theta).$$

As you know from Chapter 11, page 540, the only real-valued function whose derivative is proportional to the function itself is the exponential function. In other words, we know that if

$$g'(x) = k \cdot g(x), \quad \text{then} \quad g(x) \approx Ce^{kx}$$

Figure B.13: Complex number represented by a point on the unit circle for some constant C. If we assume that a similar result holds for complex-valued functions, then we have

$$f'(\theta) = i \cdot f(\theta), \quad \text{so} \quad f(\theta) = Ce^{i\theta}$$

for some constant C. To find C we substitute $\theta = 0$. Now $f(0) = Ce^{i\cdot 0} = C$, and since $f(0) = \cos 0 + i \sin 0 = 1$, we must have $C = 1$. Therefore $f(\theta) = e^{i\theta}$. Thus we have

---

Euler's formula

$$e^{i\theta} = \cos\theta + i\sin\theta.$$

This elegant and surprising relationship was discovered by the Swiss mathematician Leonhard Euler in the eighteenth century, and it is particularly useful in solving second-order differential equations. Another way of obtaining Euler's formula (using Taylor series) is given in Problem 45 on page 490. It allows us to write the complex number represented by the point with polar coordinates (r, θ) in the following form:

$$z = r(\cos\theta + i\sin\theta) = re^{i\theta}.$$

Similarly, since $\cos(-\theta) = \cos\theta$ and $\sin(-\theta) = -\sin\theta$, we have

$$re^{-i\theta} = r\left(\cos(-\theta) + i\sin(-\theta)\right) = r(\cos\theta - i\sin\theta).$$

Example 5   Evaluate $e^{i\pi}$.

Solution   Using Euler's formula, $e^{i\pi} = \cos\pi + i\sin\pi = -1$.

Example 6   Express the complex number represented by the point $r = 8$, $\theta = 3\pi/4$ in Cartesian form and polar form, $z = re^{i\theta}$.

Solution   Using Cartesian coordinates, the complex number is

$$z = 8\left(\cos\left(\frac{3\pi}{4}\right) + i\sin\left(\frac{3\pi}{4}\right)\right) = \frac{-8}{\sqrt{2}} + i\frac{8}{\sqrt{2}}.$$

Using polar coordinates, we have

$$z = 8e^{i\,3\pi/4}.$$

The polar form of complex numbers makes finding powers and roots of complex numbers much easier. Writing $z = re^{i\theta}$, we find any power of z as follows:

$$z^p = (re^{i\theta})^p = r^p e^{ip\theta}.$$

To find roots, we let p be a fraction, as in the following example.

Example 7   Find a cube root of the complex number represented by the point with polar coordinates $(8, 3\pi/4)$.

Solution   In Example 6, we saw that this complex number could be written as $z = 8e^{i3\pi/4}$. So,

$$\sqrt[3]{z} = \left(8e^{i\,3\pi/4}\right)^{1/3} = 8^{1/3}e^{i(3\pi/4)\cdot(1/3)} = 2e^{\pi i/4} = 2\left(\cos(\pi/4) + i\sin(\pi/4)\right)$$

$$= 2\left(1/\sqrt{2} + i/\sqrt{2}\right) = \sqrt{2}(1 + i).$$

You can check by multiplying out that $(\sqrt{2}(1 + i))^3 = -(8/\sqrt{2}) + i(8/\sqrt{2}) = z$.

Using Complex Exponentials

Euler's formula, together with the fact that exponential functions are simple to manipulate, allows us to obtain many results about trigonometric functions easily.

The following example uses the fact that for complex z, the function e^z has all the usual algebraic properties of exponents.

Example 8   Use Euler's formula to obtain the double-angle identities

$$\cos 2\theta = \cos^2 \theta - \sin^2 \theta \qquad \text{and} \qquad \sin 2\theta = 2\cos \theta \sin \theta.$$

Solution   We use the fact that $e^{2i\theta} = e^{i\theta} \cdot e^{i\theta}$. This can be rewritten as

$$\cos 2\theta + i\sin 2\theta = (\cos \theta + i\sin \theta)^2.$$

Multiplying out $(\cos \theta + i\sin \theta)^2$, using the fact that $i^2 = -1$ gives

$$\cos 2\theta + i\sin 2\theta = \cos^2 \theta - \sin^2 \theta + i(2\cos \theta \sin \theta).$$

Since two complex numbers are equal only if the real and imaginary parts are equal, we must have

$$\cos 2\theta = \cos^2 \theta - \sin^2 \theta \qquad \text{and} \qquad \sin 2\theta = 2\cos \theta \sin \theta.$$

If we solve $e^{i\theta} = \cos \theta + i\sin \theta$ and $e^{-i\theta} = \cos \theta - i\sin \theta$ for $\sin \theta$ and $\cos \theta$, we obtain

$$\sin \theta = \frac{e^{i\theta} - e^{-i\theta}}{2i} \qquad \text{and} \qquad \cos \theta = \frac{e^{i\theta} + e^{-i\theta}}{2}.$$

By differentiating the formula $e^{ik\theta} = \cos(k\theta) + i\sin(k\theta)$, for θ real and k a real constant, it can be shown that

$$\frac{d}{d\theta}\left(e^{ik\theta}\right) = ike^{ik\theta} \qquad \text{and} \qquad \int e^{ik\theta}\, d\theta = \frac{1}{ik}e^{ik\theta} + C.$$

Thus complex exponentials are differentiated and integrated just like real exponentials.

Example 9   Use $\cos \theta = \left(e^{i\theta} + e^{-i\theta}\right)/2$ to obtain the derivative formula for $\cos \theta$.

Solution   Differentiating gives

$$\frac{d}{d\theta}(\cos \theta) = \frac{d}{d\theta}\left(\frac{e^{i\theta} + e^{-i\theta}}{2}\right) = \frac{ie^{i\theta} - ie^{-i\theta}}{2} = \frac{i\left(e^{i\theta} - e^{-i\theta}\right)}{2}$$

$$= -\frac{e^{i\theta} - e^{-i\theta}}{2i} = -\sin \theta.$$

The facts that e^z has all the usual properties when z is complex leads to

$$\frac{d}{d\theta}(e^{(a+ib)\theta}) = (a+ib)e^{(a+ib)\theta} \qquad \text{and} \qquad \int e^{(a+ib)\theta}\, d\theta = \frac{1}{a+ib}e^{(a+ib)\theta} + C.$$

Example 10   Use the formula for $\int e^{(a+ib)\theta}\, d\theta$ to obtain formulas for $\int e^{ax}\cos bx\, dx$ and $\int e^{ax}\sin bx\, dx$.

Solution   The formula for $\int e^{(a+ib)\theta}\, d\theta$ allows us to write

$$\int e^{ax}e^{ibx}\, dx = \int e^{(a+ib)x}\, dx = \frac{1}{a+ib}e^{(a+ib)x} + C = \frac{a-ib}{a^2+b^2}e^{ax}e^{ibx} + C.$$

The left-hand side of this equation can be rewritten as

$$\int e^{ax}e^{ibx}\, dx = \int e^{ax}\cos bx\, dx + i\int e^{ax}\sin bx\, dx,$$

The right-hand side can be rewritten as

$$\frac{a-ib}{a^2+b^2}e^{ax}e^{ibx} = \frac{e^{ax}}{a^2+b^2}(a-ib)(\cos bx + i\sin bx),$$

$$= \frac{e^{ax}}{a^2+b^2}\left(a\cos bx + b\sin bx + i\left(a\sin bx - b\cos bx\right)\right).$$

Equating real parts gives

$$\int e^{ax}\cos bx\, dx = \frac{e^{ax}}{a^2+b^2}\left(a\cos bx + b\sin bx\right) + C,$$

and equating imaginary parts gives

$$\int e^{ax}\sin bx\, dx = \frac{e^{ax}}{a^2+b^2}\left(a\sin bx - b\cos bx\right) + C.$$

These two formulas are usually obtained by integrating by parts twice.

Example 11   Using complex exponentials, find a formula for $\int \sin 2x \sin 3x\, dx$.

Solution   Replacing $\sin 2x$ and $\sin 3x$ by their exponential form, we have

$$\int \sin 2x \sin 3x\, dx = \int \frac{\left(e^{2ix} - e^{-2ix}\right)}{2i}\frac{\left(e^{3ix} - e^{-3ix}\right)}{2i}\, dx$$

$$= \frac{1}{(2i)^2}\int \left(e^{5ix} - e^{-ix} - e^{ix} + e^{-5ix}\right)\, dx$$

$$= -\frac{1}{4}\left(\frac{1}{5i}e^{5ix} + \frac{1}{i}e^{-ix} - \frac{1}{i}e^{ix} - \frac{1}{5i}e^{-5ix}\right) + C$$

$$= -\frac{1}{4}\left(\frac{e^{5ix} - e^{-5ix}}{5i} - \frac{e^{ix} - e^{-ix}}{i}\right) + C$$

$$= -\frac{1}{4}\left(\frac{2}{5}\sin 5x - 2\sin x\right) + C$$

$$= -\frac{1}{10}\sin 5x + \frac{1}{2}\sin x + C.$$

This result is usually obtained by using a trigonometric identity.

Exercises for Appendix B

For Exercises 1–8, express the given complex number in polar form, $z = re^{i\theta}$.

For Exercises 9–18, perform the indicated calculations. Give your answer in Cartesian form, $z = x + iy$.

1. $2i$ **2.** -5 **3.** $1 + i$ **4.** $-3 - 4i$ **9.** $(2 + 3i) + (-5 - 7i)$ **10.** $(2 + 3i)(5 + 7i)$

5. 0 **6.** $-i$ **7.** $-1 + 3i$ **8.** $5 - 12i$ **11.** $(2 + 3i)^2$ **12.** $(1 + i)^2 + (1 + i)$

13. $(0.5 - i)(1 - i/4)$ **14.** $(2i)^3 - (2i)^2 + 2i - 1$

15. $(e^{i\pi/3})^2$ **16.** $\sqrt{e^{i\pi/3}}$

17. $(5e^{i7\pi/6})^3$ **18.** $\sqrt[4]{10e^{i\pi/2}}$

By writing the complex numbers in polar form, $z = re^{i\theta}$, find a value for the quantities in Exercises 19–28. Give your answer in Cartesian form, $z = x + iy$.

19. $\sqrt{i}$ **20.** $\sqrt{-i}$ **21.** $\sqrt[3]{i}$

22. $\sqrt{7i}$ **23.** $(1 + i)^{100}$ **24.** $(1 + i)^{2/3}$

25. $(-4 + 4i)^{2/3}$ **26.** $(\sqrt{3} + i)^{1/2}$ **27.** $(\sqrt{3}+i)^{-1/2}$

28. $(\sqrt{5} + 2i)^{\sqrt{2}}$

29. Calculate i^n for $n = -1, -2, -3, -4$. What pattern do you observe? What is the value of i^{-36}? Of i^{-41}?

Solve the simultaneous equations in Exercises 30–31 for A_1 and A_2.

30. $A_1 + A_2 = 2$
$(1 - i)A_1 + (1 + i)A_2 = 3$

31. $A_1 + A_2 = 2$
$(i - 1)A_1 + (1 + i)A_2 = 0$

32. (a) Calculate a and b if $\dfrac{3 - 4i}{1 + 2i} = a + bi$.
(b) Check your answer by calculating $(1 + 2i)(a + bi)$.

33. Check that $z = \dfrac{ac + bd}{c^2 + d^2} + \dfrac{bc - ad}{c^2 + d^2}i$ is the quotient $\dfrac{a + bi}{c + di}$ by showing that the product $z \cdot (c + di)$ is $a + bi$.

34. Let $z_1 = -3 - i\sqrt{3}$ and $z_2 = -1 + i\sqrt{3}$.

(a) Find $z_1 z_2$ and z_1/z_2. Give your answer in Cartesian form, $z = x + iy$.
(b) Put z_1 and z_2 into polar form, $z = re^{i\theta}$. Find $z_1 z_2$ and z_1/z_2 using the polar form, and verify that you get the same answer as in part (a).

35. Let $z_1 = a_1 + b_1 i$ and $z_2 = a_2 + b_2 i$. Show that $\overline{z_1 z_2} = \bar{z}_1 \bar{z}_2$.

36. If the roots of the equation $x^2 + 2bx + c = 0$ are the complex numbers $p \pm iq$, find expressions for p and q in terms of b and c.

Are the statements in Exercises 37–42 true or false? Explain your answer.

37. Every nonnegative real number has a real square root.

38. For any complex number z, the product $z \cdot \bar{z}$ is a real number.

39. The square of any complex number is a real number.

40. If f is a polynomial, and $f(z) = i$, then $f(\bar{z}) = i$.

41. Every nonzero complex number z can be written in the form $z = e^w$, where w is another complex number.

42. If $z = x + iy$, where x and y are positive, then $z^2 = a + ib$ has a and b positive.

For Exercises 43–47, use Euler's formula to derive the following relationships. (Note that if a, b, c, d are real numbers, $a + bi = c + di$ means that $a = c$ and $b = d$.)

43. $\sin^2 \theta + \cos^2 \theta = 1$ **44.** $\sin 2\theta = 2 \sin \theta \cos \theta$

45. $\cos 2\theta = \cos^2 \theta - \sin^2 \theta$ **46.** $\dfrac{d}{d\theta} \sin \theta = \cos \theta$

47. $\dfrac{d^2}{d\theta^2} \cos \theta = -\cos \theta$

48. Use complex exponentials to show that
$$\sin(-x) = -\sin x.$$

49. Use complex exponentials to show that
$$\sin(x + y) = \sin x \cos y + \cos x \sin y.$$

50. For real t, show that if $z_1(t) = x_1(t) + iy_1(t)$ and $z_2(t) = x_2(t) + iy_2(t)$ then
$$(z_1 + z_2)' = z_1' + z_2' \quad \text{and} \quad (z_1 z_2)' = z_1' z_2 + z_1 z_2'.$$

C NEWTON'S METHOD

Many problems in mathematics involve finding the root of an equation. For example, we might have to locate the zeros of a polynomial, or determine the point of intersection of two curves. Here we will see a numerical method for approximating solutions which cannot be calculated exactly.

One such method, bisection, is described in Appendix A. Although it is very simple, the bisection method has two major drawbacks. First, it cannot locate a root where the curve is tangent to, but does not cross, the x-axis. Second, it is relatively slow in the sense that it requires a considerable number of iterations to achieve a desired level of accuracy. Although speed may not be important in solving a single equation, a practical problem may involve solving thousands of equations as a parameter changes. In such a case, any reduction in the number of steps can be important.

Using Newton's Method

We now consider a powerful root-finding method developed by Newton. Suppose we have a function $y = f(x)$. The equation $f(x) = 0$ has a root at $x = r$, as shown in Figure C.14. We begin with an initial estimate, x_0, for this root. (This can be a guess.) We will now obtain a better estimate x_1. To do this, construct the tangent line to the graph of f at the point $x = x_0$, and extend it until it crosses the x-axis, as shown in Figure C.14. The point where it crosses the axis is usually much closer to r, and we use that point as the next estimate, x_1. Having found x_1, we now repeat the process starting with x_1 instead of x_0. We construct a tangent line to the curve at $x = x_1$, extend it until it crosses the x-axis, use that x-intercept as the next approximation, x_2, and so on. The resulting sequence of x-intercepts usually converges rapidly to the root r.

Let's see how this looks algebraically. We know that the slope of the tangent line at the initial estimate x_0 is $f'(x_0)$, and so the equation of the tangent line is

$$y - f(x_0) = f'(x_0)(x - x_0).$$

At the point where this tangent line crosses the x-axis, we have $y = 0$ and $x = x_1$, so that

$$0 - f(x_0) = f'(x_0)(x_1 - x_0).$$

Solving for x_1, we obtain

$$x_1 = x_0 - \frac{f(x_0)}{f'(x_0)}$$

provided that $f'(x_0)$ is not zero. We now repeat this argument and find that the next approximation is

$$x_2 = x_1 - \frac{f(x_1)}{f'(x_1)}.$$

Summarizing, for any $n = 0, 1, 2, \ldots$, we obtain the following result.

Newton's Method to Solve the Equation $f(x) = 0$

Choose x_0 near a solution and compute the sequence $x_1, x_2, x_3 \ldots$ using the rule

$$x_{n+1} = x_n - \frac{f(x_n)}{f'(x_n)}$$

provided that $f'(x_n)$ is not zero. For large n, the solution is well-approximated by x_n.

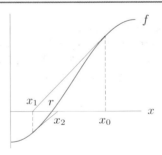

Figure C.14: Newton's method: successive approximations $x_0, x_1, x_2, \ldots$ to the root, r

---

Example 1   Use Newton's method to find the fifth root of 23. (By calculator, this is 1.872171231, correct to nine decimal places.)

Solution   To use Newton's method, we need an equation of the form $f(x) = 0$ having $23^{1/5}$ as a root. Since $23^{1/5}$ is a root of $x^5 = 23$ or $x^5 - 23 = 0$, we take $f(x) = x^5 - 23$. The root of this equation is between 1 and 2 (since $1^5 = 1$ and $2^5 = 32$), so we will choose $x_0 = 2$ as our initial estimate. Now $f'(x) = 5x^4$, so we can set up Newton's method as

$$x_{n+1} = x_n - \frac{x_n^5 - 23}{5x_n^4}.$$

In this case, we can simplify using a common denominator, to obtain

$$x_{n+1} = \frac{4x_n^5 + 23}{5x_n^4}.$$

Therefore, starting with $x_0 = 2$, we find that $x_1 = 1.8875$. This leads to $x_2 = 1.872418193$ and $x_3 = 1.872171296$. These values are in Table C.4. Since $f(1.872171231) > 0$ and $f(1.872171230) < 0$, the root lies between 1.872171230 and 1.872171231. Therefore, in just four iterations of Newton's method, we have achieved eight-decimal accuracy.

Table C.4 *Newton's method: $x_0 = 2$*

n	x_n	$f(x_n)$
0	2	9
1	1.8875	0.957130661
2	1.872418193	0.015173919
3	1.872171296	0.000004020
4	1.872171231	0.000000027

Table C.5 *Newton's method: $x_0 = 10$*

n	x_n	n	x_n
0	10	6	2.679422313
1	8.000460000	7	2.232784753
2	6.401419079	8	1.971312452
3	5.123931891	9	1.881654220
4	4.105818871	10	1.872266333
5	3.300841811	11	1.872171240

As a general guideline for Newton's method, once the first correct decimal place is found, each successive iteration approximately doubles the number of correct digits.

What happens if we select a very poor initial estimate? In the preceding example, suppose x_0 were 10 instead of 2. The results are in Table C.5. Notice that even with $x_0 = 10$, the sequence of values moves reasonably quickly toward the solution: we achieve six-decimal place accuracy by the eleventh iteration.

Example 2

Find the first point of intersection of the curves given by $f(x) = \sin x$ and $g(x) = e^{-x}$.

Solution

The graphs in Figure C.15 make it clear that there are an infinite number of points of intersection, all with $x > 0$. In order to find the first one numerically, we consider the function

$$F(x) = f(x) - g(x) = \sin x - e^{-x}$$

whose derivative is $F'(x) = \cos x + e^{-x}$. From the graph, we see that the point we want is fairly close to $x = 0$, so we start with $x_0 = 0$. The values in Table C.6 are approximations to the root. Since $F(0.588532744) > 0$ and $F(0.588532743) < 0$, the root lies between 0.588532743 and 0.588532744. (Remember, your calculator must be set in radians.)

Table C.6 *Successive approximations to root of $\sin x = e^{-x}$*

n	x_n
0	0
1	0.5
2	0.585643817
3	0.588529413
4	0.588532744
5	0.588532744

Figure C.15: Root of $\sin x = e^{-x}$

When Does Newton's Method Fail?

In most practical situations, Newton's method works well. Occasionally, however, the sequence x_0, x_1, x_2, ... fails to converge or fails to converge to the root you want. Sometimes, for example, the sequence can jump from one root to another. This is particularly likely to happen if the magnitude of the derivative $f'(x_n)$ is small for some x_n. In this case, the tangent line is nearly horizontal and so x_{n+1} will be far from x_n. (See Figure C.16.)

If the equation $f(x) = 0$ has *no* root, then the sequence will not converge. In fact, the sequence obtained by applying Newton's method to $f(x) = 1 + x^2$ is one of the best known examples of *chaotic behavior* and has attracted considerable research interest recently. (See Figure C.17.)

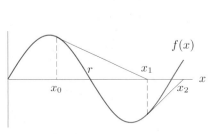

Figure C.16: Problems with Newton's method: Converges to wrong root

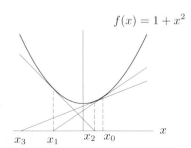

Figure C.17: Problems with Newton's method: Chaotic behavior

Exercises for Appendix C

1. Suppose you want to find a solution of the equation

   $$x^3 + 3x^2 + 3x - 6 = 0.$$

   Consider $f(x) = x^3 + 3x^2 + 3x - 6$.

   (a) Find $f'(x)$, and use it to show that $f(x)$ increases everywhere.
   (b) How many roots does the original equation have?
   (c) For each root, find an interval which contains it.
   (d) Find each root to two decimal places, using Newton's method.

For Exercises 2–4, use Newton's method to find the given quantities to two decimal places:

2. $\sqrt[3]{50}$ 3. $\sqrt[4]{100}$ 4. $10^{-1/3}$

For Exercises 5–8, solve each equation and give each answer to two decimal places:

5. $\sin x = 1 - x$ 6. $\cos x = x$

7. $e^{-x} = \ln x$

8. $e^x \cos x = 1$, for $0 < x < \pi$

9. Find, to two decimal places, all solutions of $\ln x = 1/x$.

10. How many zeros do the following functions have? For each zero, find an upper and a lower bound which differ by no more than 0.1.
    (a) $f(x) = x^3 + x - 1$ (b) $f(x) = \sin x - \frac{2}{3}x$
    (c) $f(x) = 10xe^{-x} - 1$

11. Find the largest zero of

    $$f(x) = x^3 + x - 1$$

    to six decimal places, using Newton's method. How do you know your approximation is as good as you claim?

12. For any positive number, a, the problem of calculating the square root, $\sqrt{a}$, is often done by applying Newton's method to the function $f(x) = x^2 - a$. Apply the method to obtain an expression for x_{n+1} in terms of x_n. Use this to approximate $\sqrt{a}$ for $a = 2, 10, 1000$, and π, correct to four decimal places, starting at $x_0 = a/2$ in each case.

READY REFERENCE

This section contains a concise summary of the main definitions, theorems, and ideas presented throughout the book.

READY REFERENCE

A Library of Functions

Linear functions (p. 4) have the form $y = f(x) = b + mx$, where m is the **slope**, or rate of change of y with respect to x (p. 5) and b is the **vertical intercept**, or value of y when x is zero (p. 5). The slope is

$$m = \frac{\text{Rise}}{\text{Run}} = \frac{\Delta y}{\Delta x} = \frac{f(x_2) - f(x_1)}{x_2 - x_1} \quad \text{(p. 5)}.$$

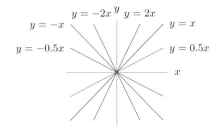

Figure R.1: The family $y = mx$
(with $b = 0$) (p. 6)

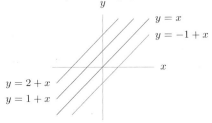

Figure R.2: The family $y = b + x$
(with $m = 1$) (p. 6)

Exponential functions have the form $P = P_0 a^t$ (p. 11) or $P = P_0 e^{kt}$ (p. 13), where P_0 is the initial quantity (p. 11), a is the growth (decay) factor per unit time (p. 11), $|k|$ is the continuous growth (decay) rate (pp. 13, 541), and $r = |a - 1|$ is the growth (decay) rate per unit time (p. 11).

Suppose $P_0 > 0$. If $a > 1$ or $k > 0$, we have **exponential growth**; if $0 < a < 1$ or $k < 0$, we have **exponential decay** (p. 12). The **doubling time** (for growth) is the time required for P to double (p. 12). The **half-life** (for decay) is the time required for P to be reduced by a factor of one half (p. 12). The continuous growth rate $k = \ln(1 + r)$ is slightly less than, but very close to, r, provided r is small (p. 541).

Figure R.3: Exponential
growth: $P = a^t$, for $a > 1$
(p. 12)

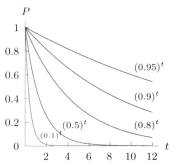

Figure R.4: Exponential decay:
$P = a^t$, for $0 < a < 1$ (p. 12)

Common Logarithm and Natural Logarithm
$\log_{10} x = \log x =$ power of 10 that gives x (p. 24)
$\log_{10} x = c$ means $10^c = x$ (p. 24)
$\ln x =$ power of e that gives x (p. 24)
$\ln x = c$ means $e^c = x$ (p. 24)
$\log x$ and $\ln x$ are not defined if x is negative or 0 (p. 24).

Properties of Logarithms (p. 25)
1. $\log(AB) = \log A + \log B$ 4. $\log(10^x) = x$

2. $\log\left(\frac{A}{B}\right) = \log A - \log B$ 5. $10^{\log x} = x$

3. $\log(A^p) = p \log A$ 6. $\log 1 = 0$

The natural logarithm satisfies properties 1, 2, and 3, and $\ln e^x = x$, $e^{\ln x} = x$, $\ln 1 = 0$ (p. 25).

Trigonometric Functions The sine and cosine are defined in Figure R.5 (see also p. 30). The tangent is $\tan t = \frac{\sin t}{\cos t} =$ slope of the line through the origin $(0, 0)$ and P if $\cos t \neq 0$ (p. 33). The period of sin and cos is 2π (p. 30), the period of tan is π (p. 33).

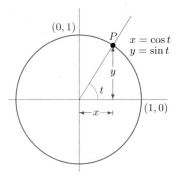

Figure R.5: The definitions of $\sin t$
and $\cos t$ (p. 30)

A **sinusoidal function** (p. 31) has the form
$y = C + A\sin(B(t + h))$ or $y = C + A\cos(B(t + h))$.

The **amplitude** is $|A|$, (p. 31), and the **period** is $2\pi/|B|$ (p. 31).

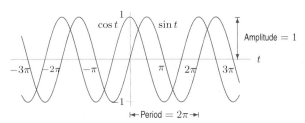

Figure R.6: Graphs of $\cos t$ and $\sin t$ (p. 30)

Trigonometric Identities

$$\sin^2 x + \cos^2 x = 1$$
$$\sin(2x) = 2\sin x \cos x$$
$$\cos(2x) = \cos^2 x - \sin^2 x = 2\cos^2 x - 1 = 1 - 2\sin^2 x$$
$$\cos(a+b) = \cos a \cos b - \sin a \sin b$$
$$\sin(a+b) = \sin a \cos b + \cos a \sin b$$

Inverse trigonometric functions: $\arcsin y = x$ means $\sin x = y$ with $-(\pi/2) \le x \le (\pi/2)$ (p. 34), $\arccos y = x$ means $\cos x = y$ with $0 \le x \le \pi$ (Problem 47, p. 36), $\arctan y = x$ means $\tan x = y$ with $-(\pi/2) < x < (\pi/2)$ (p. 34). The domain of arcsin and arccos is $[-1, 1]$ (p. 34), the domain of arctan is all numbers (p. 34).

Power Functions have the form $f(x) = kx^p$ (p. 37). Graphs for positive powers:

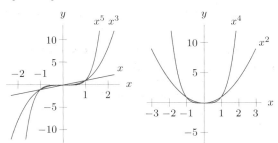

Figure R.7: Odd integer powers of x: "Seat" shaped for $k > 1$ (p. 37)

Figure R.8: Even integer powers of x: $\bigcup$-shaped (p. 37)

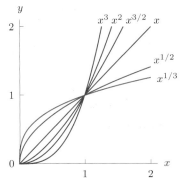

Figure R.9: Comparison of some fractional powers of x

Graphs for zero and negative powers:

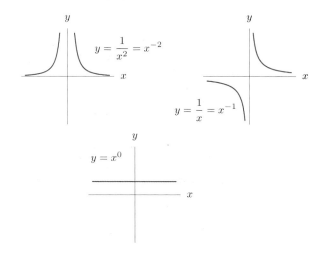

Figure R.10: Comparison of zero and negative powers of x

Polynomials have the form

$$f(x) = a_n x^n + a_{n-1} x^{n-1} + \cdots + a_1 x + a_0,\ a_n \ne 0 \quad \text{(p. 38)}.$$

The **degree** is n (p. 38) and the **leading coefficient** is a_n (p. 38).

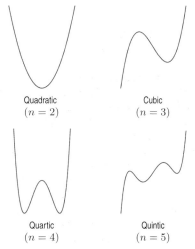

Quadratic ($n = 2$)

Cubic ($n = 3$)

Quartic ($n = 4$)

Quintic ($n = 5$)

Figure R.11: Graphs of typical polynomials of degree n (p. 38)

Rational functions have the form $f(x) = \dfrac{p(x)}{q(x)}$, where p and q are polynomials (p. 40). There is usually a vertical asymptote at $x = a$ if $q(a) = 0$ and a horizontal asymptote at $y = L$ if $\lim_{x \to \infty} f(x) = L$ or $\lim_{x \to -\infty} f(x) = L$ (p. 40).

Hyperbolic functions:

$$\cosh x = \frac{e^x + e^{-x}}{2} \qquad \sinh x = \frac{e^x - e^{-x}}{2} \qquad \text{(p. 146)}.$$

Relative Growth Rates of Functions

Power functions As $x \to \infty$, higher powers of x dominate, as $x \to 0$, smaller powers dominate.

Figure R.12: For large x: Large powers of x dominate

Figure R.13: For $0 \le x \le 1$: Small powers of x dominate

Power functions versus exponential functions Every exponential growth function eventually dominates every power function (p. 37).

Figure R.14: Exponential function eventually dominates power function

Power functions versus logarithmic functions The power function x^p dominates $A \log x$ for large x for all values of $p > 0$ and $A > 0$.

Figure R.15: Comparison of $x^{1/3}$ and $\log x$

Figure R.16: Comparison of $x^{1/3}$ and $100 \log x$

Numerical comparisons of growth rates:

Table R.1 *Comparison of $x^{0.001}$ and $1000 \log x$*

x	$x^{0.001}$	$1000 \log x$
10^{5000}	10^5	$5 \cdot 10^6$
10^{6000}	10^6	$6 \cdot 10^6$
10^{7000}	10^7	$7 \cdot 10^6$

Table R.2 *Comparison of x^{100} and 1.01^x*

x	x^{100}	1.01^x
10^4	10^{400}	$1.6 \cdot 10^{43}$
10^5	10^{500}	$1.4 \cdot 10^{432}$
10^6	10^{600}	$2.4 \cdot 10^{4321}$

Operations on Functions

Shifts, stretches, and composition Multiplying by a constant, c, stretches (if $c > 1$) or shrinks (if $0 < c < 1$) the graph vertically. A negative sign (if $c < 0$) reflects the graph about the x-axis, in addition to shrinking or stretching (p. 17). Replacing y by $(y - k)$ moves a graph up by k (down if k is negative) (p. 17). Replacing x by $(x - h)$ moves a graph to the right by h (to the left if h is negative) (p. 17). The composite of f and g is the function $f(g(x))$; f is the outside function, g the inside function (p. 18).

Symmetry We say f is an **even** function if $f(-x) = f(x)$ (p. 19) and f is an **odd** function if $f(-x) = -f(x)$ (p. 19).

Inverse functions A function f has an inverse if (and only if) its graph intersects any horizontal line at most once (p. 20). If f has an inverse, it is written f^{-1}, and $f^{-1}(x) = y$ means $f(y) = x$ (p. 20). Provided the x and y scales are equal, the graph of f^{-1} is the reflection of the graph of f about the line $y = x$ (p. 21).

Limits and Continuity

Idea of Limit (p. 49) If there is a number L such that $f(x)$ is as close to L as we please whenever x is sufficiently close to c (but $x \ne c$), then $\lim_{x \to c} f(x) = L$.

Definition of Limit (p. 50) If there is a number L such that for any $\epsilon > 0$, there exists a $\delta > 0$ such that if $|x - c| < \delta$ and $x \ne c$, then $|f(x) - L| < \epsilon$, then $\lim_{x \to c} f(x) = L$.

One-sided limits (p. 51) If $f(x)$ approaches L as x approaches c through values greater than c, then $\lim_{x \to c^+} f(x) = L$. If $f(x)$ approaches L as x approaches c through values less than c, then $\lim_{x \to c^-} f(x) = L$.

Limits at Infinity (p. 53) If $f(x)$ gets as close to L as we please when x gets sufficiently large, then $\lim_{x \to \infty} f(x) = L$. Similarly, if $f(x)$ approaches L as x gets more and more negative, then $\lim_{x \to -\infty} f(x) = L$.

Theorem: Properties of Limits (p. 51) Assuming all the limits on the right hand side exist:

1. If b is a constant, then $\lim_{x \to c} (bf(x)) = b \left(\lim_{x \to c} f(x) \right)$.
2. $\lim_{x \to c} (f(x) + g(x)) = \lim_{x \to c} f(x) + \lim_{x \to c} g(x)$.
3. $\lim_{x \to c} (f(x)g(x)) = \left(\lim_{x \to c} f(x) \right) \left(\lim_{x \to c} g(x) \right)$.
4. $\lim_{x \to c} \dfrac{f(x)}{g(x)} = \dfrac{\lim_{x \to c} f(x)}{\lim_{x \to c} g(x)}$, provided $\lim_{x \to c} g(x) \ne 0$.
5. For any constant k, $\lim_{x \to c} k = k$.
6. $\lim_{x \to c} x = c$.

Idea of Continuity (p. 44) A function is **continuous on an interval** if its graph has no breaks, jumps, or holes in that interval.

Definition of Continuity (p. 54) The function f is **continuous** at $x = c$ if f is defined at $x = c$ and

$$\lim_{x \to c} f(x) = f(c).$$

The function is **continuous on an interval** if it is continuous at every point in the interval.

Theorem: Continuity of Sums, Products, Quotients (p. 54) Suppose that f and g are continuous on an interval and that b is a constant. Then, on that same interval, the following functions are also continuous: $bf(x)$, $f(x) + g(x)$, $f(x)g(x)$. Further, $f(x)/g(x)$ is continuous provided $g(x) \neq 0$ on the interval.

Theorem: Continuity of Composite Functions (p. 55) Suppose f and g are continuous and $f(g(x))$ is defined on an interval. Then on that interval $f(g(x))$ is continuous.

Intermediate Value Theorem (p. 46) Suppose f is continuous on a closed interval $[a, b]$. If k is any number between $f(a)$ and $f(b)$, then there is at least one number c in $[a, b]$ such that $f(c) = k$.

The Extreme Value Theorem (p. 186) If f is continuous on the interval $[a, b]$, then f has a global maximum and a global minimum on that interval.

The Derivative

The slope of the secant line of $f(x)$ over an interval $[a, b]$ gives:

$$\text{Average rate of change of } f \text{ over } [a, b] = \frac{f(b) - f(a)}{b - a} \quad \text{(p. 72)}.$$

Figure R.17: Visualizing the average rate of change of f (p. 75)

The **derivative** of f at a is the slope of the line tangent to the graph of f at the point $(a, f(a))$:

$$f'(a) = \lim_{h \to 0} \frac{f(a + h) - f(a)}{h},$$

and gives the **instantaneous rate of change** of f at a (p. 74). The **second derivative** of f, denoted f'', is the derivative of f' (p. 93).

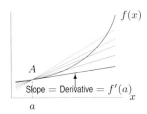

Figure R.18: Visualizing the instantaneous rate of change of f (p. 75)

The **units** of $f'(x)$ are: $\dfrac{\text{Units of } f(x)}{\text{Units of } x}$ (p. 89). If $f' > 0$ on an interval, then f is **increasing** over that interval (p. 82). If $f' < 0$ on an interval, then f is **decreasing** over that interval (p. 82). If $f'' > 0$ on an interval, then f is **concave up** over that interval (p. 93). If $f'' < 0$ on an interval, then f is **concave down** over that interval (p. 93).

The **tangent line** at $(a, f(a))$ is the graph of $y = f(a) + f'(a)(x - a)$ (p. 150). The **tangent line approximation** says that for values of x near a, $f(x) \approx f(a) + f'(a)(x - a)$. The expression $f(a) + f'(a)(x - a)$ is called the **local linearization** of f near $x = a$ (p. 150).

Derivatives of elementary functions

$$\frac{d}{dx}(x^n) = nx^{n-1} \text{ (p. 112)} \quad \frac{d}{dx}(e^x) = e^x \text{ (p. 118)}$$

$$\frac{d}{dx}(a^x) = (\ln a)a^x \text{ (p. 119)} \quad \frac{d}{dx}(\ln x) = \frac{1}{x} \text{ (p. 138)}$$

$$\frac{d}{dx}(\sin x) = \cos x \quad \frac{d}{dx}(\cos x) = -\sin x \text{ (p. 134)}$$

$$\frac{d}{dx}(\arctan x) = \frac{1}{1 + x^2} \text{ (p. 140)}$$

$$\frac{d}{dx}(\arcsin x) = \frac{1}{\sqrt{1 - x^2}} \text{ (p. 140)}$$

Derivatives of sums, differences, and constant multiples

$$\frac{d}{dx}[f(x) \pm g(x)] = f'(x) \pm g'(x) \text{ (p. 111)}$$

$$\frac{d}{dx}[cf(x)] = cf'(x) \text{ (p. 110)}$$

Product and quotient rules

$$(fg)' = f'g + fg' \text{ (p. 122)}$$

$$\left(\frac{f}{g}\right)' = \frac{f'g - fg'}{g^2} \text{ (p. 123)}$$

Chain rule

$$\frac{d}{dx}f(g(x)) = f'(g(x)) \cdot g'(x) \text{ (p. 127)}$$

Implicit differentiation (p. 143) If y is implicitly defined as a function of x by an equation, then, to find dy/dx, differentiate the equation (remembering to apply the chain rule).

Applications of the Derivative

A function f has a **local maximum** at p if $f(p)$ is greater than or equal to the values of f at points near p, and a **local minimum** at p if $f(p)$ is less than or equal to the values of f at points near p (p. 167). It has a **global maximum** at p if $f(p)$ is greater than or equal to the value of f at any point in the interval, and a **global minimum** at p if $f(p)$ is less than or equal to the value of f at any point in the interval (p. 181).

A **critical point** of a function $f(x)$ is a point p in the domain of f where $f'(p) = 0$ or $f'(p)$ is undefined (p. 167).

Theorem: Local maxima and minima which do not occur at endpoints of the domain occur at critical points (pp. 168, 186).

The First-Derivative Test for Local Maxima and Minima (p. 168):

- If f' changes from negative to positive at p, then f has a local minimum at p.

- If f' changes from positive to negative at p, then f has a local maximum at p.

The Second-Derivative Test for Local Maxima and Minima (p. 170):

- If $f'(p) = 0$ and $f''(p) > 0$ then f has a local minimum at p.

- If $f'(p) = 0$ and $f''(p) < 0$ then f has a local maximum at p.

- If $f'(p) = 0$ and $f''(p) = 0$ then the test tells us nothing.

To find the **global maximum and minimum** of a function on an interval we compare values of f at all critical points in the interval and at the endpoints of the interval (or $\lim\limits_{x \to \pm\infty} f(x)$ if the interval is unbounded) (p. 182).

An **inflection point** of f is a point at which the graph of f changes concavity (p. 170); f'' is zero or undefined at an inflection point (p. 170).

L'Hopital's rule (p. 214) If f and g are continuous, $f(a) = g(a) = 0$, and $g'(a) \neq 0$, then

$$\lim_{x \to a} \frac{f(x)}{g(x)} = \frac{f'(a)}{g'(a)}.$$

Theorems about Derivatives

Theorem: Local Extrema and Critical Points (pp. 168, 186) Suppose f is defined on an interval and has a local maximum or minimum at the point $x = a$, which is not an endpoint of the interval. If f is differentiable at $x = a$, then $f'(a) = 0$.

The Mean Value Theorem (p. 155) If f is continuous on $[a, b]$ and differentiable on (a, b), then there exists a number c, with $a < c < b$, such that

$$f'(c) = \frac{f(b) - f(a)}{b - a}.$$

The Increasing Function Theorem (p. 156) Suppose that f is continuous on $[a, b]$ and differentiable on (a, b).

- If $f'(x) > 0$ on (a, b), then f is increasing on $[a, b]$.

- If $f'(x) \geq 0$ on (a, b), then f is nondecreasing on $[a, b]$.

The Constant Function Theorem (p. 157) Suppose that f is continuous on $[a, b]$ and differentiable on (a, b). If $f'(x) = 0$ on (a, b), then f is constant on $[a, b]$.

The Racetrack Principle (p. 157) Suppose that g and h are continuous on $[a, b]$ and differentiable on (a, b), and that $g'(x) \leq h'(x)$ for $a < x < b$.

- If $g(a) = h(a)$, then $g(x) \leq h(x)$ for $a \leq x \leq b$.

- If $g(b) = h(b)$, then $g(x) \geq h(x)$ for $a \leq x \leq b$.

Theorem: Differentiability and Local Linearity (p. 151) Suppose f is differentiable at $x = a$ and $E(x)$ is the error in the tangent line approximation, that is: $E(x) = f(x) - f(a) - f'(a)(x - a)$. Then $\lim\limits_{x \to a} \dfrac{E(x)}{x - a} = 0$.

Theorem: A Differentiable Function is Continuous (p. 101) If $f(x)$ is differentiable at a point $x = a$, then $f(x)$ is continuous at $x = a$.

The Definite Integral

The **definite integral of f from** a **to** b (p. 249), denoted $\int_a^b f(x)\,dx$, is the limit of the left and the right sums as the width of the rectangles is shrunk to 0, where

Left-hand sum $= \sum\limits_{i=0}^{n-1} f(x_i)\Delta x$ (p. 244)

$$= f(x_0)\Delta x + f(x_1)\Delta x + \cdots + f(x_{n-1})\Delta x$$

Right-hand sum $= \sum\limits_{i=1}^{n} f(x_i)\Delta x$ (p. 248)

$$= f(x_1)\Delta x + f(x_2)\Delta x + \cdots + f(x_n)\Delta x$$

Figure R.19: Left-hand sum (p. 245)

Figure R.20: Right-hand sum (p. 245)

If f is nonnegative, $\int_a^b f(x)\,dx$ represents the area under the curve between $x = a$ and $x = b$ (p. 250). If f has any sign, $\int_a^b f(x)\,dx$ is the sum of the areas above the x-axis, counted positively, and the areas below the x-axis, counted negatively (p. 251). If $F'(t)$ is the rate of change of some quantity $F(t)$, then $\int_a^b F'(t)\,dt$ is the **total change** in $F(t)$ between $t = a$ and $t = b$ (p. 256). The **average value** of f on the interval $[a, b]$ is given by $\dfrac{1}{b - a}\displaystyle\int_a^b f(x)\,dx$ (p. 260).

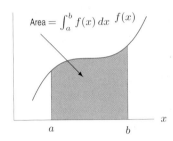

Figure R.21: The definite integral $\int_a^b f(x)\,dx$ (p. 250)

Figure R.22: Area and average value (p. 260)

The **units** of $\int_a^b f(x)\,dx$ are Units of $f(x) \times$ Units of x (p. 255).

The Fundamental Theorem of Calculus If f is continuous on $[a, b]$ and $f(x) = F'(x)$, then

$$\int_a^b f(x)\,dx = F(b) - F(a) \quad \text{(p. 256)}.$$

Properties of Definite Integrals (pp. 264, 266)
If a, b, and c are any numbers and f, g are continuous functions, then

$$\int_b^a f(x)\,dx = -\int_a^b f(x)\,dx$$

$$\int_a^b (f(x) \pm g(x))\,dx = \int_a^b f(x)\,dx \pm \int_a^b g(x)\,dx$$

$$\int_a^c f(x)\,dx + \int_c^b f(x)\,dx = \int_a^b f(x)\,dx$$

$$\int_a^b cf(x)\,dx = c\int_a^b f(x)\,dx$$

Antiderivatives

An **antiderivative** of a function $f(x)$ is a function $F(x)$ such that $F'(x) = f(x)$ (p. 282). There are infinitely many antiderivatives of f since $F(x) + C$ is an antiderivative of f for any constant C, provided $F'(x) = f(x)$ (p. 282). The **indefinite integral** of f is the family of antiderivatives $\int f(x)\,dx = F(x) + C$ (p. 288).

Construction Theorem (Second Fundamental Theorem of Calculus) If f is a continuous function on an interval and a is any number in that interval, then $F(x) = \int_a^x f(t)\,dt$ is an antiderivative of f (p. 299).

Properties of Indefinite Integrals (p. 290)

$$\int (f(x) \pm g(x))\,dx = \int f(x)\,dx \pm \int g(x)\,dx$$

$$\int cf(x)\,dx = c\int f(x)\,dx$$

Some antiderivatives:

$$\int k\,dx = kx + C \text{ (p. 288)}$$

$$\int x^n\,dx = \frac{x^{n+1}}{n+1} + C, \quad n \neq -1 \text{ (p. 289)}$$

$$\int \frac{1}{x}\,dx = \ln|x| + C \text{ (p. 290)}$$

$$\int e^x\,dx = e^x + C \text{ (p. 290)}$$

$$\int \cos x\,dx = \sin x + C \text{ (p. 290)}$$

$$\int \sin x\,dx = -\cos x + C \text{ (p. 290)}$$

$$\int \frac{dx}{1 + x^2} = \arctan x + C \text{ (p. 336)}$$

$$\int \frac{dx}{\sqrt{1 - x^2}} = \arcsin x + C \text{ (p. 334)}$$

Substitution (p. 312) For integrals of the form $\int f(g(x))g'(x)\,dx$, let $w = g(x)$. Choose w, find dw/dx and substitute for x and dx. Convert limits of integration for definite integrals.

Integration by Parts (p. 320) Used mainly for products; also for integrating $\ln x$, $\arctan x$, $\arcsin x$.

$$\int uv'\,dx = uv - \int u'v\,dx$$

Partial Fractions (p. 331) To integrate a rational function, $P(x)/Q(x)$, express as a sum of a polynomial and terms of the form $A/(x-c)^n$ and $(Ax+B)/q(x)$, where $q(x)$ is an unfactorable quadratic.

Trigonometric Substitutions (p. 334) To simplify $\sqrt{x^2 - a^2}$, try $x = a\sin\theta$ (p. 335). To simplify $a^2 + x^2$ or $\sqrt{a^2 + x^2}$, try $x = a\tan\theta$ (p. 336).

Numerical approximations for definite integrals (p. 339) Riemann sums (left, right, midpoint) (p. 340), trapezoid rule (p. 340), Simpson's rule (p. 346)

Approximation errors (pp. 341, 342)

f increasing: left underestimates, right overestimates (p. 341)
f decreasing: right underestimates, left overestimates (p. 341)
f concave up: midpoint underestimates, trapezoid overestimates (p. 342)
f concave down: trapezoid underestimates, midpoint overestimates (p. 342)

Evaluating improper integrals (p. 348)

- Infinite limit of integration: $\int_a^\infty f(x)\,dx = \lim_{b\to\infty}\int_a^b f(x)\,dx$ (p. 349)

- For $a < b$, if integrand is unbounded at $x = b$, then: $\int_a^b f(x)\,dx = \lim_{c\to b^-}\int_a^c f(x)\,dx$ (p. 352)

Testing improper integrals for convergence by comparison (p. 357):

- If $0 \le f(x) \le g(x)$ and $\int_a^\infty g(x)\,dx$ converges, then $\int_a^\infty f(x)\,dx$ converges

- If $0 \le g(x) \le f(x)$ and $\int_a^\infty g(x)\,dx$ diverges, then $\int_a^\infty f(x)\,dx$ diverges

Applications of Integration

Total quantities can be approximated by slicing them into small pieces and summing the pieces. The limit of this sum is a definite integral which gives the exact total quantity.

Applications to Geometry
To calculate the volume of a solid, slice the volume into pieces whose volumes you can estimate (pp. 368, 374). Use this method to calculate volumes of revolution (p. 375) and volumes of solids with known cross sectional area (p. 377). Curve $f(x)$ from $x = a$ to $x = b$ has

$$\text{Arc length} = \int_a^b \sqrt{1 + (f'(x))^2}\,dx \quad \text{(p. 378)}.$$

Mass and Center of Mass from Density, δ

$$\text{Total mass} = \int_a^b \delta(x)\,dx \text{ (p. 390)}$$

$$\text{Center of mass} = \frac{\int_a^b x\delta(x)\,dx}{\int_a^b \delta(x)\,dx} \text{ (p. 395)}$$

To find the center of mass of two- and three-dimensional objects, use the formula separately on each coordinate (p. 395).

Applications to Physics

$$\text{Work done} = \text{Force} \times \text{Distance (p. 399)}$$
$$\text{Pressure} = \text{Density} \times g \times \text{Depth (p. 403)}$$
$$\text{Force} = \text{Pressure} \times \text{Area (p. 403)}$$

Applications to Economics Present and future value of income stream, $P(t)$ (p. 409); consumer and producer surplus (p. 412).

$$\text{Present value} = \int_0^M P(t)e^{-rt}\,dt \text{ (p. 409)}$$

$$\text{Future value} = \int_0^M P(t)e^{r(M-t)}\,dt \text{ (p. 409)}$$

Polar Coordinates

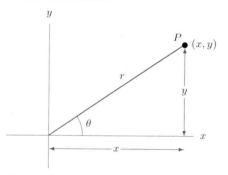

Figure R.23: Cartesian and polar coordinates for the point P

The **polar coordinates** (p. 382) of a point are related to its Cartesian coordinates by

- $x = r\cos\theta$ and $y = r\sin\theta$
- $r = \sqrt{x^2 + y^2}$ and $\tan\theta = \dfrac{y}{x}$, $x \neq 0$

For a constant a, the equation $r = a$ gives a circle of radius a, and the equation $\theta = a$ gives a ray from the origin making an angle of θ with the positive x-axis. The equation $r = \theta$ gives an **Archimedean spiral** (p. 384).

Area in Polar Coordinates (p. 387)
For a curve $r = f(\theta)$, with $\alpha \le \theta \le \beta$, which does not cross itself,

$$\text{Area of region enclosed} = \frac{1}{2}\int_\alpha^\beta f(\theta)^2\,d\theta.$$

Slope and Arclength in Polar Coordinates (p. 388)
For a curve $r = f(\theta)$, we can express x and y in terms of θ as a parameter, giving

$$x = r\cos\theta = f(\theta)\cos\theta \quad \text{and} \quad y = r\sin\theta = f(\theta)\sin\theta.$$

Then

$$\text{Slope} = \frac{dy}{dx} = \frac{dy/d\theta}{dx/d\theta}$$

and

$$\text{Arc length} = \int_{\alpha}^{\beta} \sqrt{\left(\frac{dx}{d\theta}\right)^2 + \left(\frac{dy}{d\theta}\right)^2}\, d\theta.$$

Alternatively (p. 390),

$$\text{Arc length} = \int_{\alpha}^{\beta} \sqrt{(f'(\theta))^2 + (f(\theta))^2}\, d\theta.$$

Sequences and Series

A **sequence** $s_1, s_2, s_3, \ldots, s_n, \ldots$ has a **limit** L, written $\lim_{n \to \infty} s_n = L$, if we can make s_n as close to L as we please by choosing a sufficiently large n. The sequence **converges** if a limit exists, **diverges** if no limit exists (see p. 440). Limits of sequences satisfy the same properties as limits of functions stated in Theorem 1.2 (p. 51) and

$$\lim_{n \to \infty} x^n = 0 \quad \text{if } |x| < 1 \qquad \lim_{n \to \infty} 1/n = 0 \quad \text{(p. 440)}$$

A sequence s_n is **bounded** if there are constants K and M such that $K \le s_n \le M$ for all n (p. 440). A convergent sequence is bounded. A sequence is monotone if it is either increasing, that is $s_n < s_{n+1}$ for all n, or decreasing, that is $s_n > s_{n+1}$ for all n (p. 441).

Theorem: Convergence of a Monotone, Bounded Sequence (p. 441): If a sequence s_n is bounded and monotone, it converges.

A **series** is an infinite sum $\sum a_n = a_1 + a_2 + \cdots$. The nth **partial sum** is $S_n = a_1 + a_2 + \cdots + a_n$ (p. 450). If $S = \lim_{n \to \infty} S_n$ exists, then the series $\sum a_n$ **converges**, and its sum is S (p. 450). If a series does not converge, we say that it **diverges** (p. 450). The sum of a **finite geometric series** is (p. 446):

$$a + ax + ax^2 + \cdots + ax^{n-1} = \frac{a(1-x^n)}{1-x}, \qquad x \ne 1.$$

The sum of an **infinite geometric series** is (p. 446):

$$a + ax + ax^2 + \cdots + ax^n + \cdots = \frac{a}{1-x}, \qquad |x| < 1.$$

The p-**series** $\sum 1/n^p$ converges if $p > 1$ and diverges if $p \le 1$ (p. 453). The **harmonic series** $\sum 1/n$ diverges (p. 452), the **alternating harmonic series** $\sum (-1)^{n-1}(1/n)$ converges (p. 460).

Convergence Tests

Theorem: Convergence Properties of Series (p. 451)

1. If $\sum a_n$ and $\sum b_n$ converge and if k is a constant, then
   - $\sum (a_n + b_n)$ converges to $\sum a_n + \sum b_n$.
   - $\sum k a_n$ converges to $k \sum a_n$.
2. Changing a finite number of terms in a series does not change whether or not it converges, although it may change the value of its sum if it does converge.

3. If $\lim_{n \to \infty} a_n \ne 0$ or $\lim_{n \to \infty} a_n$ does not exist, then $\sum a_n$ diverges.

4. If $\sum a_n$ diverges, then $\sum k a_n$ diverges if $k \ne 0$.

Theorem: Integral Test (p. 453) Suppose $c \ge 0$ and $f(x)$ is a decreasing positive function, defined for all $x \ge c$, with $a_n = f(n)$ for all n.

- If $\int_c^{\infty} f(x)\, dx$ converges, then $\sum a_n$ converges.

- If $\int_c^{\infty} f(x)\, dx$ diverges, then $\sum a_n$ diverges.

Theorem: Comparison Test (p. 455) Suppose $0 \le a_n \le b_n$ for all n.

- If $\sum b_n$ converges, then $\sum a_n$ converges.
- If $\sum a_n$ diverges, then $\sum b_n$ diverges.

Theorem: Limit Comparison Test (p. 457) Suppose $a_n > 0$ and $b_n > 0$ for all n. If

$$\lim_{n \to \infty} \frac{a_n}{b_n} = c \qquad \text{where } c > 0,$$

then the two series $\sum a_n$ and $\sum b_n$ either both converge or both diverge.

Theorem: Convergence of Absolute Values Implies Convergence (p. 458): If $\sum |a_n|$ converges, then so does $\sum a_n$.

We say $\sum a_n$ is **absolutely convergent** if $\sum a_n$ and $\sum |a_n|$ both converge and **conditionally convergent** if $\sum a_n$ converges but $\sum |a_n|$ diverges (p. 461).

Theorem: The Ratio Test (p. 458) For a series $\sum a_n$, suppose the sequence of ratios $|a_{n+1}|/|a_n|$ has a limit:

$$\lim_{n \to \infty} \frac{|a_{n+1}|}{|a_n|} = L.$$

- If $L < 1$, then $\sum a_n$ converges.
- If $L > 1$, or if L is infinite, then $\sum a_n$ diverges.
- If $L = 1$, the test does not tell us anything about the convergence of $\sum a_n$.

Theorem: Alternating Series Test (p. 460) A series of the form

$$\sum_{n=1}^{\infty} (-1)^{n-1} a_n = a_1 - a_2 + a_3 - a_4 + \cdots + (-1)^{n-1} a_n + \cdots$$

converges if $0 < a_{n+1} < a_n$ for all n and $\lim_{n \to \infty} a_n = 0$.

Theorem: Error Bounds for Alternating Series (p. 461) Let $S_n = \sum_{i=1}^{n} (-1)^{i-1} a_i$ be the nth partial sum of an alternating series and let $S = \lim_{n \to \infty} S_n$. Suppose that $0 < a_{n+1} < a_n$ for all n and $\lim_{n \to \infty} a_n = 0$. Then $|S - S_n| < a_{n+1}$.

Power Series

$$P(x) = C_0 + C_1(x-a) + C_2(x-a)^2 + \cdots$$
$$+ C_n(x-a)^n + \cdots$$
$$= \sum_{n=0}^{\infty} C_n(x-a)^n \quad \text{(p. 463)}.$$

The **radius of convergence** is 0 if the series converges only for $x = a$, ∞ if it converges for all x, and the positive number R if it converges for $|x-a| < R$ and diverges for $|x-a| > R$ (p. 465). The **interval of convergence** is the interval between $a - R$ and $a + R$, including any endpoint where the series converges (p. 465).

Theorem: Method for Computing Radius of Convergence (p. 466) To calculate the radius of convergence, R, for the power series $\sum_{n=0}^{\infty} C_n(x-a)^n$, use the ratio test with $a_n = C_n(x-a)^n$.

- If $\lim_{n \to \infty} |a_{n+1}|/|a_n|$ is infinite, then $R = 0$.
- If $\lim_{n \to \infty} |a_{n+1}|/|a_n| = 0$, then $R = \infty$.
- If $\lim_{n \to \infty} |a_{n+1}|/|a_n| = K|x-a|$, where K is finite and nonzero, then $R = 1/K$.

Approximations

The **Taylor polynomial of degree** n approximating $f(x)$ for x near a is:

$$P_n(x) = f(a) + f'(a)(x-a) + \frac{f''(a)}{2!}(x-a)^2 + \cdots\cdots$$
$$+ \frac{f^{(n)}(a)}{n!}(x-a)^n \quad \text{(p. 481)}$$

The **Taylor series** approximating $f(x)$ for x near a is:

$$f(x) = f(a) + f'(a)(x-a) + \frac{f''(a)}{2!}(x-a)^2 + \cdots$$
$$+ \frac{f^{(n)}(a)}{n!}(x-a)^n + \cdots \quad \text{(p. 487)}$$

Theorem: The Lagrange Error Bound for $P_n(x)$ **(p. 498)** Suppose f and all its derivatives are continuous. If $P_n(x)$ is the nth Taylor polynomial for to $f(x)$ about a, then

$$|E_n(x)| = |f(x) - P_n(x)| \leq \frac{M}{(n+1)!}|x-a|^{n+1},$$

where $\max \left| f^{(n+1)} \right| \leq M$ on the interval between a and x.

Taylor series for $\sin x$**,** $\cos x$**,** e^x **(p. 486):**

$$\sin x = x - \frac{x^3}{3!} + \frac{x^5}{5!} - \frac{x^7}{7!} + \frac{x^9}{9!} - \cdots$$
$$\cos x = 1 - \frac{x^2}{2!} + \frac{x^4}{4!} - \frac{x^6}{6!} + \frac{x^8}{8!} - \cdots$$
$$e^x = 1 + x + \frac{x^2}{2!} + \frac{x^3}{3!} + \frac{x^4}{4!} + \cdots$$

Taylor series for $\ln x$ **about** $x = 1$ converges for $0 < x \leq 2$ (p. 487):

$$(x-1) - \frac{(x-1)^2}{2} + \frac{(x-1)^3}{3} - \frac{(x-1)^4}{4} + \cdots$$

The Binomial Series for $(1+x)^p$ converges for $-1 < x < 1$ (p. 488):

$$1 + px + \frac{p(p-1)}{2!}x^2 + \frac{p(p-1)(p-2)}{3!}x^3 + \cdots$$

The **Fourier series** of $f(x)$ is (p. 505)

$$f(x) = a_0 + a_1 \cos x + a_2 \cos 2x + a_3 \cos 3x + \cdots$$
$$+ b_1 \sin x + b_2 \sin 2x + b_3 \sin 3x + \cdots,$$

where

$$a_0 = \frac{1}{2\pi} \int_{-\pi}^{\pi} f(x)\,dx,$$
$$a_k = \frac{1}{\pi} \int_{-\pi}^{\pi} f(x)\cos(kx)\,dx \quad \text{for } k \text{ a positive integer,}$$
$$b_k = \frac{1}{\pi} \int_{-\pi}^{\pi} f(x)\sin(kx)\,dx \quad \text{for } k \text{ a positive integer.}$$

Differential Equations

A **differential equation** for the function $y(x)$ is an equation involving x, y and the derivatives of y (p. 522). The **order** of a differential equation is the order of the highest-order derivative appearing in the equation (p. 524). A **solution** to a differential equation is any function y that satisfies the equation (p. 523). The **general solution** to a differential equation is the **family of functions** that satisfies the equation (p. 523). An **initial value problem** is a differential equation together with an **initial condition**; a solution to an initial value problem is called a **particular solution** (p. 523). An **equilibrium solution** to a differential equation is a particular solution where y is constant and $dy/dx = 0$ (p. 545).

First-order equations: methods of solution. A **slope field** corresponding to a differential equation is a plot in the xy-plane of small line segments with slope given by the differential equation (p. 526). **Euler's method** approximates the solution of an initial value problem with a string of small line segments (p. 532). Differential equations of the form $dy/dx = g(x)f(y)$ can be solved analytically by **separation of variables** (p. 535).

First-order equations: applications. The differential equation for **exponential growth and decay** is of the form

$$\frac{dP}{dt} = kP.$$

The **solution** is of the form $P = P_0 e^{kt}$, where P_0 is the initial value of P, and positive k represents growth while negative k represents decay (p. 540). Applications of growth and decay include **continuously compounded interest (p. 541), population growth (p. 558),** and **Newton's law of heating and cooling (p. 543).** The **logistic equation** for population growth is of the form

$$\frac{dP}{dt} = kP\left(1 - \frac{P}{L}\right),$$

where L is the **carrying capacity** of the population (p. 561). The **solution to the logistic equation** is of the form $P = \dfrac{L}{1 + Ae^{-kt}}$, where $A = \dfrac{L - P_0}{P_0}$ (p. 563).

Systems of differential equations. Two interacting populations, w and r, can be modeled by two equations

$$\frac{dw}{dt} = aw - cwr \quad \text{and} \quad \frac{dr}{dt} = -br + kwr \quad \text{(p. 571)}.$$

Solutions can be visualized as **trajectories** in the wr-**phase plane** (p. 571). An **equilibrium point** is one at which $dw/dt = 0$ and $dr/dt = 0$ (p. 571). A **nullcline** is a curve along which $dw/dt = 0$ or $dr/dt = 0$ (p. 576).

Second-order equations. A **linear second-order differential equation with constant coefficients** is of the form

$$\frac{d^2y}{dt^2} + b\frac{dy}{dt} + cy = 0 \quad \text{(p. 588)}.$$

The solution (given on p. 593) depends on the sign of the quantity $b^2 - 4c$ (see pp. 589, 590). Solutions are calculated using the **characteristic equation (p. 589)** $r^2 + br + c = 0$.

- if $b^2 - 4c > 0$, then $y = C_1 e^{r_1 t} + C_2 e^{r_2 t}$, where r_1 and r_2 are the real solutions to the characteristic equation;

- if $b^2 - 4c = 0$, then $y = (C_1 t + C_2)e^{-bt/2}$;

- if $b^2 - 4c < 0$, then $y = C_1 e^{\alpha t}\cos\beta t + C_2 e^{\alpha t}\sin\beta t$, where $\alpha \pm i\beta$ are the complex solutions to the characteristic equation.

The **equation for oscillations of a mass on a spring** is of the form $\dfrac{d^2 s}{dt^2} = -\dfrac{k}{m}s$ (p. 582). The general solution is of the form $s(t) = C_1\cos\omega t + C_2\sin\omega t$, where $\omega = \sqrt{k/m}$, and C_1 and C_2 are arbitrary constants (p. 583). The **amplitude** is $A = \sqrt{C_1^2 + C_2^2}$ and the **phase shift**, ϕ, is given by $\tan\phi = C_1/C_2$ (p. 585). We choose ϕ in $(-\pi, \pi]$ such that if $C_1 > 0$, ϕ is positive, and if $C_1 < 0$, ϕ is negative.

ANSWERS TO ODD NUMBERED PROBLEMS

Section 1.1

3   V (thousand dollars)

$(5, 6)$

a (years)

5   $y = (1/2)x + 2$

7   Slope: $-12/7$
Vertical intercept: $2/7$

9   Slope: 2
Vertical intercept: $-2/3$

11 (a) (V)
   (b) (VI)
   (c) (I)
   (d) (IV)
   (e) (III)
   (f) (II)

13 Parallel: $y = -4x + 9$
Perpendicular:
   $y = 0.25x + 4.75$

15 Domain: $0 \leq x \leq 5$
Range: $0 \leq y \leq 4$

17 Domain: $1 \leq x \leq 5$
Range: $1 \leq y \leq 6$

19 Domain: all x
Range: $0 < y \leq 1/2$

21 All $x \neq 0, -1$
   $x = \pm 2$

23 $E = kv^3$

25 (a) (II)
   (b) (I)
   (c) (IV)

27   d (distance)

t (time)

29 (a) $C = 12 + 0.2w$
   (b) 0.2 \$/kg
   (c) \$12

31 (a) Slope = 1.8
   (b) $^\circ$F $= 1.8(^\circ$C$) + 32$
   (c) 68°F
   (d) -40°

33   $y = -(950/7)x + 950$

35 (a) (i) $q = 320 - (2/5)p$
       (ii) $p = 800 - (5/2)q$
   (b) p (dollars)

800

550

100 320

q (items)

37 (a) $l = al_0(t - t_0) + l_0$
       Slope $= al_0$
       $y_{\text{int}} = -al_0 t_0 + l_0$
   (b) $l = 0.001t + 99.94$
   (c) As temperature rises:
        slope positive, metal expands
        slope negative, metal contracts

Section 1.2

1   5; 7%

3   3.2; 3% (continuous)

5 (a) $P = 1000 + 50t$
  (b) $P = 1000(1.05)^t$

7   Concave up

9   Neither

11 (a) x-interval: D to E, H to I
   (b) x-interval: A to B, E to F
   (c) x-interval: C to D, G to H
   (d) x-interval: B to C, F to G

13 (a) 125%
   (b) 9 times

15 (a)   customers

time

   (b) Concave down

17   $y = 3(2^x)$

19   $y = 2(3^x)$

21 (a) 1.5
   (b) 50%

23   $P = 15(1.2840)^t$; growth

25   $P = P_0(1.2214)^t$; growth

27 (a) $P = 10^6(e^{0.02t})$
   (b) P

1,000,000

t

29 (a) 47.3%
   (b) 11,199.4%

31   48.8%

33   2.3 years

35 (a) $k(t)$
   (b) $h(t)$
   (c) $g(t)$

37   $f(s) = 2(1.1)^s$
     $g(s) = 3(1.05)^s$
     $h(s) = (1.03)^s$

39 (a) $P = 2.5t + 50$
   (b) $P = 50(1.035)^t$
   (c)   price (\$1000s)

200
160
120
80
40

Exponential
Linear

0 10 20 30 40 t (years
                    since 1970)

   (d) Exponential

Section 1.3

1 (a) 4
  (b) 2
  (c) $(x + 1)^2$
  (d) $x^2 + 1$
  (e) $t^2(t + 1)$

3 (a) e
  (b) e^2
  (c) e^{x^2}
  (d) e^{2x}
  (e) $e^t t^2$

5

$n(t)$

-4 4   6 t

7

$k(t)$

-4 4   6 t

9

10
5

-4 -2 2 x

11

10
5

-4 -2 2 x

13

10
5

-4 -2 2 x

15 (a) $t^2 + 2t + 2$
   (b) $t^4 + 2t^2 + 2$
   (c) 5
   (d) $2t^2 + 2$
   (e) $t^4 + 2t^2 + 2$

17   $2z + 1$

19   $2zh - h^2$

23 Length of column of mercury when temperature
    is 75°F

25 Not invertible

27 (a) -1

(b)

29 $y = (x - 2)^3 - 1$

31 Neither

33 Neither

35 Even

37 Neither

39 Not invertible

41 Invertible

43 0.4

45 -0.9

47

49 $f(x) = x^3$
$g(x) = x + 1$

51 $2(y - 1)^3 - (y - 1)^2$

53 (a) $k = f(p) = (1/2.2)p$

(b) $p = 2.2k$; Weight in pounds given mass in kilograms

Section 1.4

1 $1/2$

3 $5A^2$

5 $-1 + \ln A + \ln B$

7 $(\log 11)/(\log 3) = 2.2$

9 $(\log(2/5))/(\log 1.04) = -23.4$

11 -3.26

13 -0.347

15 0.515

17 0.382, 1.815

19 $(\log P - \log P_0)/(\log a)$

21 $\log(Q_0/P_0)/\log(a/b)$

23 $\ln(P/P_0)/k$

25 $P = 10e^{0.5306t}$

27 $P = 4e^{-0.5978t}$

29 $f^{-1}(t) = 10\ln\left(\frac{t}{50}\right)$

31 A: e^x
B: x^2
C: $x^{1/2}$
D: $\ln x$

33 About 14.21 years

35 16 kg

37 $78e^{0.555t}$
55.5%

39 $y = 3.3219x + 1$

41 (a) $A = 100e^{-0.17t}$

(b) $t \approx 4$ hours

(c) $t = 4.077$ hours

43 (a) 81%

(b) 32.9 hours

(c)

45 96.34 years

Section 1.5

1 Negative
0
Undefined

3 Positive
Positive
Positive

5 Positive
Positive
Positive

7 Positive
Negative
Negative

9 Negative
Positive
Negative

11 0.588

13 (a) 1

(b) $2\pi/3$

(c)

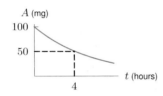

15 $8\pi; 3$

17 $2; 0.1$

19 $f(x) = 2\sin(x/4)$

21 $f(x) = -4\sin(2x)$

23 $f(x) = 5\sin((\pi/3)x)$

25 $f(x) = 2 - \sin x$

27 $f(x) = 2\sin(x/4) + 2$

29 $(\sin^{-1}(2/5))/3 \approx 0.1372$

31 $(\tan^{-1} 2)/5 = 0.221$

33 No solution

35 0.3 seconds

37 $2/\pi$ or $-2/\pi$

39 (a) Maximum displacement from equilibrium

(b) ω

41 US: 156 volts max, 60 cycles/sec
Eur: 339 volts max, 50 cycles/sec

43 (a)

(b) $P = 800 - 100\cos(\pi t/6)$

45 (a) $(2.3, -4.4)$
(b) Once around circle and back
47 (b)

$$y = \arccos x$$

(c) The domain of arccos and arcsin are the same because their inverses (sine and cosine) have the same range

Section 1.6

1 $10 \cdot 2^x$

3 As $x \to \infty, y \to \infty$
   As $x \to -\infty, y \to -\infty$

5 (I) (a) 3 (b) Negative
  (II) (a) 4 (b) Positive
  (III) (a) 4 (b) Negative
  (IV) (a) 5 (b) Negative
  (V) (a) 5 (b) Positive

7 $f(x) =$
   $k(x + 3)(x - 1)(x - 4)$
   (for $k < 0$)

9 $f(x) = k(x + 2)(x - 1)(x - 3)(x - 5)$
   (for $k > 0$)

11 (a)

(b)

(c)

(d)

13 $y = -\frac{1}{2}(x + 2)^2(x - 2)$

15 (a) 1.3 m^2
   (b) 86.8 kg
   (c) $h = 112.6s^{4/3}$

17 44.25 ft and 708 ft

19 (a) $R = kr^4$
       (k is a constant)
   (b) $R = 4.938r^4$
   (c) $3086.42 \text{ cm}^3/\text{sec}$

21 Hits the ground at $t = 4$ sec
    Maximum height at $t = 2$ sec
    Maximum height = 64 feet

23 $h(t) = ab^t$
    $g(t) = kt^3$
    $f(t) = ct^2$

25 3 zeros:
    $x \approx -1, x \approx 3,$
    $x > 10$

27 (a) $x = -c/2$
   (b)

(c) x-coordinate moves left and y-coordinate moves up

29 (a) III
   (b) IV
   (c) I
   (d) II

31 (a) $0 \le x \le 5,$
        $0 \le y \le 9$
   (b) $0 \le x \le 5,$
        $0 \le y \le 625$
   (c) $0 \le x \le 10,$
        $0 \le y \le 50,000$

Section 1.7

1 Yes
3 Yes
5 Yes
7 No
9 No

15 $f(t) = \begin{cases} 6, & 0 < t \le 7 \\ 12, & 7 < t \end{cases}$
    Not continuous on any interval around 7

17 $k = 6$
19 20
21

23 $Q = \begin{cases} 1.2t & 0 \le t \le 0.5 \\ 0.6e^{0.001}e^{-.002t} & 0.5 < t \end{cases}$

Section 1.8

1 (a) 3
  (b) 7
  (c) Does not exist
  (d) 8

3 2

5 $0.01745\ldots$

7 (b) 1
  (c)

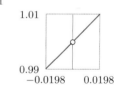

(d) $-0.0033 < x < 0.0033,$
     $0.99 < y < 1.01$

9 (b) 0
  (c)

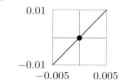

(d) $-0.005 < x < 0.005,$
     $-0.01 < y < 0.01$

11 (b) 2
   (c)

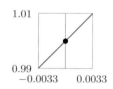

(d) $-0.0865 < x < 0.0865,$
     $1.99 < y < 2.01$

13 (b) 1
   (c)

(d) $-0.0198 < x < 0.0198,$
$0.99 < y < 1.01$

15 (b) 4

(c)

(d) $1.99 < x < 2.01,$
$3.99 < y < 4.01$

17 (b) 0

(c)

(d) $1.55 < x < 1.59,$
$-0.01 < y < 0.01$

19 $\lim_{x\to 4^+} f(x) = 1, \lim_{x\to 4^-} f(x) = -1, \lim_{x\to 4} f(x)$ does not exist

21 $\lim_{x\to 3^+} f(x) = \lim_{x\to 3^-} f(x) = \lim_{x\to 3} f(x) = 7$

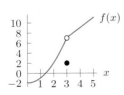

29 The limit appears to be 1

31 $1/3$

33 $2/3$

35 $3/2$

37 4 or -4

39 4

41 Any k

43 $k \le 2$

45 $0.1, 0.05, 0.01, 0.005, 0.001, 0.0005$

Chapter 1 Review

1 (a) $[0, 7]$
(b) $[-2, 5]$
(c) 5
(d) $(1, 7)$
(e) Concave up
(f) 1
(g) No

3 $(\log(25/2))/(\log 5) \approx 1.57$

5 1.68

7 Amplitude: 2
Period: $2\pi/5$

9 $y = e^{0.4621x}$ or $y = (1.5874)^x$

11 $y = -k(x^2 + 5x)$
$(k > 0)$

13 $z = 1 - \cos\theta$

15 $x = k(y^2 - 4y)$
$(k > 0)$

17 $y = -(x+5)(x+1)(x-3)^2$

19 Simplest is $y = 1 - e^{-x}$

21 $f(x) = \sin(2(\pi/5)x)$

23 $f(x) = x^3$
$g(x) = \ln x$

25 Not continuous

27 Not continuous

29 (b) 0

(c)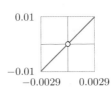

(d) $-0.0029 < x < 0.0029,$
$-0.01 < y < 0.01$

31 (b) $1/2$
(c)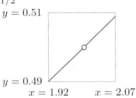

(d) $1.92 < x < 2.07,$
$0.49 < y < 0.51$

33 $\lim_{x\to 0^+} f(x) = \lim_{x\to 0^-} f(x) = \lim_{x\to 0} f(x) = 1$

35 (a)

37 (a) $S = 113 - 0.94t$
(b) During 2038

39 13,500 bacteria

41 8.46%

43 $Q(m) = T + L + Pm$
$T =$ fuel for take-off
$L =$ fuel for landing
$P =$ fuel per mile in the air
$m =$ length of the trip (miles)

45 (a) $S = 2\pi r^2 + 2V/r$
(b) $S \to \infty$ as $r \to \infty$
(c)

47 Depth $= 7 + 1.5\sin(\pi t/3)$

49 (a) f is invertible
(b) $f^{-1}(400) \approx 1979$ is year in which 400 million motor vehicles were registered
(c)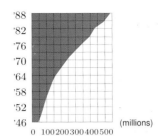

51 (a) 2π

53 (a) $r(p) = kp(A - p)$
$(k > 0)$
(b) $p = A/2$

55 $0.1, 0.05, 0.00007$

57 (a) $f(x) = (x-a)(x+a)(x+b)(x-c)$
(b)

59 (a) $f(x) \to \infty$ as $x \to \infty$
$f(x) \to 16$ as $x \to -\infty$
(b) $(e^x+1)(e^{2x}-2)(e^x-2)(e^{2x}+2e^x+4)$
Two zeros
(c) $(\ln 2)/2, \ln 2$
One twice other

61 (a) $p(x) = x^2 + 3x + 9$
$r(x) = -3, q(x) = x - 3$
(b) $f(x) \approx -3/(x-3)$ for x near 3
(c) $f(x) \approx x^2 + 3x + 9$ as $x \to \pm\infty$

63 (a) $1 - 8\cos^2 x + 8\cos^4 x$
(b) $1 - 8\sin^2 x + 8\sin^4 x$

Ch. 1 Understanding

1 False

3 True

5 True

7 True

9 False

11 False

13 True

15 False

17 False

19 False

21 True

23 True

25 True

27 True

29 False

31 Possible answer:
$$f(x) = \begin{cases} 1 & x \le 2 \\ x & x > 2 \end{cases}$$

33 Possible answer $f(x) = 1/(x + 7\pi)$

35 Possible answer:
$$f(x) = (x - 1)/(x - 2)$$

37 $f(x) = x, g(x) = -2x$

39 $f(x) = e^x, g(x) = e^{-2x}$

41 False; $f(x) = x/(x^2 + 1)$

43 False; $y = x + 1$ at points
$(1, 2)$ and $(2, 3)$

45 False; $y = 4x + 1$ starting at $x = 1$

47 False

49 True; $f(x) = 0$

51 False

53 True

55 False

57 False

59 True

61 False

Section 2.1

1 $265/3$ km/hr

3 (a) (i) 8.4 m/sec
(ii) 8.04 m/sec
(iii) 8.004 m/sec
(b) 8 m/sec

5

Slope	-3	-1	0	$1/2$	1	2
Point	F	C	E	A	B	D

7 27

9 1.9

11

13

15 $F < B < E < 0 < D < A < C$

17 $v_{avg} = 2.5$ ft/sec
$v(0.2) = 4.5$ ft/sec

19 3

21 6

Section 2.2

1 12

3 (b) 0.24
(c) 0.22

5 negative

7 $f'(2) \approx 9.89$

9

11 41

13 (a) $g(2) = 5$
(b) $g'(2) = -0.4$

15 (a) $f(4)$
(b) $f(2) - f(1)$
(c) $(f(2) - f(1))/(2 - 1)$
(d) $f'(1)$

17 (a) Line from (0,0) to $(4, f(4))$
(b) $f(3)/3$

19 (a) $(f(b) - f(a))/(b - a)$
(b) Slopes same
(c) Yes

21 $f'(2) \approx 6.77$

23 $f'(1) \approx -1.558$
$f'(\frac{\pi}{4}) \approx -1$

25 (a) They are the same

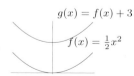

(b) $f'(x) = g'(x)$ for all x

27 -6

29 -1

31 $1/4$

33 100

35 -1

37 $-1/4$

39 $y = 100x - 500$

41 $y = x$

Section 2.3

1

3

5

7

9

11 (a) 3
(b) Positive: $0 < x < 4$
Negative: $4 < x < 12$

13 $-1/x^2$

15 $4x$

17

19

21 $f'(x)$ positive: $4 \leq x \leq 8$
$f'(x)$ negative: $0 \leq x \leq 3$
$f'(x)$ greatest: at $x \approx 8$

23

25

27

29

31

33 (a) x_3
(b) x_4
(c) x_5
(d) x_3

35 (a) Graph II
(b) Graph I
(c) Graph III

37 (a) $t = 3$
(b) $t = 9$
(c) $t = 14$
(d)

41

Section 2.4

1 (a) Negative
(b) Degrees/min

3 (a) Costs $350 for 200 gallons
(b) Costs $1.40 for 201st gallon

5 Dollars/year

7 dollars/year; negative

9 Feet/mile; negative

11 (b) Pounds/(Calories/day)

13 1.25 billion people; growing at
0.0174 billion people per year

17 mpg/mph

19 (a) $f'(a)$ is always positive
(c) $f'(100) = 2$: more
$f'(100) = 0.5$: less

21 (a) Liters per centimeter
(b) About 0.042 liters per centimeter
(c) Cannot expand much more

Section 2.5

1 (a) Negative
(b) Negative
(c) Positive

3

Point	f	f'	f''
A	$-$	0	$+$
B	$+$	0	$-$
C	$+$	$-$	$-$
D	$-$	$+$	$+$

5

7 $f'(x) > 0$
$f''(x) > 0$

9 $f'(x) < 0$
$f''(x) = 0$

11 $f'(x) > 0$
$f''(x) < 0$

13 $v(t) = 10t$
$a(t) = 10$

15 $0 \leq t \leq 1$:
acceleration $= 30$ ft/sec^2
$1 \leq t \leq 2$:
acceleration $= 22$ ft/sec^2

17 (a)

utility

quantity

(b) Derivative of utility is positive
2nd derivative of utility is negative

19 (a) x_4, x_5
(b) x_3, x_4
(c) x_3, x_4
(d) x_2, x_3
(e) x_1, x_2, x_5
(f) x_1, x_4, x_5

21 (a) x_6
(b) x_1

(c) x_3
(d) x_2
(e) x_6
(f) x_1

23 (a)

(b) One
(c) Between $x = 1$ and $x = 5$
(d) $\lim_{x \to -\infty} f(x) = -\infty$
(e) Yes
(f) No

Section 2.6

1 (a) $x = 1$
(b) $x = 1, 2, 3$

3 (a) No
(b) Yes, $x = -1$

5 Yes

7 Yes

9 Yes

11 (b) Yes
(c) No

13 (a) $B = B_0$ at $r = r_0$
B_0 is max

(b) Yes
(c) No

15 (a) Yes
(b) Yes

17 (a) No; yes.
(b) $f'(x)$ not differentiable at $x = 0$;
$f''(x)$ neither differentiable nor continuous
at $x = 0$

Chapter 2 Review

1

3

5

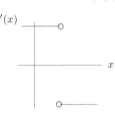

7 $10x + 1$

9 $2a$

11 $-2/a^3$

13 $-1/(2(\sqrt{a})^3)$

15 (a) $3, -1$
  (b) $-3, 1$

17 357

19 (a) $f(7) = 3$
  (b) $f'(7) = 4$

21

25 (a) $f'(0.6) \approx 0.5$
     $f'(0.5) \approx 2$
  (b) $f''(0.6) \approx -15$
  (c) Maximum: near $x = 0.8$
     minimum: near $x = 0.3$

27 (a) At $(0, \sqrt{19})$: slope $= 0$
     At $(\sqrt{19}, 0)$: slope is undefined
  (b) slope $\approx 1/2$
  (c) At $(-2, \sqrt{15})$: slope $\approx 1/2$
     At $(-2, -\sqrt{15})$: slope $\approx -1/2$
     At $(2, \sqrt{15})$: slope $\approx -1/2$

29 (a) Period 12 months

  (b) Max of 4500 on June 1st
  (c) Min of 3500 on Feb 1st
  (d) Growing fastest:
     April 1st
     Decreasing fastest:
     July 15 and Dec 15
  (e) About 400 deer/month

31 (a) Concave down

(b) $120° < T < 140°$
(c) $135° < T < 140°$
(d) $45 < t < 50$

33 (a) $f'(0) = 1.00000$
     $f'(0.3) = 1.04534$
     $f'(0.7) = 1.25521$
     $f'(1) = 1.54314$
  (b) They are about the same

35 (a) $2\cos^2 x - 2\sin^2 x$
     Answers may vary
  (b) $(d/dx)\sin(2x) = 2\cos(2x)$

37 (a) $1/x$
  (b) Graphs same shape, shifted vertically

39 (a) $\cos x, -\sin x, \cos^2 x - \sin^2 x$
     Form of answers may vary
  (b) No

Ch. 2 Understanding

1 False

3 True

5 True

7 True

9 False

11 True

13 False

15 True

17 True

19 False

21 True; $f(x) = x^2$

23 True; $f(x) = \begin{cases} 1 & x \geq 0 \\ -1 & x < 0 \end{cases}$

25 (a) Not a counterexample
  (b) Counterexample
  (c) Not a counterexample
  (d) Not a counterexample

Section 3.1

3 $11x^{10}$

5 $-12x^{-13}$

7 $3.2x^{2.2}$

9 $3x^{-1/4}/4$

11 $-4x^{-5}$

13 $6.1z^{-7.1}$

15 $(1/2)x^{-1/2}$

17 $-(1/3)\theta^{-4/3}$

19 ex^{e-1}

21 $6t - 4$

23 $2z - \frac{1}{2}z^{-2}$

25 $6/w^4 + 3/(2\sqrt{w})$

27 $15t^4 - \frac{5}{2}t^{-1/2} - 7t^{-2}$

29 $(3/2)x^{1/2} + (1/2)x^{-1/2}$

31 $-3t^{-2} - 8t^{-3}$

33 $1 - x^{-2}$

35 $-2t^{-3} - t^{-2} + 4t^{-5}$

37 $3x^2/a + 2ax/b - c$

39 a/c

41 $(8\pi rb)/3$

43 $2ax + b$

45 Do not apply

47 $\pi x^{(\pi-1)} + \pi x^{-(\pi+1)}$

49 Do not apply

51 $f'(t) = 6t^2 - 8t + 3$
   $f''(t) = 12t - 8$

53 $y = 2x - 1$

55 $x \geq 1$ or $x \leq -2$

57 For $x < 0$ or $2 < x < 3$

59 $x = -1$ and $x = 5$

61 Height $= 625$ cm,
   Changing (eroding) at -30 cm/year

63 (a) 15.2 m/sec
  (b) 5.4 m/sec
  (c) -9.8 m/sec^2
  (d) 34.9 m
  (e) 5.2 sec

65 (a) $dT/dl = \pi/\sqrt{gl}$
  (b) Positive, so period increases
     as length increases

67 $V = 4\pi r^3/3$
   $dV/dr = 4\pi r^2 = $ Surface area of a sphere

69 $n = 4, a = 3/32$

Section 3.2

1 $2e^x + 2x$

3 $(\ln 5)5^x$

5 $10x + (\ln 2)2^x$

7 $4(\ln 10)10^x - 3x^2$

9 $(\ln 2)2^x - 6x^{-4}$

11 $(\ln 4)e^x$

13 $(\ln(\ln 3))(\ln 3)^t$

15 $(\ln(\ln 2))(\ln 2)^z$

17 $3x^2 + 3^x \ln 3$

19 $(\ln \pi)\pi^x$

21 $(\ln k)k^x$

23 e^{t+2}

25 $a^x \ln a + ax^{a-1}$

27 $2x + (\ln 2)2^x$

29 Rules do not apply

31 e^{x+5}

33 Rules do not apply

35 Rules do not apply

37 ≈ -444.3 people/year

39 $\approx 22.5(1.35)^t$

41 (a) $f'(0) = -1$
  (b) $y = -x$
  (c) $y = x$

43 $g(x) = x^2/2 + x + 1$

45 $x = 1$ and $x = 2$

Section 3.3

1. $5x^4 + 10x$
3. $e^x(x+1)$
5. $2^x/(2\sqrt{x}) + \sqrt{x}(\ln 2)2^x$
7. $3^x[(\ln 3)(x^2 - x^{\frac{1}{2}}) + (2x - 1/(2\sqrt{x}))]$
9. $4^y((\ln 4)(2 - y^2) - 2y)$
11. $(1-x)/e^x$
13. $(1 - (t+1)\ln 2)/2^t$
15. $6/(5r+2)^2$
17. $1/(5t+2)^2$
19. $(t^2 + 2t + 2)/(t+1)^2$
21. $2y - 6, \ y \neq 0$
23. $\sqrt{z}(3 - z^{-2})/2$
25. $2r(r+1)/(2r+1)^2$
27. $17e^x(1 - \ln 2)/2^x$
29. $\dfrac{-4x^2 - 8x - 1}{(2+3x+4x^2)^2}$
31. (a) 4
    (b) Does not exist
    (c) -4
33. (a) -2
    (b) Does not exist
    (c) 0
35. 0.35
37. 0.75
39. -21.22
41. $f'(t) = -e^{-t}$
43. $f'(x) = 3e^{3x}$
45. $-1/\sqrt{3} < x < 1/\sqrt{3}$
47. (a) $\dfrac{d}{dx}\left(\dfrac{e^x}{x}\right) = \dfrac{e^x}{x} - \dfrac{e^x}{x^2}$
    $\dfrac{d}{dx}\left(\dfrac{e^x}{x^2}\right) = \dfrac{e^x}{x^2} - \dfrac{2e^x}{x^3}$
    $\dfrac{d}{dx}\left(\dfrac{e^x}{x^3}\right) = \dfrac{e^x}{x^3} - \dfrac{3e^x}{x^4}$
    (b) $\dfrac{d}{dx}\left(\dfrac{e^x}{x^n}\right) = \dfrac{e^x}{x^n} - \dfrac{ne^x}{x^{n+1}}$
51. (a) 19
    (b) -11
53. (a) $f(140) = 15{,}000$:
    If the cost \$140 per board then 15,000 skateboards are sold
    $f'(140) = -100$:
    Every dollar of increase from \$140 will decrease the total sales by 100 boards
    (b) $dR/dp|_{p=140} = 1000$
    (c) Positive
    Increase by \$1000
55. (c) $B'(10) \approx -3776.63$
59. $f'(x) = f(x)\left(\dfrac{1}{x - r_1} + \dfrac{1}{x - r_2} + \cdots + \dfrac{1}{x - r_n}\right)$

Section 3.4

1. $99(x+1)^{98}$
3. $300t^2(t^3+1)^{99}$
5. $-x/\sqrt{1-x^2}$
7. $50(\sqrt{t}+1)^{99}/\sqrt{t}$

9. $2r^3/\sqrt{r^4+1}$
11. $e^{2x}\left[2x^2 + 2x + (\ln 5 + 2)5^x\right]$
13. $\pi e^{\pi x}$
15. $(\ln \pi)\pi^{(x+2)}$
17. $e^{5-2t}(1-2t)$
19. $(2t - ct^2)e^{-ct}$
21. $\frac{3}{2}e^{\frac{3}{2}w}$
23. $3s^2/(2\sqrt{s^3+1})$
25. $e^{-t^2}(1-2t^2)$
27. $(\ln 2)2^x/\left(3\sqrt[3]{(2^x+5)^2}\right)$
29. $(3/2)x^2\sqrt{5^{3x}}(2 + x\ln 5)$
31. $e^{-z}/2\sqrt{z} - \sqrt{z}e^{-z}$
33. $4x\left(x^2+2\right)/9$
35. $2e^{2x}(x^2 - x + 1)/(x^2+1)^2$
37. $-8b^4 z/(a+z^2)^5$
39. $(2t+3)(1 - e^{-2t}) + (t^2+3t)(2e^{-2t})$
41. $30e^{5x} - 2xe^{-x^2}$
43. $2we^{w^2}(5w^2 + 8)$
45. $-3te^{-3t^2}/\sqrt{e^{-3t^2}+5}$
47. $2ye^{[e^{(y^2)}+y^2]}$
49. $6ax(ax^2+b)^2$
51. $ae^{-bx} - abxe^{-bx}$
53. (a) 2
    (b) Does not exist
    (c) -2
55. (a) Does not exist
    (b) Does not exist
    (c) Does not exist
57. 1/2
59. -1
61. $y = 3x - 5$
63. $-1/\sqrt{2} < x < 1/\sqrt{2}$
65. $f'(x) = [(2x+1)^9(3x-1)^6] \times (102x+1)$
    $f''(x) = [9(2x+1)^8(2)(3x-1)^6 + (2x+1)^9(6)(3x-1)^5(3)] \times (102x+1) + (2x+1)^9(3x-1)^6(102)$
67. (a) μ
    (b) σ^2
69. Yes
71. (a) $H(4) = 1$
    (b) $H'(4) = 30$
    (c) $H(4) = 4$
    (d) $H'(4) = 56$
    (e) $H'(4) = -1$
73. $d; 0$
75. Decreasing
77. Value of $h(x)$ decreases from d to $-d$
79. (a) $P(1 + r/100)^t \ln(1 + r/100)$
    (b) $Pt(1 + r/100)^{t-1}/100$
81. (a) For $t < 0, I = 0$
    For $t > 0$
    $I = -(Q_0/RC)e^{-t/RC}$
    (b) No

    (c) No
85. $(-1/b)H(t)$
87. x^2

Section 3.5

3. $\cos^2\theta - \sin^2\theta = \cos 2\theta$
5. $3\cos(3x)$
7. $3\pi\sin(\pi x)$
9. $2x\cos x - x^2\sin x$
11. $-\sin xe^{\cos x}$
13. $e^{\cos\theta} - \theta(\sin\theta)e^{\cos\theta}$
15. $\cos(\tan\theta)/\cos^2\theta$
17. $\sin x/2\sqrt{1-\cos x}$
19. $\cos x/\cos^2(\sin x)$
21. $2\sin(3x) + 6x\cos(3x)$
23. $e^{-2x}[\cos x - 2\sin x]$
25. $5\sin^4\theta\cos\theta$
27. $-3e^{-3\theta}/\cos^2(e^{-3\theta})$
29. $\cos t - t\sin t + 1/\cos^2 t$
31. $5\sin^4\alpha\cos^4\alpha - 3\sin^6\alpha\cos^2\alpha$
33. $-2(\cos w\sin w + w\sin(w^2))$
35. $2\cos(2x)\sin(3x) + 3\sin(2x)\cos(3x)$
37. $\dfrac{\sqrt{1-\cos x}(1-\cos x - \sin x)}{2\sqrt{1-\sin x}(1-\cos x)^2}$
39. $(6\sin x\cos x)/(\cos^2 x + 1)^2$
41. $d^{50}y/dx^{50} = -\cos x$
43. $F(x) = -(1/4)\cos(4x)$
45. (a) $v(t) = 2\pi\cos(2\pi t)$
    (b)

$y = 15 + \sin 2\pi t$

$v = 2\pi \cos 2\pi t$

47. (a) $t = (\pi/2)(m/k)^{1/2}$;
    $t = 0$;
    $t = (3\pi/2)(m/k)^{1/2}$
    (b) $T = 2\pi(m/k)^{1/2}$
    (c) $dT/dm = \pi/\sqrt{km}$;
    Positive sign means an increase in mass causes the period to increase
49. $k = 7.46, (3\pi/4, 1/\sqrt{2})$

Section 3.6

1. $2t/(t^2+1)$
3. 2
5. $e^{-x}/(1-e^{-x})$
7. $e^x/(e^x+1)$
9. $ae^{ax}/(e^{ax}+b)$
11. 7
13. $-\tan(w-1)$
15. $2y/\sqrt{1-y^4}$

17 1

19 $-\sin(\ln t)/t$

21 $\arcsin w + \dfrac{w}{\sqrt{1-w^2}}$

23 $2/\sqrt{1-4t^2}$

25 $-3\sin(\arctan 3x)/(1+9x^2)$

27 $-1/z(\ln z)^2$

29 $2\ln(2x)$

31 $1/(t\ln t)$

33 $8/\sin t$

35 0.8

37 -0.4

39 $-1 < x < 1$

41 $\dfrac{d}{dx}(\log x) = \dfrac{1}{(\ln 10)x}$

43 (a) $y = x - 1$
   (b) 0.1; 1
   (c) Yes

45 (a) $f'(x) = 0$
   (b) f is a constant function

47 Any x with $25 < x < 50$

49 Any x with $75 < x < 100$

51 225

53 1/3

55 (a) $10x^4 + 9x^2 + 1$
   (b) $f' > 0$ so f increasing everywhere
   (c) 6
   (d) 20
   (e) 1/20

57 1/5

59 (a) 200 m. veh
   (b) 15 m. veh/yr
   (c) 21 yrs
   (d) 0.0667 yr/m. veh

61 $(f^{-1})'(5) \neq 1/f'(10)$

63 $(f^{-1})'$ does not change sign

Section 3.7

1 $dy/dx = -x/y$

3 $-(1+y)/(1+x)$

5 1/25

7 $(1-y)/(x-3)$

9 ax/by

11 $dy/dx = (y(1 - x\ln y))/(x(x + 3y^3))$

13 $dy/dx = 0$

15 $(y^2 + x^4y^4 - 2xy)/(x^2 - 2xy - 2x^5y^3)$

17 $(a-x)/y$

19 0

21 $-5/4$

23 $y = x/2 - 3/2$

25 $y = 2$

27 $y = 0$

29 (a) $dy/dx = -9x/25y$
   (b) Slope not defined along $y = 0$

31 (a) $dy/dx = (y^2 - 3x^2)/(3y^2 - 2xy)$
   (c) $y \approx 1.9945$
   (d) Horizontal:
     (1.1609, 2.0107)
     and
     $(-0.8857, 1.5341)$
    Vertical:
     (1.8039, 1.2026)
     and
     $(\sqrt[3]{5}, 0)$

33 (a) $-(1/2P)f(1 - f^2)$

35 $dy/dx = (m/n)x^{m/n-1}$

Section 3.8

1 $2\sinh(2x)$

3 $\sinh(\sinh t) \cdot \cosh t$

5 $2\cosh t \cdot \sinh t$

7 $2te^{t^2}\sinh(e^{t^2})$

9 $3\cosh(3y) \cdot \cosh(\sinh(3y))$

11 0

15 $(t^2 - 1)/2t$

19 $\cosh(2x) = \cosh^2 x + \sinh^2 x$

23 2

25 $|k| \leq 2$

27 (a)

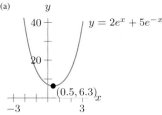

   (b) $A = 6.325$
     (stretch factor)
     $c = 0.458$
     (horizontal shift)

29 (a) $0.54\, T/w$

31 (a) 0
   (b) Positive for $x > 0$
     Negative for $x < 0$
     Zero for $x = 0$
   (c) Increasing everywhere
   (d) $1, -1$

   (e) Yes; derivative positive everywhere

Section 3.9

1 $\sqrt{1+x} \approx 1 + x/2$

3 $1/x \approx 2 - x$

7 $|\text{Error}| < 0.2$
   Overestimate, $x > 0$
   Underestimate, $x < 0$

9 (a) $\cos x \approx (-x + (\pi/4) + 1)/\sqrt{2}$
   (b) Overestimate
   (c) $|\text{Error}| < 0.3$.

11 $a = 1; f(a) = 1$
   Underestimate
   $f(1.2) \approx 1.4$

13 $E(x) = x^4 - (1 + 4(x - 1))$
   $k = 6; f''(1) = 12$
   $E(x) \approx 6(x - 1)^2$

15 $E(x) = e^x - (1 + x)$
   $k = 1/2; f''(0) = 1$
   $E(x) \approx (1/2)x^2$

17 $E(x) = \ln x - (x - 1)$
   $k = -1/2; f''(1) = -1$
   $E(x) \approx -(1/2)(x - 1)^2$

19 (c) 0

21 (c) -0.135%

23 (a) $\Delta T \approx -T\Delta g/(2g)$
   (b) 0.5% decrease

25 (a) $f'(5)$
   (b) 0
   (c) $f(5) + f'(5)\Delta x$

Section 3.10

1 False

3 True

5 False

7 Yes; yes

9 No; yes

15 Racetrack

17 Mean Value

Chapter 3 Review

1 $2e^t + 2te^t + 1/(2t^{3/2})$

3 $x^2 \ln x$

5 $(1/2)\theta^{-1/2} + \theta^{-2}$

7 $3/(y\ln(2y^3))$

9 0

11 $-6\cos(3t + 5) \cdot \sin(3t + 5)$

13 $6\cos(3\theta - \pi)\sin(3\theta - \pi)$

15 $-\theta^{-3}(\theta\cos(5 - \theta) + 2\sin(5 - \theta))$

17 $\frac{1}{2}x^{-1/2} - x^{-2} - \frac{3}{2}x^{-5/2}$

19 $-1/(1 + (2 - x)^2)$

21 $e^n\cos(e^n)$

23 $\cos(\sqrt{t}e^t) - t\sin(\sqrt{t}e^t) \cdot (\sqrt{t}e^t + e^t/(2\sqrt{t}))$

25 $e^{\tan x} + xe^{\tan x}/\cos^2 x$

27 $6x/\left(9x^4 + 6x^2 + 2\right)$

29 a

31 $ke^{k\theta}$

33 $(-cat^2 + 2at - bc)e^{-ct}$

35 $5\ln(a)a^{5x}$

37 $(2abr - ar^4)/(b + r^3)^2$

39 $-2/(x^2 + 4)$

41 $20w/(a^2 - w^2)^3$

43 $ae^{au}/(a^2 + b^2)$

45 $(2\ln(kt) - 2)/(\ln(kt) - t)^2$

47 $e^t\cos\sqrt{e^t + 1}/(2\sqrt{e^t + 1})$

49 $-(5x^4 + 2)/2$

51 $5z^4 + 20z^3 - 1$

53 $2 + 1/(3x^{4/3}) + 3^x\ln 3$

55 $\theta^2\cos\theta$

57 $6(3\theta - \pi)\cos[(3\theta - \pi)^2]$

59 k

61 $(-4 - 6x)(6x^e - 3\pi) + (2 - 4x - 3x^2)(6ex^{e-1})$

63 0

65 $4x - 2 - 4x^{-2} + 8x^{-3}$

67 $(8xy - 3x^2)/(3y^2 - 4x^2)$

69 2/3

71 $(y-1) = -(1/5)(x-3),$
   $(y+4) = -(4/5)(x-3)$

73 4

75 $x \approx 1, -0.4,$ or 2.4

77 $y = -4x + 6$

79 1.6

81 Proportional to r^2

83 (a) $H'(2) = 11$
   (b) $H'(2) = -1/4$
   (c) $H'(2) = r'(1) \cdot 3$
      (we don't know $r'(1)$)
   (d) $H'(2) = -3$

85 (a) $y = 20x - 48$
   (b) $y = 11x/9 - 16/9$
   (c) $y = -4x + 20$
   (d) $y = -24x + 57$
   (e) $y = 8.06x - 15.84$
   (f) $y = -0.94x + 6.27$

87 1.909 radians $(109.4°)$ or 1.231 radians $(70.5°)$

89 Not perpendicular; $x \approx 1.3$

91 0

93 1

95 Mexico, since $dM/dt > dU/dt$ at $t = 0$

97 (a)

$P = 30e^{-3.23 \times 10^{-5}h}$

   (b) $y = (-9.69 \times 10^{-4})h + 30$
   (c) $P = 30 - 0.001h$
   (d) Both have P intercepts of 30, and slopes are almost the same
   (e) Too small because it has a slightly smaller slope

99 $b = 1/40$ and $a = 169.36$

101 (a) 1,000,000 people
   (b) No. Max number of people to fall sick in a day is 25,000

103 (b)

105 (a) $y^{(n)} = (-1)^{n+1}(n-1)!x^{-n}$
   (b) $y^{(n)} = xe^x + ne^x$

107 $(f/g)'/(f/g) = (f'/f) - (g'/g)$

109 (a) 1
   (b) 1
   (c) $\sin(\arcsin x) = x$

111 (a) 0
   (b) 0
   (c) $\ln(1 - 1/t) + \ln(t/(t-1)) = \ln 1$

Ch. 3 Understanding

1 True

3 True

5 True

7 False

9 True

11 False

13 False

15 False; $f(x) = |x|$

17 False; $\cos t + t^2$

19 False; $f(x) = 6, g(x) = 10$

21 False; $f(x) = 5x + 7, g(x) = x + 2$

23 False; $f(x) = x^2, g(x) = x^2 - 1$

25 False; $f(x) = e^{-x}, g(x) = x^2$

27 (a) Not a counterexample
   (b) Not a counterexample
   (c) Not a counterexample
   (d) Counterexample

29 False

31 False

33 Possible answer
$$f(x) = \begin{cases} x & \text{if } 0 \le x < 2 \\ 19 & \text{if } x = 2 \end{cases}$$

Section 4.1

1

3 (a) Critical point $x \approx 0$;
      Inflection points between -1 and 0 and between 0 and 1
   (b) Critical point at $x = 0$,
      Inflection points at $x = \pm 1/\sqrt{2}$

5 $x = -1, 1/2$

7 Local maximum $x = -1$
   Local minimum $x = 1$

9 Increasing for all x, no critical point

11 Alternately incr/decr

13 Oscillating: $x < 0$
   Increasing: $x > 0$

15 Local min: $(0.37, -0.37)$

17

19

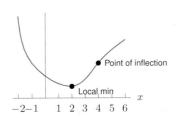

x-values of these points give inflection points of f

21 $x = 0$: not max/min
   $x = 3/7$: local max
   $x = 1$: local min

23 One: quadratic
   None: $f(x) = x^3 + x + 1$
   Infinitely many: $f(x) = \sin x$

Critical points

25 (a) Local maxima at $x \approx 1, x \approx 8,$
      Local minimum at $x \approx 4$
   (b) Local maxima at $x \approx 2.5, x \approx 9.5$
      Local minimum at $x \approx 6.5,$

27 $a = -12, b = 31$

29 $a = 3e, b = -3$

31 (a) Yes, at 2000 rabbits
   (b) 1787
      1000 rabbits

33

depth of water

time at which water reaches corner of vase

time

35 $x = 2$: local min
   $x = 4$: point infl

Point of inflection

Local min

37 (a)

(b)

39 $B = f, A = f', C = f''$

41 III even; I, II odd
   I is f'', II is f, III is f'

644

43

y

$y' < 0$ everywhere

$y'' = 0$ $y'' = 0$ $y'' = 0$

$y'' > 0$ $y'' < 0$ $y'' > 0$ $y'' < 0$

$y = f(x)$

x_1 x_2 x_3 x

45

y

$y = f(x)$

$y' = 2$ $y' > 0$
$y'' = 0$ $y'' > 0$

x_1 x

47 (a) Decreasing
(b) Local minimum at x_1
(c) Concave up at x_2

Section 4.2

1 $y = 2x - 10$

3 $y = -3x^2 + 3$

5 $y = -\sqrt{2 - x^2}$

7 $y = 5(1 - e^{-bx})$

9 $y = 0.25 \sin(\pi x/10) + 1.75$

11 $y = e^{-(x-2)^2/2}$

13 $y = (x - 1)(x - 5)(x - 7)$

15 $y = -x^4 + 2x^2 + 3$

17 (a) Local maximum:
$$p(-\sqrt{a/3}) = +2a\sqrt{a}/(3\sqrt{3})$$
Local minimum:
$$p(\sqrt{a/3}) = -2a\sqrt{a}/(3\sqrt{3})$$
(b) Further from x- and y-axis
(c)

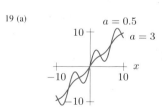

$a = 1$ $a = 3$ $a = 5$

19 (a)

$a = 0.5$

$a = 3$

10

-10 10 x

-10

(b) $-1 \le a \le 1$

---

21 $a = 200{,}000$
$k = -\ln 0.9 \approx 0.105$

25 $(1/b, 1/be)$

27 (a) $f'(x) = abe^{-bx} > 0$
(b) $f''(x) = -ab^2 e^{-bx} < 0$

29 (a) $x = e^a$
(b) $a = -1$:

1

1 2 3 x

-1

$a = 1$:

1

1 2 3 x

-1

(c) Max at (e^{a-1}, e^{a-1}) for any a

33 (a) Local max: $x = 1/b$
No local minima
Inflection point: $x = 2/b$
(b) Varying a stretches or flattens the graph vertically.
Incr b shifts critical, inflection points to left; lowers max
(c)

y Varying a

$\frac{4}{be}$ $a = 4$
$\frac{3}{be}$ $a = 3$
$\frac{2}{be}$ $a = 2$
$\frac{1}{be}$ $a = 1$

1 2 x

Varying b

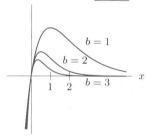

$b = 1$

$b = 2$

1 2 $b = 3$ x

37 (a) Intercept: $x = a$
Asymptotes: $x = 0$,
$U = 0$
(b) Local min: $(2a, -b/4)$
Local max: none
(c)

U

a $2a$ x

$(2a, -b/4)$

---

39 (a) Zero: $r = B/A$
Vertical asymptote:
$r = 0$
Horizontal asymptote:
$f(r) = 0$
(b) Minimum:
$$\left(3B/(2A), -4A^3/(27B^2)\right)$$
Point of inflection:
$$\left(2B/A, -A^3/(8B^2)\right)$$
(c)

$f(r)$

r

$\frac{B}{A}$

$\left(\frac{3B}{2A}, \frac{-4A^3}{27B^2}\right)$ $\left(\frac{2B}{A}, \frac{-A^3}{8B^2}\right)$

41 (a)

y

$A = B = 1$ x

y

$A = -B = 1$ x

y

$A = 2$
$B = 1$ x

y

$A = 2$
$B = -1$ x

y

$A = -2$
$B = -1$ x

y

$A = -2$
$B = 1$ x

(b) U-shaped
(c) Incr $(A > 0)$
   or Decr $(A < 0)$
(d) Max: $A < 0, B < 0$
   Min: $A > 0, B > 0$

43 (a) Vertical intercept: $W = Ae^{-e^b}$, Horizontal
   asymptote $W = A$
(b) No critical points, inflection point at $t = b/c$, $W = Ae^{-1}$
(c)

(d) Yes

Section 4.3

1

3 (a) $1/5, 1$
(b) Local max: $p = 1/5$
   Local min: $p = 1$
(c) Max: $245/3125$
   Min: 0

5 (a) $f(1)$ local minimum;
   $f(0.1), f(2)$ local maxima
(b) $f(0.1)$ global maximum
   $f(1)$ global minimum

7 Global max $= -1$ at $x = 2$
   No global min

9 Global max $= 1/e$ at $t = 1$
   No global min

11 Global max $= 1/2$ at $t = 1$
   Global min $= -1/2$ at $t = -1$

13 $0 \leq y \leq 16$

15 $0 \leq x^3 e^{-x} \leq 3^3 e^{-3}$

17 $0 \leq y$; No upper bound

19 44.1 feet

21 (a) $D = C$
(b) $D = C/2$

23 (a) $0 \leq y \leq a$

(b) $y = a/2$

25 (a) $dR/dr_1 = r_2^2/(r_2 + r_1)^2 > 0$
(b) $r_1 = b$

27 (b) 2 hours

(c) Equal

29 $(-1, -1/2)$; $(1, 1/2)$

31 (b) $f(v) = v \cdot a(v)$
(c) When $a(v) = f'(v)$
(d) $a(v)$

33 (a) $g(v) = f(v)/v$
(b) 220 mph
(c) 300 mph

35 (b) Yes, at $x = 0$
(c) Max: $x = -2$, Min: $x = 2$
(d) $5 > g(0) > g(2)$

37 False

39 False

Section 4.4

1 $5000, $2.40, $4

3 $\pi(q) = 490q - q^2 - 150$
   Max at $q = 245$

5 $5.5 < q < 12.5$ positive;
   $0 < q < 5.5$ and $q > 12.5$ negative;
   Maximum at $q \approx 9.5$

7 Above 2000

9 (a) $q = 2500$
(b) $3 per unit
(c) $3000

11

13 (a) No
(b) Yes

15 (a) $0
(b) $96.56
(c) Raise the price by $5

17 $L = [\beta p c K^{\alpha}/w]^{1/(1-\beta)}$

19 (a) and (b)

21 (a) $C(q) = 0.01q^3 - 0.6q^2 + 13q$
(b) $1
(c) $q = 30, a(30) = 4$
(d) Marginal cost is 4

23 (a) Positive
(b) Cost $= wx$, Revenue $= pf(x)$
(c) w/p
(d) Negative
(e) $1/(pf''(x^*))$, negative
(f) Less

Section 4.5

1 $x = L/2$

3 Minimum: -2 amps
   Maximum: 2 amps

5 (a) $0 \leq y \leq a$

(b) $y = 0$

7 $\theta = -\mu + \sqrt{\mu^2 + 1}$

9 Minimum: $x = -r_0/\sqrt{2}$
   Maximum: $x = r_0/\sqrt{2}$

11 Dimensions: 6 by $\sqrt{3}$

13 $(\pm 1, 0)$, $(\pm 1, 1/2)$

15 Radius $= \sqrt{2/3}$
   Height $= \sqrt{1/3}$

17 1250 square feet

19 about $4583

21 $x = \sqrt[3]{2V}$
   $y = \sqrt[3]{V/4}$

23 $(0.59, 0.35)$

25 $r = \sqrt{2A/(4 + \pi)}$;
   $h = \frac{A}{2} \cdot \frac{\sqrt{4+\pi}}{\sqrt{2A}} - \frac{\pi}{4} \cdot \frac{\sqrt{2A}}{\sqrt{4+\pi}}$

27 Maximum revenue $= $27,225$
   Minimum $= $0

29 (a) $x = e$
(b) $n = 3$
(c) $3^{1/3} > \pi^{1/\pi}$

31 (a)

(b) 7 worms
(c) increases

33 (a) $E = 500e\left(\dfrac{200 - \cos\theta}{\sin\theta}\right) + 2000e$

  $\left(\arctan\left(\dfrac{500}{2000}\right) \le \theta \le \pi/2\right)$

  (b) $\theta = \pi/3$

  (c) Independent of e, but dependent on $\overline{AB}/\overline{AL}$

35 (a) $x = 1, P = (1, 1)$

39 (a)

$Q = 100$
$Q = 200$
$Q = 300$
$C = x + 2y = 10$

  (c) 125

Section 4.6

1 69.8 m/yr

3 $-81/R^2$

5 (a) 0.3 cm/hr; 0.424 cm/hr
  (b) $t = 3; t = 3$

7 g; $ge^{-k/m}$; acceleration decreases from g at $t = 0$

9 (a) $2A/r^3 - 3B/r^4$; units of force/units of distance
  (b) $k(2A/r^3 - 3B/r^4)$; units of force/units of time

11 -0.072 ohms/sec

13 (a) $(-0.99, -0.16)$
  (b) $v_x = 0.32$
    $v_y = -1.98$

15 (a) $k \approx 0.067$
  (b) $t \approx 10.3$ hours
  (c) Formula:
    $T(24) \approx 74.1^\circ$F,
    rule of thumb: 73.6°F

17 (a) $r = \sqrt{20h - h^2}$ cm
  (b) $-1/(2\sqrt{75}) = 0.0577$ cm/hr

19 $180\pi = 565.487$ cm^3/hr

21 -0.01 cm/sec

23 (a) Increasing at 4 mph
  (b) Decreasing at 4 mph

25 $2/\cos^2\theta$

27 2513.3 cm^3/sec

29 2 cm/sec

31 (a) 25.133 meters/min
  (b) 23.335 meters/min; 0 meters/min

33 (a) $h(t) = 300 - 30t$
    $0 \le t \le 10$
  (b) $\theta = \arctan((200 - 30t)/150)$
    $d\theta/dt =$
    $-\left(\dfrac{1}{5}\right)\left(\dfrac{150^2}{150^2 + (200 - 30t)^2}\right)$
  (c) When elevator is at level of observer

35 (a) 900
  (b) 34 ships
  (c) End of battle
  (d) 30 ships
  (e) $y'(t) = kx$; negative
  (g) $k = 2$, 68 ships/hour

Section 4.7

1 Negative

3 Negative

5 0

7 0

9 $0.1x^7$

11 $x^{0.2}$

13 0

15 -2

17 $1/2$

19 2

21 0

23 $-1/2$

25 e^3

27 Does not exist

29 (a) $a\ln x + b\ln y$
  (b) $x^a y^b$

31 e

33 e^{kt}

35 $e^{-\lambda}$

37 5

39 (a)

  (b) $1/(1 - x)$

41 (a) $\ln x$
  (b)

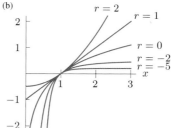

$r = 2$
$r = 1$
$r = 0$
$r = -2$
$r = -5$

Section 4.8

1 The particle moves on straight lines from $(0, 1)$ to $(1, 0)$ to $(0, -1)$ to $(-1, 0)$ and back to $(0, 1)$

3 Two diamonds meeting at $(1, 0)$

5 Clockwise for all t

7 Clockwise: $t < 0$,
  Counter-clockwise: $t > 0$

9 Counterclockwise: $t > 0$

11 Speed $= |t| \cdot \sqrt{(4 + 9t^2)}$,
  Particle stops when $t = 0$

13 Speed $= \sqrt{4\sin^2(2t) + \cos^2 t}$,
  Particle stops when
  $t = (2n + 1)\pi/2$, for
  any integer n

15 $x = 4 + 4(t - 2)$
  $y = 8 + 12(t - 2)$

17 $x = -2, y = t$

19 $x = -2\cos t, y = 2\sin t$,
  $0 \le t \le 2\pi$

21 $x = 5\cos t, y = 7\sin t$,
  $0 \le t \le 2\pi$

23 $(y - 4) = (4/11)(x - 6)$

25 $y = -(4/3)x$

27 (a) Right of $(2, 4)$
  (b) $(-1, -3)$ to $(2, 4)$
  (c) $t < -2/3$

31 $(I) = l_2, (II) = l_3$

33 (a)

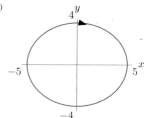

  (b) $v \approx 2.2$
  (c) $v = 2.2363$

35 (a) Yes, $t = 1, (x, y) = (-2, -1)$
  (b) Yes, $t = -1, (x, y) = (2, 3)$
  (c) Never

37 (a)

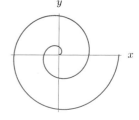

  (b) $x(\pi/4) = 5$ and $y(\pi/4) = 0$;
    $x'(\pi/4) = 0$ and $y'(\pi/4) = -8$
  (c) Twice
  (d) Parallel to y-axis, negative y direction
  (e) 10

39 (a) No
  (b) $k = 1$
  (c) Particle B

41 (a) $y + \frac{1}{2} = -\frac{\sqrt{3}}{3}(x - \pi)$
  (b) $t = \pi$
  (c) 0.291, concave up

43 (a) $t = \pi/4$; at that time, speed $= \sqrt{9/2}$
  (b) Yes, when $t = \pi/2$ or $t = 3\pi/2$
  (c) Concave down everywhere

45

47

49 (b) For example:
$R = 12, t = \pi$

(c)

Chapter 4 Review

1

Local and global max
Local max
$f(x)$
Local max
Local min
Local and global min

3 (a) $f'(x) = 3x(x - 2)$
$f''(x) = 6(x - 1)$
(b) $x = 0$
$x = 2$
(c) Inflection point: $x = 1$
(d) Endpoints:
$f(-1) = -4$ and $f(3) = 0$
Critical Points:
$f(0) = 0$ and $f(2) = -4$
Global max:
$f(0) = 0$ and $f(3) = 0$
Global min:
$f(-1) = -4$ and $f(2) = -4$
(e) f increasing:
for $x < 0$ and $x > 2$
f decreasing:
for $0 < x < 2$
f concave up:
for $x > 1$
f concave down:
for $x < 1$

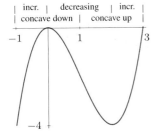

| incr. | decreasing | incr. |
| concave down | concave up |

5 (a) $f'(x) =$
$- e^{-x} \sin x + e^{-x} \cos x$
$f''(x) = -2e^{-x} \cos x$
(b) Critical points:
$x = \pi/4$ and $5\pi/4$
(c) Inflection points:
$x = \pi/2$ and $3\pi/2$
(d) Endpoints:
$f(0) = 0$ and $f(2\pi) = 0$
Global max:
$f(\pi/4) = e^{-\pi/4}(\sqrt{2}/2)$
Global min:
$f(5\pi/4) = -e^{-5\pi/4}(\sqrt{2}/2)$
(e) f increasing:
$0 < x < \pi/4$ and

$5\pi/4 < x < 2\pi$
f decreasing:
$\pi/4 < x < 5\pi/4$
f concave down:
for $0 \le x < \pi/2$
and $3\pi/2 < x \le 2\pi$
f concave up:
for $\pi/2 < x < 3\pi/2$

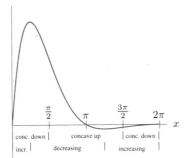

$\frac{\pi}{2}$ π $\frac{3\pi}{2}$ 2π x

conc. down | concave up | conc. down
incr. | decreasing | increasing

7 $\lim_{x \to \infty} f(x) = \infty$
$\lim_{x \to -\infty} f(x) = -\infty$
(a) $f'(x) = 6(x - 2)(x - 1)$
$f''(x) = 6(2x - 3)$
(b) $x = 1$ and $x = 2$
(c) $x = 3/2$
(d) Critical points:
$f(1) = 6$, $f(2) = 5$
Local max: $f(1) = 6$
Local min: $f(2) = 5$
Global max and min: none
(e) f increasing: $x < 1$ and $x > 2$
Decreasing: $1 < x < 2$
f concave up: $x > 3/2$
f concave down: $x < 3/2$

$f(x) = 2x^3 - 9x^2 + 12x + 1$

6
5

1   2 x

9 $\lim_{x \to -\infty} f(x) = -\infty$
$\lim_{x \to \infty} f(x) = 0$
(a) $f'(x) = (1 - x)e^{-x}$
$f''(x) = (x - 2)e^{-x}$
(b) Only critical point is at $x = 1$
(c) Inflection point: $f(2) = 2/e^2$
(d) Global max: $f(1) = 1/e$
Local and global min: none
(e) f increasing: $x < 1$
f decreasing: $x > 1$
f concave up: $x > 2$
f concave down: $x < 2$

$f(x) = xe^{-x}$
x
1   2

11 Global max $= 1$ at $t = 0$
No global min

13 Local max: $f(-3) = 12$
Local min: $f(1) = -20$
Inflection pt: $x = -1$
Global max and min: none

12
$f(x) = x^3 + 3x^2 - 9x - 15$
x
-3 1
-20

15 Global and local min: $x = 2$
Global and local max: none
Inflection pts: none

$f(x) = x - \ln x$
x
2

17 Global and local max:
$f(0) = 1$
Local minimum: none
Inflection pts: $x = \pm 1/\sqrt{2}$

$f(x) = e^{-x^2}$
x

19 $-0.014 \le e^{-x} \sin x \le 0.322$

21 $5 \le x^3 - 6x^2 + 9x + 5 \le 25$

25

$f'(x)$
f has a local min. | f has crit. pt. Neither max or min

27 (a) Increasing for $x > 0$
Decreasing for $x < 0$
(b) Local and global min: $f(0)$

29 (a) Increasing for $0 < x < 4$
Decreasing for $x < 0$ and $x > 4$
(b) Local max: $f(4)$
Local min: $f(0)$

31 $a = 5e, b = 1/2$

33 $a = 363.23$, and $b \approx 4.7665$

35 Max: 0.552; Min: 0.358

37 $x = y = \sqrt[3]{V}$

39 About 331 meters, Maintain course

41 $r \approx 2.7$ cm

43 (a) $g(e)$ is a global maximum; there is no minimum
   (b) There are exactly two solutions
   (c) $x = 5$ and $x \approx 1.75$

45 (a) Min at $x = \pi L/(4 + \pi)$
      Max at $x = L$
   (c) Yes

47 (a)

   (b) $-1/x$

49 $-1/6$

51 100π cm^3/sec

53 (a) Angular velocity
   (b) $v = a(d\theta/dt)$

55 10 ohms/min

57 $-k/V^2$

61 (a) $1/\sqrt{1 + x^2}$

63 (a) Min at $x = a$
   (c) 4.6477

65 (b) No

Ch. 4 Understanding

1 True

3 False

5 False

7 False

9 One possibility:
   $f(x) = ax^2, a \neq 0$

11 (a) True
   (b) False
   (c) True
   (d) False
   (e) True

13 True

15 True

17 False

19 False

21 $f(x) = x^2 + 1$

23 $f(x) = -x^2 - 1$

25 Impossible

Section 5.1

1 (a) Lower estimate $= 122$ ft
      Upper estimate $= 298$ ft
   (b)

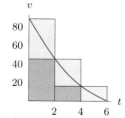

3 (a) (i) 480
       (ii) 492

(b) $n = 4$
(c) 447

5 (a) 4; 0, 4, 8, 12, 16; 25, 23, 22, 20, 17
   (b) 360; 328
   (c) 8; 0, 8, 16; 25, 22, 17
   (d) 376; 312

7 Upper estimate $= 0.75$ m;
   Lower estimate $= 0.65$ m;
   Average $= 0.70$ m

9 250 meters

11 0.09

13 0.086

15 (a) 430 ft
   (b) (ii)

17 15 cm to the left

19 25 cm to the right

21 8 cm to the right

23 (a)

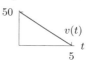

   (b) 125 feet
   (c) 4 times as far

25 (a) Car A
   (b) Car A
   (c) Car B

Section 5.2

1 (a) 224

   (b) 96

   (c) 200

   (d) 136

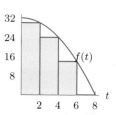

3 350

5 93.47

7 543

9 10.0989

11 1.4936

13 14,052.5

15 1.106

17 24.694

19 (a) $(\ln 1.5)0.5$

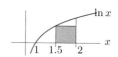

   (b) $(\ln 1.5)0.5 + (ln2)0.5$

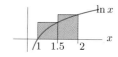

   (c) Right: overestimate
       left: underestimate

21 Left-hand sum $= 1.96875$
   Right-hand sum $= 2.71875$
   The most the estimate could be off is 0.375

23 (a)

   (b) 3.084
   (c) 2.250

25 Positive

27 (a) 13
   (b) -2
   (c) 11
   (d) 15

29 (a) -2
   (b) $-A/2$

31 $\int_0^{4\pi}(2 + \cos x)\, dx < $ LHS = RHS

33 $a = 2$, $b = 6$, $f(x) = x^2$; other answers possible

Section 5.3

1 Dollars

3 Foot-pounds

5 Change in velocity; km/hr

7 Change in salinity; gm/liter

9 2

11 15

13 Total amount $= \int_0^{60} f(t)\, dt.$

15 (a) Overestimate $= 7$ tons
   Underestimate $= 5$ tons
   (b) Overestimate $= 74$ tons
   Underestimate $= 59$ tons
   (c) Every 2 days

17 (b) 750.5 million metric tons

19 About $13,800

21 $2392.87

23 (a) (i) 1/4

   (ii) 1/4

   (iii) 0
   (b) Not true

25 (a) -4
   (b) (ii)

27 (a) III
   (b) I
   (c) II and IV

29 (a) 0.79

31 $485.80

33 (a) 0.375 thousand/hour
   (b) 1.75 thousands

35 0.732 gals, 1.032 gals

Section 5.4

1 (a) $f(1), f(2)$
   (b) $2, 2.31, 2.80, 2.77$

3 $F(0) = 0$
   $F(1) = 1$
   $F(2) = 1.5$
   $F(3) = 1$
   $F(4) = 0$
   $F(5) = -1$
   $F(6) = -1.5$

5 14.688

7 0.0833

9 2.828

11 $\int_1^5 1/t\, dt = \ln 5$

13 $\int_{-1}^1 4t^3\, dt = 0$

15 $8c$

17 -52

19 8

21 8

23 2

25 (a) Less
   (b) 0.856

27 (a) Positive, since $e^{x^2} > 0$

31 (a) 2
   (b) 12, 2
   (c) 0, 0

33 (a) Yes
   (b) No; left $n = 20$ approx to $\int_1^3 f(x)\, dx$

35 $f(3) - f(2),$
   $[f(4) - f(2)]/2,$
   $f(4) - f(3)$

37 Slope $= \frac{f(b)-f(a)}{b-a}$

39

$\frac{F(b)-F(a)}{b-a}$

Chapter 5 Review

1 (a) 570 m^3
   (b) Every 2 minutes

3 60

5 396

7 10.667

9 2

11 5.167

13 2.762

15 ≈ 455 feet or 0.086 miles

17 0.40 miles

19 30

21 42

23 0

25 (b) (i) 2

   (ii) 2

   (iii) 4

27 (a) Upper estimate: 340 liters
   Lower estimate: 240 liters
   (b)

29 (a) Yes; 20 minutes
   (b) 40 minutes; about 25 km/hr; away
   (c) 20 minutes; 3 km
   (d) 60 min; about 13.8 km

31 (a) $22°C$
   (b) $183°C$
   (c) Smaller

33 741.6 liters

35 $V < IV < II < III < I$
   I, II, III positive
   IV, V negative

37 (a) $\int_0^5 f(x)\, dx - \frac{1}{2}\int_{-2}^2 f(x)\, dx$
   (b) $\int_{-2}^5 f(x)\, dx - 2\int_{-2}^0 f(x)\, dx$
   (c) $\frac{1}{2}\left(\int_{-2}^5 f(x)\, dx - \int_2^5 f(x)\, dx \right)$

39 $(1/2)w(b_1 + b_2)$

41 (a) Approx 1.9 seconds
   (b) Approx 103.5 newton-seconds
   (c) Newton-seconds
   (d) A

(e) B

43 (a) Positive: $0 \le t < 40$
      and just before $t = 60$;
      negative: from $t = 40$
      to just before $t = 60$
   (b) About 500 feet
      at $t = 42$ seconds
   (c) Just before $t = 60$
   (d) Just after $t = 40$
   (e) A catastrophe
   (f) Total area under curve is positive
      About 280 feet

45 (a) 14.73 minutes
      15.93 minutes
   (b) 0.60 minutes

47 (a) $< (c) < (b) < (d)$

49 (a) $\sum_{i=1}^n i^4/n^5$
   (b) $(6n^4 + 15n^3 + 10n^2 - 1)/30n^4$
   (c) 1/5

51 (a) $\sum_{i=0}^{n-1} (n+i)/n^2$
   (b) $3/2 - 1/(2n)$
   (c) 3/2
   (d) 3/2

53 (a) $\sum_{i=1}^n \sin(i\pi/n)\pi/n$
   (b) $\pi \cos(\pi/2n)/n \sin(\pi/2n)$
   (c) 2
   (d) 2

55 (a) $-\ln(1 + a^2 b)/2b + \ln(1 + c^2 b)/2b$
   (b) $\ln(1 + bx^2)/2b$

Ch. 5 Understanding

1 False

3 True

5 False

7 False

9 True

11 True

13 False

15 False

17 False

19 True

21 True

23 False

25 False

27

Section 6.1

1

$F(0) = 1$

$F(0) = 0$

3

$F(0) = 1$

$F(0) = 0$

5 128, 169, 217

7 (a) -19
  (b) 6

9 (a) $x = 1, x = 3$
  (b) Local min at $x = 1$, local max at $x = 3$
  (c)
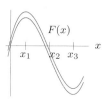

11 x_1 local max;
   x_2 inflection pt;
   x_3 local min

13 x_2, x_3 inflection pts

15 (a)
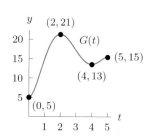

  (b) $f(0) = f(7) = -1$
  (c) 0

17 Critical points: $(0, 5), (2, 21), (4, 13), (5, 15)$

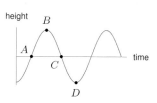

19 Acceleration is zero at points A and C

21 (a) (I) volume; (II) flow rate
  (b) (I) is an antiderivative of (II)

23 (a) $f(x)$ is greatest at x_1
  (b) $f(x)$ is least at x_5
  (c) $f'(x)$ is greatest at x_3
  (d) $f'(x)$ is least at x_5
  (e) $f''(x)$ is greatest at x_1
  (f) $f''(x)$ is least at x_5

25 (b) Maximum in July 1993
     Minimum in Jan 1994
  (c) Increasing fastest in May 1993
     Decreasing fastest in Oct 1993

Section 6.2

1 $5x$

3 $x^3/3$

5 $\sin t$

7 $\ln |z|$

9 $-1/2z^2$

11 $-\cos t$

13 $t^4/4 - t^3/6 - t^2/2$

15 $\sin t + \tan t$

17 $-\cos 2\theta$

19 $(t + 1)^3/3$

21 $5x^2/2 - 2x^{3/2}/3$

23 $x^4/4 - x^2/2 + C$

25 $t^4/4 + 5t^2/2 - t + C$

27 $2t^{3/2}/3 + C$

29 $x^4 - 7x + C$

31 $2t^{1/2} + C$

33 $F(x) = 3x$
   (only possibility)

35 $F(x) = -7x^2/2$
   (only possibility)

37 $F(x) = x^3/3$
   (only possibility)

39 $F(x) = 2x + 2x^2 + (5/3)x^3$
   (only possibility)

41 $5x^2/2 + C$

43 $-\cos \theta + C$

45 $t^3/3 - 1/t + C$

47 $x^3/3 + 5x^2/2 + 8x + C$

49 $2t^2 + 7t + C$

51 $5e^z + C$

53 $-\cos t + C$

55 $2t^{5/2}/5 - 2t^{-1/2} + C$

57 $e^{2r}/2 + C$

59 $y^3/3 - 2y - 1/y + C$

61 $\ln 3 \approx 1.0986$

63 $3e^2 - 1 = 19.167$

65 $1 - \cos 1 \approx 0.460$

67 $16/3 \approx 5.333$

69 $2e - 2 \approx 3.437$

71 $\tan(\pi/4) = 1$

73 36

75 $1/20$

77 $e - 1 - \sin 1$

79 $b = 7$

81 $103/3$

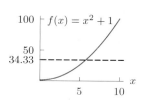

83 (a) 0
  (b) $2/\pi$

87 (a)

  (b) 7 years
  (c) 69.333 cubic yards

Section 6.3

1 $x^4/4 + 5x + C$

3 $8t^{3/2}/3 + C$

7 $10e^t + 15$

9 $2z - \cos z + 6$

11 (a) $a(t) = -9.8$ m/sec^2
      $v(t) = -9.8t + 40$ m/sec
      $h(t) = -4.9t^2 + 40t + 25$ m
  (b) 106.6 m; 4.08 sec
  (c) 8.75 sec

13 $y = 2kt^{3/2}/3$

15 19.55 ft/sec^2

17 (a)

  (b) Area $= A$
  (c) 200 ft
  (d) 200 ft

19 (a) 6 seconds
  (b) Left sum: 97.5 ft
      overestimate
      Right sum: 82.5 ft
      underestimate
  (c) 90 ft

(d) $s(t) = 30t - (5/2)t^2$; $s(6) = 90$ ft
Distance by antidiff = Average of left and right hand sums

21 (a) 4 seconds
(b) 576 ft
(c) 10 seconds
(d) 192 ft/sec downward

23 (a) 80 ft/sec
(b) 640 ft

25 -33.56 ft/sec^2

Section 6.4

1

$F(x)$
x

3
$F(x)$
x

7 (a) $Si(4) \approx 1.76$
$Si(5) \approx 1.55$
(b) $(\sin x)/x$ is negative on that interval

9 $f(x) = 5 + \int_1^x (\sin t)/t \, dt$

11

x_2 $F(x)$
x_1 x_3

13 2.747

15 $\cos(x^2)$

17 $\arctan(x^2)$

19 $-\ln x$

21 (a) 0
(b) -2
(c) $1 \le x \le 6$
(d) 8

23 (a) $F'(x) = 1/(\ln x)$
(b) Increasing, concave down

25 $\cos(\sin^2 t)(\cos t)$

27 $3t^2\sqrt{1+t^6} - e^t\sqrt{1+e^{2t}}$

29 $s(b) - s(a)$

31 $(1/2)f(x)$

33 $\text{erf}(x) + (2/\sqrt{\pi})xe^{-x^2}$

35 $3x^2 e^{-x^6}$

Section 6.5

1 (a) 1.5 m
(b) 7 m/sec
(c) 9.8 m/sec^2 downward

3 $t = 5$; $v = -160$ ft/sec

5 400 feet

9 (a) First second: $-g/2$
Second: $-3g/2$
Third: $-5g/2$
Fourth: $-7g/2$
(b) Galileo seems to have been correct

Chapter 6 Review

1

$F(0) = 1$
$F(0) = 0$
x
1

3 $(5/2)x^2 + 7x + C$

5 $2t + \sin t + C$

7 $3e^x - 2\cos x + C$

9 $16\sqrt{x} + C$

11 $e^x + 5x + C$

13 $\tan x + C$

15 $(x+1)^3/3 + C$

17 $(1/10)(x+1)^{10} + C$

19 $(1/2)x^2 + x + \ln|x| + C$

21 $3\sin x + 7\cos x + C$

23 $2e^x - 8\sin x + C$

25 $-1/x + C$

27 $-\cos x + C$

29 $5e^x + C$

31 $t^2/2 + \ln|t| + C$

33 $x^4/4 + +2x^3 - 4x + 4.$

35 $e^x + 3$

37 $\sin x + 4$

39 9

41 $\sqrt{3} - \pi/9$

43 (a) 253/12
(b) $-125/12$

45 $c = 3/4$

47 2

49 (a) $t = 2, t = 5$
(b) $f(2) \approx 55, f(5) \approx 40$
(c) -10

51

Point of inflection
$f(x)$
x
x_1 x_2 x_3 x_4

53 $\arccos x^7$

55 $e^x \cos(e^{3x})$

57 $-(\sin t)4^{7\cos t} - 5t^4 4^{7t^5}$

59 Global max: at $x = \pi$
Global min: at $x = \pi/2$

61 (a) $f(x) = 3 + \int_0^x e^{t^2} \, dt$
(b) $f(x) = 5 + \int_{-1}^x e^{t^2} \, dt$

63

$h(t)$
t

65 (a) 14,000 revs/min^2
(b) 180 revolutions

67 (b) Highest pt: $t = 2.5$ sec
Hits ground: $t = 5$ sec
(c) Left sum:
136 ft (an overest.)
Right sum:
56 ft (an underest.)
(d) 100 ft

69 (b) $t = 6$ hours
(c) $t = 11$ hrs

bacteria
B D
time (hours)
5 10 15 20

bacteria
B D
N
time (hours)
5 ≈ 11.5 20

71 Positive, zero, negative, positive, zero

73 (a) $(1/2)e^{2x}, (1/3)e^{3x}, (1/3)e^{3x+5}$
(b) $(1/a)e^{ax+b}$

75 (a) $x - \ln|x-1|, x - 2\ln|x-1|,$
$x + \ln|x-2|$
(b) $x + (b-a)\ln|x-b|$

Ch. 6 Understanding

1 True

3 True

5 False

7 True

9 True

11 True

13 False

15 True

17 True

19 True

21 True

23 True

25 True

Section 7.1

1 (a) $(\ln 2)/2$
(b) $(\ln 2)/2$

3 $(1/3)e^{3x} + C$

5 $-125e^{-0.2t} + C$

7 $-0.5\cos(2x) + C$

9 $-(1/2)e^{-x^2} + C$

11 $(1/33)(t^3 - 3)^{11} + C$

13 $(1/6)(x^2 + 3)^3 + C$

15 $(1/5)y^5 + (1/2)y^4 + (1/3)y^3 + C$

17 $\ln|y + 5| + C$

19 $(1/5)x^5 + 2x^3 + 9x + C$

21 $-(1/8)(\cos\theta + 5)^8 + C$

23 $(1/7)\sin^7\theta + C$

25 $(1/35)\sin^7 5\theta + C$

27 $(1/3)(\ln z)^3 + C$

29 $(1/2)\ln(y^2+4)+C$

31 $2e^{\sqrt{y}}+C$

33 $\ln(2+e^x)+C$

35 $(1/6)\ln(1+3t^2)+C$

37 $t+2\ln|t|-1/t+C$

39 $\sinh x+C$

41 $e^{\cosh z}+C$

43 $\frac{1}{2}\sinh x^2+C$

45 $(\pi/4)t^4+2t^2+C$

47 $\sin x^2+C$

49 $(1/5)\cos(2-5x)+C$

51 $(1/2)\ln(x^2+1)+C$

53 0

55 $1-(1/e)$

57 $3(e^2-e)$

59 $2(\sin 2-\sin 1)$

61 40

63 $\ln 3$

65 $14/3$

67 $(2/5)(y+1)^{5/2}-(2/3)(y+1)^{3/2}+C$

69 $(2/5)(t+1)^{5/2}-(2/3)(t+1)^{3/2}+C$

71 $(2/7)(x-2)^{7/2}+(8/5)(x-2)^{5/2}+$
$(8/3)(x-2)^{3/2}+C$

73 $(2/3)(t+1)^{3/2}-2(t+1)^{1/2}+C$

75 (a) Yes; $-0.5\cos(x^2)+C$
(b) No
(c) No
(d) Yes; $-1/(2(1+x^2))+C$
(e) No
(f) Yes; $-\ln|2+\cos x|+C$

77 (a) $3/2$
(b) 3
(c) $3/2$

79 $\ln 3$

81 $2V_0/\omega$

83 (a) x^4+2x^2+C
(b) $(x^2+1)^2+C$
(c) Both correct but differ by a constant

87 (a) 50 thousand liters/minute
15.06 thousand liters/minute
(b) 1747 thousand liters

89 $(mg/k)t-(m^2g/k^2)(1-e^{-kt/m})+h_0$

91 (a) First case: $\approx 23,211$
Second case: ≈ 3.5 billion
(b) In both cases, 6.7 yrs
(c) About 3.5 yrs

Section 7.2

1 $x\cdot\arctan x-\frac{1}{2}\ln(1+x^2)+C$

3 $-t^2\cos t+2t\sin t+2\cos t+C$

5 $\frac{1}{5}t^2e^{5t}-\frac{2}{25}te^{5t}+\frac{2}{125}e^{5t}+C$

7 $(1/4)(2z+1)e^{2z}+C$

9 $(x^4/4)\ln x-(x^4/16)+C$

11 $(1/3)\theta^2\sin 3\theta+(2/9)\theta\cos 3\theta$
$-(2/27)\sin 3\theta+C$

13 $(1/6)\cos(3\alpha+1)\sin(3\alpha+1)$
$+\alpha/2+C$

15 $(2/3)y(y+3)^{3/2}$
$-(4/15)(y+3)^{5/2}+C$

17 $-(\theta+1)\cos(\theta+1)+\sin(\theta+1)+C$

19 $-x^{-1}\ln x-x^{-1}+C$

21 $-2t(5-t)^{1/2}-(4/3)(5-t)^{3/2}$
$-14(5-t)^{1/2}+C$

23 $w\arcsin w+\sqrt{1-w^2}+C$

25 $\frac{1}{2}x^2\arctan x^2-\frac{1}{4}\ln(1+x^4)+C$

27 $(1/3)x^3\sin x^3+(1/3)\cos x^3+C$

29 $(x-1)\sinh x-\cosh x+C$

31 $\cos 5+5\sin 5-\cos 3-3\sin 3\approx -3.944$

33 $(9/2)\ln 3-2\approx 2.944$

35 $6\ln 6-5\approx 5.751$

37 $(1/2)(\pi/2-1)\approx 0.285$

39 π

41 Integration by parts gives:
$(1/2)\sin\theta\cos\theta+(1/2)\theta+C$
The identity for $\cos^2\theta$ gives:
$(1/2)\theta+(1/4)\sin 2\theta+C$

43 $(1/2)e^\theta(\sin\theta+\cos\theta)+C$

45 $(1/2)\theta e^\theta(\sin\theta+\cos\theta)-$
$(1/2)e^\theta\sin\theta+C$

47 Integrate by parts choosing
$u=x^n,v'=\cos ax$

49 Integrate by parts choosing $u=\cos^{n-1}x$,
$v'=\cos x$

51 Approximately 77

53 (a) $-a^2e^{-a}-2ae^{-a}-2e^{-a}+2$
(b) Increasing
(c) Concave up

55 (a) V_0: increases,
ω: none, ϕ:none
(b) V_0: increases,
ω: increases, ϕ:none
(c) V_0: increases,
ω: none, ϕ:none

57 (a) $C_1=\sqrt{2}$
(b) $C_n=\sqrt{2}$

Section 7.3

1 $(1/10)e^{(-3\theta)}(\sin\theta-3\cos\theta)+C$

3 $-(1/5)x^3\cos 5x+(3/25)x^2\sin 5x+$
$(6/125)x\cos 5x-(6/625)\sin 5x+C$

5 $(1/7)x^7+(5/2)x^4+25x+C$

7 $-(1/4)\sin^3 x\cos x$
$-(3/8)\sin x\cos x+(3/8)x+C$

9 $(\frac{1}{2}x^3-\frac{3}{4}x^2+\frac{3}{4}x-\frac{3}{8})e^{2x}+C$

11 $(1/4)\arcsin(4x/5)+C$

13 $(5/16)\sin 3\theta\sin 5\theta$
$+(3/16)\cos 3\theta\cos 5\theta+C$

15 $((1/3)x^2-(2/9)x+2/27)e^{3x}+C$

17 $((1/3)x^4-(4/9)x^3+(4/9)x^2-(8/27)x$
$+(8/81))e^{3x}+C$

19 $\frac{1}{2}\frac{\sin x}{\cos^2 x}+\frac{1}{4}\ln\left|\frac{\sin x+1}{\sin x-1}\right|+C$

21 $-(1/2)x^2\cos x^2+(1/2)\sin x^2+C$

23 $-(1/2)y^2\cos 2y+(1/2)y\sin 2y$
$+(1/4)\cos 2y+C$

25 $\frac{1}{4}\frac{\sin x}{\cos^4 x}+\frac{3}{8}\frac{\sin x}{\cos^2 x}+$
$\frac{3}{16}\ln\left|\frac{(\sin x)+1}{(\sin x)-1}\right|+C$

27 $-\frac{1}{6}\frac{\cos 3\theta}{\sin^2 3\theta}+\frac{1}{12}\ln\left|\frac{\cos(3\theta)-1}{\cos(3\theta)+1}\right|+C$

29 $\frac{1}{2}(\ln|x+1|-\ln|x+3|)+C$

31 $-(1/3)(\ln|z|-\ln|z-3|)+C$

33 $\arctan(z+2)+C$

35 $-1/(x+2)+C$

37 $-(1/9)(\cos^3 3\theta)+(1/15)(\cos^5 3\theta)+C$

39 $\frac{1}{5}\cosh^5 x-\frac{1}{3}\cosh^3 x+C$

41 $\frac{1}{3}(1-\frac{\sqrt{2}}{2})$

43 $(\ln 2)/2$

47 (a) 0
(b) $V_0/\sqrt{2}$
(c) 156 volts

49 0

Section 7.4

1 $2/(5-x)+2/(5+x)$

3 $-2/y+1/(y-2)+1/(y+2)$

5 $1/(2(s-1))-1/(2(s+1))-1/(s^2+1)$

7 $1/(w-1)-1/w-1/w^2-1/w^3$

9 $(1/6)(\ln|x|+5\ln|6+x|)+C$

11 $\ln|s|-\ln|s+2|+C$

13 $\ln|y-1|+\arctan y-\frac{1}{2}\ln\left|y^2+1\right|+C$

15 $\ln|x-2|+\ln|x+1|+\ln|x-3|+K$

17 $2\ln|x-5|-\ln\left|x^2+1\right|+K$

19 $x^3/3+\ln|x+1|-\ln|x+2|+C$

21 $\arctan(x+2)+C$

23 (a) $w=x^2+10$
(b) $x=\sqrt{10}\tan\theta$

25 $x=(4\tan\theta)-3$

27 $w=(x+1)^2$

29 $w=1-(z-1)^2$

31 $w=(\theta-2)^2-4$

33 $\ln|x+2|-\ln|x+3|+C$

35 $-\ln|x-1|+2\ln|x-2|+C$

37 $(\ln|x+1|-\ln|x+4|)/3+C$

39 $-4\ln|x-1|+7\ln|x-2|+C$

41 $\ln|x|-(1/2)\ln\left|x^2+1\right|+\arctan x+K$

43 $-x\sqrt{9-x^2}/2+(9/2)\arcsin(x/3)+C$

45 $-(\sqrt{1+t^2})/t+C$

47 $2\ln|s+2|-\ln\left|s^2+1\right|+4\arctan s+C$

49 $(1/3)(\ln|e^x-1|-\ln|e^x+2|)+C$

51 $2\ln 2-\pi/4$

53 $8(2/3-5/(6\sqrt{2}))$

55 $(1/6)\ln((3-2\sqrt{2})(7+4\sqrt{3}))$

57 $\ln|x+1|+\ln|x-1|+C$

59 (b) $-\sqrt{5-y^2}/(5y)+C$

61 (a) $(a\ln|x-a|-b\ln|x-b|)/(a-b)+C$
(b) $\ln|x-a|-a/(x-a)+C$

63 (a) $t(p)=b\ln(p(1-a)/(a(1-p)))$
(b) $a=0.01$
(c) $b=0.218$
(d) 1.478

Section 7.5

1 (a) Underestimate

(b) Overestimate

(c) Overestimate

(d) Underestimate

3 (a) Underestimate

(b) Overestimate

(c) Underestimate

(d) Overestimate

5 (a) Underestimate

(b) Overestimate

(c) Overestimate

(d) Underestimate

7 (a) 27
(b) 135
(c) 81
(d) 67.5

9 (a) MID(2)= 24;
TRAP(2)= 28
(b) MID(2) underestimate;
TRAP(2) overestimate

Area shaded = MID(2)

$f(x) = x^2 + 1$

Area shaded = TRAP(2)

$f(x) = x^2 + 1$

11 (a) (i) 0.8160
(ii) 0.7535
(iii) 0.7847

13 RIGHT < TRAP < Exact < MID < LEFT

15 RIGHT: over; LEFT: under

17 LEFT, TRAP: over; RIGHT, MID: under

19 (a) $f(x)$
(b) $f(x)$

21 (a) 0
(c) MID(n) = 0 for all n

23 445 lbs. of fertilizer

Section 7.6

1 72

5 (a) ≈ 53.598
(b) LEFT(2)= 16.778;
error= 36.820
RIGHT(2)= 123.974;
error= −70.376
TRAP(2)= 70.376;
error= −16.778
MID(2)= 45.608;
error= 7.990
SIMP(2)= 53.864;
error= −0.266
(c) LEFT(4)= 31.193;
error=22.405
RIGHT(4)= 84.791;
error=−31.193
TRAP(4)= 57.992;
error=−4.394
MID(4)= 51.428;
error=2.170
SIMP(4)= 53.616;
error=−0.018

7 (a) 4 places: 2 seconds
8 places: ≈ 6 hours
12 places: ≈ 6 years
20 places: ≈ 600 million years
(b) 4 places: 2 seconds
8 places: ≈ 3 minutes
12 places: ≈ 6 hours
20 places: ≈ 6 years

Section 7.7

1 (a)

3 (a) 0.9596, 0.9995, 0.99999996
(b) 1.0

5 Diverges

7 Converges to 1/2

9 1

11 ln 2

13 $\pi/2$

15 Does not converge

17 $\pi/4$

19 Does not converge

21 Does not converge

23 0.01317

25 $2 - 2e^{-\sqrt{\pi}}$

27 $\pi/2$

29 $(1/2)\ln(5/3)$

31 Converges; 0

33 1

35 Converges for $p < -1$
to $-pe^{p+1}/(p+1)^2$

37 (a) 1
(b) α
(c) $2\alpha^2$

39 0

41 0

43 (a) $\Gamma(1) = 1$
$\Gamma(2) = 1$
(c) $\Gamma(n) = (n-1)!$

45 9×10^9 joules

Section 7.8

1 Converges; behaves like $1/x^2$

3 Diverges; behaves like $1/x$

5 Diverges; behaves like $1/x$

7 Converges; behaves like $5/x^3$

9 Converges; behaves like $1/x^2$

11 Does not converge

13 Converges

15 Does not converge

17 Converges

19 Converges

21 Converges

23 Converges

25 Does not converge

27 0.606

29 Converge

31 Converges for $p < 1$
Diverges for $p \geq 1$

33 (a) $a = 0.399$
(b) They are the same

35 (a) e^t is concave up for all t

Chapter 7 Review

1 $-\cos t + C$

3 $(1/5)e^{5z} + C$

5 $(-1/2)\cos 2\theta + C$

7 $2x^{5/2}/5 + 3x^{5/3}/5 + C$

9 $(r+1)^4/4 + C$

11 $x^2/2 + \ln|x| - x^{-1} + C$

13 $(1/2)e^{t^2} + C$

15 $((1/2)x^2 - (1/2)x + 1/4)e^{2x} + C$

17 $(1/2)x^2 \ln x - (1/4)x^2 + C$

19 $x(\ln x)^2 - 2x \ln x$
$+ 2x + C$

21 $-e^{0.5-0.3t}/0.3 + C$

23 $-(1/3)(4-x^2)^{3/2} + C$

25 $2\sin\sqrt{y} + C$

27 $(1/4)\sin 2\theta + (1/2)\theta + C$

29 $-(1/2)\ln|\cos(2x-6)| + C$

31 $\ln|t| - 4/t - 2/t^2 + C$

33 $\ln|t| - 1/t + C$

35 $-\ln|\cos\theta| + C$

37 $(1/2)\ln|x^2+1| + C$

39 $(1/2)\arctan 2z + C$

41 $(-1/20)\cos^4 5\theta + C$

43 $(1/12)(t-10)^{12}$
$+ (10/11)(t-10)^{11} + C$

45 $xe^x - e^x + C$

47 $(1/142)(10^{71} - 2^{71})$

49 $(1/3)\arctan(u/3) + C$

51 $-\ln(|\cos(\ln x)|) + C$

53 $(1/2)\arcsin(2x) + C$

55 $y - (1/2)e^{-2y} + C$

57 $\ln|\ln x| + C$

59 $\sin\sqrt{x^2+1} + C$

61 $ue^{ku}/k - e^{ku}/k^2 + C$

63 $(1/\sqrt{2})e^{\sqrt{2}+3} + C$

65 $(1/2)e^{2x} + 2xe^x - 2e^x$
$+ (x^3/3) + C$

67 $(5/2)\ln|x^2+4| + 3\arctan(x/2) + C$

69 $(1/20)[\ln|r-10| - \ln|r+10|] + C$

71 $-e^{-ct}(k\cos kt + c\sin kt)/(k^2+c^2) + C$

73 $x^{\sqrt{k}+1}/(\sqrt{k}+1)$
$+ (\sqrt{k})^x/(\ln\sqrt{k}) + C$

75 $\ln\left|\left(x - \frac{3}{2}\right) + \sqrt{x^2-3x+2}\right| + C$

77 $x + 5\ln|x-2| - 2\ln|x-1| + C$

79 $(1/2)\ln|ax^2 + 2bx + c| + C$

81 $(1/\ln 2)\ln|2^t + 1| + C$

83 $(1/7)x^7 + 3x^5 + 25x^3 + 125x + C$

85 $(1/2)x\sqrt{4-x^2} + 2\arcsin(x/2) + C$

87 $-(1/2)\ln|1 + \cos^2 w| + C$

89 $x\tan x + \ln|\cos x| + C$

91 $(2/3)(\sqrt{x+1})^3 - 2\sqrt{x+1} + C$

93 $(1/2)\ln|e^{2y} + 1| + C$

95 $-1/(z-5) - 5/(2(z-5)^2)$
$+ C$

97 $e^{x^2-x} + C$

99 $-(2/9)(2 + 3\cos x)^{3/2} + C$

101 $\sin^3(2\theta)/6 - \sin^5(2\theta)/10 + C$

103 $(x + \sin x)^4/4 + C$

105 $\frac{1}{3}\sinh^3 x + C$

107 $49932\frac{1}{6}$

109 201,760

111 $3/2 + \ln 2$

113 $e - 2 \approx 0.71828$

115 $-11e^{-10} + 1$

117 $2e(e-1)$

119 0

121 $(\ln|x-2| - \ln|x+2|)/4 + C$

123 $(\ln|x| - \ln|x+5|)/5 + C$

125 $-3\ln|x|/2 - \ln|x+2|/6 + 5\ln|x-1|/3 + C$

127 $2/(1+x) + \ln|x| + C$

129 $\arctan(x+2) + C$

131 $\ln|\sin x| - (1/2)\ln\left|\sin^2 x + 1\right| + C$

133 Converges
$\int_4^\infty t^{-3/2}\, dt = 1$

135 Converges to 1

137 Converges to 0

139 Converges to $(\ln(3/2))/4$

141 Does not converge

143 Does not converge

145 Does not converge

147 5/6

149 (a) 3.5
(b) 35

151 11/3

153 (a) (i) 0
(ii) $2/\pi$
(iii) 1/2
(b) Smallest to largest:
Average value of $f(t)$
Average value of $k(t)$
Average value of $g(t)$

155 (a) $x + C$
(b) $\ln\left|e^x + e^{-x}\right| + C$
(c) $E(x) = (x + \ln\left|e^x + e^{-x}\right|)/2 + C$
$F(x) = (x - \ln\left|e^x + e^{-x}\right|)/2 + C$

157 $\sqrt{\pi}/4$

159 (a) $(k/L)\ln 3$
(b) $T \to \infty$

161 (a) 0.5 ml
(b) 99.95%

163 (a) 360 degree-days
(b)

(c) $T_2 = 36$

165 (a) $-x + x\ln x$,
$2x - 2x\ln x + x(\ln x)^2$,
$-6x + 6x\ln x -$
$3x(\ln x)^2 + x(\ln x)^3$,
$24x - 24x\ln x + 12x(\ln x)^2 -$
$4x(\ln x)^3 + x(\ln x)^4$
(b) $\int(\ln x)^n\, dx =$
$-n\int(\ln x)^{n-1}\, dx + x(\ln x)^n$

167 (a) $-\cos(4x)/16$
(b) $\sin(4x)/4$

Ch. 7 Understanding

1 False

3 False

5 True

7 False

9 False

11 False

13 True

15 False

17 True

19 True

21 False

23 False

25 True

27 False.

Section 8.1

1 15

3 15/2

5 $(5/2)\pi$

7 1/6

9 $\int_0^9 4\pi\, dx = 36\pi$ cm^3

11 $\int_0^5 (4\pi/25)y^2\, dy = 20\pi/3$ cm^3

13 $\int_0^5 \pi(5^2 - y^2)\, dy = 250\pi/3$ mm^3

15 Triangle; $b, h = 1, 3$

17 Quarter circle $r = \sqrt{15}$

19 Hemisphere, $r = 12$

21 Cone, $h = 6, r = 3$

23 $V = \pi$

25 $V = (\pi r^2 h)/3$

27 $\int_0^{0.2} 2250h\, dh = 45$ km^3

Section 8.2

1 $\pi/5$

3 $256\pi/15$

5 $\pi(e^2 - e^{-2})/2$

7 $\pi/2$

9 $2\pi/15$

11 3.526

13 $\sinh 2$

15 $e - 1$

17 (a) $16\pi/3$
(b) 1.48

19 $V = (16/7)\pi \approx 7.18$

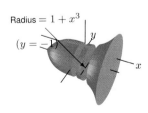

Radius = $1 + x^3$

$(y = -1)$

21 $V = (\pi^2/2) \approx 4.935$

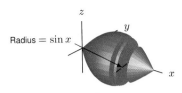

Radius = $\sin x$

23 $4\pi/5$

25 8/15

27 $\sqrt{3}/8$

29 $V \approx 42.42$

31 $V = (e^2 - 1/2) \approx 3.195$

33 (a) Volume ≈ 152 in^3
(b) About 15 apples

35 $40,000LH^{3/2}/(3\sqrt{a})$

37 (a) $dh/dt = -6/\pi$
(b) $t = \pi/6$

39 (a) $4\int_0^r \sqrt{1 + (-x/y)^2}\, dx$
(b) $2\pi r$

41 $e - e^{-1}$

Section 8.3

1 $(-1/2, \sqrt{3}/2)$

3 $(3, -\sqrt{3})$

5 $(\sqrt{2}, \pi/4)$

7 $(2\sqrt{2}, -\pi/6)$

9 (b)

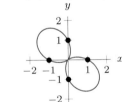

(c) Cartesian:
$(\sqrt{3}/4, 1/4)$;
$(-\sqrt{3}/4, 1/4)$ or polar:
$r = 1/2, \theta = \pi/6$ or $5\pi/6$

(d)

11 Looks the same

13 Rotated by $90°$ clockwise

15 $\pi/4 \le \theta \le 5\pi/4$;
$0 \le \theta \le \pi/4$ and $5\pi/4 \le \theta \le 2\pi$

17 $\sqrt{8} \le r \le \sqrt{18}$ and $\pi/4 \le \theta \le \pi/2$

19 $0 \le \theta \le \pi/2$ and $1 \le r \le 2/\cos\theta$

21 $8\pi^3/6$

23 $3\pi/2$

25 $2 - \pi/4$

27 $\pi/2$

29 $a = \sqrt[3]{6}$

31 $y = (\sqrt{3}/5)x + 9/5$

33 Horiz: $\theta = 0, 2.029, 4.913$
Vert: $\theta = 0.860, 3.426$

35 1.760

37 0.712

Section 8.4

1 $1 - e^{-10}$ gm

3 (a) $\sum_{i=1}^{N} (2 + 6x_i)\Delta x$
(b) 16 grams

5 (b) $\sum_{i=1}^{N}[600 + 300\sin(4\sqrt{x_i + 0.15})](20/N)$
(c) ≈ 11513

7 2 cm to right of origin

9 1 gm

11 (a) $\int_0^5 2\pi r(0.115e^{-2r})\, dr$
(b) 181 cubic meters

13 (a) $\pi r^2 l/2$
(b) $2klr^3/3$

15 $\int_0^{60} (1/144)g(t)\, dt$ ft^3

17 $x = 2$

19 $\pi/2$

21 (a) Right
(b) $2/(1 + 6e - e^2) \approx 0.2$

23 (a) 2/3 gm
(b) Greater than 1/2

x^2

(c) $\overline{x} = 3/4$ cm

27 $\overline{x} = (1 - 3e^{-2})/(2 - 2e^{-2}) \approx 0.343$;
$\overline{y} = 0$

Section 8.5

1 9/2 joules

3 (a) 1.5 joules,
13.5 joules
(b) For $x = 4$ to $x = 5$
Force larger

5 $1.489 \cdot 10^{10}$ joules

7 3437.5 ft-lbs

9 6860 joules

11 3,088,800 ft-lbs

13 2,822,909.50 ft-lbs

15 518,363 ft-lbs

17 354,673 ft-lbs

19 (a) 187.2 lbs
(b) 2808 ft-lbs

21 (a) Bottom: 187,200 lbs
(b) 15×10 side: 70,200 lbs
(c) 15×20 side: 140,400 lbs

23 Bottom: 1497.6 lbs
Front and back: 499.2 lbs
Both sides: 374.4 lbs

25 (a) 21,840 lb/ft^2; 151.7 lb/in^2
(b) (i) 546,000 pounds
(ii) 542,100 pounds

27 $9800 \int_0^{100} h(3600 - 6h)\, dh$
$= 1.6 \cdot 10^{11}$ newtons

29 60 joules

31 $\dfrac{GmM}{a(a+l)}$

33 $GMmy(a^2 + y^2)^{-3/2}$ toward center

Section 8.6

1 $15,319.30

3 $8,242, $12,296

5 $/year

t (years from present)

7 (a) $5820 per year
(b) $36,787.94

9 Installments

11 $46,800

13 (a) 10.6 years
(b) 624.9 million dollars

15 $85,750,000

19 (a) Less
(b) Can't tell
(c) Less

Section 8.7

1 % of population
per dollar of income

income

% of population having
at least this income

income

3 % of population
per dollar of income

income

% of population having
at least this income

income

5 pdf; 1/2

7 pdf; 2/3

9 pdf; 2

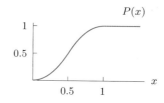

11 (a) 0.9 m–1.1 m

15 (a) Cumulative distribution
increasing
(b) Vertical 0.2,
horizontal 2

17 (a) 22.1%
(b) 33.0%
(c) 30.1%
(d) $C(h) = 1 - e^{-0.4h}$

19 (b) About 3/4

Section 8.8

5 (a) 0.684 : 1
(b) 1.6 hours
(c) 1.682 hours

7 (a) $P(t) =$ Fraction of population who survive
up to t years after treatment
(b) $S(t) = e^{-Ct}$
(c) 0.178

9 (a) $p(x) = \dfrac{1}{15\sqrt{2\pi}} e^{-\frac{1}{2}\left(\frac{x-100}{15}\right)^2}$
(b) 6.7% of the population

11 (c) μ represents the mean of the distribution,
while σ is the standard deviation

13 (b)

15 (a) $p(r) = 4r^2 e^{-2r}$

(b) Mean: 1.5 Bohr radii
Median: 1.33 Bohr radii
Most likely:
1 Bohr radius

Chapter 8 Review

1

3 $2 \int_{-r}^{r} \sqrt{r^2 - x^2}\, dx = \pi r^2$

5 $3772\pi/15 = 790.006$

7 $\pi(1 - e^{-4})/4 = 0.771$

9 $27\pi = 84.823$

11 $36\pi = 113.097$ m^3

13 $36\pi = 113.097$ m^3

15 $2 \int_{-a}^{a} \sqrt{1 + b^2 x^2/(a^2(a^2 - x^2))}\, dx$

17 45.230

19 (a)

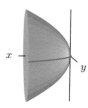

(b) $\sum_{i=1}^{N} \pi x_i \Delta x$
(c) Volume $= \pi/2$

21 (a) $8a/3$
(b) $32a^2\pi/5$

23 (a) 1.669
(b) 11.550

25 (a) $423\pi/5 = 265.778$
(b) $657\pi/5 = 412.805$

27 $\pi^2 + 2\pi/3 = 11.964$

29 2/3

31 1/3

33 (a) $a = b/l$
(b) $(1/3)\pi b^2 l$

35 Volume $= 6\pi^2$

37 πa^2

39 (a) $(\sin^2\theta - \cos^2\theta)/(2\cos\theta\sin\theta)$
(b) $\pi/4$

41 1/4

43 4400 ft-lb

45 661,619.41 ft-lb

47 1,170,000 lbs

49 (a) Future value $= 6389.06
Present value $= 864.66
(b) 17.92 years

51 49 kg

53 $y = x^4$

55 B

57 0.43m^3

59 1.25 gallons

61 (a) $y = h + (2 - 2h)x^2$
   (b) Both events happen
        simultaneously

63 The hoop rotating about the cylindrical axis

65 (a) $\frac{1}{2}t\sqrt{1 + 4t^2} + \frac{1}{4}\text{arcsinh}\,(2t)$
   (b) t^2

67 (b) $(4\pi/3)(r^2 - a^2)^{3/2}$

Ch. 8 Understanding

1 True

3 False

5 False

7 False

9 True

11 True

13 False

15 False

17 False

19 True

21 False

23 False

25 False

27 True

29 False

Section 9.1

1 $0, 3, 2, 5, 4$

3 $2/3, 4/5, 6/7, 8/9, 10/11$

5 $1, -1/2, 1/4, -1/8, 1/16$

7 $s_n = 2^{n+1}$

9 $s_n = n^2 + 1$

11 $s_n = n/(2n + 1)$

13 $1, 3, 6, 10, 15, 21$

15 $0, 1/2, 3/4, 7/8, 15/16, 31/32$

17 (a) (IV)
   (b) (III)
   (c) (II)
   (d) (I)

19 (a) (II)
   (b) (III)
   (c) (I)
   (d) (IV)
   (e) (V)

21 Diverges

23 Converges to 3

25 Diverges

27 Converges to 2

29 Converges to 0

31 Diverges

33 $0.25, 0.00, 0.25, 1.00, 2.25, 4.00$

35 $0, 6, -6, 6, -6, 6, \ldots$ and
    $3, 0, 2, -2, 2, \ldots$

37 $1.5, 2, 3, 4, 5, 6, 7 \ldots$ and
    $1.75, 2.17, 3, 4, 5, 6, \ldots$

39 (a) $75.747(1.003)^n$
   (b) $0.227(1.003)^{n-1}$; incr in consumption
   (c) Total consumption 2003–2020

41 $15{,}500 - 500n$

43 (a) $20{,}000(0.12)$; $20{,}000(0.88)(0.12)$;
        $20{,}000(0.88)^2(0.12)$
        $d_n = 20{,}000(0.88)^{n-1}(0.12)$
   (b) $400, 400(1.18), 400(1.18)^2$
        $r_n = 400(1.18)^{n-1}$
   (c) 7799.52
   (d) 6923.05; two-year-old

47 $s_n = s_{n-1} + 2, s_1 = 2$

49 $s_n = s_{n-1} + n^2, s_1 = 1$

51 $s_n = (1/s_{n-1}) + 1, s_1 = 1$

55 (a) k rows; $a_n = n$
   (b) $T_n = T_{n-1} + n, n > 1; T_1 = 1$

57 Converges to 0.7391

Section 9.2

1 Yes, $a = 2$, ratio $= 1/2$

3 No

5 No

7 Yes, $a = 1$, ratio $= 2z$

9 Yes, $a = 1$, ratio $= -x$

11 $1/(1 - 2z), |z| < 1/2$

13 $1/(1 + x), |x| < 1$

15 26 terms; 2.222

17 9 terms; 0.0000222

19 $3(2^{11} - 1)/2^{10}$

21 $(3^{17} - 1)/(2 \cdot 3^{20})$

23 Yes

25 (a) $\approx 7\%$
   (b) $Q_n = 50(1 - (0.07)^{n+1})/(1 - 0.07)$
   (c) $P_n = 0.07(50)(1 - (0.07)^n)/(1 - 0.07)$

27

q (quantity, mg)

31 $400 million

Section 9.3

1 Converges

3 Converges

7 Terms not decreasing

9 Terms neither positive nor decreasing

11 Converges

13 Diverges

15 Converges

17 Diverges

19 Diverges

23 (b) $0 < r < 1/e$

31 If $a_n = 1$, converges
    If $a_n = n$, diverges

35 (d) 0.58

37 (a) $65/24 = 2.708$
   (c) 3

Section 9.4

5 Converges

7 Converges

9 Diverges

11 Converges

13 Converges

15 Diverges

17 Alternating

19 Not alternating

21 Converges

23 Converges

25 Converges

27 Diverges

29 Converges

31 Converges

33 Terms not positive

35 Limit undefined

37 Not alternating series

39 Does not converge

41 Converges to approximately 0.3679

43 Diverges

45 Diverges

47 Diverges

49 Converges

51 Diverges

53 Diverges

55 11 or more terms

57 Absolutely convergent

59 Divergent

61 $\sum c_n$ converges, $\sum a_n$ diverges

69 Converges

Section 9.5

1 Yes

3 No

5 $1 \cdot 3 \cdot 5 \cdots (2n - 1)x^n/(2^n \cdot n!); n \geq 1$

7 $(-1)^k(x - 1)^{2k}/(2k)!; k \geq 0$

9 $(x - a)^n/(2^{n-1} \cdot n!); n \geq 1$

11 $1/5$

13 2

15 1

17 ∞

19 $1/4$

21 1

23 Diverges when $x = \frac{1}{2}$,
    converges when $x = -\frac{1}{2}$

25 $2 \leq x < 4$

27 $3 \leq x \leq 7$

31 $7 \leq R \leq 10$

33 Diverges, converges, converges

35 (a) 1
   (b) All real numbers
   (c) $\sum_{n=0}^{\infty} x^n/n!$
   (d) e^x

37 (a) 1
   (b) $p(x) = \cos x, q(x) = \sin x$

Chapter 9 Review

1 Converges 4/7

3 Diverges

5 $b^5(1 - b^6)/(1 - b)$ when $b \neq 1$;
   6 when $b = 1$

7 $32/3$

9 Diverges

11 Converges

15 Converges

17 Converges

19 Converges

21 Converges

23 Converges

25 Converges

27 Diverges

29 Diverges

31 Converges

33 Diverges

35 Diverges

37 Converges

39 Diverges

41 Diverges

43 $1/4$

45 ∞

47 Divergent

49 $-3 \le x \le 3$

51 $-1 < x \le 1$

53 (a) All real numbers
   (b) Even

55 $926.40

57 (a) $1000
   (b) When the interest rate is 5%, the present value equals the principal
   (c) The principal
   (d) Because the present value is more than the principal

59 (a)   (i) $16.43 million
      (ii) $24.01 million
   (b) $16.87 million

61 (a) $1250
   (b) 12.50

63 $5/6$

67 Diverges

69 Converges

71 Diverges, converges, yes

73 32

Ch. 9 Understanding

1 False

3 True

5 False

7 False

9 False

11 True

13 True

15 True

17 True

19 False

21 False

23 True

25 False

27 True

29 True

31 True

33 True

35 False

37 True

39 False

41 True

43 False

45 False

Section 10.1

1 $P_3(x) = 1 + x + x^2 + x^3$
$P_5(x) = 1 + x + x^2 + x^3$
$\quad + x^4 + x^5$
$P_7(x) = 1 + x + x^2 + x^3$
$\quad + x^4 + x^5 + x^6 + x^7$

3 $P_2(x) = 1 + (1/2)x - (1/8)x^2$
$P_3(x) = 1 + (1/2)x - (1/8)x^2 + (1/16)x^3$
$P_4(x) = 1 + (1/2)x - (1/8)x^2$
$\quad + (1/16)x^3 - (5/128)x^4$

5 $P_2(x) = 1 - x^2/2!$
$P_4(x) = 1 - x^2/2! + x^4/4!$
$P_6(x) = 1 - x^2/2! + x^4/4! - x^6/6!$

7 $P_3(x) = P_4(x) = x - (1/3)x^3$

9 $P_2(x) = 1 - (1/2)x + (3/8)x^2$
$P_3(x) = 1 - (1/2)x + (3/8)x^2$
$\quad - (5/16)x^3$
$P_4(x) = 1 - (1/2)x + (3/8)x^2$
$\quad - (5/16)x^3 + (35/128)x^4$

11 $P_4(x) = e[1 + (x - 1)$
$\quad + (1/2)(x - 1)^2 + (1/6)(x - 1)^3$
$\quad + (1/24)(x - 1)^4]$

13 $P_4(x) = 1 - (1/2!)(x - \pi/2)^2$
$\quad + (1/4!)(x - \pi/2)^4$

15 $P_4(x) = 2(x - 1) - (x - 1)^2 + (2/3)(x - 1)^3 - (1/2)(x - 1)^4$

17 $c < 0, b > 0, a > 0$

19 $a < 0, b > 0, c > 0$

21 (a) 5
   (b) -7
   (c) 16

23 (a) $3 - 2(x - 5)$
   $\quad + 0.5(x - 5)^2$
   $\quad - 0.5(x - 5)^3$
   (b) $3.205, 3.2055$

25 $P_2(x) = 4x^2 - 7x + 2$
$f(x) = P_2(x)$

27 (a) If $f(x)$ is a polynomial of degree n, then $P_n(x)$, the nth degree Taylor polynomial for $f(x)$ about $x = 0$, is $f(x)$ itself

31 (a) $3/7$
   (b) 0

33 (a) $P_4(x) = 1 + x^2 + (1/2)x^4$
   (b) If we substitute x^2 for x in the Taylor polynomial for e^x of degree 2, we will get $P_4(x)$, the Taylor polynomial for e^{x^2} of degree 4
   (c) $P_{20}(x) = 1 + x^2/1! + x^4/2!$
      $\quad + \cdots + x^{20}/10!$
   (d) $e^{-2x} \approx 1 - 2x + 2x^2$
      $\quad - (4/3)x^3 + (2/3)x^4 - (4/15)x^5$

35 (a) $0.94444 \cdots$
   (b) $0.94611 \cdots$

Section 10.2

1 $f(x) = 1 + 3x/2 + 3x^2/8 - x^3/16 + \cdots$

3 $f(x) = -x + x^3/3! - x^5/5! + x^7/7! + \cdots$

5 $f(x) = 1 + x + x^2$
$\quad + x^3 + \cdots$

7 $f(x) = 1 + x/2 - x^2/8$
$\quad + x^3/16 + \cdots$

9 $\sin x = \sqrt{2}/2 + (\sqrt{2}/2)(x - \pi/4)$
$\quad - (\sqrt{2}/4)(x - \pi/4)^2$
$\quad - (\sqrt{2}/12)(x - \pi/4)^3$
$\quad - \cdots$

11 $\sin \theta = -\sqrt{2}/2 + (\sqrt{2}/2)(\theta + \pi/4)$
$\quad + (\sqrt{2}/4)(\theta + \pi/4)^2$
$\quad - (\sqrt{2}/12)(\theta + \pi/4)^3 + \cdots$

13 $1/x = 1 - (x - 1) +$
$\quad (x - 1)^2 - (x - 1)^3 + \cdots$

15 $\frac{1}{x} = -1 - (x + 1)$
$\quad - (x + 1)^2 - (x + 1)^3$
$\quad - \cdots$

17 $(-1)^n x^n; n \ge 0$

19 $(-1)^{n-1} x^n/n; n \ge 1$

21 $(-1)^k x^{2k+1}/(2k + 1); k \ge 0$

23 $(-1)^k x^{4k+2}/(2k)!; k \ge 0$

25 (a) $\ln(1 + 2x)$
      $= 2x - 2x^2 + (8/3)x^3 + \cdots$
   (b) To get the expression for $\ln(1 + 2x)$ from the series for $\ln(1 + x)$, substitute $2x$ for x in the series
   (c) $-1/2 < x < 1/2$

27 $-1 < x < 1$

29 1

31 e^2

33 $4/3$

35 $\ln(3/2)$

37 e^3

39 $e^{-0.1}$

41 $e^{0.2} - 1$

43 $\frac{d}{dx}(x^2 e^{x^2})|_{x=0} = 0$
$\frac{d^6}{dx^6}(x^2 e^{x^2})|_{x=0} = \frac{6!}{2} = 360$

Section 10.3

1 $e^{-x} = 1 - x + x^2/2!$
$\quad - x^3/3! + \cdots$

3 $\sqrt{1 - 2x} = 1 - x - x^2/2$
$\quad - x^3/2 - \cdots$

5 $\arcsin x = x + (1/6)x^3$
$\quad + (3/40)x^5 + (5/112)x^7 + \cdots$

7 $1/\sqrt{1 - z^2} = 1 + (1/2)z^2 + (3/8)z^4$
$\quad + (5/16)z^6 + \cdots$

9 $3t^2 - (9/2)t^4 + (81/40)t^6 - (243/560)t^8 + \cdots$

11 $\sqrt{(1 + t)} \sin t = t + (1/2)t^2$
$\quad - (7/24)t^3 - (1/48)t^4 + \cdots$

13 $1 + 3x + 3x^2 + x^3$
$0 \cdot x^n$ for $n \ge 4$

15 $1 + (1/2)y^2 + (3/8)y^4 + \cdots + ((1/2)(3/2) \cdots (1/2 + n - 1)y^{2n})/n! + \cdots;$
$n \ge 1$

17 $1/(a - r) = (1/a)(1 + (r/a)$
$\quad + (r/a)^2 + (r/a)^3 + \cdots)$

19 $\sqrt[3]{P + t} = \sqrt[3]{P}(1 + (1/3)(t/P)$
$\quad - (1/9)(t/P)^2 + (5/81)(t/P)^3 + \cdots)$

21 $a/\sqrt{a^2 + x^2} = 1 - (1/2)(x/a)^2 + (3/8)(x/a)^4$
$\quad - (5/16)(x/a)^6 + \cdots$

23 (a) $2x + x^3/3 + x^5/60 + \cdots$
   (b) $P_3(x) = 2x + x^3/3$

25 $1/(1 - x)^2$

27 $1 - \cos y < \ln(1 + y^2) < \sin(y^2)$

29 (a) First two terms are the same in both series
  (b) $1/(1+x^2)$
  (c) Even, because all terms are of even degree

31 $\sinh 2x = \sum_{m=0}^{\infty} (2x)^{2m+1}/(2m+1)!$
   $\cosh 2x = \sum_{m=0}^{\infty} (2x)^{2m}/(2m)!$

33 $E \approx kQ/R^2 \left(4/R + 8/R^3\right)$

37 (a) $((l_1 + l_2)/c) \cdot$
    $(v^2/c^2 + (5/4)v^4/c^4)$
  (b) $v^2, (l_1 + l_2)/c^3$

39 (a) $GMm/h^2 \left(1 - (3/4)(a^2/h^2)\right)$
  (b) $F \approx (GMm)/(h^2)$
  (c) 0.03%

41 (a) Left = 0.8076
    Right = 0.6812
  (b) $1 - x^2 + x^4/2 - x^6/6$
  (c) 0.74286

Section 10.4

1 $|E_3| \le 0.00000460$

3 $|E_3| \le 0.000338$

5 $|E_4| \le 0.016$

7 $|E_3| \le 16.5$

9 (a) Underestimate
  (b) 1

11 (a) Overestimate:
    $0 < \theta \le 1$
    Underestimate:
    $-1 \le \theta < 0$
  (b) $|E_2| \le 0.17$

13 (a) (i) 4, 0.2
     (ii) 1
  (b) $-4 \le x \le 4$

15 Four decimal places: $n = 7$
   Six decimal places: $n = 9$

17 (a)

  (b)

Section 10.5

1 Not a Fourier series

3 Fourier series

5 $F_1(x) = F_2(x) = (4/\pi)\sin x$
  $F_3(x) = (4/\pi)\sin x + (4/3\pi)\sin 3x$

$F_1(x) = F_2(x) = \frac{4}{\pi}\sin x$

$F_3(x) = \frac{4}{\pi}\sin x + \frac{4}{3\pi}\sin 3x$

7 99.942% of the total energy

9 $H_n(x) =$
  $\pi/4 + \sum_{i=1}^{n}((-1)^{i+1}\sin(ix))/i +$
  $\sum_{i=1}^{[n/2]}(-2/((2i-1)^2\pi))\cos((2i-1)x),$
where $[n/2]$ denotes the biggest integer smaller than or equal to $n/2$

11 (a) $F_3(x) =$
    $1/2 + (2/\pi)\cos x - (2/3\pi)\cos 3x$

  (b) There are cosines instead of sines, but the energy spectrum remains the same

13 $F_4(x) = 1 - (4/\pi)\sin(\pi x)$
  $- (2/\pi)\sin(2\pi x) - (4/3\pi)\sin(3\pi x)$
  $- (1/\pi)\sin(4\pi x)$

19 (a) 15.9155%, 0.451808%
  (b) $(4\sin^2(k/2))/(k^2\pi^2)$
  (c) The constant term and the first five harmonics are needed
  (d) $F_5(x) = 1/2\pi + (2\sin(1/2)/\pi)\cos x$
    $+ (\sin 1/\pi)\cos 2x$
    $+ (2\sin(3/2)/(3\pi))\cos 3x$
    $+ (\sin 2/(2\pi))\cos 4x$
    $+ (2\sin(5/2)/(5\pi))\cos 5x$

---

21 (a) 31.83%, 76.91%
  (b) 90.07%
  (c) $F_3(x) = 1/\pi + (2\sin 1/\pi)\cos x +$
    $(\sin 2/\pi)\cos 2x + (2\sin 3/(3\pi))\cos 3x$

Chapter 10 Review

1 $e^x \approx 1 + e(x-1) + (e/2)(x-1)^2$

3 $\sin x \approx -1/\sqrt{2} + (1/\sqrt{2})(x+\pi/4)$
  $+ (1/2\sqrt{2})(x+\pi/4)^2$

5 $P_3(x) = 4 + 12(x-1)$
  $+ 10(x-1)^2 + (x-1)^3$

7 $1 - (9/2)y^2 + (27/8)y^4 - (81/80)y^6 + \cdots$

9 $\sin t^2 =$
  $t^2 - t^6/3! + t^{10}/5! - t^{14}/7! + \cdots$

11 $1/(1-4z^2) =$
  $1 + 4z^2 + 16z^4 + 64z^6 + \cdots$

13 $z^2 + (1/2)z^4 + (3/8)z^6 + (15/16)z^8 + \cdots$

15 $a/(a+b) = 1 - b/a + (b/a)^2 - (b/a)^3 \cdots$

17 $1/(a+r)^{3/2} = (1/a^{3/2})(1 - (3/2)(r/a)$
  $+ (15/8)(r/a)^2 - (35/16)(r/a)^3 + \cdots)$

19 3/4

21 e^{-2}

23 $3e$

25 (a) 2
  (b) -1
  (c) $-2/3$
  (d) 12

27 (a) $P_2(x) = b - bx^2/(2a^2)$
  (b) 0 because graph horizontal
  (c) $y = b - bx^2/(2a^2)$
    Intercepts $x = \pm\sqrt{2}a$
  (d) $3 \cdot 10^{-7}$

29 $-1 < x < 1; R = 1$

31 (a) 2
  (b) Near $\theta = 0$, we make the approximation
    $(\sin 2\theta)/\theta \approx 2 - (4/3)\theta^2$

33 (a) $2 - (x^2/4) - (x^4/64)$
  (b) 1.9135
  (c) $(\sqrt{3}/2) + (\pi/3)$
  (d) Three decimal places

35 (a) $V(x) \approx V(0) + V''(0)x^2/2$
    $V''(0) > 0$
  (b) Force $\approx -V''(0)x$
    $V''(0) > 0$
    Toward origin

37 (a) Set $dV/dr = 0$, solve for r. Check for max or min.
  (b) $V(r) = -V_0$
    $+ 72V_0 r_0^{-2} \cdot (r - r_0)^2(1/2)$
    $+ \cdots$
  (d) $F = 0$ when $r = r_0$

41 (a) $g(x) \approx P_n(x) =$
    $g(0) + (g''(0)/2!)x^2 + (g'''(0)/3!)x^3$
    $+ \cdots + (g^{(n)}(0)/n!)x^n$
  (b) If $g''(0) > 0$:
    0 is a local minimum
    If $g''(0) < 0$:
    0 is a local maximum

45 (b) If the amplitude of the kth harmonic of f is A_k, then the amplitude of the kth harmonic of f' is kA_k

(c) The energy of the kth harmonic of f' is k^2 times the energy of the kth harmonic of f

49 (a) $P_7(x) = x - x^3/6 + x^5/120 - x^7/5040$
$Q_7(x) = x - 2x^3/3 + 2x^5/15 - 4x^7/315$

(b) For n odd, ratio of coefficients of x^n is 2^{n-1}

51 (a) $P_{10}(x) = 1 + x^2/12 - x^4/720 + x^6/30240 - x^8/1209600 + x^{10}/47900160$

(b) All even powers

(c) f even

Ch. 10 Understanding

1 False

3 False

5 False

7 True

9 False

11 False

13 True

15 False

17 True

19 True

21 False

23 True

Section 11.1

1 (a) (III)
(b) (V)
(c) (I)
(d) (II)
(e) (IV)

5 12, 18, 27, 40.5

7 $k = -0.03$ and C is any number, or $C = 0$ and k is any number

13 (a) $C = 0$ for any n, or $n = 3$
(b) $C = 5$

15 (a) (IV)
(b) (III)
(c) (III), (IV)
(d) (II)

Section 11.2

1 Possible curves:

3 (a)

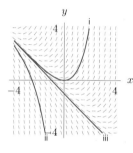

(b) $y = -x - 1$

5 (a)

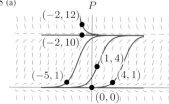

(b) Incr: $0 < P < 10$; decr: $P > 10$; P tends to 10

7 (a)

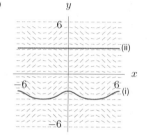

(b) $y = n\pi$

9 (a) II
(b) VI
(c) IV
(d) I
(e) III
(f) V

Section 11.3

1 (a) $y(0.4) \approx 1.5282$
(b) $y(0.4) = -1.4$

3 (a)

5 (a) $\Delta x = 0.5$, $y(1) \approx 1.5$
$\Delta x = 0.25$, $y \approx 1.75$
(b) $y = x^2 + 1$, so $y(1) = 2$
(c) Yes

9 (a) $B \approx 1050$
(b) $B \approx 1050.63$
(c) $B \approx 1050.94$

Section 11.4

1 $P = e^{-2t}$

3 $L = 100e^{p/2}$

5 $y = 10e^{-x/3}$

7 $m = 5e^{3t-3}$

9 $z = 5e^{5t-5}$

11 $u = 1/(1 - (1/2)t)$

13 $y = 200 - 150e^{0.5}$

15 $y = 2 + 3e^{2x-4}$

17 $m = 3000e^{0.1t} - 2000$

19 $B = 25 + 75e^{2-2t}$

21 $z = -\ln(1 - t^2/2)$

23 $y = -2/(t^2 + 2t - 4)$

25 $w = 2/(\cos\theta^2 + 1)$

27 $u = -1/(\ln|2x/(x+1)| - 1)$

29 $R = Ae^{kt}$

31 $P = a + Ae^t$

33 $P = a + Ae^{kt}$

35 $P = (1/a)(Ae^{at} - b)$

37 $R = \tan(ax + C)$

39 $y = 2/(1 + e^{-2t})$

41 $x = e^{At}$

43 (a) $y(t) = 100 - Ae^{-t}$
(b)

(c) $y(t) = 100 - 75e^{-t}$
$y(t) = 100 + 10e^{-t}$
(d) $y(t) = 100 - 75e^{-t}$

45 (a) and (b)

(c) $y^2 - x^2 = 2C$

Section 11.5

1 (a) (I)
  (b) (IV)
  (c) (II) and (IV)
  (d) (II) and (III)

3 (a) $H = 200 - 180e^{-kt}$
  (b) $k \approx 0.027$ (if t is in minutes)

5 (a) $y = 2$: stable; $y = -1$: unstable
  (b)

7 (a) $dS/dt = kS$
  (b) $S = Ce^{kt}$
  (c) $C = 5$
  (d) $S = 5e^{0.1576t}$

9 (a) $dQ/dt = kQ$
  (b) $Q = Ce^{kt}$
  (c) $k \approx -0.182$
  (d) $Q(12) \approx 1.126$ mg

11 (a) $dH/dt = k(68 - H)$
   (b) $H = 68 - Ce^{-kt}$
   (c) $57.8°$F

13 2.21 hours

15 (b) $dQ/dt = -0.347Q + 2.5$
   (c) $Q = 7.2$ mg

17 Michigan: 72 years
   Ontario: 18 years

19 (a) 69,300 barrels/year
   (b) 25.9 years

21 (a) $dT/dt = -k(T - 68)$
   (b) $T = 68 + 22.3e^{-0.06t}$;
       3:45 am.

23 (a) $k \approx 0.000121$
   (b) 779.4 years

Section 11.6

1 (a) $dB/dt = 0.10B + 1000$
  (b) $B = 10000e^{0.1t} - 10000$

3 (a) $H' = k(H - 50)$; $H(0) = 90$
  (b) $H(t) = 50 + 40e^{-0.05754t}$
  (c) 24 minutes

5 (a) $dy/dt = ky$
  (b) 0.2486 grams

7 $dD/dt = -0.75(D - 4)$
  Equilibrium = 4 g/cm^2

9 about 3 days

11 (a) $dW/dt = (1/3500)(I - 20W)$

   (b) $W = I/20 + (W_0 - I/20)e^{-(2/350)t}$

(c)

100 days

13 5.7 days, 126.32 liters

15 $y = A\sqrt{x}$

17 (a) $B_{60} = \sum_{k=0}^{59} 100e^{(0.1)k/12}$
   (b) $B_{60} = 7752.26$
   (c) Smaller because deposits are made later

19 (a) $dy/dt = -k(y - a)$
   (b) $y = (1 - a)e^{-kt} + a$
   (c) a: fraction remembered in the long run
       k: rate material is forgotten

21 (a) $Q = (r/\alpha)(1 - e^{-\alpha t})$
       $Q_\infty = r/\alpha$

   (b) Doubling r doubles Q_∞;
       Altering r doesn't alter the time it takes to
       reach $(1/2)Q_\infty$.
   (c) Both Q_∞ and the time to reach $(1/2)Q_0$
       are halved by doubling α

23 $t \approx 11$ hr and 13 min

25 (c) $v^2 = v_0^2 + (2gR^2/(R + h)) - 2gR$
   (d) $v_0 = \sqrt{2gR}$

Section 11.7

1 $P = (6.6 \cdot 10^6)e^{0.002t}$

3 (a)

  (b) $P = 1$
  (c) Increasing: $0 < P < 1$
      Decreasing: $P > 1$
      In the long run:
        $P \to 1$ (unless $P_0 = 0$)
      Inflection points:
        $P = 1/2$ (for $P_0 < 1/2$)
      At $P = 1/2$, population growing fastest
  (d) $dP/dt > 0$ for $0 < P < 1$
      $dP/dt < 0$ for $P > 1$
      $dP/dt = 0$ for $P = 0$
        and $P = 1$
      dP/dt max at $P = 1/2$

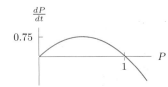

5 (b) $P = (1986, 36)$; $L = 72\%$
  (c) 75%

7 dP/dt is largest in 1920.
  Using this population to estimate the 1990 population, we get 211.4 million

9 (a) $P = 5000/(1 + 499e^{-1.78t})$
  (b)

  (c) $t \approx 3.5$; $P \approx 2500$

11 (a) $dI/dt = k(M - I)$
       $(k > 0)$

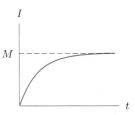

   (b) $dI/dt = kI(M - I)$
       $(k > 0)$

13 (a)

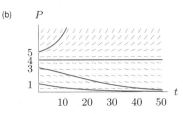

   (b)

   (c) $P = 0$ (stable)
       $P = 4$ (unstable)

15 (a)

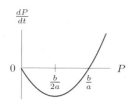

(b) $dP/dt < 0$ for $P < b/a$
$dP/dt > 0$ for $P > b/a$

(c) $P > b/a$: increase
$P < b/a$: extinction

Section 11.8

3 $r = r_0 e^{-t}$, $w = w_0 e^t$

5 Worms decrease, robins increase. Long run: populations oscillate

7

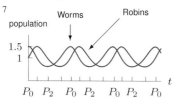

9 Robins increase;
Worms constant, then decrease;
Both oscillate in long run

11 x and y increase, about same rate

13 x decreases quickly while y increases more slowly

15 (a) Symbiosis
(b) Both $\to \infty$ or both $\to 0$

17 (a) Predator-prey
(b) x, y tend to ≈ 1

19 (a) $dy/dx = bx/ay$

21 (b) $dx/dt = -xy$, $dy/dt = -x$
(c) $dy/dx = 1/y$
soln: $y^2/2 = x + C$
(d) If $C > 0$, y wins
If $C < 0$, x wins
If $C = 0$, mutual annihilation
(e)

23 (a) $dy/dx = y(x+3)/(x(y+2))$
$y^2 e^y = Ax^3 e^x$
(d) $y^2 e^y = x^3 e^{x+4}$
(e) $y = x = e^{-4} \approx 0.0183$
(f) $y \approx e^{-13}$

Section 11.9

1 (a) $(0, 0)$; $(5, 2)$
(b) dx/dt and dy/dt not both 0

3 $dw/dt = 0$ where $w = 0$ or $r = 1$
$dr/dt = 0$ $r = 0$ or $w = 1$
Equilibrium points: $(0, 0)$ and $(1, 1)$

5 Horizontal nullclines; $x = 1/2$, $y = 0$
Vertical nullclines; $y = (1/3)(2 - x)$, $x = 0$
Equilibrium points; $(0, 0)$, $(1/2, 1/2)$, $(2, 0)$

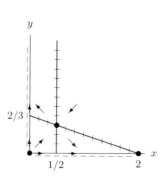

7 Horizontal nullclines; $y = 0$, $y = 2(1 - x)$
Vertical nullclines; $x = 0$, $y = 1 - x/3$
Equilibrium points; $(0, 0)$, $(0, 2)$, $(3, 0)$, $(3/5, 4/5)$

9 $dx/dt = 0$ when $x = 0$ or $y + x/2 = 1$
$dy/dt = 0$ when $y = 0$ or $x + y/3 = 1$
Equilibrium points:
$(2, 0)$, $(0, 3)$, $(0, 0)$, $(4/5, 3/5)$

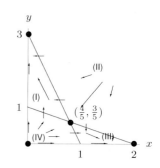

11 (a) $dP_1/dt = 0$ where $P = 0$ or
$P_1 + 3P_2 = 13$
$dP_2/dt = 0$ where $P = 0$ or
$P_2 + 0.4P_1 = 6$

13 (a) $dx/dt = 0$ where
$x = 23.3$
$dy/dt = 0$ where
$8.2x - 0.8y - 142 = 0$
Equilibrium point: $(23.3, 61.7)$
(b) y (US)

Section 11.10

7 (a)

$s = 4\cos t + 3\sin t$

(c) $A = 5, \phi = 0.93$

9 $\sqrt{58}$

11 $\omega = 2, A = 13, \psi = \tan^{-1} 12/5$

13 (a) goes with (II)
(b) goes with (I) and (IV)
(c) goes with (III)
(I) $x = 2\sin 2t$
(II) $x = -\sin t$
(III) $x = \cos 4t$
(IV) $x = -3\sin 2t$

15 Highest point, at rest

17 Lowest point, at rest

19 (a) $y(t) = A\sin(3t) + B\cos(3t)$
(b) (i) $y(t) = (1/3)\sin(3t)$
(ii) $y(t) = \cos(3t)$
(iii) $y(t) = (-\cos 3/\sin 3)\sin(3t) + \cos(3t)$
(iv) $y(t) = (1/\sin 3)\sin(3t)$

(c)(i)

(ii)

(iii)

(iv)

21 (a) $x'' + (g/2)x = 0, x(0) = 5, x'(0) = 0$

(b) $x = 5\cos((\sqrt{g/2})t)$

23 (a) Starts higher; period same
(b) Period increases

25 (a) $Q = 36\sin(1/18)t$
(b) $Q = 6\cos(1/18)t$

27 Oscillate, same frequency, out of phase

Section 11.11

1 $y(t) = C_1 e^{-t} + C_2 e^{-3t}$

3 $y(t) = C_1 e^{-2t}\cos t + C_2 e^{-2t}\sin t$

5 $s(t) = C_1 \cos\sqrt{7}t + C_2 \sin\sqrt{7}t$

7 $z(t) = C_1 e^{-t/2} + C_2 e^{-3t/2}$

9 $p(t) = C_1 e^{-t/2}\cos\frac{\sqrt{3}}{2}t + C_2 e^{-t/2}\sin\frac{\sqrt{3}}{2}t$

11 $y(t) = A + Be^{-2t}$

13 $y(t) = -2e^{-3t} + 3e^{-2t}$

15 $y(t) = \frac{1}{5}e^{4t} + \frac{4}{5}e^{-t}$

17 $y(t) = \frac{5}{4}e^{-t} - \frac{1}{4}e^{-5t}$

19 $y(t) = 2e^{-3t}\sin t$

21 $y(t) = \frac{1}{1-e}e^{-2t} + \frac{-e}{1-e}e^{-3t}$

23 $p(t) = 20e^{(\pi/2)-t}\sin t$

25 (a) (IV)
(b) (II)
(c) (I)
(d) (III)

27 $k = 6$
$y(t) = C_1 e^{2t} + C_2 e^{3t}$

29 Overdamped if:
$b > 2\sqrt{5}$ or $b < -2\sqrt{5}$
Critically damped if:
$b = \pm 2\sqrt{5}$
Underdamped if:
$-2\sqrt{5} < b < 2\sqrt{5}$

31 (iii)

33 (iv)

35 (ii)

39 (a) $d^2 y/dt^2 - y = 0$
(b) $y = C_1 e^t + C_2 e^{-t}$,
so $x = C_2 e^{-t} - C_1 e^t$

41 (a) $Q(t) = 2te^{-t/2}$
(b) $Q(t) = (2+t)e^{-t/2}$

43 As $t \to \infty, Q(t) \to 0$

Chapter 11 Review

1 $P(t) = t^2/2 + C$

3 $P = 5 + Ae^{-2t}$

5 $R = Ae^{2t}/(1 + 3Ae^{2t})$

7 $y = 2/(x^2 + 1)$

9 $y = \tan x$

11 $k = t^t$

13 $500\ln|y| + \frac{5}{2}y = 18\ln|x| + \frac{1}{10}x + 482\ln 10 + 24$

15 $y = 10/(1 + 9e^{-10t})$

17 $f(x) = (\frac{1}{3}x^{3/2} + \frac{2}{3})^2$
(for $x \geq 0$)

19 $y = -\ln(1 + e - e^x)$

21 $-y - \ln|1 - y| = \arctan t - \frac{\pi}{4}$

23 $z(t) = A\cos\pi t + B\sin\pi t$

25 $y(t) = C_1 \cos(t/3) + C_2 \sin(t/3)$

27 $y(t) = e^{-t}(A\sin\sqrt{2}t + B\cos\sqrt{2}t)$

29 (I) = (b); (II) = (d) and (f); (III) = (f);
(IV) = (c); (V) = (a) and (c); (VI) = (e)

31 (a) $y = 3$ and $y = -2$
(b) $y = 3$ unstable;
$y = -2$ stable

33 (a) $dB/dt = (r/100)B$
Constant $= r/100$
(b) $B = Ae^{(r/100)t}$
(c)

35 (a) $dV/dt = -kV^{2/3}$
(b) $V = ((-kt/3) + V_0^{1/3})^3$
(c) $3V_0^{1/3}/k$

37 (a) $dx/dt = k(a-x)^2$
(b) $x = a^2 kt/(akt + 1)$

39 Overdamped: $c < 2$
Critically damped: $c = 2$
Underdamped: $c > 2$

41 (a) $dp/dt = kp(B - p)$
$(k > 0)$
(b) Half of the tin

43 (b) $d^2 r/dt^2 = -ABr$
$r(t) = C_1 \cos\sqrt{AB}t + C_2 \sin\sqrt{AB}t$
(c) $r(t) = \cos\sqrt{AB}t$
$j(t) = \sqrt{A/B}\sin\sqrt{AB}t$
(d) One quarter of the time

45 (a) $P = 0, 1, 2$
(c) $1/2, 3/2, 0, 2$

47 (a)

Ch. 11 Understanding

1 True

3 False

5 True

7 False

9 True

11 True

13 False

15 True

17 True

19 True

21 False

23 False

25 True

27 False

29 False

31 True

33 True

35 $dy/dx = y - x^2$

37 $dy/dx = x/y$

Appendix A

1 (a) $y \leq 30$
  (b) two zeros

3 -1.05

5 2.5

7 $x = -1.1$

9 0.45

11 1.3

13 (a) $x = -1.15$
  (b) $x = 1$, $x = 1.41$,
    and $x = -1.41$

15 (a) $x \approx 0.7$
  (b) $x \approx 0.4$

17 (a) 4 zeros
  (b) $[0.65, 0.66]$, $[0.72, 0.73]$,
    $[1.43, 1.44]$, $[1.7, 1.71]$

19 (b) $x \approx 5.573$

21 Bounded $-5 \leq f(x) \leq 4$

23 Not bounded

Appendix B

1 $2e^{i\pi/2}$

3 $\sqrt{2}e^{i\pi/4}$

5 $0e^{i\theta}$, for any θ.

7 $\sqrt{10}e^{i\arctan(-3)}$

9 $-3 - 4i$

11 $-5 + 12i$

13 $1/4 - 9i/8$

15 $-1/2 + i\sqrt{3}/2$

17 $-125i$

19 $\sqrt{2}/2 + i\sqrt{2}/2$

21 $\sqrt{3}/2 + i/2$

23 -2^{50}

25 $2i\sqrt[3]{4}$

27 $(1/\sqrt{2})\cos(-\pi/12) + (i/\sqrt{2})\sin(-\pi/12)$

29 $-i, -1, i, 1$
  $i^{-36} = 1$, $i^{-41} = -i$

31 $A_1 = 1 + i$
  $A_2 = 1 - i$

37 True

39 False

41 True

Appendix C

1 (a) $f'(x) = 3x^2 + 6x + 3$
  (b) At most one
  (c) $[0, 1]$
  (d) $x \approx 0.913$

3 $\sqrt[4]{100} \approx 3.162$

5 $x \approx 0.511$

7 $x \approx 1.310$

9 $x \approx 1.763$

11 $x \approx 0.682328$

Differentiation Formulas

1. $(f(x) \pm g(x))' = f'(x) \pm g'(x)$

2. $(kf(x))' = kf'(x)$

3. $(f(x)g(x))' = f(x)g'(x) + g(x)f'(x)$

4. $\left(\dfrac{f(x)}{g(x)}\right)' = \dfrac{g(x)f'(x) - f(x)g'(x)}{(g(x))^2}$

5. $(f(g(x)))' = f'(g(x)) \cdot g'(x)$

6. $\dfrac{d}{dx}(x^n) = nx^{n-1}$

7. $\dfrac{d}{dx}(e^x) = e^x$

8. $\dfrac{d}{dx}(a^x) = a^x \ln a \quad (a > 0)$

9. $\dfrac{d}{dx}(\ln x) = \dfrac{1}{x}$

10. $\dfrac{d}{dx}(\sin x) = \cos x$

11. $\dfrac{d}{dx}(\cos x) = -\sin x$

12. $\dfrac{d}{dx}(\tan x) = \dfrac{1}{\cos^2 x}$

A Short Table of Indefinite Integrals

I. Basic Functions

1. $\displaystyle\int x^n \, dx = \dfrac{1}{n+1}x^{n+1} + C, \quad n \neq -1$

2. $\displaystyle\int \dfrac{1}{x} \, dx = \ln|x| + C$

3. $\displaystyle\int a^x \, dx = \dfrac{1}{\ln a}a^x + C, \quad a > 0$

4. $\displaystyle\int \ln x \, dx = x\ln x - x + C$

5. $\displaystyle\int \sin x \, dx = -\cos x + C$

6. $\displaystyle\int \cos x \, dx = \sin x + C$

7. $\displaystyle\int \tan x \, dx = -\ln|\cos x| + C$

II. Products of e^x, $\cos x$, and $\sin x$

8. $\displaystyle\int e^{ax}\sin(bx) \, dx = \dfrac{1}{a^2 + b^2}e^{ax}[a\sin(bx) - b\cos(bx)] + C$

9. $\displaystyle\int e^{ax}\cos(bx) \, dx = \dfrac{1}{a^2 + b^2}e^{ax}[a\cos(bx) + b\sin(bx)] + C$

10. $\displaystyle\int \sin(ax)\sin(bx) \, dx = \dfrac{1}{b^2 - a^2}[a\cos(ax)\sin(bx) - b\sin(ax)\cos(bx)] + C, \quad a \neq b$

11. $\displaystyle\int \cos(ax)\cos(bx) \, dx = \dfrac{1}{b^2 - a^2}[b\cos(ax)\sin(bx) - a\sin(ax)\cos(bx)] + C, \quad a \neq b$

12. $\displaystyle\int \sin(ax)\cos(bx) \, dx = \dfrac{1}{b^2 - a^2}[b\sin(ax)\sin(bx) + a\cos(ax)\cos(bx)] + C, \quad a \neq b$

III. Product of Polynomial $p(x)$ with $\ln x$, e^x, $\cos x$, $\sin x$

13. $\displaystyle\int x^n \ln x \, dx = \dfrac{1}{n+1}x^{n+1}\ln x - \dfrac{1}{(n+1)^2}x^{n+1} + C, \quad n \neq -1$

14. $\displaystyle\int p(x)e^{ax} \, dx = \dfrac{1}{a}p(x)e^{ax} - \dfrac{1}{a}\int p'(x)e^{ax} \, dx$

$\qquad = \dfrac{1}{a}p(x)e^{ax} - \dfrac{1}{a^2}p'(x)e^{ax} + \dfrac{1}{a^3}p''(x)e^{ax} - \cdots$

$\qquad (+ - + - \ldots)$

(signs alternate)